GEOLOGY AND OFFSHORE RESOURCES OF PACIFIC ISLAND ARCS—SOLOMON ISLANDS AND BOUGAINVILLE, PAPUA NEW GUINEA REGIONS

Circum-Pacific Council for Energy and Mineral Resources

Earth Science Series, Volume 12

GEOLOGY AND OFFSHORE RESOURCES OF PACIFIC ISLAND ARCS—SOLOMON ISLANDS AND BOUGAINVILLE, PAPUA NEW GUINEA REGIONS

Edited By

J. G. Vedder and T. R. Bruns

Published by the Circum-Pacific Council
for Energy and Mineral Resources
Houston, Texas, U.S.A.
1989

Published September, 1989

ISBN: 0-933687-13-3

Cover Design: Ben Servino, U.S. Geological Survey

Cover Illustration: Schematic tectonic map of the Solomon Islands—Bougainville, Papua New Guinea region. Color layout by Phyllis Swenson.

Circum-Pacific Council for Energy and Mineral Resources Earth Science Series
F. L. Wong and H. G. Greene, Editors

1. Tectonostratigraphic Terranes of the Circum-Pacific Region
   edited by D. G. Howell
2. Geology and Offshore Resources of Pacific Island Arcs—Tonga Region
   compiled and edited by D. W. Scholl and T. L. Vallier
3. Investigations of the northern Melanesian Borderland
   edited by T. M. Brocher
4. Geology and Offshore Resources of Pacific Island Arcs—Central and Western Solomon Islands
   edited by J. G. Vedder, K. S. Pound, and S. Q. Boundy

5A. The Antarctic Continental Margin: Geology and Geophysics of Offshore Wilkes Land
   edited by S. L. Eittreim and M. A. Hampton

5B. The Antarctic Continental Margin: Geology and Geophysics of the Western Ross Sea
   edited by A. K. Cooper and F. J. Davey

6. Geology and Resource Potential of the Continental Margin of Western North American and Adjacent Ocean Basins—Beaufort Sea to Baja California
   edited by D. W. Scholl, A. Grantz, and J. G. Vedder
7. Marine Geology, Geophysics, and Geochemistry of the Woodlark Basin—Solomon Islands
   edited by B. Taylor and N. F. Exon
8. Geology and Offshore Resources of Pacific Island Arcs—Vanuatu Region
   edited by H. G. Greene and F. L. Wong
9. Geology and Offshore Resources of Pacific Island Arcs—New Ireland and Manus Region, Papua New Guinea
   edited by M. S. Marlow, S. V. Dadisman, and N. F. Exon
10. Petroleum Resources of China and Related Subjects
   edited by H. C. Wagner, L. C. Wagner, Frank F. H. Wang, and F. L. Wong
11. Geology of the Andes and its Relation to Hydrocarbon and Mineral Resources
   edited by G. E. Ericksen, M. T. Cañas Pinochet, and J. A. Reinemund
12. Geology and Offshore Resources of Pacific Island Arcs—Solomon Islands and Bougainville, Papua New Guinea Regions
   edited by J. G. Vedder and T. R. Bruns

Circum-Pacific Council publications are available from:
The AAPG Bookstore • P.O. Box 979 • Tulsa, Oklahoma 74101-0979 • USA
and its international distributors.

CIRCUM-PACIFIC COUNCIL FOR ENERGY AND MINERAL RESOURCES

FOREWORD

The Earth Science Series of the Circum-Pacific Council for Energy and Mineral Resources (CPCEMR) is designed to convey the results of geologic research in and around the Pacific Basin. Topics of interest include framework geology, petroleum geology, hard minerals, geothermal energy, environmental geology, volcanology, oceanography, tectonics, geophysics, geochemistry, and applications of renewable energy. The CPCEMR supports and publishes results of scientific research that will advance the knowledge of energy and mineral resource potential in the circum-Pacific region. The Earth Science Series is specifically designed to publish papers that include new data and new maps, report on CPCEMR-sponsored symposia and workshops, and describe the results of onshore and marine geological and geophysical explorations.

This volume reports the results of one of fourteen internationally sponsored surveys to investigate the energy and mineral resources in the Southwest Pacific. The 1982, 1984 and 1986-1987 surveys were fostered by the Australia-New Zealand-United States Tripartite Agreement. Geophysical and geological data were collected aboard the U.S. Geological Survey's (USGS) R/V *S.P. Lee* and the Hawaii Institute of Geophysics' (HIG) R/V *Kana Keoki* and R/V *Moana Wave* in the waters of Fiji, Kiribati, Papua New Guinea, Solomon Islands, Tonga, Western Samoa, and Vanuatu.

Funding for ship time was made available through the U.S. Agency for International Development, the USGS, the U.S. Office of Naval Research (for HIG's 1982 work), the Australian Development Assistance Bureau, the Australian Bureau of Mineral Resources (BMR), the New Zealand Ministry of Foreign Affairs, the New Zealand Department of Scientific and Industrial Research (DSIR), the New Zealand Geological Survey, and the New Zealand Oceanographic Institute (NZOI). Coordination of the program was provided by the U.S. Department of State and the United Nations-sponsored Committee for Co-ordination of Joint Prospecting for Mineral Resources in South Pacific Offshore Areas (CCOP/SOPAC). Over 150 scientists and technicians participated in the cruises and represented the South Pacific island nations and funding countries mentioned above, as well as CCOP/SOPAC, New Caledonia, the United Kingdom and France.

Michel T. Halbouty

Chairman and President

PREFACE

The scientific results presented in this volume are part of a major collaboration of the Australian Bureau of Mineral Resources (BMR), the New Zealand Department of Scientific and Industrial Research (DSIR), the U.S. Geological Survey (USGS) and nations in the South Pacific to investigate the geology and to evaluate potential oil, gas, and mineral resources of the circum-Pacific region.

This collaboration began in 1982 with SOPAC Tripartite I which investigated the potential for offshore oil and gas resources in Solomon Islands, Tonga, and Vanuatu and was responsible for spearheading renewed resource-oriented marine geological and geophysical exploration in the Pacific. The joint surveys were continued in 1984-1987 as SOPAC Tripartite II, designed in part to investigate the Exclusive Economic Zones of the Southwest Pacific and to study other selected parts of the Pacific with an international team composed primarily of scientists from the Tripartite countries.

The studies reported in this volume are beneficial to the Australia, New Zealand, and United States geoscientific research partners and the member countries of the Committee for Co-ordination of Joint Prospecting for Mineral Resources in South Pacific Offshore Areas (CCOP/SOPAC). The benefits to the participating groups include obtaining a comprehensive update and evaluation of the Pacific Basin energy and mineral resources; assisting developing nations of the Pacific in assessing their potential resources; providing scientists of the participating countries an opportunity to study analogous geologic provinces and then develop models, hypotheses and theories that will be useful in unravelling complex geologic problems in their home waters; and bringing together a team of international scientists that work cooperatively in sharing information and ideas that are germane to the scientific understanding of the circum-Pacific geology. The benefits of this work to the general geoscientific community are also great as new information about remote areas of the Pacific is being added to our knowledge and understanding of geology and geologic processes worldwide. Results of other investigations that are part of Operation Deep Sweep are, or will be, reported in other volumes of the Earth Science Series of the Circum-Pacific Council for Energy and Mineral Resources.

Roye Rutland
Director, Australian Bureau of Mineral Resources, Geology and Geophysics

David Kear
Formerly Director-General, New Zealand Department of Scientific and Industrial Research

Dallas Peck
Director, United States Geological Survey

The research recorded here is a major contribution by the countries in the CCOP/SOPAC region and by the wider international community to the understanding of the natural resource potential and the geological origin and significance of the seabed of national Exclusive Economic Zones (EEZ) and the oceanic areas of the South Pacific. At the same time, it is an outstanding example of international partnership and co-operation, with mutual benefits to all concerned.

CCOP/SOPAC warmly welcomes the success of this collaborative enterprise, and, on behalf of the member governments and its Technical Secretariat, I wish to place on record our sincere appreciation for the very generous contribution by the Tripartite Partners—the governments of Australia, New Zealand, and the United States—and by the Circum-Pacific Council for Energy and Mineral Resources.

Jioji Kotobalavu

Director, CCOP/SOPAC Technical Secretariat

EDITORS' FOREWORD

The scientific papers presented in this volume are based upon data gathered during the 1984 CCOP/SOPAC Australia-New Zealand-United States Tripartite marine geologic investigation of the Bougainville, Papua New Guinea and Solomon Islands regions. The amount of data collected nearly doubled that obtained from a similar cruise in 1982 which resulted in the publication of Volume 4 of the Circum-Pacific Council for Energy and Mineral Resources (CPCEMR) Earth Science Series. Because the area of coverage was expanded and significant new findings were made in 1984, the scientific staff agreed that preparation of a second volume would be an appropriate way to augment the material presented in the first volume. Furthermore, a second volume would link the region covered in 1982 in the central and western Solomon Islands to the New Ireland-Manus region, which also was investigated in 1984 (Volume 9 of the CPCEMR Earth Science Series). Subsequent approval by the CCOP/SOPAC member countries and the Tripartite nations and financial support by the Government of Australia and the CPCEMR allowed the work to proceed as planned.

This second collection of papers differs from the four-part format used in the first in that introductory reports on regional geologic setting, shipboard operations, and analytical methods are omitted and a summary chapter is not appended. To include such parts in a second volume would be redundant; therefore, the papers are limited to topical studies. Because each contribution is intended to stand alone as a complete and separate article, some of the background information is repeated chapter by chapter, particularly in those that deal with aspects of regional tectonics.

The editorial assistance of a number of individuals has been indispensable. Without the expert advice and technical supervision by Florence Wong and the formatting of camera-ready copy of texts and illustrations by Susan Boundy-Sanders, publication of this volume in its present style would not have been possible. A number of the papers by U.S. Geological Survey writers received thoughtful and careful editing by Catherine Campbell.

Listed below are the sponsoring organizations for the Tripartite work. In addition, many scientists, technicians, and administrators devoted much time and effort to this investigation. To these individuals, our grateful thanks.

J.G. Vedder
T.R. Bruns

Sponsors:

Australian Development Assistance Bureau, Department of Foreign Affairs, Australia
Bureau of Mineral Resources, Geology and Geophysics, Australia
Committee for Co-ordination of Joint Prospecting for Mineral Resources in South Pacific Offshore Areas
Department of Geology, Mines and Rural Water Supplies, Vanuatu
Department of Scientific and Industrial Research, New Zealand
Geological Survey, Papua New Guinea
Ministry of Lands, Energy and Natural Resources, Solomon Islands
Ministry of Lands, Survey and Natural Resources, Kingdom of Tonga
New Zealand Ministry of Foreign Affairs, External Aid Division
U.S. Agency for International Development (Office of Energy), Department of State
U.S. Geological Survey, Department of the Interior

TABLE OF CONTENTS

Vedder, J.G., and Bruns, T.R., editors, 1989, Geology and offshore resources of Pacific island arcs—Solomon Islands and Bougainville, Papua New Guinea Regions: Houston, Texas, Circum-Pacific Council for Energy and Mineral Resources Earth Science Series, v. 12.

INTRODUCTION TO THE GEOLOGY AND OFFSHORE RESOURCES OF THE CENTRAL AND WESTERN SOLOMON ISLANDS AND EASTERN PAPUA NEW GUINEA

J.G. Vedder
U.S. Geological Survey, Menlo Park, CA 94025, USA

J.B. Colwell
Bureau of Mineral Resources, Geology, and Geophysics
GPO Box 378, Canberra 2601, Australia

PURPOSE AND SCOPE

As part of the Tripartite Australia-New Zealand-United States CCOP/SOPAC series of cruises in 1984, scientists aboard the U.S. Geological Survey research vessel *S.P. Lee* collected geological and geophysical data in the central and western Solomon Islands and in eastern Papua New Guinea (Figures 1 and 2). The purpose of this reconnaissance survey was to supplement similar work done in 1982 and to provide CCOP/SOPAC member nations with additional information about the offshore petroleum and mineral resource potential as well as about geologic hazards.

A published volume consisting of reports on shipboard operations and topical research resulted from the 1982 cruise in the central and western Solomon Islands (Vedder, Pound, and Boundy, 1986). This second volume of research papers is intended to complement those published in the first volume and to deal with an expanded area of investigation that includes part of eastern Papua New Guinea.

Personnel from the Solomon Islands, Australia, and the United States formed the research contingent (Table 1) for the 1984 Solomon Islands-Papua New Guinea segment (cruise identifier L6-84-SP) of the Tripartite cruises. Specific objectives of the 1984 cruise included: 1) identification of structural and stratigraphic traps (including deeply buried reefs) suitable for entrapment of petroleum, 2) correlation of known onshore rock units with offshore seismic stratigraphy, 3) detection and analysis of shallowly buried gas accumulations, and 4) assessment of subsea geologic hazards such as nascent volcanoes and incipient slumps. Problems that were partly or completely resolved by the 1984 cruise included: 1) delineation of the southern and western margins of the Shortland basin, 2) structure and composition of the crust beneath the Central Solomons Trough and bordering ridges, 3) thickness of the sedimentary sequence in the area of the New Georgia wedge, 4) configuration and structure of a deep basin between New Ireland and Bougainville, 5) nature of deformation along the Kilinailau Trench, and 6) origin and composition of seafloor sediments and shallow gas pockets beneath the Central Solomons Trough.

Although most of the objectives of the 1984 cruise were successfully met, some of the problems remain unsolved, particularly the times of inception and completion of arc-polarity reversal. Attempts at sampling and dating seafloor bedrock outcrops in order to correlate seismic stratigraphy with onshore stratigraphic sequences rocks were disappointing. Coring for Quaternary samples, however, provided favorable results.

EQUIPMENT AND DATA ACQUIRED

Navigation equipment, geophysical systems, and sampling gear used during the cruise are listed in Table 2. Throughout this volume, the use of trade names is for purposes of identification only and does not imply endorsement by the U.S. Geological Survey, the New Zealand Geological Survey, or the Bureau of Mineral Resources, Geology, and Geophysics, Australia. The quantity of data collected during the survey is summarized in Table 3.

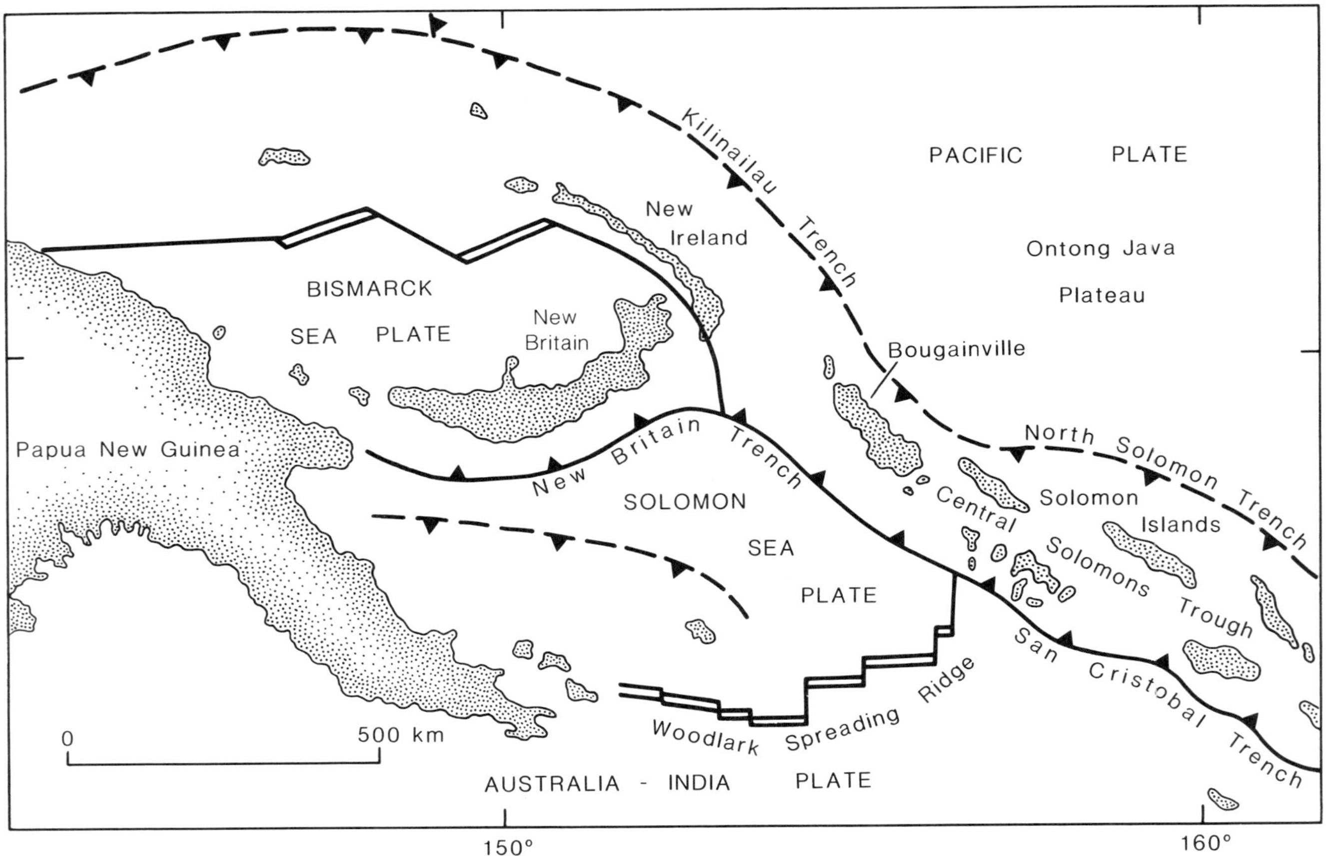

Figure 1. Tectonic elements in the Solomon Islands—Papua New Guinea segment of the Melanesian Borderlands. Other features noted in the text are shown on Figure 2.

Table 1. List of Shipboard Scientific Participants

B.J. Blubaugh	Navigator	USGS
T.R. Bruns	Geophysicist	USGS
J.B. Colwell	Co-chief scientist	BMR
A.K. Cooper	Geophysicist	USGS
D.A. Dinter	Geologist	USGS
J.C. Erikson	Marine technician	USGS
J.A. Fife	Geophysicist	USGS
K.L. Kinoshita	Navigator	USGS
C. Midobatu	Geochemist	SIGS
A.R. Niem	Geologist	OSU
G.E. O'Brien	Marine technician	USGS
H.J. Ryan	Geophysicist	USGS
S. Shafik	Micropaleontologist	BMR
R.W. Vail	Electronics technician	USGS
J.R. Vaughn	Electronics technician	USGS
J.G. Vedder	Co-chief scientist	USGS
R.E. Wells	Geologist	USGS
C.R. Wilson	Navigator	USGS

USGS - U.S. Geological Survey
BMR - Bureau of Mineral Resources, Geology, and Geophysics, Australia
OSU - Oregon State University
SIGS - Solomon Islands Geological Survey

PREVIOUS WORK AND PRELIMINARY REPORTS

Solomon Islands Area

During the 1960s and 1970s, several reconnaissance offshore surveys were made in the Central Solomons Trough and adjacent areas by exploration companies and research institutions. These organizations (and year of their survey) included: Magellan Petroleum Ltd. (1969), Shell Oil Co. and Teledyne (1971), Gulf and Mobil Corporations (1972), Western Geophysical Co. (1973), Pacific Energy and Minerals (1978), and CCOP/SOPAC (1979). Reports based upon an analysis of the results of these surveys were prepared by CCOP/SOPAC for the Solomon Islands government by Katz (1980) and Maung and Coulson (1983).

In 1982, two cruises were made in the Solomon Islands area by U.S. ships under the terms of the Tripartite 1 Agreement. The first, by the Hawaii Institute of Geophysics research vessel *Kana Keoki*, involved the collection of approximately 6,860 km of underway geophysical data and substantial geological data from the Woodlark Basin-New Georgia islands area (Exon and Taylor, 1984; Taylor and Exon, 1987). Supplemental geological and geophysical data were acquired in the

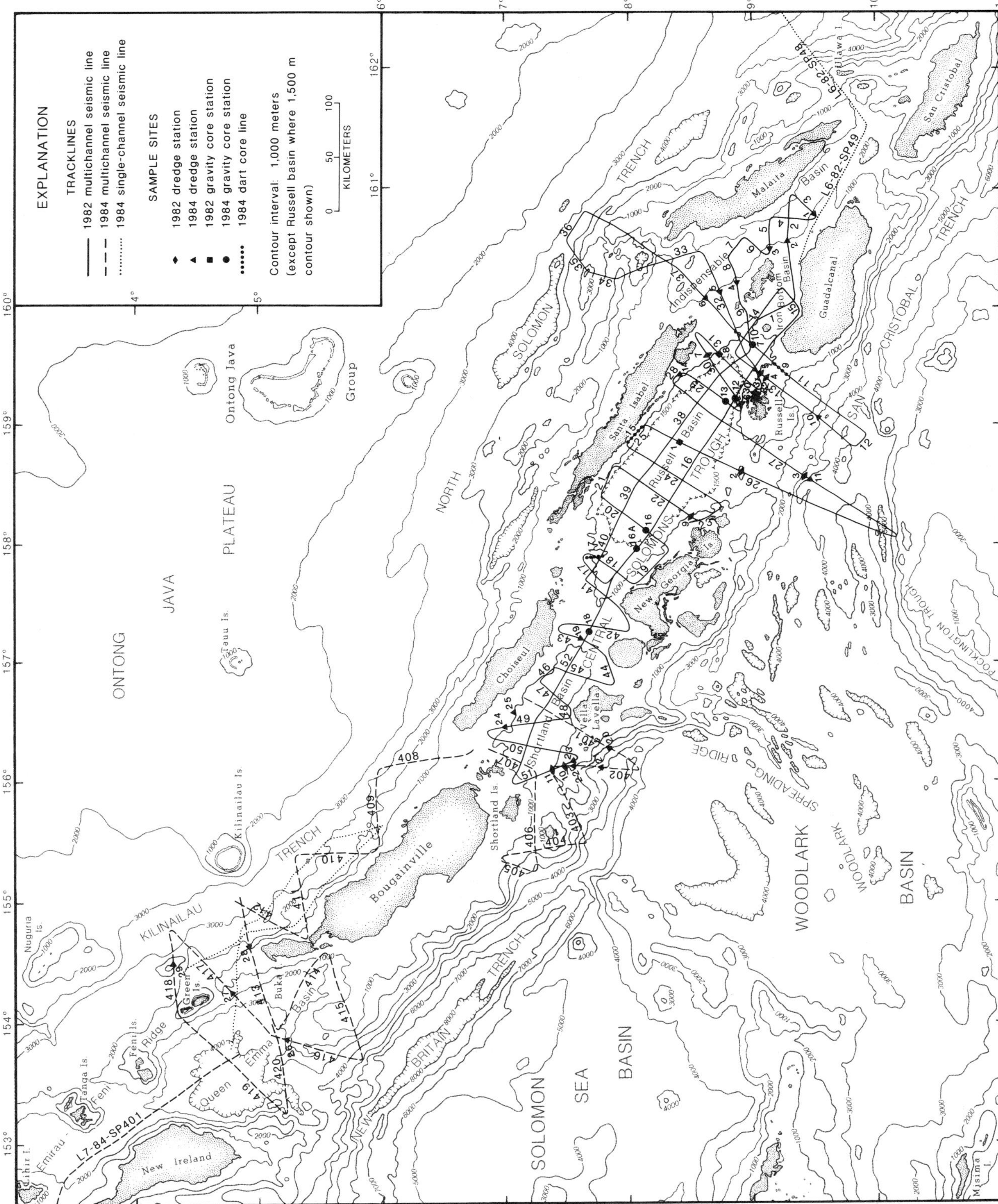

Figure 2. Ship tracklines and sample sites, 1982 and 1984 Tripartite cruises of the R/V *S.P. Lee.*

Central Solomons Trough by the same ship. The second, cruise L7-82-SP by the U.S. Geological Survey research vessel *S.P. Lee*, consisted of seafloor sampling and recording of approximately 3,700 km of geophysical data (Figure 2) from the Central Solomons Trough and adjacent areas (Vedder and Pound, 1984; Vedder, Pound, and Boundy, eds., 1986). Both cruises provided significant new information on the structure, tectonic history, and resource potential of the area.

A preliminary account of the results of the 1984 *S.P. Lee* cruise (L6-84-SP) was given by Colwell and Vedder (1984), and the sampling phases of the cruise were summarized by Colwell et al (1985). These two reports covered both the Solomon Islands and the Bougainville-Buka areas.

Table 2. Geophysical and Sampling Equipment

1. Integrated Magnavox[1] MX702A dual-channel satellite navigation system supplemented by radar ranges and visual bearings*
2. La Coste and Romberg S-53 gravimeter*
3. Geometrics proton-precession magnetometer*
4. 12 kHz echo sounder with a Raytheon CESP III correlator*
5. 3.5-kHz high-resolution sub-bottom profiler with a Raytheon CESP III correlator*
6. EG&G Uniboom high-resolution seismic profiler with hull-mounted transducer array
7. Single-channel seismic-reflection system
8. Multichannel seismic-reflection system consisting of five Bolt airguns with a combined capacity of approximately 1,200 in^3, a 22-group SEI streamer, a Litton airgun triggering system, four air compressors, and a GUS 4300 digital recording system
9. Six Raytheon line-scan recorders (two as single-channel monitors)
10. Fairfield sonobuoy refraction/wide-angle reflection system
11. Cesium clock and NBS synchronizing digital timing equipment
12. Deep-sea winch with approximately 7,000 m of 1/2" wire
13. Four chain-bag dredges
14. Three time-depth recorders
15. Five 3-m-long gravity corers
16. Forty 30-cm-long dart corers
17. Carle Model 311 gas chromatograph

*Routine recording

[1]Use of trade names is for purposes of identification only and does not imply endorsement by the U.S. Geological Survey, the New Zealand Geological Survey, or the Bureau of Mineral Resources, Geology, and Geophysics, Australia.

Bougainville-Buka Area

Until 1984, knowledge of the offshore area directly west of Bougainville and Buka Island was limited largely to data collected along reconnaissance geophysical lines by the Gulf Corporation, Office de la Recherche Scientifique et Technique Outre-Mer (ORSTOM), and the Australian Bureau of Mineral Resources (BMR) in the early to middle 1970s. A preliminary report, as noted above, was issued on the new data and results of the 1984 *S.P. Lee* cruise segment in the Bougainville-Buka area (Colwell and Vedder, 1984). At the Marine Studies Workshop on the petroleum geology of south Pacific island countries, 14th BMR Symposium, Canberra, Australia, short papers were presented on new findings in offshore areas of Bougainville (Shaw, 1985; Colwell et al, 1985; Vedder, Colwell, and Bruns, 1985).

Interpretations of the structure and seismic stratigraphy of the Kilinailau Trench, southern Emirau-Feni Ridge, and the deep basin between New Ireland and Buka Island were reported and exhibited at the 1986 Circum-Pacific Meeting in Singapore (Vedder et al, 1986). Results from modeling of sonobuoy refraction data from the Kilinailau Trench were described at the 1986 American Geophysical Union meeting in San Francisco (Cochrane, Cooper, and Bruns, 1986). The hydrocarbon potential of the platform southwest of Bougainville was outlined by Stewart, Francis, and Pederson (1987).

Table 3. Summary of Collected Data

Data	Amount	Unit	km
Multichannel seismic data	968	nmi	(1,792 km)
Single-channel seismic data	155	nmi	(287 km)
Sonobuoy refraction data, 40 lines	14-22	nmi	(25-40 km)
Gravity data	1,620	nmi	(3,000 km)
Magnetic data	1,250	nmi	(2,315 km)
Bathymetric data, 3.5 kHz	2,088	nmi	(3,868 km)
Bathymetric data, 12 kHz	2,127	nmi	(3,940 km)
Uniboom data	324	nmi	(600 km)
Gravity cores	6		
Dart cores (3 lines)	19		
Gas samples from gravity cores	16		
Dredge hauls (19 stations)	12		

Joint Cruise Report

In August 1987, a collection of 16 preliminary research papers and maps, which were prepared in partial fulfillment of the Tripartite 2 agreement, were transmitted as an administrative report to the CCOP/SOPAC Technical Secretariat and to the participating institutions. This compilation (Vedder, Bruns, and Colwell, 1987) included reports that resulted directly from the 1984 *S.P. Lee* cruise, together with topics of broader scope such as regional biostratigraphy. These preliminary papers formed the nucleus from which this volume was produced.

REGIONAL GEOLOGIC SETTING

Aspects of Cenozoic tectonism throughout the island arcs of the southwest Pacific Ocean were synthesized by Kroenke (1984), and the regional geologic setting outlined here was treated in more detail by Vedder and Coulson (1986) and Coulson and Vedder (1986). The central and western Solomon Islands and Bougainville regions lie within the northwestern Melanesian Borderlands described by Coleman and Packham (1976) and form a complex island-arc system that includes an intra-arc basin and remnants of both forearc and back-arc depositional wedges. These features developed in response to episodic Cenozoic tectonism along the convergent boundary between the Pacific and Australia-India plates (Kroenke, 1984). Although the timing and nature of successive tectonic events are debatable, presumed early Tertiary southwest-directed subduction of the Pacific plate and associated arc magmatism were curtailed by impingement of the leading edge of the Ontong Java Plateau (Coleman and Kroenke, 1981). Aprons of back-arc and forearc sediment were derived from high-standing parts of the arc during the late Oligocene and early Miocene. Late Tertiary arc-polarity reversal and northeastward-directed subduction of the Woodlark spreading-ridge system (Taylor and Exon, 1987) at the northeastern edge of Woodlark Basin along the New Britain-San Cristobal Trench caused a renewal of island-arc magmatism that enclosed the Central Solomons Trough to form an intra-arc basin.

The modern Central Solomons Trough is a composite intra-arc basin consisting of aligned sub-basins, some of which contain more than 5.0 km of late Oligocene(?) and younger sedimentary rocks (Vedder and Bruns, 1986). Five lenticular seismic-stratigraphic units are discernible in most parts of this composite basin; these units are differentiated by unconformities and abrupt velocity changes on multichannel seismic-reflection profiles (Bruns et al, 1986). Late Miocene and younger folds and faults deform the northeastern and southwestern flanks of the trough.

Profiles across the Kilinailau Trench recorded during the 1984 *S.P. Lee* cruise show Ontong Java Plateau rocks covered by 2 to 3 km of Miocene(?) and younger trench sediment. The inner trench wall consists of folded, upfaulted, and rotated blocks of trench and forearc strata that are juxtaposed with Paleogene volcanic rocks of Bougainville (Bruns, Vedder, and Culotta, this volume).

Queen Emma basin, the deep-water intra-arc basin west of Buka Island has a southern margin that apparently is formed by a subsided part of the early Cenozoic arc that crosses Bougainville (Vedder, Bruns, and Cooper, this volume). This basin contains Oligocene and younger strata as much as 8 km thick. Pre-Pliocene faults and folds deform the lower part of the sedimentary section and Quaternary faults cut the flanking ridges. The part of the Emirau-Feni Ridge that extends northwestward from Buka Island to form the northeastern flank of the basin is believed to be a segment of the late Cenozoic arc that cuts diagonally across early Cenozoic arc rocks on Bougainville. This basin may have initially developed during the process of arc reversal in the Miocene (Vedder, Bruns, and Cooper, this volume).

ACKNOWLEDGMENTS

The expert seamanship of Captain A.J. McClenaghan and the mechanical proficiency of Chief Engineer J. Dobranski together with the support of their respective crews resulted in superior ship operation. Much of the credit for the success of the cruise is attributable to the complement of research scientists and their assistants as well as to the expert technicians (listed in Table 1).

Shore-based logistical support in Honiara, Solomon Islands, was kindly provided by S. Danitofea, Chief Geologist, Solomon Islands Geological Survey, and R.D. Walshaw, Principal Geologist for Permanent Secretary, Ministry of Natural Resources. R.J. Ramstad was instrumental in expediting the transport of equipment, records, and personnel in Kieta, Papua New Guinea.

REFERENCES CITED

Bruns, T.R., Cooper, A.K., Mann, D.M., and Vedder, J.G., 1986, Seismic stratigraphy and structure of sedimentary basins in the Solomon Islands region, *in* Vedder, J.G., Pound, K.S., and Boundy, S.Q., eds., Geology and offshore resources of Pacific island arcs—central and western Solomon Islands: Houston, Texas, Circum-Pacific Council for Energy and Mineral Resources, Earth Science Series, v. 4, p. 177-214.

Cochrane G.R., Cooper, A.K., and Bruns, T.R., 1986, Modeling of sonobuoy refraction data from the Kilinailau Trench, Bougainville, Papua New Guinea [abs.], EOS, v. 67, n. 44, p. 1196.

Coleman, P.J., and Kroenke, L.W., 1981, Subduction without volcanism in the Solomon Islands arc: Geomarine Letters, v. 1, p. 129-134.

Coleman, P.J., and Packham, G.H., 1976, The Melanesian Borderlands and India-Pacific plates boundary: Earth-Science Reviews, v. 12, p. 197-223.

Colwell, J.B., and Vedder, J.G., 1984, Initial report on 1984 R/V *S.P. Lee* cruise L6-84-SP in western Solomon Islands and eastern Papua New Guinea: UN ESCAP CCOP/SOPAC Cruise Report n. 91, 22 p.

Colwell, J.B., Vedder, J. G., Crook, K. A. W., and Shafik, S., 1985, Offshore sampling in the Solomon Islands and Bougainville—Buka areas: Extended abstracts, Petroleum geology of south Pacific island countries, marine studies workshop, 14th BMR Symposium, Canberra, Australia, p. 43-47.

Coulson, F.I., and Vedder, J.G., 1986, Geology of the central and western Solomon Islands, *in* Vedder, J.G., Pound, K.S., and Boundy, S.Q., eds., Geology and offshore resources of Pacific island arcs—central and western Solomon Islands: Houston, Texas, Circum-Pacific Council for Energy and Mineral Resources, Earth Science Series, v. 4, p. 59-87.

Exon, N.F., and Taylor, B., eds., 1984, Sea-floor spreading, ridge subduction, volcanism, and sedimentation in the offshore Woodlark-Solomons region and Tripartite cruise report for *Kana Keoki* cruise 82-03-16, leg 4: UN ESCAP, CCOP/SOPAC Technical Report n. 34, 386 p.

Katz, H.R., 1980, Basin development in the Solomon Islands, their petroleum potential: UN ESCAP, CCOP/SOPAC Technical Bulletin n. 3, p. 59-75.

Kroenke, L.W., 1984, Cenozoic tectonic development of the southwest Pacific: UN ESCAP, CCOP/SOPAC Technical Bulletin n. 6, 126 p.

Maung, T.U., and Coulson, F.I., 1983, Assessment of the petroleum potential of the central Solomons Basin: UN ESCAP CCOP/SOPAC Technical Report n. 26, 68 p.

Shaw, R.D., 1985, Hydrocarbon potential of the Bougainville Basin, Papua New Guinea: Extended abstracts, Petroleum geology of south Pacific island countries, marine studies workshop, 14th BMR Symposium, Canberra, Australia, p. 19-29.

Stewart, W.D., Francis, G., and Pederson, S.L., 1987, Hydrocarbon potential of Papua New Guinea's Bougainville, southeastern New Ireland basins: Oil and Gas Journal, v. 85, n. 47, p. 83-87.

Taylor, B., and Exon, N.F., 1987, An investigation of ridge subduction in the Woodlark-Solomons region, *in* Taylor, B., and Exon, N.F., eds., Marine geology, geophysics, and geochemistry of the Woodlark Basin-Solomon Islands: Houston, Texas, Circum-Pacific Council for Energy and Mineral Resources, Earth Science Series, v. 7, 363 p.

Vedder, J.G., and Bruns, T.R., 1986, Offshore petroleum potential, Solomon Islands, *in* Vedder, J.G., Pound, K.S., and Boundy, S.Q., eds., Geology and offshore resources of Pacific island arcs—central and western Solomon Islands: Houston, Texas, Circum-Pacific Council for Energy and Mineral Resources, Earth Science Series, v. 4, p. 277-291.

Vedder, J.G., Bruns, T.R., and Colwell, J.B., 1987, Compilers, Aspects of regional geology, Solomon Islands and Papua New Guinea, Joint Cruise Report, R/V *S.P. Lee*, 1984: Administrative Report to UN CCOP/SOPAC Technical Secretariat, 258 p.

Vedder, J.G., Colwell, J.B., and Bruns, T.R., 1985, New findings in offshore areas of western Solomon Islands and Buka-Bougainville, Papua New Guinea: Extended abstracts, Petroleum geology of south Pacific island countries, marine studies workshop, 14th BMR Symposium, Canberra, Australia, p. 37-41.

Vedder, J.G., Colwell, J.B., Bruns, T.R., and Cooper, A.K., 1986, Regional offshore geology of the central and western Solomon Islands and Bougainville, Papua New Guinea [abs.]: American Association of Petroleum Geologists Bulletin, v. 70, n. 7, p. 937-938.

Vedder, J.G., and Coulson, F.I., 1986, Regional offshore geology of the Solomon Islands, *in* Vedder, J.G., Pound, K.S., and Boundy, S.Q., eds., Geology and offshore resources of Pacific island arcs—central and western Solomon Islands: Houston, Texas, Circum-Pacific Council for Energy and Mineral Resources, Earth Science Series, v. 4, p. 99-108.

Vedder, J.G., and Pound, K.S., 1984, Compilers, Geology and offshore resources of the Solomon Islands, Joint Cruise Report, R/V *S. P. Lee*, 1982: Administrative Report to UN CCOP/SOPAC Technical Secretariat, 226 p.

Vedder, J.G., Pound, K.S., and Boundy, S.Q., eds., 1986, Geology and offshore resources of Pacific island arcs—central and western Solomon Islands: Houston, Texas, Circum-Pacific Council for Energy and Mineral Resources, Earth Science Series, v. 4, 306 p.

Vedder, J.G., and Bruns, T.R., editors, 1989, Geology and offshore resources of Pacific island arcs—Solomon Islands and Bougainville, Papua New Guinea Regions: Houston, Texas, Circum-Pacific Council for Energy and Mineral Resources Earth Science Series, v. 12.

THE OCEANIC BASALT BASEMENT OF THE SOLOMON ISLANDS ARC AND ITS RELATIONSHIP TO THE ONTONG JAVA PLATEAU—INSIGHTS FROM CENOZOIC PLATE MOTION MODELS

R.E. Wells
U.S. Geological Survey, Menlo Park, California 94025, USA

ABSTRACT

Cenozoic global plate-motion models based on a hotspot reference frame may provide a useful framework for analyzing the tectonic evolution of the Solomon Islands convergent margin. A postulated late Miocene collision of the Ontong Java Plateau (OJP) with a northeast-facing arc is consistent with the predicted path of the OJP across the Pacific basin and its early(?) Miocene arrival at the trench. Late-stage igneous activity (65-30 Ma) predicted for the OJP as it rode over the Samoan hotspot occurred in correlative stratigraphic sections on Malaita, the supposed accreted flake of OJP in the Solomon Islands arc. Convergence similar to the present velocities between the Australia-India and Pacific plates were characteristic of the last 43 million years. Prior to 43 Ma, Pacific/Australia plate motions were divergent, seemingly at odds with geologic evidence for early Tertiary convergence, particularly in Papua New Guinea. A postulated south Pacific plate may have existed between the Australia and Pacific plates and would have allowed northward subduction along the northeastern Australia-India plate boundary that lasted into the early Eocene. Subsequent reorganization of plate motions in the middle Eocene correlates with middle Eocene marginal basin formation along ridges oblique to the main plate boundary. Cessation of spreading on the Pacific-South Pacific ridge and its subsequent subduction beneath Asia followed the change in Pacific plate motion at 43 Ma. A trapped remnant of the extinct, northwest-trending ridge may still lie beneath the western Philippine Sea. The terminal deformation, metamorphism and ophiolite obduction in the Eocene orogen of the southwest Pacific also correlates with the major change in Pacific plate motion at 43 Ma and the subsequent compression of the dying Eocene arc against outlying continental and oceanic crustal blocks of the Australian plate. The Solomon Islands oceanic basement may represent juxtaposition of oceanic plateaus of the Australian plate beneath overthrust, dismembered ophiolite derived from adjacent marginal basin crust.

INTRODUCTION

The double island chain of the Solomon Islands arc occupies a central position astride a long-lived convergent boundary between the Australia-India and Pacific plates, which stretches from New Guinea south to New Zealand. Recorded in the submarine topography of numerous marginal basins, inactive ridges and troughs, and adjacent oceanic plateaus is a complex tectonic history of episodic arc volcanism, multiple collisions of arcs, continents and oceanic plateaus, and subsequent changes in the polarity of subduction (Figure 1). At present, northeast-directed subduction of the Australia-India plate is occurring along the New Britain-San Cristobal Trench system, and volcanism in the Solomon Islands is restricted to the southwest-facing part of the arc. An inactive trench borders the island chain to the north and separates it from the Ontong Java Plateau (OJP), a broad high-standing region of Cretaceous oceanic basalt mantled by 2 km of pelagic sediments (Kroenke, 1972; Packham and Andrews, 1975). Based on seismic-reflection profiles across the arc (Bruns et al, 1986), the region is undergoing southwest-northeast compression, and stratal tilts on the island platforms generally dip toward the Central Solomons Trough, which consists of a series of en echelon basins filled with more than 5 km of late Oligocene and younger sediments.

The geology of the Solomon Islands has been summarized by Coleman (1965, 1970), Hackman (1973; 1980), Coleman and Packham (1976), Coulson (1985), and Coulson and Vedder (1986). Coleman (1970) originally subdivided the Solomon Islands into three geologic

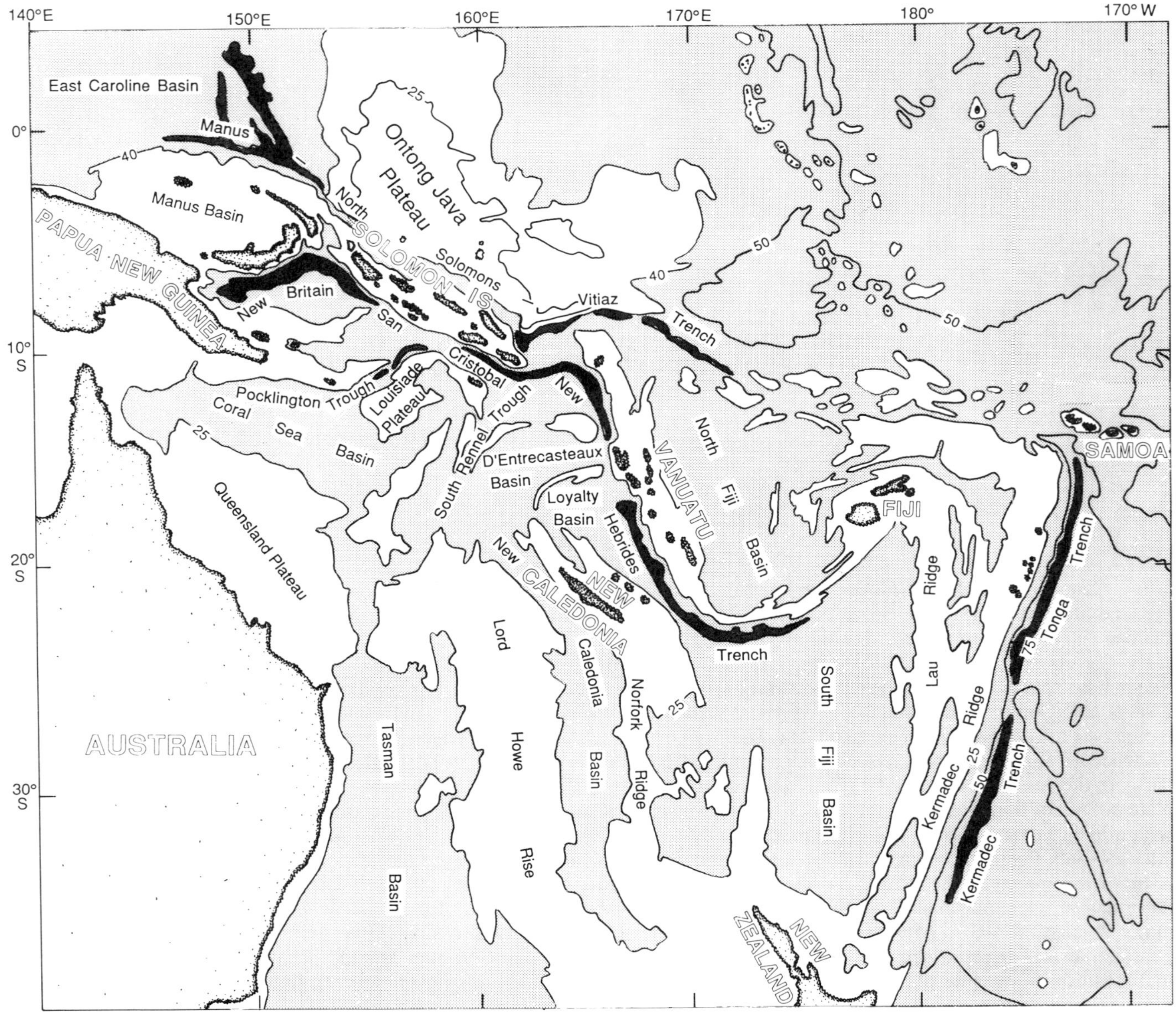

Figure 1. A) Bathymetric map of the southwest Pacific region showing major geographic features from Kroenke (1984) Contour values x 100 m. **B)** Tectonic map modified from plate tectonic map of the Circum-Pacific region (Drummond et al, 1985). Fine straight lines are selected Mesozoic magnetic anomalies; barbed lines are zones of convergence—solid are presently active, open symbols are Oligocene-Miocene in age, dashed zones are Eocene in age. Double lines are spreading

provinces based on gross geologic differences (Figure 2). The Volcanic Province is essentially the locus of active volcanism, an area which extends northwestward from Guadalcanal through the Russell Islands, New Georgia island group, Shortland Islands, and on to Bougainville. The Central Province (Choiseul, southwestern Santa Isabel, Florida Islands, Guadalcanal, and San Cristobal) is composed of a variably tectonized platform of Cretaceous and early Tertiary(?) oceanic basalts, locally metamorphosed to greenschist and amphibolite facies and interleaved with ultramafic tectonites. Oligocene and early Miocene arc volcanics overlie the basement and are in turn overlain by Miocene and younger shallow-water carbonates and volcaniclastic sandstone. The Pacific Province of Malaita, Ulawa, and northeastern Santa Isabel consists of folded, unmetamorphosed Cretaceous and Paleocene oceanic basalt and an overlying thick pelagic carbonate sequence ranging from Late Cretaceous to Pliocene in age. The stratigraphic sequence is strikingly similar to that described for the nearby Ontong Java Plateau (OJP) (Hughes and Turner, 1977).

Kroenke (1972), Coleman and Kroenke (1981), Kroenke (1984), and many others consider the folded pelagic sediments and oceanic basalt of the Pacific

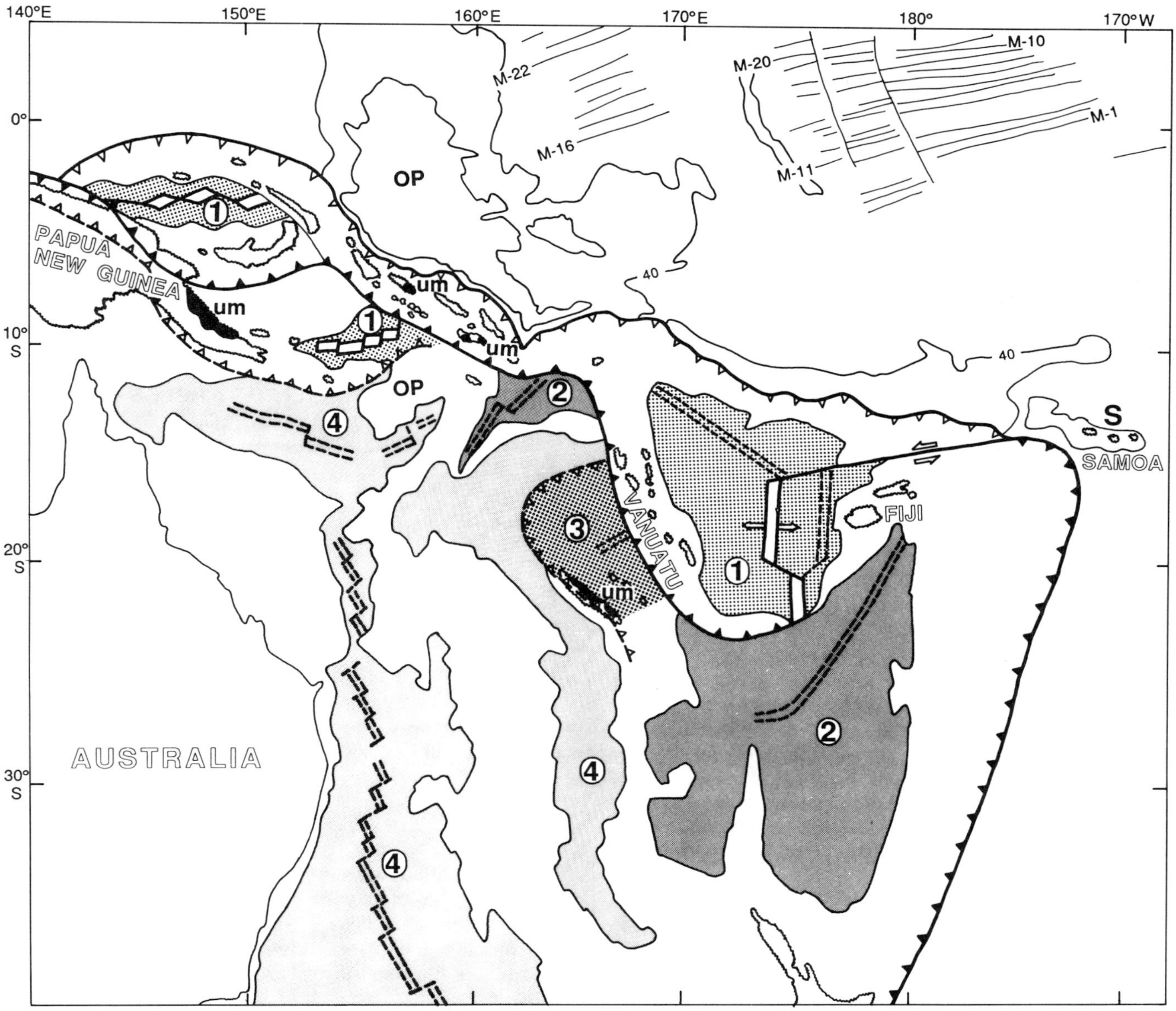

ridges, dotted where inactive; age of spreading indicated as (1) presently active, (2) Oligocene, (3) Eocene, (4) Late Cretaceous-Paleocene; S is location of Samoan hotspot; um represents ultramafic rocks obducted in late Eocene time; op represents oceanic plateaus underlain by thickened oceanic crust.

Province to be the exposed edge of the OJP, obducted onto the arc platform in late Miocene time. They proposed that subduction along the North Solomon Trench ceased upon collision of the OJP, thus causing subduction polarity to reverse, with the Australia-India plate now subducting beneath the Solomon Islands along the New Britain-San Cristobal trench system. A modification to this idea was proposed by Bruns, Cooper, and Vedder (1986). They suggest, based on seismic-reflection profiles across the arc, that the present terrane boundary along the Kia-Korigole-Kaipito fault zone and its offshore extension may be the former subduction zone, rather than the North Solomon Trench. Alternatively, Ramsay (1978, 1982) argued that stratigraphic and petrochemical similarities between locally unmetamorphosed oceanic basement of the Central Province and the oceanic basalt of the Pacific Province obviate the need for major convergence between the two regions. He proposed that the entire Solomon Island arc is underlain by oceanic basement with affinities to the OJP, thus implying either a long history of northward subduction beneath the arc as it rode with the OJP across the Pacific, or a separate local history for the OJP with respect to the Pacific plate. Similarly, Turner and Hughes (1982) state that similarities in the sedimentation history between the two regions do not require large transport between the Central and Pacific provinces. On the basis of

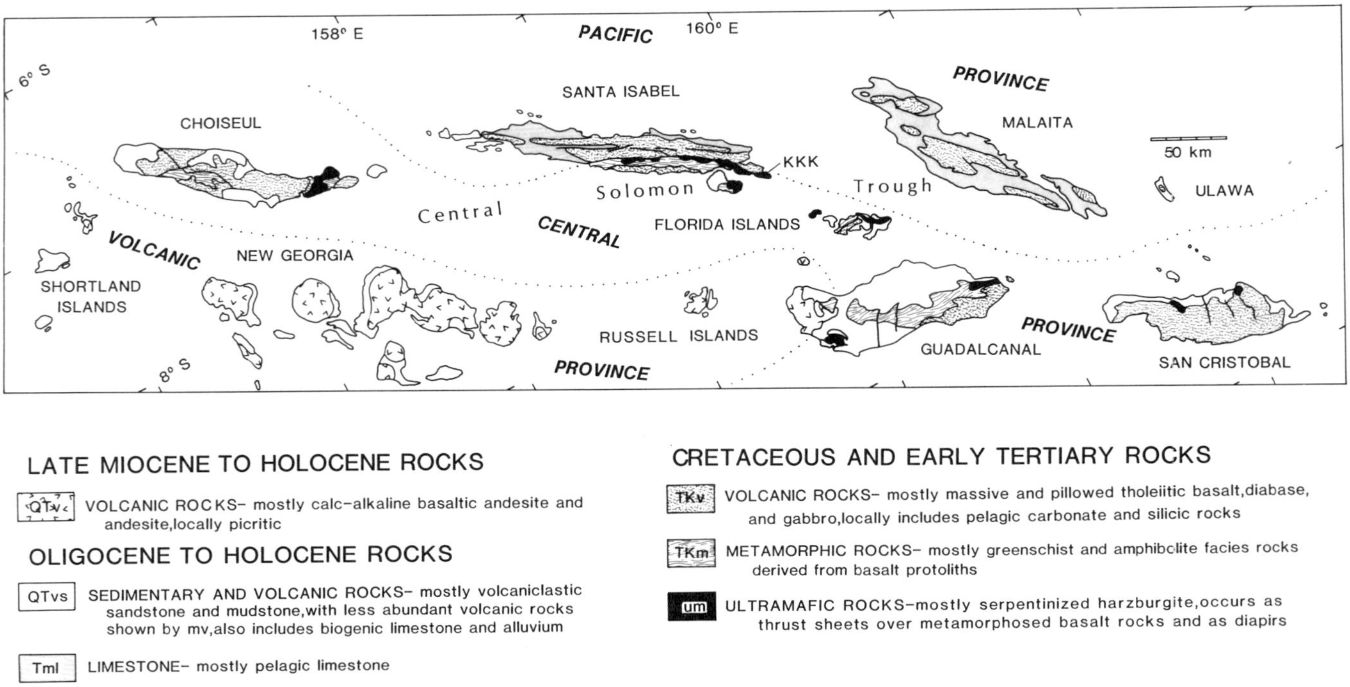

Figure 2. Geologic map of the Solomon Islands, modified from Coulson and Vedder (1986); province boundaries (dotted) modified from Coleman (1970).

paleomagnetic studies in early Tertiary pelagic sediments and basalts of Malaita, Musgrave (1985) concluded that continuity between Malaita and the Australia-India plate persisted for most of Tertiary time, and he suggested that the upraised pelagic sediments and oceanic basalt are an autochthonous forearc assemblage, rather than a part of the OJP.

The purpose of this paper is to examine these apparently conflicting ideas for the origin and deformation of the oceanic basement of the Solomons arc in the context of recent plate-motion models for the Pacific basin (Engebretson, Cox, and Gordon, 1985; Gordon and Jurdy, 1986). These models provide a quantitative framework for the past position of Pacific basin oceanic features, including oceanic plateaus, with respect to both paleomagnetic and hotspot reference frames, and thus provide some constraint on the path of the OJP across the Pacific basin from its probable point of origin to its ultimate time of arrival at the North Solomon Trench. The plate models can also be useful in evaluating the origin of the Cretaceous-Tertiary oceanic basalt of the Central Province and the cause of locally intense Paleogene deformation, metamorphism, and ophiolite obduction in both the Central Province and adjacent regions of Papua New Guinea and New Caledonia. Results so far indicate that there is a good correlation between Tertiary plate motions in a hotspot reference frame and tectonic models for Miocene collision of the OJP with the Solomon Islands arc. Moreover, provisional correlations can be made between early Tertiary plate motions and geologically less certain episodes of early Tertiary convergence, marginal basin formation, and ophiolite obduction.

PLATE MODEL

The relative motion model used in this paper is essentially that of Engebretson, Cox, and Gordon (1985) and Gordon and Jurdy (1986), in which global plate motions are determined relative to a hotspot reference frame. Gordon and Jurdy (1986) summarized evidence showing that the hotspot reference frame is in reasonable accord (±5°) with the no-net-torque mean lithosphere frame and the paleomagnetic axis over Cenozoic time. This first order accordance is reassuring and suggests that gross errors in reconstructions for the larger plates are unlikely for Cenozoic time, although the possibility of significant hotspot drift still remains (Gordon, 1987; Molnar and Stock, 1987). Reconstructions of Pacific plate motion with respect to hotspots for the last 125 Ma by Henderson and Gordon (1981) Henderson, Gordon, and Engebretson (1984), and Duncan and Clague (1985) suggested that most of the linear volcanic chains and oceanic plateaus of the Pacific basin could be explained by hotspot volcanism. These reconstructions account for known age progressions and paleomagnetically derived paleolatitudes, as well as more complex patterns of renewed volcanism and uplift in some oceanic plateaus as a result of passing over multiple hotspots. Because these reconstructions are part of an internally consistent global plate-motion model, they have predictive value in assessing the interactions between oceanic plateaus and convergent margins (see for example, Henderson, Gordon, and Engebretson, 1984).

South Pacific plate reconstructions in the hotspot reference frame for selected intervals of Cretaceous and Cenozoic time are shown in Figure 3. The Cenozoic time scale used is that of Berggren and others (1985). The reconstructions are modified from Gordon and Jurdy (1986), Engebretson, Cox, and Gordon (1985), Hilde, Uyeda, and Kroenke (1977), Duncan and Clague (1985), Stock and Molnar (1982, 1987), and Duncan (1981) and show major plate boundaries, selected active hotspots, and the location of the OJP, presuming that it is part of the main Pacific plate. Given the close ties of the OJP to the Phoenix magnetic anomalies and the lack of a structural break, there is no known reason to think otherwise. Although the errors for the Cretaceous reconstructions are probably large (see discussion by Engebretson, Cox, and Gordon, 1985, and Molnar and Stock, 1987), and almost no constraints are known about the location and number of plate boundaries northeast of Australia at that time, I have included speculative figures to allow illustration of the approximate starting position of the Ontong Java Plateau. Plate boundaries shown in these reconstructions are the minimum number required to explain calculated relative motions between the Australia-India and Pacific plates.

Velocity space diagrams were calculated from the angular velocity vectors of Gordon and Jurdy (1986) for a point on the northeastern edge of the Australian plate, now located at the Solomon Islands (Figure 4). Details of the relative plate motions are subject to revisions and are probably within the uncertainties of the plate reconstructions, but some major trends can be identified. A consistent pattern of Pacific/Australia convergence much like the present is indicated for the last 43 Ma, and is a result of the rapid west-northwest motion of the Pacific plate combined with the rapid northward motion of the Australian away from Antarctica since 43 Ma. Prior to 43 Ma, several reorganizations of relative plate motion probably occurred, although the Pacific plate was generally moving northward away from a slow-moving Australian plate, resulting in divergence between the two plates. If the Australian and Pacific plates were in contact along the north side of Australia in the early Tertiary, then a spreading ridge must have separated them (Figure 3). The existence of such a spreading ridge contradicts tectonic models that require early Tertiary convergence between the Pacific and Australian plates and will be discussed in a later section, along with a possible convergent boundary. This spreading ridge was anticipated by Hilde, Uyeda, and Kroenke (1977), who imagined that it was the eastern extension of a Tethys ridge that was subducted beneath Eurasia following the breakup of Gondwanaland. They suspected that the Australia-Pacific ridge segment may have been subducted by 53 Ma, although in the hotspot reference frame, much of the ridge appears to have been far from any Eurasian subduction zone (Figure 3). I suggest that the now partly subducted ridge may have continued spreading at least until 43 Ma when a gradual shut down followed the change in Pacific plate motion.

Seno (1984) proposed a similar spreading ridge between the Pacific and a southward-subducting North New Guinea plate, which he deemed a requirement for his model of early Tertiary back-arc spreading in the Philippine Sea. He suggested that the subduction of this ridge beneath Australia and the Philippine Sea precipitated the change in Pacific plate motion at 43 Ma. In the model presented here, Seno's South Pacific ridge is retained but for different reasons. I prefer a model involving trapped oceanic crust rather than early Tertiary backarc spreading as the origin of the Philippine basin. Hilde and Lee (1984) summarized evidence that requires the Paleogene spreading center in the Philippine basin to be a remnant of a northwest-trending Pacific spreading ridge that met its demise following initiation of subduction on the Palau-Kyushu ridge at about 45 Ma. Following Uyeda and Ben Avraham (1972), Hilde and Lee (1984) that suggested the trapped ridge was a remnant of the Kula-Pacific ridge. A more likely possibility is that the trapped ridge is a remnant of the Pacific-South Pacific ridge, which is required by plate motions to be present between Australia and the Pacific plate.

Plate Motions and the Formation of Oceanic Basement of the Solomon Islands

Kroenke (1972, 1984), Coleman and Kroenke (1981), and Kroenke, Resig, and Cooper (1986) are the foremost proponents of a tectonic model in which the Ontong Java Plateau (OJP) collided with the Solomon Islands in the late Miocene (at about 8 Ma), thus halting southwest-directed subduction of the Pacific plate and forcing a reversal of subduction polarity. They envisioned the OJP to be of oceanic origin, probably generated on a slow spreading ridge to produce its 40-km-thick crust. Although five times thicker than normal oceanic crust, the velocity structure is compatible with oceanic crustal velocities (Hussong, Wipperman, and Kroenke, 1979). Alternatively, Nur and Ben-Avraham (1977, 1982) argued that the 6.1 km/s velocities of the OJP at intermediate crustal depths are characteristic of continental crust, and that the OJP might be a fragment of a now-dispersed continental mass they called Pacifica, which was originally marginal to Gondwanaland.

Based on a correlation between the stratigraphy of Deep Sea Drilling Project (DSDP), sites 287 and 288 on the OJP and rocks on Malaita, Kroenke (1972) and Hughes and Turner (1977) interpreted the oceanic basalts and pelagic sediments of the Pacific Province of the Solomon Islands to be part of the OJP obducted onto the arc platform. Both sequences are characterized by unmetamorphosed Cretaceous oceanic pillow basalt overlain locally by siliceous deep-ocean sediments and a thick section of pelagic carbonates ranging in age from Early Cretaceous to Pliocene. Prominent sedimentary reflectors on seismic-

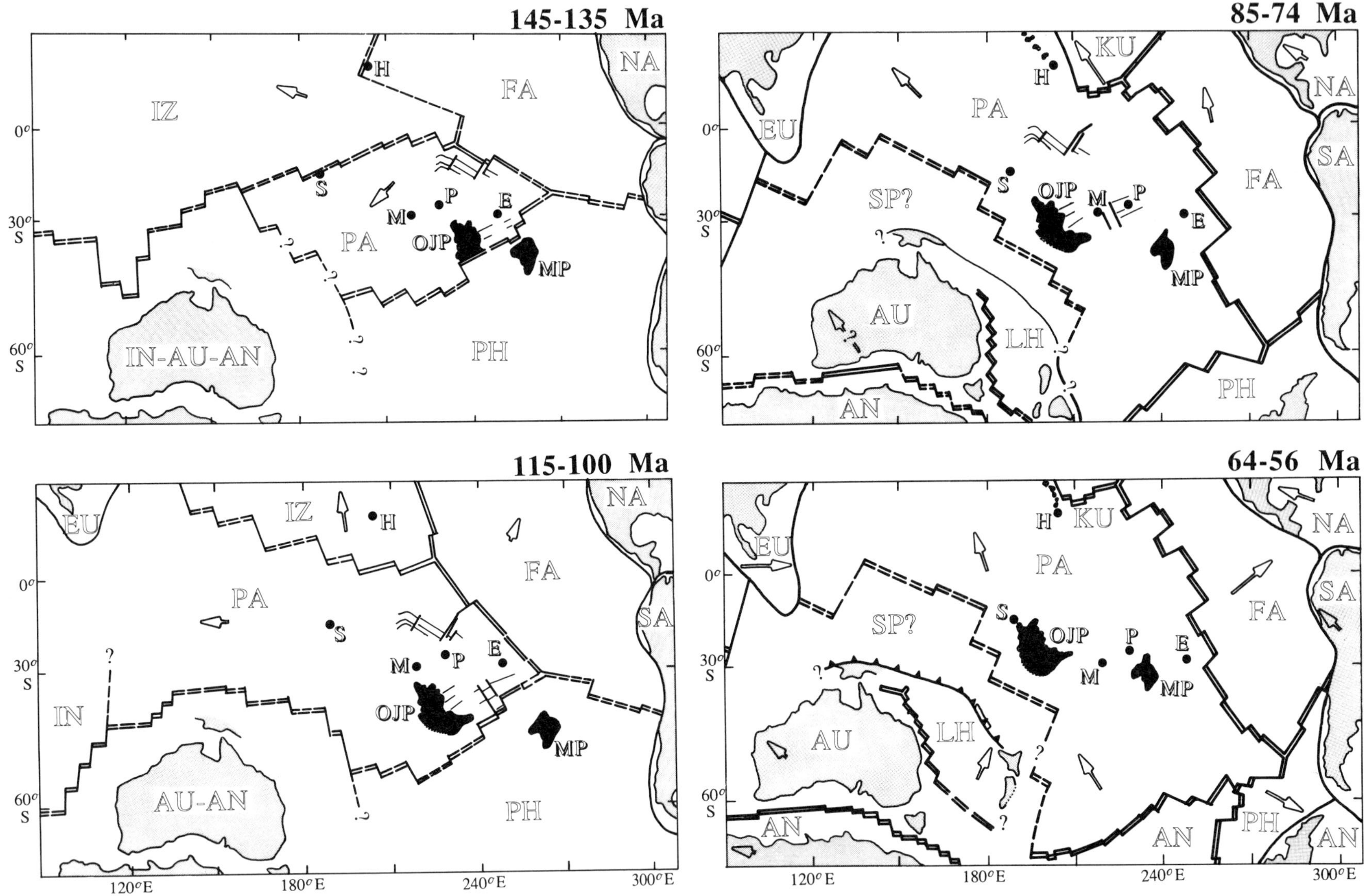

Figure 3. Plate reconstructions for the south Pacific for selected time intervals, modified from Gordon and Jurdy (1986), Engebretson, Cox, and Gordon (1985), Hilde, Uyeda, and Kroenke (1977), Duncan and Clague (1985), and Duncan (1981). Reconstructions prior to 65 Ma are not well constrained. Spreading ridges shown by double lines, dashed where speculative, dotted where extinct. Selected Mesozoic magnetic anomalies on Pacific plate shown by fine lines. Probable convergent margins in Solomon Islands region shown by barbed lines with barbs on upper plate (region of ophiolite emplacement shown

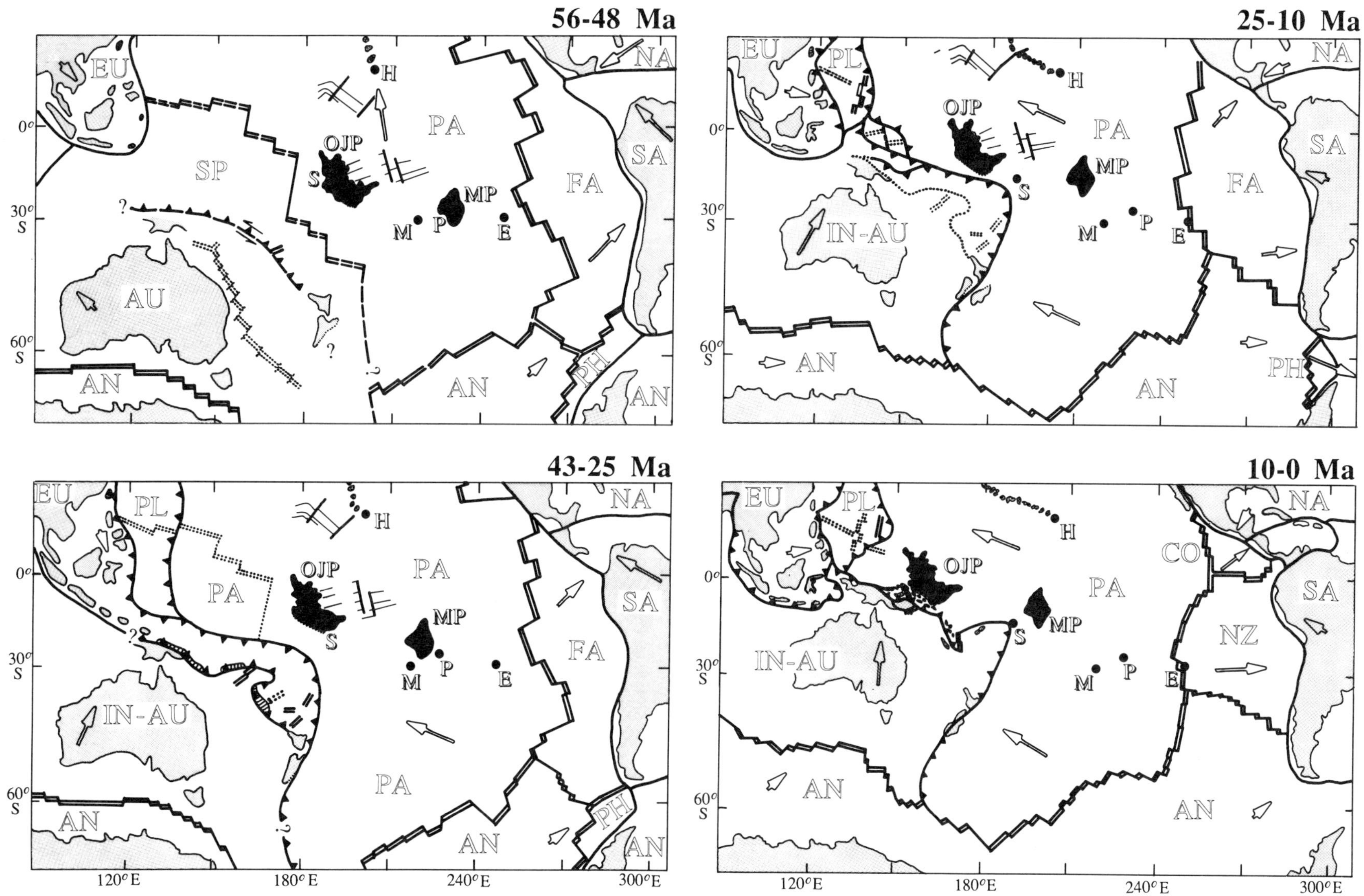

by diagonal line pattern); all other plate boundaries are generalized by solid lines. Motions of plates with respect to hotspots shown by arrows; selected hotspots include: H = Hawaii, E = Easter Island, M = Macdonald, P = Pitcairn, S = Samoa (from Duncan and Clague, 1985). Plate symbols include: AN = Antarctica, AU = Australian, EU = Eurasia, FA = Farallon, IN = India, IZ =Izanagi, KU = Kula, LH = Lord Howe Rise, NA = North America, NZ = Nazca, PA = Pacific, PH = Phoenix, PL = Philippine, SA = South America, SP = South Pacific. OJP represents Ontong Java Plateau, MP represents Manihiki Plateau.

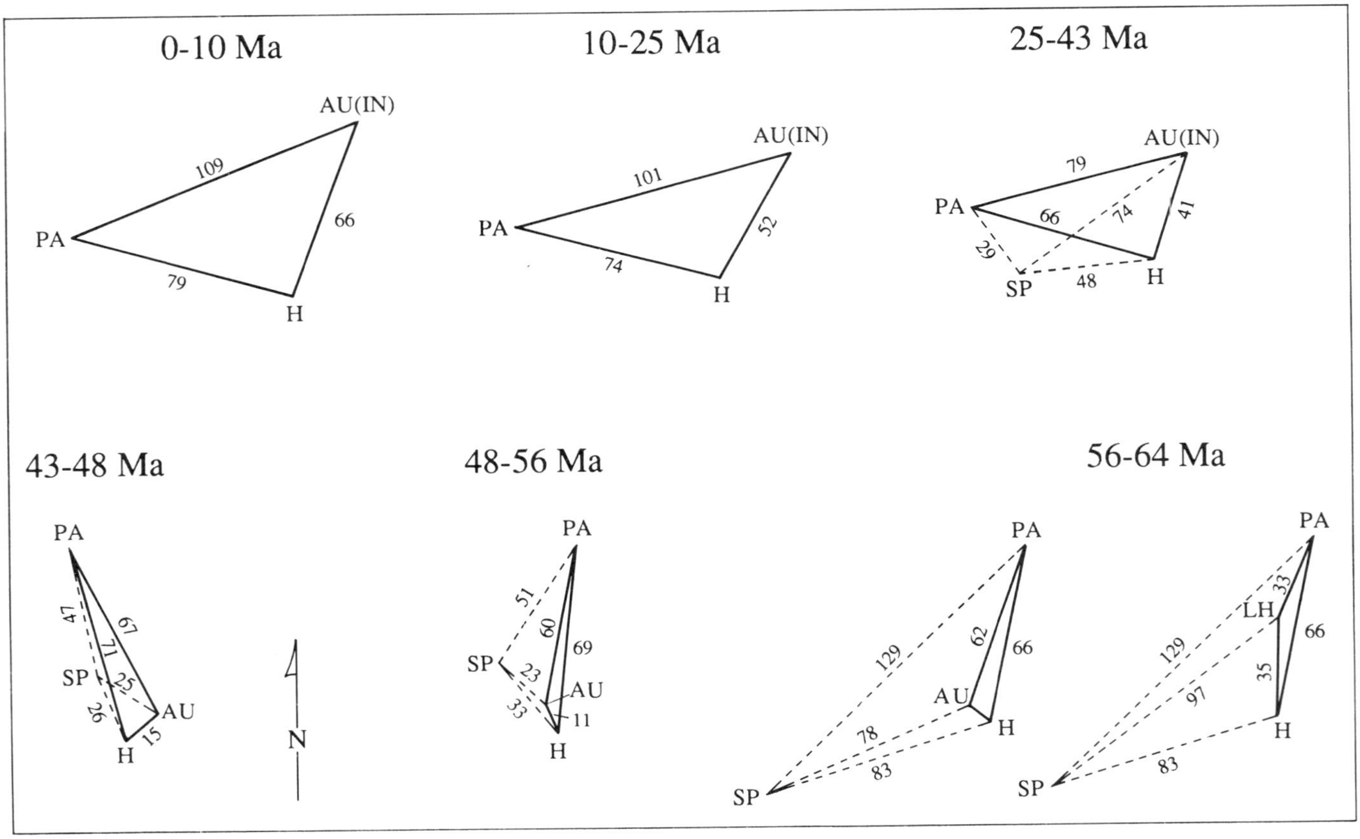

Figure 4. Cenozoic velocity space diagram, calculated from model of Gordon and Jurdy (1986), for location on edge of Australian plate, presently in the Solomon Islands. Plate symbols as in Figure 3, with H = hotspot reference frame; Rates are in km/m.y. Possible relative motions for South Pacific plate calculated assuming continuity with Chatham Rise region.

reflection profiles of the OJP can be traced southwestward beneath ponded sediments in the North Solomon Trench onto the Malaita platform where they become increasingly folded and locally overturned (Kroenke, 1972; Coleman and Kroenke, 1981; Kroenke, Resig, and Cooper, 1986). Presumably this Pliocene folding, which is also characteristic of exposures on Malaita, Santa Isabel, and other islands of the Pacific Province, occurred during obduction of part of the plateau onto the arc.

Objections to the collision hypothesis center around perceived similarities across the supposed suture between the oceanic basalt basement of the Pacific and Central provinces of the Solomon Islands (Ramsay 1978, 1982). Ramsay suggested continuity between the metamorphosed oceanic basalt basement of the Central Province, such as the Voza Lavas of Choiseul and the Mbirao Group of Guadalcanal, and the unmetamorphosed oceanic tholeiites of the Malaita Volcanics and Sigana Volcanics on Malaita and Santa Isabel in the Pacific Province. Both groups of igneous rocks are apparently similar in age (Late Cretaceous and early Tertiary) and petrochemistry. He argued that the metamorphism in the Central Province is localized along large fault zones and that otherwise the units are the same. Likewise, Turner and Hughes (1982) contended that there is continuity or, at least, proximity between the provinces based on similarities in the overlying pelagic sequences of the Central and Pacific provinces. They also asserted that distal mudstone turbidites interbedded in the Oligocene pelagic carbonates of Malaita could have been derived from Guadalcanal, which was emergent at that time. However, if one accepts the unbroken stratigraphic link between the Ontong Java Plateau and the Pacific Province (Kroenke, Resig, and Cooper, 1986), this continuity would require that the entire Solomon Islands platform be attached to the OJP, with little or no convergence between them. An attachment of this nature, however, is contrary to geomorphic evidence for a supposed partly filled trench along the north side of the Solomon Islands and evidence for seismicity on a southwest-dipping relict slab extending at least 200 km beneath the northern trench (Cooper, Kroenke, and Resig, 1986). Alternatively, Musgrave (1985) suggested that paleomagnetic poles from Eocene to Miocene sediments and volcanic rocks on Malaita are more consistent with the Australian polar-wander path rather than a path characteristic of the Pacific plate. He preferred a forearc origin for the Malaitan assemblage and argued against correlation with the OJP. The data are not yet published, however, and the Pacific polar-wander path is currently under revision (Tarduno, McWilliams, and Sliter, 1986). It is also possible that because paleolatitudes of the OJP and Australia are similar (Figure 3), the apparent differences in direction between that expected for the Pacific plate and observed on Malaita could have resulted from local tectonic rotation of Malaita. In any case, the mostly pelagic sedimentary section of Malaita would be unusual for a forearc basin adjacent to an active arc.

In the plate reconstruction of Figure 3, the OJP is considered to be part of the Pacific plate based on its relationship to the Phoenix magnetic anomalies, which strike westward across the plate into the plateau, and on the lack of any structural break between the OJP and the rest of the Pacific plate. Thus, the Ontong Java Plateau moved with the Pacific plate, which travelled north and west across the Pacific basin during the last 100 Ma. The position of the OJP with respect to anomalies M29 to M1 suggests that the plateau may have formed along the Phoenix-Pacific ridge sometime between 160 and 120 Ma (time scale of Kent and Gradestein, 1985). This is consistent with DSDP results, which suggest oceanic basement is older than 110 Ma, and the observation that the sediment thickness decreases southward, consistent with the southward decrease in age of the anomalies (Packham and Andrews, 1975; Coleman and Kroenke, 1981). The thick crust of the OJP implies the presence of a hotspot on or near the ridge to produce such large volumes of basalt, an origin suggested for many oceanic plateaus (Henderson and Gordon, 1981; Henderson, Gordon, and Engebretson, 1984). In this scenario, the "other half" of the OJP alluded to by Coleman, McGowan, and Ramsay (1978), Coleman and Kroenke (1981) and Coulson (1985) would have been on the Phoenix plate and subsequently subducted beneath South America or Antarctica. The likely point of origin of the OJP at about 140 Ma was many thousands of kilometers east of the Australia-India plate, in the vicinity of the presently active south Pacific hotspot cluster near Tahiti. The paleolatitude of the OJP in the hotspot model is compatible with the paleomagnetically determined paleolatitude of 33°S from Aptian sediments and basalt recovered from the plateau at DSDP site 289 (Hammond and others, 1975). If the assumption of slow-moving hotspots is correct, then the position of the plateau suggests an oceanic origin. Because the plateau is far removed from the margins of Gondwanaland, a continental origin as part of "Pacifica" (Nur and Ben-Avraham, 1977, 1982) seems unlikely.

The path of the OJP across the Pacific traversed the region of several presently active hotspots that may have affected the plateau (Figure 3). At around 85 Ma, the OJP passed close to the MacDonald hotspot on a northwest trajectory that eventually approached the Samoan hotspot (Duncan and Clague, 1985) by about 65 Ma. The OJP passed north-northwest directly over the Samoan hotspot between 65 and 43 Ma, essentially along its northern limb. From 43 Ma to about 30 Ma, the OJP moved west-northwest over the hotspot along its eastern limb, thus producing a path congruent with the Hawaiian-Emperor bend in the north Pacific. If the hotspots were active during passage of the OJP, it may explain: 1) continued high elevation of the plateau through its history (Packham and Andrews, 1975) and high topography that mimics the Hawaiian-Emperor bend; 2) subsequent tholeiitic and alkalic igneous activity of Late Cretaceous and early Tertiary age on the plateau (Rea and Vallier, 1983) and in correlative rocks on Malaita and Santa Isabel (Hughes and Turner, 1976, 1977; Coleman, McGowan, and Ramsay 1978); and 3) a major unconformity spanning late Paleocene to early Oligocene time recorded at DSDP 288 on the OJP (Packham and Andrews, 1975).

By 20 Ma, the OJP was less than a few hundred kilometers from the trench north of the Solomon Islands, if 500+ km of Australia-India plate, which were subducted beneath the islands since arc-polarity reversal, are restored. This inferred position agrees well with the proposed time of arrival based on the early Miocene hiatus in Solomon Islands arc volcanism, which is interpreted to be the result of partial subduction of the plateau (Kroenke, 1984). In this reconstruction, the OJP arrived somewhat later than implied by Coleman and Kroenke (1981) and Kroenke (1984), who regarded the 34-Ma alnoitic intrusive breccias on Malaita (Nixon and Coleman, 1978) to be related to flexure of the OJP as it entered the trench. As described here, the breccia pipes and associated deep-mantle xenoliths are more likely to be related to passage of the plateau over the Samoan hotspot. In late Miocene and Pliocene time, a flake of the OJP was folded and obducted onto the arc platform, possibly overthrusting and covering any forearc basin that may have existed along the North Solomon-Vitiaz trench system during the Oligocene and Miocene. Arc volcanism did not resume in the Solomon Islands until the establishment of northeast-directed subduction in the late Miocene (Kroenke, 1984).

SPECULATIONS ON THE EARLY TERTIARY TECTONICS OF THE SOLOMON ISLANDS REGION

Several workers have suggested that prior to late Eocene-early Oligocene initiation of southwest-directed subduction along the Manus-North Solomon-Vitiaz trench system, there was an earlier Eocene period of northward convergence of Australia beneath the Pacific plate (Davies and Smith, 1971; Hamilton, 1979; Parrot and Dugas, 1980; Kroenke, 1984). The locus of this subduction is thought to be the Aure-Moresby-Pocklington trough system which curves eastward from Papua New Guinea (Figure 1b) to the Rennell Trough where it swings southward toward New Caledonia (Kroenke, 1984). However, considerable uncertainty remains regarding the tectonic significance of the various submarine troughs and ridges in the northeastern Coral Sea region, and there are major gaps in the continuity of these features, particularly across the D'Entrecasteaux Basin. The best evidence for Eocene north-directed subduction is in Papua New Guinea where 55-50 Ma tonalites intruded the Papuan ultramafic belt (Davies and Smith, 1971; D'Addario, Dow, and Swoboda, 1976; Davies, 1980) before its southward emplacement over an accretionary wedge containing an Eocene high pressure-low temperature metamorphic assemblage. Hamilton (1979) interpreted the ophiolite obduction to be the result of collision of an Eocene island arc with northern Australia in the Miocene, whereas Kroenke (1984) envisioned collision of the arc as occurring in the Eocene. Late Eocene blueschist metamorphism also accompanied southwest-directed thrusting of peridotite over the accretionary wedge in New Caledonia (Blake, Brothers, and Lanphere, 1977), tectonic events which Kroenke related to collision of the western part of the Norfolk Ridge with the New Caledonia Arc to cause obduction of the Loyalty back-arc basin floor over the arc. Parrot and Dugas (1980) pointed out the similarities in the radiometric age of convergence-related metamorphism in New Guinea, the Solomon Islands Central Province, and New Caledonia (Table 1) and argued for original continuity between the regions and nearly simultaneous regional emplacement of the ophiolite assemblage in the late Eocene. They suggested that the Solomon Islands portion of the belt was displaced northwestward from its original position between New Guinea and New Caledonia by several episodes of transform faulting. Although details of this model present problems, the basic idea may be useful in the context of the plate-motion model presented here.

Paleocene-Eocene Tectonics

Local plate-tectonic reconstructions derived from an analysis of geology and submarine morphology all seem to require Eocene convergence between a southwest-facing arc on the Pacific plate and Australia. As discussed above, there is an apparent conflict between these models and plate motions in the hotspot reference frame which show about 70 km/my *divergence* between the Pacific and Australian plates during most of Eocene time (Figure 4). This conflict may be resolved if another plate is inferred to have been present between the Australian and Pacific plates.

It is noteworthy that in some previous attempts to determine plate motions for the south Pacific, the Chatham Rise region was considered a possible separate plate from the Pacific plate (Gordon and Jurdy, 1986). The plate motions for this region (as originally determined by Stock and Molnar, 1982) permit Paleocene-Eocene convergence between the Chatham Rise region and the Lord Howe Rise, which is consistent with geologic evidence in New Guinea for early Eocene subduction. If we assume that the Chatham Rise region was part of our postulated South Pacific plate in the early Tertiary, then possible plate motions for the south Pacific plate could be calculated (dotted lines on Figure 4). The plate motions are compatible with a northwest-trending spreading ridge between the Pacific and the postulated south Pacific plate and with convergence between the south Pacific plate and Australia-Lord Howe Rise. These are intriguing results but are, of course, speculative. Stock and Molnar (1987) subsequently suggested that their earlier reconstructions (Stock and Molnar, 1982) indicating Paleogene convergence between the Lord Howe Rise and the Pacific plate may have resulted from failure to account for an additional plate boundary in the system. They proposed the existence of Bellinghausen-Antarctica spreading center to explain the subtle bight in Pacific plate anomalies 31 to 18; and in the new reconstructions, significant convergence between the Pacific plate and Lord Howe Rise is not required. Recent work has also shown that the Louisville Ridge, which lies a few hundred kilometers north of the Chatham Rise, can be modelled as a hotspot track parallel to the Hawaiian-Emperor chain, and thus is probably part of the Pacific plate (Watts and others, 1988; Lonsdale, 1988). Any plate boundary between the Chatham Rise and the Pacific plate would have to lie just north of the rise. Nonetheless, the existence of a south Pacific plate, with or without a connection to the Chatham Rise, is a distinct possibility, given that much of the southwest Pacific sea floor has been subducted since 43 Ma. I prefer to think that in early Tertiary time, the southwest Pacific region contained two oceanic plates: the Pacific plate, which was moving away from Australia as indicated by the Emperor Seamount trend, and a south Pacific plate, which was converging with Australia-Lord Howe Rise during the Paleocene and early Eocene. In the Eocene, convergence between the south Pacific and northeast Australia became more oblique, possibly turning into sinistral slip during middle Eocene plate reorganization of the Pacific basin. This relationship is geometrically compatible with the observed formation of marginal basins by northeast-trending ridge segments, such as in the Loyalty Basin, between 56 and 43 Ma (Figure 1).

Meanwhile, spreading continued between the Lord Howe Rise and Australia into the early Eocene. The formation of oceanic plateaus or continental-margin flood-basalt provinces during this rifting (e.g., Louisiade Plateau of Kroenke, 1984) at the north end of the Lord Howe Rise may have been facilitated by the slow motion of Australia over a cluster of hotspots whose existence is suggested by several southward-younging volcanic chains along the margin of the Tasman Basin (Kroenke and Chun, 1987). The Milne Volcanics of the Papuan peninsula, the Poya Formation of New Caledonia, and part of the Voza-Sigana-Mbirao lavas of the central Solomon Islands may be the exposed part of these tholeiitic volcanics which now occupy lower plate positions beneath ophiolite slabs obducted from the north in the late Eocene.

Late Eocene-Early Oligocene Tectonics

At about 43 Ma, major convergence was reestablished along the northeastern segment of the Australian plate boundary, as the Pacific plate shifted to a more westerly course and the Australian plate began to move more rapidly northward away from Antarctica (Figures 3 and 4). This convergence was contemporaneous with a slow-down or cessation of spreading on the Pacific-

Table 1. K-Ar Ages (in Ma) and Stratigraphic Relationships Relevant to Tertiary Ophiolite Emplacement, S.W. Pacific.

	Papua New Guinea Papuan Peninsula	Solomon Islands Central Province	New Caledonia
VOLCANISM			
Pre-existing arc	55-50 wr tonalite[1]		middle Eocene[4]
Lower plate basalt (oceanic plateau and rift volcanics of Australian and Lord Howe plates)	Late Cretaceous-middle Eocene Milne tholeiites[2]	66±3 wr gabbro Sta. Isabel[3] 92±20 hb gabbro Guadalcanal[3]	Late Cretaceous-early Eocene Poya Formation tholeiitic volcanics
Obducted oceanic crust	150 wr gabbro[1] 147 wr gabbro[1]	Late Cretaceous-early Eocene(?)[3]	49±1.5 wr gabbro[4] 54±2.8 wr gabbro
ACCRETIONARY WEDGE			
Protolith age	Cretaceous-Eocene[5]	No accretionary wedge (all basalt)	Eocene[4]
METAMORPHISM			
Accretionary wedge			36.2±1.1 pa[6] 36.7±1.4 ph 36.8±1.1 ph 39.3±0.7 wr metagraywacke 41.1±2.5 ph 41.7±2.6 ph
"Basal" amphibolite	52-42 wr[7]	Small Nggela[8] 35.2±1.4 hb amphibolite 36.7±0.4 pl and diorite 38.4±0.7 hb tremolite 44.7±2.1 hb schist Choiseul[9] 32.4±6.8 pl 44.3±6.8 hb 40.6±6.8 pl amphibolite 43.6±6.8 hb and schists 49.5±6.8 pl 51.5±6.8 hb	
ONLAP SEQUENCE	Early Miocene	Early Miocene[3]	25-Ma granite intrusion[4]

wr = whole rock; hb = hornblende; pa = paragonite; ph = phengite; pl = plagioclase

[1]Page (1976); [2]Smith (1982); [3]Hackman (1980); [4]Paris (1981); [5]Kroenke(1984); [6]Blake, Brothers, and Lanphere (1977); [7]Davies (1980); [8]Neef and McDougall (1976); Neef and Plimer (1978); [9]Richards et al (1966)

South Pacific ridge, a remnant of which is now trapped in the Philippine plate, and with the extinction of spreading in the middle Eocene marginal basins along the plate boundary with Australia (e.g., Loyalty Basin). During the change in plate motion, the Eocene arc and its marginal basins were compressed, telescoped, and bent around outlying buttresses of the Australian plate such as New Guinea and the Louisiade and West Torres plateaus (Kroenke, 1984), and ophiolites were emplaced in New Guinea, the Solomon Islands Central Province, New Caledonia, and possibly Vanuatu at about 40 Ma (Figure 5 and Table 1). The oceanic basement of the Solomon Islands Central Province may, in part, be composed of two juxtaposed oceanic terranes (see for example, Hackman, 1973; Hughes, 1981). The thick oceanic basalts and scraps of overlying pelagic carbonate may represent tectonized

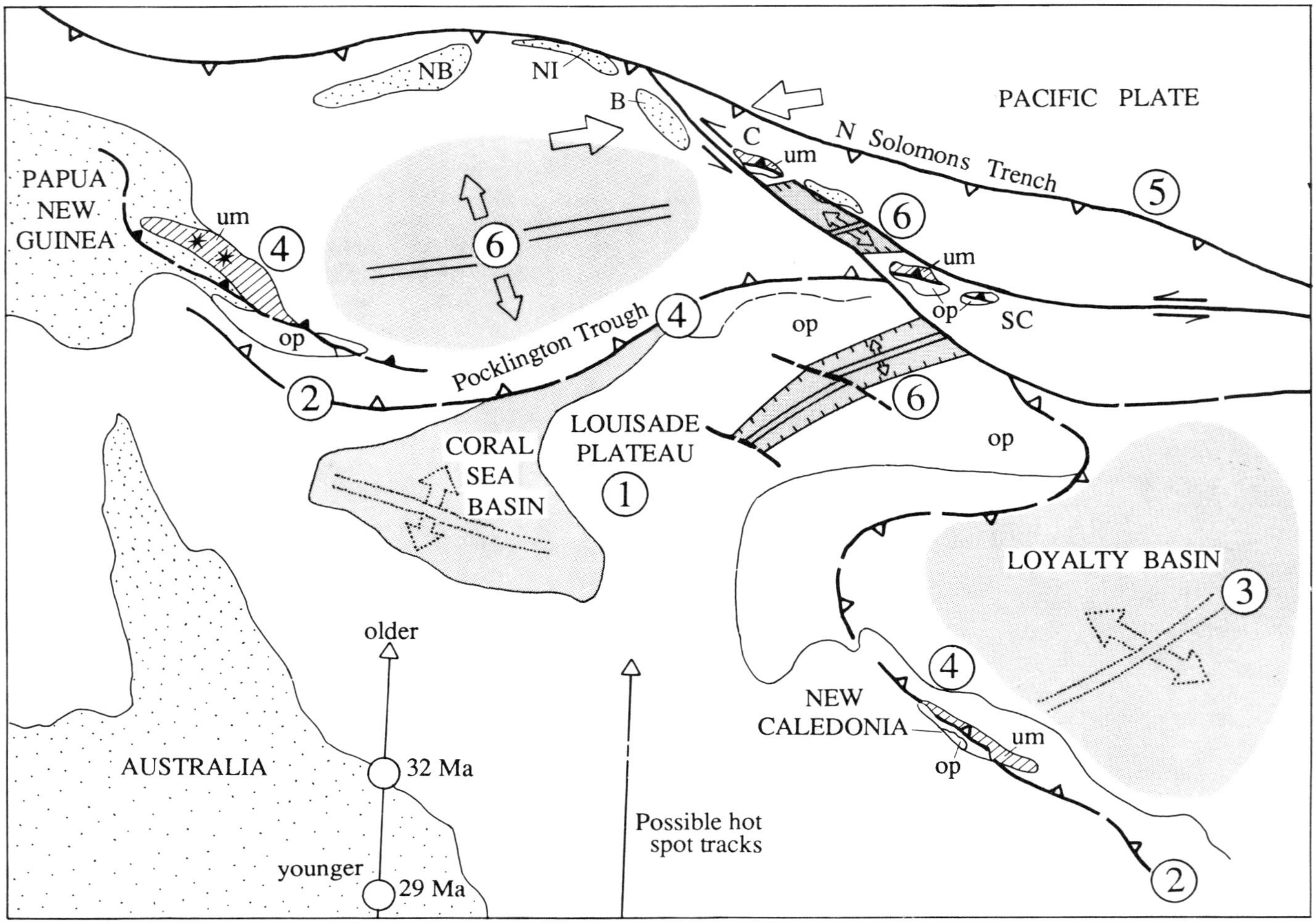

Figure 5. A possible early Cenozoic tectonic history for the Solomon Islands region: (1) Creation of oceanic plateaus by hotspot volcanism on slow moving Australian plate and opening of Coral Sea and Tasman Basins (Late Cretaceous-early Tertiary). (2) Northward subduction of Australian and Lord Howe Rise plates along Pocklington Trough and formation of 55-50 Ma arc tonalites in Papua New Guinea, shown by stars (Paleocene-Eocene). (3) Oblique spreading in Loyalty back-arc basin following change to predominantly transcurrent motion between Australia and the south Pacific plate (Eocene). (4) Change in Pacific and Australian plate motion. Collision of Australian oceanic plateau with arc, formation of arc reentrant, and obduction of ophiolite (um) over oceanic plate (op) rocks of lower plate (late Eocene). (5) Beginning of subduction along North Solomon Trench (late Eocene). (6) Oblique opening of South Rennel Trough, Solomon Sea, Central Solomons Trough and related sinistral faulting in the arc (late Eocene or early Oligocene)

oceanic plateaus of the Australian plate that are now thrust beneath a tectonic slab of dismembered ophiolite derived from arc basement or marginal basin crust. The situation is analogous to that in Papua New Guinea and New Caledonia where ophiolite slabs are thrust over outboard tholeiitic basalt piles such as the Milne Volcanics. However, the intervening terrigenous accretionary wedge, which is best developed in New Guinea adjacent to the Australian continent, is missing in the Solomon Islands. The entire collisional welt was trapped as part of the Australian plate by the subsequent jump in subduction. Shortly after collision, southwest-directed subduction probably began along the Manus-North Solomon-Vitiaz trench system north of the Solomon Islands to produce the late Eocene and Oligocene arc volcanics of New Britain, New Ireland, the Solomon Islands, Vanuatu, and Fiji (Dunkley, 1983; Kroenke, 1984).

Renewed oblique rifting and marginal basin formation along the edge of the Australian plate probably accompanied the beginning of late Eocene-Oligocene subduction, which had a significant sinistral slip component in the Solomon Islands region (Figures 4 and 5). The South Fiji Basin opened at this time (Malahoff, Feden, and Fleming 1982) and possibly also the South Rennell Trough and part of the Santa Cruz Basin along a northeast-trending rift (Larue et al, 1977). This rifting

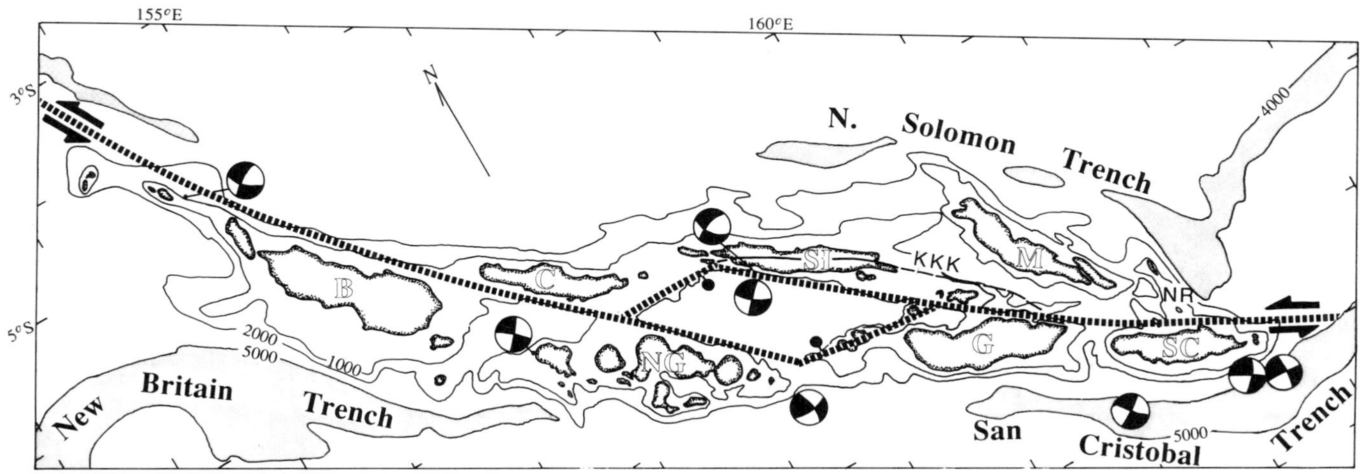

Figure 6. Interpretation of Central Solomons Trough as early Tertiary pull-apart basin formed during oblique subduction of the Pacific plate. Dotted lines indicate possible late Eocene or Oligocene(?) sinistral faults; basin-margin normal faults indicated by ball on downthrown side. Selected fault-plane solutions for shallow earthquakes (50 km depth; Cooper and Taylor, 1984) show that some present-day sinistral slip may be associated with these older structures. B = Bougainville, C = Choiseul, G = Guadalcanal, M = Malaita, NG = New Georgia, CS = San Cristobal, SI = Santa Isabel, NR = Nudha Ridge, KKK = Kia-Korigole-Kaipito fault system. Water depth in meters.

may have transported the Solomon Islands platform to the northwest, away from its position in the Eocene collisional orogen. Sinistral faulting within the arc may have been contemporaneous with formation of the marginal basin and may have caused the initial opening of the Central Solomons Trough as a sinistral pull-apart basin in the late Eocene or early Oligocene.

The Russell basin segment of the Central Solomons Trough appears to be a nearly ideal rhombic depression, oriented such that the linear northeastern and southwestern flanks could be left-lateral faults, and the northern and southern ends could be normal faults (Figure 6). Steep boundary faults along the basin margins are suggested by the gravity data (Beyer, 1986; Cooper, Marlow, and Bruns, 1986), but major offsets of acoustic basement are only locally resolvable in seismic-reflection lines that cross the major topographic breaks (Bruns et al, 1986). This absence of well-defined offsets could result from the fact that acoustic basement locally includes both the late Oligocene basin fill of volcaniclastic and reefal debris and the tectonized volcanic basement of the arc (Cooper, Bruns, and Wood, 1986; Bruns, et al, 1986). In addition, vertical strike-slip faults are difficult to image with seismic-reflection data. Strike-slip faulting along the inter-arc basin margins apparently has not disrupted post-Oligocene sediments. A possible sinistral slip fault with major displacements is the Kia-Korigole-Kaipito fault system on Santa Isabel (Coulson and Vedder, 1986), which may extend offshore to the southeast where it cuts the eastern flank of the Florida Islands platform and Nudha ridge on marine seismic-reflection profiles (Vedder and Coulson, 1986; Bruns et al, 1986). This fault system may be a reactivated collisional suture between Pacific Province oceanic rocks to the north, and tectonized Solomon Islands platform to the south (Bruns, Cooper, and Vedder, 1986). The available fault- plane solutions for recent earthquakes in the Solomon Islands indicate that left-lateral slip in the arc, oriented parallel to the trend of these presumed strike-slip faults, is occurring now (Cooper and Taylor, 1984).

SUMMARY

Global plate models provide some useful constraints on possible local tectonic reconstructions for the Solomon Islands and the southwest Pacific in general. A tectonic model involving origin of the Ontong Java Plateau at an oceanic spreading center and the subsequent early(?) Miocene collision with a northeast-facing Solomon Islands arc is compatible with plate reconstructions in the hotspot reference frame. The plate model also provides a hotspot source at the right time to explain early Tertiary alkalic volcanism in strata on Malaita that correlate with the Ontong Java Plateau. Models that require continuity between the Central Province of the Solomon Islands and the Ontong Java Plateau are less appealing because they require migration of the entire arc-trench complex westward across the Pacific in accompaniment with the plateau, and they also require a very long transform separating the Australia-India plate on the south from the Pacific plate on the north.

A previously postulated episode of earliest Tertiary Australia-Pacific convergence is consistent with a plate model in which an additional south Pacific plate was separated from the Pacific plate by a spreading ridge, and in which the south Pacific plate overrode the Australian and

Lord Howe Rise plates at least during the Paleocene and earliest Eocene. Subsequent middle Eocene back-arc basin formation along oblique spreading centers was terminated by the major reorganization of the plate boundary at 43 Ma. A remnant of the mostly-subducted Pacific-South Pacific ridge was trapped as part of the Philippine plate at this time. Renewal of compression across the Pacific/Australian boundary and regional late Eocene-early Oligocene obduction of marginal basin ophiolites over oceanic plateaus in New Guinea, the Solomon Islands, and New Caledonia closely followed the major change in Pacific plate motion with respect to hotspots from northward to west-northwest. Development of southwest-directed subduction north of the Solomon Islands in the latest Eocene may have trapped the collisional welt as Solomon Islands basement. New marginal basin formation in response to somewhat oblique early Oligocene subduction may have dismembered the Eocene orogen and transported the Solomon Islands to the northwest along sinistral faults. The rhomb-shaped Russell basin in the Central Solomons Trough may be a relict pull-apart basin associated with dismemberment of the Eocene arc. Oblique convergence has been characteristic of the Solomon Island region since 43 Ma and is still driving left-lateral faulting in the arc today.

ACKNOWLEDGMENTS

I thank J. Vedder, R. Gordon, J. Tarduno, and T. Bruns for discussions which helped clarify my thoughts, although I accept the primary responsibility for the opinions expressed in this paper. Reviews by J. Vedder, W. Harbert, and an anonymous reviewer are also appreciated.

REFERENCES CITED

Berggren, W.A., Kent, D.V., Flynn, J.J., and VanCouvering, J.A., 1985, Cenozoic geochronology: Geological Society of America Bulletin, v. 96, p. 1407-1418.

Beyer, L.A., 1986, Gravity anomalies of the Solomon Islands between 156° and 161° E. longitude, *in* Vedder, J.G., Pound, K.S., and Boundy, S.Q., eds., Geology and offshore resources of Pacific island arcs—central and western Solomon Islands: Houston, Texas, Circum-Pacific Council for Energy and Mineral Resources Earth Science Series, v. 4, p. 123-135.

Blake, M.C., Brothers, R.N., and Lanphere, M.A., 1977, Radiometric ages of blueschists in New Caledonia, *in* International Symposium on the Geodynamics of the Southwest Pacific, New Caledonia, August-September 1976: Paris, Editions Technip, p. 279-281.

Bruns, T.R., Cooper, A.K., Mann, D.M., and Vedder, J.G., 1986, Seismic stratigraphy and structure of sedimentary basins in the Solomon Islands region, *in* Vedder, J.G., Pound, K.S., and Boundy, S.Q., eds., Geology and offshore resources of Pacific island arcs—central and western Solomon Islands: Houston, Texas, Circum-Pacific Council for Energy and Mineral Resources Earth Science Series, v. 4, p. 177-214.

Bruns, T.R., Cooper, A.K., and Vedder, J.G., 1986, A multichannel seismic- reflection profile across the Solomon Islands arc, *in* Vedder, J.G., Pound, K.S., and Boundy, S.Q., eds., Geology and offshore resources of Pacific island arcs—central and western Solomon Islands: Houston, Texas, Circum-Pacific Council for Energy and Mineral Resources Earth Science Series, v. 4, p. 215-224.

Coleman, P.J., 1965, Stratigraphical and structural notes on the British Solomon Islands with reference to the first geological map: British Solomon Islands Geological Record (1959-1962), v. 2, p. 17-31.

Coleman, P.J., 1970, Geology of the Solomon and New Hebrides Islands, as part of the Melanesian re-entrant, Southwest Pacific: Pacific Science, v. 24, p. 289-314.

Coleman, P.J., and Kroenke, L. W., 1981, Subduction without volcanism in the Solomon Islands arc: Geo-Marine Letters, v. 1, p. 129-134.

Coleman, P.J., McGowran, B. and Ramsay, W.R.H., 1978, New, early Tertiary, ages for basal pelagites, northeastern Santa Isabel, Solomon Islands (central southwest flank, Ontong Java Plateau): Bulletin of the Australian Society of Exploration Geophysicists, v. 9, n. 3, p. 110-114.

Coleman, P.J., and Packham, G.H., 1976, The Melanesian Borderlands and India-Pacific plates' boundary: Earth-Science Reviews, v. 12, p. 197-233.

Cooper, A.K., Bruns, T.R., and Wood, R.A., 1986, Crustal structure of the Solomon Islands intra-arc basins from sonobuoy seismic studies, *in* Vedder, J.G., Pound, K.S., and Boundy, S.Q., eds., Geology and offshore resources of Pacific island Arcs—central and western Solomon Islands: Houston, Texas, Circum-Pacific Council for Energy and Mineral Resources Earth Science Series, v. 4, p. 135-156.

Cooper, P.A., Kroenke, L.W., and Resig, J.M., 1986, Tectonic implications of seismicity northeast of the Solomon Islands, *in* Vedder, J.G., Pound, K.S., and Boundy, S.Q., eds., Geology and offshore resources of Pacific island arcs—central and western Solomon Islands: Houston, Texas, Circum-Pacific Council for Energy and Mineral Resources Earth Science Series, v. 4, p. 117-122.

Cooper, P.A., and Taylor B., 1984, The spatial distribution of earthquakes, focal mechanisms, and subducted lithosphere in the Solomon Islands, *in* Exon, N.F., and Taylor, B., eds., Seafloor spreading, ridge subduction, volcanism, and sedimentation in the offshore Woodlark-Solomons region and Tripartite cruise report for Kana Keoki cruise 82-03-16, leg 4: UN ESCAP, CCOP/SOPAC Technical Report n. 34, p. 43-76.

Coulson, F.I., 1985, The Solomon Islands, *in* Nairn, A.E.M., Stehli, F.G., and Uyeda, S., eds., The ocean basins and margins, v. 7A, The Pacific Ocean: New York, Plenum Press, p. 607-682.

Coulson, F.I., and Vedder, J.G., 1986, Geology of the central and western Solomon Islands, *in* Vedder, J.G., Pound, K.S., and Boundy, S.Q., eds., Geology and offshore resources of Pacific island arcs—central and western Solomon Islands: Houston, Texas, Circum-Pacific Council for Energy and Mineral Resources Earth Science Series, v. 4, p. 59-88.

D'Addario, G.W., Dow, D.B., Swoboda, R., 1976, Geology of Papua New Guinea, 1:2,500,000: Bureau of Mineral Resources, Canberra, Australia.

Davies, H.L., 1980, Folded thrust fault and associated metamorphics in the Suckling-Dayman Massif, Papua New

Guinea: American Journal of Science, v. 280A, p. 171-191.

Davies, H.L., and Smith, I.E., 1971, Geology of eastern Papua: Geological Society of America Bulletin: v. 82, p. 3299-3312.

Drummond, K.J., Nishiwaki, C., Corvalan, J., Doutch, H.F., and Craddock, C. 1985, Plate tectonic map of the Circum-Pacific region, Pacific basin sheet: Houston, Texas, Circum-Pacific Council for Energy and Mineral Resources, scale 1:17,000,000.

Duncan, R.A., 1981, Hotspots in the southern oceans—an absolute frame of reference from motion of the Gondwanda continents: Tectonophysics, v. 74, p. 29-42.

Duncan, R.A., and Clague, D.A., 1985, Pacific plate motion recorded by linear volcanic chains, *in* Nairn, A.E.M., Stehli, F.G., and Uyeda, S., eds., The ocean basins and margins, v. 7A, The Pacific Ocean: New York, Plenum Press, p. 89-122.

Dunkley, P.M., 1983, Volcanism and the evolution of the ensimatic Solomon Islands arc, *in* Shimozuru, D., and Yokoyama, I., eds., Arc volcanism: physics and tectonics: Tokyo, Terrapub, p. 225-241.

Engebretson, D.C., Cox, A.V., and Gordon, R.G., 1985, Relative motion between oceanic and continental plates in the Pacific basin: Geological Society of America Special Paper 206, 59 p.

Gordon, R.G., 1987, Circum-Pacific plate motions and paleomagnetic data: EOS, Transactions of the American Geophysical Union, v. 68, p. 290.

Gordon, R.G., and Jurdy, D.M., 1986, Cenozoic global plate motions: Journal of Geophysical Research, v. 91, p. 12,389-12,406.

Hackman, B.D., 1973, The Solomon Islands fractured arc, *in* Coleman, P.J., ed., The western Pacific: island arcs, marginal seas, geochemistry: Perth University of Western Australia Press, p. 179-191.

Hackman, B.D., 1980, The geology of Guadalcanal, Solomon Islands: Overseas Memoir of the Institute of Geological Sciences, n. 6, London, 115 p.

Hamilton, W., 1979, Tectonics of the Indonesian region: U.S. Geological Survey Professional Paper 1078, 345 p.

Hammond, S.R., Kroenke, L.W., Thaeyer, F., and Keeling, D.L., 1975, Late Cretaceous and Palaeogene palaeolatitudes of the Ontong Java Plateau: Nature, v. 255, p. 46-47.

Henderson, L.J., and Gordon, R.G., 1981, Oceanic plateaus and the motion of the Pacific plate with respect to the hotspots [abs.]: EOS, Transactions of the American Geophysical Union, v. 62, p. 1028.

Henderson, L.J., Gordon, R.G., and Engebretson, D.C., 1984, Mesozoic aseismic ridges of the Farallon plate and southward migration of shallow subduction during the Laramide orogeny: Tectonics, v. 3, p. 121-132.

Hilde, T.W.C., and Lee, C.-S., 1984, Origin and evolution of the West Philippine Basin: a new interpretation: Tectonophysics, v. 102, p. 85-104.

Hilde, T.W.C., Uyeda, S., and Kroenke, L.,W. 1977, Evolution of the western Pacific and its margin: Tectonophysics, v. 38, p. 145-165.

Hughes, G.W., 1981, The geology of the Oaka and Siruka Bay area, Choiseul, an explanation of the 1:50,000 Geological Sheet CH 10: British Technical Cooperation Western Solomons Mapping Project, Geology Division, Solomon Islands Ministry of Natural Resources Report, n. 10, 92 p.

Hughes, G.W., and Turner, C.C., 1976, Geology of south Malaita: Solomon Islands Geological Survey Bulletin 2, 80 p.

Hughes, G.W., and Turner, C.C., 1977, Upraised Pacific Ocean floor, southern Malaita, Solomon Islands: Geological Society of America Bulletin, v. 88, p. 412-424.

Hussong, D.M., Wipperman, L.K., and Kroenke, L.W., 1979, The crustal structure of the Ontong Java and Manihiki oceanic plateaus: Journal of Geophysical Research, v. 84, p. 6003-6010.

Kent, D.V., and Gradestein, F.M., 1985, A Cretaceous and Jurassic geochronology: Geological Society of America Bulletin, v. 96, p. 1419-1427.

Kroenke, L.W., 1972, Geology of the Ontong Java Plateau: Hawaii Institute of Geophysics, Report n. HIG-72-5, University of Hawaii, 119 p.

Kroenke, L.W., 1984, Cenozoic tectonic development of the southwest Pacific: UN ESCAP, CCOP/SOPAC Technical Bulletin 6, 122 p.

Kroenke, L.W., and Chun, Y.Y., 1987, Indo-Australian plate motions: the origin of hotspots: Geological Society of America Abstracts with Programs, v. 19, p. 396.

Kroenke, L.W., Resig, J.M., and Cooper, P.A., 1986, Tectonics of the southeastern Solomon Islands: Formation of the Malaita Anticlinorium, *in* Vedder, J.J., Pound, K.S., and Boundy, S.Q., eds., Geology and offshore resources of Pacific island arcs—central and western Solomon Islands: Houston, Texas, Circum-Pacific Council for Energy and Mineral Resources Earth Science Series, p. 109-116.

Larue, B.M., Daniel, J., Jouannic, C., and Recy, J, 1977, The South Rennell Trough, evidence for a fossil spreading zone, *in* International Symposium on Geodynamics in the Southwest Pacific: Paris, Editions Technip, p. 51-62.

Lonsdale, P., 1988, Geography and history of the Louisville hotspot chain in the southwest Pacific: Journal of Geophysical Research, v. 93, p. 3078-3104.

Malahoff, A., Feden, R.H., and Fleming, H.F, 1982, Magnetic anomalies and tectonic fabric of marginal basins north of New Zealand: Journal of Geophysical Research, v. 87, B5, p. 4109-4125.

Molnar, P., and Stock, J., 1987, Relative motions of hotspots in the Pacific, Atlantic and Indian Oceans since Late Cretaceous time: Nature, v. 237, p. 587-591.

Musgrave, R.J., 1985, Paleomagnetism and tectonics of Malaita, Solomon Islands: a cautionary tale for terrane seekers: Third Circum-Pacific terrane conference extended abstracts, Geological Society of Australia Publication 14.

Neef, G., and McDougall, I., 1976, Potassium-argon ages on rocks from Small Nggela Island, British Solomon Islands, S.W. Pacific: Pacific Geology, v. 11, p. 81-86.

Neef, G., and Plimer, I.R. 1979, Ophiolite complexes on Small Nggela Island, Solomon Islands; summary: Geological Society of America Bulletin, pt. 1, v. 90, p. 136-138.

Nixon, P.H., and Coleman, P.J., 1978, Garnet-bearing lherzolites and discrete nodule suites from the Malaita alnoite, Solomon Islands, and their bearing on the nature and origin of the Ontong Java Plateau: Australian Society of Exploration Geophysicists Bulletin, v. 9, n. 3, p. 103-107.

Nur, A., and Ben-Avraham, Z., 1977, Lost Pacifica continent: Nature, v. 270, p. 41-43.

Nur, A., and Ben-Avraham, Z., 1982, Oceanic plateaus, the fragmentation of continents, and mountain building: Journal of Geophysical Research, v. 87, p. 3644-3661.

Packham, G.H., and Andrews, J.E., 1975, Results of Leg 30 and the geologic history of the southwest Pacific arc and marginal sea complex, *in* Andrews, J.E., Packham, G.H., et al, Initial Reports of the Deep Sea Drilling Project : Washington, U.S. Government Printing Office, v. 30, p. 691-705.

Page, R.W., 1976, Geochronology of igneous and metamorphic rocks in the New Guinea Highlands: Bulletin of the Bureau of Mineral Resources, Geology and Geophysics of Australia 162.

Paris, J.P., 1981, Geologie de la Nouvelle Caledonie: un essai de synthese: Memoire Bureau de Recherches Geologiques et Minieres, v. 113, 279 p.

Parrot, J.F., and Dugas, F., 1980, The disrupted ophiolitic belt of the southwest Pacific: Evidence of an Eocene subduction zone: Tectonophysics, v. 66, p. 349-372.

Ramsay, W.R.H., 1978, Field, mineralogical, and structural observations on some basement rocks, southeast Choiseul, Solomon Islands: Australian Society of Exploration Geophysicists Bulletin, v. 9, p. 107-110.

Ramsay, W.R.H., 1982, Crustal strain phenomena in the Solomon Islands, constraints from field evidence, and relationship to the India-Pacific plates' boundary: Tectonophysics, v. 87, p. 109-126.

Rea, D.K., and Vallier, T.L., 1983, Two Cretaceous volcanic episodes in the western Pacific Ocean: Geological Society of America Bulletin, v. 94, p. 1430-1437.

Richards, J.R., Cooper, J.A., Webb, A.W., and Coleman, P.J., 1966, Potassium-argon measurements of the age of basal schists in the British Solomon Islands: Nature, v. 211, p. 1251-1252.

Smith, I.E.M., 1982, Volcanic evolution in eastern Papua: Tectonophysics, v. 87, p. 315-333.

Seno, T., 1984, Was there a North New Guinea plate?: Geological Survey of Japan Report 263, p. 29-42.

Stock, J., and Molnar, P., 1982, Uncertainties in the relative positions of the Australia, Antarctica, Lord Howe, and Pacific plates since the Late Cretaceous: Journal of Geophysical Research, v. 87, p. 4697-4714.

Stock, J. and Molnar, P., 1987, Revised history of early Tertiary plate motion in the southwest Pacific: Nature, v. 325, p. 495-499.

Tarduno, J.A., McWilliams, M., and Sliter, W.V., 1986, Paleomagnetic analysis of Pacific oceanic rises [abs.]: EOS, Transactions of the American Geophysical Union, v. 67, p. 926.

Turner, C.C., and Hughes, G.W., 1982, Distribution and tectonic implications of Cretaceous-Quaternary sedimentary facies in the Solomon Islands: Tectonophysics, v. 87, p. 127-146.

Uyeda, S., and Ben-Avraham, Z., 1972, Origin and development of the Philippine Sea: Nature, v. 240, p. 176-178.

Vedder, J.G., and Coulson, R.I., 1986, Regional offshore geology of the Solomon Islands, *in* Vedder, J.G., Pound, K.S., and Boundy, S.Q., eds., Geology and offshore resources of Pacific island arcs—central and western Solomon Islands: Houston, Texas, Circum-Pacific Council for Energy and Mineral Resources Earth Science Series, v. 4, p. 99-108.

Watts, A.B., Weissel, J.K., Duncan, R.A., and Larson, R.L., 1988, Origin of the Louisville Ridge and its relationship to the Eltanin fracture system: Journal of Geophysical Research, v. 93, p. 3051-3077.

A version of this paper also appears in Tectonophysics.

Vedder, J.G., and Bruns, T.R., editors, 1989, Geology and offshore resources of Pacific island arcs—Solomon Islands and Bougainville, Papua New Guinea Regions: Houston, Texas, Circum-Pacific Council for Energy and Mineral Resources Earth Science Series, v. 12.

VELOCITY-STRUCTURE OF THE UPPER CRUST BENEATH THE SOLOMON ISLANDS-BOUGAINVILLE ISLAND ARC

A.K. Cooper, G.R. Cochrane, T.R. Bruns
U.S. Geological Survey, Menlo Park, CA 94025, USA

ABSTRACT

During 1984, 39 seismic sonobuoy stations were recorded across the island platforms, intra-arc basins, and trenches of the Solomon Islands-Bougainville region to measure rock velocities within the upper crust (0-10 km) using refraction-slope-intercept and ray-tracing techniques.

Basement rocks have velocities in excess of 4.4 km/s. At shallow depths, basement (4.5-5.5 km/s) is probably composed of volcanic rocks associated with Paleogene and late Cenozoic volcanic centers, especially near Bougainville and the New Georgia island group, where thick flows appear to cover the island flanks. High-velocity basement rocks (5.8-6.3 km/s) may indicate the presence of local mafic intrusions, arc-basement rocks, or deep-seated mafic igneous rocks of oceanic origin. Velocities in excess of 7.5 km/s were measured near Choiseul and Santa Isabel, indicating possible small ultramafic bodies.

Sedimentary rocks are at least 6.5 km thick in the deep parts of Queen Emma basin and the Central Solomons Trough. Island platforms and arc flanks are draped with high-velocity (2.9-3.9 km/s) strata that generally are 1 to 2 km thick and as much as 3.5 km thick near Vella Lavella. These strata conform with the dipping basement surface and are the deepest sedimentary units in the basins. The deep sedimentary units are likely to be Oligocene and early Miocene in age and were deposited prior to late Cenozoic volcanism and major uplift of the island platforms. Shallow flat-lying strata (1.6 to 2.6 km/s) cover the region and are probably Pliocene and Pleistocene in age.

The 1984 refraction data support the premise that the acoustically concealed New Georgia wedge northeast of New Georgia is composed of as much as 6.5 km of interbedded high-velocity (3.9-5.5 km/s) layers of volcanic rocks and low-velocity (2.3-3.1 km/s) basin strata. Refraction data near Bougainville suggest the absence of an analogous sedimentary wedge northeast of the island. Sonobuoy measurements are consistent with the concept that the Solomon Islands-Bougainville segment of the Melanesian arc system and the associated intra-arc basins are underlain by mafic oceanic rocks intruded and subsequently uplifted during the collision of the Pacific and Australia-India plates in Paleogene to Holocene times.

INTRODUCTION

The Solomon Islands and Bougainville, Papua New Guinea, form a nearly linear segment of the extensive island-arc chain lying along the boundary between the Pacific and Australia-India plates (Figure 1). The Solomon Islands are a double island chain separated by an intra-arc Miezitis, 1967).

The geologically complex island arc incorporates forearc and back-arc depositional structures as well as clearly-defined intra-arc basins. The geologic history of these structures has been dominated by major oceanic-plate interaction in the Cenozoic. The island arc probably developed on oceanic crust in early Tertiary time as the basin. The northeastern set of islands is believed to be floored by an early Cenozoic island arc, whereas the southwestern island chain is underlain principally by late Cenozoic volcanic rocks (Coulson and Vedder, 1986). In the Bougainville region, the late Cenozoic arc crosses to the northeastern side of the island and the early Tertiary arc lies on the southwestern edge of the island (Blake and result of southwest-directed subduction of the Pacific plate along the arc (Coleman and Packham, 1976; Katz, 1980; Kroenke, 1984; Coulson and Vedder, 1986; Vedder, 1986). In the Solomon Islands region of the arc, large back-arc debris fans formed along the flanks of the early Tertiary arc (northeastern island chain) and were later buried within

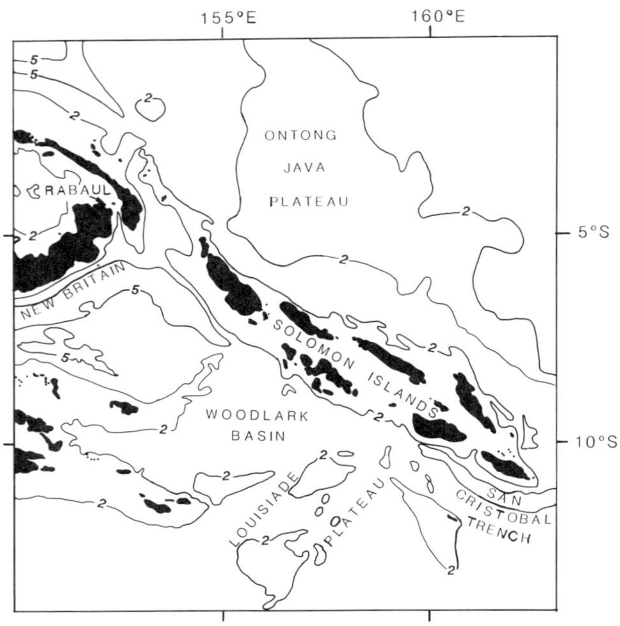

Figure 1. The Solomon Islands-Papua New Guinea region. A. Interpretation of the microplate configurations (from Cooper, Kroenke, and Resig, 1986); B. Physiographic map with bathymetry in km (from Cooper, Bruns, and Wood, 1986).

intra-arc basins by the development of the late Cenozoic volcanic centers (southwestern island chain).

The arc-forming rocks of the region have been uplifted and eroded along the northeastern and southwestern sides of the arc to create the exposed islands and the adjacent wave-planed basement platforms. These platforms are underlain by basinward-dipping strata and high-velocity volcanic and basement rocks (Cooper, Bruns, and Wood, 1986; Cooper, Marlow, and Bruns, 1986; Bruns et al, 1986).

The Solomon Islands-Bougainville segment of the Melanesian arc system is bordered on the northeast by the North Solomon-Kilinailau trench system and on the southwest by the San Cristobal-New Britain trench system. Trenches on the northeast side of the arc are believed to have been the site of active subduction from early Oligocene to early Miocene time (Kroenke, 1984). With the impingement of the Ontong Java Plateau against the arc in early Miocene time, subduction ceased, leading to a quiescent period that extended until late Miocene time. In late Miocene time, subduction shifted to the southwest side of the arc, shortly before the Pliocene opening of the Woodlark basin (Luyendyk, McDonald, and Bryan , 1973; Weissel, Taylor, and Karner, 1982) and Manus basin (Kroenke, 1984). Northeast-directed underthrusting of Woodlark Basin crust beneath the arc probably began soon after this opening began (Coleman and Kroenke, 1981; Luyendyk, McDonald, and Bryan, 1973; Weissel, Taylor and Karner, 1982; Kroenke, 1984).

Previous sonobuoy and two-ship seismic-refraction studies in the region (Furumoto et al, 1970, 1973, 1976; Ewing, Houtz, and Ludwig, 1970; Murauchi et al, 1973; Hussong, Wipperman, and Kroenke, 1979; Ravenne, deBroin, and Aubertin, 1982; Cooper and Wood, 1986; Cooper, Bruns, and Wood, 1986; Cooper, Marlow, and Bruns, 1986) were located principally in the intra-arc basins of the Solomon Islands and over the Ontong Java Plateau and nearby abyssal ocean floors. Few refraction stations are recorded in the areas around Bougainville. Marlow et al (1988) and Childs and Marlow (1988) list sonobuoy measurements for the New Ireland forearc basin, which lies northwest of Queen Emma basin, a deep-water basin that separates New Ireland and Bougainville. Previous studies have shown that basement rocks of the composite Melanesian island arc lie at shallow depths, with the exception of the intra-arc basins, where basement locally is covered by as much as 6 km of sedimentary rocks of Oligocene(?) to Holocene age (Katz, 1980; Vedder, Colwell, and Bruns, 1985; Bruns et al, 1986). Velocities of the basement rocks are indicative of volcanogenic and/or highly lithified sedimentary units. The basement rocks are believed to be underlain by mafic and ultramafic rocks of the 13- to 25-km-thick crust of the arc (Cooper, Bruns, and Wood, 1986).

During 1984, the U.S. Geological Survey (R/V *S.P. Lee* cruise L6-84-SP) recorded unreversed sonobuoy seismic stations at 39 sites on the Solomon Islands-Bougainville island-arc platform, on the flanks of the arc, in the intra-arc basins around Bougainville, and across the trenches adjacent to the northeastern and southwestern sides of the island arc (Figure 2). In this paper, we describe the new sonobuoy refraction-velocity results and use these results to address four general topics: 1) the shallow (0-10 km) velocity structure of the basement rocks underlying the island arc, 2) the thickness and velocity of the extensive volcanogenic units surrounding the late Cenozoic islands and extending into the intra-arc basins, 3) velocities of sedimentary units filling the intra-arc basins, especially around Bougainville, and 4) the structure of the trenches adjacent to the island arc.

METHODS

Refraction velocities for the 1984 sonobuoy stations (Figure 2) were computed using slope-intercept techniques (Childs and Cooper, 1978). Table 1 includes results from all sonobuoys except one; sonobuoy station 76 is excluded because, although good refraction data were recorded, the slope-intercept solutions could not be determined because of complex geologic structure. Parameters for the field-data collection and the processing procedures for analysis of the 1984 sonobuoy field data are given in Tables 2 and 3. The field techniques for the 1984 survey were similar to those used during a 1982 sonobuoy-seismic survey of the Solomon Islands region (Figure 3; Cooper and Wood, 1986).

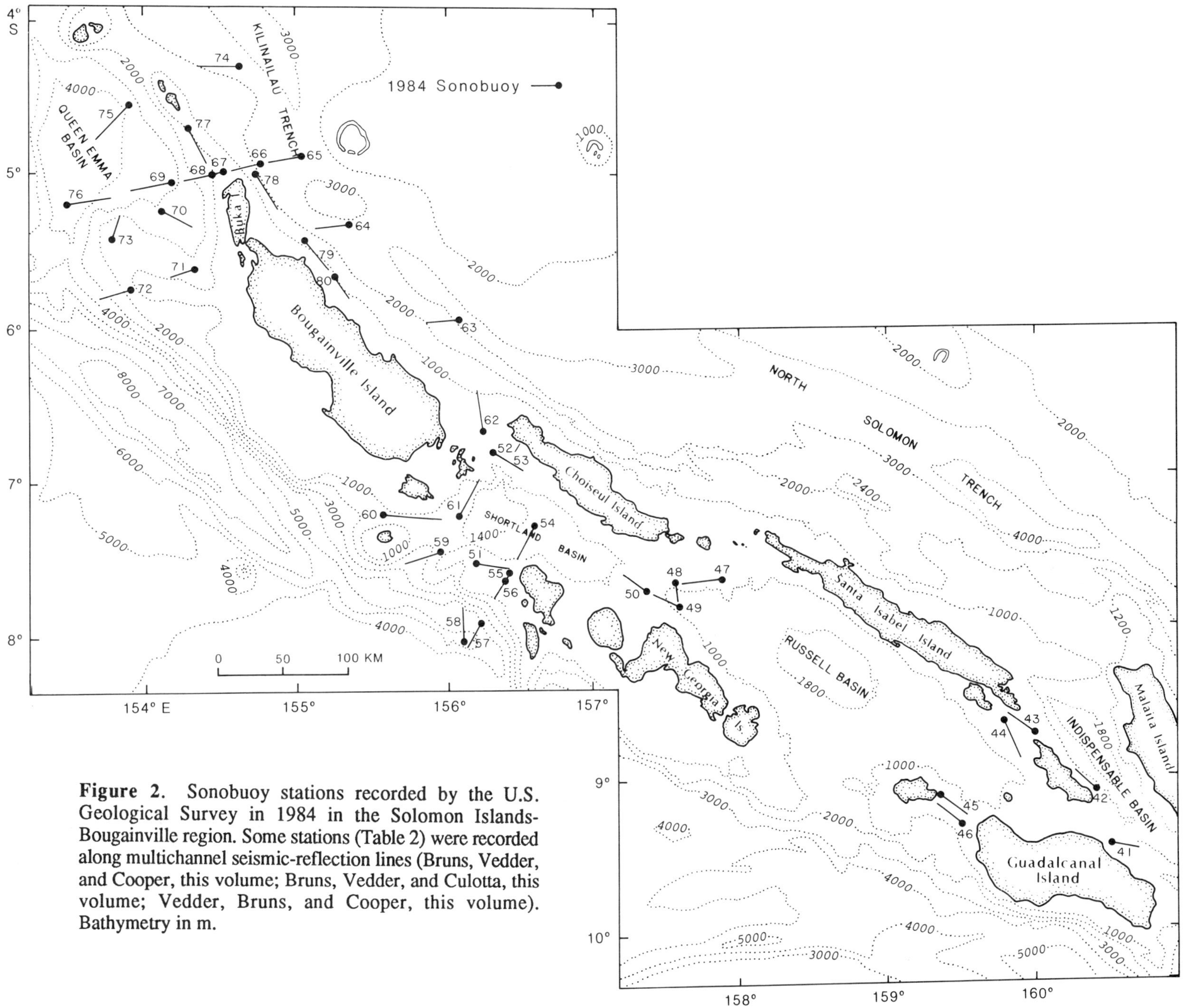

Figure 2. Sonobuoy stations recorded by the U.S. Geological Survey in 1984 in the Solomon Islands-Bougainville region. Some stations (Table 2) were recorded along multichannel seismic-reflection lines (Bruns, Vedder, and Cooper, this volume; Bruns, Vedder, and Culotta, this volume; Vedder, Bruns, and Cooper, this volume). Bathymetry in m.

Table 1. Seismic Refraction Solutions for Sonobuoy Measurements by the U.S. Geological Survey in the Solomon Islands during 1984.

Sonobuoy No.	Lat (S) Line no.	Long (E) Shotpoint	Water depth (m)	Layer 1 V_c (V_0) h D	Layer 2 V_c (V_0) h D	Layer 3 V_c (V_0) h D	Layer 4 V_c (V_0) h D	Layer 5 V_c (V_0) h D	Layer 6 V_c (V_0) h D	Layer 7 V_c (V_0) h D
41	9° 24.02′	160° 25.88′	373	1.60 (1.60*)	2.02 (2.02)	3.70(3.70)	6.67 (6.67)			
	-	-		0.41	0.34	0.11				
				0.0	0.0	0.0	0.0			
42	8° 59.44′	160° 16.42′	100	1.60 (1.60*)	2.00 (2.00)	3.87(3.87)				
	-	-		0.22	0.08					
				0.0	0.0	0.0				
43	8° 44.16′	159° 59.52′	183	1.60 (1.60*)	1.94 (1.94)	2.26(2.26)	4.54 (4.54)	6.26 (6.26)		
	-	-		0.04	0.28	0.45	1.48			
				0.0	0.0	0.0	0.0	0.0		
44	8° 38.09′	159° 51.84′	160	1.60 (1.60*)	2.30 (2.30)	4.63(4.63)	6.40 (6.40)			
	-	-		0.04	0.67	2.46				
				0.0	0.0	0.0	0.0			
45	9° 17.31′	159° 23.58′	440	1.60 (1.60*)	2.36 (2.36)	4.86(4.86)				
	-	-		0.18	0.69					
				0.0	0.0	0.0				
46	9° 17.27′	159° 31.17′	281	1.61 (1.60*)	2.28 (2.24)	3.46(3.37)	5.67 (5.51)			
	-	-		0.13	0.83	1.01				
				-0.74	-0.74	-0.74	0.0			
47	7° 39.97′	157° 50.25′	695	1.60 (1.60*)	2.29 (2.29)	4.10(4.10)	5.18 (5.18)			
	-	-		0.49	1.62	0.84				
				0.0	0.0	0.0	0.0			
48	7° 43.14′	157° 36.47′	945	1.60 (1.60*)	3.86 (3.86)	4.60(4.60)				
	-	-		0.33	5.00					
				0.0	0.0	0.0				
49	7° 50.54′	157° 38.14′	1085	1.60 (1.60*)	2.74 (2.74)	4.34(4.34)	5.30 (5.30)			
	-	-		0.17	0.18	3.30				
				0.0	0.0	0.0	0.0			
50	7° 44.70′	157° 22.84′	910	1.60 (1.60*)	3.58 (3.58)	3.87(3.87)	5.07 (5.07)			
	-	-		0.31	2.64	2.15				
				0.0	0.0	0.0	0.0			
51	7° 33.15′	156° 12.71′	735	1.60 (1.60*)	3.35 (3.35)	4.04(4.04)	6.63 (6.63)			
	-	-		0.09	3.46	1.82				
				0.0	0.0	0.0	0.0			
52	6° 53.86′	156° 26.37′	39	1.60 (1.60*)	2.27 (2.27)	5.32(5.32)				
	-	-		0.04	0.34					
				0.0	0.0	0.0				
53	6° 54.54′	156° 27.52′	85	1.92 (1.92)	2.34 (2.34)	4.42(4.42)	5.81 (5.81)			
	-	-		0.23	0.19	0.76				
				0.0	0.0	0.0	0.0			
54	7° 20.95′	156° 36.69′	1348	1.60 (1.60*)	1.81 (1.81)	3.37(3.37)	4.96 (4.96)			
	401	445		0.75	0.23	1.03				
				0.0	0.0	0.0	0.0			
55	7° 39.50′	156° 24.57′	650	1.60 (1.60*)	2.70 (2.70)	3.80(3.80)				
	401	1000		0.28	0.73					
				0.0	0.0	0.0				
56	7° 40.21′	156° 24.06′	679	1.60 (1.60*)	4.40 (5.56)					
	401	1236		0.65						
				1.3	4.5					
57	7° 56.29′	156° 14.07′	4012	1.60 (1.60*)	2.62 (2.43)	4.57(4.12)				
	401	1943		1.26	0.46					
				-4.4	-3.5	-1.2				
58	7° 59.28′	156° 07.11′	4398	4.37 (4.29)	5.12 (5.01)	6.23(6.01)				
	402	270		3.46	1.07					
				-0.43	-0.43	-0.43				
59	7° 29.30′	155° 52.44′	842	1.60 (1.60*)	2.00 (2.01)	3.29(3.24)	4.67 (4.26)			
	402	547		0.20	0.48	2.05				
				0.4	0.4	-0.9	-5.0			
60	7° 14.39′	155° 37.96′	1129	1.60 (1.60*)	4.80 (4.80)					
	406	737		0.51						
				0.0	0.0					

V_0 is the observed velocity in km/s.
* indicates assumed velocity.
V_c is the velocity corrected for layer dip.
h is the layer thickness in km.
D is the layer dip in degrees (negative sign indicates increasing depth along shotline).

Table 1. (continued)

Sonobuoy No.	Lat (S) Line no.	Long (E) Shotpoint	Water depth (m)	Layer 1 V_c (V_0) h D	Layer 2 V_c (V_0) h D	Layer 3 V_c (V_0) h D	Layer 4 V_c (V_0) h D	Layer 5 V_c (V_0) h D	Layer 6 V_c (V_0) h D	Layer 7 V_c (V_0) h D
61	7° 13.11′	156° 07.46′	1373	1.60 (1.60*)	1.92 (1.91)	2.78 (2.83)	4.68 (4.87)			
	407	28		0.58	0.70	0.84				
				0.0	-0.5	1.1	1.1			
62	6° 37.19′	156° 14.13′	491	1.60 (1.60*)	1.90 (1.90)	3.66 (3.66)	5.01 (5.01)			
	408	73		0.32	0.29	1.56				
				0.0	0.0	0.0	0.0			
63	5° 58.07′	156° 05.10′	3036	1.59 (1.60*)	1.86 (1.88)	2.15 (2.18)	4.72 (4.46)	5.17 (4.84)	6.13 (5.66)	
	409	62		0.59	0.15	0.20	1.13	3.14		
				0.9	0.8	0.7	-2.2	-2.2	-2.2	
64	5° 18.20′	155° 16.78′	3304	1.60 (1.60*)	2.12 (2.12)	2.71 (2.71)	5.73 (5.73)			
	411	240		0.61	0.40	1.44				
				0.0	0.0	0.0	0.0			
65	4° 50.09′	155° 02.40′	3340	1.59 (1.60*)	2.14 (2.17)	2.30 (2.41)	2.54 (2.74)	3.58 (3.84)	4.17 (4.52)	
	413	57		0.19	0.43	0.31	0.74	0.44		
				0.9	0.9	5.1	5.1	1.3	1.3	
66	4° 53.59′	154° 46.73′	2140	1.60 (1.60*)	2.20 (2.20)	2.94 (2.92)	3.22 (3.20)	4.29 (5.62)		
	413	586		0.48	0.57	0.49	1.14			
				-0.4	-0.1	-0.1	-0.1	13.9		
68	4° 57.26′	154° 30.71′	680	1.64 (1.60*)	2.13 (2.02)	3.34 (3.04)	5.64 (4.78)			
	413	1146		0.27	0.65	1.14				
				-3.15	-3.15	-2.77	-2.77			
69	5° 02.42′	154° 11.50′	2850	1.61 (1.60*)	2.53 (2.38)	3.75 (3.21)	4.61 (3.78)			
	413	1829		0.87	0.62	2.23				
				-1.0	-3.0	-6.0	-6.0			
70	5° 15.50′	154° 10.39′	2727	2.31 (2.35)	3.15 (3.49)	3.73 (4.02)	5.78 (8.56)			
	414	551		1.06	1.67	4.10				
				-2.3	1.9	-3.2	7.8			
71	5° 36.65′	154° 21.39′	1567	1.60 (1.60*)	2.18 (2.18)	3.93 (3.93)				
	415	494		0.15	0.43					
				0.0	0.0	0.0				
72	5° 46.60′	153° 53.85′	2250	1.69 (1.60*)	3.07 (2.53)	5.08 (3.64)				
	415	1507		1.19	1.13					
				-7.1	-7.1	-7.1				
73	5° 24.71′	153° 48.59′	2021	1.60 (1.60*)	5.15 (5.15)					
	416	835		0.34						
				0.0	0.0					
74	4° 17.35′	154° 29.71′	859	1.70 (1.60*)	1.97 (1.80)	3.22 (2.61)	4.81 (3.48)	5.98 (4.02)		
	418	630		0.12	0.04	1.19	1.18			
				-7.4	-6.4	-7.6	-7.6	-7.6		
75	4° 38.31′	153° 46.84′	4145	1.60 (1.60*)	1.73 (1.73)	2.7 (2.70)	3.91 (3.91)			
	419	910		0.14	0.43	1.80				
				0.0	0.0	0.0	0.0			
77	4° 55.72′	154° 32.01′	700	1.61 (1.60*)	2.63 (2.58)	3.56 (3.46)	5.21 (4.98)	6.08 (5.76)	6.18 (5.86)	8.05 (7.50**)
	-	-		0.28	0.74	0.97	0.45	0.60	2.30	
				-0.8	-0.8	-0.8	-0.8	-0.8	-0.8	-0.8
78	4° 58.35′	154° 47.08′	2414	1.60 (1.60*)	1.95 (1.98)	2.76 (2.83)	5.25 (5.66)			
	-	-		0.28	0.29	0.97				
				0.0	1.0	1.0	1.6			
79	5° 23.38′	154° 58.52′	1720	1.60 (1.60*)	2.16 (2.16)	2.88 (2.88)	5.04 (5.04)			
	-	-		0.25	0.54	1.32				
				0.0	0.0	0.0	0.0			
80	5° 34.30′	155° 12.90′	2459	1.59 (1.60*)	2.66 (2.71)	3.60 (3.71)	4.58 (4.77)	7.48 (8.03**)		
	-	-		0.07	0.80	0.62	3.63			
				0.8	0.8	0.8	0.8	0.8		

V_0 is the observed velocity in km/s.
* indicates assumed velocity.
V_c is the velocity corrected for layer dip.
h is the layer thickness in km.
D is the layer dip in degrees (negative sign indicates increasing depth along shotline).
** see text.

Table 2. Field Recording Parameters

Sonobuoy type: U.S. Navy type 41-B (176 MHz)

Normal maximum offset-range (ship to sonobuoy): 20-30 km

Station geometry: All stations unreversed

Seismic source:
- Stations 41-53 : One 1,000 in^3 airgun
- Stations 49-50 : Five airguns totaling about 2,350 in^3.
- Stations 54-76 : Five airguns totaling about 1,300 in^3.
- Stations 77-80 : One 1,000 in^3 airgun

Shot interval:
- Stations 41-53 : 30 s (about 100 m)
- Stations 54-76 : 50 m
- Stations 77-80 : 30 s (about 100 m)

Seismic-reflection data:
- Stations 41-53 : 3.5 kHz and 12 kHz bathymetry only
- Stations 54-76 : Multichannel seismic-reflection data
- Stations 77-80 : Single-channel seismic-reflection data

Table 3. Processing Procedures

I. Slope-intercept analysis: (All sonobuoys)

A. Select and digitize water arrival (D-wave) and refractions (linear segments) from subsurface rock units
B. Compute trial velocities assuming uniformly thick layers with no dip
C. Determine layer dips from seismic-reflection profiles
D. Correct trial velocities for dip
E. Evaluate effects of geologic structure on velocities and discard velocities for layers that violate assumptions of the slope-intercept technique

II. Ray-tracing procedure: (Sonobuoys 49 and 50, this paper; sonobuoys 63, 65, and 66, Cochrane, Cooper, and Bruns, this volume)

A. Construct a trial geologic model based on geometry from seismic-reflection data and velocities from slope-intercept analysis.
B. Adjust the geometry and velocities of the model to obtain a good fit between the observed refraction and reflection arrival times and the arrival times computed using the model.
C. Evaluate the model to insure that final velocities and structures are realistic in terms of regional geologic framework.

Several sonobuoys in the present study were interpreted using a ray-tracing technique (McMechan and Mooney, 1980) to more accurately determine the crustal velocities required to match the observed seismic arrival times, especially in areas of complex geologic structure. The ray-tracing models are discussed in the following sections.

VELOCITY STRUCTURE OF THE ISLAND ARC

Several sonobuoy stations were recorded across the shallow basement platforms adjacent to the islands. These sonobuoys were sited in areas where acoustic basement in seismic-reflection data was known to be shallow and where good refractions in sonobuoy data were expected from high-velocity intrusive, metasedimentary, or ultramafic rocks close to the sea floor. Sonobuoy-seismic data from these areas generally confirmed that acoustic basement rocks have velocities greater than 4.4 km/s and that locally, intrusive bodies of higher velocity rock are present.

Central Solomon Islands

Six sonobuoy stations (41-46) and concurrent high-resolution (3.5 kHz and 12 kHz) bathymetric data were recorded on the basement platforms near Santa Isabel, the Florida Islands, and Guadalcanal (Figures 2 and 4). Because seismic-reflection data were not acquired, the geometry of subsurface bodies deeper than a few tens of meters below the sea floor is unknown. Multichannel seismic-reflection data collected nearby in 1982 (Bruns et al, 1986; Bruns, Cooper, and Vedder, 1986; Cooper, Bruns, and Wood, 1986; Cooper, Marlow, and Bruns, 1986), show little penetration into the arc-framework basement rocks. Velocities in excess of 4.4 km/s, for likely basement rocks, were measured at all stations except sonobuoy 42, where complex bathymetry and noisy sonobuoy records yielded only shallow, low-velocity refractions (1.6 km/s to 3.9 km/s) (Figure 4). On the north side of Guadalcanal (sonobuoy 41), about 850 m of probable sedimentary rocks, with velocities less than 3.8 km/s overlie a high-velocity, 6.7 km/s basement. Velocities of 6.3 km/s and 6.4 km/s were also recorded near Santa Isabel (sonobuoys 43 and 44). However, in both of these areas, and elsewhere in the central Solomon Islands region, velocities in excess of 6.0 km/s occur beneath acoustic basement recorded on seismic-reflection profiles. The absence of rocks with velocities of about 5.0 km/s near Guadalcanal (sonobuoy 41) suggests that the highly-indurated sedimentary or volcanic units, typical of basement in other parts of the Solomon region, are not present. Rather, the 6.7 km/s rocks may be equivalent to mafic volcanic and intrusive rocks, possibly similar to the upper Mesozoic Mbirao Volcanics and older intrusives on Guadalcanal (Hackman, 1980).

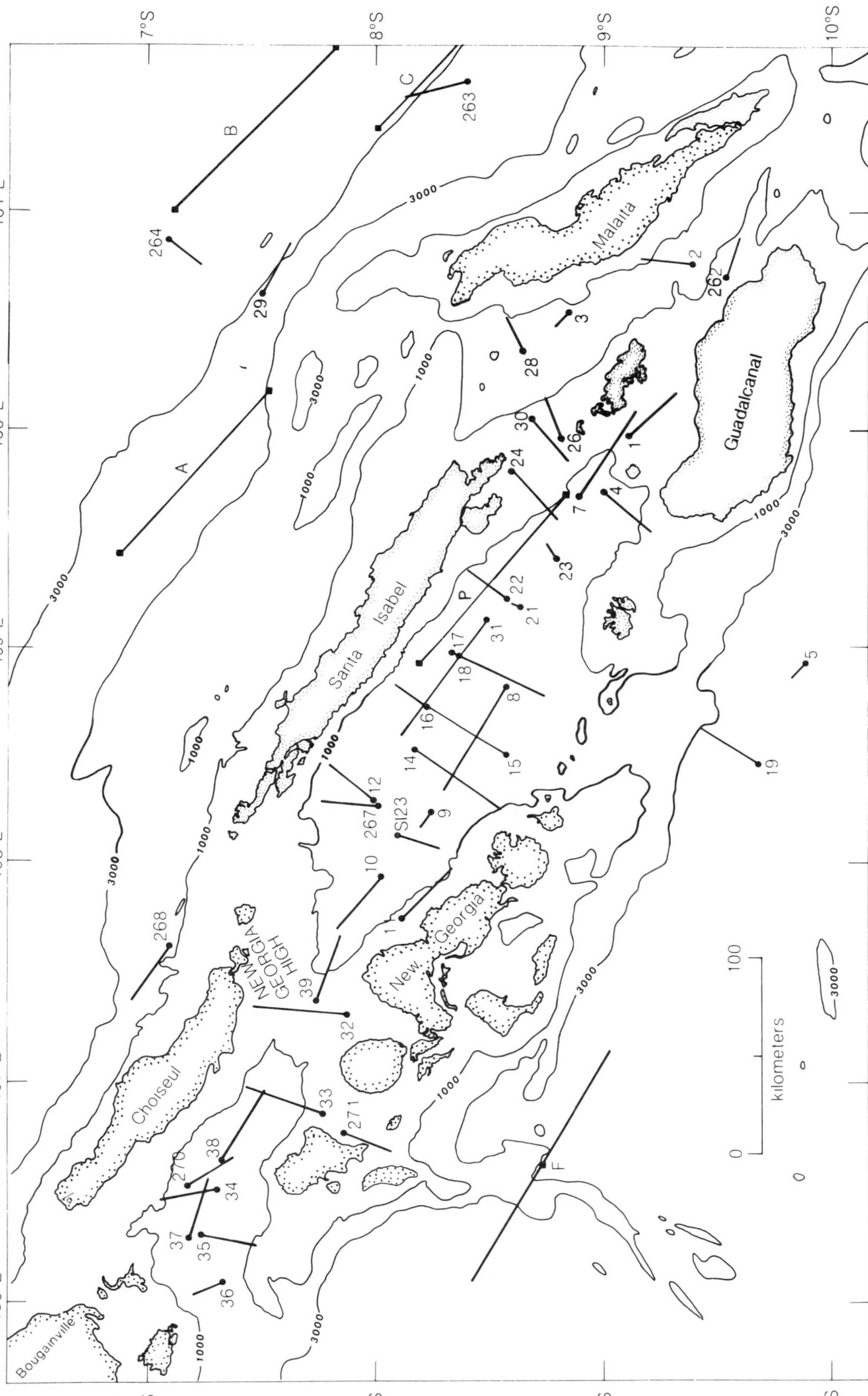

Figure 3. Pre-1984 sonobuoy and 2-ship refraction stations in the central and western Solomon Islands (from Cooper, Bruns, and Wood, 1986). Bathymetry in m.

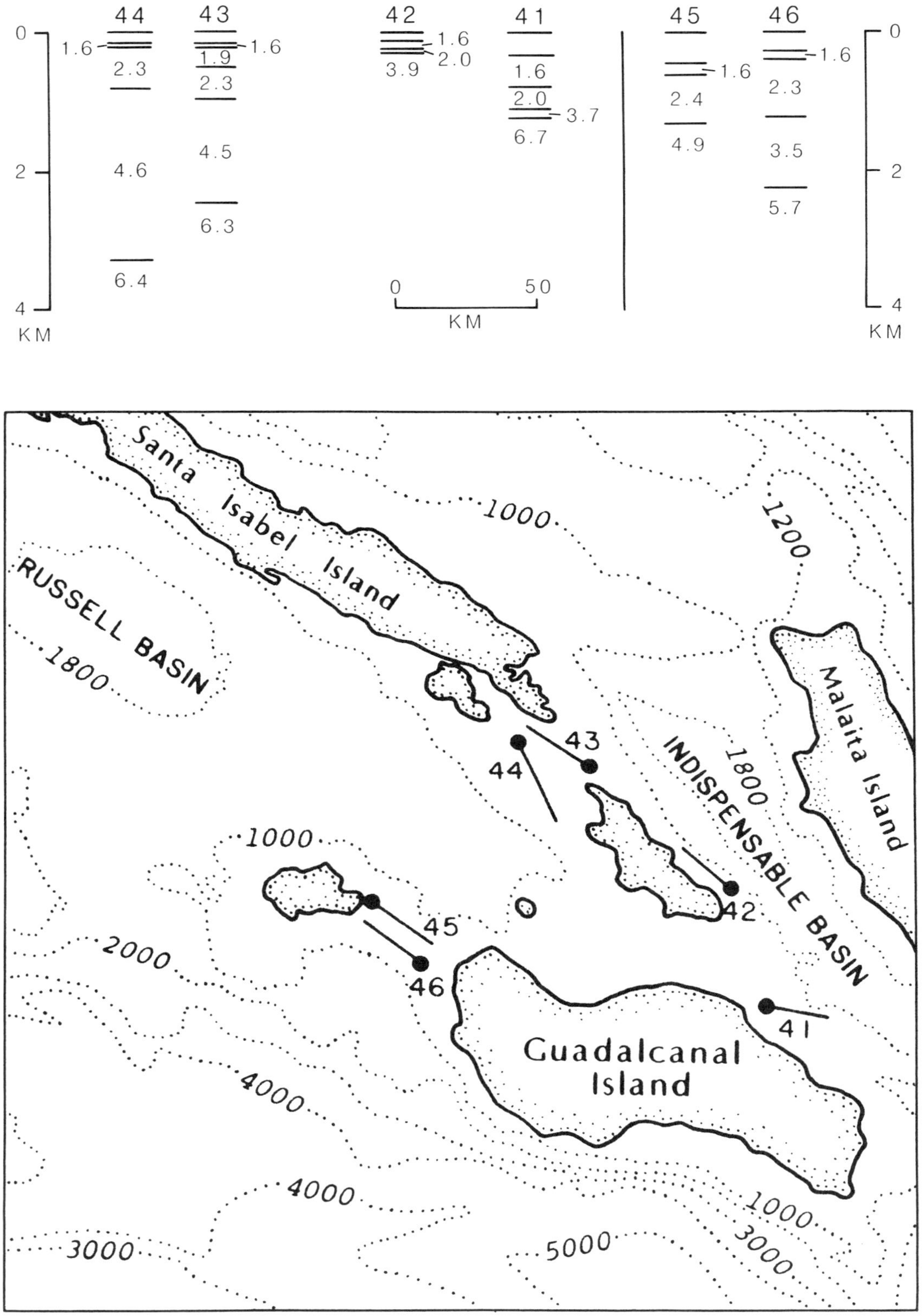

Figure 4. Refraction-velocity results for sonobuoy stations over shallow arc-framework basement rocks in the Santa Isabel and Guadalcanal areas. See Table 1 for numerical listing and Figures 1 and 2 for regional location. Velocities in km/s; water depths in m.

Multichannel seismic-reflection profiles and sonobuoy-refraction data across the southeastern end of the Santa Isabel platform show that shallow ultramafic rocks may lie within a fault zone that trends northwest into onshore structures (Cooper, Marlow, and Bruns, 1986). Sonobuoys 43 and 44 were recorded on the southwest (intra-arc basin) side of the fault zone, parallel with regional structure, to verify and delineate the existence of possible offshore ultramafic bodies similar to those on Santa Isabel (Pound, 1986). Refraction arrivals from the two sonobuoys are disrupted and antiformal in character, showing clear evidence of a subsurface structure, possibly an intrusive body. However, sonobuoy 44 also recorded a refraction arrival (a short, second-arrival segment) from the body with an apparent velocity between 7.5 and 8.0 km/s. This velocity is not included on Table 1 because the depth and geometry of the body cannot be determined using slope-intercept analysis. The existence of the antiformal structure and the high apparent velocity suggests that local ultramafic bodies may lie at shallow (several hundred meters) subsurface depths in the offshore areas.

Sonobuoys 45 and 46 were recorded across a bathymetric saddle lying between Guadalcanal and the Russell Islands. Late Cenozoic volcanic rocks and Quaternary limestone are exposed on the islands near both sonobuoy stations (Coulson and Vedder, 1986). The sonobuoy results show a 0.9- to 2.0-km-thick sedimentary section with velocities less than 3.6 km/s overlying probable volcanogenic basement rocks with velocities of 4.9-5.7 km/s. Although there is some variance in the basement velocities, the good continuity of refraction arrivals for the entire length of both sonobuoys suggests that there is no major structural discontinuity in the region between Guadalcanal and the Russell Islands.

Western Solomon Islands

Several sonobuoy stations were sited along the southwestern (51, 55, 56, 59, 60) and northeastern (52, 53, 62) sides of the Central Solomons Trough near Bougainville (Shortland basin) (Figures 2 and 5). Five of these stations lie along multichannel seismic-reflection profiles that cross the regional northwest-southeast structural grain. The other three sonobuoy stations (51-53) were recorded without vertical-incidence seismic-reflection control but are parallel to regional structure.

Sonobuoy stations on the southwestern side of the Solomon Islands arc between the New Georgia island group and the Shortland Islands show that the acoustic basement in multichannel seismic-reflection records (Bruns et al, 1986; Bruns, Vedder, and Cooper, this volume) has velocities ranging from 4.0 km/s (51) to 4.8 km/s (60). These velocities are slightly lower than those observed for basement elsewhere in the Solomon Islands (4.5-5.5 km/s, Cooper, Bruns, and Wood, 1986; Cooper, Marlow, and Bruns, 1986), except near the central New Georgia island group where thick volcanic units are likely to be present. The 4.0-4.8 km/s velocities may indicate the occurrence of massive limestone beds, but we suspect that these velocities represent volcaniclastic debris because late Cenozoic volcanic rocks are abundant and limestone occurs only as Quaternary reefs on the nearby islands. Relatively thin (tens of meters) limestone beds probably lie close to the sea floor near Vella Lavella because calcareous rocks were dredged nearby (Colwell and Vedder, 1986) and because a relatively-high-velocity (3.4 km/s) thin layer was detected near the sea floor in refraction data at sonobuoy station 51.

From refraction data, the sedimentary section beneath Vella Ridge is at least 3.5 km thick near Vella Lavella (station 51, Figure 5) and thins northwestward to about 0.5 km near Bougainville (station 60). The section near station 51 may be thinner, however, based on multichannel seismic-reflection interpretation (3.0 km, Bruns et al, 1986) and on our interpretation that undetected low-velocity layers may lie within the sedimentary section beneath station 51. The low-velocity layers are suggested by the abrupt termination of shallow refraction arrivals. The presence of low-velocity sediment layers would yield smaller sediment-layer thicknesses than those shown in Table 1. The New Georgia wedge, northeast of the New Georgia island group, has interbedded low-velocity layers (volcanics and sediments, Figure 6) and is a possible analog for the sedimentary section near Vella Lavella.

Near Bougainville, the acoustic basement is shallow (500 m sub-seafloor) and is marked by a continuous, nearly linear, high-amplitude refraction arrival with a velocity of 4.8 km/s. The acoustic character and velocity of the refraction arrival suggest the presence of thick volcanic units without interbedded sedimentary layers. These volcanic units are probably the offshore continuation of the volcanic rocks that occur in southeastern Bougainville. At station 60, we interpret the shallow volcanic(?) basement rocks as part of the arc framework, rather than as shallow volcanic flows that cover an older and low-velocity sedimentary section.

Two sonobuoy stations (52, 53) were recorded on the northeastern side of the Solomon Islands arc in the shallow-water platform area south of Choiseul. The location of the sonobuoys is in an area where previous gravity surveys mapped a very large (210 mgal) positive gravity anomaly, which was thought to be caused by a large mass of relatively dense intrusive and (or) metamorphic rocks at shallow depth (Beyer, 1986). Sonobuoy station 52 recorded a 5.3 km/s refraction arrival from basement. Nearly continuous but wavy refraction arrivals were recorded to a sonobuoy-offset range of 22 km at station 53 and are interpreted as indicating that an antiformal structure with high-velocity rocks (5.8 km/s) underlies this part of the platform. The 5.8 km/s velocity may signify the presence of mafic intrusive or volcanic rocks, but the velocity is too low to correspond to ultramafic rocks similar to those exposed on Choiseul.

A thin or narrow ultramafic body may lie beneath station 62 near the northwestern end of Choiseul. Here, as at the southeastern end of the island (sonobuoy 44), a short second-arrival in the refraction data yields a velocity greater

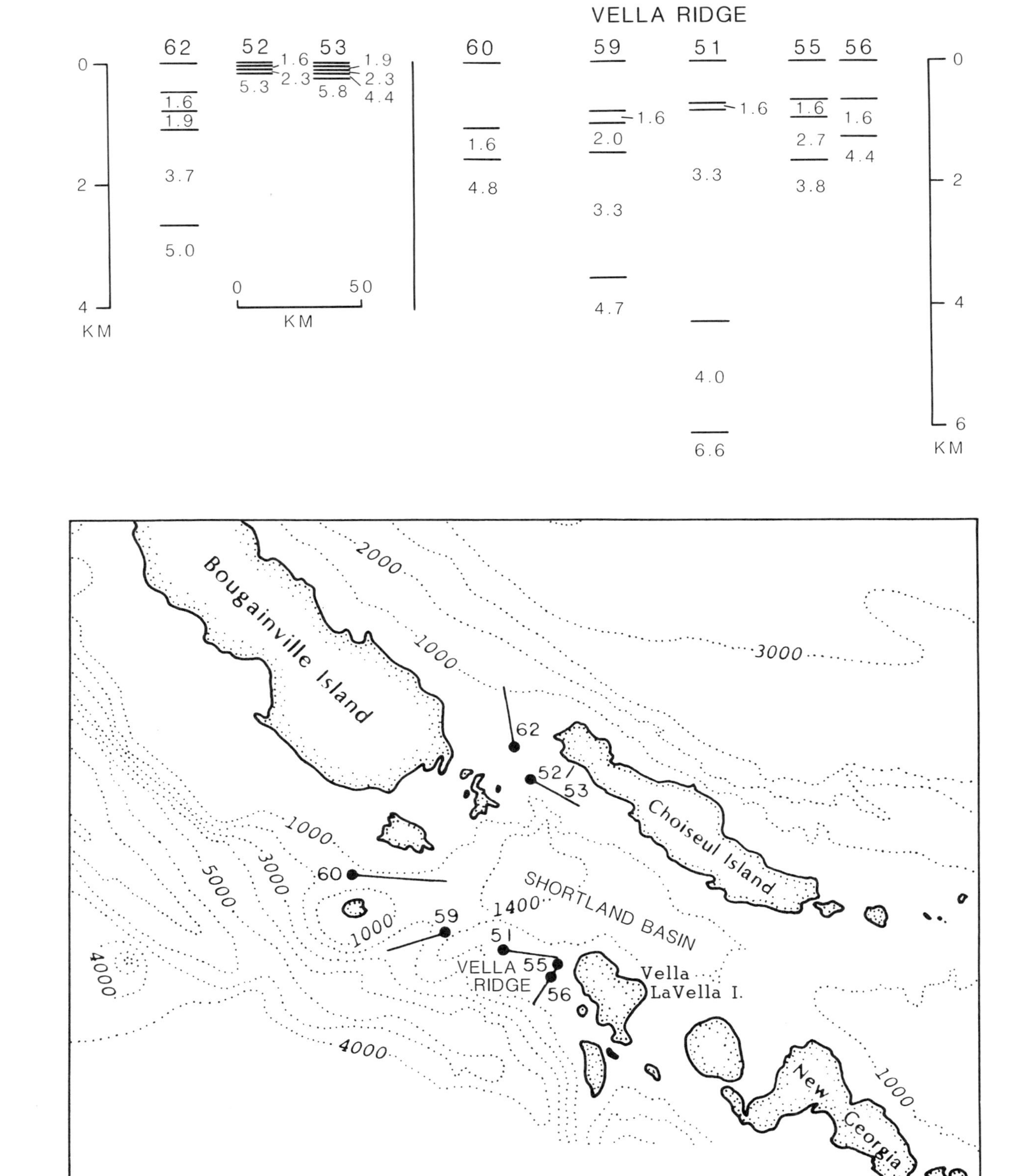

Figure 5. Refraction-velocity results for sonobuoy stations over shallow arc-framework basement rocks at the northwestern end of the Solomon Islands. See Table 1 for numerical listing and Figures 1 and 2 for regional location. Velocities in km/s; water depths in m.

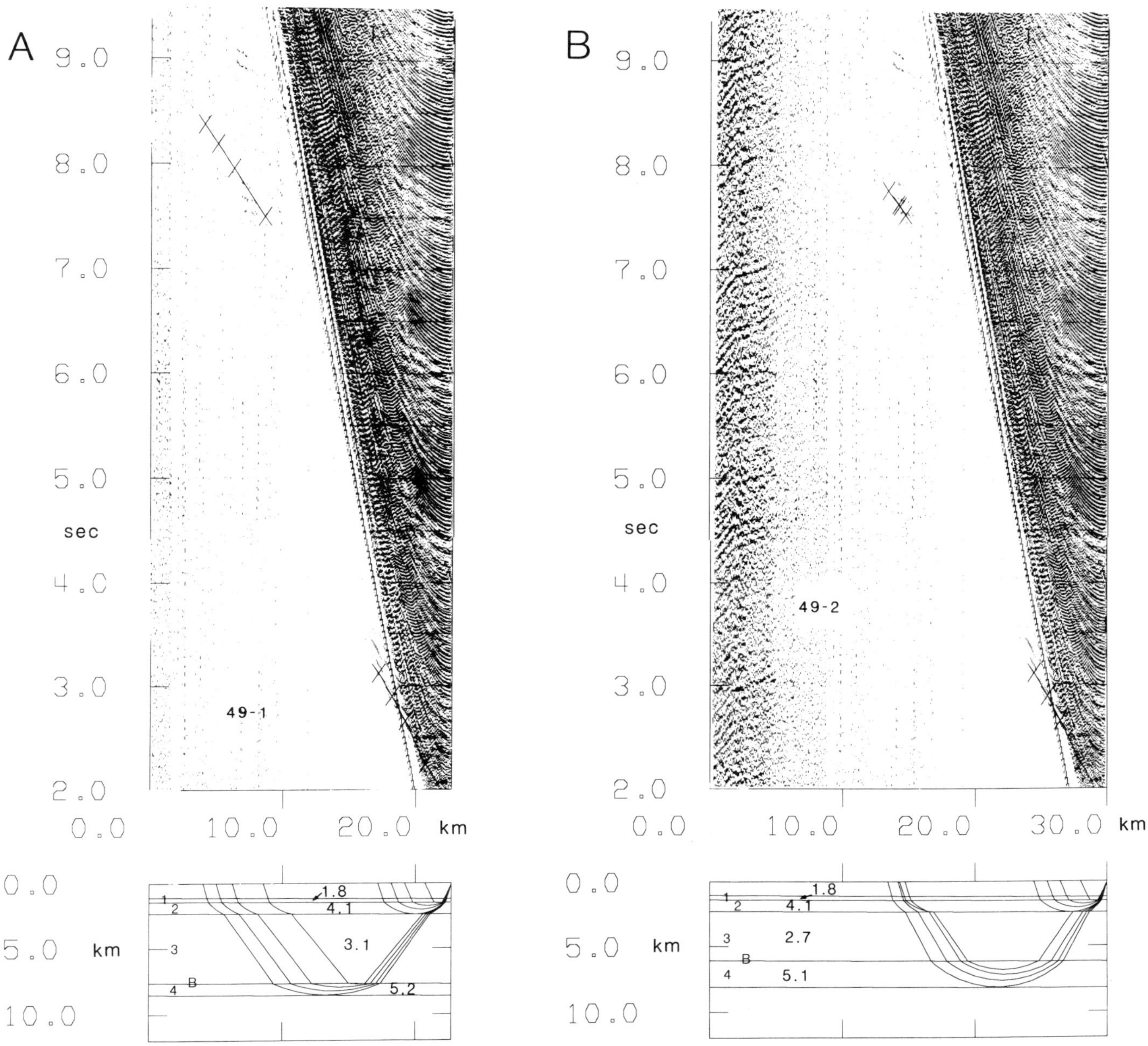

Figure 6. Seismic ray-tracing models for sonobuoy station 49 over the New Georgia wedge, Central Solomons Trough (See Figures 2 and 7 for location). A. Model 49-1 and field data showing higher sedimentary velocities (3.1 km/s) and deeper basement (5.6 km) than model 49-2; B. Model 49-2 and field data showing lower sedimentary velocities (2.7 km/s) and shallower basement (5.1 km) than model 49-1. These models illustrate the abrupt termination and time-offset of refraction arrivals caused by the presence of low-velocity sedimentary(?) rocks (layer 3) underlying a shallow, thin high-velocity volcanic(?) layer (layer 2). Layer velocities and thicknesses cannot be uniquely determined where low velocity zones are present. In general, high sedimentary velocities (for layer 3) require deeper basement depths (to layer 4) to give correct refraction arrival times. Xs are computed times. Numbers are average layer velocities in km/s. See Table 4 for specific layer velocities and thicknesses. "B" is top of basement.

than 7.5 km/s near the top of a feature that may represent an antiformal intrusive body. Deformation is not apparent in multichannel seismic-reflection data at station 62, indicating that the suspect high-velocity body does not extend to the sea floor. However, the body's depth and geometry cannot be determined from slope-intercept methods. Alternatively, the high-velocity refraction arrivals could be explained by large basement dips or by wide-angle reflections or reflected refractions from nearby basement structures.

Bougainville, Buka, and the Green Islands

Six sonobuoy stations were recorded along the crestal platform (68, 77) and the east flank (74, 78, 79, 80) of Bougainville, Buka, and the Green Islands to measure velocities for arc-framework rocks (Figures 2 and 7). The central and northern parts of Bougainville, as well as Buka and the Green Islands, are believed to be underlain in large part by late Cenozoic volcanic rocks that cut diagonally across the early Cenozoic arc beneath Bougainville (Blake and Miezitis, 1967; Kroenke, 1984).

Two sonobuoys (77, 68) were recorded between Buka and the Green Islands. These islands lie at the southeastern end of Emirau-Feni Ridge, which, farther northwest, forms an outer high bounding the New Ireland basin (Marlow et al, 1988). Sonobuoy 77 recorded high-amplitude refraction arrivals along the crest of Emirau-Feni Ridge and gave rock velocities to subsurface depths of 5.5 km. Rocks within 2.0 km of the sea floor are likely to be non-indurated sedimentary and/or volcanic units (1.6-3.6 km/s), and they overlie basement (5.2 km/s) and intra-basement layers (6.1-6.2 km/s) that have a combined thickness of about 3.4 km. A refraction velocity of 7.5 km/s was also recorded; however, this velocity could result from relief on intra-basement layers. Vertical-incidence seismic-reflection data from the single-channel profiles that parallel the ridge crest could not resolve structures deeper than 1 to 2 km beneath the surface.

Sonobuoy 68 was recorded close to station 77 but down the southwestern flank of Emirau-Feni Ridge. Multichannel seismic-reflection data at station 68 show chaotic reflectors directly below the sea floor (Vedder, Bruns, and Cooper, this volume), yet on the lower flanks of the ridge, these reflectors are increasingly evenly layered. The sonobuoy data indicate that this chaotic-to-layered unit is about 2.1-km-thick and has velocities of 2.1-3.3 km/s, suggesting a sedimentary origin, even on the ridge crest. Part of this layered unit is interpreted as being correlative with the Oligocene and early Miocene(?) strata of the onshore Buka Formation (Vedder, Bruns, and Cooper, this volume). Rocks with a velocity of 5.6 km/s lie directly below the sedimentary units and correspond with acoustic basement that is likely to consist of volcanogenic rocks beneath the ridge flanks.

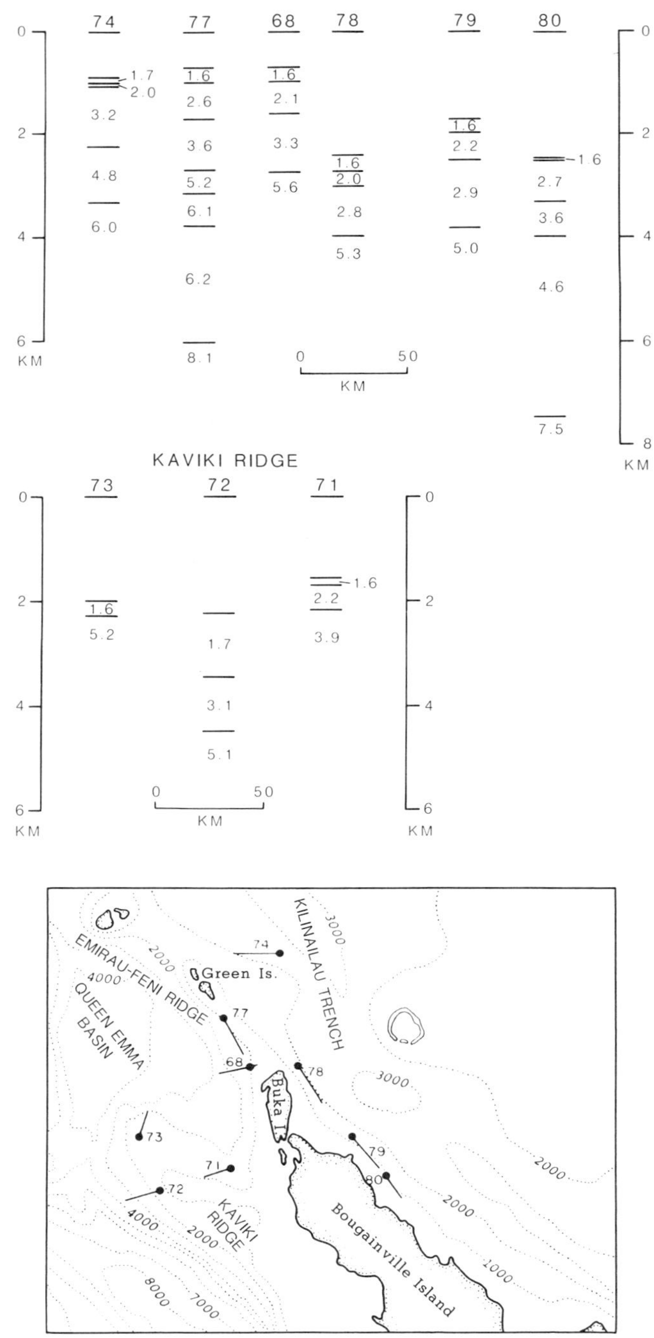

Figure 7. Refraction-velocity results for sonobuoy stations over shallow arc-framework basement rocks near the northwestern end of Bougainville. See Table 1 for numerical listing and Figures 1 and 2 for regional location. Velocities in km/s; water depths in m.

Sedimentary rocks overlie the basement ridge that lies about 25 km east of the Green Islands (Figure 7). Sonobuoy 74 was recorded down the western flank of this ridge, where multichannel seismic-reflection data show well-layered strata about 160 meters thick overlying disrupted reflectors that delineate the acoustic basement. The refraction data indicate that the upper 1.2 km of acoustic basement may be sedimentary units (3.2 km/s) that overlie deeper volcanogenic basement layers (4.8-6.0 km/s). The acoustic basement rocks (3.2 km/s) may be equivalent to the deformed and uplifted sedimentary units that occur 100 km to 250 km farther southeast on the landward side of the Kilinailau Trench (Cochrane, Cooper, and Bruns, this volume; Bruns, Vedder and Culotta, this volume).

Three successive sonobuoy stations (78, 79, 80) were recorded about midslope and parallel with the coast of northern and central Bougainville. The sonobuoys were located within 30 km of the Kilinailau Trench, with the hope that refraction arrivals might be recorded from old oceanic crust that presumably is underthrust beneath Bougainville. However, the combination of insufficient ship-to-sonobuoy offset range, irregular canyon bathymetry along the sonobuoy stations, and the likely presence of thick volcanic flows covering the arc flank, prevented acoustic penetration beneath the top of the volcanic flows (4.6-5.3 km/s) at subsurface depths of 1.5-2.1 km. Sonobuoy 80 recorded a weak refraction second-arrival with a 8.0 km/s velocity at a depth of 5.1 km; however, this apparent velocity (and depth) is suspect and may result from structural relief on basement rocks at great depth.

Three sonobuoys (71-73, Figure 7) were recorded across Kaviki Ridge, the structural high that extends westward from Bougainville. This feature lies along the structural trend of onshore Paleogene volcanogenic rocks and may be the offshore continuation of the early Tertiary Solomon Islands volcanic arc (Vedder, Colwell, and Bruns, 1985; Vedder, Bruns, and Cooper, this volume).

Multichannel seismic-reflection data suggest that a thick sedimentary section underlies station 71. Buried strata beneath the station are folded; consequently, refraction velocities could only be determined for the upper few hundred meters of the section (1.6-3.9 km/s). At station 72, on the western flank of the ridge, refraction data indicate that the acoustically disrupted strata may be composed of deformed or well-indurated sedimentary strata. A reliable velocity of 5.2 km/s was measured for basement rocks beneath Kaviki Ridge at station 73, where basement was flat lying. Based on limited refraction data, the ridge is probably underlain by Paleogene basement rocks that have higher velocities than the younger(?) basement rocks beneath the northeastern flank of Bougainville.

VELOCITY STRUCTURE OF THE INTRA-ARC BASINS

Several intra-arc basins lie along the island arc in the Solomon Islands-Bougainville region. These basins contain Cenozoic sedimentary units that may have been deposited as overlapping debris fans originating from the adjacent uplifted island areas (Katz, 1980; Ravenne, deBroin, and Aubertin, 1982; Bruns et al, 1986). Distinct velocity discontinuities occur in the sedimentary sequence that fills the Central Solomons Trough, a structural depression lying between the double island chain of the Solomon Islands (Cooper, Bruns, and Wood, 1986). These discontinuities generally delineate the tops of sedimentary packages that were derived from high-standing platform areas on opposite sides of the basin (Bruns et al, 1986). The layered sedimentary units in the Central Solomons Trough are believed to be Oligocene and younger. Velocities of 1.6 to 2.6 km/s (Pliocene and younger) occur above a prominent velocity discontinuity beneath which velocities are 2.8 to 4.4 km/s (Cooper, Bruns, and Wood, 1986). The 1984 refraction data for the intra-arc basins are located in the Central Solomons Trough near the New Georgia island group (47, 48, 49, 50), near the Shortland Islands (54, 61), and in Queen Emma basin (69, 70, 75) (Figure 8).

Central Solomons Trough, New Georgia-Shortland Islands Area

Previous seismic studies showed conflicting results regarding the thickness of the sedimentary section beneath the Central Solomons Trough directly north of the New Georgia island group. Multichannel seismic-reflection data were interpreted as showing 1.5-2.0 km of strata over acoustic basement (Figure 6 of Bruns et al, 1986), whereas sonobuoy-seismic data show more than 5.0 km of 1.6-4.4 km/s rock overlying basement (Figure 6 of Cooper, Bruns, and Wood, 1986). This apparent discrepancy was explained by the likely existence of a sedimentary body of lower velocity strata, named the New Georgia wedge, underlying a near-surface, high-velocity layer, which forms acoustic basement on the multichannel seismic-reflection data (Cooper, Bruns, and Wood, 1986).

During the 1984 survey, a larger seismic-energy source (2,350 in^3 rather than 1,300 in^3 used in 1982) was used to attain greater penetration through the New Georgia wedge. The two new sonobuoy stations (stations 49 and 50) show only slightly better penetration than the 1982 sonobuoy station 39 (between stations 49 and 50). The similarity between the 1982 and 1984 sonobuoy records confirms that the New Georgia wedge is a broad feature comprised of thick highly dispersive acoustic units of probable interlayered sedimentary and volcanic origin.

Ray-tracing models for sonobuoy stations 49 and 50 are included in Figures 6 and 9 to illustrate possible layer velocities and thicknesses within the wedge. Two models for station 49 show that unique velocities and thicknesses cannot be determined where lower velocity strata (layer 3) underlie high-velocity rocks (layer 2), in effect causing low-velocity zones (Figure 6). However, the possible combinations of layer velocities and layer thicknesses is limited because seismic rays in the models must arrive at

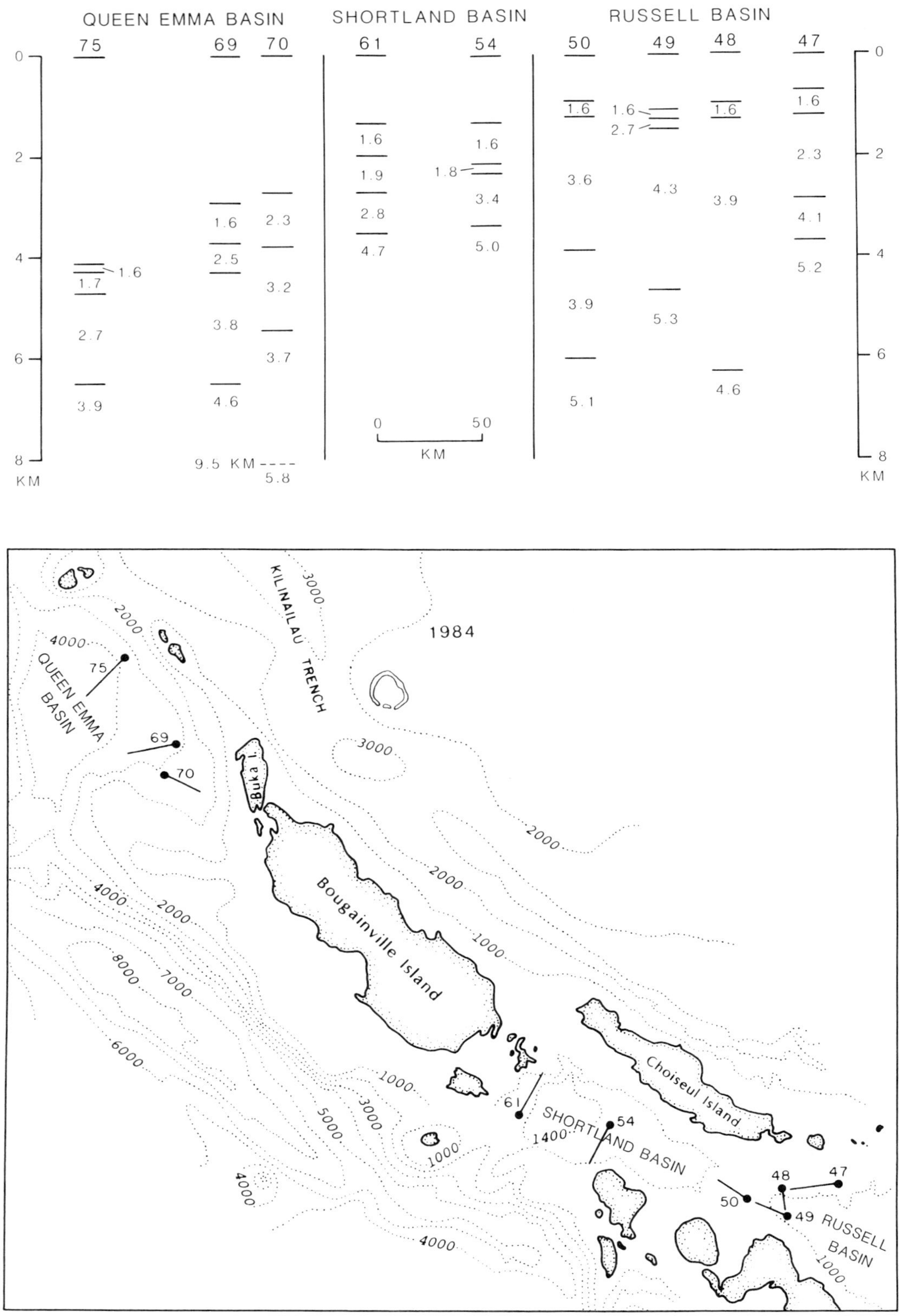

Figure 8. Refraction-velocity results for sonobuoy stations over intra-arc basins in the Solomon Islands-Bougainville region. See Table 1 for numerical listing and Figures 1 and 2 for regional location. Velocities in km/s; water depths are in m.

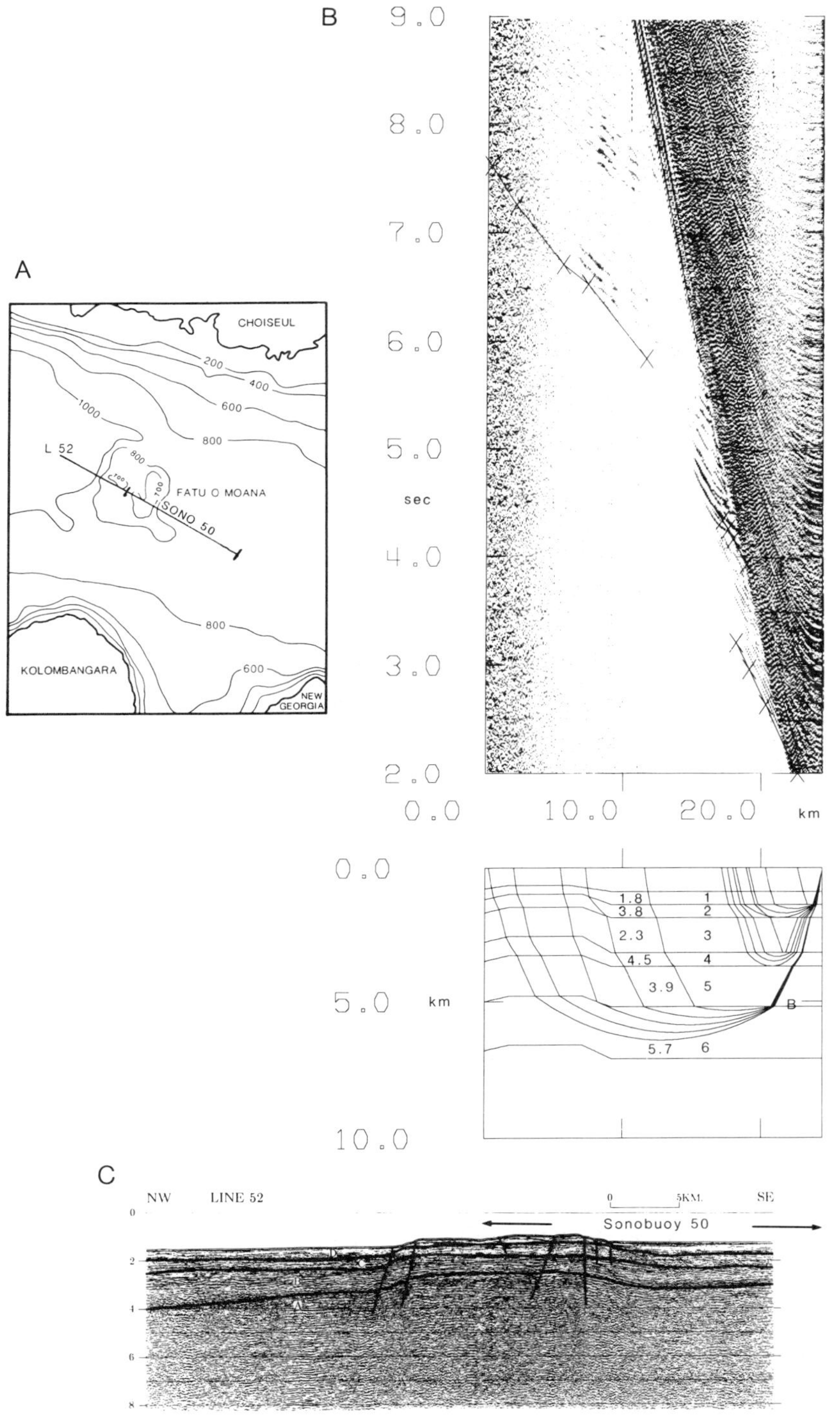

Figure 9. Seismic ray-tracing model and multichannel seismic-reflection (MCS) data for sonobuoy station 50 over the New Georgia wedge and Fatu O Moana, Central Solomons Trough (see Figures 2 and 7 for location). A. Location map; B. Field data for sonobuoy station 50 superimposed with computed arrival times from ray-tracing model; C. MCS line 52 showing seismic stratigraphy interpreted by Bruns et al (1986) and sonobuoy 50 ray-tracing results (Table 4). The abrupt termination and time-offset of refraction arrivals caused by interbedded high-velocity (layers 2 and 4) and low-velocity (layers 1,3, and 5) layers in the sedimentary section overlying basement (layer 6) are illustrated. The high-amplitude arrivals at 16 km offset probably are caused by focusing of seismic energy from uplifted basement rocks beneath Fatu O Moana. Xs are computed times. Numbers are average layer velocities in km/s. See Table 4 for specific layer velocities and thicknesses. "B" is top of basement.

Table 4. Seismic Refraction Solutions for Sonobuoys in the Central Solomons Trough Modeled with the Ray-tracing Method.

Sonobuoy Model	Lat (S)	Long (E)	Water depth (m)	Layer 1 V_t - V_b h t_i	Layer 2 V_t - V_b h t_i	Layer 3 V_t - V_b h t_i	Layer 4 V_t - V_b h t_i	Layer 5 V_t - V_b h t_i	Layer 6 V_t - V_b h t_i
49-1	7° 50.54′	157° 38.14′	1085	1.8 - 1.8	3.5 - 4.8	3.1 - 3.1	4.9 - 5.4		
				0.29	0.90	5.30	0.90		
				0.32	0.43	3.42	0.35		
49-2	7° 50.54′	157° 38.14′	1085	1.8 - 1.8	3.5 - 4.8	2.7 - 2.7	4.3 - 5.9		
				0.29	0.90	3.80	2.00		
				0.32	0.43	2.81	0.78		
50	7° 44.70′	157° 22.84′	910	1.8 - 1.8	3.6 - 3.9	2.3 - 2.3	4.1 - 4.8	3.9 - 3.9	5.5 - 5.9
				0.41	0.50	1.10	0.70	1.50	1.80
				0.46	0.27	0.96	0.31	0.77	0.63

Directly beneath the sonobuoy:
V_t is the velocity at the top of the layer in km/s.
V_b is the velocity at the bottom of the layer.
h is the layer thickness in km.
t_i is the interval two way travel time for the layer.

the correct horizontal-offset distances and refraction travel-times. The abrupt termination of refraction arrivals at shallow depths in the sonobuoy data (Figures 6 and 9) is a result of shallow, thin, high-velocity and high-velocity-gradient layers (causing the refractions to terminate abruptly) and of underlying low-velocity strata (causing refraction arrivals to be abruptly shifted to greater arrival times).

Models 49-1 and 49-2 (Figure 6; Table 4) show the least complex case, in which the deep sedimentary section within the New Georgia wedge (layer 3) has a uniform velocity. These models show that the total thickness of strata above basement (layer 4) can range from 6.5 km (model 49-1) to 5.0 km (model 49-2) depending on the chosen velocity. As higher velocities are assigned to the buried sedimentary section, greater sediment thicknesses are required for model ray-path travel times to match the observed refraction-arrival data.

Ray-tracing results for station 50 are similar to those for station 49, but we show a more complex model that incorporates two high-velocity layers within the sedimentary section (Figure 9). The additional layer is needed to explain the high-amplitude refraction arrival at about 4 seconds of travel time. This model is based on the assumption that high-velocity rocks, probably volcanic flows and debris, have at times covered the sedimentary sequence in parts of the Central Solomons Trough. Model 50 indicates that there is a total stratal thickness (layers 1-5) of about 4.5 km. This estimate is a minimum total sedimentary thickness because low sedimentary-rock velocities have been assumed. Higher assumed velocities at depth would cause the total thickness to increase, possibly by 1-2 km.

Station 50 was sited near a large bathymetric feature, Fatu O Moana (Figure 9), which is interpreted as an uplifted structural feature (Taylor and Exon, 1984; Bruns et al, 1986). Model 50 indicates that the increased amplitudes and bending of seismic arrivals at about 17 km offset may be caused by the focusing of rays at the edge of a basement horst, or basement intrusion, that uplifted the sedimentary section beneath Fatu O Moana. A vertical basement-offset of about 500 m is adequate to explain observed arrival times, based on model 50.

Velocity versus depth curves determined by slope-intercept and ray-tracing methods for sonobuoy stations 49 and 50 (Figure 10) illustrate the uncertainty and range in layer thicknesses and depths caused by buried low-velocity layers. Sonobuoy 49 (Figure 6) shows a broad range in basement depths (3.5 to 6.5 km) because the short and weak refraction arrivals cannot be modeled as accurately as the longer and higher amplitude arrivals of sonobuoy 50 (Figure 9). Slope-intercept thicknesses and velocities were used in the initial ray-trace model for station 49 (not shown); however, these values were changed so that the computed seismic arrival times matched the observed arrival times. The ray-tracing models illustrate that if low-velocity layers are present, then layer-thickness determinations are non-unique. However, refraction velocities and layer thicknesses for layers below the low-velocity zone can be accurately determined.

Two other sonobuoys (47, 48) were recorded across the perimeter of the New Georgia wedge on the northeastern side of the Central Solomons Trough (Figure 8). Excellent refraction arrivals were recorded at station 47 giving continuously-increasing velocities of 1.6-4.1 km/s throughout the 3.0-km-thick sedimentary section overlying basement (5.2 km/s). In the center of the trough at station 48, poor refraction arrivals were recorded from strata beneath a 0.3-km-deep, high-velocity (3.9 km/s) layer similar to that observed at stations 49 and 50. The 5.0 km

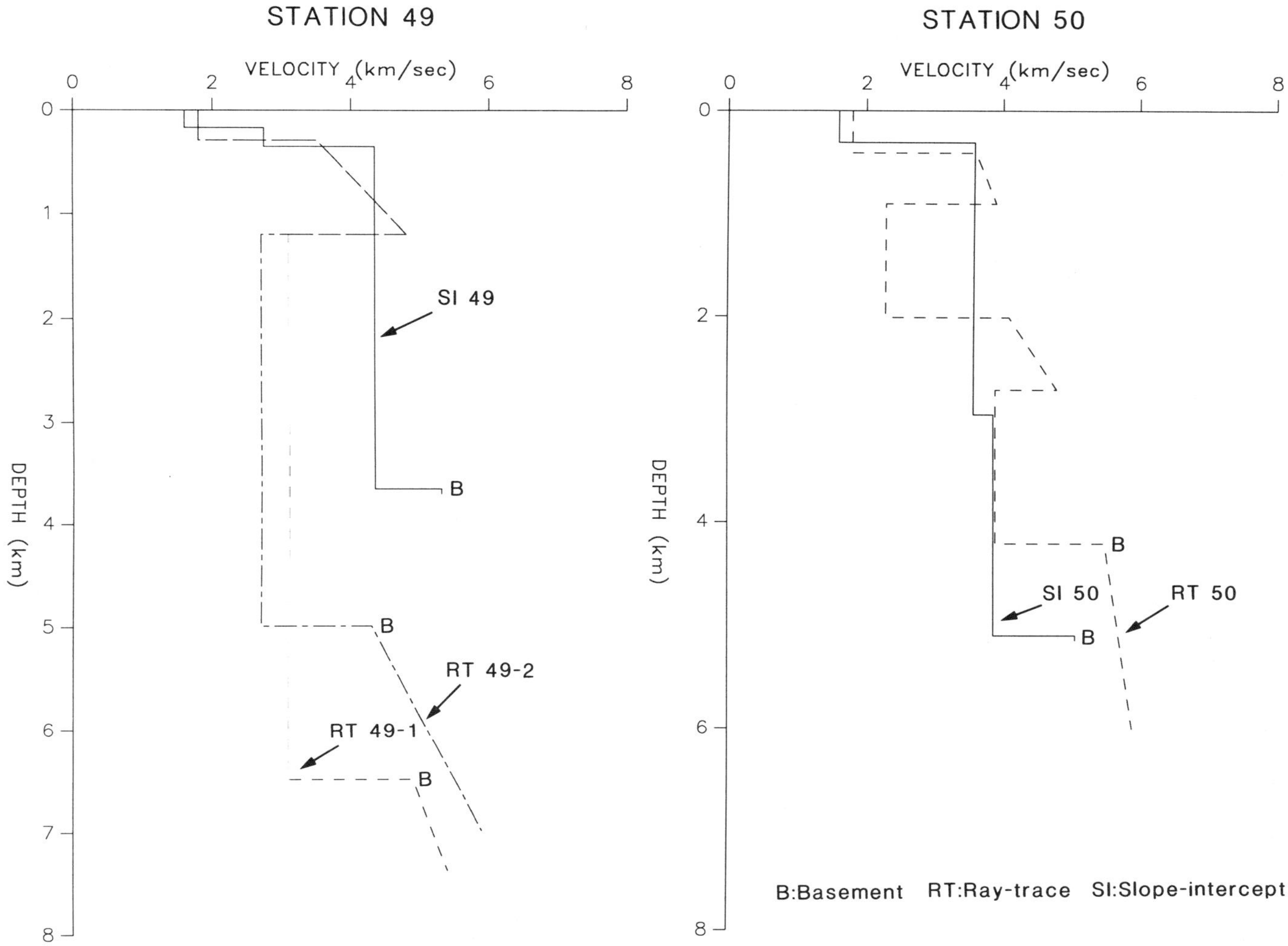

Figure 10. Velocity-versus-depth profiles derived by slope-intercept and ray-tracing methods for sonobuoy stations 49 and 50 over the New Georgia wedge, Central Solomons Trough (See Figures 2 and 7 for location). The large range in depths to the arc basement rocks (B) at station 49 illustrates the uncertainty in depths caused by interbedded high- velocity and low-velocity rocks.

thickness of this 3.9 km/s layer is suspect because lower-velocity strata probably occur below the 3.9 km/s layer, as discussed above.

At the northwestern end of the Central Solomons Trough, two sonobuoy stations (54, 61, Figure 8) were recorded on the perimeter of Shortland basin in order to measure velocities in the overlapping depositional bodies originating from the island platforms. The velocity-versus-reflection time profiles from these sonobuoys are incorporated with the 1982 results of Cooper and Wood (1986) and Cooper, Bruns, and Wood (1986) and shown in Figure 11. Both stations measured sedimentary thicknesses of about 2 km overlying basement with velocities of 4.7-5.0 km/s. Thick low-velocity layers probably do not exist at these two sites based on the long and continuous refraction arrivals that were recorded.

Queen Emma Basin

Three sonobuoy stations (69, 70, 75) were recorded across Queen Emma basin (defined by Vedder, Bruns, and Cooper, this volume), which lies northwest of Bougainville (Figure 8).

Multichannel seismic-reflection data recorded near the sonobuoy stations in Queen Emma basin show three distinct, layered sequences overlying acoustic basement: two almost flat-lying units and a thick basinward-dipping unit (Vedder, Bruns, and Cooper, this volume). In the basin, these units are folded above an antiformal structure that parallels Emirau-Feni Ridge. Two of the three sonobuoys (69, 70) were located over the thick sedimentary section lying between the antiform and the ridge; these sonobuoys recorded good refraction arrivals from the deep

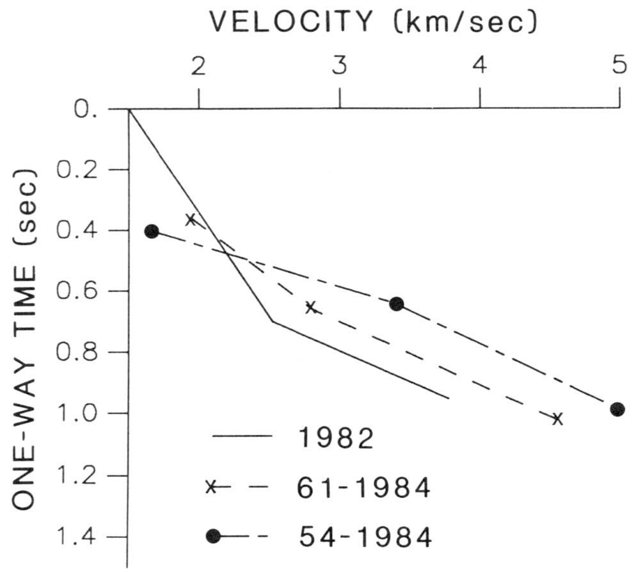

Figure 11. Velocity versus one-way reflection-time curves for sonobuoy stations 54 and 61 in the Shortland basin at the northwestern end of the Central Solomons Trough. See Figures 2 and 7 for location. The data are compared with a curve from Cooper, Bruns, and Wood (1986) for the Shortland basin.

sedimentary section. The third sonobuoy station (75) crossed the antiform, and consequently, refraction arrivals from great depths are distorted and cannot be used to determine velocities using slope-intercept methods (Table 1).

The sonobuoys show that the shallow, flat-lying, layered section in Queen Emma basin has velocities of 1.6-2.6 km/s. The dip-corrected velocities for the deep, dipping section are 3.7 km/s at stations 69 and 70. Acoustic basement has velocities of 4.6 and 5.8 km/s at two stations. The large, layer-dip angles at stations 69 and 70 result in large corrections to the observed velocities, especially for basement (uncorrected velocity of 8.5 km/s) at station 70. Consequently, the velocities for the deeper units are likely to have large errors, because of uncertainties in layer dips.

The total sedimentary section in Queen Emma basin may be thicker than 6.5 km, although accurate measurement of sediment thickness requires additional refraction stations parallel with the axis of the basin. Based upon interpretation of seismic-reflection data, Vedder, Bruns, and Cooper (this volume) estimate local thicknesses of as much as 8.0 km.

VELOCITY STRUCTURE OF THE TRENCHES

In this section, we discuss the velocity-structure of the Kilinailau Trench along the northeastern side of Bougainville and a small segment of the New Britain Trench along the southwestern side of Vella Lavella (Figure 12).

Kilinailau Trench

Four sonobuoy stations (63-66) were recorded along multichannel seismic-reflection lines across the Kilinailau Trench. Stations 63-65 were sited in the trench and station 66 was across a small lower-slope basin directly west of the trench (Figure 12).

The trench is a bathymetric depression underlain by two layered sedimentary units, an upper flat-lying unit and a lower arcward-dipping unit that overlies acoustic basement. The layered sections are abruptly folded at the base of the arc flank (Bruns, Vedder, and Culotta, this volume; Cochrane, Cooper, and Bruns, this volume). The folded units become laterally more disrupted beneath the lower flank of the arc and comprise (or structurally merge with) acoustic basement. Ray-tracing indicates that the acoustic basement landward of the trench probably is composed of uplifted trench sediment that now forms an accretionary wedge (Cochrane, Cooper, and Bruns, this volume).

Slope-intercept velocities from three sonobuoys (63, 64, and 65) across the Kilinailau Trench show that the flat-lying sedimentary sequence in the trench has velocities of 1.6-2.7 km/s (Table 1). The underlying dipping strata that rest on acoustic basement have velocities of 3.6-4.7 km/s. Refraction arrivals from acoustic basement were recorded from only one station (63) where velocities of 5.2-6.2 km/s were derived. A high velocity of 5.7 km/s was measured at station 64 for the deep, layered unit that dips into the trench, but this velocity is uncorrected for steep dips that are likely to occur in the folded sedimentary section.

At station 66, on the landward flank of the trench, velocities of 1.6-2.9 km/s were measured in a perched, layered sedimentary sequence that is acoustically similar to the upper layered unit in the nearby trench. The perched unit rests on a highly disrupted sedimentary unit, which in turn has a velocity of 3.2 km/s. A strong reflection event and a high-velocity layer of 4.3 km/s is present about 1.1 s below this disrupted unit, suggesting that deformed sedimentary units underlie the landward wall of the trench (Cochrane, Cooper, and Bruns, this volume).

The flat-lying sedimentary sequence filling the Kilinailau Trench is composed of distinct layers with small velocity gradients (i.e., layers of nearly constant velocity), which give nearly straight refraction arrivals (Cochrane, Cooper, and Bruns, this volume). Such layers are typical in sequences where distinct lithologic changes result from rapid sedimentation, abrupt alterations in depositional patterns, or variation in subsidence rates. In contrast, in other sequences where uniform deposition and compaction occur, layers are commonly characterized by large velocity gradients with curved refraction arrivals (see discussion in Cooper, Bruns, and Wood, 1986). The 1984 sonobuoy

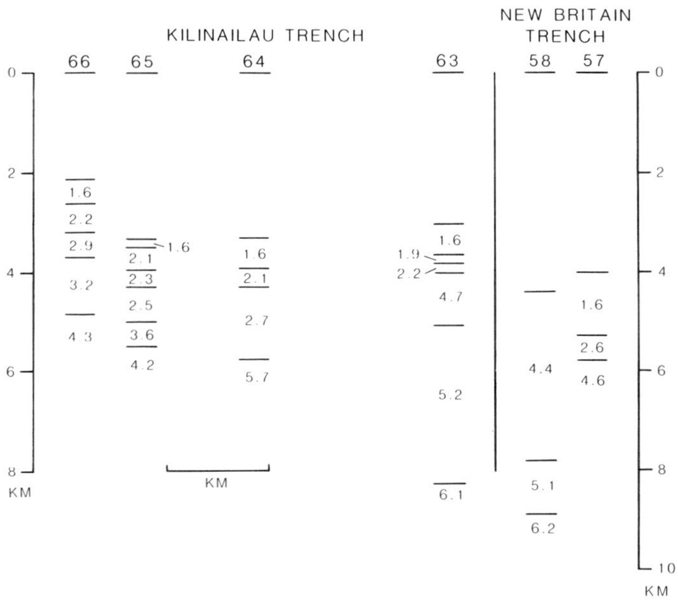

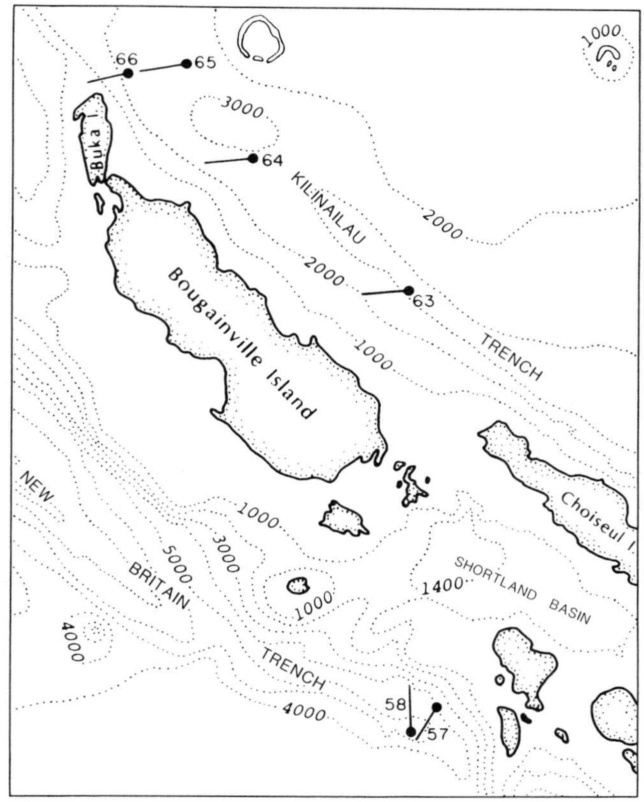

Figure 12. Refraction-velocity results for sonobuoy stations over the Kilinailau and New Britain trenches near Bougainville. See Table 1 for numerical listing and Figures 1 and 2 for regional location. Velocities in km/s; water depths in m.

data suggest that the sedimentary section in the Kilinailau Trench was deposited rapidly and that the sediment supply may have changed abruptly. These observations may reflect the episodic active uplift and rapid erosion of the nearby island arc.

New Britain Trench

Two sonobuoys, 57 and 58, were recorded across the inner wall of the New Britain Trench south of Vella Lavella where the trench is a steep-sided bathymetric trough bordered on both sides by irregular seafloor relief (Figure 12). About 100 m (0.2 s) of flat-lying sediment is present in the trench and is underlain by acoustically disrupted reflectors. A strong continuous reflection is conspicuous in multichannel seismic-reflection data about 1-2 s sub-bottom under the arc flank (Bruns et al, this volume). This reflection may be from the top of the Woodlark Basin oceanic crust, which has been subducted beneath the arc (Weissel, Taylor, and Karner, 1982). Similar deep reflections from oceanic crust are evident across the lower, southwestern flank of the Solomon Island arc near the Russell Islands, 300 km to the southeast (Cooper, Marlow and Bruns, 1986).

Strong sub-seafloor refraction arrivals were recorded at stations 57 and 58; however, these refractions are distorted in places because of seafloor and deeper structure. At station 57, the deepest refraction (4.6 km/s) corresponds with the strong oceanic-plate(?) reflection in multichannel seismic-reflection data (Bruns, Vedder, and Cooper, this volume). This subsurface reflection is less distinct at station 58 but appears to correspond with the 5.1 km/s sonobuoy refraction arrival that can be traced landward about 17 km from the trench. The subsurface reflection, which is associated with the 5.1 km/s refraction arrival, is traceable in multichannel seismic-reflection data for an additional 8 km beneath the arc flank. The measured velocities of the overlying rocks are 2.6 and 4.4 km/s, but a near-surface, low-velocity layer, may occur at station 57. The variable velocities, together with the disrupted acoustic character of rocks near the sea floor along this part of the arc flank suggest the presence of a young accretionary wedge at the base of the arc.

DISCUSSION

Basement Rocks

Sonobuoy-refraction data from the 1984 and previous surveys show that rocks with velocities greater than 4.5 km/s form the basement framework for the Solomon Islands-Bougainville segment of the Melanesian arc system (Cooper, Bruns, and Wood, 1986). However, high-amplitude reflection events, in some cases identified as acoustic basement on seismic-reflection profiles, have lower velocities (3.6-4.2 km/s especially beneath late

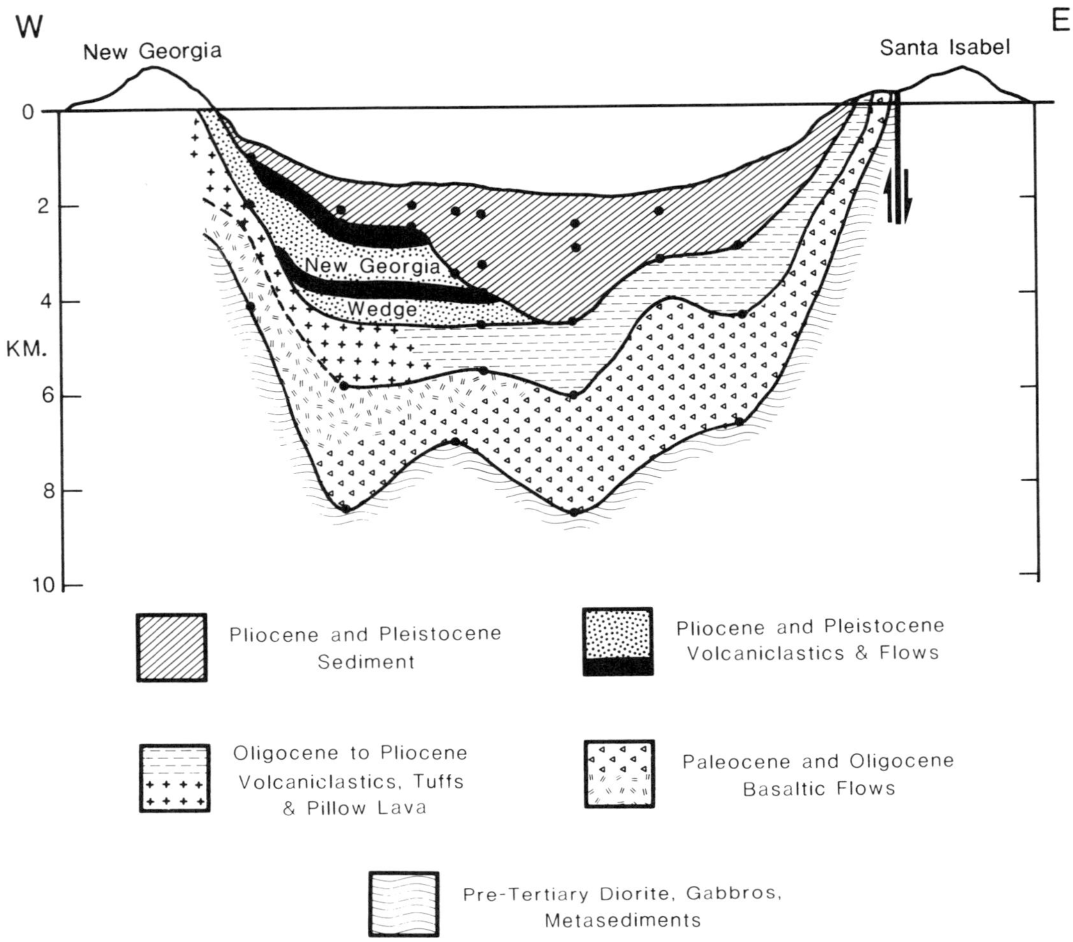

Figure 13. Schematic cross-section showing the New Georgia wedge between the New Georgia island group and central Santa Isabel based on Cooper, Bruns, and Wood (1986). Ray-tracing model studies show evidence for interbedded high-velocity layers within the thick sequence of Pliocene-Pleistocene volcaniclastic and flow rocks. Although the exact thickness is unknown, the total sedimentary section may be as much as 6.0 km thick. The large dots indicate the top of velocity layers computed from sonobuoy-refraction profiles.

Cenozoic volcanic rocks near the New Georgia wedge. We suspect, as has been suggested by Cooper, Bruns, and Wood (1986), that in these areas thin layers of volcanic flows and coarse volcaniclastic rocks (3.6-4.2 km/s) are interbedded within a thick lower-velocity sedimentary section (Figure 13). The volcanic rocks reflect and disperse the seismic energy to effectively hide the underlying sedimentary section. The true thickness of the hidden sedimentary section cannot be determined accurately with refraction-arrival data and may vary by 1 to 2 km, as estimated by ray-tracing techniques.

In other areas where late Cenozoic volcanics occur, such as near the Russell Islands (Figure 4) and near the southeastern and northeastern flanks of Bougainville, sonobuoy data do not indicate the presence of thick sedimentary sections beneath the shallow volcanic rocks (4.6-5.6 km/s). Instead, these areas appear, from refraction-velocity gradient information, to be underlain by massive flows that overlie the deeper metasedimentary and igneous core rocks of the arc. The best example is from the southeastern end of Bougainville where refraction arrivals at station 61 indicate that at least the upper 2-3 km of rocks beneath the 4.7 km/s basement layer have velocities higher than 4.7 km/s. A deeper, intra-basement layer with velocities greater than 6.0 km/s is not, however, discernible at this site.

Velocities of arc-framework rocks beneath the Solomon Islands-Bougainville segment of the Melanesian arc are best determined in the Santa Isabel-Choiseul area and near Buka Island and the Green Islands where these rocks are near the surface. Near Santa Isabel and Choiseul, sonobuoy data show evidence for mantle-type velocities from small or thin bodies of ultramafic rocks, probably the offshore equivalent of ultramafic bodies exposed onshore. Local intrusions of possible mafic, high-velocity (5.8 km/s; 6.3-6.4 km/s) rocks are revealed in the 1984 sonobuoy data from this area. These arc-framework rocks have velocities that are commonly observed for rocks of oceanic origin (Spudich and Orcutt, 1980). The presence of these high-velocity rocks within 1-5 km of the sea floor and on nearby islands indicates extensive deformation and uplift of likely old oceanic crust.

Near Buka Island and the Green Islands, only two sonobuoys recorded velocities greater than 5.9 km/s within the arc basement rocks. These intra-basement velocities of 6.0-6.2 km/s are not common in oceanic rocks, although the overlying basement layer at nearby stations has velocities of 4.6-5.6 km/s, which are characteristic of oceanic volcanic rocks. We believe that the sonobuoy 77 velocities of 6.1-6.2 km/s are reliable because of the refraction line's orientation parallel with the arc and the strong refraction arrivals. Without additional sonobuoy information, we cannot establish whether the low sub-crustal velocities are typical of the region. If they are typical, then metamorphic rocks or extensive intrusive rocks may compose the intra-basement framework of this part of the arc.

Sedimentary Section

At places where sonobuoys were recorded, the data show that the intra-arc basins and flanks of the Solomon Islands-Bougainville segment of the Melanesian arc system are covered by as much as 6.5 km of sedimentary rocks with velocities of 1.6-4.3 km/s, as illustrated by the isopach map on Figure 14. The thickest strata underlie the axial troughs of the intra-arc basins, usually areas of rapid subsidence and sediment infilling. Other areas of thick (1-3 km) sediment include: 1) many of the volcanic ridges underlying the southwestern chain of the Solomon Islands, especially near Vella Ridge, 2) the Kaviki and Emirau-Feni ridges that border Queen Emma basin, and 3) the northeastern flank of the Bougainville platform, especially along the inner wall of the Kilinailau Trench.

The sedimentary sequences throughout the region are typically composed of two distinct units. The lower unit, which overlies basement, is well-layered and conformable with the basement (Bruns et al, 1986; Bruns, Vedder, and Cooper, this volume; Vedder, Bruns, and Cooper, this volume) and has velocities of 2.9-4.3 km/s. This lower unit consists of distinctly layered sedimentary rocks that drape conformably over the flanks of Emirau-Feni Ridge and thicken into Queen Emma basin. The upper unit is composed of flat-lying strata that onlap the lower unit, thicken basinward, and have velocities of 1.6-2.6 km/s. This upper unit is best exemplified in the Central Solomons Trough. Correlation with onshore rocks (Cooper, Bruns, and Wood, 1986; Bruns et al, 1986; Vedder, Bruns, and Cooper, this volume) suggests that the lower unit may be Oligocene and Miocene in age and that the upper unit is Pliocene and Pleistocene in age. The offshore geometry and velocities of the two units are similar throughout the region, supporting the concept that these coeval rocks may have a common structural and depositional origin.

The area surrounding the New Georgia island group is anomalous because interbedded high-velocity rocks mask the underlying sedimentary strata as shown in Figure 13. Thick sedimentary sequences may be present elsewhere beneath the southwestern island chain, especially at places where the thick intra-arc basin sequence occurs nearby but is either abruptly or gradually hidden beneath shallow high-amplitude reflections. The older part of the sedimentary sequence may be composed of Paleogene marine strata that are quite different from the overlying late Cenozoic volcanic flows and debris.

SUMMARY

The major results of sonobuoy studies conducted during 1984 in the central and western Solomon Islands and eastern Papua New Guinea are:

1) The sedimentary sequences in the intra-arc basins and adjacent trenches of the region are characterized by an upper 1.6-2.6 km/s unit of flat-lying strata and by a lower 2.9-4.3 km/s layered unit that conformably overlies the basement surface.

2) Where sonobuoy data were recorded, the thickest parts of the sedimentary sequence, which are as much as 6.5 km thick, underlie the axial troughs of the intra-arc basins, and the bordering ridges are covered by 1-4 km of strata. Layered strata in the adjacent Kilinailau Trench and deformed sedimentary units beneath the inner trench wall may have a composite thickness of 3-4 km.

3) The sedimentary sequence in the area northeast of the New Georgia island group is characterized by shallow and thin high-velocity units that effectively mask a 2-5-km-thick underlying sedimentary section. These shallow units are probably volcanic flows and highly-lithified volcaniclastic debris. Wide-angle seismic-refraction data resolve the masked strata where vertical incidence reflections did not penetrate the volcanic units. The deep, masked section may be composed of low-velocity sedimentary rocks.

4) Acoustic basement generally corresponds to rocks that have velocities of 4.5-5.5 km/s. Near Bougainville, igneous basement rocks occur at all stations but, unlike elsewhere, were not penetrated to detect deeper intra-basement layers. Shallow basement rocks probably consist of volcanic and metasedimentary rocks.

5) Sub-basement layers with velocities of 5.8-6.6 km/s are present beneath or near the island platforms. These velocities suggest mafic igneous and metamorphic rocks of oceanic affinity. Velocities of 6.0-6.2 km/s from Emirau-Feni Ridge are not, however, typical of oceanic rocks and may be related to island-arc intrusive igneous rocks.

6) Small and shallow ultramafic bodies with velocities greater than 7.5 km/s may underlie the southeastern end of Santa Isabel and the northwestern end of Choiseul. These bodies may be similar to nearby onshore ultramafic exposures. Velocities in excess of 7.5

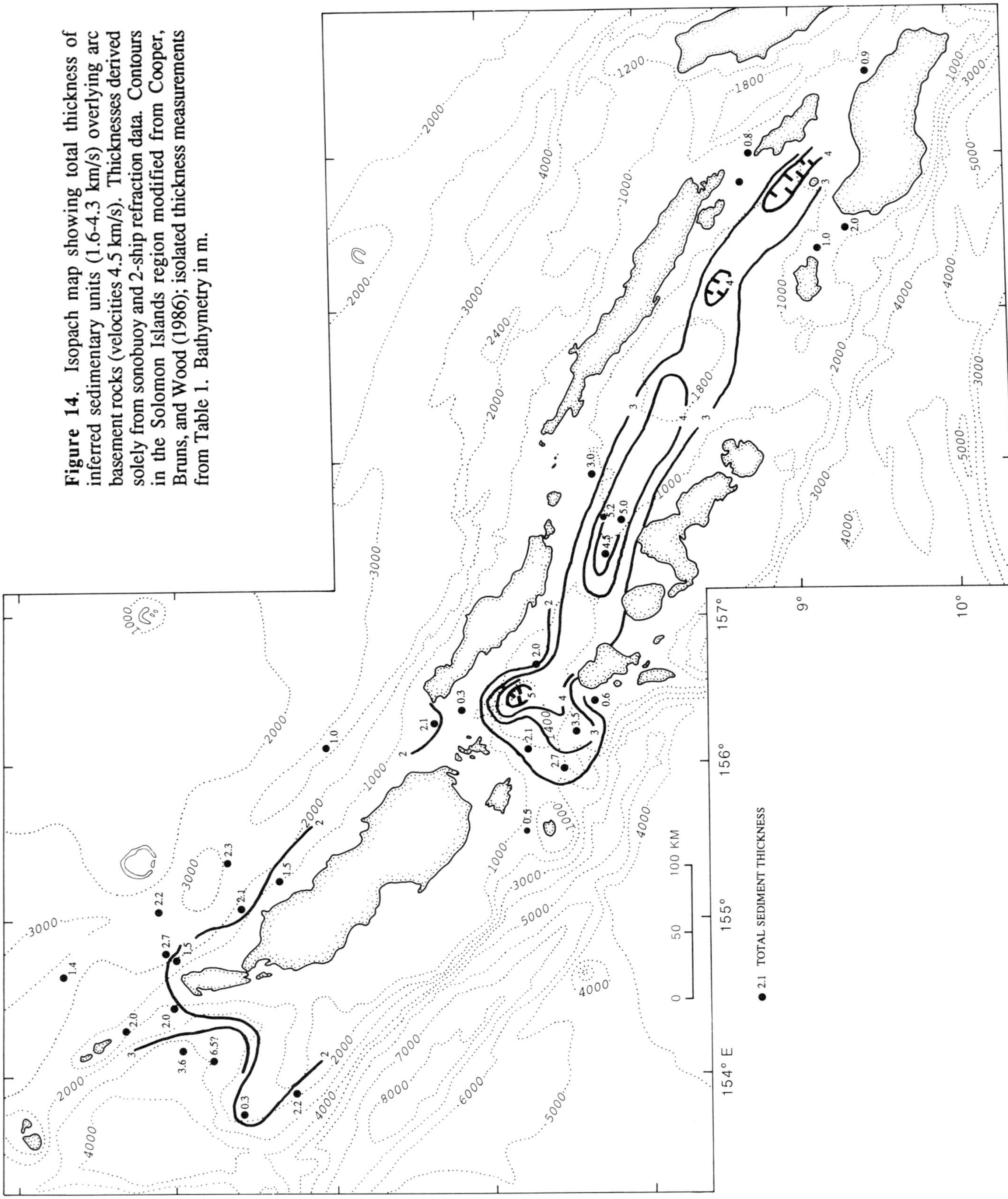

Figure 14. Isopach map showing total thickness of inferred sedimentary units (1.6-4.3 km/s) overlying arc basement rocks (velocities 4.5 km/s). Thicknesses derived solely from sonobuoy and 2-ship refraction data. Contours in the Solomon Islands region modified from Cooper, Bruns, and Wood (1986); isolated thickness measurements from Table 1. Bathymetry in m.

km/s recorded at depths of 6-8 km could, however, be caused by locally steep dips on basement structures.

7) The velocity structure of sedimentary and basement rocks of the Solomon Islands-Bougainville region support the concept (Vedder and Coulson, 1986; Cooper, Bruns, and Wood, 1986; Cooper, Marlow, and Bruns, 1986) that the island arc and intra-arc basins are underlain by mafic oceanic rocks emplaced and subsequently uplifted during the collision of the Pacific and Australia-India plates in Paleogene to Holocene time.

ACKNOWLEDGMENTS

We wish to thank the U.S. Navy for providing the sonobuoys used in this study. Helpful reviews of the manuscript were provided by H.J. Ryan and M.S. Marlow.

REFERENCES CITED

Beyer, L.A., 1986, Gravity anomalies of the Solomon Islands between 156° and 161° E longitude, *in* Vedder, J.G., Pound, K.S., and Boundy, S.Q., eds., Geology and offshore Resources of Pacific island arcs—central and western Solomon Islands: Houston, Texas, Circum-Pacific Council for Energy and Mineral Resources Earth Science Series, v. 4, p. 123-134.

Blake, D.H., and Miezitis, Y., 1967, Geology of Bougainville and Buka Islands, New Guinea: Department of National Development, Bureau of Mineral Resources, Geology and Geophysics Bulletin 93 and Bulletin PNG 1, Canberra, A.C.T., Australia, 56 p.

Bruns, T.R., Cooper, A.K., Mann, D.M., and Vedder, J.G., 1986, Seismic stratigraphy and structure of sedimentary basins in the Solomon Islands region, *in* Vedder, J.G., Pound, K.S., and Boundy, S.Q., eds., Geology and offshore resources of Pacific island arcs—central and western Solomon Islands: Houston, Texas, Circum-Pacific Council for Energy and Mineral Resources Earth Science Series, v. 4, p. 177-214.

Bruns, T.R., Cooper, A.K., and Vedder, J.G., 1986, A multichannel seismic-reflection profile across the Solomon Islands arc, *in* Vedder, J.G., Pound, K.S., and Boundy, S.Q., eds., Geology and offshore resources of Pacific island arcs—central and western Solomon Islands: Houston, Texas, Circum-Pacific Council for Energy and Mineral Resources Earth Science Series, v. 4, p. 215-224.

Childs, J.R., and Cooper, A.K., 1978, Collection, reduction, and interpretation of marine seismic sonobuoy data: U.S. Geological Survey Open-File Report 78-442, 219 p.

Childs, J.R., and Marlow, M.S., 1988, Geophysical data near Manus and New Ireland islands, Papua New Guinea, *in* Marlow, M.S., Dadisman, S.V., and Exon, N.F., eds., Geology and offshore resources of Pacific island arcs—New Ireland and Manus region, Papua New Guinea: Houston, Texas, Circum-Pacific Council for Energy and Mineral Resources Earth Science Series, v. 9, p. 273-288.

Coleman, P.J., and Kroenke, L.W., 1981, Subduction without volcanism in the Solomon Islands arc: Geo-Marine Letters v. 1, p. 129-134.

Coleman, P.J., and Packham., G.H., 1976, The Melanesian borderlands and India-Pacific plates' boundary: Earth Science Reviews, n. 12, p. 197-233.

Colwell, J.B., and Vedder, J.G., 1986, Description and interpretation of dredged rocks, Solomon Islands, *in* Vedder, J.G., Pound, K.S., and Boundy, S.Q., eds., Geology and offshore resources of Pacific island arcs—central and western Solomon Islands: Houston, Texas, Circum-Pacific Council for Energy and Mineral Resources Earth Science Series, v. 4, p. 225-236.

Cooper, A.K., Bruns, T.R., and Wood, R.A., 1986, Crustal structure of the Solomon Islands intra-arc basins from sonobuoy seismic studies, *in* Vedder, J.G., Pound, K.S., and Boundy, S.Q., eds., Geology and offshore resources of Pacific island arcs—central and western Solomon Islands: Houston, Texas, Circum-Pacific Council for Energy and Mineral Resources Earth Science Series, v. 4, p. 135-156.

Cooper, A.K., Marlow, M.S., and Bruns, T.R., 1986, Deep structure of the central and southern Solomon Islands region—implications for tectonic origin, *in* Vedder, J.G., Pound, K.S., and Boundy, S.Q., eds., Geology and offshore resources of Pacific island arcs—central and western Solomon Islands: Houston, Texas, Circum-Pacific Council for Energy and Mineral Resources Earth Science Series, v. 4, p. 157-176.

Cooper, A.K., and Wood, R.A., 1986, Wide-angle seismic-reflection and -refraction data from the Solomon Islands, *in* Vedder, J.G., Pound, K.S., and Boundy, S.Q., eds., Geology and offshore resources of Pacific island arcs—central and western Solomon Islands: Houston, Texas, Circum-Pacific Council for Energy and Mineral Resources Earth Science Series, v. 4, p. 41-48.

Cooper, P.A., Kroenke, L.W., and Resig, J.M., 1986, Tectonic implications of seismicity northeast of the Solomon Islands, *in* Vedder, J.G., Pound, K.S., and Boundy, S.Q., eds., Geology and offshore resources of Pacific island arcs—central and western Solomon Islands: Houston, Texas, Circum-Pacific Council for energy and Mineral Resources Earth Science Series, v. 4, p. 117-122.

Coulson, F.I., and Vedder, J.G., 1986, Geology of the central and western Solomon Islands, *in* Vedder, J.G., Pound, K.S., and Boundy, S.Q., eds., Geology and offshore resources of Pacific island arcs—central and western Solomon Islands: Houston, Texas, Circum-Pacific Council for Energy and Mineral Resources Earth Science Series, v. 4, p. 59-88.

Ewing, J.I., Houtz, R.E., and Ludwig, W.J., 1970, Sediment distribution in the Coral Sea: Journal of Geophysical Research, v. 75, n. 11, p. 1963-1974.

Furumoto, A.S., Hussong, D.M., Campbell, J.F., Sutton, G.H., Malahoff, A., Rose, J.C., and Woolard, G.P., 1970, Crustal and upper mantle structure of the Solomon Islands as revealed by seismic refraction of November-December, 1966: Pacific Science, v. 24, p. 315-332.

Furumoto, A.S., Webb, J.P., Odegard, M.E., and Hussong, D.M., 1976, Seismic studies on the Ontong Java Plateau:

Furumoto, A.S., Wiebenga, W.A., and Sutton, G.H., 1973, Crustal structure of the Hawaiian Archipelago, northern Melanesia, and the central Pacific basin by seismic refraction methods: Tectonophysics, v. 20, p. 153-164.

Hackman, B.D., 1980, The geology of Guadalcanal, Solomon Islands: London, Overseas Memoir of the Institute of Geological Science, n. 6, 115 p.

Hussong, D.M., Wipperman, L.K., and Kroenke, L.W., 1979, The crustal structure of the Ontong Java and

Manihiki oceanic plateaus: Journal of Geophysical Research, v. 84, n. B11, p. 6003-6010.

Katz, H.R., 1980, Basin development in the Solomon Islands and their petroleum potential: UN ESCAP, Committee for the Coordination of Joint Prospecting for Mineral Resources in south Pacific offshore areas UN CCOP/SOPAC), Technical Bulletin, n. 3, p. 59-75.

Kroenke, L.W., 1984, The Solomon Islands: San Cristobal to Bougainville and Buka, Chapter 4, *in* Kroenke, L.W., ed., Cenozoic Tectonic Development of the Southwest Pacific: UN CCOP/SOPAC Technical Bulletin n. 6, p. 47-62.

Luyendyk, B.P., McDonald, K.C., and Bryan, W.B., 1973, Rifting history of the Woodlark Basin in the South Pacific: Geological Society of America Bulletin, v. 84, p. 1125-1134.

Marlow, M.S., Exon, N.F., Ryan, H.F., and Dadisman, S.V., 1988, Offshore structure and stratigraphy of New Ireland basin in northern Papua New Guinea, *in* Marlow, M.S., Dadisman, S.V., and Exon, N.F., eds., Geology and offshore resources of Pacific island arcs—New Ireland and Manus region, Papua New Guinea: Houston, Texas, Circum-Pacific Council for Energy and Mineral Resources Earth Science Series v. 9, p. 137-155.

McMechan, G.A., and Mooney, W.D., 1980, Asymptotic ray theory and synthetic seismograms for laterally varying structures—theory and application to the Imperial Valley, California: Bulletin of Seismological Society of America, v. 70, p. 2021-2035.

Murauchi, S., Ludwig, W.J., Den, N., Hotta, H., Asanuma, T., Yoshii, T., Kubotera, A., and Hagiwara, K., 1973, Seismic refraction measurements on the Ontong Java Plateau northeast of New Ireland: Journal of Geophysical Research, v. 78, n. 35, p. 8653-8662.

Pound, K.S., 1986, Correlation of rock units in the central and western Solomon Islands, *in* Vedder, J.G., Pound, K.S. and Boundy, S.Q., eds., Geology and offshore resources of Pacific island arcs—Central and western Solomon Islands: Houston, Texas, Circum-Pacific Council for Energy and Mineral Resources Earth Science Series, v. 4, p. 89-98.

Ravenne, C., deBroin, C.E. and Aubertin, F., 1982, Structure et Histoire de la region Solomon Nouvelle-Irlande, *in* Contribution à l'etude Géodynamique du Sud-ouest Pacifique, Equipe Géologie-Geophysique Nouméa: Paris, O.R.S.T.O.M., p. 327-340.

Spudich, P., and Orcutt, J., 1980, A new look at the seismic velocity structure of the oceanic crust: Reviews of Geophysics and Space Physics, v. 18, n. 3, p. 627-645.

Taylor, B., and Exon, N.F., 1984, An investigation of ridge subduction and related phenomena in the Woodlark-Solomons region—introduction and background, *in* Exon, N.F., and Taylor, B., eds., Sea floor spreading, ridge subduction, volcanism and sedimentation in the offshore Woodlark - Solomons region and Tripartite cruise report for KANA KEOKI Cruise 82-03-16 leg 4: UN ESCAP, CCOP/SOPAC Technical Report n. 34, p. 1-12.

Vedder, J.G., 1986, Summary of the geology and offshore resources of the Solomon Islands, *in* Vedder, J.G., Pound, K.S., and Boundy, S.Q., eds., Geology and offshore resources of Pacific island arcs—central and western Solomon Islands: Houston, Texas, Circum-Pacific Council for Energy and Mineral Resources Earth Science Series, v. 4, p. 295-306.

Vedder, J.G., Colwell, J.B., and Bruns, T.R., 1985, New findings in offshore areas of western Solomon Islands and Buka—Bougainville, Papua New Guinea, *in* Extended abstracts—Petroleum geology of south Pacific island countries: Marine workshop, 14th BMR Symposium, Canberra, Australia, October 1985, p. 37-41.

Vedder, J.G., and Coulson, F.I., 1986, Introduction to geology and offshore resources of the Solomon Islands, *in* Vedder, J.G., Pound, K.S., and Boundy, S.Q., eds., Geology and offshore resources of Pacific island arcs—central and western Solomon Islands: Houston, Texas, Circum-Pacific Council for Energy and Mineral Resources Earth Science Series, v. 4, p. 1-18.

Weissel, J.K., Taylor, B., and Karner, G.D., 1982, The opening of the Woodlark Basin, subduction of the Woodlark spreading system, and the evolution of northern Melanesia since mid-Pliocene time: Tectonophysics, v. 87, p. 253-277.

Vedder, J.G., and Bruns, T.R., editors, 1989, Geology and offshore resources of Pacific island arcs—Solomon Islands and Bougainville, Papua New Guinea Regions: Houston, Texas, Circum-Pacific Council for Energy and Mineral Resources Earth Science Series, v. 12.

MODELING OF SONOBUOY REFRACTION DATA FROM THE KILINAILAU TRENCH, BOUGAINVILLE, PAPUA NEW GUINEA

G.R. Cochrane, A.K. Cooper, T.R. Bruns
U.S. Geological Survey, Menlo Park, CA 94025, USA

ABSTRACT

Sonobuoy seismic-refraction data from the Kilinailau Trench, an Oligocene(?) to early Miocene subduction zone northeast of Bougainville, Papua New Guinea, are modeled with two-dimensional ray tracing. The sonobuoys cross the trench and part of the inner trench wall, an area where structural complexity made two-dimensional techniques necessary for interpretation of the data.

Modeling suggests that beneath the partly filled trench, southwestward-dipping acoustic basement, which consists of Cretaceous-Paleocene layered basalts of the Ontong Java Plateau, has a velocity of 4.5 km/s; this velocity increases to 6.0 km/s at a depth of about 10 km. Pelagic strata as much as 500 meters thick conformably overlie acoustic basement and have a velocity of about 3.2 km/s. Miocene and younger trench-fill sediments 2-3 km thick unconformably overlie the pelagic unit. Velocities increase from about 1.7 km/s at the sea floor to 2.5 km/s at the base of the trench fill. Anticlinal folds deform the entire sedimentary section. A slight horizontal velocity gradient (velocities increasing toward the trench axis) occurs in these strata. A zone of highly deformed sedimentary rock underlies the inner trench wall and extends about 10 km toward the arc where it grades into a high-velocity sediment apron. The rocks below this apron have a velocity of 4.5 km/s and are interpreted to be Neogene volcanogenic rocks of the Bougainville Group and possibly older arc-basement rocks.

Continuity of refraction arrivals across the boundary between the slightly folded Neogene-Quaternary(?) trench-fill sediments and deformed strata beneath the inner trench wall suggests that these two sediment bodies are coeval. Therefore, at least some of the horizontal stress caused by convergence of the Pacific and Australia-India plates has been accommodated in the Kilinailau Trench after the Miocene reversal in arc polarity. Our results support a model of a rootless arc riding atop the boundary between the Australia-India and Pacific plates.

INTRODUCTION

In 1984, the U.S. Geological Survey acquired marine geological and geophysical data in the Solomon Islands and eastern Papua New Guinea on a reconnaissance cruise aboard the R/V *S.P. Lee*. These data include in part 1,800 km of multichannel seismic-reflection profiles (Bruns, Vedder, and Culotta, this volume) and 40 sonobuoy refraction stations (Cooper, Cochrane, and Bruns, this volume).

The study area straddles the convergent boundary between the Pacific and Australia-India plates (Figure 1). Tectonism since the Eocene can be roughly divided into two periods of oppositely directed subduction (Kroenke, 1984). During the Eocene to early Miocene, southwest-directed subduction occurred along the northeastern side of the study area (Figure 1). Subduction ceased in the early Miocene. In the late Miocene, northeast-directed subduction began along the New Britain Trench. This sequence of events indicates that the northern trench system should be inactive. Our study focuses on this older trench, the Kilinailau Trench, northeast of Bougainville.

The tracklines for the multichannel seismic-reflection data collected in the Kilinailau Trench are shown on Figure 2. The three sonobuoys discussed in this paper cross the trench and part of the inner trench wall, a structurally complex area where plane-layer solutions of the refraction data are inadequate for interpretation of the geology. We therefore developed two-dimensional ray-trace models to delineate the distribution of sedimentary sequences and igneous basement. Our interpretation shows that compressive deformation has continued in the trench during the late Cenozoic after the arc polarity reversal.

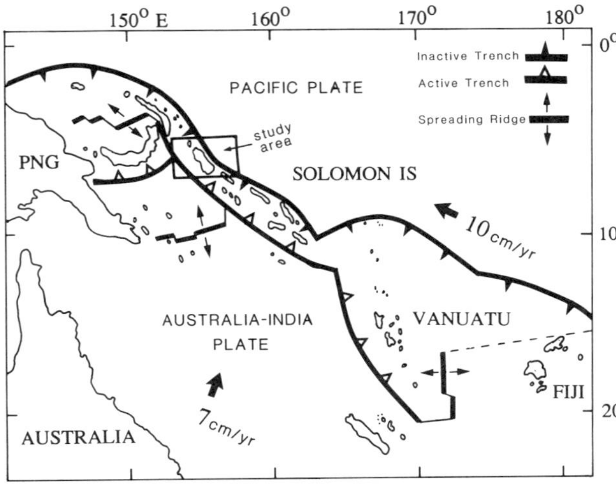

Figure 1. Map showing study area situated on the convergent boundary between Pacific plate and Australia-India plate. From Drummond et al (1985).

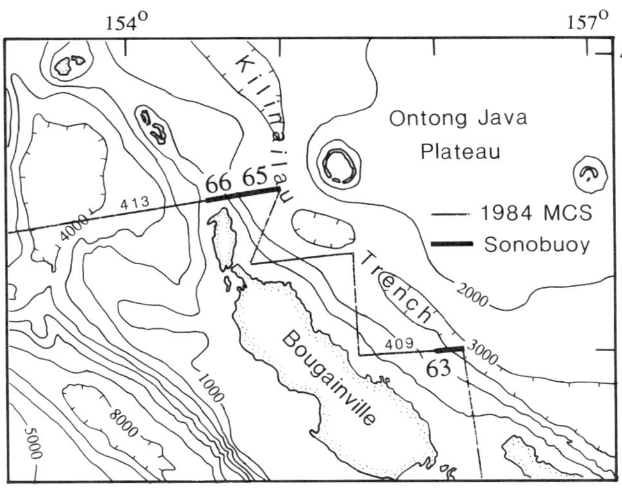

Figure 2. Locations of 1984 U.S. Geological Survey sonobuoy stations and multichannel seismic-reflection lines across the Kilinailau Trench.

RAY-TRACE MODELS

Two-dimensional ray-trace modeling of the sonobuoy data was required because of significant variation in rock physical properties, structure, and bathymetry in the Kilinailau Trench. The procedure followed was to generate an initial laterally-homogeneous model using vertical travel times to structural and stratigraphic boundaries interpreted from multichannel seismic-reflection records and using slope-intercept refraction-velocity stations of Cooper, Cochrane, and Bruns (this volume). Structural and stratigraphic boundaries are represented in the models by lines between which velocities are determined by two-dimensional linear interpolation. P-wave travel times for raypaths through the model are computed by the two-dimensional ray-tracing algorithm described by McMechan and Mooney (1980). Successive models were produced to improve the fit of the synthetic data to the observed sonobuoy data. The final models satisfy the condition that vertical travel times computed using model boundaries are equal to vertical travel times measured in seismic-reflection profiles.

Sonobuoy 63 began above the sediment-filled trench (Figure 3). The trench-fill strata unconformably overlie westward-dipping reflectors, interpreted to be pelagites, which in turn conformably overlie igneous rocks of the Ontong Java Plateau (Vedder, Colwell, and Bruns, 1985; Bruns, Vedder, and Culotta, this volume). The sonobuoy data were recorded across the edge of the inner trench wall where undeformed strata terminate near a westward-dipping reflector (20-30 km, Figure 3), which we believe is a thrust fault because of its similarity to structure observed on other convergent margins (von Huene, 1988). Modeling of sonobuoy data was done to test our hypothesis that the material to the west of this thrust consists of low-velocity accreted sedimentary rocks rather than high-velocity basement.

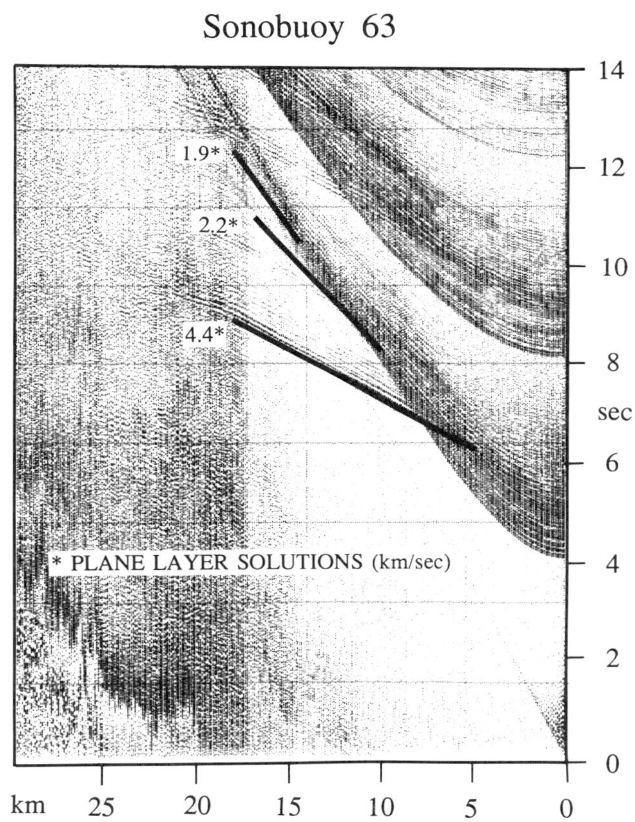

Figure 4. Example of sonobuoy seismic-refraction data (sonobuoy 63). Plane layer velocities shown are initial trial velocities for ray-trace model in Figure 5.

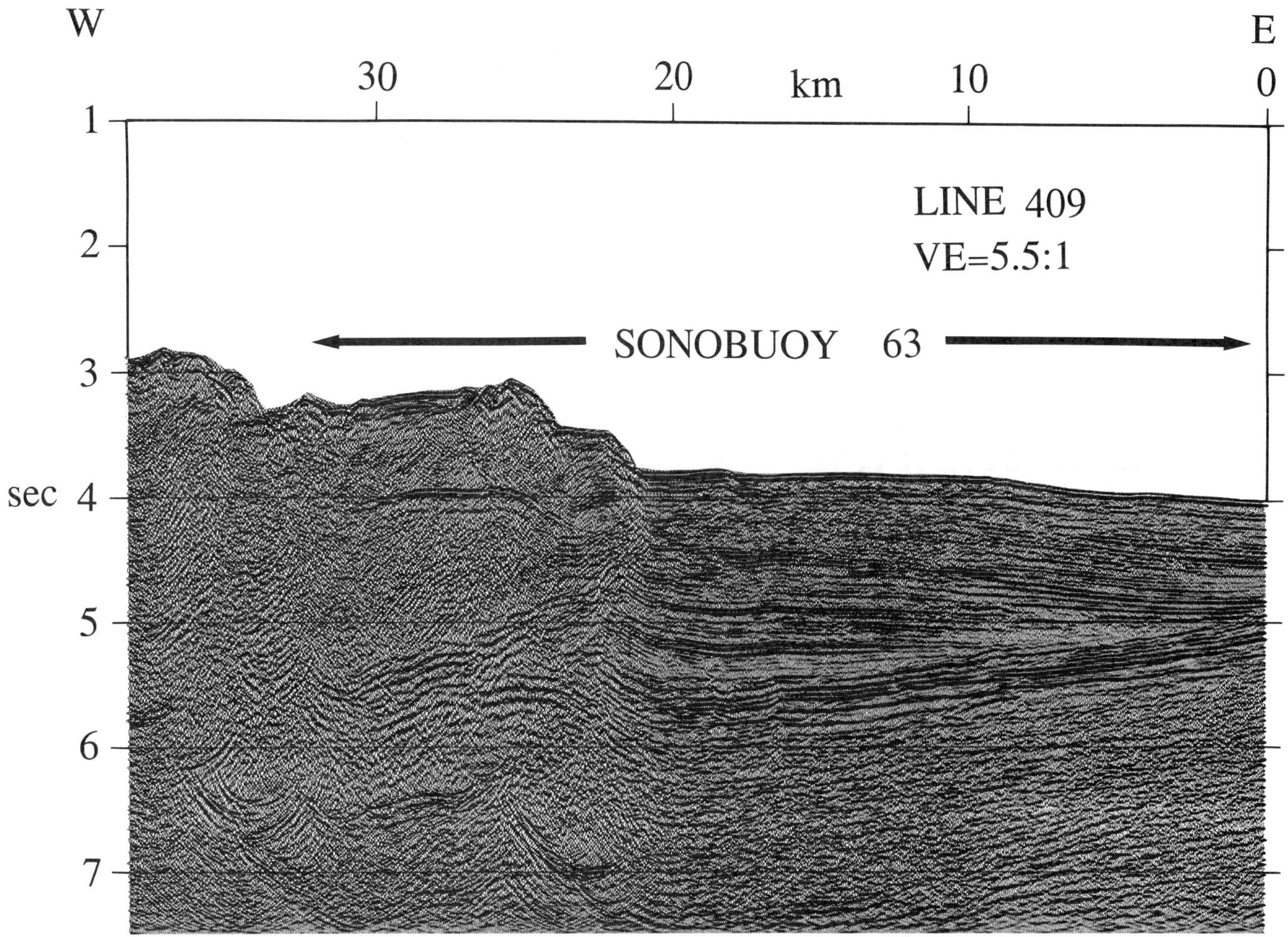

Figure 3. Migrated time section of multichannel seismic-reflection line 409 showing location of sonobuoy 63.

The initial trial velocities for the model of sonobuoy 63 are shown on Figure 4. An enlarged section of the sonobuoy 63 data and our best ray-tracing model for this part of line 409 are shown on Figure 5. Layer 1 represents the trench-fill strata. An average velocity of 1.8 km/s and a thickness of about 1 km produces arrival times that match the data. The average vertical velocity of this layer increases to 2.1 km/s at a range of 18 km, indicating a slight horizontal velocity gradient in this layer possibly owing to lateral stress compaction or overburden. Similarly, layer 2 (pelagites) average velocity increases from 2.55 to 2.8 km/s in the direction of the arc. A deflection in refraction arrivals from the layer 3 (basalt) at a range of 20 km is modeled by passing raypaths through the bathymetric high associated with the postulated thrust.

The second slope area modeled lies to the northwest along line 413 (Figure 6) where sonobuoys 65 and 66 were recorded. Sonobuoy 65 crosses the area that is geologically equivalent to the previous model. Sonobuoy 66 continues up the slope and over the submerged crest of the arc. The most recent trench-fill sediments in this area show more deformation than those to the southeast. The layered sequence lying above the accreted sediment of the inner trench wall (20-40 km distance) consists of levee deposits along a depositional channel which originates on the flank of Bougainville. Beneath the accreted sediment (5.6 s at a distance of 30 km, Figure 6), eastward-dipping reflectors are interpreted to be the top of igneous rocks of the Bougainville Group of Blake and Miezitis (1967) and possibly older arc-basement rocks.

The model for sonobuoy 65 (Figure 7, Table 2) produces a good fit to refractions through the two principal sediment layers. Velocities for both layers are higher than for sonobuoy 63, suggesting greater compaction because of the greater thickness. Velocities increase from the trench-fill sediments to the more deformed accreted sediments across a nearly vertical boundary at a range of about 15 km (Figure 6). Rays that pass through this boundary area have arrival times and ranges approximately coinciding with diffractions in the sonobuoy data. Prominent diffractions are also evident on unmigrated multichannel seismic-reflection data in this area. (Bruns, Vedder, and Culotta, this volume) The velocity of the accreted sediments and the geometry of the top of the igneous crust at ranges greater

Table 1. Velocity and depth data for raytrace model of sonobuoy 63 (Figure 5).

Range (km)	Depth (km)	V_u	V_1
30.0	2.4	1.5	1.9
30.0	5.1	2.4	2.4
30.0	5.101	2.4	2.75
30.0	5.4	2.85	4.95
30.0	10.0	6.5	-
26.4	2.5	1.5	2.0
26.4	4.98	2.4	2.4
26.4	4.981	2.4	2.75
26.4	D_i	2.85	4.95
26.4	10.0	6.5	-
22.2	2.2	1.5	2.2
22.2	D_i	2.2	2.2
22.2	D_i	2.4	2.75
22.2	D_i	2.85	4.95
22.2	10.0	6.5	-
18.0	2.8	1.5	1.7
18.0	2.801	1.7	1.7
18.0	D_i	2.0	2.54
18.0	D_i	2.59	4.95
18.0	10.0	6.5	-
0.0	3.0	1.5	1.65
0.0	3.001	1.65	1.65
0.0	4.1	1.9	2.54
0.0	4.4	2.59	4.95
0.0	10.0	6.5	-

V_u = velocity above boundary in km/s.
V_1 = velocity below the in km/s.
D_i indicates depth interpolated by raytracing program.

Table 2. Velocity and depth data for raytrace model of sonobuoy 65 (Figure 7).

Range (km)	Depth (km)	V_u	V_1
27.25	2.21	1.5	1.9
27.25	3.65	2.1	3.8
27.25	5.0	3.8	4.5
27.25	10.0	6.0	-
20.24	2.62	1.5	1.9
20.24	3.15	1.9	3.4
20.24	D_i	3.8	4.5
20.24	10.0	6.0	-
16.5	D_i	1.5	1.9
16.5	D_i	1.9	2.9
16.5	6.7	3.8	4.5
16.5	10.0	6.0	-
15.56	3.15	1.5	1.9
15.56	3.151	1.9	2.4
15.56	D_i	3.8	4.5
15.56	10.0	6.0	-
15.0	D_i	1.5	1.85
15.0	5.7	2.7	3.7
15.0	D_i	3.8	4.9
15.0	10.0	6.0	-
0.0	3.32	1.5	1.85
0.0	5.0	2.5	3.7
0.0	5.7	3.8	4.9
0.0	10.0	6.0	-

V_u = velocity above boundary in km/s.
V_1 = velocity below boundary in km/s.
D_i indicates depth interpolated by raytracing program.

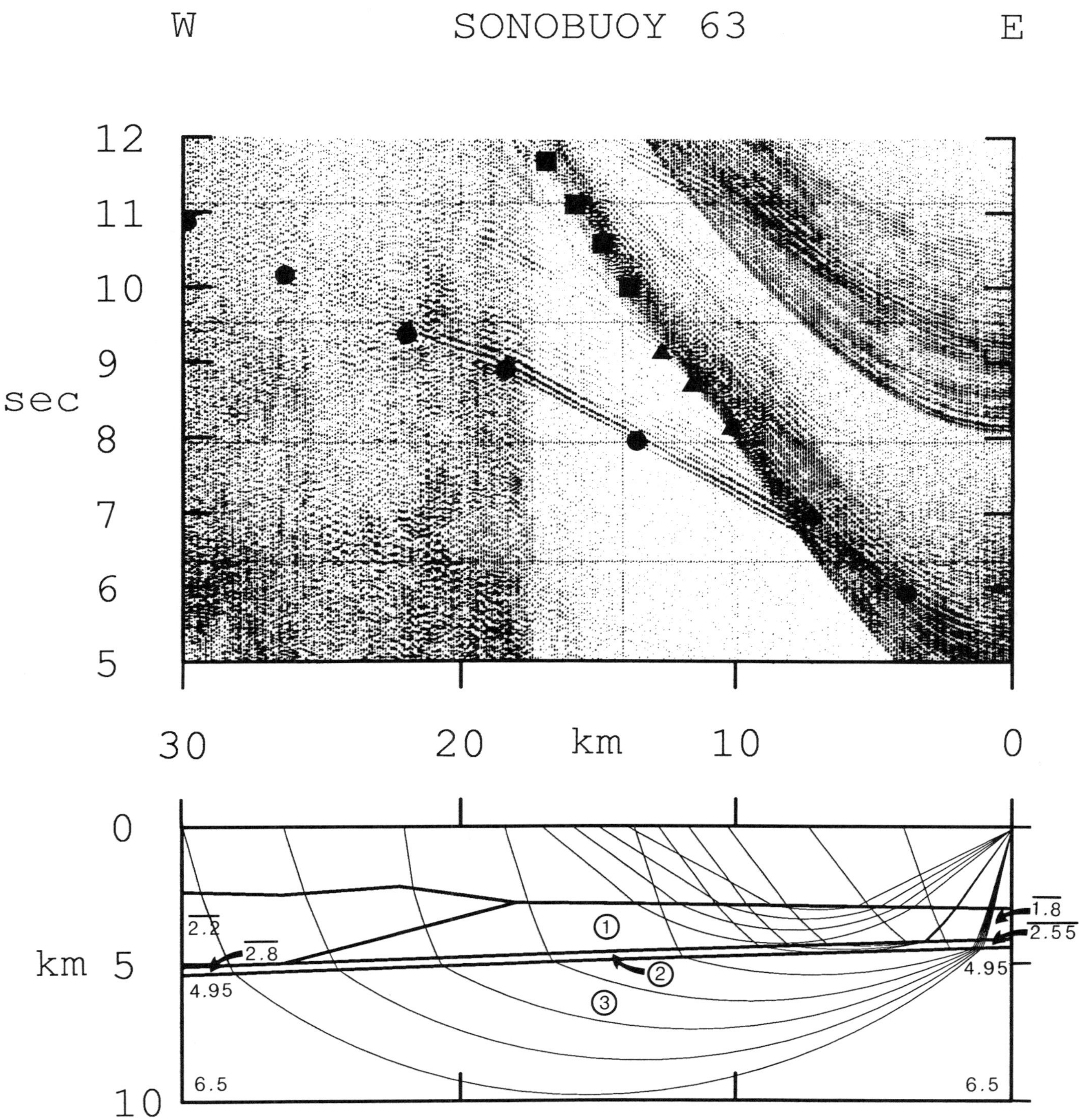

Figure 5. Data for sonobuoy 63 (top). Computed arrival times for the two-dimensional ray-trace model (bottom) and plotted on the data as squares, triangles, and dots for layers 1, 2, and 3, respectively). Velocities shown for layer 1 (1.8 km/s) and 2 (2.6 km/s) are average velocities. Velocities for the top and bottom of layer 3 are shown. See Table 1 for complete model-input data.

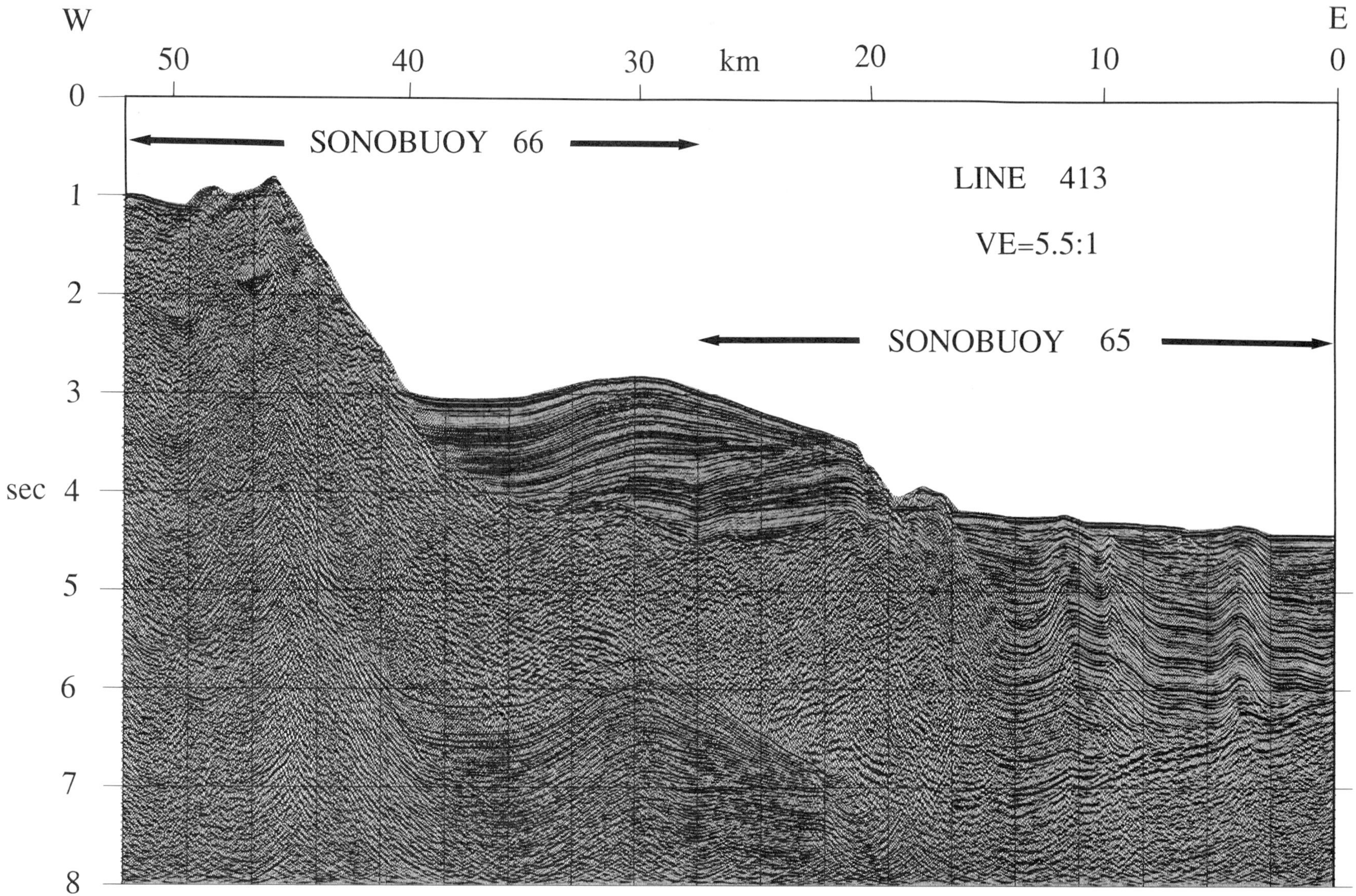

Figure 6. Migrated time section of multichannel seismic-reflection line 413 showing location of sonobuoys 65 and 66.

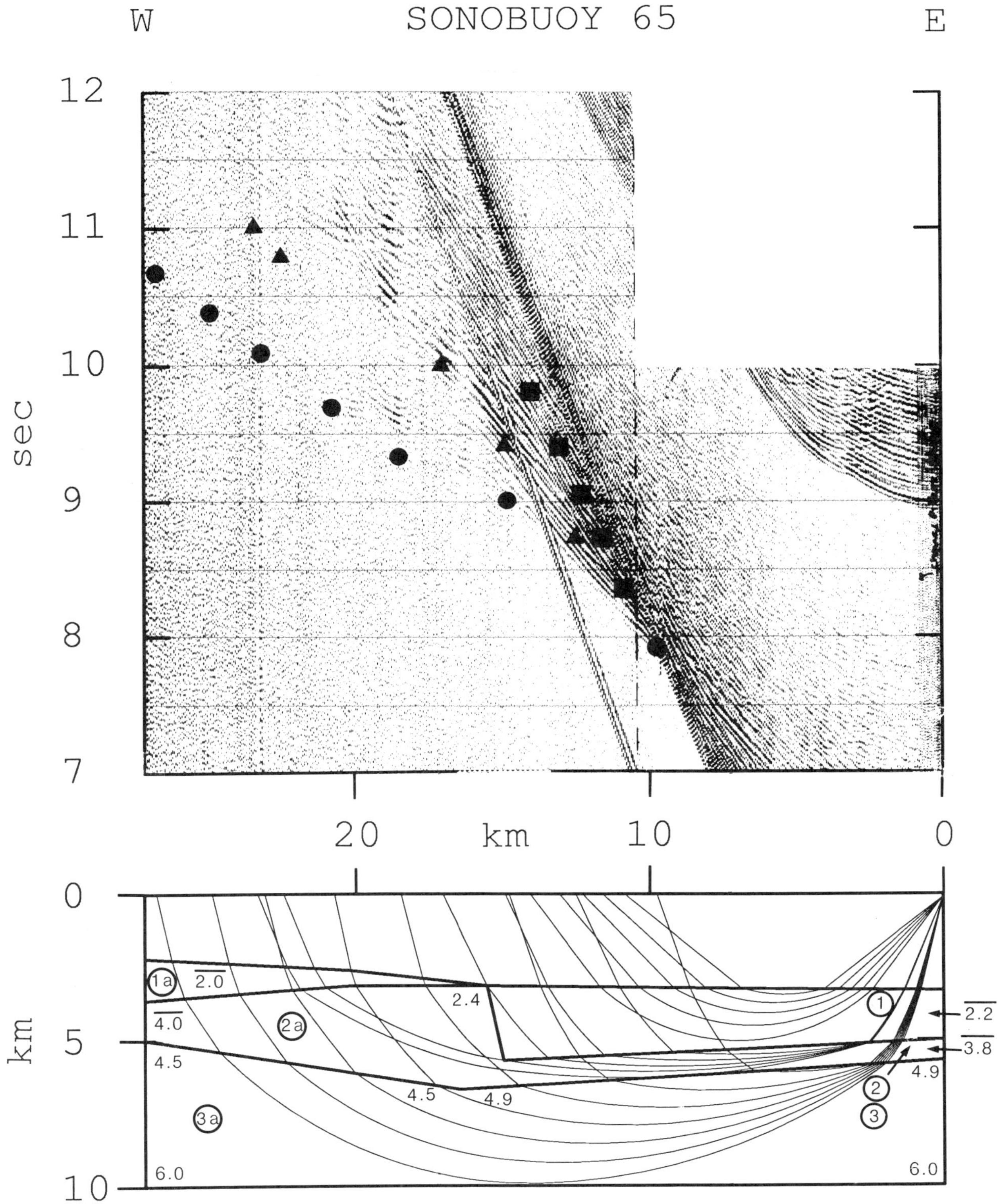

Figure 7. Data for sonobuoy 65 (top). Computed arrival times for the two-dimensional ray-trace model (bottom) are plotted on the data as squares, triangles and dots for layers 1, 2, and 3, respectively. Velocities shown for layer 1 (2.2 km/s) and 2 (3.8 km/s) are average velocities. Velocities for the top and bottom of layer 3 are shown. See Table 2 for complete model-input data.

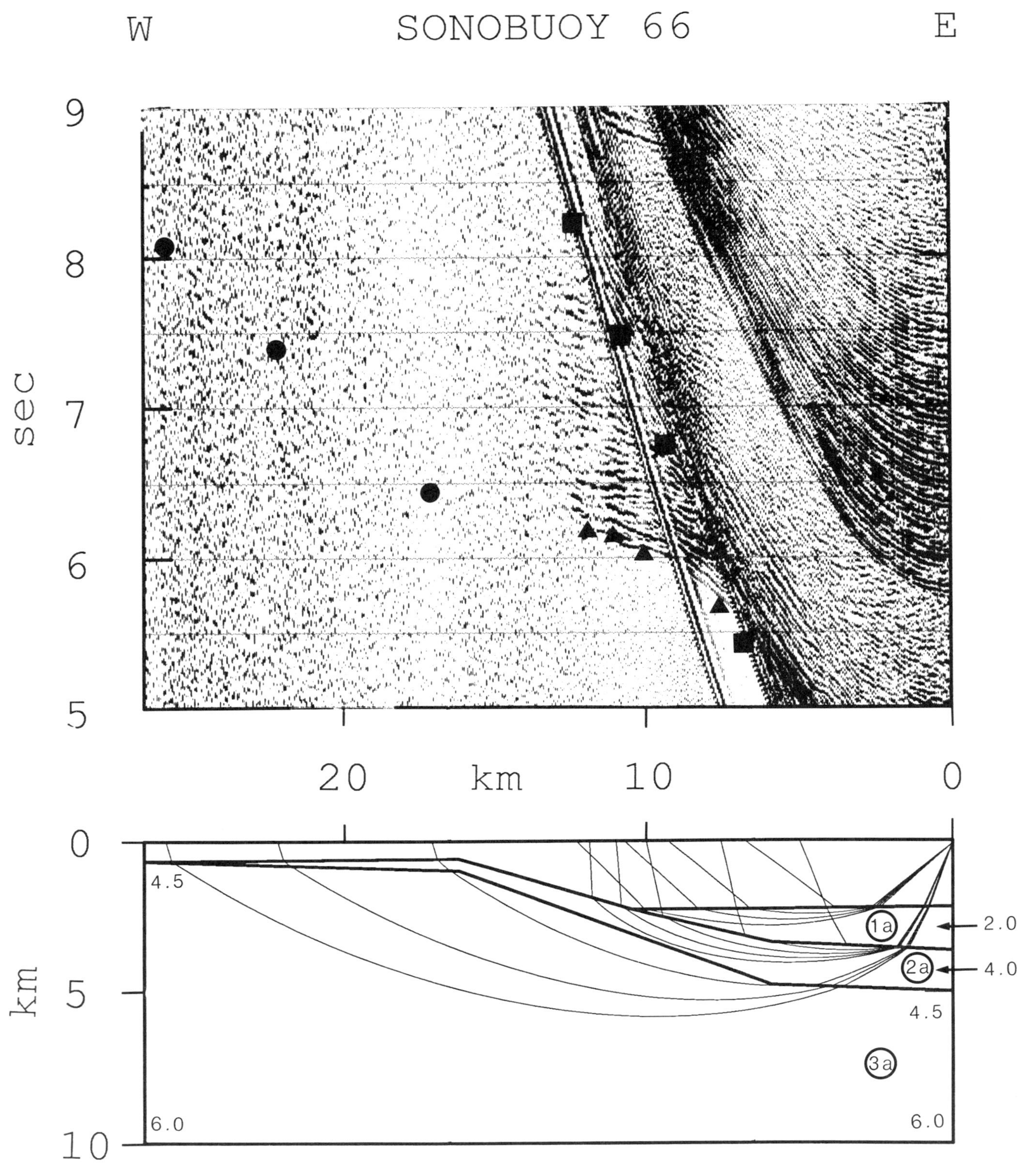

Figure 8. Data for sonobuoy 66 (top). Computed arrival times for the two-dimensional ray-trace model (bottom) are plotted on the data as squares, triangles and dots for layers 1a, 2a, and 3a, respectively. Velocities shown for layer 1a (2.0 km/s) and 2a (4.0 km/s) are average velocities. Velocities for the top and bottom of layer 3a are shown. See Table 3 for complete model-input data.

than 20 km are based principally on the results of the adjacent sonobuoy 66 model.

Sonobuoy 66 crosses the slope and the submerged crest of the arc (Figure 6). In the model (Figure 8, Table 3), layer 1 represents channel levee deposits. Using an average vertical velocity of 2.0 km/s and no horizontal gradient in the layer, a good fit of synthetic to real data resulted. Layer 2 represents accreted sediments. The curvature and abrupt attenuation of the refraction arrivals through layer 2 is reproduced by a combination of bathymetry, vertical velocity gradient, and thinning in the direction of the arc. Layer 3 represents the basement rocks of the arc, although arrivals from this layer are not well defined in the data from sonobuoy 66. However, the velocities of 4.5-6.0 km/s are constrained at the eastern edge of the model by the results of the adjacent sonobuoy 65.

DISCUSSION

Velocity profiles for Kilinailau Trench sonobuoys 63 and 65 (Figure 9) are similar to profiles from other areas of oceanic crust with high sedimentation and slow convergence (Cochrane, Lewis, and McClain, in press; von Huene, Shor, and Wageman, 1979; Spudich and Orcutt, 1980). Velocities in the upper trench-fill sediments are higher in the more deformed northwestern part of the trench, suggesting more consolidation of sedimentary strata that may have resulted from greater horizontal compression or overburden. Layer 2 has a higher velocity and a lower vertical velocity gradient than the overlying sediment. The velocity-depth curves for Ontong Java Plateau basalts for the two sonobuoys are similar. The results of this analysis of the Kilinailau Trench sonobuoys confirm the initial interpretation of the seismic-reflection data seaward of the trench.

Our modeling suggests that the upper layer of the lower slope along the northeastern side of Bougainville is composed of accreted sediments. The calculated velocities (2.2 and 2.4 km/s) for deformed rocks beneath the inner trench wall at the base of the arc slope (18 km, Figure 5; 16 km, Figure 7) are too low to be igneous or metasedimentary rocks. The small layer-1 horizontal velocity gradient used in the models for sonobuoys 63 and 65 are similar to horizontal gradients in deformed sediments accreted at the base of the Oregon-Washington slope (Cochrane, Lewis, and McClain, in press), and other convergent margins (Bray and Karig, 1985).

Table 3. Velocity and depth data for raytrace model of sonobuoy 66 (Figure 8).

Range (km)	Depth (km)	V_u	V_l
26.6	0.68	1.5	4.5
26.6	0.681	4.5	4.5
26.6	0.682	4.5	4.5
26.6	10.0	6.0	-
16.3	0.6	1.5	2.1
16.3	1.601	2.1	3.8
16.3	1.0	4.3	4.5
16.3	10.0	6.0	-
10.6	2.29	1.5	2.1
10.6	2.291	2.1	3.9
10.6	D_i	4.3	4.5
10.6	10.0	6.0	-
6.0	2.29	1.5	1.9
6.0	3.39	2.1	3.9
6.0	4.8	4.3	4.5
6.0	10.0	6.0	-
0.0	2.21	1.5	1.9
0.0	3.65	2.1	3.8
0.0	5.0	4.3	4.5
0.0	10.0	6.0	-

V_u = velocity above boundary in km/s.
V_l = velocity below boundary in km/s.
D_i indicates depth interpolated by raytracing program.

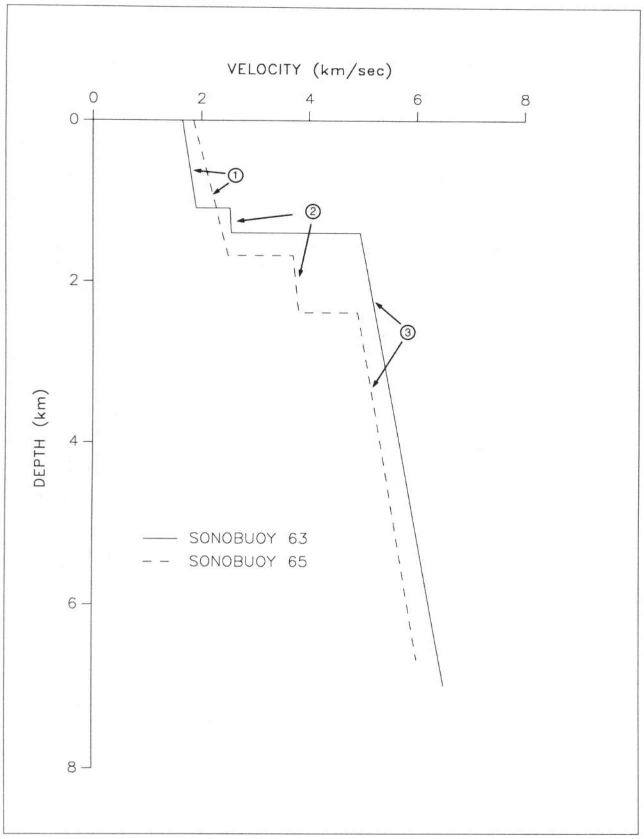

Figure 9. Velocity-depth profiles for Kilinailau Trench sonobuoys 63 and 65. Layer 1: trench-fill sediments. Layer 2: pelagite. Layer 3: Ontong Java Plateau basalt.

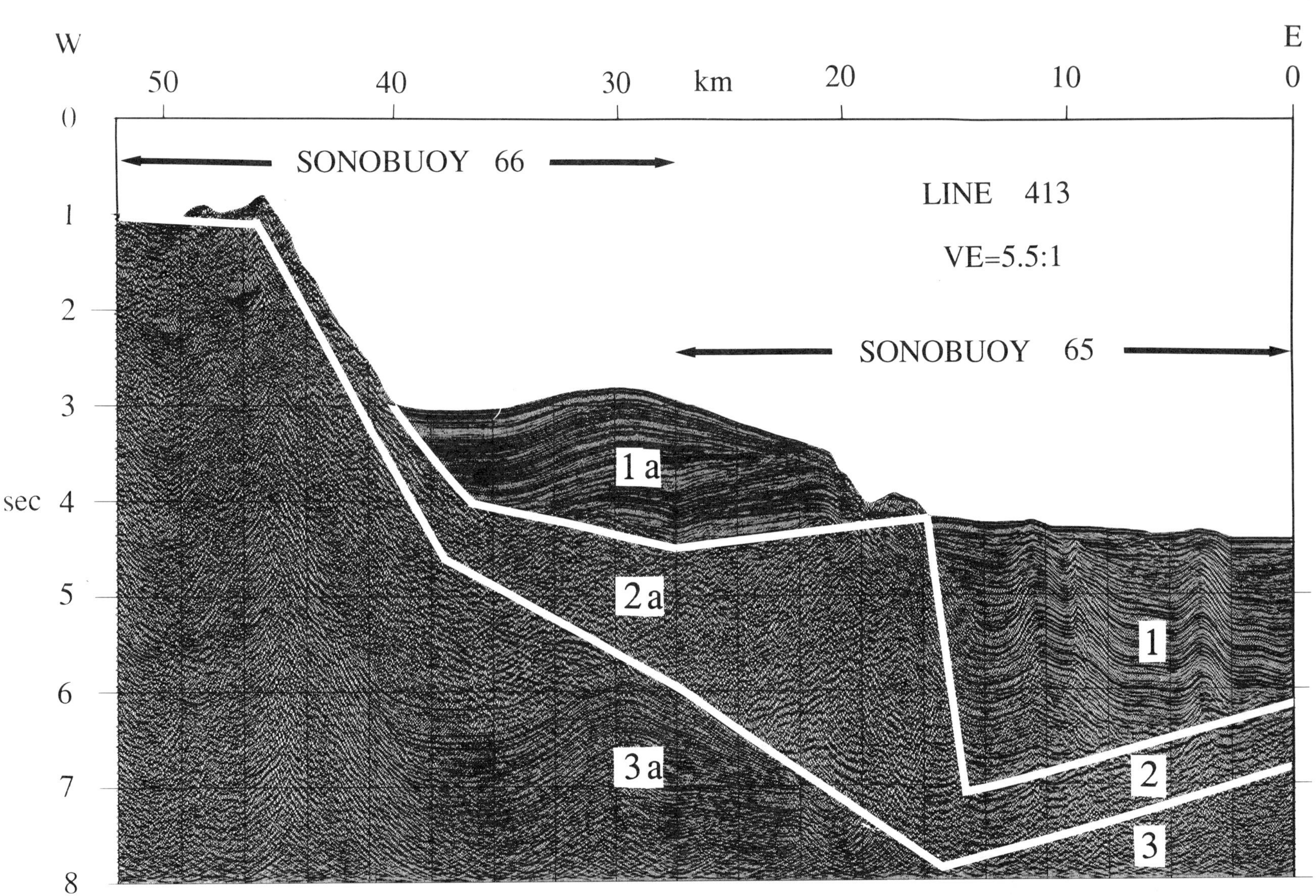

Figure 10. Migrated time-section of multichannel seismic-reflection line 413 with boundaries from ray-trace model (Figures 7 and 8) superimposed. Layer 1: Paleogene(?)-Neogene trench-fill sediments. Layer 2: Paleogene and Neogene(?) pelagite. Layer 3: Ontong Java Plateau basalt. Layer 1a: Neogene channel levee deposits. Layer 2a: Paleogene-Neogene(?) accreted sediments. Layer 3a: Bougainville Group and older volcanogenic arc-basement rocks.

Accreted sediments may grade into slope-apron sediment at the base of the arc flank (25-35 km, Figure 10). No structural contact between the two sedimentary bodies can be resolved in seismic-reflection or refraction data. Recent channel levee deposits that overlie accreted sediments and onlap the slope-sediment apron (unit 2a, Figure 10), suggest that the two lower sedimentary bodies are roughly coeval. Alternatively, the apron on the arc flank may be composed of uplifted accreted sediments. Uplift of Bougainville occurred during the late middle to early late Miocene (Blake and Miezitis, 1967) and was associated with incipient northeast-directed subduction.

Basement rocks of the Ontong Java Plateau can be distinguished from basement rocks of Bougainville on the basis of velocity. Both trench sonobuoys show similar velocity profiles for oceanic basalt (Figure 9) and an upper velocity of 4.9 km/s. Arc-flank sonobuoy 66 shows a lower velocity at the top of basement (4.5 km/s). Structure separating Bougainville basement, arc rocks, and Ontong Java Plateau basalts could not be resolved in the seismic-reflection data. The transition from basalt (unit 3, Figure 10) to volcanogenic arc basement (unit 3a, Figure 10) occurs beneath the base of the slope (18 km, Figure 10) based on seismic-refraction velocities from sonobuoy 65.

The topmost strata of unit 1 (Figure 10) are interpreted to be recent trench-fill sediments. These strata show significant compressional deformation. If our interpretation is correct, some of the horizontal stress resulting from oblique convergence of the Pacific and Australia-India plates has been accommodated in the Kilinailau Trench after the late Miocene reversal in arc polarity and the establishment of the New Britain Trench. This accommodation of stress implies that at the plate boundary between the arc, in the vicinity of Bougainville, the Pacific plate and arc continue to converge, contrary to the interpretation represented by Figure 1. Convergence on both the northern and southern margins of Bougainville supports a model of a rootless arc riding atop the boundary between the Australia-India and Pacific plates as suggested by Cooper, Marlow, and Bruns (1986) for the central and western Solomon Islands area.

ACKNOWLEDGMENTS

R.C. Culotta provided migrated multichannel seismic-reflection profiles for this study. The U.S. Navy supplied the sonobuoys used by the U.S. Geological Survey in this and other projects. M.S. Marlow and L.A. Beyer reviewed the manuscript.

REFERENCES CITED

Blake, D. H., and Miezitis, Y., 1967, Geology of Bougainville and Buka Islands, New Guinea: Australia Department of National Development, Bureau of Mineral Resources, Geology and Geophysics Bulletin N. 96 and Bulletin N. PNG 1, Canberra, Australia, 56 p.

Bray, C. J., and Karig, D. E., 1985, Porosity of sediments in accretionary prisms and some implications for dewatering processes: Journal of Geophysical Research, v. 90, p. 768-778.

Cochrane, G. R., Lewis, B. T. R., and McClain, K. J., in press, Structure and subduction processes along the Oregon-Washington margin, *in* Ruff, L. J., and Kanamori, H., eds., Subduction zones, Pure and Applied Geophysics, Special Issue.

Cooper, A. K., Marlow, M. S., and Bruns, T. R., 1986, Deep structure of the central and southern Solomon Island region: implications for tectonic origin, *in* Vedder, J. G., Pound, K.S., and Boundy, S. Q., eds., Geology and offshore resources of Pacific island arcs—central and western Solomon Islands: Houston, Texas, Circum-Pacific Council for Energy and Mineral Resources, v. 4, p. 157-176.

Drummond, K.J., Nishiwaki, C., Corvalan, J., Doutch, H.F., and Craddock, C., 1985, Plate tectonic map of the circum-Pacific region, Pacific basin sheet: Houston, Texas, Circum-Pacific Council for Energy and Mineral Resources, scale 1:17,000,000.

Kroenke, L. W., 1984, Solomon Islands: San Cristobal to Bougainville and Buka, *in* Kroenke, L.W., ed., Cenozoic tectonic development of the southwest Pacific: UN ESCAP, CCOP/SOPAC Technical Bulletin, n. 6, p. 47-61.

McMechan, G. A., and Mooney, W.D., 1980, Asymptotic ray theory and synthetic seismograms for laterally varying structures—theory and application to the Imperial Valley, California: Bulletin of the Seismological Society of America, v. 70, p. 2021-2035.

Spudich, P., and Orcutt, J., 1980, A new look at the seismic velocity structure of the oceanic crust: Reviews of Geophysics and Space Physics, v. 18, n. 3, p. 627-645.

Vedder, J.G., Colwell, J.B., and Bruns, T.R., 1985, New findings in offshore areas of western Solomon Islands and Buka-Bougainville, Papua New Guinea: 14th BMR Symposium, Extended Abstracts—Petroleum Geology of South Pacific Island countries, Marine Studies Workshop, Canberra, Australia, p. 37-41.

von Huene, R., 1988, Seismic images of modern convergent margin tectonic structure: American Association of Petroleum Geologists Studies in Geology, n. 26, 60 p.

von Huene, R., Shor, G. G., and Wageman, J., 1979, Continental margins of the eastern Gulf of Alaska and boundaries of tectonic plates, *in* Watkins, J. S., Montadert, L., and Dickerson, P.W., eds., Geological and geophysical investigations of continental margins: American Association of Petroleum Geologists Memoir 29, p. 273-290.

Vedder, J.G., and Bruns, T.R., editors, 1989, Geology and offshore resources of Pacific island arcs—Solomon Islands and Bougainville, Papua New Guinea Regions: Houston, Texas, Circum-Pacific Council for Energy and Mineral Resources Earth Science Series, v. 12.

GEOLOGIC FRAMEWORK OF QUEEN EMMA BASIN, EASTERN PAPUA NEW GUINEA

J.G. Vedder, T.R. Bruns, A.K. Cooper
U.S. Geological Survey, Menlo Park, CA 94025, USA

ABSTRACT

Queen Emma basin (new name) is a northwest-trending structural and depositional intra-arc basin that lies between New Ireland and Bougainville in eastern Papua New Guinea. The southwestern margin of the basin is formed by the combination of Kaviki Ridge (new name), a subsided segment of the early Cenozoic North Solomon Island Arc, and southeastern New Ireland, an uplifted segment of the same arc. The northeastern margin of the basin is marked by the continuation of Emirau-Feni Ridge through the Green Islands and Buka Island, which together are part of the late Cenozoic South Solomon Island Arc that cuts northwestward across the west-trending early Cenozoic arc in central Bougainville. Oligocene and younger volcaniclastic strata derived from flanking ridges were deposited as forearc aprons of sediment that now underlie the intra-arc basin. During the Oligocene and early Miocene, these forearc sediments came primarily from sources along the high-standing Kaviki Ridge and were transported northeastward while subordinate amounts of detrital material may have moved southward into the same area from ancestral Nuguria Ridge, a possible forearc high at that time. The approach and arrival of the Ontong Java Plateau may have checked subduction-related magmatism as early as the early Miocene, and after a period of relative tectonic quiescence in the middle Miocene, arc polarity reversed in the late Miocene. Uplift along the newly formed southwest-facing magmatic arc resulted in accumulation of late Miocene and younger basin sediments that were derived mostly from sources on ancestral Bougainville to the southeast. The maximum composite thickness of Oligocene and younger strata in Queen Emma basin may be as much as 8.2 km.

Northwest-trending structures prevail in the region between the southeastern New Ireland-Kaviki Ridge culmination and the Kilinailau Trench. Folds include a buried late Miocene(?) anticlinal feature in Queen Emma basin and the Pliocene to Holocene broad antiform that underlies the southern continuation of Emirau-Feni Ridge and Buka Island. Faults occur in three main zones: 1) the southeastern offshore continuation of the Weitin fault zone on New Ireland, 2) along the northeast side of Kaviki Ridge, and 3) along the buried anticlinal structure in Queen Emma basin. The old, buried structures may be related to enigmatic processes that were operative during arc-polarity reversal near the end of the Miocene. The younger structures that deform the seafloor may be the result of either transpression associated with the transform zone that separates New Britain and New Ireland or convergence accompanying back-arc thrusting along the Kilinailau Trench.

INTRODUCTION

Purpose and Scope

Investigation of the offshore area between New Ireland and Bougainville was part of the internationally sponsored marine surveys conducted under the Australia-New Zealand-United States Tripartite Agreement of 1984 and supervised by the Committee for Coordination of Joint Prospecting for Mineral Resources in South Pacific Offshore Areas (CCOP/SOPAC). The marine data were collected during a cruise segment of the U.S. Geological Survey's R/V *S.P. Lee* in June 1984 (Vedder and Colwell, this volume) with funding largely from the U.S. Agency for International Development.

The primary purpose of this report is to describe the stratigraphy, structure, and inferred tectonic history of a poorly known intra-arc basin that adjoins the northern Solomon Sea and to relate the geology of this basin to that of adjacent segments of the Melanesian arc system (Melanesian Borderlands of Coleman and Packham, 1976) as described by Kroenke (1984). The western and northern perimeter of the basin is delimited geographically by the southeastern coast of New Ireland together with the Feni

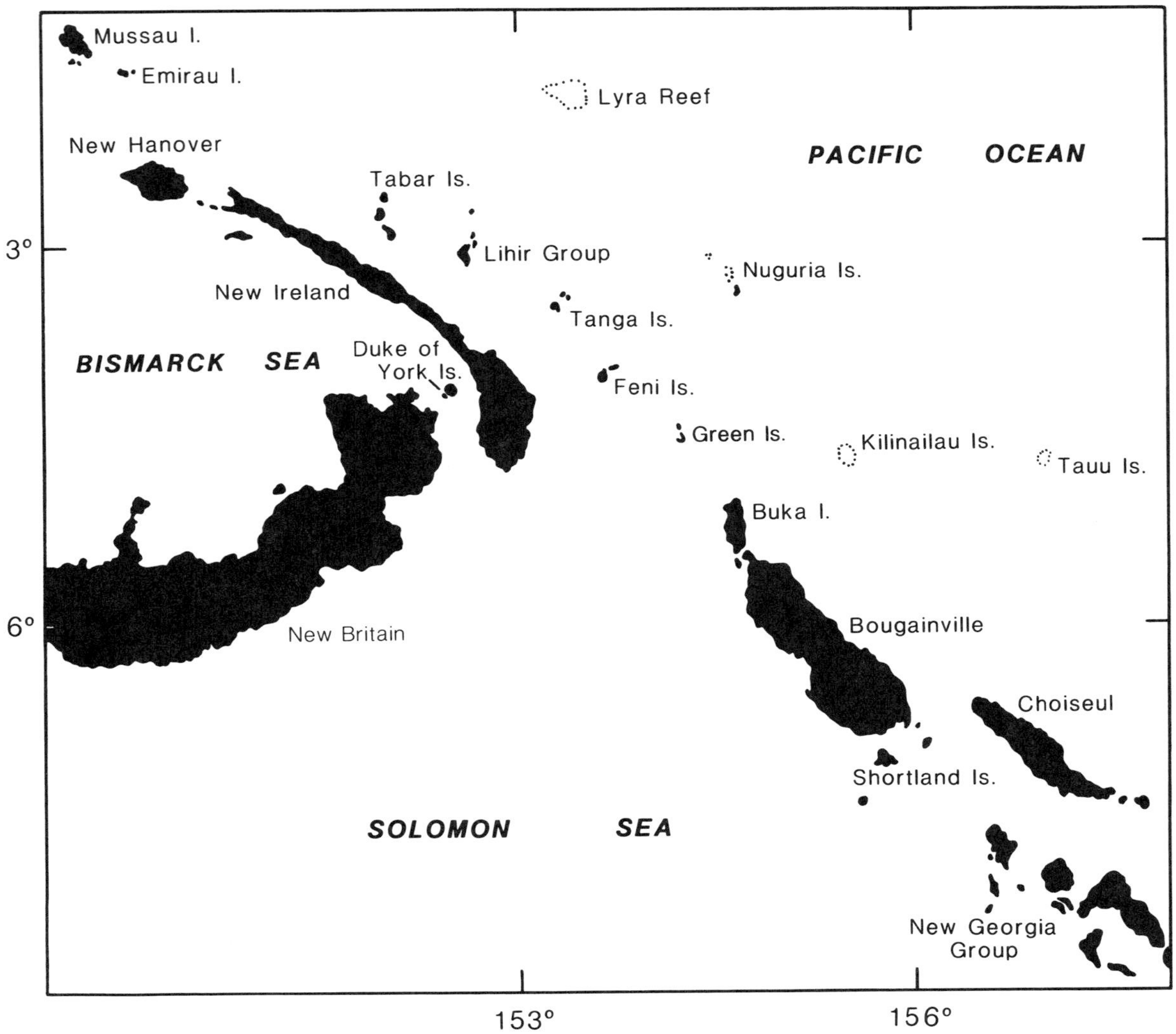

Figure 1. Regional map of eastern Papua New Guinea and the western Solomon Islands showing conventional names for large islands, small island groups, and reefs.

and Green island groups (Figures 1 and 2). The eastern and southeastern edge of the basin is rimmed by Buka Island and Bougainville, and the southern part is open to the Solomon Sea.

The Geologic Feature and Derivation of Its Name

In accordance with the application of names to structural sub-basins and depocenters in the Central Solomons Trough (Tiffin, Vedder, and Cooper, 1983), the name Queen Emma basin is herewith assigned to the structural and depositional feature that underlies the northern reentrant of the Solomon Sea between New Ireland and Bougainville (Figures 1 and 2). The basin is named for Emma Coe Forsayth Kolbe (1850-1913), who was instrumental in the early development of the copra trade in the region (Robson, 1973). She was the eldest daughter of the U.S. Commercial Agent and Consul in Samoa, J.M. Coe, and his Samoan wife, Le'utu Taletale. Emma, her trader companion, Thomas Farrell, and other partners, became successful planters in the Duke of York Islands and neighboring areas in the late 1870s and 1880s and expanded their holdings into New Britain, New Ireland, Bougainville, and other island groups. To her contemporaries, colleagues, and commercial rivals alike, she was known as "Queen Emma" throughout the southwest Pacific.

Previous Work

Although the islands surrounding Queen Emma basin have been relatively thoroughly investigated by geologists, offshore studies have been sporadic, few, and reconnaissance in nature. Significant among the

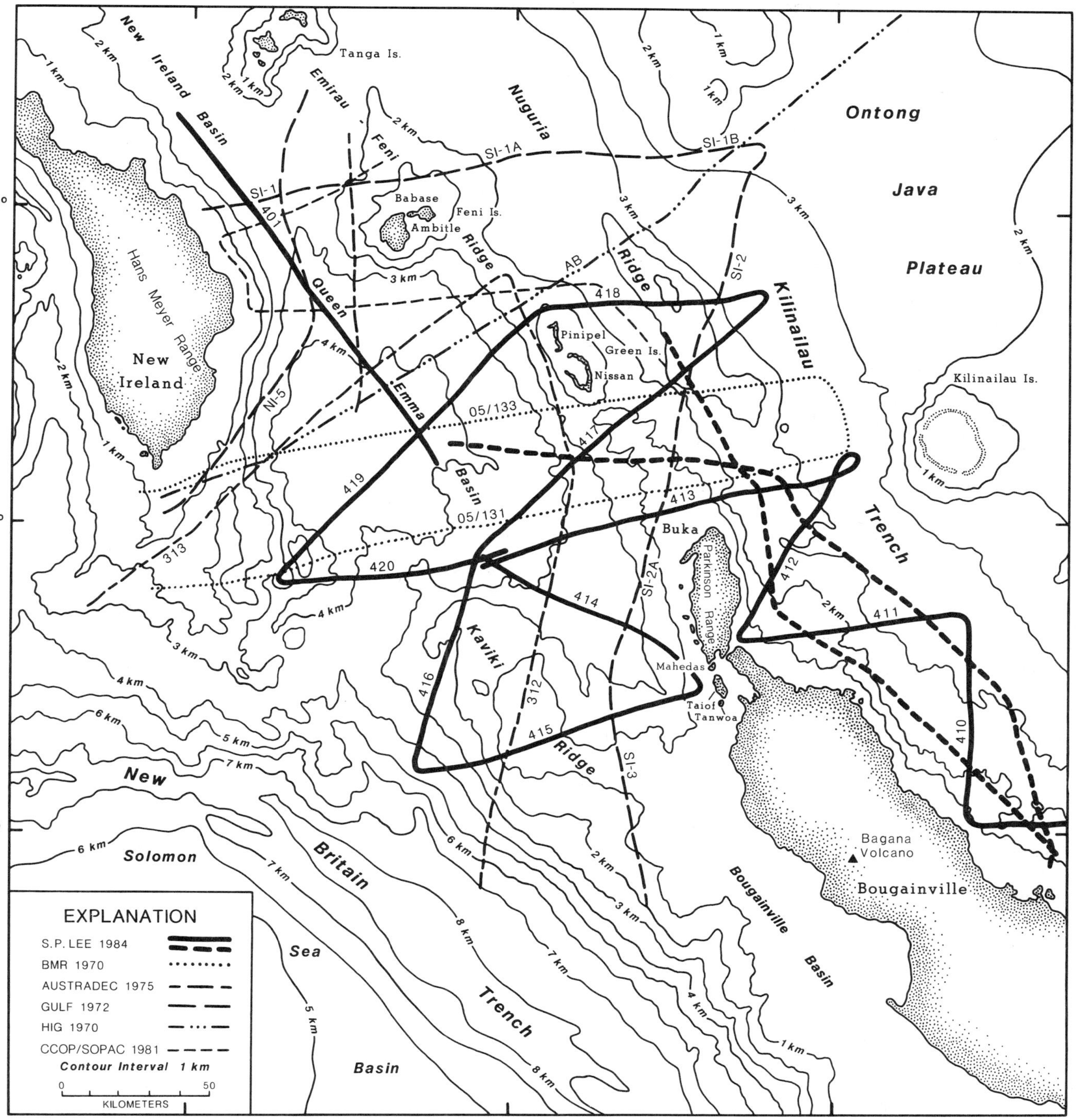

Figure 2. Ship tracklines and line designations, Queen Emma basin. Map includes generalized bathymetric contours and names of seafloor features and depositional basins used in the text. R/V *S.P. Lee* multichannel seismic-reflection trackline shown by heavy solid lines; single-channel tracklines, by heavy dashed lines.

publications on island geology are those for New Ireland by Hohnen (1978) and Stewart and Sandy (1988), the Tabar to Feni island chain by Wallace et al (1983), and Buka and Bougainville by Blake and Miezitis (1967). In this report, the authors rely heavily upon the onshore geology described in the papers cited above.

Regional syntheses include Kroenke's (1984) overview, Taylor's (1979) tectonic analysis, Johnson's

(1979) history of magmatism, and Falvey and Pritchard's (1984) reconstructions. Adjoining offshore areas in the New Ireland basin recently have been described by Exon and Tiffin (1984), Exon et al (1986), and Marlow et al (1988). Bougainville basin, which lies southeast of Queen Emma basin (Figure 2), has been studied in detail by Shaw (1985) and Stewart, Francis, and Pederson (1987).

Widely spaced single-channel seismic-reflection profiling across the basin was done in 1970 by the Hawaii Institute of Geophysics (HIG) and in the same year by Compagnie General de Geophysique for the Australian Bureau of Mineral Resources (BMR) (Figure 2). Gulf Research and Development Company (Gulf) ran two multichannel seismic-survey lines along the eastern and western margins of the basin in 1972. The consortium of Institut Français du Pétrole, Comite d'Études Pétrolières Marines, and Office de la Recherche Scientifique et Technique Outre-Mer (IFP-CEPM-ORSTOM) collected multichannel data on their AUSTRADEC III cruise, which crossed the deep part of the basin, paralleled the northeast flank, and traversed the southeastern segment in 1975. A CCOP/SOPAC cruise using a single-channel seismic-reflection system crossed the northern end of the basin in 1981.

Geomorphology and Names of Subsea Features

The bathymetric configuration of Queen Emma basin is roughly elliptical below the 3,500-m isobath, but above the 2,500-m isobath, the shape is increasingly elongate and almost triangular with the apex at the southeastern end (Chase, Seekins, and Young, this volume). Bathymetric profiles across the longest (~200 km) and widest (~125 km) segments of the basin are asymmetric. The deepest seafloor is in the western part of the basin where maximum water depths are approximately 4,800 m. Water depth to the sill at the open southern margin is about 3,500 m and nearly 2,900 m in the passage into the New Ireland Basin; the Emirau-Feni Ridge on the northeast is less than 1,900 m deep in its deepest part. The western, northern, and northeastern flanks of the bathymetric feature steepen above the 3,000 m isobath. In general, the thickest sedimentary sequences underlie the areas of deepest water.

New names applied to subsea features by Exon et al (1986) have been adopted for this paper. Because some of the informal names assigned by earlier workers were confusing or inconsistent, additional new names are proposed herein (Figure 2). These adopted and newly proposed names and their informal precursors are as follows:

1. Emirau-Feni Ridge (Exon et al, 1986) = North East ridge (de Broin et al, 1977) and Tabar-Feni-South Solomon ridge (Kroenke, 1984).
2. Nuguria Ridge (Exon et al, 1986) = no precursor.
3. Kaviki Ridge (new name) = South West ridge (de Broin et al, 1977) and New Ireland ridge (Kroenke, 1984).
4. Queen Emma basin (new name) = South New Ireland basin and Solomon basin (de Broin et al, 1977) and Buka basin (Vedder, Colwell, and Bruns, 1985).

Other names shown on Figure 2 are those used by most workers in the region and are depicted on the bathymetric maps of the southwest Pacific by Kroenke, Jouannic, and Woodward (1983).

Regional Tectonic Setting

Queen Emma basin lies within the Melanesian Borderlands island-arc system, a broad zone of complex tectonism between the convergent Pacific and Australia-India plates. Various earlier interpretations of the configuration of microplates, spreading-ridge systems, and magmatic-arc elements in the Papua New Guinea-western Solomon Islands region (Figure 3) were reviewed by Johnson (1979), who asserted that at least six (and perhaps as many as ten) plate boundaries are present in this zone of convergence. Taylor (1979), however, rejected some aspects of this concept, in particular, the idea that the North Bismarck plate is part of the Pacific plate. He proposed back-arc spreading as the primary tectonic mechanism for development of the Manus Basin.

Exon and Tiffin (1982) suggested that the New Ireland basin formed as a late Eocene forearc basin between the West Melanesian arc and trench, the southwestern flank of which was truncated by a transform system that was initiated in the Pliocene as Manus Basin began to open. Reconstructions by Falvey and Pritchard (1984), based largely upon paleomagnetic data, indicate that the Solomon Islands, Bougainville, New Ireland, and New Britain were segments of a continuous, northwest-trending magmatic arc during the Oligocene and that much of the northern Solomon Islands arc was part of the Australia-India plate before the late Miocene, but thereafter became part of the Pacific plate following arc reversal.

Most workers in the region invoke three episodes of Cenozoic tectonism to explain the geologic relations in the Bougainville-Solomon Islands segment of the Melanesian arc system. Not included here is the obscure development of the Paleocene(?) to Eocene(?) Rennell arc (Kroenke, 1984). The first well-documented episode was the late Eocene(?) and early Oligocene southwest-directed subduction of the Pacific plate that resulted in island-arc magmatism. The second episode began with the collision of the Ontong Java Plateau with the arc and was characterized by waning volcanism, uplift, and rapid sedimentation during the late Oligocene and early Miocene, and subsequent Miocene deposition of volcanogenic detritus and subordinate carbonates on upwarped shelves and basin slopes. The third episode, which began in the late Miocene and continues to the present, included arc-polarity reversal and northward-directed subduction of

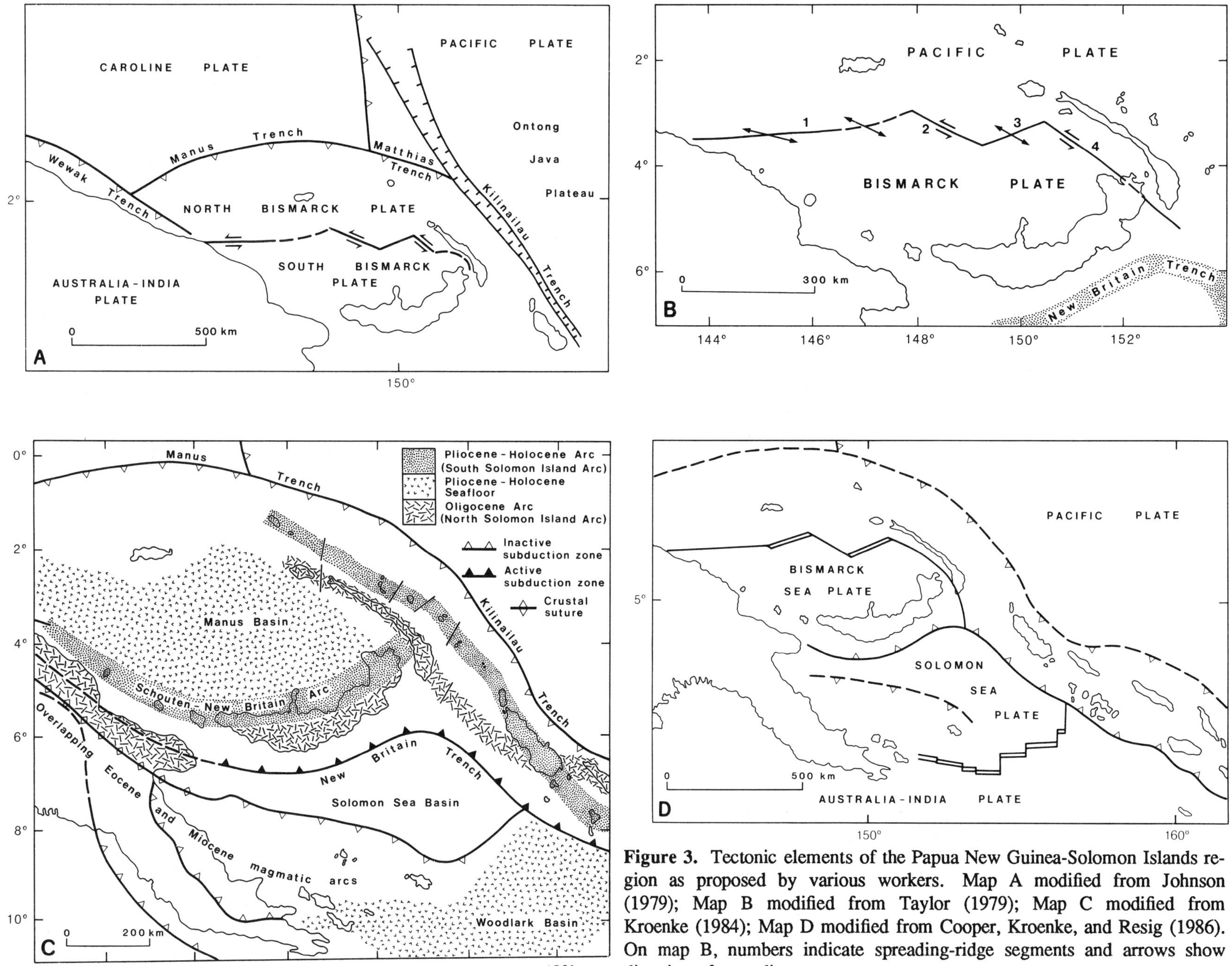

Figure 3. Tectonic elements of the Papua New Guinea-Solomon Islands region as proposed by various workers. Map A modified from Johnson (1979); Map B modified from Taylor (1979); Map C modified from Kroenke (1984); Map D modified from Cooper, Kroenke, and Resig (1986). On map B, numbers indicate spreading-ridge segments and arrows show direction of spreading.

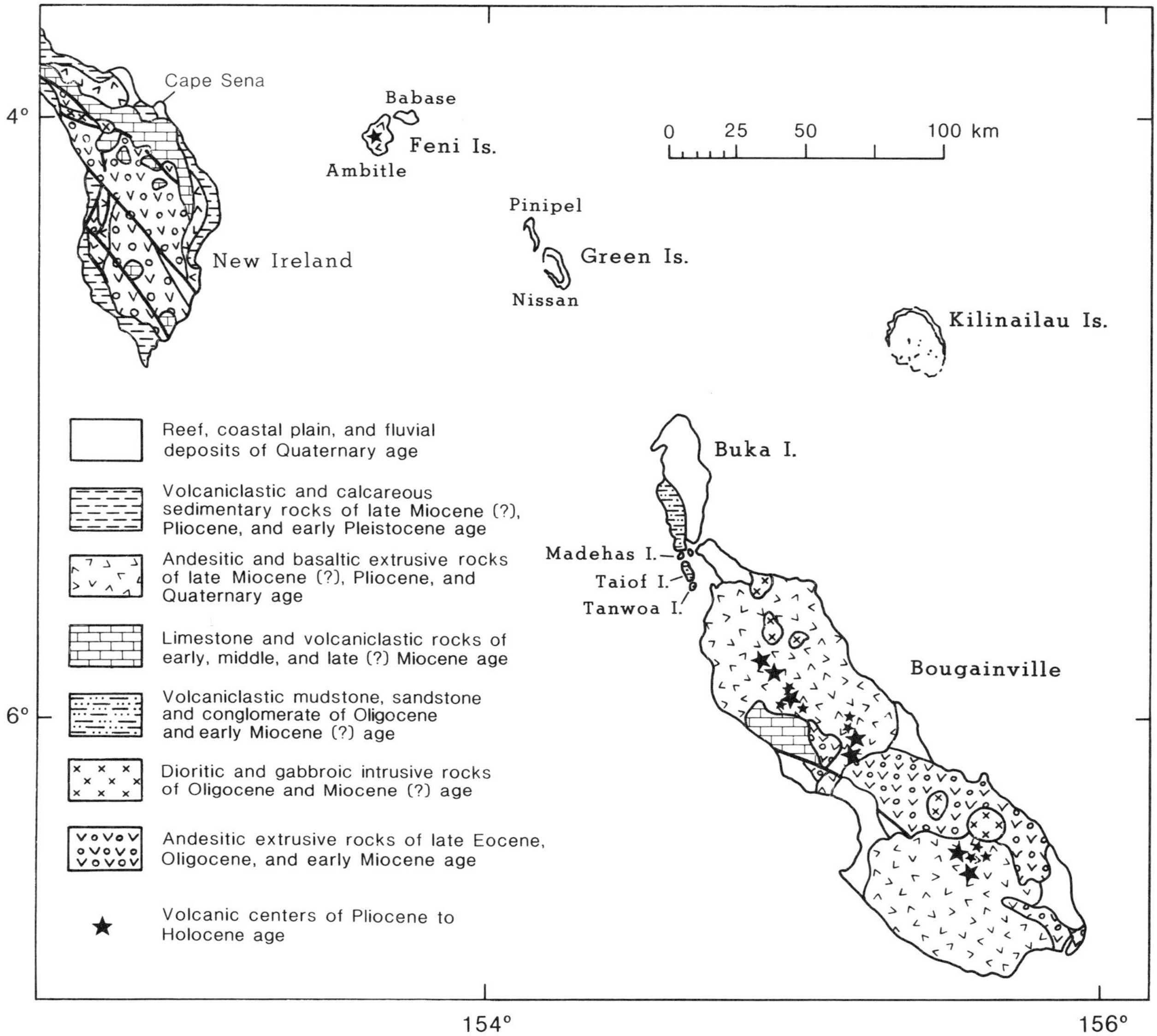

Figure 4. Generalized geologic map of islands adjacent to Queen Emma basin. Modified from Bain et al (1972). See Figure 5 for formation names on New Ireland, Buka Island, and Bougainville and their inferred stratigraphic relations and correlations. A revised geologic map of Bougainville is included in Hilyard and Rogerson (this volume).

Solomon Sea plate crust and nascent oceanic crust of the Woodlark spreading-ridge system, which resulted in the creation of a new magmatic arc. Direct effects of the third episode were continued uplift of New Georgia and Bougainville and subsidence accompanied by high rates of sedimentation in the intra-arc basins. In his synthesis of the Melanesian arc systems, Kroenke (1984, Figures 3.3 and 4.3) showed the early Cenozoic arc (North Solomon Island Arc) passing through Choiseul and New Ireland, and the late Cenozoic arc (South Solomon Island Arc) through New Georgia and the Tabar-to-Feni island chain with an apparent cross-of the new arc through the old arc at Bougainville (Figure 3C). If these events and regional relations are accepted, Queen Emma basin represents a subsided intra-arc basin that received sediment in the form of overlapping forearc aprons, first from the old arc rocks beneath Kaviki Ridge to the southwest and subsequently from the new arc to the northeast. Sediment sources concentrated along high-standing parts of the old arc may have been supplemented by subordinate sources along a possibly emergent old forearc structural high, the Nuguria Ridge, which is situated between the Kilinailau Trench and Emirau-Feni Ridge (Figure 2).

GEOLOGY OF ISLANDS CONTIGUOUS TO THE BASIN

Before making provisional correlations of exposed rock units and seismic-stratigraphic units, a review of the geology of the islands that lie on the perimeter of Queen Emma basin is essential. This review is based principally on the work of others, except for a few observations made by Vedder during brief visits to Buka Island and Bougainville in 1985 and 1987. A highly generalized geologic map and simplified stratigraphic columns for New Ireland, Buka Island, and central Bougainville are shown in Figures 4 and 5 (After this paper was completed, the stratigraphy of Buka Island and Bougainville was revised by Hilyard and Rogerson, this volume). Inferred correlations of the exposed rock with the basin sequences are discussed in the section on seismic stratigraphy.

Southeastern New Ireland

Descriptions of the geology of southeastern New Ireland used herein are based entirely upon the work of Hohnen (1978), Exon et al (1986), and Stewart and Sandy (1988). The oldest exposed rocks are the Jaulu Volcanics which have an age range of Eocene(?), Oligocene, and early Miocene(?). These volcanogenic rocks, which may be more than 2,000 m thick, consist largely of massive calc-alkaline andesitic agglomerate and flows, lapilli and crystal-lithic tuff, and ignimbrite. Subordinate strata in the Jaulu Volcanics are volcaniclastic sandstone and small lenses of tuffaceous limestone. Bodies of gabbro, dioritic rocks, and leucocratic rocks of variable composition known as the Lemau Intrusive Complex have yielded a K/Ar age of 37.4 ± 1.7 Ma (central New Ireland) and a range of 32.6 ± 1.0 to 14.2 ± 0.5 Ma (southern New Ireland) (Hohnen, 1978; Stewart and Sandy, 1988).

The Lelet Limestone of probable early to late Miocene age unconformably overlies the Jaulu Volcanics and Lemau Intrusive Complex and may be as much as 1,000 m thick at places. Rock types in the limestone are mainly massive to thick-bedded foraminiferal-algal biomicrite and microsparite. Fossils, including foraminifera, echinoids, corals, mollusks, and calcareous algae, indicate that these calcareous strata accumulated in neritic depositional environments. Fine-grained clastic sedimentary rocks that unconformably overlie the Jaulu Volcanics in southeastern New Ireland recently were named the Tamiu Siltstone by Stewart and Sandy (1988). The lower part of the unit consists chiefly of massive calcareous siltstone; the upper part consists of tuffaceous siltstone and sandy mudstone. These strata have a probable age range of early into middle Miocene and may represent a regressive clastic facies deposited during early phases of progradation by the platform carbonates of the Lelet Limestone. Deposition of the Tamiu Siltstone took place in increasingly shallow water depths; the lower part (early? and middle Miocene) is upper bathyal to outer neritic, and the upper part (middle Miocene) probably is inner neritic (Stewart and Sandy, 1988). The unit is locally 300 m or more thick.

The Punam Limestone of probable late Miocene and early Pliocene age occurs mostly along the northeast coast and extends southeastward as far as Cape Sena. The formation is composed largely of massive, friable foraminiferal biomicrite that contains variable amounts of tuffaceous material. Middle to lower bathyal depositional environments are inferred from the faunal content. A minimum thickness of 200 m is estimated for the sequence.

Early Pliocene to early Pleistocene epiclastic beds assigned to the Uluputur unit of the Rataman Formation by Stewart and Sandy (1988) are distributed in fault blocks along the east coast of southeastern New Ireland. Interbedded strata in the unit have diverse compositions that include calcareous silty sandstone, siltstone, biomicrite, conglomerate, and pebbly mudstone. Clasts in the coarser grained beds were derived from underlying formations. Bedding characteristics and foraminiferal content suggest that most of the strata in the Uluputur unit represent deposition in submarine-fan environments at water depths of 500 m or more. However, conglomerates in the uppermost part may be channel deposits that accumulated in upper bathyal and outer neritic depths.

Coarse clastic deposits that unconformably overlie the Uluputur unit, Lelet Limestone, and Jaulu Volcanics are designated the Maton Conglomerate. This Pleistocene conglomeratic sequence consists mainly of volcanolithic conglomerate that contains lenticular sandstone beds. Most of the clasts in the conglomerate were derived from the Jaulu Volcanics and deposited in alluvial-fan environments. Uplifted reef limestone in the easternmost part of the island is a lateral equivalent of the Maton Conglomerate (Hohnen, 1978; Stewart and Sandy, 1988).

Feni and Green Islands

The geology of the Feni Islands, which include Ambitle and Babase, was described by Wallace et al (1983). According to these workers, Ambitle is a Pliocene and Pleistocene stratovolcano that is constructed, at least in part, on middle or late Oligocene marine limestone. This deep-water limestone consists of unconsolidated bedded calcilutite that is known to be exposed only along a short tract of stream bed in the northern part of the island. Most of the overlying thick volcanic succession on the flanks of the composite central-vent structure is made up of interbedded mafic and intermediate lava, lahar deposits, and tuff.

Babase is formed by two eroded Pliocene and Pleistocene volcanoes, the eastern remnant of which is composed mainly of basaltic and trachybasaltic flows and the western remnant of which is a dissected trachytic cumulodome that is overlain by Pleistocene reef limestone similar to uplifted reef deposits on northern Ambitle.

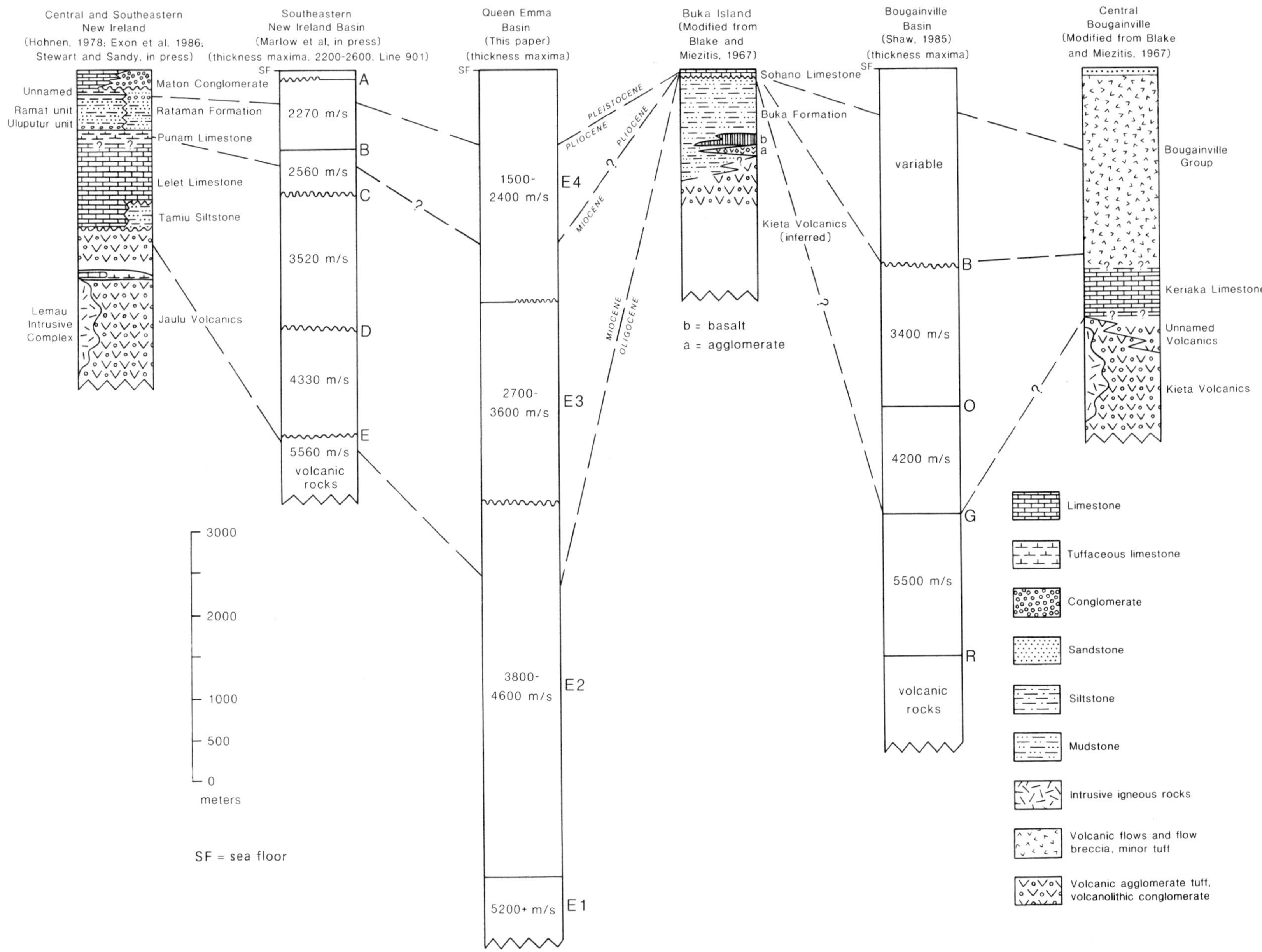

Figure 5. Generalized stratigraphic columns based upon composite island sections and interpretations of offshore seismic-reflection profiles. Correlations are inferred. Sources of data are given at the top of each column (Revised stratigraphy for central Bougainville is given in Hilyard and Rogerson, this volume).

Hornblende concentrates of Babase trachyte gave a radiometric age of 1.53 ± 0.15 Ma, and age determinations on biotite from Ambitle trachyte provided ages of 0.68 ± 0.10 and 0.49 ± 0.10 Ma (Wallace et al, 1983). The trachytic composition of rocks on the Feni Islands contrasts markedly with that of the Pliocene to Holocene andesitic volcanic rocks on Bougainville (Wallace et al, 1983). According to Wallace et al (1983), the trachytic rocks may represent the final stages of volcanism in the Tabar-to-Feni chain where contamination by intracrustal melting may have changed the original andesitic composition of the magmas.

The Green Islands, which include Nissan and Pinipel, are atoll-like and are composed of uplifted Quaternary reef deposits (Bain et al, 1972). Because stratified rocks about 1-km thick are evident on seismic-reflection profiles across the ridge crest northwest and southeast of the islands, the rocks underlying these islands presumably consist of late Cenozoic volcanogenic rocks and possibly older strata that may correlate with the Buka Formation and rocks on the Feni Islands.

Buka Island

The oldest known rocks on Buka Island consist mainly of volcaniclastic mudstone and sandstone turbidites of the Buka Formation that Blake and Miezitis (1967) inferred to be of late Oligocene to early Miocene age. This westward-dipping homoclinal sequence forms most of the north-trending Parkinson Range and extends southward onto Madehas, Taiof, and Tanwoa islands (Figures 2 and 4). Blake and Miezitis (1967) reported subordinate amounts of agglomerate, tuff, and basaltic lava in the Buka Formation and implied, but did not make a direct correlation with the Kieta Volcanics on Bougainville (See Hilyard and Rogerson, this volume, for revised stratigraphy).

Although Blake and Miezitis (1967) stated that the Buka Formation apparently is unfossiliferous, samples of tuffaceous clayey mudstone from the southern part of the Parkinson Range have yielded calcareous nannofossils (Table 1) of probable middle Oligocene age (P.J. Quinterno, written communication, 1987). These fossil assemblages indicate direct correlation with parts of both the basement volcanics on Bougainville and the Jaulu Volcanics on New Ireland. Because the fossiliferous samples are from the middle of a subaerially exposed sequence that may be more than 1,000-m thick, outcrops of the Buka Formation could range in age from Eocene to early Miocene. Moreover, a conformably overlying sequence offshore is as much as 2,000-m thick and is likely to include strata of early Miocene age. Inferred stratigraphic position and compositional similarity of some beds strongly suggest that the Buka Formation is a lateral facies of the upper part of the basement volcanics on Bougainville. On Madehas, Taiof, and Tanwoa islands, lenticular beds of poorly sorted volcaniclastic conglomerate are interlayered with finer grained strata. Some of these conglomerate beds are strikingly like fanglomerate sequences described by Blake and Miezitis (1967) in the vicinity of Kieta.

Pliocene(?) and Pleistocene reef deposits that make up the Sohano Limestone occur in an uplifted and tilted platform around the northern and eastern parts of the island. This limestone platform dips very gently westward and southward from its culmination in sea cliffs nearly 100-m high at the northern end of the island. The Sohano Limestone extends southeastward from Buka Island about 45 km along the northern coast of Bougainville. Uplifted reefs in the Green Islands presumably are correlatives.

Bougainville

Rocks exposed on Bougainville were mapped and described by Blake and Miezitis (1967); their nomenclature and age assignments are used herein except as noted (See Hilyard and Rogerson, this volume, for revised stratigraphy of the units noted below). The oldest outcropping rocks are the late Oligocene(?) and early Miocene(?) Kieta Volcanics, which form a broad, roughly west-trending belt through the midsection of the island. Calc-alkaline andesitic and basaltic flows, agglomerate, tuff, and water-laid volcaniclastic strata make up most of the formation, which is more than 1,500-m thick. The flows are largely subaerial. Dioritic intrusions are locally present in the Kieta Volcanics as dike- and stock-like bodies surrounded by hornfelsed volcanic and volcaniclastic rocks. Imbrication of cobbles and boulders in volcanic conglomerate exposed near Kieta indicates transport directions of north to northwest (Vedder, unpublished data). If the basement volcanics on Bougainville are assumed to correlate directly with the Jaulu Volcanics on New Ireland and the upper part of the Voza Lavas on Choiseul, the age assigned by Blake and Miezitis (1967) should be extended to include all of the Oligocene and possibly late Eocene. Furthermore, the age and composition of some of the strata in the Buka Formation lead us to believe that it is a lateral facies of the upper part of the Kieta Volcanics.

In the west-central part of Bougainville, coralline-algal reef deposits of the Keriaka Limestone unconformably overlie the basement volcanics and contain a shallow-water foraminiferan assemblage of early Miocene age. Whether these limestone beds represent fringing reefs or platform deposits is uncertain. According to Blake and Miezitis (1967), these massive limestones form a plateau that dips west-southwest about 5 degrees. They further state that the Keriaka Limestone apparently is overlain at places by unnamed Miocene(?) and Pliocene(?) sequences of andesitic, basaltic, and dacitic flows, agglomerate, and tuff that are locally intruded by microdiorite. Elsewhere, the formation is overlain by various volcanogenic units in the Pliocene(?) to Holocene Bougainville Group.

By far the largest areas of the island are underlain by calc-alkaline andesitic flows, agglomerate, tuff, and associated detrital material of the Bougainville Group. These Pliocene(?) to Holocene volcanogenic rocks are

Table 1. Ages and Ranges of Calcareous Nannoplankton Species from the Buka Formation. Identifications and age assignments furnished by P.J. Quinterno.

Locality	Identified species	Range (calcareous nanno-fossil zones of Okada and Bukry, 1980)	Age
1A	*Sphenolithus distentus*	CP18 - 19a	Middle - late Oligocene
	Dictyococcites scrippsae	CP13 - CP19b	Middle Eocene - late Oligocene
	D. bisectus	CP14b - CP19b	Middle Eocene - late Oligocene
	Discoaster deflandrei[1]	CP10? - CN5b?	Eocene - early Miocene
			Assemblage suggests middle - late Oligocene
1B	*Sphenolithus predistentus*	CP12b - CP18	Middle Eocene - middle Oligocene
	Discoaster deflandrei[1]	CP10? - CN5b?	
	Dictyococcites bisectus	CP14b - CP19b	Middle Eocene - late Oligocene
			Assemblage suggests middle Eocene - middle Oligocene
			In combination, assemblages from 1A and 1B suggest Oligocene, possibly middle Oligocene

[1]According to Perch-Neilson (1985), *D. deflandrei* is common only in the Oligocene.

Locality description: Southern Parkinson Range, Buka Island, North Solomons Province, Papua New Guinea; approximately 5 km west and 3 km north of lodge on Sohano Island in Buka Passage. In clayey mudstone float from cut bank about 3 m high on east side of West Coast Road a short distance north of the point where the road turns northward after skirting the west side of Ramun Bay.

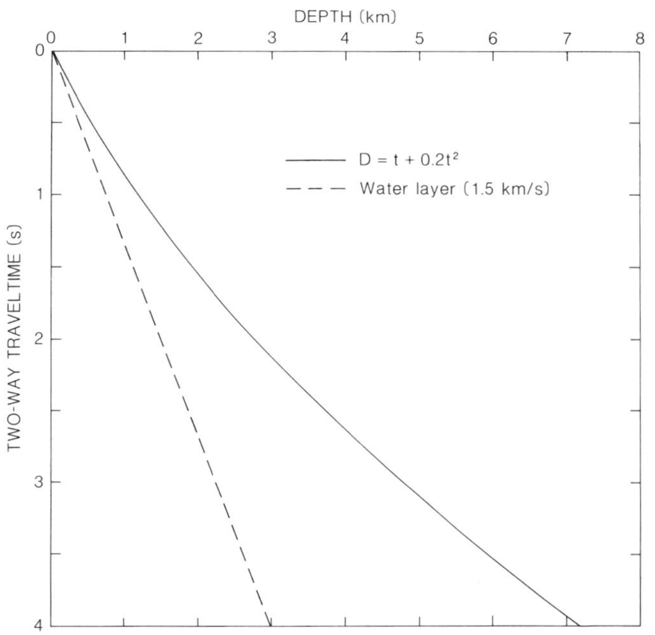

Figure 6. Time-depth curves for converting two-way sub-seafloor traveltime on seismic-reflection sections to depth. Water-layer thickness must be added to obtain total depth.

concentrated in the northwestern and southeastern parts of the island where they consist of aligned effusive centers, one of which, Bagana volcano, is intermittently active.

Uplifted and tilted Pleistocene reef deposits of the Sohano Limestone fringe the northernmost end of the island where they dip about 1° southward. Some of the lowest exposed beds in these reef deposits may be as old as Pliocene.

SEISMIC STRATIGRAPHY

Introduction

Data used as the principal source for interpreting the stratigraphy of Queen Emma basin and its flanking ridges are multichannel seismic-reflection profiles collected over about 700 km of trackline by the U.S. Geological Survey's R/V *S.P. Lee* in June 1984 (Figure 2). These data were acquired by means of a 1,311 in^3 airgun array, a Globe Universal Sciences (GUS)[1] seismic recording system, and a 2,400 m streamer. A shooting geometry of 50 m shotpoint intervals and a 100 m group intervals provided 24-fold data. Processing of the records was done on a DISCO[1] (trademark of Digicon, Inc.) system. The main steps included demultiplexing, velocity analysis using a coherence-estimation technique combined with constant-velocity stacks, normal moveout correction, stacking, and filtering. Migration velocities for the migrated sections were selected from different post-stack constant velocity migrations. Additional details on the methods of collection and processing are given in Mann (1986). Supplementary materials examined include profiles recorded by Compagnie General Geophysique for the Australian Bureau of Mineral Resources in 1970, by Gulf Research and Development Company in 1972, and by Institut Français du Pétrole-Comite d'Études Pétrolières Marines-Office de la Recherche Scientifique et Technique Outre-Mer (IFP-CEPM-ORSTOM) in 1975 (Figure 2).

Positions of the tracklines were fixed by satellite navigation supplemented by doppler sonar, a speed log, and a gyrocompass to calculate dead-reckoning positions between satellite fixes.

Time-to-Depth Conversion

Because most refraction sonobuoys deployed in the Queen Emma basin/Emirau-Feni Ridge area were recorded where the strata are deformed, velocities were determined from seismic-reflection records across thick, essentially undeformed basin sequences. On profiles 413, 415, 417, and 419 (Figure 2), interval velocities were calculated from stacking velocities for seismic stratigraphic units between selected seismic horizons. These interval velocities were averaged for each seismic unit, and a time-depth conversion curve was constructed for each profile using the average interval velocities. Curves for the four profiles are similar and were combined to give a basin-wide time-depth curve.

This curve was then fit with a least-squares quadratic equation, yielding a time-depth function:

$$D = t + 0.2t^2$$

where D = approximate subbottom depth in kilometers and t = two-way sub-seafloor travel time in seconds (Figure 6). Depths to reflection horizons and isopachs were ascertained using this curve.

Seismic-Stratigraphic Units and Inferences on Their Origin

In this report, the rocks that form Queen Emma basin are divided into four seismic-stratigraphic units (Figure 7). Each unit is bounded by a surface that is either an unconformity or an abrupt seismic-velocity change, or both. In order of decreasing age, the four units are designated E1 through E4. Seismic-reflection surfaces (horizons) between units are indicated by hyphenated symbols; for example, horizon E1-E2 separates seismic units E1 and E2. Selected interpreted profiles, structure contours, and isopachs depict the configuration of the seismic units in the basin on Figures 8 through 13. Each unit, inferred correlative onshore rocks, and the causes of

[1]Any use of trade names is for purposes of identification only and does not imply endorsement by the U.S. Geological Survey, the New Zealand Geological Survey, or the Bureau of Mineral Resources, Australia.

SEISMIC UNIT DESCRIPTION			
SEISMIC UNIT	VELOCITY RANGE (KM/S)	THICKNESS RANGE (KM)	PROVISIONAL AGES AND ONSHORE CORRELATIVES
UNIT E4	1.5 - 2.4	1 - 2.5	PLEISTOCENE, PLIOCENE, AND LATE MIOCENE (?) (BOUGAINVILLE GROUP; RATAMAN FORMATION, AND UPPER PUNAM LIMESTONE?)
UNIT E3	2.7 - 3.6	0 - 2.5	EARLY (?) TO LATE (?) MIOCENE; CHIEFLY MIDDLE (?) MIOCENE (LELET LIMESTONE AND TAMIU SILTSTONE)
UNIT E2	3.8 - 4.6 IN BASIN 2.1 - 3.6 ON FLANKS	1 - 5	EARLY MIOCENE (?) AND OLIGOCENE (KERIAKA LIMESTONE (?), BUKA FORMATION, AND UPPER KIETA VOLCANICS)
UNIT E1	3.9 - 6.1 COMMONLY GREATER THAN 5.2 KM/SEC	UNKNOWN	OLIGOCENE AND OLDER (KIETA VOLCANICS, JAULU VOLCANICS AND OLDER IGNEOUS ROCKS, INCLUDING LEMAU COMPLEX)

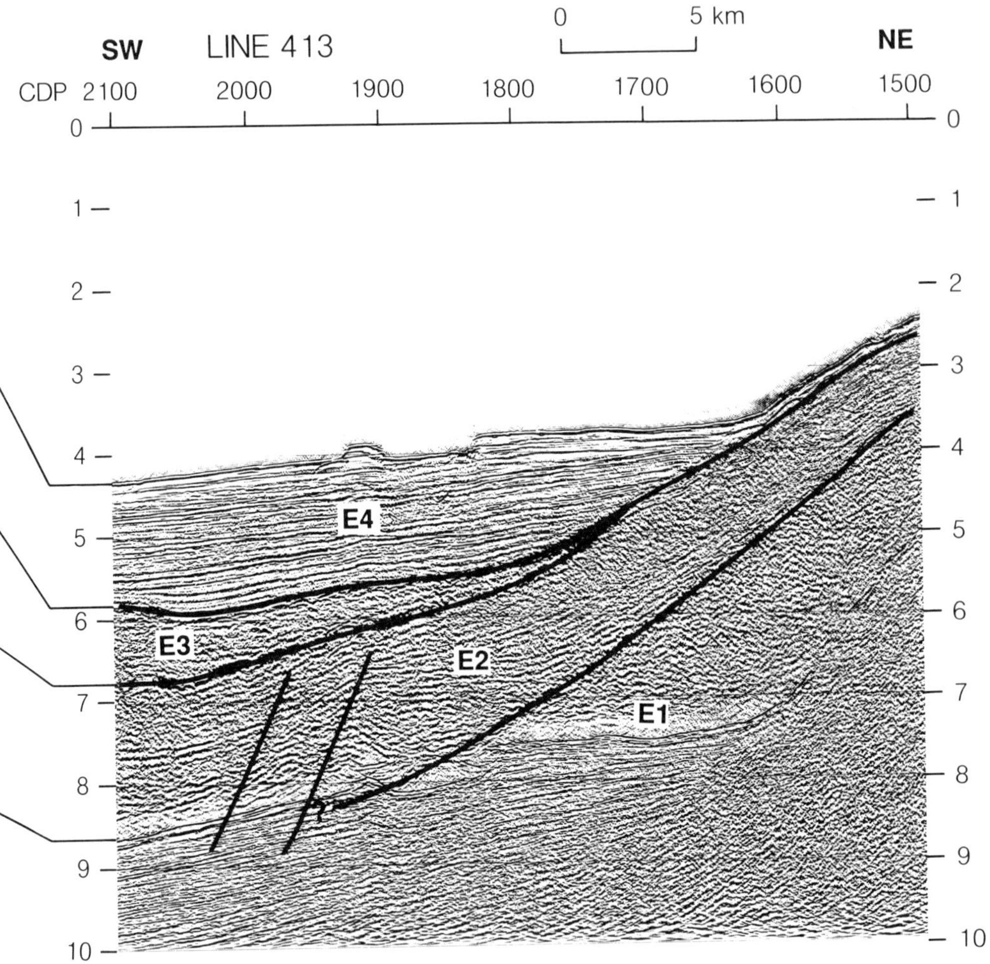

Figure 7. Correlation and description of seismic-stratigraphic units in Queen Emma basin. Velocity range is based upon seismic-refraction and reflection data of Cooper, Bruns, and Wood (1986) and on interval velocities obtained from multichannel seismic data. CDP - Common Depth Point; 100 CDPs equal 5 km. Correlations with onshore rocks are provisional (see text).

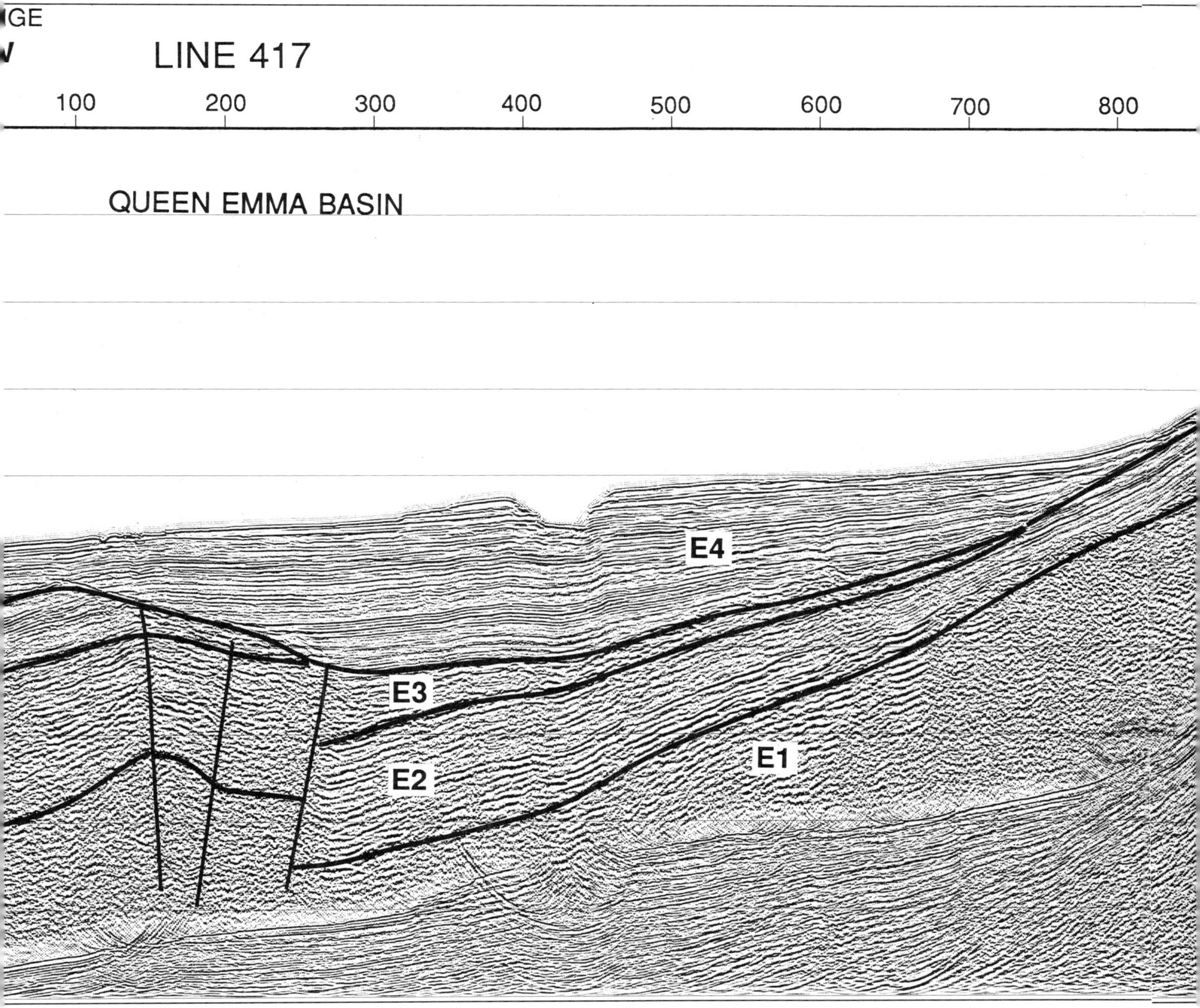
GE
LINE 417
100
200
300
400
500
600
700
800
QUEEN EMMA BASIN
E4
E3
E2
E1

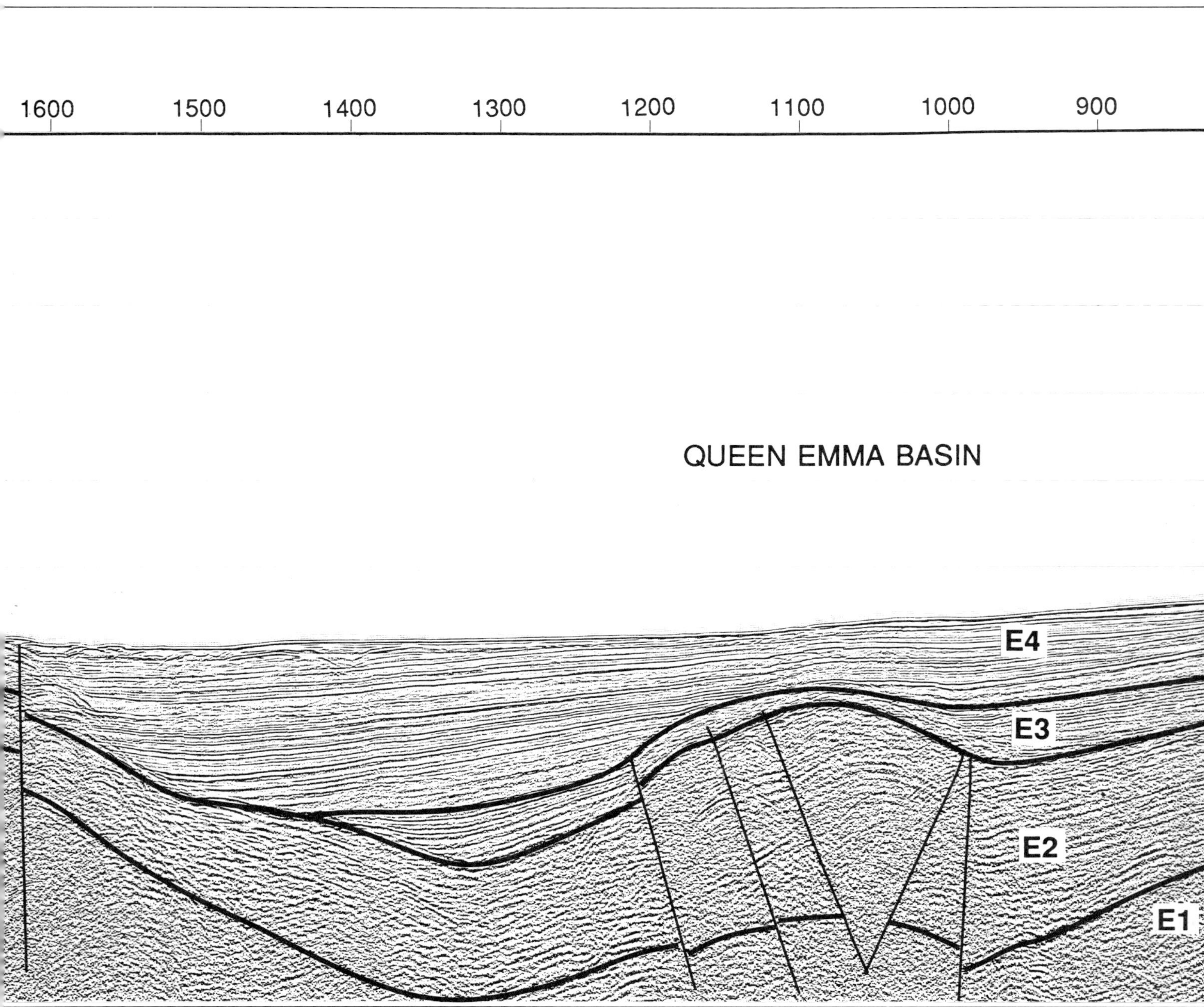

1600
1500
1400
1300
1200
1100
1000
900
QUEEN EMMA BASIN
E4
E3
E2
E1

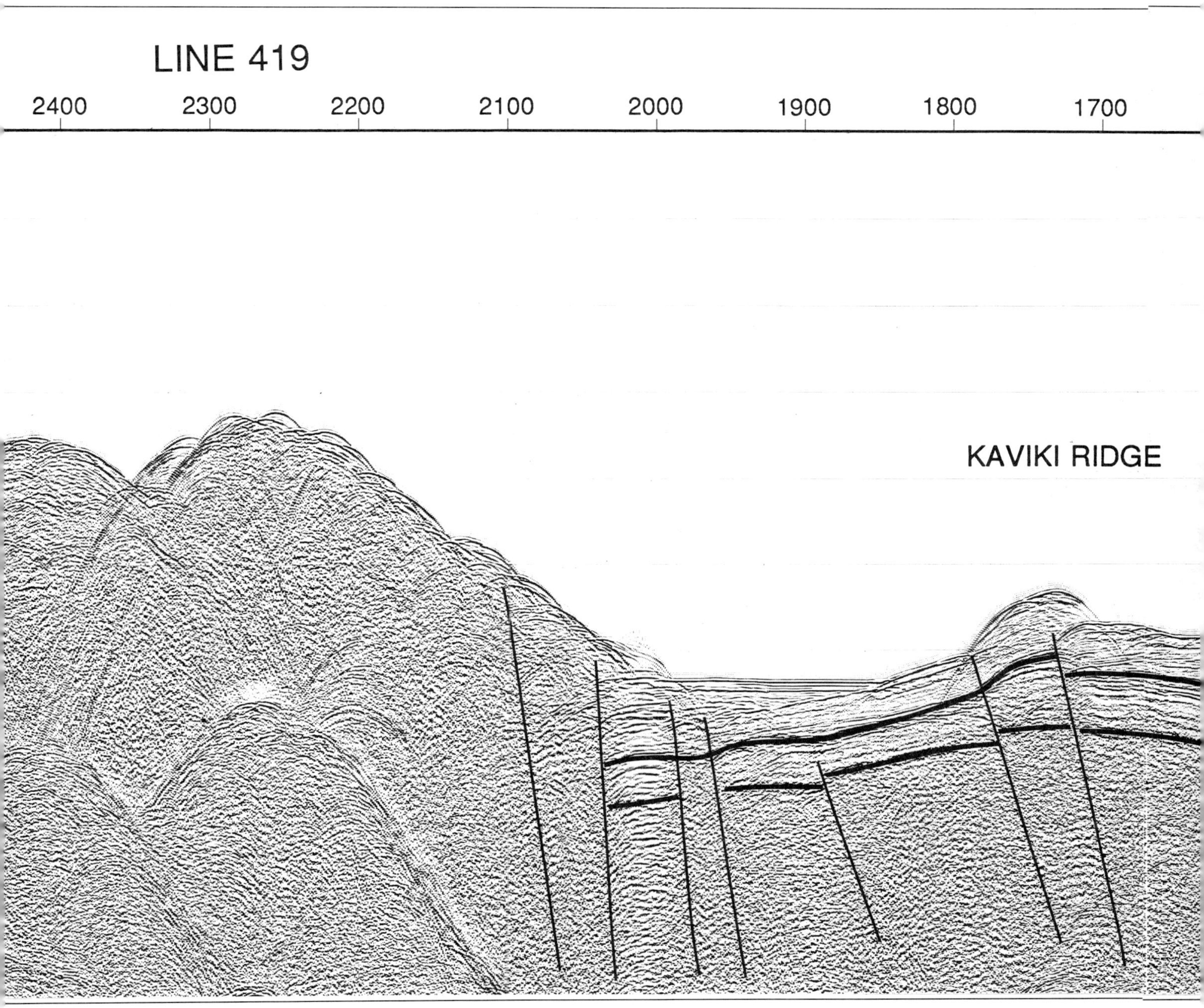

e 10. Seismic-reflection profile 419, unmigrated time section. Seismic-stratigraphic units as defined in Figure 7 and

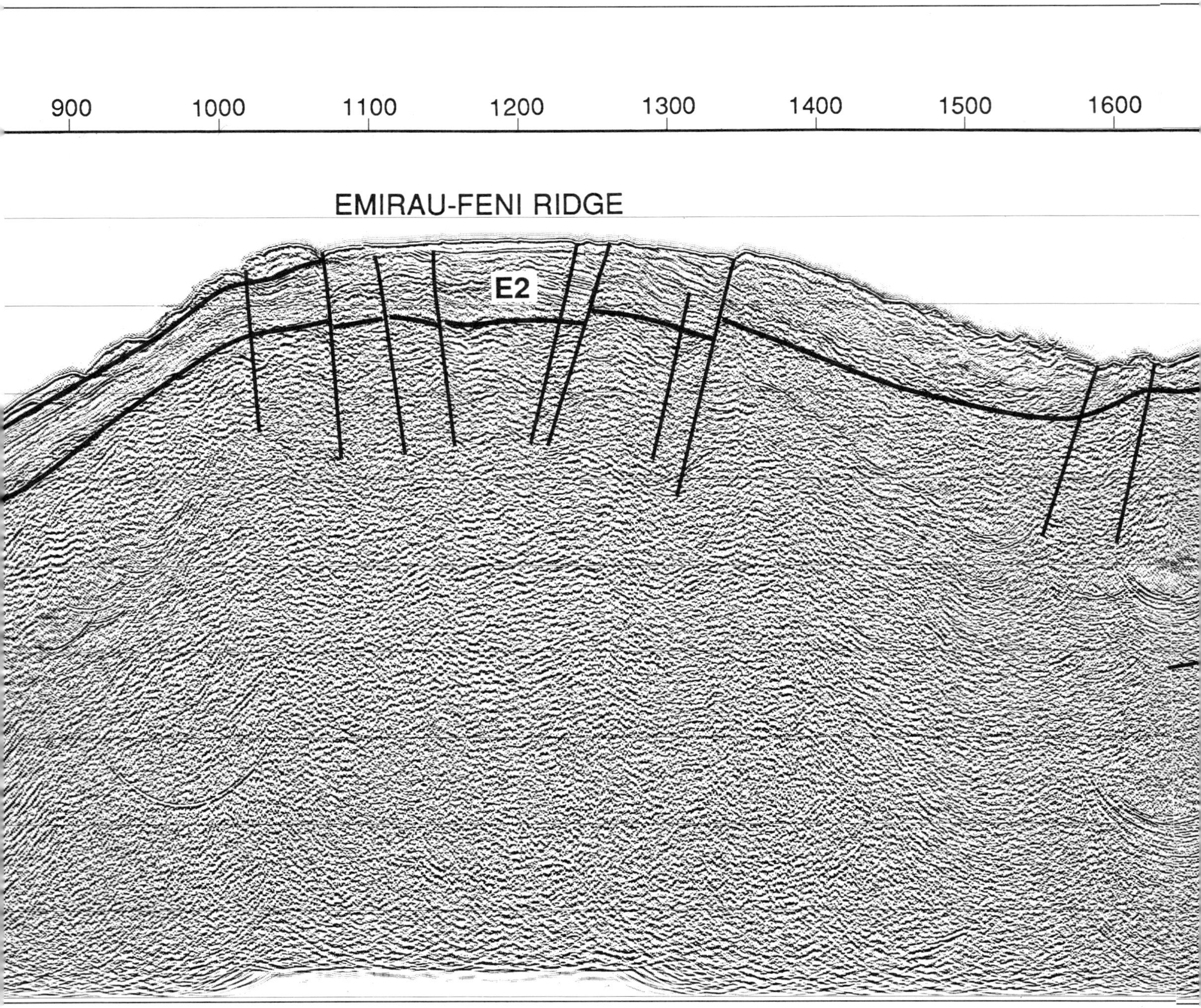
900
1000
1100
1200
1300
1400
1500
1600
EMIRAU-FENI RIDGE
E2

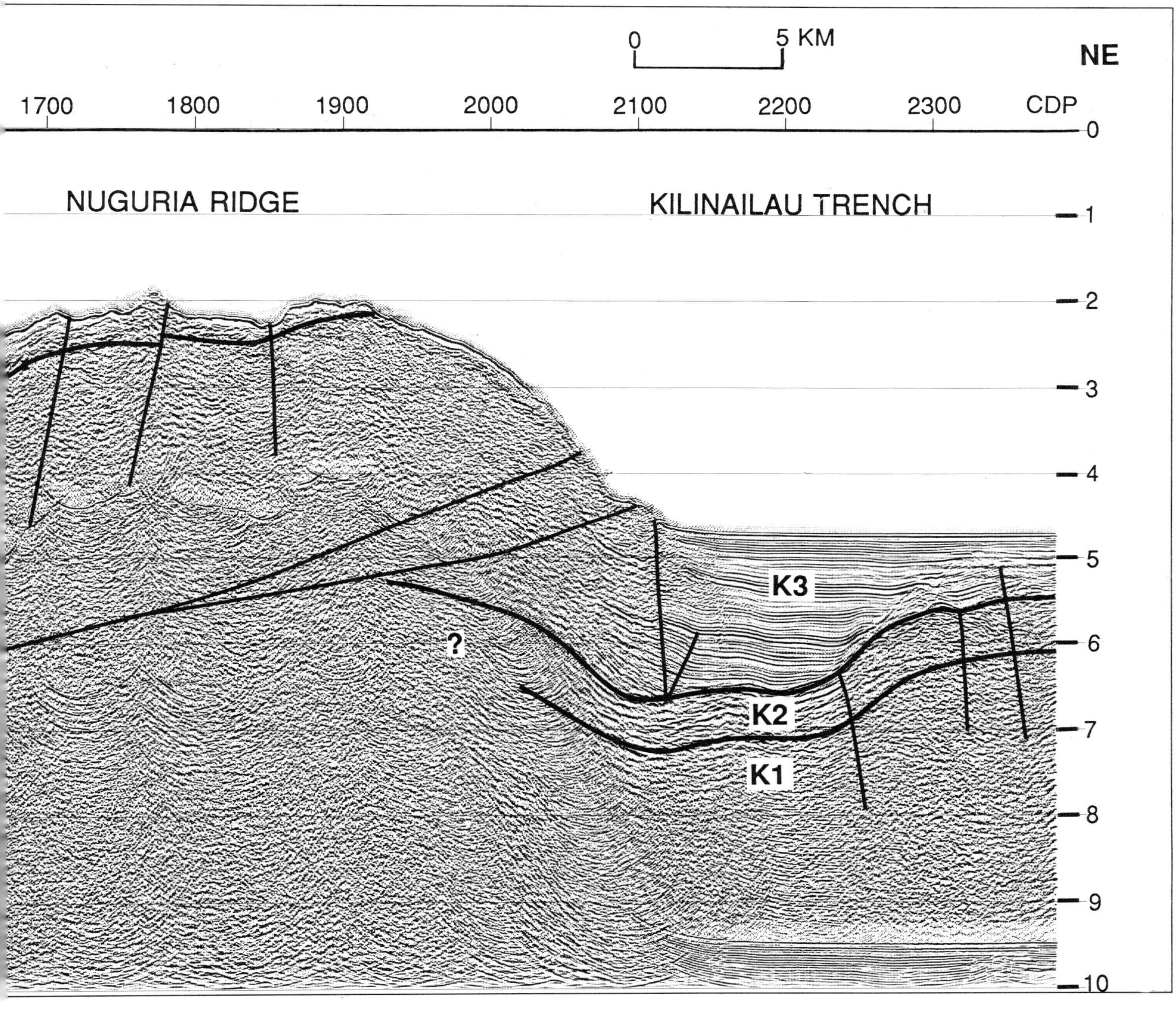
0
5 KM
NE
1700
1800
1900
2000
2100
2200
2300
CDP
0
1
2
3
4
5
6
7
8
9
10
NUGURIA RIDGE
KILINAILAU TRENCH
K3
?
K2
K1

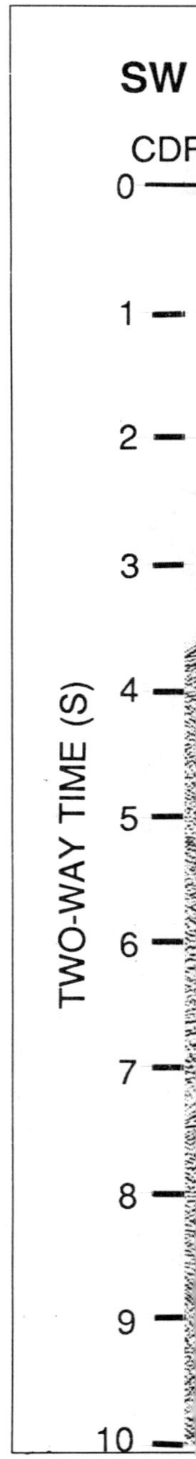

Fig
tex

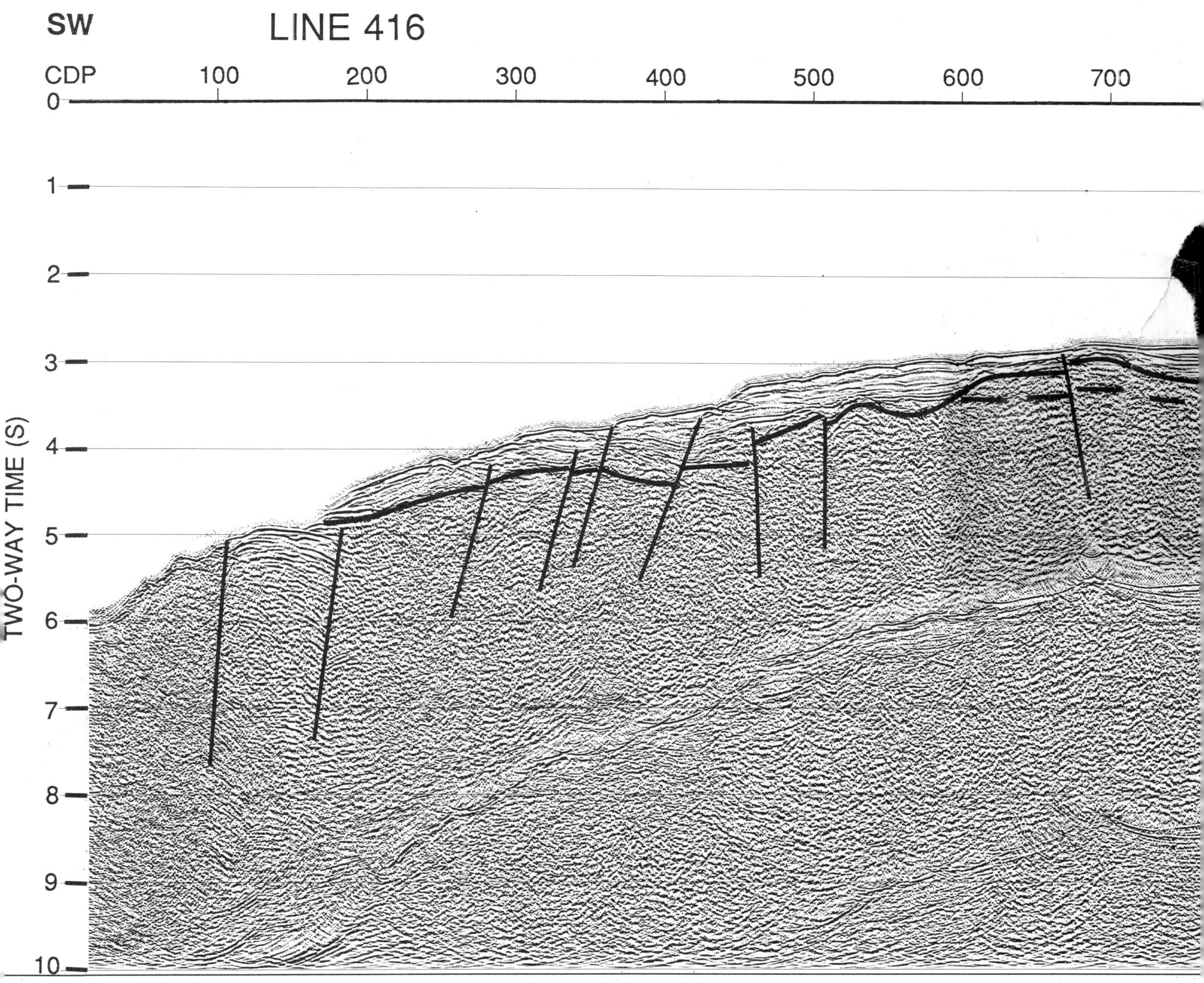

Figure 8. Seismic-reflection profiles 416 and 417, migrated time sections, through Queen Emma basin. Interpreted seismic-stratigraphic units are the same as those defined in Figure 6 and text; for Kilinailau Trench units, see Bruns, Vedder, and Culotta (this volume). CDP - Common Depth Point; 100 CDPs equal 5 km. Vertical exaggeration at the seafloor on these and subsequent sections is 4:1 unless otherwise stated.

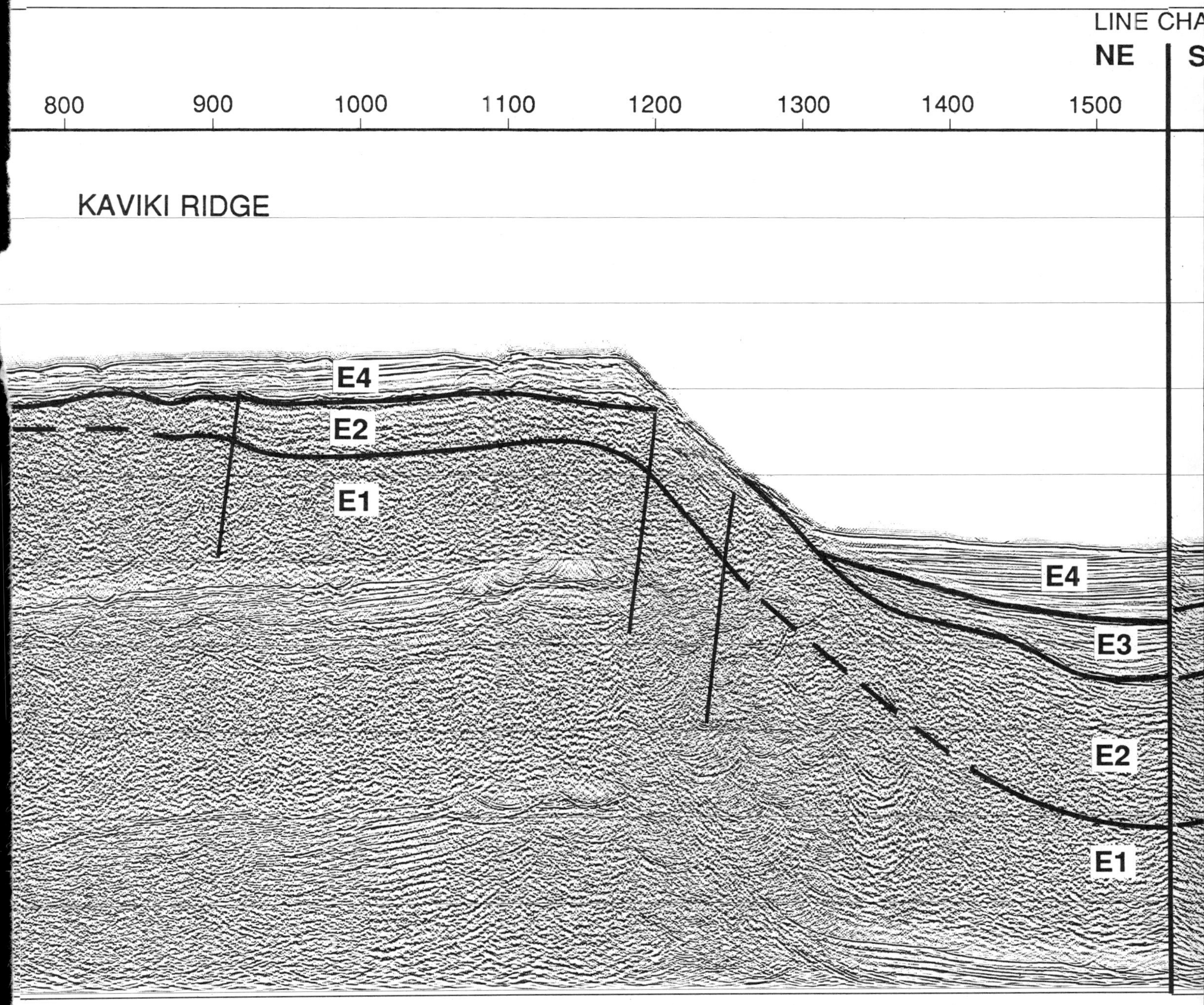

LINE CHA
NE
S
800
900
1000
1100
1200
1300
1400
1500
KAVIKI RIDGE
E4
E2
E1
E4
E3
E2
E1

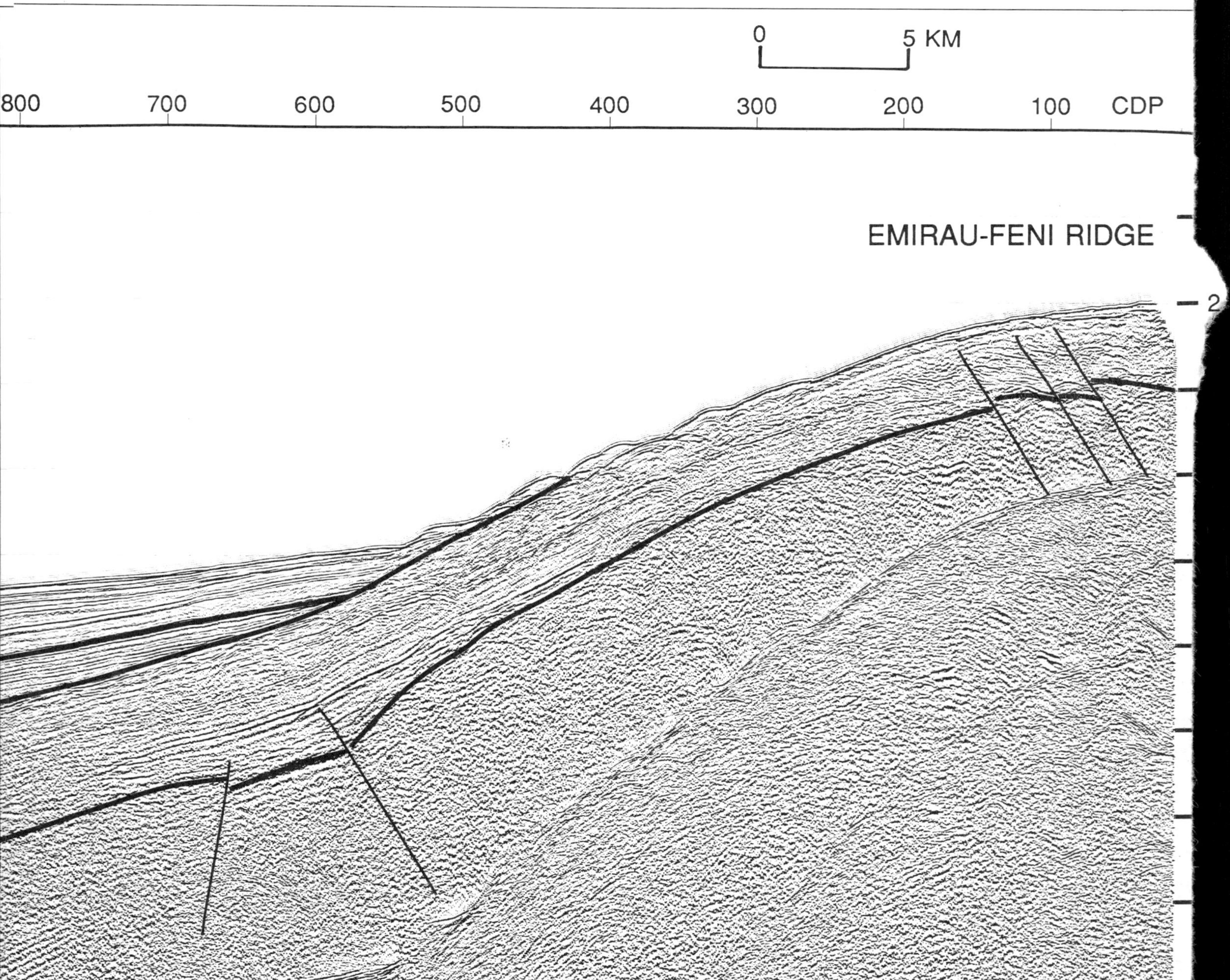
0
5 KM
800
700
600
500
400
300
200
100
CDP
EMIRAU-FENI RIDGE
2

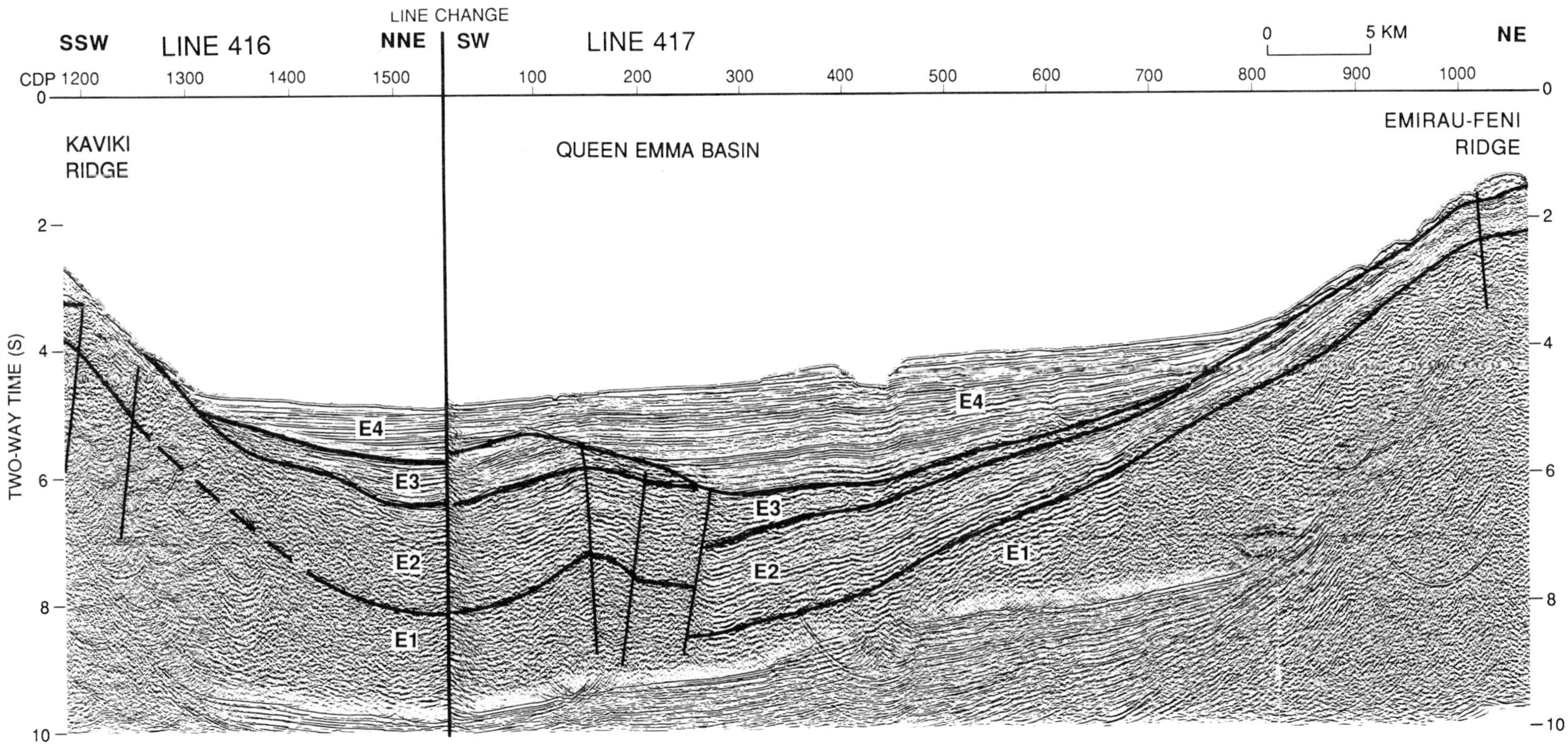

Figure 8. Seismic-reflection profiles 416 and 417, migrated time sections, through Queen Emma basin. Interpreted seismic-stratigraphic units are the same as those defined in Figure 6 and text; for Kilinailau Trench units, see Bruns, Vedder, and Culotta (this volume). CDP - Common Depth Point; 100 CDPs equal 5 km. Vertical exaggeration at the seafloor on these and subsequent sections is 4:1 unless otherwise stated.

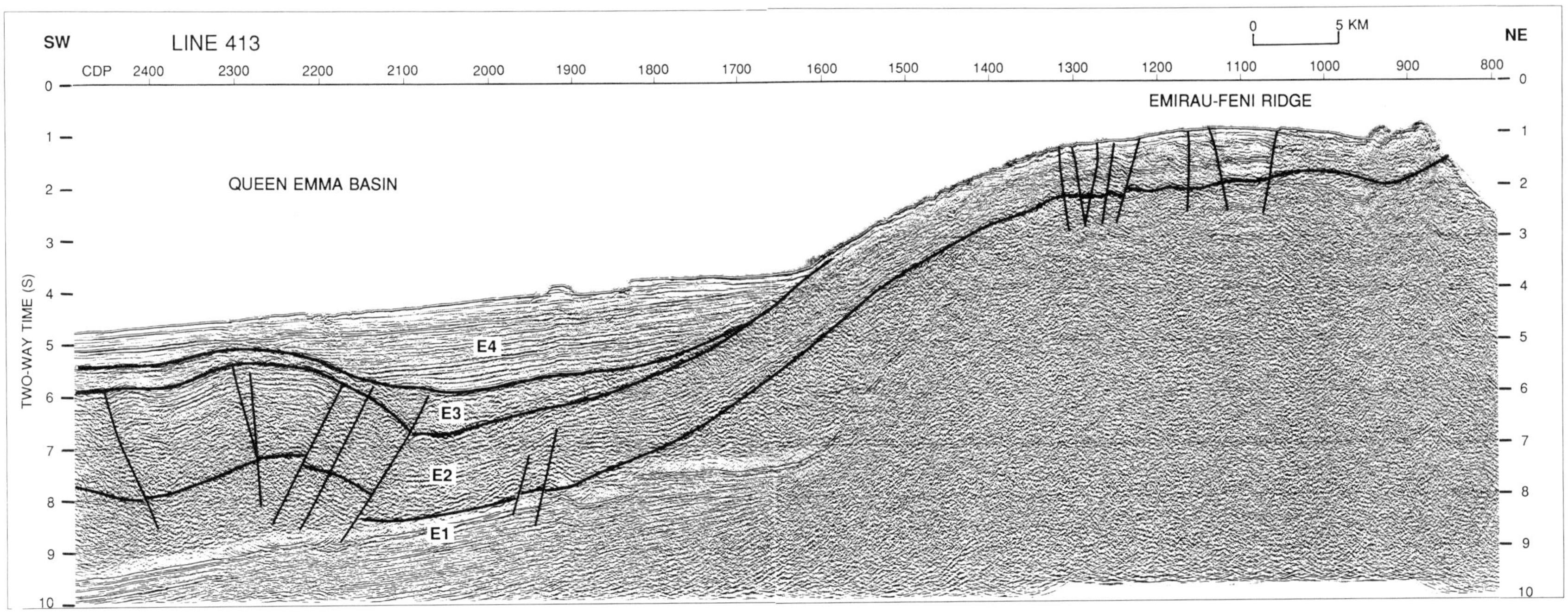

Figure 9. Seismic-reflection profile 413, migrated time section. Seismic-stratigraphic units are the same as those defined in Figure 7 and text.

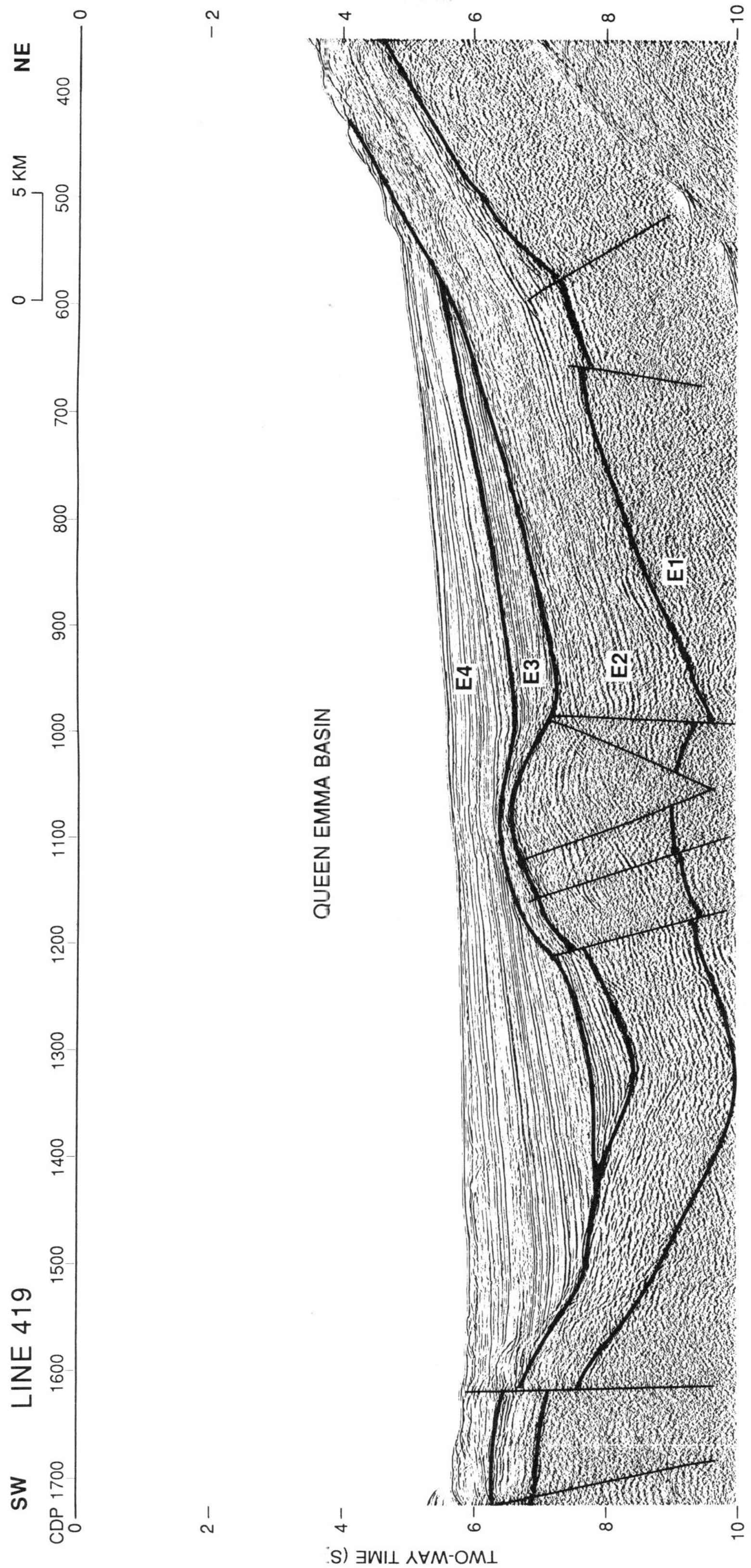

Figure 10. Seismic-reflection profile 419, unmigrated time section. Seismic-stratigraphic units as defined in Figure 7 and text.

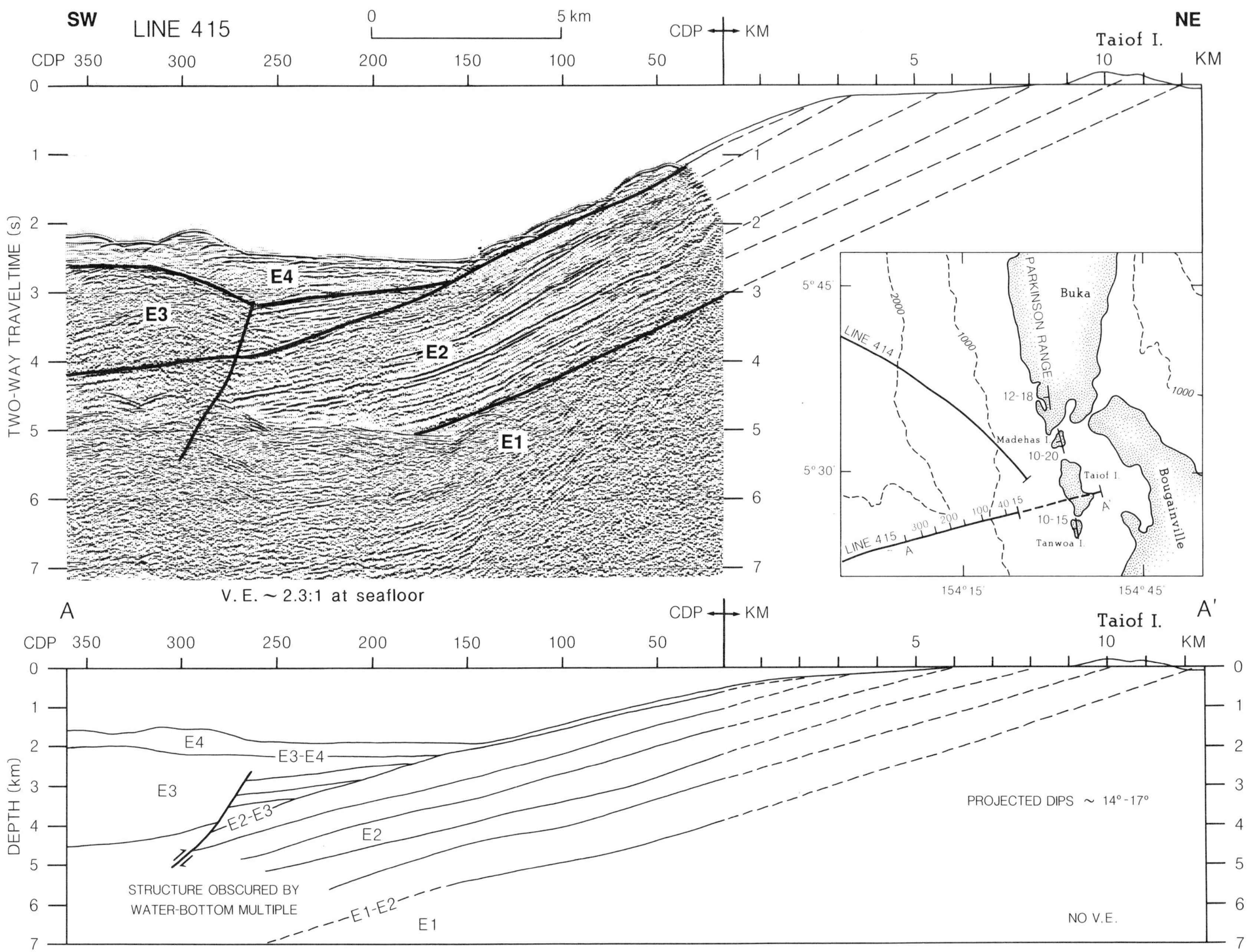

Figure 11. Segment of seismic-reflection profile 415 (eastern end), unmigrated, and eastward projection of the section through Taiof Island where the Buka Formation is exposed. Dips and strikes on the islands are composites from estimated readings.

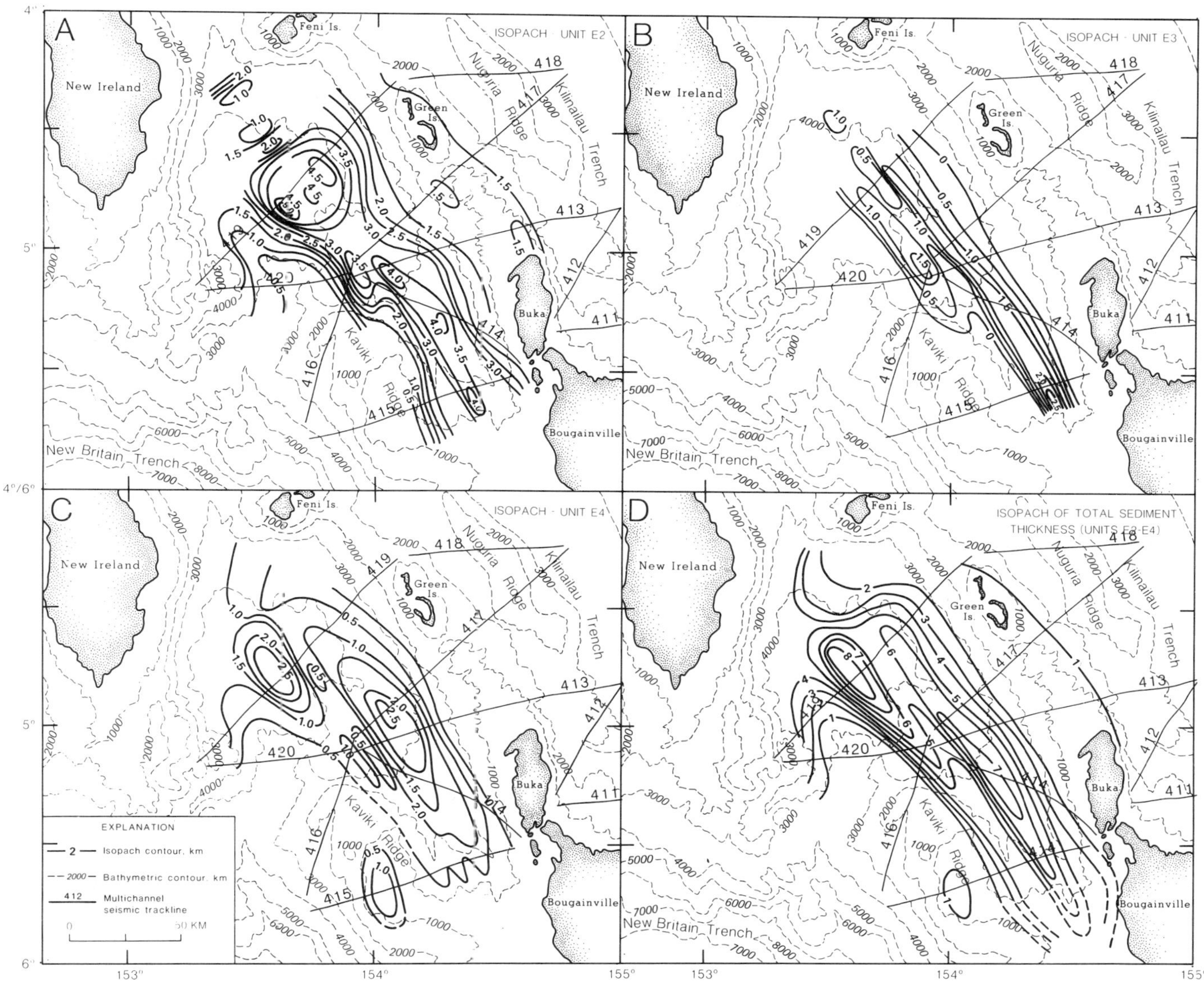

Figure 12. Isopach maps of sediment thickness for units E2(A), E3(B), E4(C), and for total sediment thickness above acoustic basement (between seafloor and top of unit E4(D)), Queen Emma basin.

structural deformation are described below. However, the limited number of offshore samples that can be tied directly to seismic-reflection profiles inhibits correlation of seismic units with onshore rock units and hinders interpretations of the timing of tectonic events.

Seismic-stratigraphic Unit E1

Unit E1 is acoustic basement. This basement generally is characterized by discontinuous to chaotic seismic reflections and locally by diffractions at the top. On the northeast flank of the basin, local poorly defined sub-parallel reflections within unit E1 may indicate stratification (Figures 8 and 9). Refraction data show that this unit commonly has a seismic velocity of >5 km/s, but the velocity range is 3.9 to 6.2 km/s (Cooper, Cochrane, and Bruns, this volume). These velocities indicate that the unit is likely to consist chiefly of igneous rocks, particularly on the ridge crests and flanks, although local intrusive bodies of dense subcrustal igneous and metamorphic rocks may be included where velocities are 6.0 to 6.2 km/s.

Beneath Kaviki Ridge, unit E1 presumably consists of volcanic and intrusive rocks emplaced along the Paleogene magmatic arc and, therefore, would correlate with the basement volcanics on Bougainville and the Jaulu Volcanics and Lemau Complex on New Ireland. The nature and correlation of the unit are uncertain directly under the deep part of Queen Emma basin; there unit E1 conceivably could include pre-Eocene oceanic crust trapped between the arcs during reversal but is more likely to be

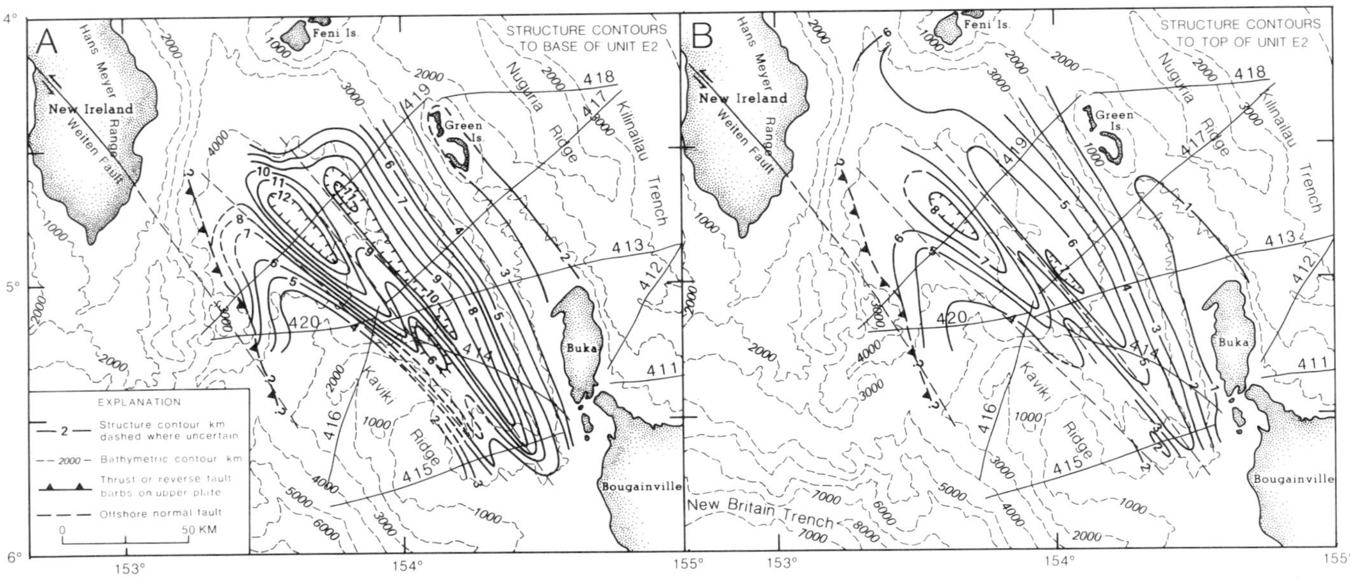

Figure 13. Structure contour maps for base and top of unit E2, Queen Emma basin. The offshore extension of the Weitin fault presumably is offset in a left lateral sense as indicated on New Ireland.

subsided Eocene(?) and Oligocene arc crust similar to that which forms the southern perimeter of the basin. Beneath the southeastern continuation of Emirau-Feni Ridge between the Feni Islands and Buka, the unit may be composed of uplifted remnants of these old arc rocks that have been invaded by Pliocene and Quaternary igneous intrusions related to late Cenozoic arc magmatism. Magnetic anomalies that occur along the ridge crest support this interpretation (Bruns and Cooper, unpublished data), although the older igneous (and metamorphic?) rocks cannot be differentiated from the younger intrusive rocks on the basis of the available geophysical data. If present, these late Cenozoic arc rocks correlate with the Bougainville Group on Bougainville and the Pliocene-Pleistocene volcanic rocks on the Feni Islands.

Seismic-stratigraphic Unit E2

Unit E2 is characterized by seismic reflections that vary from persistent and acoustically well-defined on the northeast side of Queen Emma basin (Figures 8 and 9) to chaotic and poorly defined on the south and southwest sides (Figures 8, 9, 10, and 11). The unit is the thickest of the three basin-fill sequences and generally ranges from 1.0 to 3.0 km thick on the northeast to as much as 5.0 km in the west-central part (Figure 12A). On Kaviki Ridge, a group of reflections questionably assigned to unit E2 occur in a sequence less than 1.5 km thick (Figures 8, 10, and 12A). Although seismic velocities within the unit range from about 3.8 to 4.5 km/s where the sedimentary cover is thick, they decrease to 2.1 to 3.6 km/s on the basin flanks where there is little or no sedimentary cover and no evidence for compaction. Reflection horizon E1-E2, which separates unit E2 from unit E1, is recognized at most places by an abrupt upward change from chaotic reflections to fairly well-defined subparallel reflections. This surface is not discernible where the overlying strata are thickest (>3.0 km). Unit E2 is deformed by faults and folds on all of the profiles.

Most of unit E2 may represent volcaniclastic strata that were deposited as a northeastward-sloping forearc sedimentary apron derived from a formerly high-standing, active segment of the early Cenozoic magmatic arc southwest of the basin. This subsided old arc segment is now manifested by Kaviki Ridge. The thin sequence of unit E2 on Kaviki Ridge may consist of remnant coarse epiclastic rocks of the old arc that are interlayered with subordinate carbonate zones; however, the lack of samples precludes identification of rock types.

Figure 11 shows the projection of seismic reflections in the lower part of unit E2 into westward-dipping strata in the Buka Formation on Taiof Island. As explained in the section on island geology, the Buka Formation in the Parkinson Range includes Oligocene beds and may span a time range of late Eocene to early Miocene. Because the paleontologically dated beds on Buka Island strike directly into the sequence exposed on Taiof and Tanwoa islands, and the dips of the seismic reflections and the island strata are nearly the same, an unqualified correlation seems warranted.

The well-defined, westward-dipping reflections along the northeast flank of the basin probably represent fine-grained turbidite beds that originated mainly as northeastward-dipping distal forearc deposits. Some of the coarser strata, such as those exposed at places on southern Buka, Madehas, and Taiof islands, may have had nearby sources in the vicinity of easternmost Kaviki Ridge.

Subordinate amounts of material shed southwestward from an old forearc structural high(?), the ancestral Emirau-Feni and Nuguria ridges, may be intermixed in the forearc sequence (Vedder, Colwell, and Bruns, 1985). Until field studies on the islands confirm sediment-transport directions, source areas will remain suspect.

The following correlations of unit E2 are proposed: 1) unit E2(?) on Kaviki Ridge is equivalent to the upper part of the basement volcanics and Keriaka Limestone on Bougainville and to the upper, carbonate-bearing part of the Jaulu Volcanics and/or Lelet Limestone on New Ireland, 2) the lower part of unit E2 in the deep central part of Queen Emma basin is equivalent to volcanic agglomerate and epiclastic beds that occur stratigraphically high in the Jaulu Volcanics and the basement volcanics on Bougainville, and 3) the lower part of unit E2 on the northeastern flank of basin is equivalent to the Buka Formation in the Parkinson Range and Taiof Island (Figure 11) and to the unnamed Oligocene limestone on Ambitle. Although the upper 2,000 m of unit E2 has no known correlatives in the Parkinson Range, more distant correlatives of the upper part may be the lower beds of the Lelet Limestone and Tamiu Siltstone of New Ireland.

Initial uplift and tilt of unit E2 in the distal forearc area may have begun as early as the early Miocene during the impaction of the Ontong Java Plateau against the early Cenozoic arc in the Bougainville region. However, most of the deformation of unit E2 probably was caused by late Miocene(?) regional compression along the late Cenozoic arc axis during the process of arc reversal.

Seismic-stratigraphic Unit E3

Unit E3 is characterized by reflections that are predominantly parallel and moderately well defined, except near the structural high in the center of the basin where they are poorly defined and disordered. Unit E3 generally shows better defined and more closely spaced layering than unit E2 as exemplified on Figure 10. Where it laps out on the northwestern flank of the basin, unit E3 dips basinward less than unit E2. Unit E3 not only onlaps and thins over the anticlinal structure in unit E2 in the central basin but also is folded, faulted, and slumped along the flanks of the structure (Figures 8, 9, 10 and 14). The apparent truncation of unit E3 on the north limb of the anticlinal structure (Figure 8) may have been caused by slumping along the growing fold. Reflection horizon E2-E3, which separates unit E3 from E2, is marked by an unconformity and a velocity increase from about 2.7-3.6 km/s to 3.8-4.6 km/s. Southwest of Taiof Island, unit E3 is as much as 2.5 km thick; elsewhere, it generally is less than 2.0 km thick (Figure 12B). Although unit E3 is primarily confined to the central part of Queen Emma basin, the unit probably is present as obscure wedges along the northern slope of Kaviki Ridge (Figure 8).

Unit E3 may have originated in a hemipelagic depositional system in which substantial amounts of fine-grained detrital volcaniclastic material were transported northwestward along the structural axis of Queen Emma basin (Figure 13B). Because the unit thins northeastward and southwestward on the basin flanks and truncates by onlap onto unit E2, both of these flanks probably were minor contributors of sediment into the dominant axial transport system of the basin.

The age of unit E3 is uncertain; although it may span much of early Miocene to late Miocene time, a middle Miocene age for most of the unit seems likely. Inasmuch as underlying unit E2 correlates with Oligocene and early Miocene(?) strata in the Buka Formation (Figures 7 and 11) and probably was derived largely from Paleogene arc rocks on Kaviki Ridge, the lowermost onlapping strata of unit E3 conceivably could be as old as early Miocene. In Bougainville basin, rocks inferred to overlie strata equivalent to unit E2 (Figure 5, G to B interval) are late early Miocene to late Miocene (Shaw, 1985), thus implying a similar age for unit E3. By inference, subsidence of Kaviki Ridge began after southwest-directed subduction ceased in the early Miocene, and complete submergence may have occurred by the middle Miocene. Such submergence would curtail sediment input from the ridge. Under these circumstances in the middle Miocene, the only remaining high-standing region in the vicinity of Queen Emma basin was ancestral Bougainville where early Cenozoic arc rocks were the sole source for the axially transported basin sediments of unit E3.

Alternatively, unit E3 could be younger than middle Miocene based upon the reasoning that deposition of unit E2 continued until the early Cenozoic arc subsided to a depth that no longer provided a source of sediment, possibly as late as the middle Miocene. Thus, if all of the uplift of unit E2 on the northeast flank of the basin did not occur until the late Cenozoic arc was constructed, onlapping unit E3 would be late Miocene and/or early Pliocene in age. This alternative age assignment is judged unlikely because unit E3 apparently was deposited before arc reversal and uplift of Emirau-Feni Ridge.

If the late early to late Miocene age for unit E3 is accepted, the unit correlates with parts of the Lelet Limestone and Tamiu Siltstone on New Ireland and with the Keriaka Limestone on Bougainville (Figure 5). If, however, the less likely late Miocene to early Pliocene age is correct, the unit correlates with the Punam Limestone and possibly the lower part of the Rataman Formation on New Ireland and with the lowermost part of the Bougainville Group.

Seismic-stratigraphic Unit E4

Unit E4 is typified by well-defined, persistent, nearly parallel reflections (Figures 8, 9, and 10). The unit is widely distributed in Queen Emma basin where it generally ranges in thickness from 1.0 to 2.5 km (Figure 12C); the unit thins on the basin margins to less than 1.0 km and laps out on the lower slopes of the flanking ridges. A thin veneer of unit E4 caps Kaviki Ridge (Figure 8) where the thickness mainly is less than 0.5 km and beds of

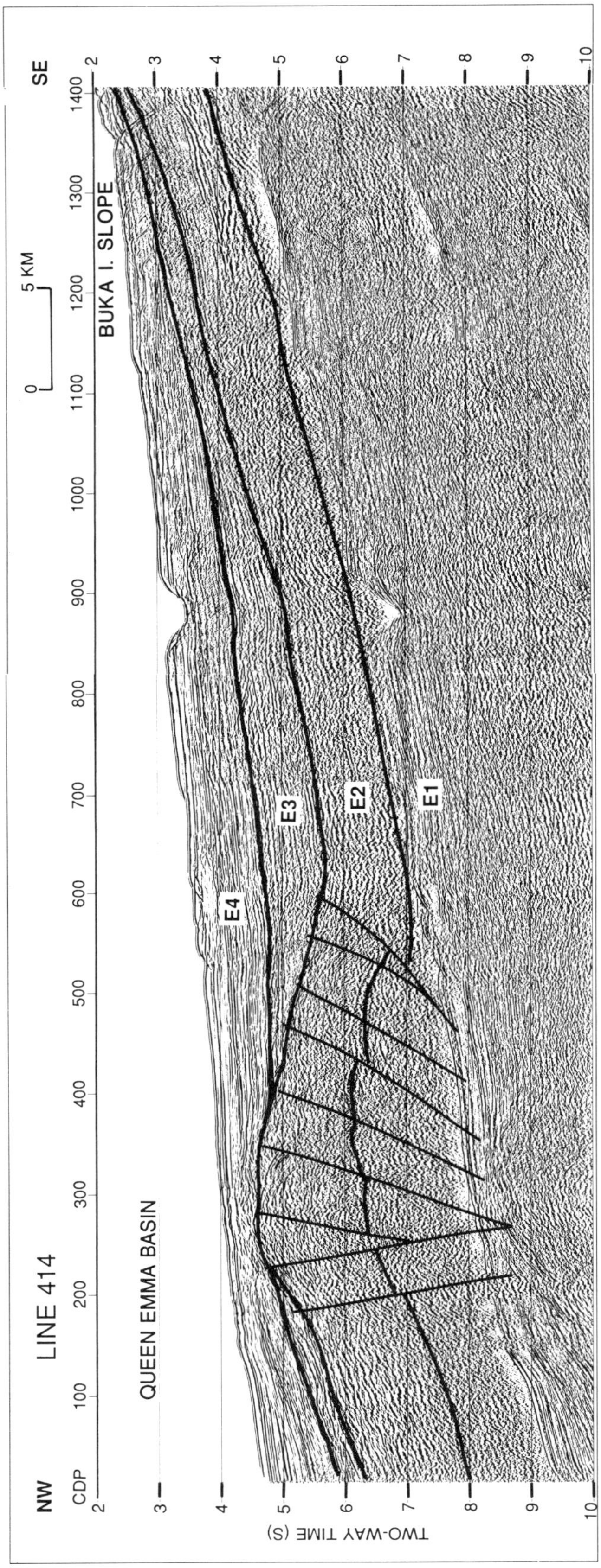

Figure 14. Segment of seismic-reflection profile 414, unmigrated, showing interpreted faults that are associated with the buried anticline underlying the long axis of Queen Emma basin.

unit E3 may be included in the lower part. Reflection horizon E3-E4, which separates unit E4 from E3, is marked by an increase in velocity from about 2.0-2.4 km/s to 2.7-3.6 km/s and separates deformed strata in unit E3 from draped strata in unit E4 in the central basin.

Unit E4 is an undeformed basin-fill sequence that accumulated in the intra-arc basin, which is flanked on the south by the subsided Kaviki Ridge segment of the early Cenozoic arc and on the northeast by the uplifted Emirau-Feni Ridge segment of the late Cenozoic arc. On the west, unit E4 laps out or is faulted against an uplifted segment of the older arc in southeastern New Ireland (Figure 10), and on the north the unit merges with equivalent strata in the New Ireland basin. Much of the sequence may have been derived from the Bougainville-Buka region by northwest-directed axial transport; however, supplementary sources of sediment presumably included other structurally high-standing areas such as New Ireland and Emirau-Feni Ridge. The well-developed leveed channel in the upper part of the unit (Figures 8 and 9) is a persistent northwest-trending feature that probably anastomoses into a fan-like pattern in the deep western part of Queen Emma basin where channels are not evident on bathymetry and seismic-reflection profiles (Figure 10). Both the orientation and slope of this channel system together with the position of the levee indicate sediment flow from southeast to northwest.

Based upon its stratigraphic position and general lack of deformation, unit E4 is provisionally assigned an age of latest Miocene(?) to Holocene. Thus, correlatives of the lowermost part of unit 4 on New Ireland would include the upper part of the Punam Limestone and lower part of the Rataman Formation. If the lowermost part is younger than early Pliocene, unit E4 is equivalent to the upper part of the Rataman Formation (late Pliocene and early Pleistocene) and the Maton Conglomerate (late Pleistocene[?] and Holocene[?]). The younger effusive rocks in the Bougainville Group on Bougainville and volcanogenic deposits on Ambitle are equivalent in age to at least the upper part of unit E4.

Total Sediment Thickness

The combined thickness of units E2, E3, and E4 is as much as 8.2 km in the central part of Queen Emma basin (Figure 12D). On the upper flanks and crest of Emirau-Feni Ridge, where the units are difficult to differentiate on seismic-reflection profiles, the total thickness of stratified rocks apparently ranges from about 1.0 to 1.5 km, although sonobuoy data suggest the local presence of as much as 2.0 km of strata. The nearly flat-lying sequence of units E2 and E4 on the broad high-standing part of Kaviki Ridge generally is less than 1.0 km thick.

Structure

Queen Emma basin is a structural depression surrounded by both uplifted and partly subsided fault-bounded ridges and constitutes an intra-arc basin that we believe formed in response to late Cenozoic arc-polarity reversal and tectonic interactions caused by development of the early Cenozoic and late Cenozoic arc systems (Falvey and Pritchard, 1984; Vedder and Colwell, this volume). The basin is analogous to the intra-arc basins in the Solomon Islands (Bruns et al, 1986; Vedder, 1986), although tectonic reconstructions indicate that the depositional elements in Queen Emma basin represent forearc rather than backarc aprons of sediment.

Structure-contour maps of the base and top of unit E2 (Figures 13A and B) show the prevailing northwest structural grain in the basin. The structural relief between the crest of the Hans Meyer Range in southeastern New Ireland and the top of the buried basement in the central deep of Queen Emma basin is about 15 km (Figure 13A).

Folds

Two northwest-trending low-amplitude folds occur within Queen Emma basin. One is a faulted anticline (or closely spaced en echelon anticlines) that deforms late Miocene(?) and older strata (units E1 through E3) in the central part of the basin; the other is a broad antiform that underlies the segment of Emirau-Feni Ridge that connects the Feni Islands and Buka Island (Figures 8, 9, and 10).

The faulted anticline (or anticlines) forms a remarkably linear feature that parallels the long axis of Queen Emma basin and lies about midway between Emirau-Feni Ridge and Kaviki Ridge (Figures 8, 10, 13A and B). Slight vergence on this fold is southwestward. The origin and duration of the folding along this structure are uncertain; the youngest deformed strata are in unit E3, which thins over and is slumped along the fold, indicating growth. This folding event terminated before deposition of unit E4, which is believed to be latest Miocene(?) and younger; thus the folding died out in the late Miocene and structural growth may have been related to processes of arc reversal.

The southwestern flank of the broad antiform that underlies the Green Islands-Buka Island segment of Emirau-Feni Ridge extends southeastward beyond Taiof Island to form the northeastern margin of Queen Emma basin (Figures 8, 10, and 11). Because the antiform has an arched carapace of unit-E2 strata (Figures 8 and 10), the age of the folding is, for the most part, younger than early Miocene(?), even though regional uplift caused by early Cenozoic arc magmatism began as early as the Oligocene. The antiform may have begun to develop during the last stages of collision of the Ontong Java Plateau with the early Cenozoic arc and continued to grow when new arc magmatism and uplift commenced near the end of Miocene time. Benthic foraminifers from a sample taken near the ridge crest about midway between the Green Islands and Buka Island indicate at least 700 m of uplift since deposition during the Pliocene or Pleistocene (Resig *in* Crook and Niem, this volume). At present, upwarping probably is locally active along Emirau-Feni Ridge.

Faults

Three major northwest-trending fault zones transect the subsea region between New Ireland and Bougainville (Figures 13A and B). In sequential order from southwest to northeast, the first zone is the offshore southeastern extension of the Weitin fault on New Ireland (Figures 4, 10, and 13). The second throughgoing zone incorporates the faults that cut the northeastern flank of Kaviki Ridge. The third zone consists of buried faults that are distributed along the anticlinal structure that underlies the long axis of Queen Emma basin. Minor normal faults are commonplace along the upper slopes of the ridges, especially Emirau-Feni Ridge (Figures 8 and 10).

The Weitin fault was described as a major left-lateral strike-slip fault by Hohnen (1978); since then, most workers, including Johnson (1979), Taylor (1979), and Stewart and Sandy (1988), considered this fault to be among the primary post-Miocene structures in the ridge-trench transform zone that separates New Ireland and New Britain (Figure 3D). Hohnen (1978) attributed late Pliocene to Holocene strike slip to the Weitin fault. A highly deformed zone that borders the southwestern side of Queen Emma basin (Figure 10) probably is the southeastern continuation of the Weitin fault zone. If projected directly southeastward, this zone falls along the Solomon Sea side of Kaviki Ridge and intersects the inner wall of the New Britain Trench. The dip of the faults within the zone cannot be determined on the basis of seismic-reflection profiles 419 and 420.

The zone of faults that lies along the northeastern flank of Kaviki Ridge (Figures 8, 9, and 10) trends northwest nearly parallel to the offshore extension of the Weiten fault. Most of the faults in this zone seem to be basin-side-down normal faults; however, the tracklines are too widely spaced to ascertain whether the traces form either parallel, anastomosing, or en echelon patterns. Although the age of these faults is uncertain, they may have resulted from differential rates of downwarp between the deep part of Queen Emma basin and its southwestern flank before regional subsidence of Kaviki Ridge.

Faults associated with the buried anticline beneath the basin occur in a belt as wide as 15 km. Because they are deeply covered by overlying strata, the faults in this zone are barely discernible on seismic-reflection profiles, and the pattern of their traces is poorly defined. Apparent normal and reverse faults on profiles that cross this zone show separations that range from less than 100 m to nearly 1 km. The complex arrangement (flower structure?) of these faults in vertical section (Figures 8, 9, 10, and 14) suggests that they may have formed as a result of transpression. Reasons for this interpretation include 1) linearity of both the zone of faults and the anticline (or anticlines), 2) the faults have opposing vergences on different seismic profiles, and 3) some of the structures seem to be arranged en echelon. The youngest of these faults deform the uppermost (late? Miocene) part of unit E3 and thus may be related to the arc-reversal process.

Evidence is lacking for the north-trending fault mapped by Blake and Miezitis (1967) along the western shoreline of Buka Island. This inferred fault is not recognizable on seismic-reflection profile 413 just north of the island (Figure 9). Furthermore, a fault is not required on the direct projection of dips on profile 415 into dips on Taiof Island (Figure 11).

CRUSTAL STRUCTURE FROM SONOBUOY DATA

Sonobuoy and two-ship seismic-refraction studies, chiefly in the Solomon Islands region, were reviewed by Cooper, Bruns, and Wood (1986), and Cooper, Marlow, and Bruns (1986). Sonobuoy data collected during the 1984 R/V *S.P. Lee* cruise is interpreted in Cooper, Cochrane, and Bruns (this volume). The earlier refraction studies in the central and western Solomon Islands indicated that basement rocks occur at shallow depths except beneath the intra-arc basins where there is as much as 6 km of Oligocene(?) to Holocene sedimentary cover (Bruns et al, 1986; Cooper, Bruns, and Wood, 1986). Basement-rock seismic velocities of 4.5 to 6.3 km/s imply the presence of plutonic, volcanic, or highly lithified or metamorphosed sedimentary rocks (Cooper, Cochrane, and Bruns, this volume). Presumably, these basement rocks are underlain by mafic and ultramafic rocks that constitute the 13- to 25-km thick crust beneath the arc (Cooper, Marlow, and Bruns, 1986).

The methods used for interpreting the sonobuoy records from north and west of Bougainville in 1984 (Figure 15) are described in Cooper, Cochrane, and Bruns (this volume). Of these 14 stations, eight are pertinent to the understanding of the kinds of rocks that underlie Queen Emma basin and are discussed below. The results are shown on Figure 15.

Green-Buka-Bougainville Islands Area

The subsea parts of southern Emirau-Feni Ridge, the Green Islands, and eastern Buka Island are believed to be underlain at shallow depth by intrusive and extrusive igneous rocks of late Cenozoic age similar to those on northern Bougainville. These rocks form a belt (Figure 3C) that diagonally transects early Cenozoic igneous and metamorphic(?) rocks beneath central Bougainville (Blake and Miezitis, 1967; Kroenke, 1984; Vedder, Colwell, and Bruns, 1985).

Interpretations of a sonobuoy record from station 77 along the crest of the southern extension of Emirau-Feni

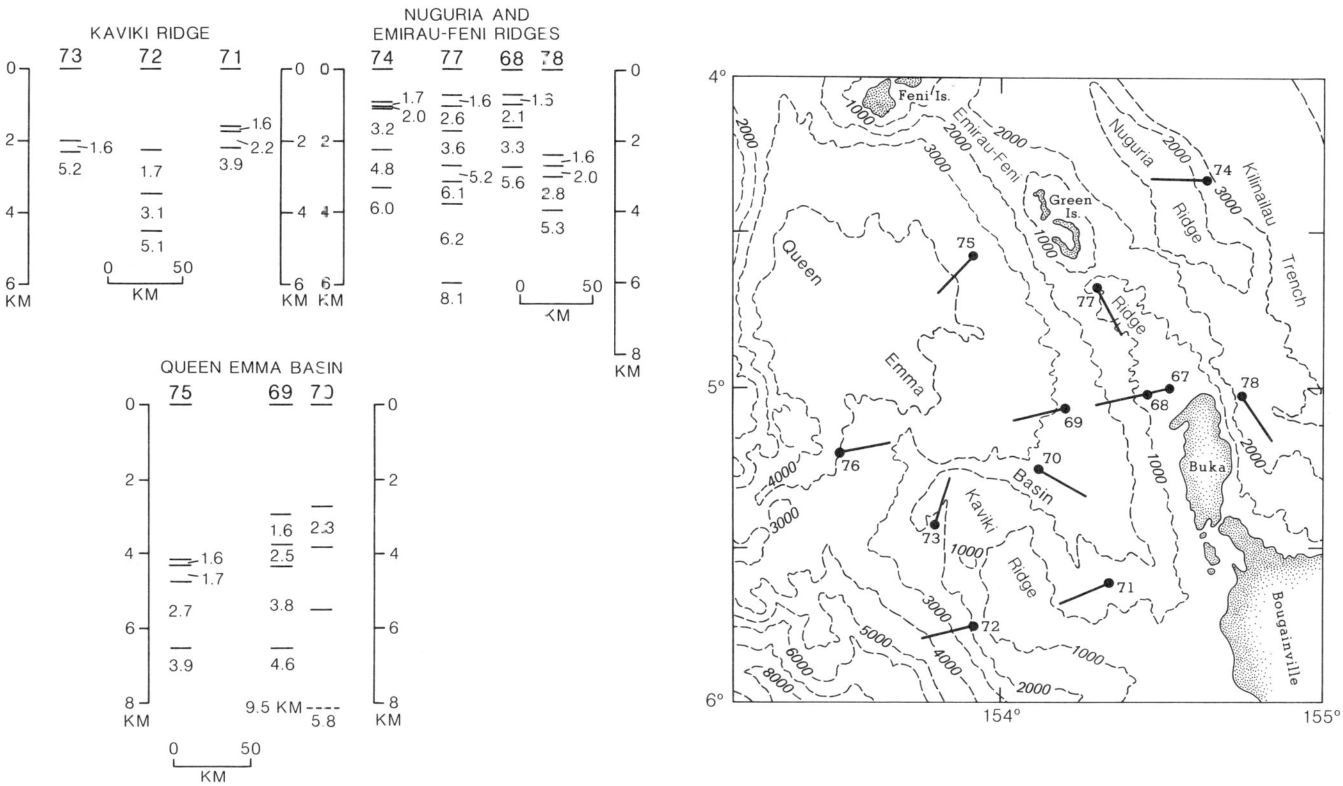

Figure 15. Refraction-velocity results and sites of sonobuoy records in Queen Emma basin area. Dots indicate location of sonobuoys; lines denote length of record. Sonobuoys 67 and 76 did not provide interpretable records. Contours are in meters and velocities are in km/s.

Ridge (Figure 15) indicate that rocks in the uppermost 2.0 km of section have seismic velocities of 1.6 to 3.6 km/s and probably are composed of volcaniclastic strata that correlate with seismic-stratigraphic unit E2. These strata are underlain by about 3.4 km of rocks that have velocities of 5.2 km/s and 6.1 to 6.2 km/s and that are equivalent to seismic-stratigraphic unit E1. At a greater depth, a recorded velocity of 7.5 km/s dip-corrected to 8.1 km/s may be caused by uplifted or intruded rock bodies from beneath unit E1.

Data from sonobuoy 68 were recorded downslope on the southwestern flank of Emirau-Feni Ridge (Figure 15) and shows seismic velocities of 2.1 to 3.3 km/s in a section about 2.1 km thick that extends upslope to the ridge crest. This sedimentary sequence, which is represented by seismic-stratigraphic unit E2, is correlated with the Oligocene and early Miocene(?) Buka Formation. Rocks directly below the sedimentary sequence have a velocity of 5.6 km/s and are inferred to consist of igneous rocks that are equivalent to seismic-stratigraphic unit E1 beneath the ridge slope and to volcanogenic rocks on northern Bougainville. On the eastern slope of Buka Island where data from sonobuoy 78 were recorded, irregular bathymetry and the likely presence of volcanic flows prevented acoustic penetration beneath the flows, the top of which occur at subsurface depths of 1.5 to 2.1 km.

The record from sonobuoy 74 on the western flank of Nuguria Ridge indicates that the upper 1.2 km of acoustic basement may be composed of sedimentary rocks. These rocks have seismic velocities of 3.2 km/s and are underlain by igneous(?) rocks that have velocities of 4.8 to 6.0 km/s.

Kaviki Ridge

Interpretations of the records from sonobuoy stations 71, 72, and 73 furnish varying results over Kaviki Ridge (Figure 15), a feature that we believe is a partly subsided segment of the early Cenozoic magmatic arc. At station 71, subsurface structure limits interpretation to only a few hundred meters of section, in which seismic velocities range from 1.6 to 3.9 km/s. On the southwestern flank of the ridge, station 72 indicates probable sedimentary rocks about 2.0 km thick, a thickness that is substantiated by seismic-reflection data (seismic-stratigraphic units E2 and E4, Figure 8). A velocity of 3.1 km/s was recorded from rocks lower in the section. At station 73, a velocity of 5.2 km/s was determined for rocks beneath a thin veneer of sedimentary cover. These high-velocity rocks are inferred to be related to the basement volcanics and associated intrusive rocks on Bougainville.

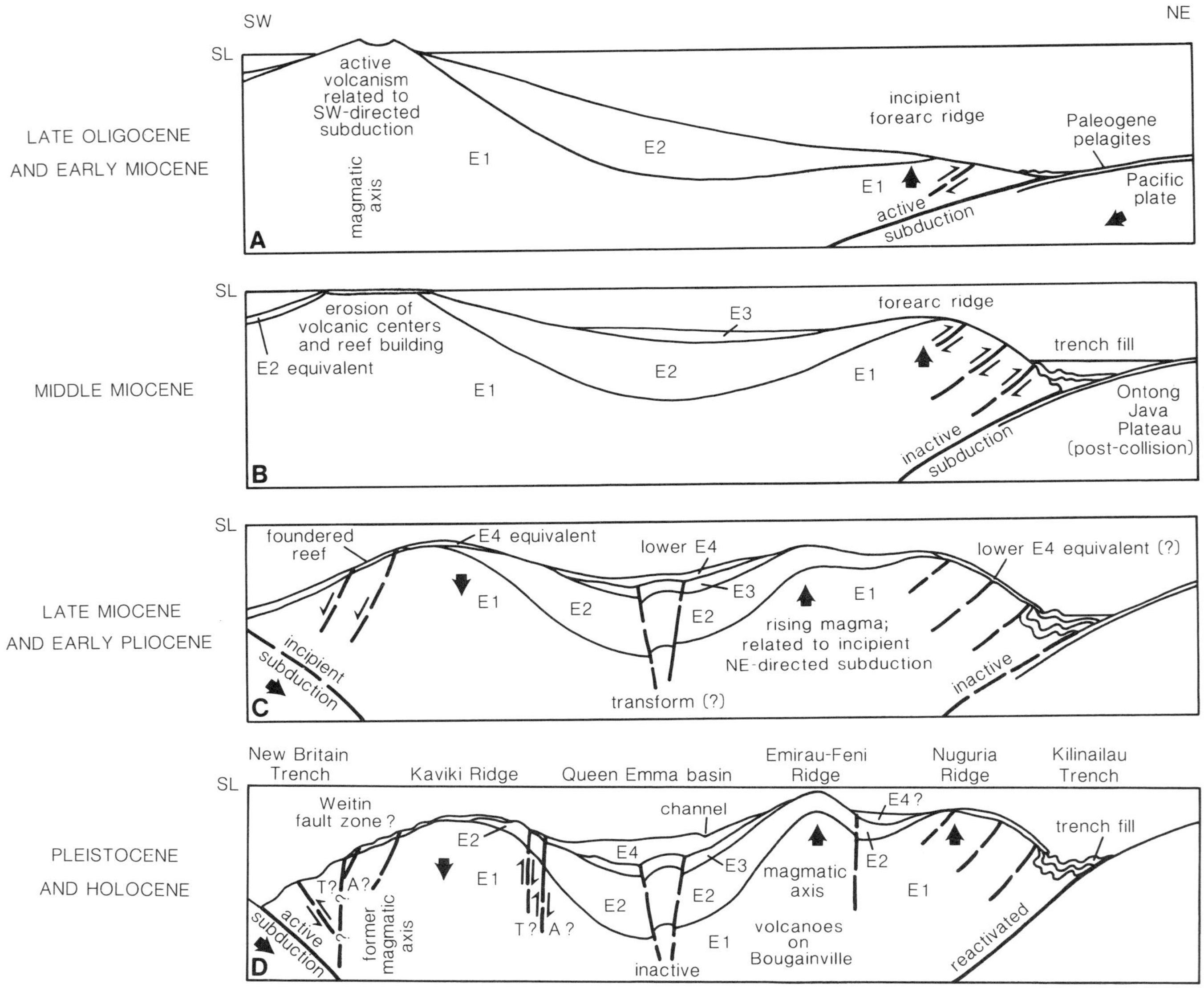

Figure 16. Inferred tectonic development of Queen Emma basin from late Oligocene to Holocene time.

Queen Emma Basin

Three sonobuoys deployed in Queen Emma basin (Figure 15) provided useful records; a fourth was unsuitable for interpretation because of complex subsurface structure. Data from sonobuoys 69 and 70 were recorded over the thickest parts of the basin's sedimentary section, and data from sonobuoy 75 was recorded across the buried, faulted anticlinal structure that trends northwest along the basin axis.

The nearly flat-lying section (seismic-stratigraphic units E4 and E3) has seismic velocities of 1.6 to 2.5 km/s, and the deformed underlying section (seismic-stratigraphic unit E2) has a velocity of 3.2 to 3.8 km/s. Where basement-refraction arrivals were recorded, velocities are 4.6 to 5.8 km/s. Velocities for all of the deeper units are likely to have large errors because of uncertainties in the dip of layers. Results from the three successful sonobuoys indicate that more than 6.5 km of sedimentary section fills the deeper parts of the basin (Cooper, Cochrane, and Bruns, this volume).

DISCUSSION AND CONCLUSIONS

Tectonic History of Queen Emma Basin and Environs

About 25 to 20 Ma, the close approach and arrival of the Ontong Java Plateau at the site of subduction along the Pacific plate/Australia-India plate boundary set off a series of tectonic events that influenced the architecture of the North and South Solomon Island arcs for the remainder of Cenozoic time. The shutdown of southwest-directed subduction and cessation of accompanying arc magmatism was followed by relatively little tectonic activity about 15 to 10 Ma. Arc polarity reversal about 10 Ma instituted northeast-directed subduction and generated a new phase of arc magmatism that continues today. These tectonic events and their effect on basin deposition are schematically depicted by a sequence of panel diagrams on Figure 16.

Interpretations of the data collected during the 1984 R/V *S.P. Lee* cruise support the concept of arc-polarity reversal and the resulting new magmatic arc which crosses an older arc axis beneath Bougainville (Blake and Miezitis, 1967; Kroenke, 1984; Vedder, Colwell, and Bruns, 1985). These data also clarify some of the effects of arc reversal on depositional patterns.

Rocks generated by and derived from early Cenozoic arc magmatism not only are exposed in a west-trending belt through central Bougainville (basement volcanics and related intrusive rocks) but also form the structurally high-standing region of southeastern New Ireland (Jaulu Volcanics and Lemau Intrusive Complex). These basement-rock correlations indicate that Kaviki Ridge is a subsided segment of the axial part of the early Cenozoic northeast-facing magmatic arc that formerly continued without interruption into the now uplifted part of the same arc northeast of the Weitin fault on New Ireland. The upper part of the acoustic basement (unit E1) on Kaviki Ridge is composed of rocks that have a seismic velocity of 5.2 km/s and no resolvable stratigraphy. Strata derived from the ridge probably make up much of the oldest basin-fill sequence (unit E2); and these strata form a lenticular northeastward-thinning sediment apron, a circumstance which implies that Kaviki Ridge was a high-standing feature at least until middle Miocene time. Thus, the deformed sedimentary sequence northeast of Kaviki Ridge (unit E2 in Queen Emma basin; Figures 8, 9, and 10) represents forearc-basin strata derived from the ridge (Figure 16, A) and correlates with seismic sequence D-E of Exon et al (1986) in southeastern New Ireland Basin.

Because the Pliocene to Holocene stratovolcanoes that form the northwest-trending spine of Bougainville (Figure 4) are in direct alignment with the southern extension of Emirau-Feni Ridge (Figure 2), much of this ridge is likely to be underlain by correlative volcanogenic rocks. The upper 3.4 km of acoustic basement beneath the submergent ridge crest between the Green Islands and Buka Island is composed of rocks that have a seismic velocity of 5.2 km/s and correlate in part with the seismic-stratigraphic basement (unit E1).

If the axial trend of the early Cenozoic magmatic arc follows the inferred path along Kaviki Ridge, the crossing of the old and new arcs must be in central Bougainville (Figure 3C). Therefore, in Queen Emma basin, the two uppermost seismic-stratigraphic units (units E3 and E4) would represent Miocene and younger forearc and intra-arc basin strata derived chiefly from Bougainville and the southern segment of Emirau-Feni Ridge between Buka Island and the Green Islands (Figure 16, B, C, and D).

Uplift of Nuguria Ridge and the Green Islands-Buka segment of the Emirau-Feni Ridge probably took place in two episodes (Figure 16, B, and C). The first episode occurred during the Oligocene and early Miocene(?) when convergence of the Ontong Java Plateau against the old forearc region caused a forearc high to develop southwest of the Kilinailau Trench (Figure 16, B) (Bruns et al, this volume). The second episode began along Emirau-Feni Ridge as a result of late Miocene and early Pliocene magmatism beneath the incipient new arc that was being created by northeast-directed subduction along the ancestral New Britain Trench (Figure 16, C). In the same region, Pliocene(?) and Quaternary back-arc thrusting along the Kilinailau Trench (Figure 16, D) deformed rocks along the eastern slope of Emirau-Feni Ridge (Bruns et al, this volume).

The buried, faulted anticlinal structure in Queen Emma basin resulted from compression that occurred after unit E3 was deposited, possibly in late middle and late Miocene time. This folding may have been related to southeastward-propagating arc reversal (Falvey and Pritchard, 1984) advancing on the dying (southeastward retreating?) early Cenozoic subduction zone about 10 Ma.

Post-Pliocene(?) normal faults along the northeastern edge of Kaviki Ridge probably are related to the Weitin fault and may have resulted from late-stage intra-arc transtension. This extensional episode presumably contributed to the subsidence of Kaviki Ridge and the southern margin of Queen Emma basin. The southeastward offshore continuation of the Weitin fault is part of the zone of faults that lies within the Pliocene to Holocene transform zone on the northeast side of Manus Basin (Taylor, 1979; Johnson, 1979; Figures 3A, B, and D). This left-lateral strike-slip fault may splay out southeastward and die before reaching Kaviki Ridge, or it may continue along the southwestern flank of Kaviki Ridge and extend diagonally into the north wall of the New Britain Trench. Alternatively, the fault may break into an en echelon pattern and step over to the zone of faults along the northeastern flank of Kaviki Ridge.

Relation to Intra-arc Basins in Adjacent Arc Segments

By using the reconstructions of Falvey and Pritchard (1984) and Kroenke (1984) for the various arc segments in the New Ireland-Bougainville region, the obvious inference is that Queen Emma basin now is attached to the Pacific plate, even though it formerly was wholly or partly detached from it. Similarly, the basin is interpreted to be one of the series of intra-arc basins that lie along the plate boundary (Vedder and Coulson, 1986). From northwest to southeast, these intra-arc basins are the New Ireland Basin of Exon and Tiffin (1984), Queen Emma basin (this paper), and the composite Central Solomons Trough of Katz (1980), which was divided into three sub-basins (Shortland basin, Russell basin, and Iron Bottom basin) by Tiffin, Vedder, and Cooper (1983). This entire set of basins was referred to as the Solomon Basin by de Broin et al (1977). Even though the old and new magmatic arcs cross at Bougainville and converge at Guadalcanal, the intervening basins are considered to be intra-arc basins; those to the southeast consisting of back-arc sediment aprons (Vedder and Coulson, 1986; Bruns et al, 1986), and those to the northwest consisting of forearc sediment aprons. The Bougainville Basin of Shaw (1985), which is situated close to the crossing of the early Cenozoic and late Cenozoic arc axes, differs from the others in that it originated as a backarc basin but now occupies a forearc position southwest of the Pliocene-Holocene volcanoes on Bougainville.

SUMMARY

Queen Emma basin is a subsided structural and depositional trough that developed initially in a forearc setting associated with the early Cenozoic New Ireland-Bougainville segment of the North Solomon Island Arc and that continued to evolve in a similar setting after late Cenozoic arc reversal formed the South Solomon Island Arc. The basin is a discrete feature that is separated from the New Ireland Basin by stratigraphic discontinuities and from Bougainville basin by a basement ridge. Queen Emma basin is one of a series of intra-arc depocenters that extend southeastward from the Bismarck Archipelago into the central Solomon Islands.

Interpretations of seismic-reflection records across Queen Emma basin indicate the presence of four seismic-stratigraphic units that range in age from Eocene(?) to Holocene. Although direct correlation with onshore rock sequences is uncertain, some of the offshore units have recognizable counterparts on the adjoining islands. Acoustic basement (unit E1) at most places probably is composed of volcanogenic rocks that correlate with the early Cenozoic Jaulu Volcanics on New Ireland and the basement volcanics on Bougainville. The lowermost bedded sequence (unit E2) is deformed, as much as 5.0 km thick, and correlates in part with the Oligocene and early Miocene(?) Buka Formation and possibly with the Tamiu Siltstone on New Ireland. The next higher unit in the succession above an unconformity is a deformed, relatively thin and areally restricted sedimentary sequence (unit E3) of late early(?) to middle(?) Miocene strata that occupies the deeper parts of the basin and that generally is less than 2.0 km thick. This sequence may correlate with the middle part of the Lelet Limestone on New Ireland, although the strata are more likely to consist of bedded clastic sediment rather than limestone in the inferred deep-water environment. The highest depositional sequence (unit E4) probably ranges in age from late Miocene(?) to Holocene and may correlate in part with the Punam Limestone and younger sedimentary units on New Ireland and with the volcanic rocks of the Bougainville Group on Bougainville. This youngest seismic-stratigraphic unit has a maximum thickness of about 2.5 km and onlaps the underlying units on the basin margins.

Folds and faults deform all of the seismic-stratigraphic units that are inferred to be older than Pliocene (units E1, E2, and E3). A buried northwest-trending anticline underlies the central trough of the basin and predates the deposition of the youngest stratigraphic sequence (unit E4). The segment of Emirau-Feni Ridge that connects Feni, Green, and Buka islands is underlain by a broad antiform that structurally affects all rocks older than late Pleistocene. Three well-defined northwest-trending fault zones transect the basin area: 1) the southeastward offshore continuation of the left-lateral(?) strike-slip Weitin fault on New Ireland, 2) a zone of apparent normal faults that cuts the northeastern flank of Kaviki Ridge, and 3) flower-structure-like faults along the buried anticline in the central part of Queen Emma basin. Small normal(?) faults are concentrated on the crests and slopes of all the structurally high-standing features on the periphery of the basin; these faults cut the youngest seismic stratigraphic unit (E4) and apparently extend to the seafloor at places.

Crustal seismic-velocity structure interpreted from sonobuoy refraction data includes regional basement-rock velocities of more than 4.5 km/s, some of which are as high as 5.5 km/s in large tracts of shallow basement, and from 5.8 to 6.3 km/s in small areas of basement under high-standing structures. Upper layers of basement on Kaviki Ridge have a velocity of 5.2 km/s. Sonobuoy refraction records from the deeper parts of Queen Emma basin suggest the presence of sedimentary sequences more than 6.5-km thick.

Stratigraphic and structural inferences derived from interpretations of the data collected during the 1984 R/V *S.P. Lee* cruise, together with older data, support the concept of crossing early Cenozoic and late Cenozoic magmatic arcs beneath Bougainville. Kaviki Ridge, which flanks Queen Emma basin on the south, is inferred to be a subsided segment of the axial part of the northeast-facing early Cenozoic arc. The southeastward extension of Emirau-Feni Ridge through the Green Islands to Buka Island and Bougainville forms the uplifted northeastern flank of Queen Emma basin and is the surface manifestation of the late Cenozoic arc. Consequently, Queen Emma basin occupies an intra-arc position and contains detrital materials derived initially from the early Cenozoic arc axis on the south and subsequently from the late Cenozoic arc axis on the northeast. The conclusion, therefore, is that both depositional episodes occurred in forearc settings.

Although the origin of the buried faulted anticline in the central part of the basin is unknown, it may have formed in response to transpression associated with the process of arc reversal; or the folding may have been an early expression of vigorous convergence and coupling along the New Britain Trench. Transform motion associated with the Pliocene to Holocene opening of the Manus Basin presumably caused uplift of southeastern New Ireland and may have been responsible for subsidence of Kaviki Ridge and Queen Emma basin. Late Cenozoic magmatism resulted in resurgent uplift and folding of the old forearc high along the Nuguria and Emirau-Feni ridges and tilting of the Parkinson Range on Buka Island. Pliocene(?) and Quaternary back-arc thrusting along the

Kilinailau Trench may have contributed to deformation along the northeastern slope of the Emirau-Feni Ridge.

ACKNOWLEDGMENTS

We wish to single out R.C. Culotta for his expert processing of the seismic-reflection data. We are grateful to J.B. Colwell for supplying copies of seismic-reflection profiles from the BMR, Gulf, and IFP-CEPM-ORSTOM cruises. We also thank N.F. Exon and M.S. Marlow for freely providing access to geophysical data and preliminary reports resulting from their 1984 Tripartite Cruise. We express our appreciation to P.J. Quinterno for furnishing the paleontologic reports on calcareous nannofossils from the Buka Formation. New marine topographic maps of part of the region were made by T.E. Chase, B.A. Seekins, and J.D. Young. Special thanks are due to G.J. Francis and D. Hilyard of the Geological Survey of Papua New Guinea for sharing their ideas and knowledge of the regional geology.

Suggestions by H.G. Greene and M.S. Marlow on an early version of this paper helped to improve its content and quality. Monique Martin and Susan Vath prepared the text and illustrations, respectively.

REFERENCES CITED

Bain, J.H.C., Davies, H.L., Hohnen, P.D., Ryburn, R.J., Smith, I.E., Grainger, R., Tingey, R.J., and M.R. Moffat, 1972, Geology of Papua New Guinea: Department of National Development in cooperation with the Geological Survey Division of the Papua New Guinea Department of Lands, Surveys and Mines, Bureau of Mineral Resources, Geology and Geophysics, scale 1:1,000,000.

Blake, D.H., and Miezitis, Y., 1967, Geology of Bougainville and Buka Islands, New Guinea: Department of National Development, Bureau of Mineral Resources, Geology and Geophysics, Bulletin 93, PNG 1, Canberra, A.C.T., 56 p.

Bruns, T.R., Cooper, A.K., Mann, D.M., and Vedder, J.G., 1986, Seismic stratigraphy and structure of sedimentary basins in the Solomon Islands region, *in* Vedder, J.G., Pound, K.S., and Boundy, S.Q., eds., Geology and offshore resources of Pacific island arcs—central and western Solomon Islands: Houston, Texas, Circum-Pacific Council for Energy and Mineral Resources Earth Science Series, v. 4, p. 177-223.

Coleman, P.J., and Packham, G.H., 1976, The Melanesian Borderlands and India-Pacific plates' boundary: Earth Science Reviews, v. 12, p. 197-233.

Cooper, A.K., Bruns, T.R., and Wood, R.A., 1986, Shallow crustal structure of the Solomon Islands intra-arc basins from sonobuoy seismic studies, *in* Vedder, J.G., Pound, K.S., and Boundy, S.Q., eds., Geology and offshore resources of Pacific island arcs—central and western Solomon islands: Houston, Texas, Circum-Pacific Council for Energy and Mineral Resources Earth Science Series, v. 4, p. 135-156.

Cooper, A.K., Marlow, M.S., and Bruns, T.R., 1986, Deep structure of the central and southern Solomon Islands region: implications for tectonic origin, *in* Vedder, J.G., Pound, K.S., and Boundy, S.Q., eds., Geology and offshore resources of Pacific island arcs—central and western Solomon Islands: Houston, Texas, Circum-Pacific Council for Energy and Mineral Resources Earth Science Series, v. 4, p. 157-175.

Cooper, P.A., Kroenke, L.W., and Resig, J.M., 1986, Tectonic implications of seismicity northeast of the Solomon Islands, *in* Vedder, J.G., Pound, K.S., and Boundy, S.Q., eds., Geology and offshore resources of Pacific island arcs—central and western Solomon Islands: Houston, Texas, Circum-Pacific Council for Energy and Mineral Resources Earth Science Series, v. 4, p. 117-122.

de Broin, C.E., Aubertin, F., and Ravenne, C., 1977, Structure and history of the Solomon-New Ireland region: International Symposium on Geodynamics in the Southwest Pacific, Noumea [New Caledonia], 27 August - 2 September 1976, Editions Technip, Paris, 1977, p. 37-50.

Exon, N.F., Stewart, W.D., Sandy, M.J., and Tiffin, D.L., 1986, Geology and offshore petroleum prospects of the eastern New Ireland Basin, northeastern Papua New Guinea: BMR Journal of Australian Geology and Geophysics, v. 10, p. 39-51.

Exon, N.F., and Tiffin, D.L., 1984, Geology and petroleum prospects of offshore New Ireland Basin in northern Papua New Guinea, *in* Watson, S.T., ed., Transactions of the Third Circum-Pacific Energy and Mineral Resources Conference 22-28 August 1982, Honolulu, Hawaii, p. 623-630.

Falvey, D.A., and Pritchard, T., 1984, Preliminary paleomagnetic results from northern Papua New Guinea: evidence for large microplate rotations, *in* Watson, S.T., ed., Transactions of the Third Circum-Pacific Energy and Mineral Resources Conference 22-28 August 1982, Honolulu, Hawaii, p. 593-599.

Hohnen, P.D., 1978, Geology of New Ireland, Papua New Guinea: Department of National Development, Bureau of Mineral Resources, Geology and Geophysics, Bulletin 194, PNG 12, Australian Government Publishing Service, 39 p.

Johnson, R.W., 1979, Geotectonics and volcanism in Papua New Guinea: a review of the late Cainozoic: BMR Journal of Australian Geology and Geophysics, v. 4, p. 181-207.

Katz, H.R., 1980, Basin development in the Solomon Islands and their petroleum potential: UN ESCAP, CCOP/SOPAC Technical Bulletin n. 3, p. 59-75.

Kroenke, L.W., 1984, Cenozoic tectonic development of the southwest Pacific: UN ESCAP, CCOP/SOPAC Technical Bulletin n. 6, 126 p.

Kroenke, L.W., Jouannic, C., and Woodward, P., 1983, CCOP/SOPAC Bathymetry of the southwest Pacific, Chart 1 of the geophysical atlas of the southwest Pacific: Survey and Mapping Division, Ministry of Lands, Energy and Natural Resources, Honiara, Solomon Islands, 1st Edition.

Mann, D.M., 1986, Multichannel seismic operations for CCOP/SOPAC cruise, leg 3, Solomon Islands, *in* Vedder, J.G., Pound, K.S., and Boundy, S.Q., eds., Geology and offshore resources of Pacific island arcs—central and western Solomon Islands: Houston, Texas, Circum-Pacific Council for Energy and Mineral Resources Earth Science Series, v. 4, p. 35-40.

Marlow, M.S., Exon, N.F., Ryan, H., and Dadisman, S., 1988, Offshore structure and stratigraphy of New Ireland Basin in northern Papua New Guinea, *in* Marlow, M.S., Dadisman, S., and Exon, N.F., eds., Geology and offshore resources of Pacific island arcs—New Ireland and Manus region, Papua New Guinea: Houston, Texas, Circum-Pacific Council for Energy and Mineral Resources Earth Science Series, v. 9, p. 137-155.

Okada, H., and Bukry, D., 1980, Supplementary modification and introduction of code numbers to the low-latitude coccolith biostratigraphic zonation (Bukry, 1973; 1975): Marine Micropaleontology, v. 5, p. 321-325.

Perch-Nielsen, K., 1985, Cenozoic calcareous nannofossils, *in* Bolli, H.M., Saunders, J.B., and Perch-Nielson, K., eds., Plankton stratigraphy: Cambridge, U.K., Cambridge University Press, p. 427-554.

Robson, R.W., 1973, Queen Emma: Sydney, Pacific Publications Pty., Ltd., 245 p.

Shaw, R.D., 1985, Hydrocarbon potential of the Bougainville Basin Papua New Guinea: 14th BMR Symposium, Extended Abstracts - Petroleum Geology of South Pacific Island Countries, Marine Studies Workshop, Canberra, 22-23 October, p. 19-29.

Stewart, W.D., Francis, G., and Pederson, S.L., 1987, Hydrocarbon potential of Papua New Guinea's Bougainville, southeastern New Ireland basins: Oil and Gas Journal, v. 85, n. 47, p. 83-87.

Stewart, W.D., and Sandy, M.J., 1988, Geology of New Ireland and Djaul Island, northeastern Papua New Guinea, *in* Marlow, M.S., Dadisman, S., and Exon, N.F., eds., Geology and offshore resources of Pacific island arcs—New Ireland and Manus region, Papua New Guinea: Houston, Texas, Circum-Pacific Council for Energy and Mineral Resources Earth Science Series, v. 9, p. 13-30.

Taylor, B., 1979, Bismarck Sea: Evolution of a back-arc basin: Geology, The Geological Society of America, v. 7, n. 4, April, p. 171-174.

Tiffin, D.L., Vedder, J.G., and Cooper, A.K., 1983, Multichannel seismic and geophysical survey of "The Slot" and adjacent areas in the Solomon Islands: United Nations Development Programme, Economic and Social Commission for Asia and the Pacific CCOP/SOPAC, Cruise Report n. 71 of PE/SI.2 and PE/SI.15.

Vedder, J.G., 1986, Summary of the geology and offshore resources of the Solomon Islands, *in* Vedder, J.G., Pound, K.S., and Boundy, S.Q., eds., Geology and offshore resources of Pacific island arcs—central and western Solomon Islands: Houston, Texas, Circum-Pacific Council for Energy and Mineral Resources Earth Science Series, v. 4, p. 295-306.

Vedder, J.G., Colwell, J.B., and Bruns, T.R., 1985, New findings in offshore areas of western Solomon Islands and Buka-Bougainville, Papua New Guinea: 14th BMR Symposium, Extended Abstracts - Petroleum Geology of South Pacific Island Countries, Marine Studies Workshop, Canberra, 22-23 October, p. 37-41.

Vedder, J.G., and Coulson, F.I., 1986, Regional offshore geology of the Solomon Islands, Vedder, J.G., Pound, K.S., and Boundy, S.Q., eds., Geology and offshore resources of Pacific island arcs—central and western Solomon Islands: Houston, Texas, Circum-Pacific Council for Energy and Mineral Resources Earth Science Series, v. 4, p. 99-108.

Wallace, D.A., Johnson, R.W., Chappell, B.W., Arculus, R.J., Perfit, M.R., and Crick, I.H., 1983, Cainozoic volcanism of the Tabar, Lihir, Tanga, and Feni Islands, Papua New Guinea: Geology, whole-rock analyses, and rock-forming mineral compositions: Bureau of Mineral Resources, Australia, Report 243, BMR Microform MF197.

Vedder, J.G., and Bruns, T.R., editors, 1989, Geology and offshore resources of Pacific island arcs—Solomon Islands and Bougainville, Papua New Guinea Regions: Houston, Texas, Circum-Pacific Council for Energy and Mineral Resources Earth Science Series, v. 12.

REVISED STRATIGRAPHY OF BOUGAINVILLE AND BUKA ISLANDS, PAPUA NEW GUINEA

D. Hilyard and R. Rogerson
Geological Survey of Papua New Guinea, Port Moresby, Papua New Guinea

ABSTRACT

The oldest exposed rocks on Bougainville belong to the late Eocene and early Oligocene Atamo Volcanics and consist of spilitized basalt-basaltic andesite, commonly pillowed, together with interbedded volcaniclastic breccia, sandstone, and mudstone. The Buka Formation, the oldest rock unit on Buka Island north of Bougainville, consists of volcanolithic sandstone and siltstone, with minor conglomerate and basalt. The oldest fossils obtained from the Buka Formation are reworked(?) specimens of late Eocene to early Oligocene age; the youngest fossils collected from the formation have a range of late early to early middle Miocene. In central and southern Bougainville, the Atamo Volcanics are unconformably overlain by the inner neritic Keriaka Limestone, which is late Oligocene to latest early Miocene in age. The Keriaka Limestone is unconformably overlain by the Toniva Formation and locally by the Aropa Andesite, a sequence of two-pyroxene andesite flows at the base of the Toniva Formation in southern Bougainville. The Toniva Formation consists of interbedded volcaniclastic sandstone, conglomerate, mudstone, siltstone, and minor micrite. Microfossils in the Toniva Formation indicate middle to late Miocene and early Pliocene ages and a bathyal environment of deposition; sedimentary facies indicate deposition by mass-flow processes. Minor amounts of andesite are scattered through the sequence. The Toniva Formation shows a gross coarsening upward toward a gradational conformable contact with the overlying latest (?) Miocene and Pliocene Arawa Conglomerate which is characterized by cliff-forming volcanolithic pebble-boulder diamictite of mass-flow origin. This conglomeratic unit contains interbedded sandstone in places, and pyroclastic rocks appear to become more abundant up section. The distribution of Arawa Conglomerate suggests that syndepositional faulting was active.

The latest Miocene Isinai Granodiorite, the Pliocene Kupei Complex, and numerous unnamed Miocene and Pliocene shallow intrusives occur throughout central and southern Bougainville, intruding all older stratigraphic units. The Isinai Granodiorite (ca. 5 Ma) consists of even-textured, medium-grained hornblende diorite-granodiorite. The Kupei Complex (ca. 2 to 6 Ma) consists of variably altered, variably textured diorite and granodiorite and hosts low-grade Au-Cu porphyry mineralization at Panguna and Kupei. The Bougainville Group consists of latest Pliocene and Quaternary andesitic stratovolcano complexes that locally interfinger with the shallow-marine Pleistocene and Holocene Sohano Limestone.

INTRODUCTION

Bougainville and Buka Island (Figure 1) form the northern part of the Solomon Island arc, which currently faces southwest toward the New Britain Trench. New geological mapping and mineral exploration on these islands was undertaken by the Geological Survey of Papua New Guinea during 1987 and 1988. Work included 1:100,000 scale mapping, as well as more than 40 new isotopic ages and 80 micropaleontological determinations. This geological note presents a preliminary summary of the revised stratigraphy of the two islands. Full details along with 1:100,000 scale geological maps and formal definitions of new stratigraphic units are being prepared for publication

STRATIGRAPHY

The stratigraphy of Bougainville and Buka Island is summarized in Figure 2. The previous regional study by Blake and Miezitis (1967) recognized an older basement unit, the Kieta Volcanics, overlain by the Miocene Keriaka Limestone and younger volcanic sequences. Our mapping has shown that their Kieta Volcanics is a composite unit that can be subdivided into the Atamo Volcanics, Keriaka

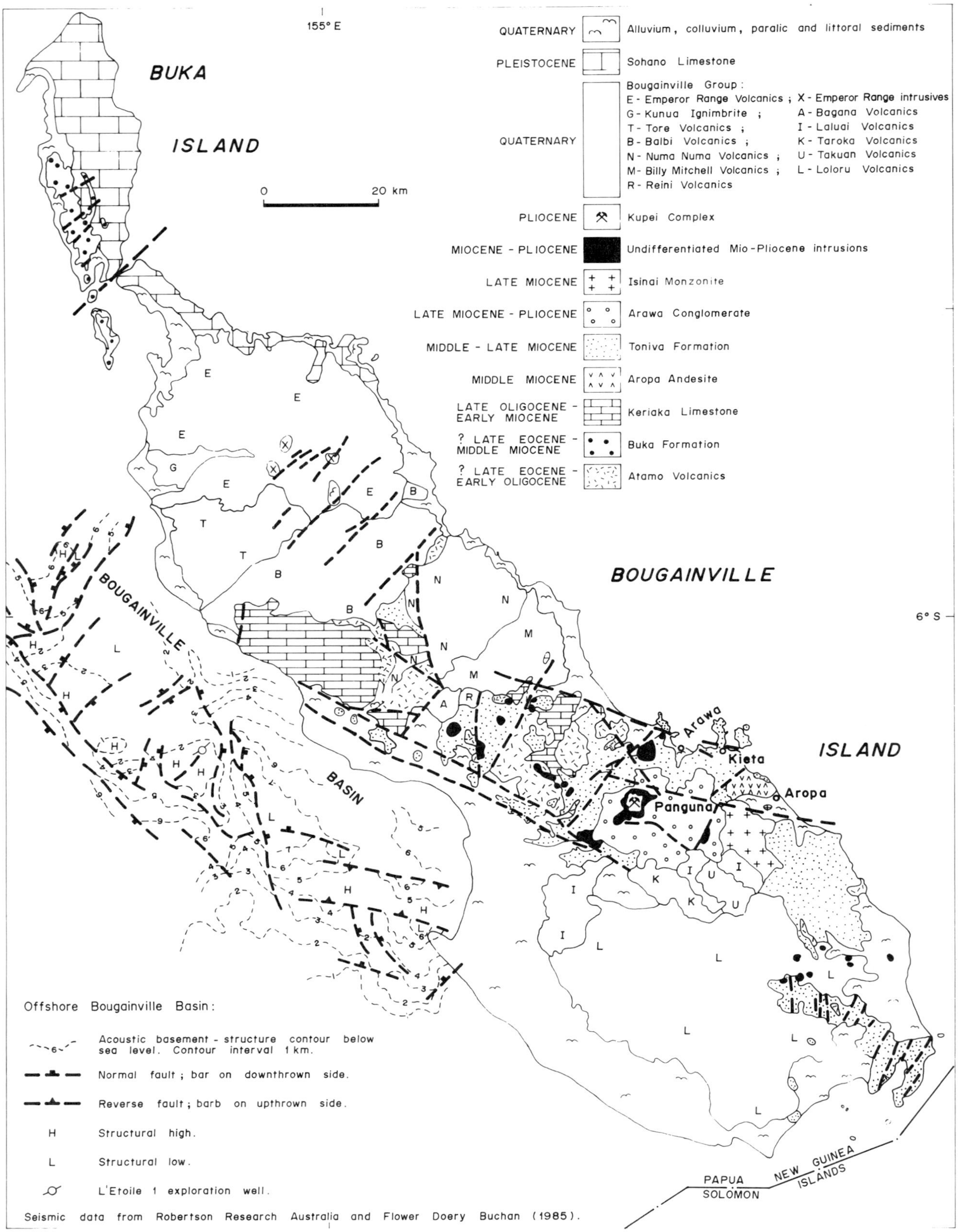

Figure 1. Geological map of Bougainville and Buka Island.

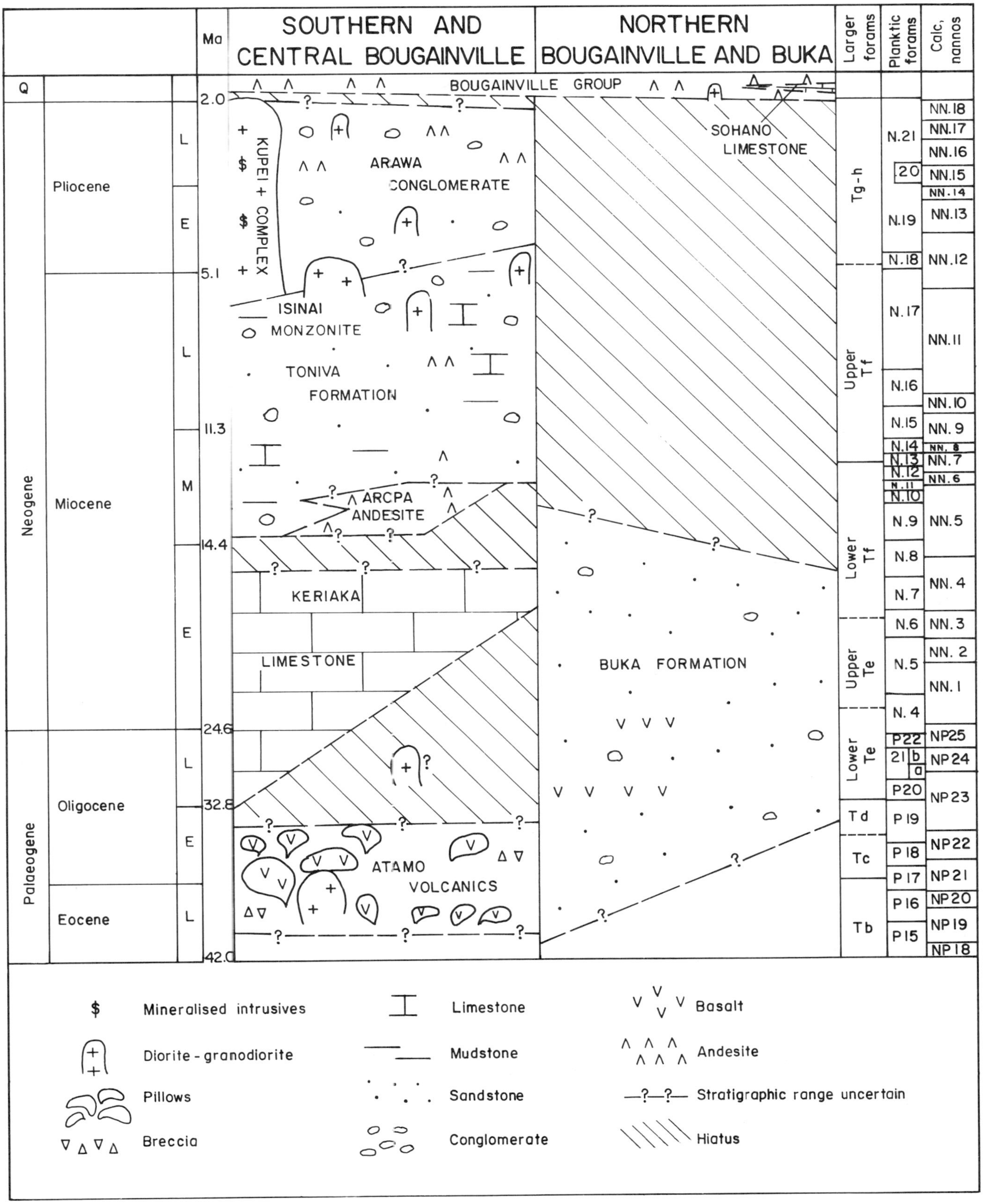

Figure 2. Stratigraphic column for Bougainville and Buka Island.

Limestone, Aropa Andesite, Toniva Formation, and Arawa Conglomerate as defined here; thus the name Kieta Volcanics is superseded.

Atamo Volcanics (new name)

The Atamo Volcanics consist of spilitized, commonly amygdaloidal and pillowed, flows of basalt-basaltic andesite. Dikes of similar material, plus altered dolerite and microdiorite, commonly intrude the flows. Interbeds of pillow breccia, volcanolithic sandstone, and minor calcareous mudstone are present. A late Eocene to early Oligocene age was obtained from foraminiferal assemblages (D. Behring, written commun., 1988).

Buka Formation

The Buka Formation is known to crop out only on Buka Island and adjacent small islands. The unit consists of medium- to thick-bedded volcanolithic sandstone and interbedded siltstone with lesser amounts of volcanolithic conglomerate and minor basalt-andesite lava. Stratigraphic relationships between the Buka Formation and other Tertiary units are not exposed. The oldest microfossil assemblage collected indicates a late Eocene to early Oligocene age, and a general Oligocene and early Miocene age has been obtained on other assemblages (see also Table 1, Vedder, Bruns, and Cooper, this volume). The degree of reworking of these faunas is unknown. The youngest age-diagnostic fauna is assigned a range of uppermost planktonic foraminiferal Zone N7 to N9 (late early to early middle Miocene) (Figure 2). The microfaunal associations indicate paleoenvironments ranging from inner neritic to open marine; at least some of the shallow-marine indicators probably were reworked into more open marine, deeper water sediments.

Keriaka Limestone

The Keriaka Limestone forms an extensive karst plateau in western Bougainville and occurs as small outliers in central Bougainville. Stratigraphic relations with the underlying Atamo Volcanics are uncertain although a hiatus is indicated by known age ranges, and an unconformity is possible. The base of the Keriaka Limestone appears to be markedly transgressive, with ages ranging from benthic foraminiferal lower Te Stage (late Oligocene) to nannofossil Zones NN4-NN5 (late early and early middle Miocene) (Figure 2). The latter ages were obtained from a lagoonal facies sequence of calcareous mudstone and sandstone containing minor coaly lenses at the base of the Keriaka Limestone in western Bougainville. This sequence is remarkably similar to the Matakan unit of the Miocene Lelet Limestone on New Ireland (Stewart and Sandy, 1988). Although the youngest ages obtained from the Keriaka Limestone are not particularly precise, a general benthic foraminiferal Tf Stage (Miocene) designation is indicated. The occurrence of Zone N9 indicators in the unconformably overlying Toniva Formation indicates that the upper part of the Keriaka Limestone may be restricted to the lower part of the lower Tf Stage (late early Miocene). The Keriaka Limestone consists mainly of algal-foraminiferal packstone that includes variable amounts of mollusk, echinoid, and bryozoan debris. The faunal assemblages indicate deposition in an inner neritic environment.

Aropa Andesite (new name)

This unit, which is restricted to the area around Aropa on the southeastern coast of Bougainville, consists of relatively fresh two-pyroxene andesite flows and very minor volcanolithic sandstone interbeds. The stratigraphic relationships of the Aropa Andesite are not well established; the unit appears to lie at or near the base of the Toniva Formation and to be overlain by Toniva Formation sediments. K/Ar dating is in progress.

Toniva Formation (new name)

The Toniva Formation is a widespread sequence of mainly volcanolithic sediments that unconformably overlies both the Atamo Volcanics and Keriaka Limestone. The distribution of sub-units in central Bougainville suggests that a major episode of block faulting occurred prior to deposition of the Toniva Formation. The unit consists largely of interbedded volcanolithic sandstone, siltstone, and conglomerate. Minor amounts of andesite lava and micritic limestone are present. The proportion of conglomerate appears to increase up section toward the contact with the conformably overlying Arawa Conglomerate.

The formation consists of mass-flow sediments including turbiditic sandstone, massive thick-bedded sandstone, and massive matrix-supported pebble-cobble diamictite. Outer neritic and bathyal faunas generally are contained in mudstone and micrite that are interbedded with the coarser strata.

Ages determined from isolated outcrops of the Toniva Formation need to be interpreted with care, as not only is block faulting pervasive, but also reworking of microfossils is common. Most samples give middle Miocene or younger ages, though specific N9 (middle Miocene) indicators to provide the oldest age-diagnostic assemblages known from the formation. Upward, the section becomes more difficult to date, as the amount of coarse volcanolithic detritus increases. Assemblages representing Zones NN11 and NN12-15 are the youngest age-diagnostic faunas obtained, indicating a latest Miocene and early Pliocene age for the upper part of the formation.

Arawa Conglomerate (new name)

The Arawa Conglomerate, which conformably overlies the Toniva Formation, is restricted to the area around the Panguna Mine. Syndepositional faulting may have helped control the extent of this unit. Cliff-forming matrix-supported pebble-boulder volcanolithic diamictite and interbedded massive, thick-bedded sandstone are the dominant rock types. Andesitic pyroclastic rocks are common in the upper part. Although the age is poorly constrained, a Pliocene assignment is indicated for most of the formation. The Arawa Conglomerate includes a number of informal stratigraphic units used by geologists in the Panguna Mine area: the Pakia Gap tuff, Karoona and Dopinpoi tuff breccias, the Binivan andesite breccia, and parts of the Panguna andesite (Bougainville Copper Ltd., 1988; Clark, in press).

Mono Siltstone

The Mono Siltstone is not known to crop out anywhere on Bougainville or on Buka Island. Strata assigned to the unit were intersected between the depths of 433 m and 1335 m in petroleum exploration well L'Etoile 1 drilled in the Bougainville basin to the west of the island (Figure 1), and seismic profiles suggest that the unit is widespread in the subsurface (Robertson Research Australia Ltd. and Flower Doery Buchan Ltd., 1985). The Mono Siltstone and equivalent units crop out on the Shortland Islands approximately 20 km southeast of Bougainville (Ridgway and Coulson, 1987). The unit consists of calcareous tuffaceous siltstone and mudstone with interbedded volcaniclastic sandstone and minor limestone which together represent a regressive sequence ranging from lower and middle bathyal at the base to outer neritic in the upper part. Inner neritic limestone beds are locally present. The Mono Siltstone in L'Etoile 1 ranges from Zones N17-N18 (latest Miocene to earliest Pliocene) to uppermost Zone N21-basal N22 (latest Pliocene? to earliest Pleistocene) (Robertson Research Australia Ltd.and Flower Doery Buchan Ltd., 1985). On the basis of age and lithology, the Mono Siltstone is regarded as the distal equivalent of the Arawa Conglomerate.

Isinai Monzonite

The Isinai Monzonite intrudes both the Toniva Formation and Arawa Conglomerate. The unit consists of relatively even-textured, largely unaltered, medium-grained hornblende monzonite, quartz monzonite, and monzodiorite. Pink K-feldspar, orthopyroxene, clinopyroxene, and biotite are minor but distinctive components. K/Ar ages of 5.4 to 5.2 Ma have been recorded (Blake and Miezitis, 1967; R.W. Page, unpublished data).

Kupei Complex (new name)

The Kupei Complex consists of a number of small multiple intrusions that host economic Au-Cu porphyry ore bodies in the Panguna-Kupei area. The intrusions consist of diorite to granodiorite, and textures range from even textured to porphyritic. Extensive hydrothermal alteration has affected most components of the Kupei Complex. The complex includes a number of informally named units in the mine area: the Karoona and Karua porphyries, Kawerong quartz diorite, Biuro granodiorite, and Nautango andesite (Bougainville Copper Ltd., 1988; Clark, in press). Published and unpublished isotopic dates (Page and McDougall, 1972; R.W. Page, unpublished data) indicate a range of ages for the Kupei Complex from 5 to 1.6 Ma. Pre-mineralization Kawerong quartz diorite falls in the 5 to 4 Ma range; mineralized, intensely altered intrusives are ca. 3.5 Ma; and post-mineralization unaltered intrusives are dated at 3.5 and 1.6 Ma. These ages indicate that the Kupei Complex is broadly coeval with the Arawa Conglomerate, and the close spatial association of these units suggests that they may be genetically related. The Arawa Conglomerate is regarded as the eruptive pyroclastic and epiclastic apron around a stratovolcano intruded by stocks of the Kupei Complex.

Undifferentiated Miocene and Pliocene Intrusions

Numerous small stocks and dikes intrude the areas of pre-Quaternary outcrop in central Bougainville. Most are too small to be shown on Figure 1. Available K/Ar dates from samples of these intrusions range from 8 to 4 Ma. The intrusives are typically diorite, microdiorite, or andesite, with common phenocrystic hornblende. Alteration of these rocks is variable. Parts of the informal Panguna andesite unit of the Panguna Mine area (Bougainville Copper Ltd., 1988; Clark, in press) may be included in this suite.

Bougainville Group

The Bougainville Group encompasses the latest Pliocene and Quaternary andesitic stratovolcanoes of Bougainville. The oldest part of the group, including shallow sub-volcanic plutons exposed in northern Bougainville, includes latest Pliocene (2.1 Ma) rocks. Blake and Miezitis (1967) subdivided the Bougainville Group into a number of constituent formations, most of which have been retained in this study. From north to south, we recognize the Emperor Range Volcanics, Kunua Ignimbrite (new name), Tore Volcanics, Balbi Volcanics, Numa Numa Volcanics, Billy Mitchell Volcanics, Reini Volcanics, Bagana volcanics, Laluai Volcanics (new name), Takuan Volcanics, Taroka Volcanics, and Loloru Volcanics (new name). The Bakanovi Volcanics described by Blake and Mietzitis (1967) comprise a residual hill of diamictite which we correlate with the Miocene Toniva Formation; the name Bakanovi Volcanics thus is superseded. Bagana Volcano is currently active, whereas Balbi and Loloru Volcanoes have active fumaroles. New isotopic dating shows that Billy Mitchell and Loloru Volcanoes are dormant with potential for moderate to large pyroclastic eruptions in the future (McKee et al., 1988).

Sohano Limestone

The Sohano Limestone crops out extensively on Buka Island and northern Bougainville. The formation consists of clean coralline limestone and appears to be, at least in part, coeval with the Bougainville Group. One specimen of limestone yielded a Pliocene or younger foraminiferal assemblage, and stratigraphic relationships suggest a Pleistocene age. Considering the general overlap of age of the Sohano Limestone and Bougainville Group, the absence of volcaniclastic detritus in the limestone is remarkable.

ACKNOWLEDGMENTS

We gratefully acknowledge the collaboration of the mapping team who worked with us on Bougainville and Buka: Steve Nion, Eric Finlayson, Lasark Joseph, John Duguman, Robert Sumaiang, and Adam Wangu from Minerals Division, Geological Survey of Papua New Guinea; Chris McKee and Herman Patia, Rabaul Volcanological Observatory; Wally Johnson, Bureau of Mineral Resources, Geology and Geophysics, Australia; and the geologists of the Bundesanstalt für Geowissenschaften und Rohstoffe, Hannover. Bougainville Copper Ltd. assisted with work around the Panguna Mine area. Heinz Hekel, Alan Lloyd, David Haig, and Paula Quinterno provided the micropaleontologic data. Alan Webb, Mark Fanning, Mike Barbetti, and Gillian Taylor undertook the isotopic dating.

REFERENCES CITED

Bougainville Copper Ltd., 1988, Geological plan, Panguna Special Mining Lease. Plan no. G-5-200A, Bougainville Copper Ltd., Panguna, Papua New Guinea (unpubl.).

Blake, D.H., and Miezitis, Y., 1967, The geology of Bougainville and Buka Islands, New Guinea: Bureau of Mineral Resources, Geology and Geophysics, Australia, Bulletin 93, and Bulletin PNG 1, 56 p.

Clark, G.H., in press, The Panguna porphyry copper/gold deposit, Papua New Guinea, *in* Hughes, F. E., ed., Geology and mineral deposits of Australia and Papua New Guinea: Parkville, Australia, Australasian Institute of Mining and Metallurgy.

McKee, C., Johnson, R.W., and Patia, H., 1988. Introduction to volcanic hazards on Bougainville Island: Geological Survey of Papua New Guinea Report 88/23.

Page, R.W., and McDougall, I., 1972, Geochronology of the Panguna porphyry copper deposit, Bougainville Island, Papua New Guinea: Economic Geology 67, p. 1065-1074.

Ridgway, J., and Coulson, F.I.E., 1987, The geology of Choiseul and the Shortland Islands, Solomon Islands: British Geological Survey Overseas Memoir 8, 134 p.

Robertson Research Australia Ltd. and Flower Doery Buchan Ltd., 1985, Petroleum potential of the Cape Vogel, Bougainville, and New Ireland Basins, Papua New Guinea. Report prepared for the Geological Survey of Papua New Guinea, 143 p.

Stewart, W.D., and Sandy, M.J., 1988, Geology of New Ireland and Djaul Islands, northeastern Papua New Guinea, *in* Marlow, M.S., Dadisman, S.V., and Exon, N.F.,eds., Geology and offshore resources of Pacific island arcs—New Ireland and Manus region, Papua New Guinea: Houston, Texas, Circum-Pacific Council Earth Science Series, v. 9, p. 13-30.

Vedder, J.G., and Bruns, T.R., editors, 1989, Geology and offshore resources of Pacific island arcs—Solomon Islands and Bougainville, Papua New Guinea Regions: Houston, Texas, Circum-Pacific Council for Energy and Mineral Resources Earth Science Series, v. 12.

STRUCTURE AND TECTONICS ALONG THE KILINAILAU TRENCH, BOUGAINVILLE-BUKA ISLAND REGION, PAPUA NEW GUINEA

T.R. Bruns, J.G. Vedder, and R.C. Culotta
U.S. Geological Survey, Menlo Park, California 94025

ABSTRACT

Multichannel seismic-reflection data show stratigraphic relations and structural features along the Kilinailau Trench and adjacent slope off Bougainville and Buka Island, Papua New Guinea. The Kilinailau Trench is a site of active back-arc thrusting. Thrust faults and large anticlinal folds along the thrust zone have affected the youngest trench-fill strata and have caused marked bathymetric relief at the sea floor.

Rocks in the trench can be separated into three seismic-stratigraphic units. These are (1) unit K1, Cretaceous oceanic crust (layered?); (2) unit K2, an overlying 700- to 900-m-thick Late Cretaceous to early Miocene(?) pelagic unit similar to that on the Ontong Java Plateau; and (3) unit K3, a 3- to 5-km-thick (axial thickness) clastic turbidite trench fill of probable middle Miocene(?) and younger age. Rocks beneath the adjacent slope on the southwest side of the trench are also divided into three seismic-stratigraphic units corresponding to: (1) unit B1, chaotically bedded Oligocene and early Miocene volcanic and derived sedimentary rocks, possibly including remnants of Late Cretaceous and younger metamorphosed oceanic crust and deformed sedimentary rocks accreted during Cenozoic subduction; (2) unit B2, bedded but markedly deformed late Miocene rocks which have a maximum thickness greater than 2 km; and (3) unit B3, an overlying well-bedded and slightly deformed to undeformed Pliocene(?) and younger unit as much as 2 km thick. The unit B1 and B2 slope rocks have been folded and faulted, presumably during early Tertiary subduction and late Cenozoic back-arc thrusting, whereas the overlying younger B3 strata are largely undeformed.

The Kilinailau Trench thrust zone is a result of closure between the Pacific plate and the island arc systems of the western Solomon Islands and Bougainville. This closure is best explained as being caused by partial coupling of the island arc to the actively subducting plate at the New Britain subduction zone on the south side of the arc. This coupling has pushed the island arc system northward relative to the Pacific plate, reactivated an early Tertiary subduction zone suture, and generated back-arc thrusting along this reactivated suture. Thus, at present, the island arcs of the Solomon Islands and Bougainville regions are moving with, but are partly decoupled from, the Pacific plate.

INTRODUCTION

During June and July, 1984, geological and geophysical investigations of the offshore area of the Solomon Islands, Bougainville, and New Ireland were carried out from the U.S. Geological Survey research vessel R/V *S.P. Lee*. This cruise was part of a series of internationally sponsored marine surveys made under the Australia-New Zealand-United States Tripartite Agreement of 1984, and coordinated by the U.S. Department of State and the United Nations Committee for Co-ordination of Joint Prospecting for Mineral Resources in South Pacific Offshore Areas (CCOP/SOPAC).

In this report, we present interpretations of 550 km of multichannel seismic-reflection data that were acquired across the Kilinailau Trench region off Buka Island and Bougainville, Papua New Guinea, and Choiseul, Solomon Islands (Figures 1-3). From these interpretations, we draw inferences on the stratigraphy, structure, and tectonic history of the trench. Sonobuoy refraction data collected along some of these lines are described and interpreted in reports by Cooper, Cochrane, and Bruns (this volume), and Cochrane, Cooper, and Bruns (this volume).

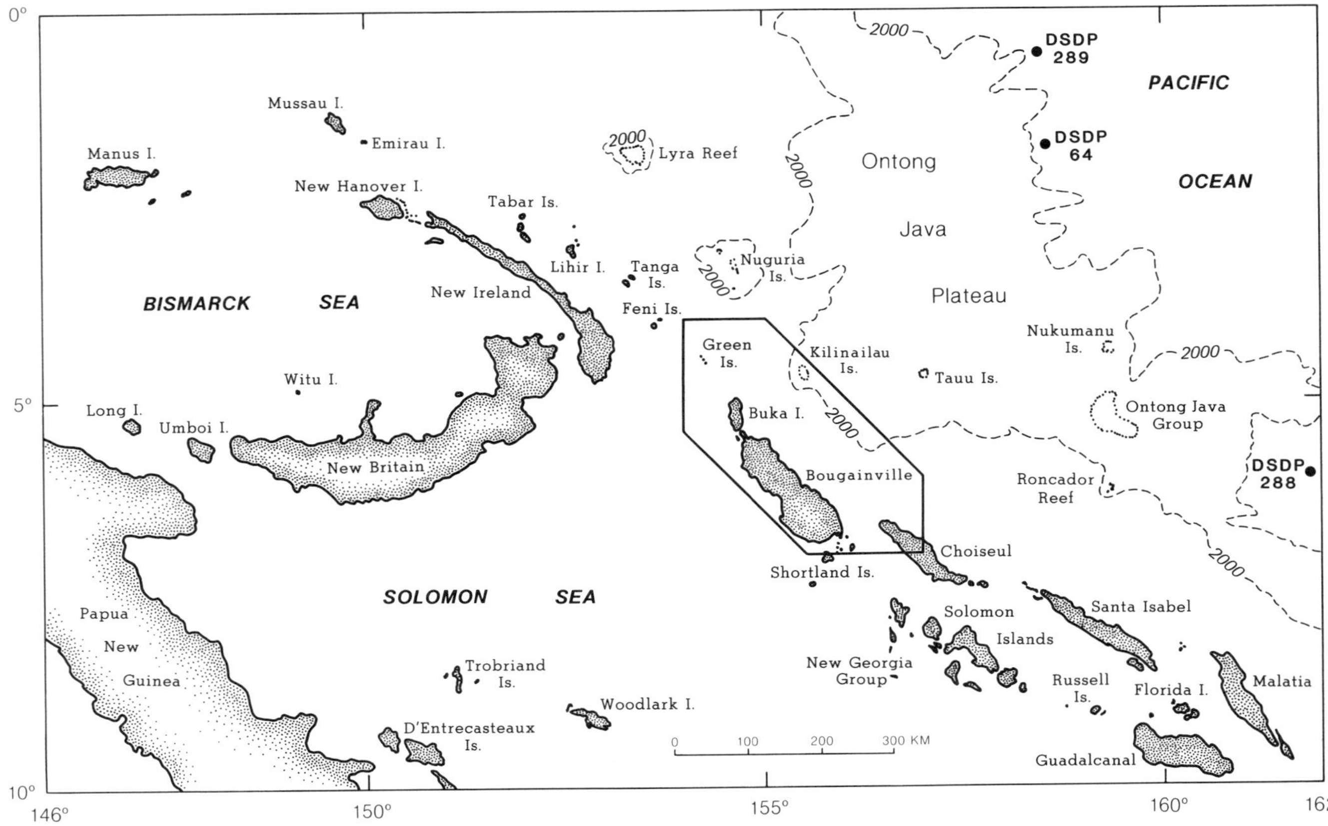

Figure 1. Location map of the Solomon Islands and Bougainville-New Ireland region of Papua New Guinea. The Ontong Java Plateau is bordered by the 2,000-m bathymetric contour. The outline shows the Kilinailau Trench area of Figures 3, 14, 15, and 16.

GEOGRAPHIC SETTING

The Kilinailau Trench is one segment of the long modern geomorphic trench system that stretches southeastward from Papua New Guinea to the eastern Solomon Islands. This system includes the Manus, Kilinailau, and North Solomon trenches (Figure 2). The Kilinailau segment of this system extends as a geomorphic feature about 650 km northwest-southeast from the Lihir Island-Lyra Reef region to the area north of Choiseul.

Within the Kilinailau Trench, water depths range from about 3,000 to 3,500 m near Choiseul and Bougainville, deepen northwest of the Nuguria Islands, and reach as much as 4,500 m adjacent to Lyra Reef. Both walls of the trench rise abruptly to shallow depths along Bougainville and Buka Island. On the southwestern side of the trench lie the islands of Choiseul, Bougainville, Buka, and the Tabar-to-Feni chain (Tabar, Lihir, Tanga, Feni, and Green Islands), which lie atop the Emirau-Feni Ridge. On the northeastern side of the trench, the Ontong Java Plateau lies beneath 1,000 to 2,000 m of water; atolls rise from the plateau to sea level adjacent to the trench at Lyra Reef and the Nuguria and Kilinailau islands.

PREVIOUS WORK

Studies of the Kilinailau Trench made before 1984 are few and are reconnaissance in nature. A pair of single-channel seismic-reflection lines were acquired across the trench in 1970 by Compagnie Genereal de Geophysique for the Australian Bureau of Mineral Resources (BMR 1970, Figure 3). Gulf Research and Development Company ran two multichannel seismic-reflection lines across the northwestern end of the trench in 1972 (Gulf 1972, Figure 3). A French consortium, Comite d'Etudes Petrolieres Marines (CEPM), acquired widely spaced lines in 1975 during the AUSTRADEC III cruise (AUSTRADEC 1975, Figure 3; de Broin, Aubertin, and Ravenne, 1977). Johnson, Egloff, and Hemler (1974) published a reflection-line run with a sparker source that crossed the trench; Hamilton (1979) reproduced the record. Furumoto et al (1976) also showed a single-channel seismic-reflection record across the trench. The high vertical exaggeration made this last record difficult to interpret, and the Kilinailau Trench and adjacent Queen Emma basin (Figure 3) were described as grabens, inferring that the region was extensional.

However, most studies, such as those of Johnson

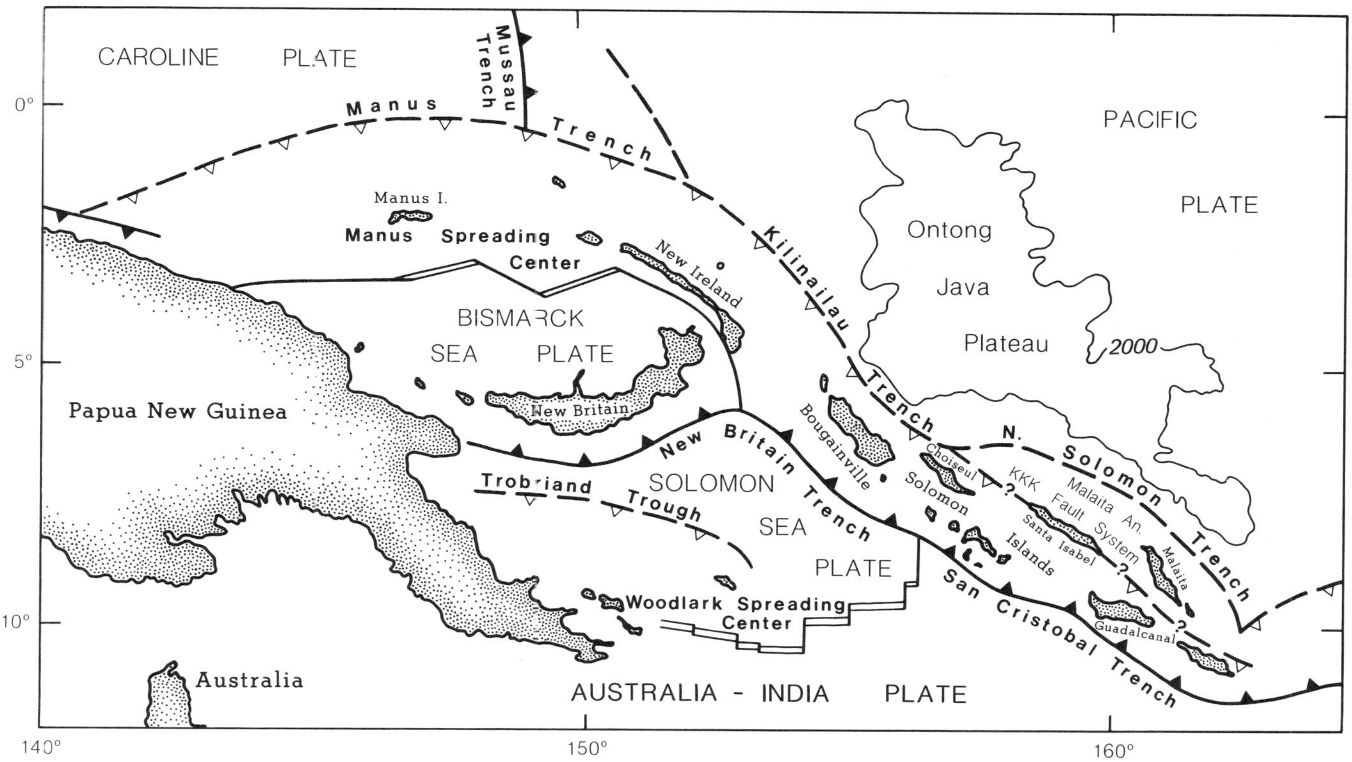

Figure 2. Plate boundaries and tectonic features in the Solomon Islands and Bougainville-New Ireland region of Papua New Guinea. Active trenches (subduction zones) —solid lines with filled barbs; inactive trenches (inactive subduction zones) — dashed lines with open barbs; spreading systems — double line for spreading ridge and single line for transform faults; Malaita An. — Malaita Anticlinorium; KKK — Kia-Korigole-Kaipito fault system; the Ontong Java Plateau is outlined by the 2,000-m bathymetric contour.

and Molnar (1972), Kroenke (1972), Coleman and Packham (1976), de Broin, Aubertin, and Ravenne (1977), Hamilton (1979), Coleman and Kroenke (1981), and Cooper and Taylor (1985, 1987a, b), suggested that the trench was a fossil feature related to early Tertiary subduction and early Miocene arc reversal. Furthermore, a compressional tectonic regime is clearly indicated by the numerous anticlines of the Malaita Anticlinorium — a broad zone of folded oceanic crust lying north of the Solomon Islands (Figure 2; Kroenke, 1972; 1984; this volume; Kroenke, Resig, and Cooper, 1986). Recent regional syntheses include overviews of the southwest Pacific by Johnson (1979), Hamilton (1979), and Kroenke (1984). Tectonic analyses of areas near the Kilinailau Trench are given by Taylor (1979, 1987), Falvey and Pritchard (1984), Exon and Tiffin (1984), Vedder, Pound, and Boundy (1986), Exon et al (1986), Cooper and Taylor (1987a, b), and Marlow et al (1988).

GEOPHYSICAL AND GEOLOGICAL DATA

The multichannel seismic-reflection data were recorded by using a 1,311 in^3 (21.5 l) airgun array, a Globe Universal Sciences (GUS)[1] seismic-recording system, and a 2,400-m streamer. A shooting geometry with a 50-m shotpoint interval and a 100-m group interval provided 24-fold data. The ship's position was determined by a satellite navigation system that used doppler sonar, speed log, and gyrocompass information to calculate dead-reckoned positions between satellite fixes.

The seismic-reflection data were processed by the U.S. Geological Survey in Menlo Park, California, on a DISCO (trademark of Digicon, Inc.) processing system. The main processing steps included demultiplexing, velocity analysis using a coherence-estimation technique combined with constant-velocity stacks, normal moveout correction, stacking, and filtering. All sections are migrated; the migration velocities were selected from a series of post-stack constant-velocity migrations. A further description of the standard U.S. Geological Survey acquisition and processing methods is given by Mann (1986).

Aside from data acquired on the R/V *S.P. Lee*, geo-

[1]Any use of trade names is for purposes of identification only and does not imply endorsement by the U.S. Geological Survey, the New Zealand Geological Survey, or the Bureau of Mineral Resources, Geology, and Geophysics, Australia.

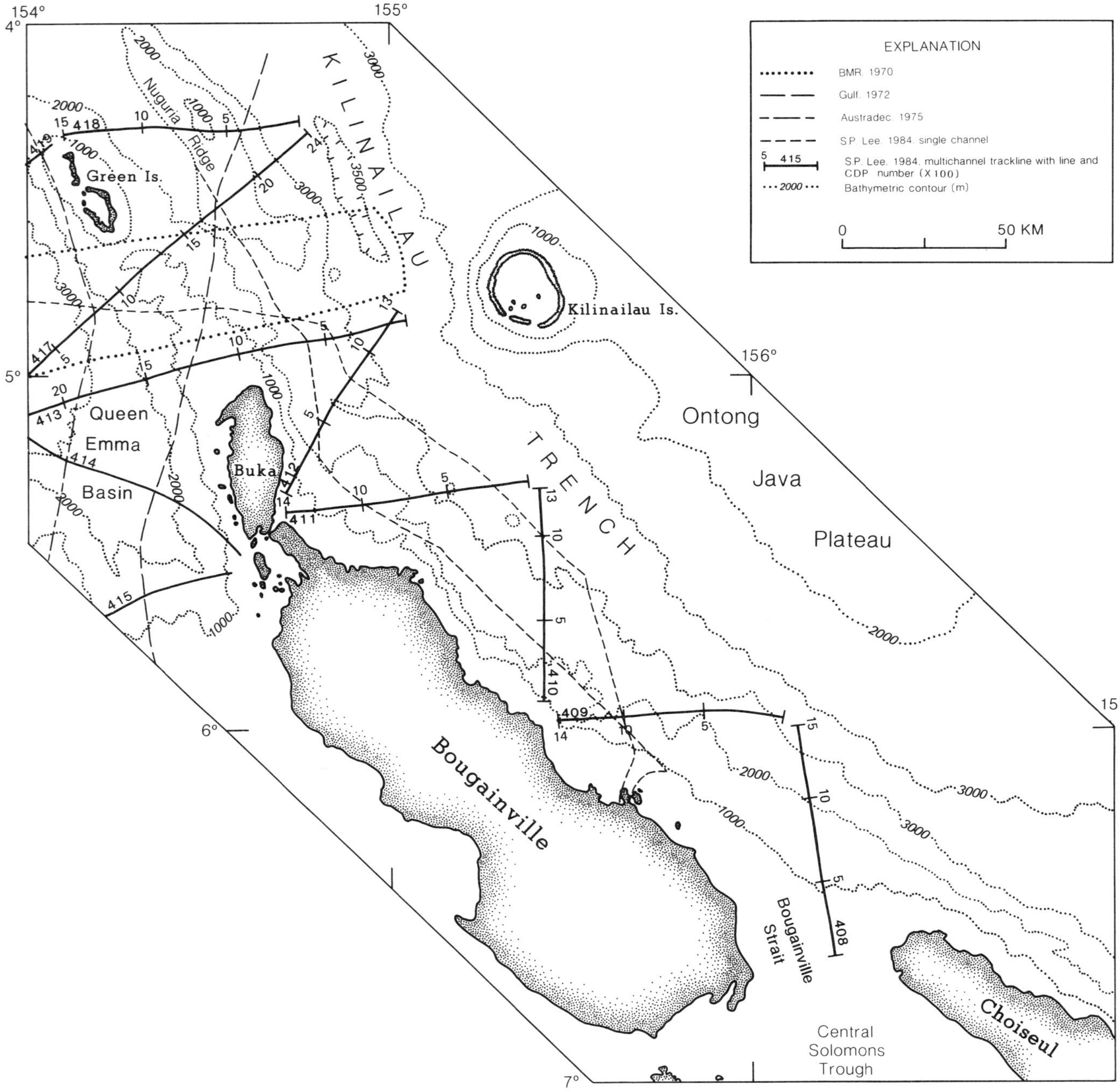

Figure 3. Trackline map showing tracklines, line numbers, and common depth point (CDP) numbers for single-channel and multichannel seismic data acquired by the U.S. Geological Survey research vessel R/V *S.P. Lee* in 1984. Tracklines also shown for seismic-reflection lines acquired prior to 1984 by the Australian Bureau of Mineral Resources (BMR), Gulf Research and Development Co. (Gulf), and the French consortium (CEPM) AUSTRADEC III cruise. R/V *S.P. Lee* seismic sections across the Kilinailau Trench are shown in Figures 4-12. Bathymetry from Chase, Seekins, and Lund (1986) and Chase, Seekins, and Young (this volume).

physical data examined include seismic reflection-profiles acquired by the Australian Bureau of Mineral Resources in 1970, by Gulf Research and Development Company in 1972, and by the CEPM group AUSTRADEC III Survey in 1975 (Figure 3; see also Vedder, Bruns, and Cooper, this volume).

Because rock samples have not been collected along the Kilinailau Trench and slope, the composition and age of these rocks can only be inferred by comparison with the geology of the adjacent islands and the Ontong Java Plateau. Rocks similar to those on the islands may underlie the slope off Choiseul, Bougainville, and Buka Island. Possible correlatives to the Pacific plate oceanic crust and sedimentary rocks beneath the trench have been drilled on the Ontong Java Plateau (Winterer, Reidel et al, 1971; Andrews, Packham, et al, 1975), and crop out to the southeast on Malaita (Kroenke, 1972, 1984; Kroenke, Resig, and Cooper, 1986; Coulson and Vedder, 1986; Pound, 1986).

REGIONAL RELATIONS

The rocks and structures exposed on New Ireland, the Tabar-to-Feni island chain, Green and Buka Islands, Bougainville, and Choiseul (Figure 1) reflect volcanic arc growth during at least two episodes. New Ireland, Bougainville, and Choiseul are all underlain by an Eocene to early Miocene volcanic arc that, at least on Choiseul, is built on partly metamorphosed oceanic crust of Cretaceous(?) and early Tertiary age. A late Miocene and younger volcanic arc underlies the southwestern island chain of the Solomon Islands (western Guadalcanal, the Russell Islands, New Georgia island group, and Shortland Islands) and crosses the early Tertiary arc through Bougainville to the Green Islands and the Emirau-Feni Ridge.

Oligocene through Quaternary volcaniclastic and carbonate rocks cover large parts of the volcanic arcs, and up to 8 km of strata lie in basins between the arcs. These basins include New Ireland basin, which lies north of Manus Island and New Ireland (Exon and Tiffin, 1984; Exon et al, 1986; Marlow et al, 1988); Queen Emma basin, which lies between New Ireland and Bougainville (Figure 3; Vedder, Bruns, and Cooper, this volume); and the Central Solomons Trough, a composite intra-arc basin, which lies beneath the double island chain of the Solomon Islands (Figure 1; Bruns et al, 1986; Cooper, Bruns, and Wood, 1986; Cooper, Cochrane, and Bruns, this volume).

Northeast of the Kilinailau Trench is the Ontong Java Plateau, which is part of the Pacific plate. The plateau is characterized by crust as much as 40 km thick, in marked contrast to the 10-km thick oceanic crust of the Pacific and Solomon Sea plates. Sedimentary rocks as much as 1.4 km thick overlie the basaltic basement (Kroenke, 1972; 1984). The crustal seismic-refraction velocities and samples from Deep Sea Drilling Project (DSDP) sites indicate that the rocks of both this thickened crust and the overlying sedimentary cover are entirely oceanic in character (Winterer, Reidel, et al, 1971; Kroenke, 1972, 1984; Andrews, Packham, et al, 1975; Furumoto et al, 1976; Hussong, Wipperman, and Kroenke, 1979).

The crust of the plateau is folded into a series of anticlines and synclines between the Solomon Islands and the North Solomon Trench in a region called the Malaita Anticlinorium (Figure 2). Malaita of the Solomon Islands is an emergent and breached crest of one of the Malaita Anticlinorium structures. Development of the anticlinorium began in about the late Miocene and continues to the present (Kroenke, 1972; 1984; this volume).

Regional Geology

Feni and Green Islands

The Feni and Green islands lie atop the Emirau-Feni Ridge northwest of Buka Island and Bougainville. The Feni Islands consist of eroded Pliocene and Pleistocene stratovolcanos overlain by uplifted Pleistocene reef limestones (Wallace et al, 1983). Ambitle, one of the Feni Islands, is constructed at least in part of an Oligocene deep-water marine limestone that crops out on the northern part of the island. The Green Islands are composed of uplifted Quaternary reef deposits (Bain et al, 1971). Seismic-reflection data near both the Feni and Green island groups (e.g. line 417, Figure 3) show that the top of the ridge is covered by about 1 km of strata that correlate with the late Eocene(?) through middle Miocene(?) Buka Island Formation (Vedder, Bruns, and Cooper, this volume; Hilyard and Rogerson, this volume). Therefore, both the island groups are presumably underlain by these Oligocene and Miocene strata.

Buka Island

The oldest rocks on Buka Island are late Eocene(?) through middle Miocene(?) volcanic agglomerates, and volcaniclastic mudstone and sandstone turbidites of the Buka Island Formation (Blake and Miezitis, 1967; Hilyard and Rogerson, this volume). These rocks are correlated with the upper Atamo Volcanics and Keriaka Limestone of Bougainville and probably are a basinal facies of these rocks (Hilyard and Rogerson, this volume; Vedder, Bruns, and Cooper, this volume). Overlying these rocks is the Pleistocene and Pliocene(?) Sohano Limestone, a reef deposit that borders the northern and eastern ends of the island and dips gently westward. This limestone platform extends southeastward about 45 km along the northern coast Bougainville.

Bougainville

On Bougainville is a succession of volcaniclastic, carbonate, and volcanic rocks initially described by Blake and Miezitis (1967), and recently remapped and described by Hilyard and Rogerson (this volume). The oldest rocks exposed are the Eocene(?) and Oligocene(?) Atamo

Volcanics, which form the basement and are part of the Paleogene (older) arc. They consist of basalt, basaltic andesite, and interbedded volcaniclastic rocks. These rocks are overlain unconformably by the shallow-water, late Oligocene to latest early Miocene Keriaka Limestone. Unconformably overlying the Keriaka Limestone are the volcaniclastic debris flows of the middle to late Miocene Toniva Formation, and locally a sequence of andesite flows of the Aropa Andesite. These rocks are distributed along the northeastern coast between Kieta and the southeastern end of the island, and probably extend well beneath the adjacent slope. The Toniva Formation is overlain conformably by the latest Miocene(?) and Pliocene Arawa Conglomerate, characterized by volcanolithic pebble-boulder beds of mass-flow origin. Various igneous rocks of latest Miocene and younger age intrude the older rocks (Isinai Granodiorite, Kupei Complex, and unnamed intrusives). The latest Pliocene and younger volcanic rocks of the Bougainville Group cover most of the northwestern and southeastern parts of the island.

Choiseul

Choiseul is underlain by an oceanic basement complex composed of the Choiseul Schists and Voza Lavas. The Choiseul Schists consist largely of metamorphosed Cretaceous(?) and early Tertiary basaltic rocks. The Voza Lavas are dated as pre-Miocene, but interpretations of geological and geochemical data indicate that the Voza Lavas are probably the protolith for the Choiseul Schists, and therefore are probably also of Cretaceous(?) and early Tertiary age (Coleman, 1965; Coulson and Vedder, 1986; Pound, 1986).

Sedimentary rocks directly above the basement schist and basalt are assigned to the Oligocene through early Pliocene Mole Formation. The formation is as much as 450 m thick and has an early Oligocene basal conglomerate — the Koloe Breccia Member — which is overlain by Oligocene to early Pliocene dominantly turbidite-derived siltstone and sandstone, and subordinate shallow-water near-shore deposits. The Pliocene Pemba Formation overlies the Mole Formation conformably or with slight discordance. The Pemba Formation has a maximum thickness of 600 m and consists of a lower calcarenite member and an upper calcisiltite member. Both the Mole and Pemba formations dip and thicken to the west and southwest toward Bougainville Strait and the Central Solomons Trough. Overlying all these units are the Quaternary reef deposits of the Nukiki Limestone Formation, which are now elevated as much as 440 m above sea level on the northwestern end of Choiseul, and up to 800 m above sea level on the southeast end of the island (Coulson and Vedder, 1986; Pound, 1986)

Malaita and Ontong Java Plateau

On Malaita, unmetamorphosed Cretaceous oceanic basalt is overlain by 1,600 to 2,000 m of pelagic and hemipelagic carbonate rocks, radiolarian chert, and siliceous mudstone. This sequence includes about 220 m of the Cretaceous Kwara'ae Mudstones, up to 1,000 m of the Cretaceous through Oligocene Alite Limestone and Apuloto Limestone and the Oligocene through Pliocene Haruta Calcisiltite and Hada Calcisiltite, and up to 825 m of the Miocene Suaba Chalk (Rickwood, 1957; Coulson and Vedder, 1986; Pound, 1986).

Rocks correlative to those on Malaita were drilled on the Ontong Java Plateau at DSDP sites 64, 288, and 289, from 500 to 600 km northeast and east of the Kilinailau Trench region (Figure 1; Winterer, Reidel, et al, 1971; Kroenke, 1972; Andrews, Packham, et al, 1975). The rocks exposed on Malaita and sampled at the DSDP sites, although far from the Kilinailau Trench, provide correlatives for what are probably similar pelagic and hemipelagic rocks lying on the Pacific plate beneath the Kilinailau Trench.

Tectonic Setting

The Kilinailau Trench lies within a broad zone of complex tectonism between the convergent Pacific and Australia-India plates (Figure 2). Major tectonic elements include the active Manus and Woodlark spreading centers, active subduction zones along the New Guinea Trench and the New Britain-San Cristobal trench system, and presumed inactive subduction zones such as the Manus-Kilinailau-North Solomon trench system. Between the major Pacific and Australia-India plates, the most significant small plates or possible plates include the Caroline, Bismarck Sea, and Solomon Sea plates. At present, the Australia-India plate is moving northward at a rate of about 7 cm/yr, and the Pacific plate is moving northwestward at a rate of 10.7 cm/yr. East of the Woodlark spreading center, the Australia-India plate subducts beneath the Pacific plate along the San Cristobal Trench, and the relative convergence rate exceeds 10 cm/yr in a direction of about N63°E. West of the Woodlark spreading center, the Solomon Sea plate is subducting beneath the New Britain Trench in a direction of about N36°E at a rate perhaps as high as 15 cm/yr (Minster and Jordan, 1978; Cooper and Taylor, 1987b; Taylor, 1987; Drummond et al, 1985; Wells, this volume).

Intense seismicity characterizes the New Britain and San Cristobal trenches, except along the subducting Woodlark spreading center, and clearly defines a northward-dipping Wadati-Benioff zone beneath the islands. In contrast, the Manus-Kilinailau-North Solomon trench system is characterized by very low levels of seismicity. This low-level seismicity, however, outlines a relic Pacific plate slab beneath Santa Isabel and Choiseul. Based on the seismicity, the Solomon Islands and Bougainville, which lie between the seismically inactive Manus-Kilinailau-North Solomon Trench system and the seismically active New Britain-San Cristobal subduction zone, are primarily attached to and moving with the Pacific plate (Johnson,

1979; Cooper, Kroenke, and Resig, 1986; Cooper and Taylor, 1985; 1987b).

Tectonic Model

Tectonism in the Solomon Islands-Bougainville region has been explained by a variety of models. The generalized model presented below is summarized from Johnson (1979); Taylor (1979, 1984); Weissel, Taylor, and Karner (1982); Hegarty, Weissel, and Hayes (1983); Taylor and Exon (1984, 1987); Kroenke (1972; 1984); Cooper and Taylor (1985, 1987a, b); Kroenke, Resig, and Cooper (1986); Cooper, Kroenke, and Resig (1986); and Wells (this volume).

Southward-directed subduction and related arc volcanism built the North Solomon Island Arc (Kroenke, 1984) — the early Tertiary island arc that lies beneath Santa Isabel, Choiseul, Bougainville, and New Ireland (Figure 1). The early Tertiary subduction zone probably lay along the present Manus and Kilinailau trenches. Southeastward, however, instead of following the North Solomon Trench, the subduction zone probably continued along or slightly north of what is now the Kia-Korigole-Kaipito (KKK) fault system on Santa Isabel (Stanton, 1961; Ramsay, 1978, 1982; Coulson and Vedder, 1986) and along a major fault zone that lies between Guadalcanal and Malaita on the north side of the Florida Islands (Figures 1 and 2; Kroenke, 1972; Bruns et al, 1986; Bruns, Cooper, and Vedder, 1986; Vedder and Bruns, this volume). This possible continuation of the early Tertiary subduction zone may now be partly overridden by obducted Ontong Java Plateau crust on Santa Isabel (Kroenke, 1984; this volume; Kroenke, Resig, and Cooper, 1986).

In the middle Miocene, volcanic and tectonic quiescence ensued along the arc, possibly caused when the thickened Pacific plate crust of the Ontong Java Plateau entered and blocked the subduction zone about 22 to 20 Ma. The absence of volcanic activity along the North Solomon Island Arc indicates that middle Miocene Australia-Pacific plate convergence was accommodated elsewhere than on the Manus-Kilinailau-Solomon trench system, perhaps north of New Guinea along the southward-dipping New Guinea Trench (Wewak Trench of Kroenke, 1984) and the Trobriand Trough subduction zone (Kroenke, 1984; Cooper and Taylor, 1985). During this arc-quiescent period, the early Cenozoic arc of the Solomon Islands could have been largely locked to, and moving with, the Ontong Java Plateau and Pacific plate.

About 10 Ma, subduction shifted to the presently active northward-dipping New Britain and San Cristobal subduction zones on the southern side of New Britain, Bougainville, and the Solomon Islands, thus completing the arc-subduction reversal in the Solomons. This subduction shift was accompanied by the development of the late Cenozoic arc that underlies and uplifted the New Georgia islands group, Bougainville, Buka Island, the Green Islands, and the Tabar-to-Feni island chain.

Additional late Cenozoic tectonic events include the development of the Malaita Anticlinorium and the Woodlark and Manus spreading centers. Compression between the Solomon Islands and the Ontong Java Plateau during late Miocene and later time created the folds of the Malaita Anticlinorium and uplifted Malaita (Kroenke 1972; 1984). The present North Solomon Trench is a late Cenozoic structural hinge line or regional synclinal trend lying at the base of the northernmost structures of the Malaita Anticlinorium (Kroenke, 1972; 1984; this volume; Kroenke, Resig, and Cooper, 1986; Bruns, Cooper, and Vedder, 1986).

The Woodlark spreading center developed on the south side of the arc prior to 5 Ma — the age of the oldest identifiable magnetic anomaly (Taylor, 1987). The spreading center is now subducting into the New Britain-San Cristobal subduction zone, leading to anomalously low seismicity, near-trench volcanism, and pronounced tectonic uplift in the New Georgia region (Weissel, Taylor, and Karner, 1982; Taylor, 1984, 1987, Colwell and Vedder, 1986). The Manus spreading center developed on the west, beginning at about 3.5 Ma, and caused an eastward translation of the New Britain island arc from an original position west of New Ireland into its present position (Taylor, 1979).

This complex tectonic history indicates that no significant subduction should have occurred along the early Tertiary subduction zone, which is composed of the Manus-Kilinailau-KKK trench/fault system, since the the early Miocene (Kroenke, 1972, 1984; Hamilton, 1979; Kroenke, Resig, and Cooper, 1986; Cooper and Taylor, 1985, 1987). However, interpretations of the 1984 R/V *S.P. Lee* cruise data indicate that the Kilinailau Trench has been tectonically active during the late Cenozoic, with compressional deformation caused by persistent convergence between the Pacific plate and the Solomon Islands arc. In the following pages, we show the evidence for deformation along the trench and discuss some of the possible causes and implications of this tectonic activity.

SEISMIC STRATIGRAPHY AND CORRELATION OF SEISMIC UNITS

The rocks that underlie the Kilinailau Trench and adjacent slopes of Buka Island, Bougainville, and Choiseul are divided into seismic-stratigraphic units that are characterized by distinctive seismic-reflection geometries, and that are in part bounded by unconformities and abrupt changes in seismic velocities. Separate designations for slope and trench units are required to distinguish varying ages and compositions of arc and ocean-plate rocks that are juxtaposed along the trench subduction zone.

The seismic units are designated by letter and number symbols in order of decreasing age (Figure 4). Units K1 through K3 (Kilinailau Trench units 1 through 3) lie beneath the trench, and units B1 through B3 (Bougainville slope units 1 through 3) lie beneath the slope. In addition, two seismic-stratigraphic units from Queen Emma basin, units E1 and E2 (Vedder, Bruns, and Cooper, this volume)

SLOPE SEISMIC UNIT DESCRIPTION

SEISMIC UNIT	VELOCITY RANGE (KM/S)	THICKNESS RANGE (KM)	PROVISIONAL AGE CORRELATION
UNIT B3 WELL-LAYERED UNDEFORMED TO MILDLY DEFORMED	1.5 AT TOP TO 3.1 AT BASE	LESS THAN 0.5 TO 2.0	PLIOCENE (?) AND QUATERNARY
UNIT B2 ACOUSTICALLY LAYERED DEFORMED	2.4 - 4.5 ESTIMATE THICKNESS WITH 4 KM/S	0 - 3 (?) HIGHLY VARIABLE	LATE MIOCENE (?)
UNIT B1 NON-LAYERED TO POORLY LAYERED DEFORMED VOLCANICS ?	2.4 - 4.5+ ESTIMATE THICKNESS WITH 4 KM/S	UNKNOWN	CENOZOIC

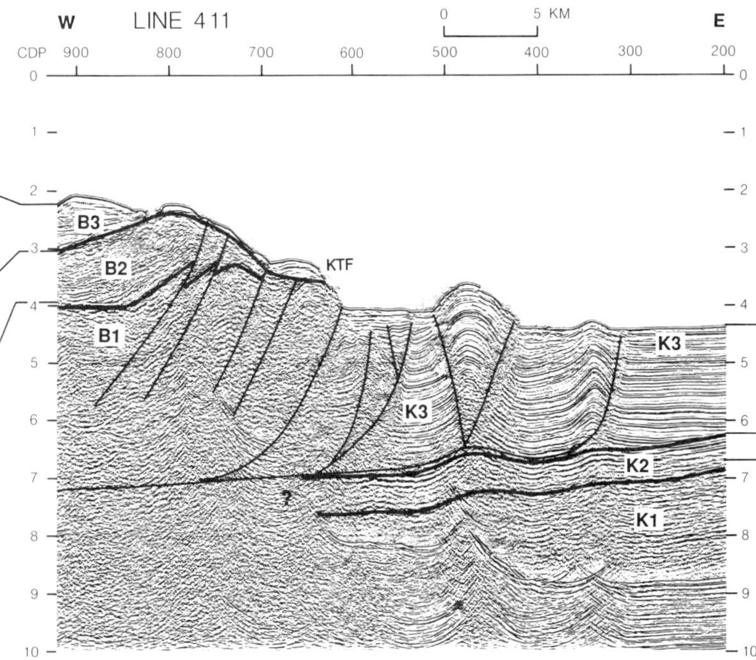

TRENCH SEISMIC UNIT DESCRIPTION

SEISMIC UNIT	VELOCITY RANGE (KM/S)	THICKNESS RANGE (KM)	PROVISIONAL AGE CORRELATION
UNIT K3 WELL-LAYERED TRENCH FILL	1.5 AT TOP TO 4.5 AT BASE OF THICKEST SECTION	0.5 - 5.5 (THINS SEAWARD)	MIDDLE MIOCENE (?) AND YOUNGER
UNIT K2 WELL-LAYERED PARALLEL TO TOP UNIT K1 PELAGIC CARBONATES AND CHERTS	3.6 - 5.1 AVERAGE ABOUT 4.6	0.7 - 0.9	EARLY MIOCENE (?) TO LATE CRETACEOUS(?)
UNIT K1 OCEANIC BASALT (LAYERED?)	4.5 - 6.5 AVERAGE ABOUT 5.5	UNKNOWN	CRETACEOUS AND OLDER

Figure 4. Description of seismic-stratigraphic units in the Kilinailau Trench and adjacent slope, shown on part of multichannel seismic profile 411. Velocity range is based on seismic refraction and reflection data of Cooper, Cochrane, and Bruns (this volume); Cochrane, Cooper and Bruns (this volume); and on interval velocities obtained from multichannel seismic data. Provisional age correlation is largely inferred and poorly constrained. Profile 411 is a migrated time section; location shown in Figures 3 and 14-16; complete line shown in Figure 8. CDP — common depth point; CDP's annotated on Figures 3, and 14-16. Vertical exaggeration at the sea floor is about 4:1.

can be traced across the Emirau-Feni Ridge between Buka Island and the Green Islands and beneath the Bougainville slope, where they form part of unit B1. Seismic-reflection horizons between units are indicated by hyphenated symbols; for example, horizon K2-K3 separates seismic units K2 and K3. The seismic units are described below and are interpreted on seismic-reflection profiles (Figures 5-12). Following the description of each unit is a discussion of the inferred ages and correlations with onshore rocks.

Trench Seismic-Stratigraphic Units

Unit K1

Description. Unit K1, the lowermost unit of the seismic-reflection data, is characterized by low-amplitude, variably discontinuous to continuous, parallel reflections that dip southwest and extend below the lowermost part of the slope. No bottom to unit K1 is evident; weak layered reflections are present to the water-bottom multiple, below which no internal details of the unit are discernible. The top of the unit (horizon K1-K2) is smooth and difficult to distinguish, merging almost imperceptibly into the overlying highly stratified and conformable unit K2 (Figures 5-12).

Unit K1 has high refraction and reflection velocities. On sonobuoy refraction records from the trench, the velocity within unit K1 ranges from 5.2 to 6.2 km/s (Cooper, Cochrane, and Bruns, this volume; Cochrane, Cooper, and Bruns, this volume). Analysis of the seismic-reflection stacking-velocity data shows considerable variation of the derived interval velocities. Some of the unit K1 internal reflection events are probably interbed multiples generated within the overlying sedimentary section, because interval velocities derived from the stacking velocities are low, within the 2 to 3 km/s range. However, high stacking velocities, and hence high interval velocities, are common for rocks within the unit. The interval velocities are mainly between 5 and 6 km/s, but range from about 4 km/s at the top of the unit to greater than 6 km/s in the lower part. These high velocities are unlikely to have originated from interbed multiples or from out-of-plane reflections; both types of events commonly have relatively low velocities. Thus, we believe the velocity of unit K1 rocks is in the 5 to 6 km/s range.

Inferred Age and Correlation of Unit K1. We correlate unit K1 with the oceanic basalt of the Ontong Java Plateau. The primary characteristics of unit K1 — the smoothness of the top of the unit, the internal acoustic stratification, and the high velocity — have been described for the igneous basement on the Ontong Java Plateau (Kroenke, 1972; Hussong, Wipperman, and Kroenke, 1979). On the plateau, the igneous upper crustal layer has refraction velocities of 5.5 to 5.6 km/s (Kroenke, 1972; Hussong, Wipperman, and Kroenke, 1979). The top of the Ontong Java Plateau oceanic basalt layer is also very smooth, quite unlike the diffractive or choatic reflection character commonly observed for oceanic basement.

The Ontong Java Plateau basalts were sampled at DSDP site 289 (Figure 1) and consist of Lower Cretaceous mid-ocean ridge tholeiite flows, lying beneath 1,262 m of Cretaceous through Pleistocene sedimentary rocks (Andrews, Packham, et al, 1975). Similar basalts are present on the breached Ontong Java Plateau structural high that forms the island of Malaita (Hughes and Turner, 1977; Coulson and Vedder, 1986; Pound, 1986) and are deep-water tholeiitic basalt of Late Cretaceous and older age.

Kroenke (1972) proposed that the Ontong Java Plateau originated from extrusion of submarine flood basalts; and Hussong, Wipperman, and Kroenke (1979) proposed that the smoothness at the top of the basement was due to interlayering of basalt flows and high-velocity sedimentary rocks at the top of the oceanic basalt. Such features would presumably also be present in ocean-plate rocks beneath the Kilinailau Trench and would account for the characteristic layered acoustic signature of unit K1. We correlate unit K1 with the Late Cretaceous and older oceanic basalt of the Ontong Java Plateau, and speculate that the unit could be composed of layered igneous rocks, overlain by interbedded basalt flows and sedimentary rocks.

Unit K2

Description. Unit K2 ranges from about 0.4 to 0.8 s thick on the time section (two-way traveltime), and is characterized by continuous high-amplitude reflections (Figures 4-12). The unit lies parallel to the ocean-plate surface (to unit K1), dips southwest, and extends beneath the island-arc slope in the footwall of the thrust-fault blocks. The velocity within unit K2 is high, ranging from 3.6 to 5.1 km/s according to seismic-refraction (Cooper, Cochrane, and Bruns, this volume) and reflection data (see below in time-to-depth conversion section), and averaging about 4.6 km/s according to the reflection data. The unit thus ranges in thickness from about 900 to 1,800 m. The lowest velocities within the unit are on line 408 (Figure 5), where the overlying unit K3 is thinnest; the highest velocities are where unit K2 is most deeply buried by unit K3. The base of the unit (horizon K1-K2) is difficult to distinguish, both because the velocity contrast between the two units is small and because unit K2 is conformable with the layered unit K1. The top of the unit (horizon K2-K3) is a distinctive onlap surface that is well defined except where it is obscured by structural deformation.

Inferred Age and Correlation of Unit K2. The characteristics of unit K2, which include the continuous reflections parallel to the top of the ocean plate, the thickness of the unit, and the moderately high velocity, are similar to the characteristics of sedimentary rocks on the Ontong Java Plateau (Kroenke, 1972). Therefore, we infer that the age of the lowermost part of unit K2 is Late Cretaceous, the same as the age of the oldest sediments on the plateau.

We assume here that the youngest age of unit K2 is about late early or early middle Miocene, but the age could range from early Miocene to Pleistocene. At one extreme,

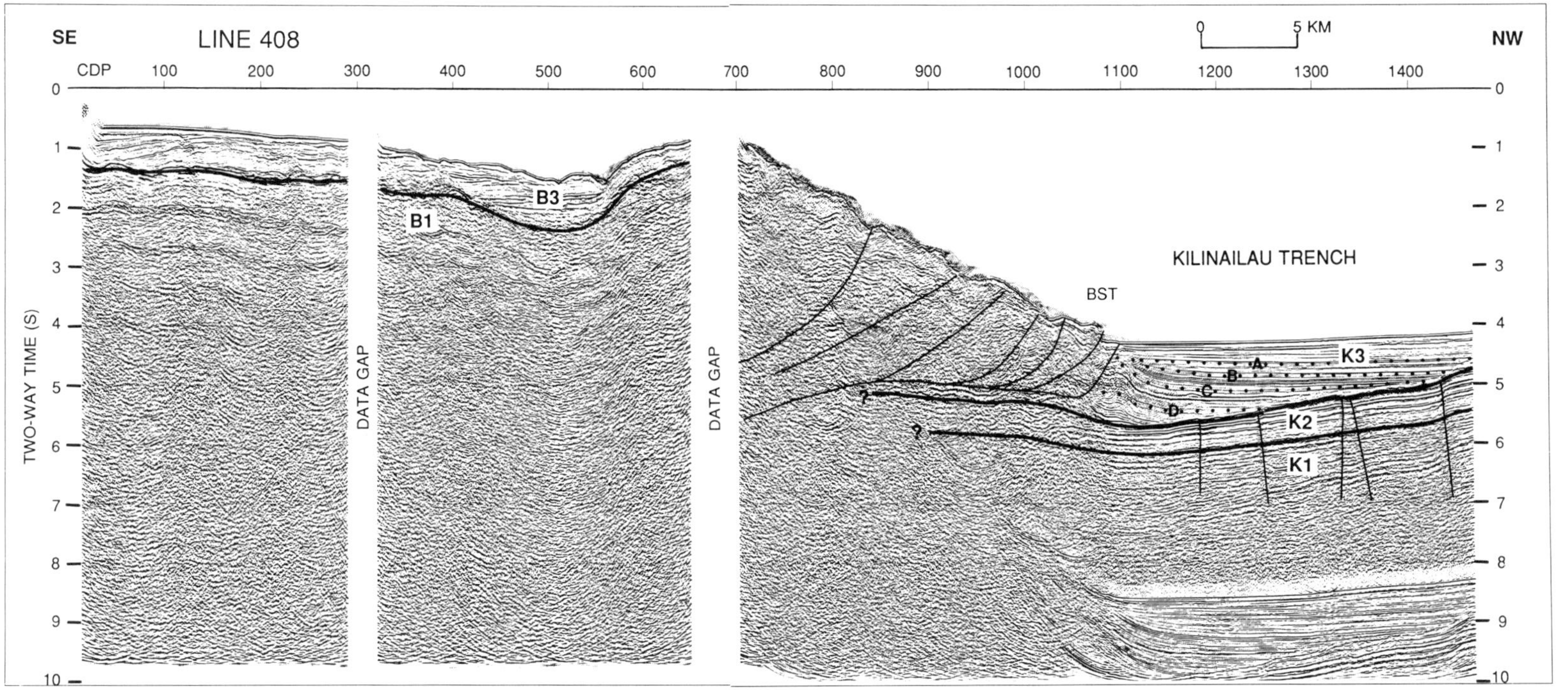

Figure 5. Multichannel seismic-reflection profile 408, migrated time section. This and subsequent seismic sections show seismic-stratigraphic units K1-K3 and B1-B3 as shown in Figure 4 and described in text, correlated reflection events A through D (dotted) in unit K3, and interpreted faults. BST-base-of-slope thrust marking underthrusting of trench strata beneath slope rocks. CDP — common depth point, annotated on seismic profile and on Figures 3 and 14-16; 100 CDP's equal 5 km. Vertical exaggeration at the sea floor about 4:1. Location of seismic profiles shown in Figures 3 and 14-16.

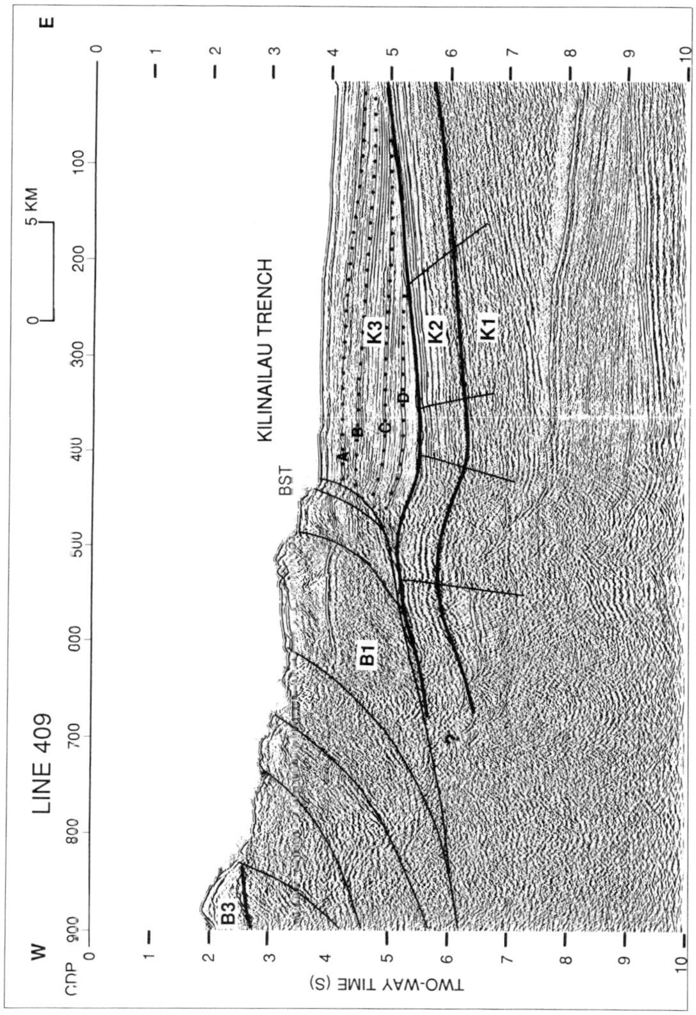

Figure 6. Profile 409, migrated time section. Explanation the same as in Figure 5.

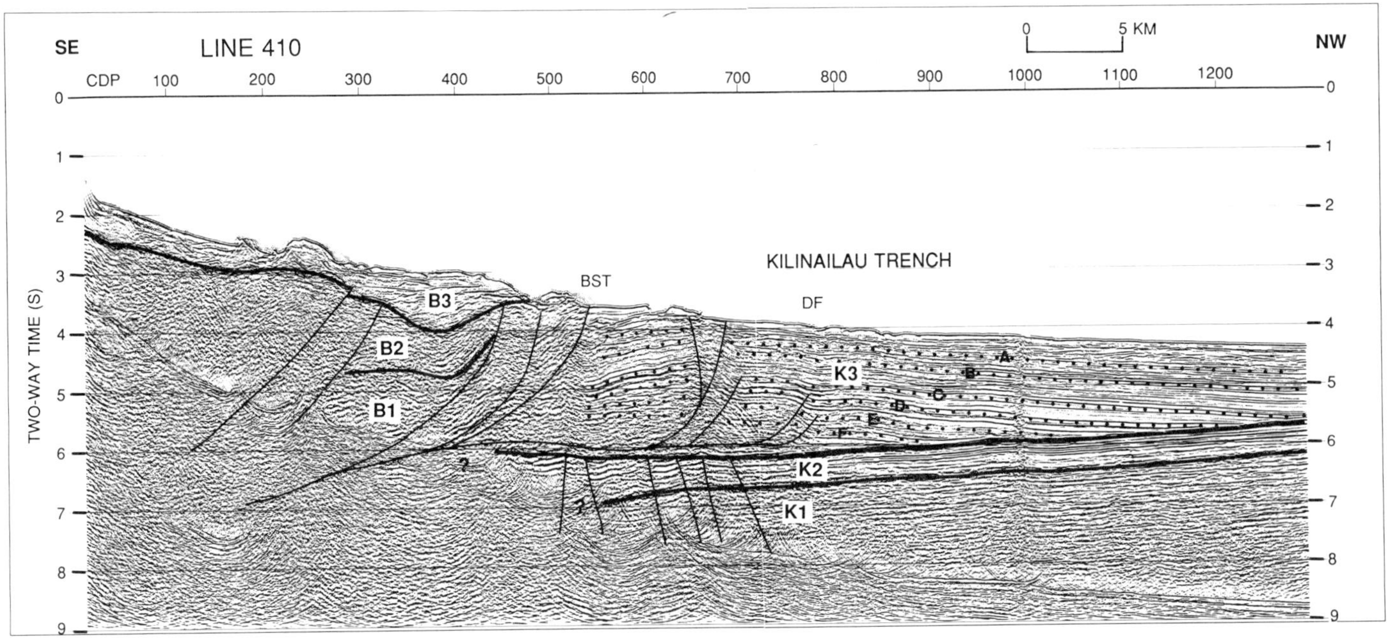

Figure 7. Profile 410, migrated time section showing seismic-stratigraphic units K1-K3 and B1-B3 as shown in Figure 4 and described in text, correlated reflection events A through F (dotted) in unit K3, and interpreted faults. Note that unit K3 thickens markedly compared to section on lines 408 and 409, and includes reflection events E and F, not present on lines 408 and 409. DF — deformation front marking transition from undeformed to deformed trench strata; BST — base-of-slope thrust marking underthrusting of trench strata beneath slope rocks. CDP — common depth point, annotated on seismic profile and on Figures 3 and 14-16; 100 CDP's equal 5 km. Vertical exaggeration at the sea floor about 4:1. Location of seismic profiles shown in Figures 3 and 14-16.

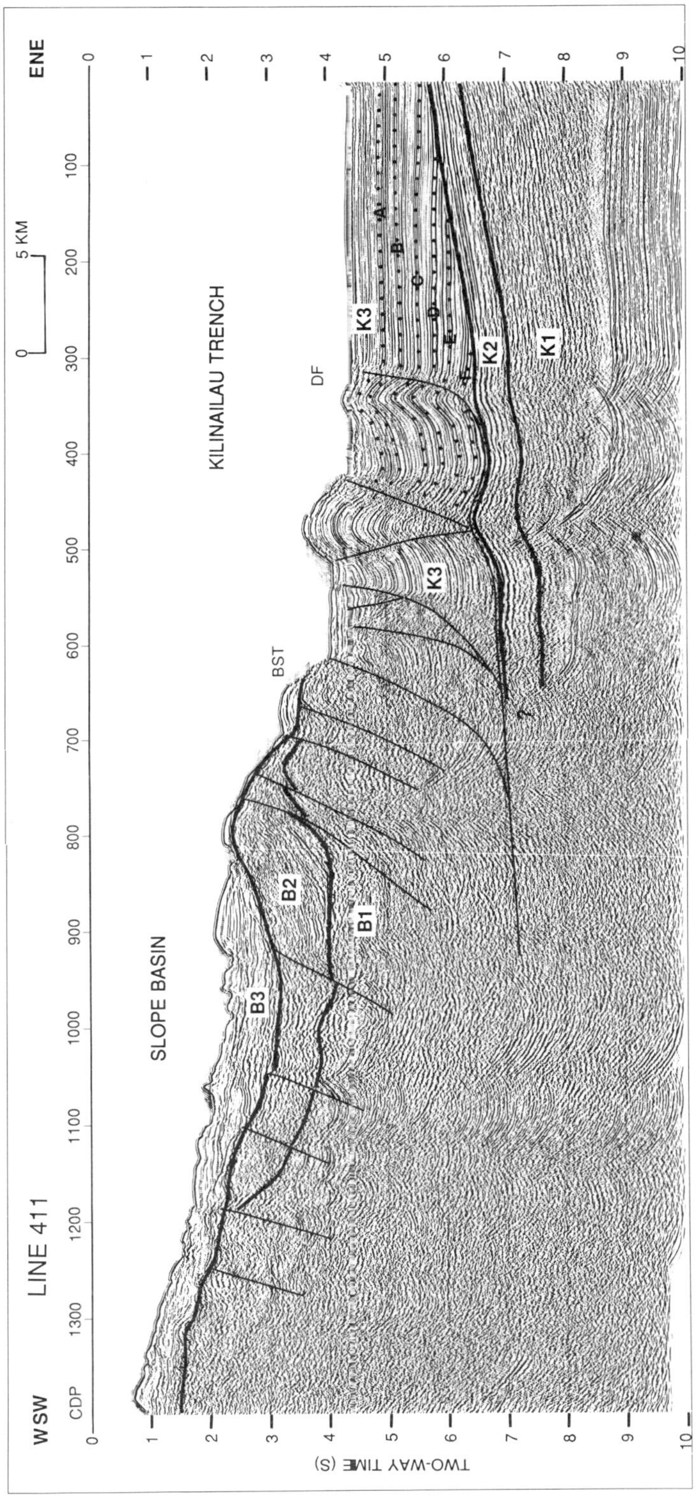

Figure 8. Profile 411, migrated time section. Explanation the same as in Figure 7.

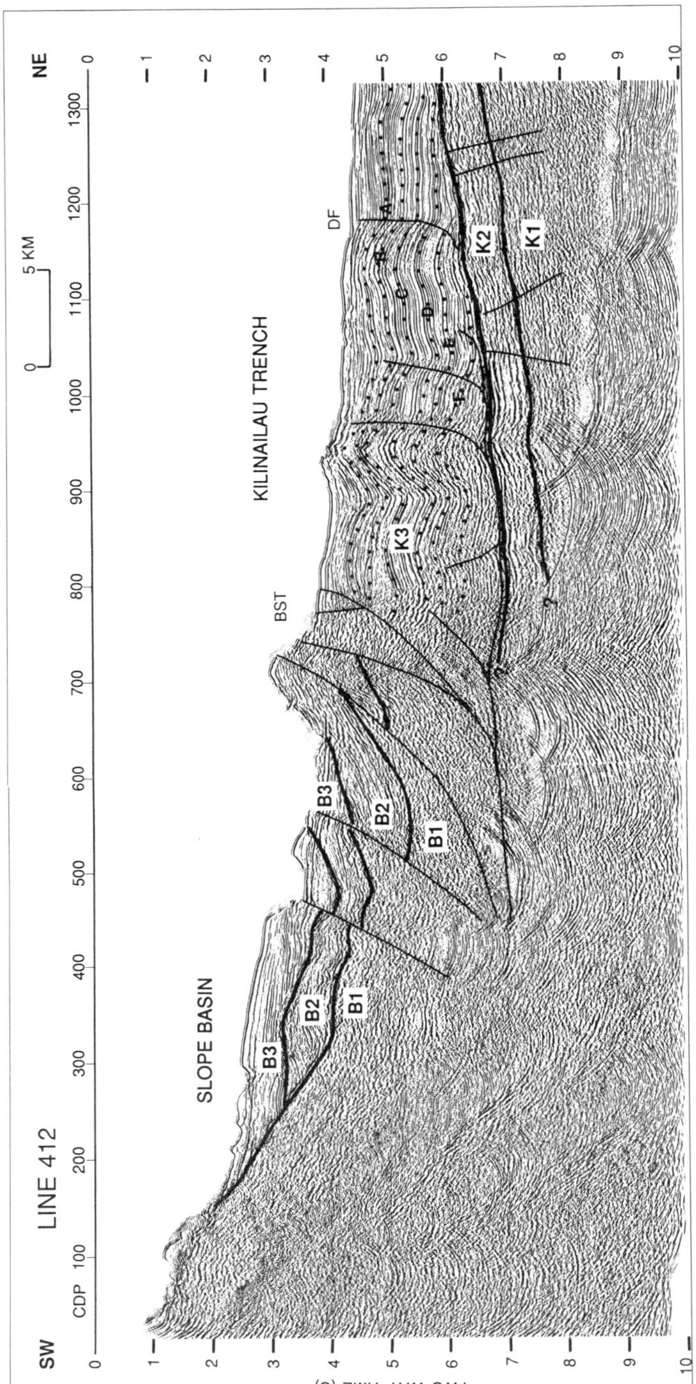

Figure 9. Profile 412, migrated time section. Explanation the same as in Figure 7. Line is located over a northeastward trending submarine channel; unit B3 is erosionally truncated by the channel.

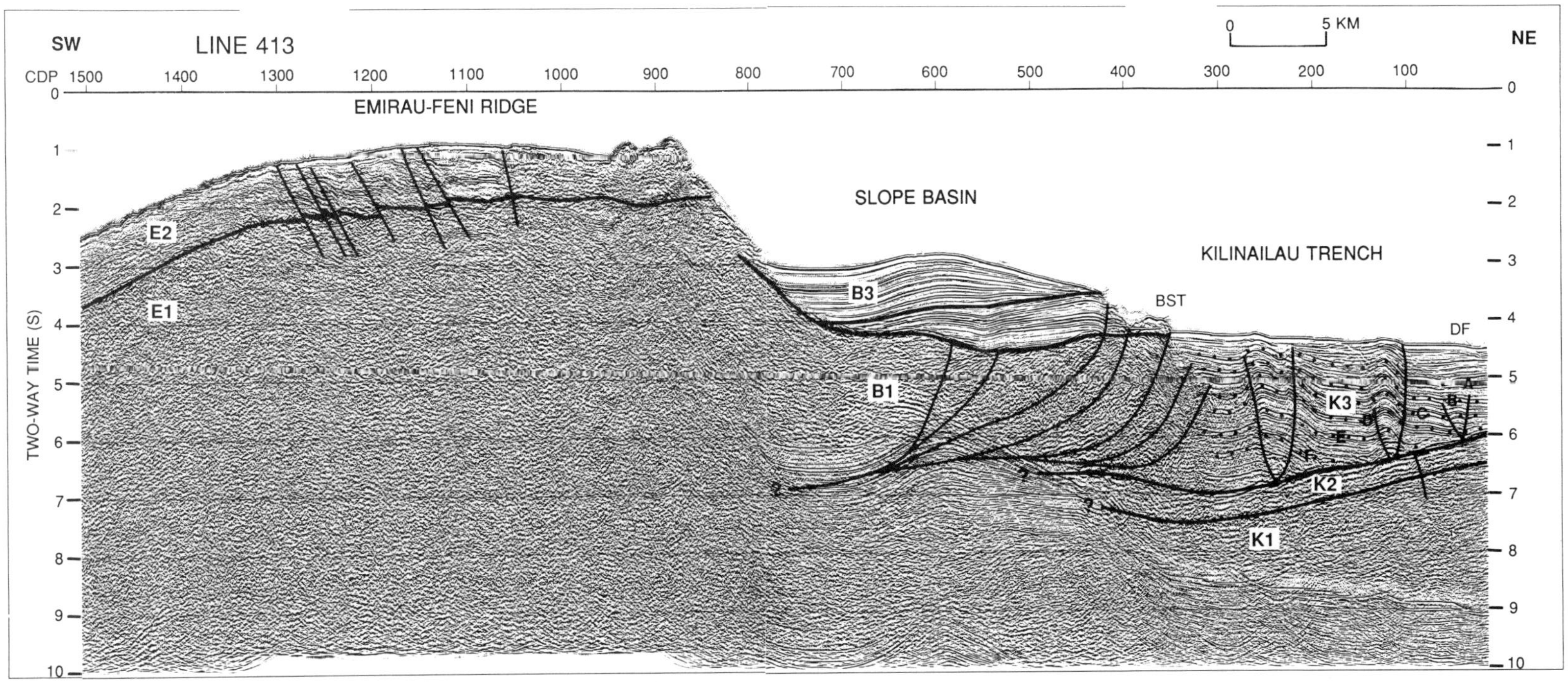

Figure 10. Profile 413, migrated time section. Seismic stratigraphic units E1 and E2 are continued from Queen Emma basin (Vedder, Bruns, and Cooper, this volume) and correspond to volcanic and metamorphic basement and Eocene(?) through middle Miocene(?) Buka Formation respectively. A submarine channel runs southeast along the southwest side of the slope basin. Further explanation given in Figure 7.

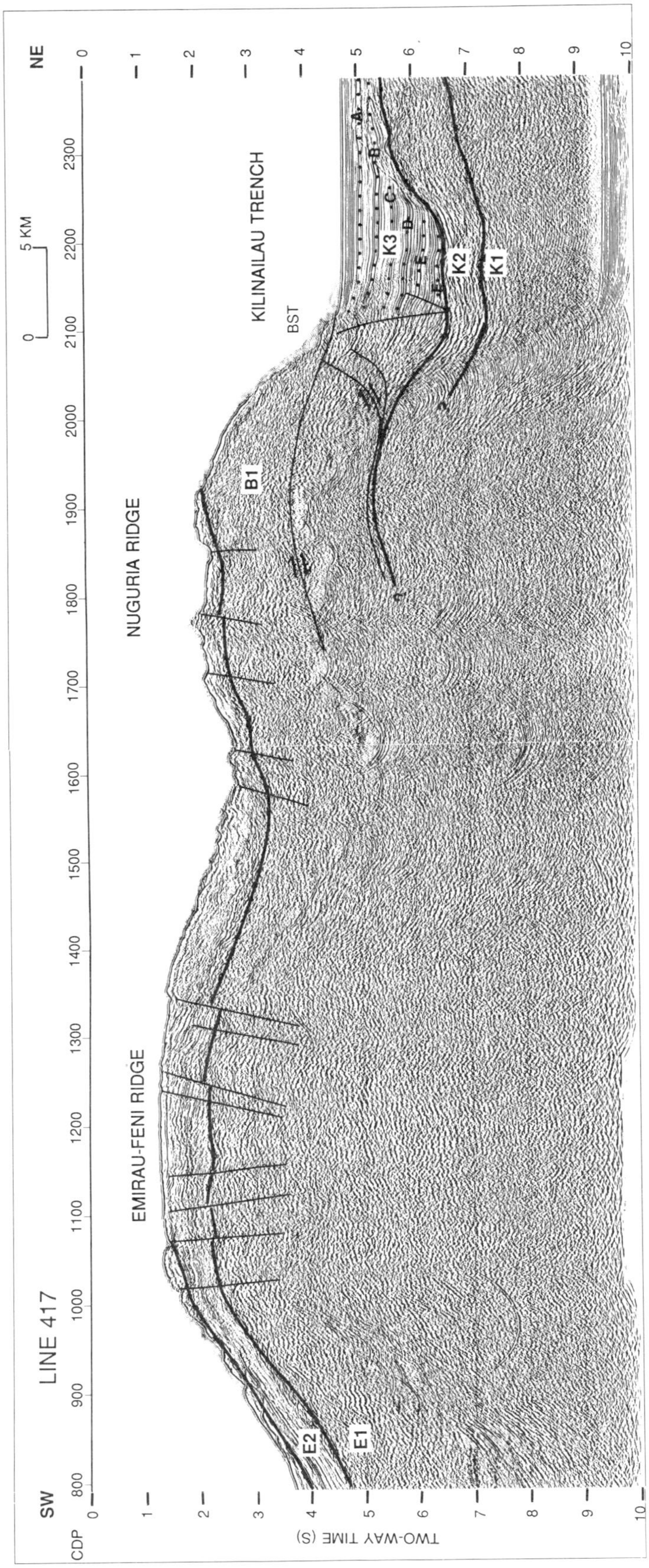

Figure 11. Profile 417 across Nuguria Ridge; migrated time section. Explanation the same as in Figure 7.

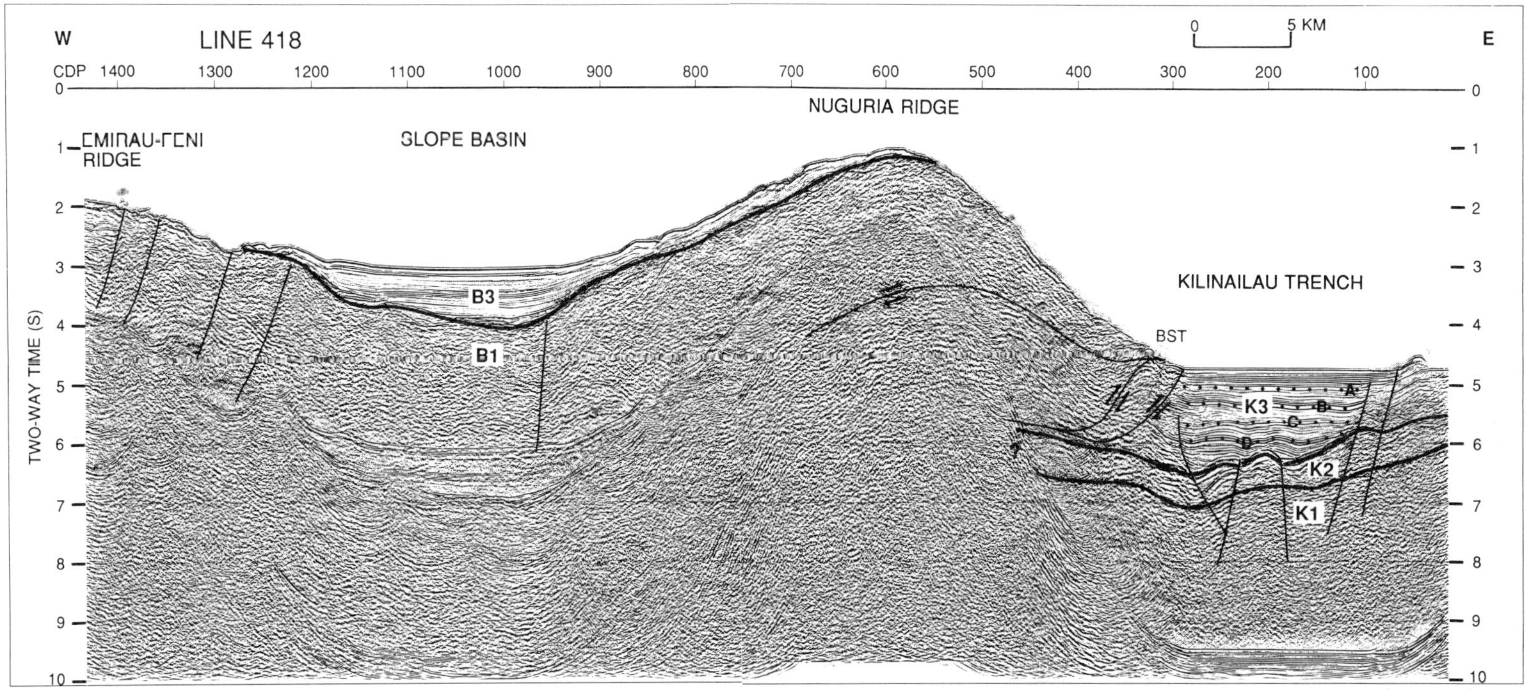

Figure 12. Profile 418 across Nuguria Ridge; migrated time section. Dipping unit B3 on southwest flank of ridge suggests that uplift of ridge is young, perhaps occurring in late Quaternary. Further explanation given in Figure 7.

if we accept the tectonic model in which subduction along the trench stopped in the early Miocene and has not been subsequently reactivated (e.g. Kroenke, 1984), then the overlying unit K3 trench-fill sequence would have begun to accumulate in about the late early or early middle Miocene, and the youngest age for unit K2 would be early Miocene. However, slow subduction could have continued until the late Miocene, when the subduction zone on the south side of the arc became active (e.g. Coleman and Kroenke, 1981); in this case, the youngest age of unit K2 could be late Miocene. The unit might even be considerably younger than Miocene, because we cannot determine an upper age limit that is independent of a tectonic model. For example, the upper 200 to 500 m at the DSDP sites consist of middle Miocene and younger sedimentary rocks. These rocks could continue beneath the trench as part of unit K2, and the age range for unit K2 could be anywhere in the interval from early Miocene to Pleistocene. We provisionally accept the model in which subduction ceased in the late early Miocene, and a trench-fill unit K3 began to accumulate over unit K2 in the early middle Miocene. However, this assumption is strongly dependent on the tectonic-model, and we cannot rule out either older or younger ages.

Unit K2 is probably composed of pelagic carbonate and chert, similar to the Late Cretaceous to Holocene cherty limestone, nannofossil and foraminiferal limestone, and chalk drilled on the Ontong Java Plateau (Winterer, Reidel, et al, 1971; Andrews, Packham, et al, 1975) and exposed on Malaita (Kroenke, 1972, 1984; Hughes and Turner, 1977; Coulson and Vedder, 1986; Pound, 1986).

Unit K3

Description. Unit K3 is characterized by parallel, strongly continuous reflections. It is a trench-filling sequence, onlapping the underlying unit K2 on the K2-K3 horizon and thinning to the northeast (Figures 4-12). The unit is flat-lying to gently northeastward dipping beneath the undeformed (plateau-ward) part of the trench, but, near the base of the slope, the unit is folded and faulted. The contact between these faulted unit K3 rocks and the acoustically incoherent rocks of slope unit B1 is never clearly resolved on the seismic records. Individual seismic reflections or reflection packages within unit K3 show distinctive reflection characteristics that can be correlated between adjacent seismic lines (dotted reflectors A through F, Figures 5-12) and show the pattern of trench infilling through time. Sonobuoy refraction velocities (Cooper, Cochrane, and Bruns, this volume; Cochrane, Cooper, and Bruns, this volume) and reflection interval velocities in the unit range from about 1.5 km/s at the top of the unit (weakly consolidated) to as high as 4.5 km/s below 2.2 s sub-bottom time (highly consolidated; see below in time-to-depth conversion section).

Inferred Age and Correlation of Unit K3. The youngest age of unit K3 is Holocene; the uppermost strata of the unit are now being deposited. The age of the lowermost strata in unit K3 is unknown. If subduction ceased in the early Miocene, then the base of unit K3 could be as old as late early or early middle Miocene, because preservation of trench fill would begin after subduction stopped. However, since the top of unit K2 could be much younger than early Miocene, the base of unit K3 could likewise be much younger. To be consistent with the assumed late early Miocene age for unit K2, we further assume that unit K3 is early middle Miocene and younger. As with unit K2, however, the assumed age depends on the model and the unit could be considerably younger.

Unit K3 is probably composed of trench turbidites derived from the surrounding islands. If it is, the unit would be primarily clastic, in contrast to the inferred pelagic to hemipelagic character of unit K2. Onshore correlative sedimentary rocks that may be lithologically similar to unit K3 rocks are the turbidites of the Miocene Mole Formation on Choiseul and possibly the calcarenite and calcisiltite of the Pliocene and Pleistocene Pemba Formation on the same island. Correlative, but lithologically dissimilar, rocks include the volcaniclastic rocks of the Bougainville Group on Bougainville, and Quaternary limestone on the Feni and Green islands, Bougainville, Buka Island, and Choiseul (Coulson and Vedder, 1986; Vedder, Bruns, and Cooper, this volume; Hilyard and Rogerson, this volume).

Slope Seismic-Stratigraphic Units

Slope seismic-stratigraphic units B1, B2, and B3, described below, lie beneath the northeast wall of the Kilinailau trench off the Green Islands, Buka Island, Bougainville, and Nuguria Ridge. Two additional seismic-stratigraphic units, E1 and E2 (Vedder, Bruns, and Cooper, this volume), extend into the slope region from the Queen Emma basin on lines 413, 417, and 418. A description of these units is summarized here.

Units E1 and E2

Unit E1 is the acoustic basement in Queen Emma basin and is characterized by discontinuous to choatic seismic reflections. Velocities within the unit commonly are greater than 5 km/s; they are as high as 6.1 km/s beneath the ridge crest and flank, but as low as 3.9 km/s beneath Queen Emma basin. The velocities indicate that beneath the ridge crests and flanks, the unit consists of igneous or metamorphic rocks (Vedder, Bruns, and Cooper, this volume; Cooper, Cochrane, and Bruns, this volume). Thus, unit E1 presumably includes volcanic and intrusive rocks emplaced along the Cenozoic magmatic arcs and is correlative with the volcanic and volcaniclastic rocks of Bougainville. The unit may include rocks similar to the Cretaceous(?) and early Tertiary Choiseul Schists and the Voza Lavas of Choiseul. Unit E1 is continuous with Bougainville slope unit B1 described below.

Unit E2 is characterized by persistent and well-defined to poorly defined seismic reflections in Queen

Emma basin and over the Emirau-Feni Ridge. These reflections are evident beneath the ridge on lines 413, 417, and 418 (Figures 10-12). On line 413 (Figure 10), unit E2 cannot be traced on the seismic section past the scarp that forms the northeastern edge of the Emirau-Feni Ridge just north of Buka Island. On lines 417 and 418 (Figures 11 and 12), unit E2 is traceable across the Emirau-Feni Ridge and onto Nuguria Ridge. However, the correlation is questionable because the unit does not display the well-bedded seismic character that typifies it in Queen Emma basin and on the Emirau-Feni Ridge. Based on a correlation between seismic-reflection data and the geology exposed on Buka Island and nearby islands, we believe unit E2 is equivalent to the late Eocene(?) through middle Miocene(?) Buka Island Formation (Vedder, Bruns, and Cooper, this volume; Hilyard and Rogerson, this volume). Beneath the Bougainville slope and the Nuguria Ridge, unit E2, if present, is included as part of unit B1.

Unit B1

Description. We include in unit B1 all acoustically choatic and non-stratified to indistinctly layered rocks beneath the Bougainville slope that are assumed to be arc-related, rather than accreted from an underthrusting Pacific plate. On the seismic records, the contact between units B1 and K3 lies at or southwest of the last resolvable thrust fault that appears to affect unit K3 rocks, and may lie well landward of the base of the slope. Beneath the Emirau-Feni Ridge and the upper slope, unit B1 is continuous with units E1 and E2 of Queen Emma basin, and velocities are as high as 6.1 km/s, indicating the presence of igneous and/or metamorphic rocks. However, beneath the middle and lower slope, seismic refraction velocities in the unit are low, ranging from about 2.4 to 4.5 km/s (Cooper, Cochrane, and Bruns this volume; Cochrane, Cooper, and Bruns, this volume); we infer that the unit is composed of highly deformed or very poorly bedded sedimentary rocks. Therefore, rocks within the unit are extremely variable in both lithology and age, ranging from volcanic and/or metamorphic rocks at the ridge crest to deformed sedimentary rocks beneath the middle and lower slope.

Inferred Age and Correlation of Unit B1. Unit B1 could be correlated with rocks of virtually all ages and lithologies on the adjacent islands. We presume that beneath the arc and upper slope, unit B1 is composed of the basement rocks on which both the early and late Cenozoic volcanic arcs were built, and of intrusive and extrusive volcanic rocks and epiclastic strata emplaced during arc building and subduction. Interbedded igneous flows and sedimentary rocks may be present, but so deformed that they show choatic reflections; or the igneous rocks could prevent acoustic penetration into, and reflections from, layered interbedded sedimentary rocks. Rocks beneath the arc and upper slope probably include the metamorphic and igneous rocks of the Cretaceous and early Tertiary Kieta Volcanics on Bougainville, the Choiseul Schist and Voza Lavas of Choiseul, and possibly deformed poorly stratified equivalents to sedimentary rocks such as the Oligocene and Miocene(?) Buka Island Formation of Buka Island, and the Oligocene and Miocene Mole Formation of Choiseul.

Beneath the middle and lower slope, the low seismic-refraction velocities of unit B1 rocks indicate that the unit must be largely composed of non-stratified sedimentary rocks. These rocks may include highly deformed slope deposits derived from erosion of high-standing parts of the arc, as well as offscraped and accreted ocean-plate rocks tectonically emplaced during subduction of the Pacific plate beneath the arc. However, at the base of the slope the unit is likely to include rocks that have been underthrust and accreted from the ocean plate — rocks similar to those that compose units K2 and K3. Therefore, the deformed rocks near the base of the slope that are included within unit B1 could be as young as Quaternary.

Units B2 and B3

Description. Units B2 and B3 are both composed of bedded sedimentary rocks. These units are distinguished by their contrasting structure — unit B2 is moderately folded and faulted, whereas unit B3 is mainly undeformed or gently dipping, the dip being largely due to depositional rather than tectonic processes.

Unit B2 can be distinguished only on lines 410-412; on lines 408, 409, 413, 417, and 418, the unit is not present as a separate recognizable unit. The base of unit B2 is either indistinct and difficult to distinguish from the underlying unit B1 (e.g., line 411, sp. 900-1,000, Figure 8) or is marked by a strong seismic reflector along which unit B2 onlaps the underlying unit B1 (e.g., line 411, sp. 800-900, Figure 8). The top of unit B2 is an unconformity, onlapped by the flat-lying to gently dipping strata of unit B3. The maximum thickness of the unit is more than 2 km.

Unit B3 is a slope-draping unit characterized by laterally continuous, high-amplitude reflections alternating with zones that are almost acoustically transparent. The unit is a thin (less than 0.5 to 0.8 km thick) channeled blanket that covers much of the northeastern slope of Bougainville and a thick (almost 2 km) basin fill in a mid-slope basin east of the Green Islands, Buka Island, and northern Bougainville (Figures 8-10). The velocity in unit B3 ranges from less than 2 km/s at the top to about 3.1 km/s at the base. Beneath the middle slope, unit B3 onlaps a distinctive unconformity at the top of units B1 and B2.

Unit B3 is most strikingly displayed on line 413 across the mid-slope basin (Figure 10), where B3 strata form an almost 2-km-thick, southwest-dipping sequence overlying acoustically chaotic unit B1. On this profile, unit B3 includes two subunits separated by a low-angle unconformity. The lower subunit thickens and is erosionally truncated to the northeast; the upper subunit thins northeast and onlaps the lower unit (profile 413, Figure 10). On lines 410 through 412, unit B3 cannot be subdivided, and the unit terminates to the northeast by onlap onto uplifted unit B1 or B2 rocks.

Inferred Age and Correlation of Units B2 and B3. The ages of units B2 and B3 are inferred largely from tectonic models. Unit B2 could be as old as late early to early middle Miocene, if the inferences are made that: (1) the unit B1 rocks of the slope, which unit B2 onlaps along the upper slope, are early Tertiary arc-volcanic rocks or rocks accreted during early Tertiary subduction; and (2) deposition of the layered unit B2 over an early Tertiary accretionary wedge began when southward-directed subduction ended in the late early Miocene.

However, uplift of the Emirau-Feni Ridge and parts of Bougainville and Choiseul occurred in the late Miocene and early Pliocene, during which time southward-directed subduction at the New Britain Trench resulted in development of the late Cenozoic volcanic arc. Geologic evidence for uplift includes the exposed Oligocene through Quaternary marine strata on Choiseul (Mole, Pemba, and Nukiki Limestone formations) and Bougainville (Buka Island Formation, Keriaka Limestone, and Sohano Limestone). In this case, unit B2 could be as young as late Miocene to early Pliocene, because the unit onlaps uplifted rocks of the ridge, and because unit B2 does not dip steeply northeastward as might be expected if the unit were uplifted with the ridge. Therefore, although the oldest age of unit B2 could be late early Miocene, we believe that the unit is more likely to be late Miocene in age.

Unit B3 also onlaps both the uplifted arc rocks and the deformed unit B2 rocks, indicating deposition after uplift of the Emirau-Feni Ridge and an age no older than about late Miocene. The youngest rocks of unit B3 are Holocene, since the unit is still being deposited. We provisionally assume that unit B3 is largely Pliocene and Quaternary in age.

The combined unit B2 and B3 section may be equivalent to unit K3 in the Kilinailau Trench, with a once-continuous unit B2-unit K3 section broken by faulting, and with the uplifted part of the section now underlying the Bougainville slope. Onshore lithologic equivalents to units B2 and B3 could be the turbidites of the Mole Formation and the calcarenites and calcisiltites of the Pemba Formation on Choiseul, and the volcaniclastic rocks of the Toniva Formation, Arawa Conglomerate, and Bougainville Group on Bougainville.

TIME-TO-DEPTH CONVERSION

Velocities vary markedly both between and within seismic-stratigraphic units (Figure 13). Velocity information for the units is primarily derived from sonobuoy seismic-refraction measurements (Cooper, Cochrane, and Bruns, this volume; Cochrane, Cooper, and Bruns, this volume) and from seismic-reflection data, with stacking velocities from areas of undeformed strata converted to interval velocities by the Dix equation (Dix, 1955). Other velocity information is obtained from refraction studies on the adjacent Ontong Java Plateau (Kroenke, 1972; Andrews, Packham, et al, 1975; Furumoto et al, 1976; Hussong, Wipperman, and Kroenke, 1979).

For unit K1, oceanic basalt, plane-layer solutions for seismic-refraction sonobuoys on lines 409 and 411 give velocities of 5.2 to 5.9 km/s (Cooper, Cochrane, and Bruns, this volume). Modeling studies of refraction data on lines 409 and 413 give basement velocities of about 4.5 km/s at the top of the unit, increasing to 6 to 6.5 km/s at a depth of 10 km (Cochrane, Cooper, and Bruns, this volume). Seismic-reflection interval velocities from the acoustically layered unit K1 are as low as about 4 km/s, but range from 4.5 to 6.8 km/s.

For the oceanic crust of the adjacent Ontong Java Plateau that we correlate with unit K1, Andrews, Packham, et al (1975) reported a velocity of around 5.5 km/s for the basalt sampled at DSDP Hole 289 (range 4.7 to 5.9 km/s). Kroenke (1972) reported a value of 5.6 km/s for the upper crustal layer, based on data from refraction sonobuoys. Hussong, Wipperman, and Kroenke (1979), likewise found upper crustal values of 5.5 to 5.6 km/s on sonobuoy refraction records, and Furumoto et al (1976) reported a 10- to 12-km-thick crustal unit with velocities ranging from 5.0 to 6.4 km/s.

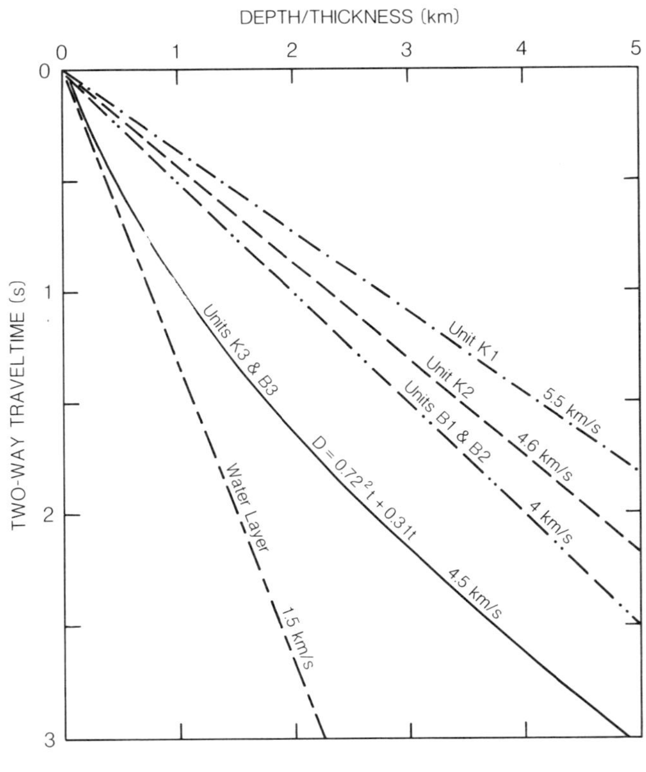

Figure 13. Time-depth conversion curves for converting two-way traveltime through each seismic-stratigraphic unit to depth for the Kilinailau Trench and adjacent slope. Equation for units K3 and B3 valid for 0 to 2.2 s; below 2.2 s, a constant interval velocity of 4.5 km/s is assumed. D=depth in km and t=two-way traveltime in seconds. Water-layer thickness must be added to cumulative sub-seafloor depth to obtain total depth. See text for discussion of derivation and usage.

Thus, the velocity of unit K1 can be slightly less than 5 km/s at the top of the unit and increase rapidly with depth to 5.5 to 6.5 km/s. The thickness of unit K1 can perhaps best be estimated by using an interval velocity of about 5.5 km/s, similar to that reported by Kroenke (1972), Andrews, Packham, et al (1975), and Cooper, Cochrane, and Bruns (this volume).

For the inferred pelagic carbonate deposits of unit K2, interval velocities were determined from seismic-reflection stacking velocities on lines 408 through 418. The velocities range from about 3.9 to 5.1 km/s and average 4.6 km/s with a standard deviation of about 10 percent. The average velocity of the unit is higher than the values of 3.3 to 3.8 km/s that were reported for the lower 500 m of the sedimentary unit overlying the Ontong Java Plateau (Hussong, Wipperman, and Kroenke, 1979); we believe the higher velocities in unit K2 are due to burial by the trench-fill unit K3. The lowest interval velocity in the unit is at the northern end of line 408, where the overlying trench fill is thinnest and unit K2 is least compacted. The highest velocity is beneath the thickest trench-fill unit K3 on lines 412 and 413 where unit K2 is most compacted as a result of burial. The seismic-reflection velocities are similar to those from refraction data, which give a range of about 3.6 to 4.7 km/s (Cooper, Cochrane, and Bruns, this volume). An approximate thickness of unit K2 can be obtained by using the average velocity of 4.6 km/s.

On lines 408 through 418, where flat-lying to gently dipping units B3 and K3 cover both the trench and slope sequences, stacking velocities were converted to interval velocities between predetermined subseafloor reflection times that lie approximately along continuous seismic-reflection horizons. The interval velocities from all lines were then averaged to obtain a regional average interval velocity for each subsurface time interval . This process is valid to a subbottom two-way travel time of 2.2 s; below this time, stacking-velocity information from the reflection data is poor. The averaged interval velocities were used to derive a time-to-depth conversion curve for t less than the 2.2 s given by the equation: $D=0.72t+0.31t^2$, where D=subbottom thickness of units K3 and B3 and t=two-way traveltime. Below 2.2 s, an interval velocity of 4.5 km/s was assumed, which is close to the velocity measured in the interval of 1.8 to 2.2 s and is similar to refraction velocities measured on lines 409 (4.7 km/s) and 413 (4.3 km/s; Cooper, Cochrane, and Bruns, this volume; Cochrane, Cooper, and Bruns, this volume). Also, from the equation given above, the interval velocity in unit K3 from subbottom to 2.2 s is given by $V=1.24t+1.44$, where V=interval velocity, and t=two-way traveltime in seconds.

For units B1 and B2, refraction sonobuoy velocities and modeling studies on line 413 give values ranging from 2.4 to 4.5 km/s. The thickness of these two units can be estimated by using a velocity of 4 km/s (Cochrane, Cooper, and Bruns, this volume).

The time-to-depth conversion curves for the various seismic-stratigraphic units and for the water layer (Figure 13) were used to construct the isopach and structure maps for the trench and slope (Figures 14, 15). The total depth to the base of a unit can be obtained by using a combination of the appropriate curves on times measured the seismic profiles, or the interval thickness of any unit can be obtained separately.

SEDIMENT THICKNESS AND DISTRIBUTION

Trench

Unit K2, the southwestward continuation of the pelagic sediment blanket which covers the Ontong Java Plateau, is commonly about 0.4 to 0.8 s or about 900 to 1,800 m thick (Figures 5-12). This thickness is greater than that reported for pelagic strata on the plateau, where the sequence is about 0.6-0.8 s or 600 to 800 m thick (Kroenke, 1972). However, the Ontong Java Plateau strata are as thick as 1,200 m at DSDP site 289. Unit K2 is relatively uniform in thickness, although some thickening occurs locally toward the slope. Thickness differences between profiles may be partly due to errors in locating the K1/K2 boundary.

The overlying unit K3 trench fill is thickest at the trench axis, thins markedly toward the Ontong Java Plateau, and thins both southeastward and northwestward along the trench from lines 411 and 412 (Figure 14). On line 412 at the trench axis, the unit is more than 5 km thick on line 412 in a structural low just south of the Kilinailau Islands. The unit thins to about 1.5 km on line 408 at the southeastern end of the study area and to 2.7 km on lines 417 and 418 at the northwestern end. The thin axial fill on line 408 lies on a bathymetric and structural high along the trench axis (Figures 14 and 15).

Slope

Unit B2 is over 2 km thick, and thins to about 1 km toward the trench axis where the unit is uplifted and truncated by faulting. Landward, unit B2 thins to about 1 km beneath the upper slope. The extent of unit B2 beneath the slope is unknown. The unit is only mappable on lines 410-412, and even on these lines the position of the base of the unit is uncertain. Elsewhere, unit B2 strata, if present, are highly deformed and cannot be differentiated from the basement unit B1.

A slope basin containing unit B3 strata as much as 1.8 km thick is present between Nuguria Ridge and the Bougainville-Buka Island slope (Figure 14). The fill in the slope basin is thickest adjacent to Buka Island (line 413, Figure 10). Northwest of the basin, unit B3 thins to less than 0.5 km at the high point of the saddle between Nuguria Ridge and the Green Islands. Little or no unit B3 sediment is present on the crest and northeastern flank of the ridge (Figures 11, 12). The southeastern extent of the slope basin lies at the northern end of Bougainville (Figure 14) and is coincident with the end of the uplifted Quaternary reefs on Bougainville. Farther southeastward, from the end

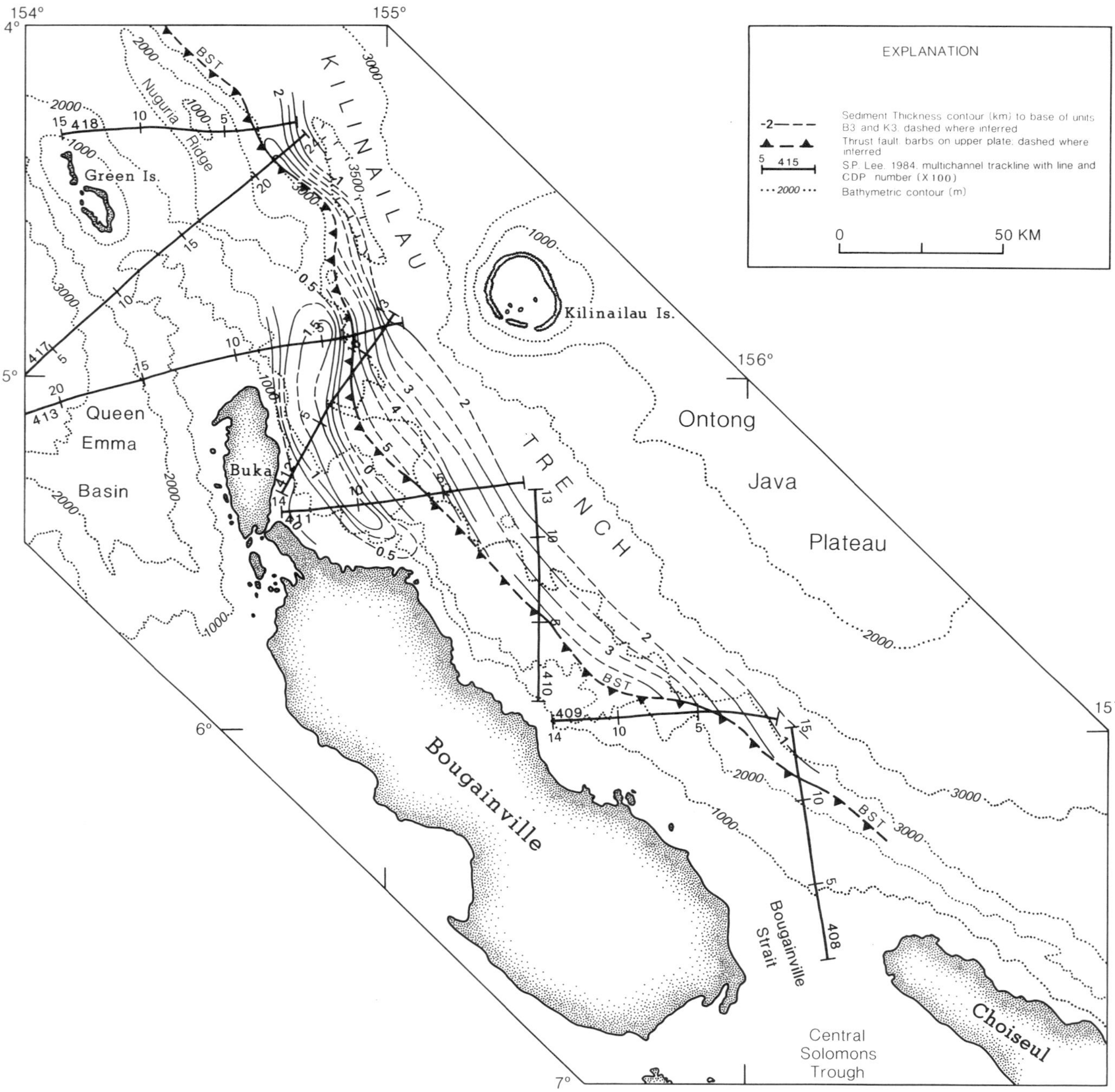

Figure 14. Isopach map of sediment thickness in seismic-stratigraphic units K3 and B3. BST — base-of-slope thrust.

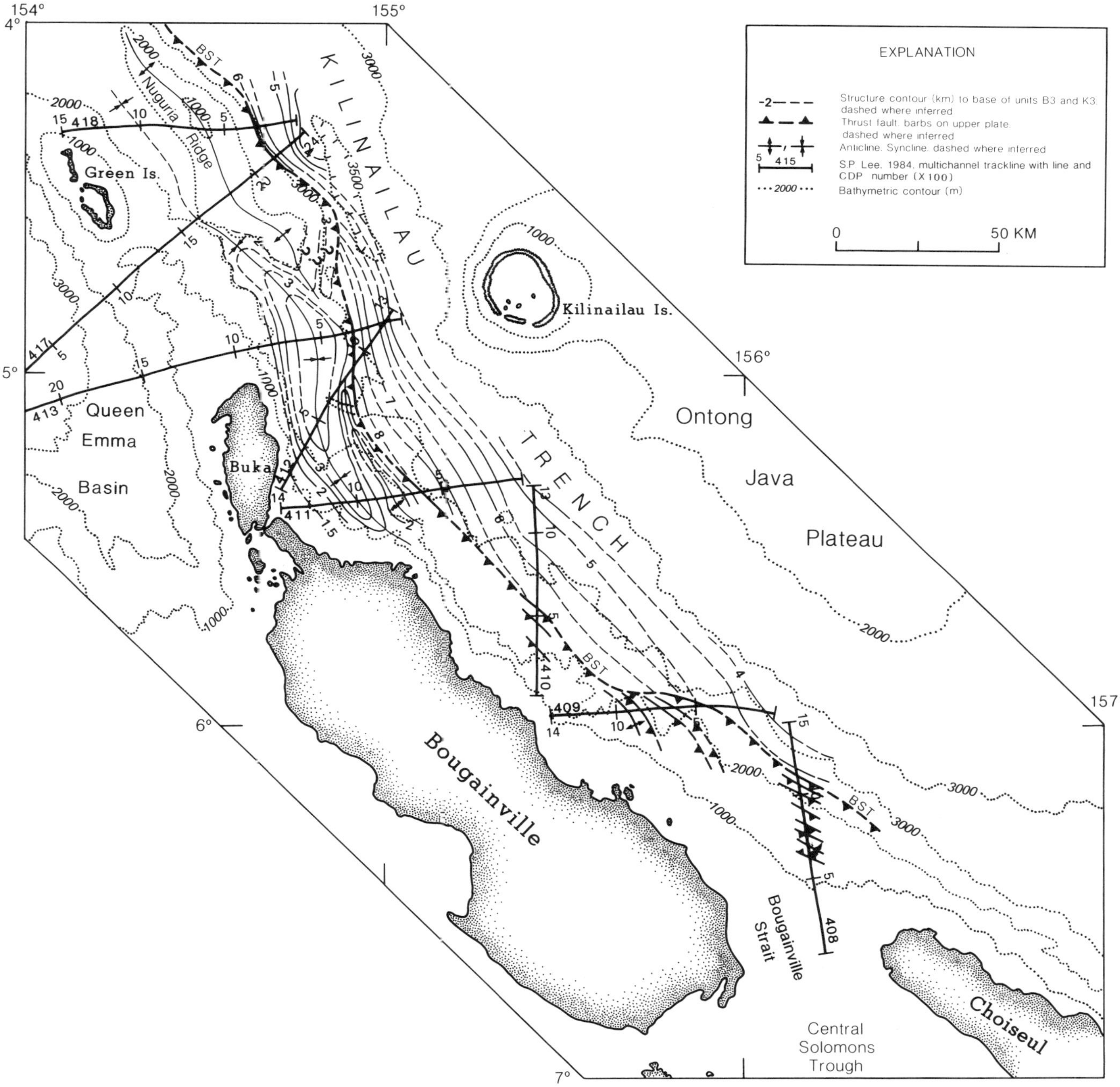

Figure 15. Structural contours at the base of seismic-stratigraphic units K3 and B3. BST — base-of-slope thrust.

of the slope basin to Choiseul, unit B3 sediment beneath the slope is mainly less than about 0.5 km thick. Limited seismic-reflection and bathymetric data suggest that these strata are dissected by channels about 5 to 10 km apart.

Sediment Dispersal Paths

The distribution of sediment in the Kilinailau Trench region is controlled by the structural and bathymetric relief along the trench. On a structure map showing the base of unit K3, the structurally lowest part of the trench is in the area around lines 411 and 412 (Figure 15) and corresponds to the region of thickest sediment accumulation (Figure 14). This structural low is also reflected in the present-day bathymetry. Both northwest and southeast of this low, the trench shallows to about 3,000 m at local structural highs near the Nuguria Islands and off the northwestern tip of Choiseul. Northwest of the Nuguria Islands, the trench abruptly deepens to 4,500 m; southeast of line 408, the trench also deepens to about 3,500 m. Sediment entering the trench from Bougainville and Buka Island is trapped in the bathymetric and structural low between Choiseul and the Nuguria Island local highs. The bathymetric and structural gradient of the trench funnels sediment from Choiseul to the southeast away from Bougainville. Similarly, sediment derived from northwest of the Green Islands is channeled northwestward. Therefore, the dominant source terrains for units K3 and B3 between seismic profiles 408 and 418 are Bougainville, Buka Island, the Green Islands, and the connecting submarine ridge.

The maximum sediment thickness in the trench is adjacent to Buka Island at the mouth of a submarine channel (Figure 14). This channel arises between Nuguria Ridge and the Emirau-Feni Ridge, trends southeastward between the ridges and along the axis of the slope basin, then turns eastward toward the trench at the southeastern end of Nuguria Ridge. The channel empties into the local bathymetric low in the trench axis near lines 412 and 413. Correlated seismic units in unit K3 (reflectors A-F, Figures 5-12) show that the oldest units are thickest in this same region, and thin and are truncated northwestward and southeastward along the trench axis. For example, below reflector D on lines 408, 409, and 418, only a thin section is present. In contrast, a relatively thick section is present below reflector D on lines 411-413 and 417. These differences in the thickness suggest that the channel has been the major conduit for transport of sediment into the trench throughout unit K3 time.

Southeast of the channel mouth, numerous small channels are present on single-channel seismic reflection lines which were shot parallel to the slope. The sediment from Bougainville is funneled into the trench along these channels, which appear on the limited seismic and bathymetric data to cut directly downslope. Relatively thin sediment cover on the slope suggests that much or most of the sediment derived from the island bypasses the slope and is deposited in the trench. Little sediment enters the trench northwest of the channel mouth because Nuguria Ridge forms a barrier to sediment transport. The ridge is covered by a young thin sediment cover (less than about 100 m thick) that has not provided large amounts of sediment to the trench through erosion, either subsea or subaerial.

STRUCTURE

The structure of the Kilinailau Trench region is typical of that of an active convergent-margin subduction zone, although the region has been in a back-arc location during the late Cenozoic. The lower slope and trench are underlain by a back-arc thrust and a foldbelt that extends from at least Choiseul to the Green Islands. Along this thrust belt, relatively undeformed ocean-plate rocks (units K1 and K2) are thrust beneath the slope, and trench-fill rocks (unit K3) are in part imbricated by faults and transferred to the upper plate, and in part thrust beneath the slope (Figures 5-12). Deformation throughout the trench region is young and ongoing, as evidenced by folding of even the youngest strata and by marked disruption of the sea floor well northeast of the base of the slope.

The thrust belt is underlain by a major detachment plane. In the region from line 408 to line 413, unit K3 strata are markedly deformed by folds and faults, but unit K2 rocks are largely unaffected, even beneath the lower slope. Thus, a major detachment is present at or near the top of unit K2. At the northwestern end of the region on lines 417 and 418, however, this detachment cuts upward and lies at the top of the unit K3 rocks, as these rocks are thrust beneath Nuguria Ridge as a relatively coherent unit, without major fault imbrication or folding.

The structure along the trench is variable both parallel and perpendicular to the trench (Figure 16). Beneath the lower slope at the southeastern end of the area (lines 408 and 409), the structural configuration is that of a thrust-imbricated accretionary wedge. In the central area, the deformed area widens markedly. Although the lower slope is still underlain by an accretionary wedge, a fold-and-thrust belt extends and deforms the trench sediments well northeast of the base of the slope. Finally, in the northwestern part of the area at Nuguria Ridge (lines 417 and 418), trench strata are thrust beneath the ridge almost undeformed, with little folding or fault-imbrication of trench strata. Each of these areas is discussed further below.

Lines 408 and 409

At the southeastern end of the study region on lines 408 and 409 adjacent to Bougainville Strait, strata northeast of the base of the slope are undeformed. But at the base of the slope, unit K2 strata are thrust beneath the slope for at least 12 to 15 km with little deformation, and strata of the overlying unit K3 are fault-imbricated into an accretionary wedge-like structure about 12 to 25 km wide. The rocks beneath the lower slope have low velocities according to models of seismic-refraction data (Cochrane, Cooper, and Bruns, this volume), thus emphasizing that the material

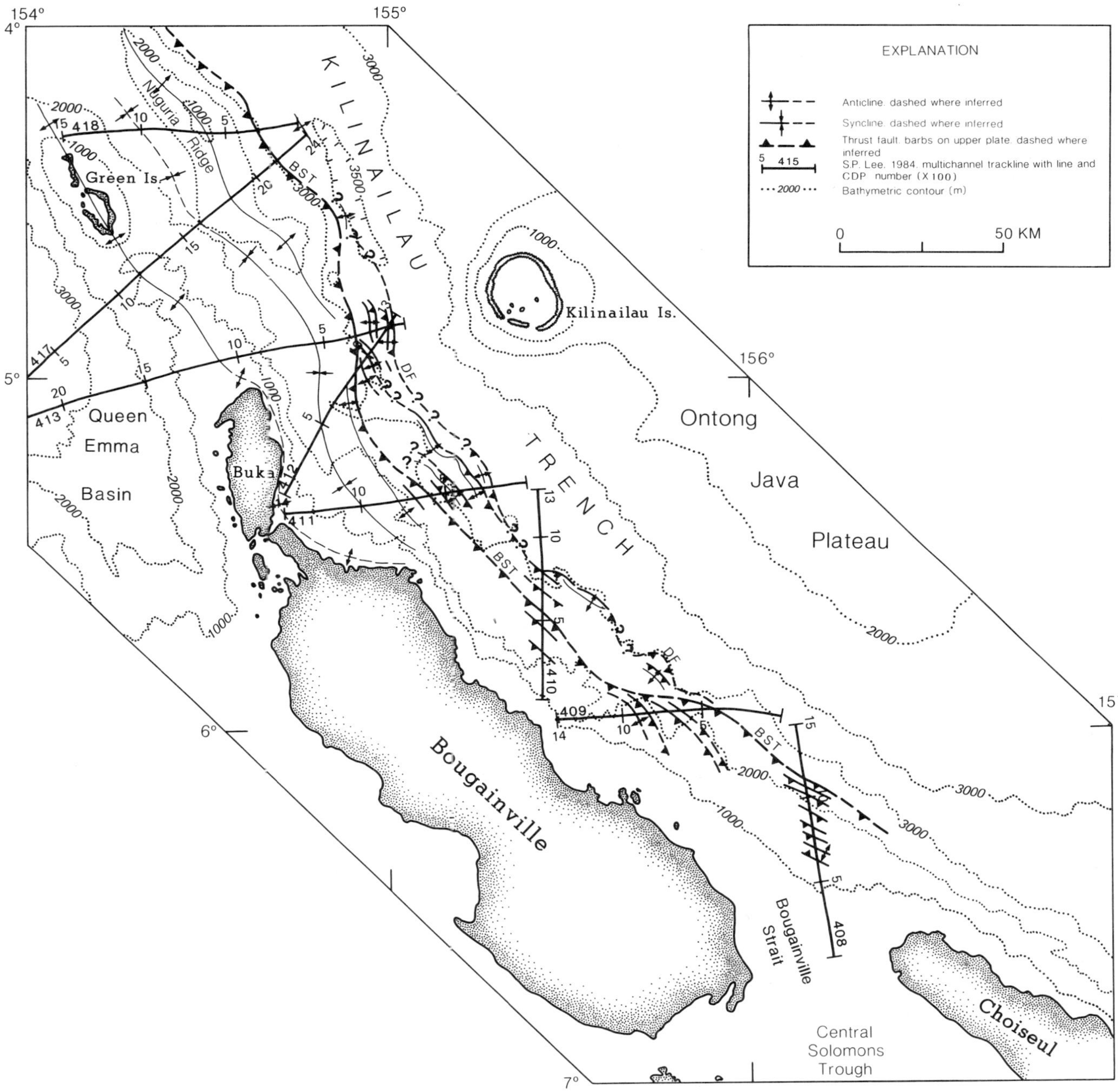

Figure 16. Structural axis and trends based on single- and multichannel seismic-reflection data and on bathymetry of Chase, Seekins, and Young (this volume). DF — deformation front; BST — base-of-slope thrust.

beneath the slope consists of low-velocity accreted sedimentary rocks rather than high-velocity basement rocks of Choiseul and Bougainville. The contact between the accreted rocks and the basement rocks is not clear on the seismic-reflection records, but, based on these records and on refraction models of data along lines 409 and 413 (Cochrane, Cooper, and Bruns, this volume), we think the contact probably lies on a southwest-dipping thrust fault that surfaces near the northeast edge of unit B3.

The unit K3 trench strata on lines 408 and 409 are tilted seaward; this tilt continues on line 410 also (see next section). While this tilt could be depositional, it could also reflect gentle uplift of the entire base-of-slope region off Bougainville and Choiseul due to convergence between the Pacific plate and the islands.

Lines 410 through 413

Between Bougainville Strait and Nuguria Ridge, the thrust fault which marks the base of the slope in the regions of line 408-409 bifurcates into the deformation front (DF) and the base-of-slope thrust (BST; Figure 16). The deformation front is the forwardmost fault or fold of the thrust zone, and marks the transition from undeformed trench strata to faulted and folded strata. The base-of-slope thrust marks the region at the base of the slope where rocks of the trench and ocean plate are thrust beneath the slope. The separation between these two breaks is as much as 20 km. The DF forms a generally straight continuation of the thrust faults at the base of the slope at the northwestern and southeastern ends of the region, whereas the BST steps southwestward of this continuation.

In between the DF and the BST structural breaks, unit K3 trench rocks are strongly deformed by thrust faults which cut upward from the detachment plane lying along, or near the top of, unit K2. The degree of deformation becomes greater from southeast to northwest along the trench, and from northeast to southwest across the trench. Deformation at the DF is variable, ranging from minor on a thrust fault which does not break through to the surface (line 410, Figure 7), to moderate on simple anticlines which barely disrupt the sea floor (lines 412 and 413, Figures 9 and 10). Within the deformed sequence between the BST and the DF, vergence of the thrust faults is variable; although most faults verge northeastward, some (lines 411 and 413, Figures 8 and 10) appear to have a southwestward vergence. Faults are steeply dipping at the sea floor and flatten to nearly horizontal only at the base of the trench fill where they merge with the underlying decollement. Based on the limited bathymetric data and seismic lines, the DF appears to be formed by a series of lobate overlapping thrust faults, and the anticlines between the DF and BST are a series of en-echelon structures.

The BST marks a region of folding and faulting, in which structural complexity increases southwestward, as the trench and slope rocks are imbricated into an accretionary wedge. The unit K3 strata are cut by thrust faults and rotated southwestward, and unit B2 strata are markedly uplifted along the base of the slope. Within the accretionary wedge rocks, thrust faults appear to be steeply dipping at the top of the deformed sequence, and to flatten into the underlying detachment surface. The spacing between individual faults which can be readily identified near the front of the wedge is from 1 to over 5 km. Numerous other thrust faults may be present but are not resolved on the seismic records.

As on lines 408 and 409, the contact between the fault-imbricated trench rocks and the basement rocks of the arc cannot be distinguished on lines 410 through 413. However, this contact probably lies near the steep slope leading to the top of the island platform, because high-velocity material is required in this region when seismic-refraction data are modeled (Cochrane, Cooper, and Bruns, this volume).

The deformed strata are overlain by the relatively undeformed unit B3 strata. The undeformed sequence is especially striking on line 413, where the imbricated rocks are overlain by almost undeformed strata about 2 km thick. A bounding fault may be present at the top of the deformed rocks in unit B1, separating unit B1 from the partly deformed unit B2 and the mainly undeformed unit B3. Such a fault would form a duplex structure, in which the accretionary wedge rocks are bounded above and below by decollements and the trench rocks are underthrust without deformation of the overlying slope strata. Such a thrust is apparently present at the top of underthrust unit K3 strata on lines 417 and 418. However, a throughgoing thrust seems unlikely on lines 410 through 413, because the overlying unit B3 strata are so little deformed, particularly on line 413. The thick unit B3 sequence is probably due to high deposition rates along the mid-slope channel.

Lines 417 and 418

At the northwestern end of the region along Nuguria Ridge, both unit K2 and unit K3 trench strata are underthrust from 10 to 15 km beneath arc rocks of the ridge. The underthrust unit K3 rocks are gently folded (line 418), but fault imbrication of the underthrust strata beneath the ridge appears to be minimal, because seismic reflectors from the unit are relatively coherent beneath the ridge. Beneath Nuguria Ridge, the fault detachment surface must lie near the top of unit K3, because there is little evidence for accretion of unit K3 strata to the base of the slope, as on lines 408 through 413; rather, the unit K3 rocks lie entirely in the underthrust plate. Therefore the decollement surface, which lies at or near the top of unit K2 to the southeast, must cut up through the unit K3 strata between lines 413 and 417.

Uplift of Nuguria Ridge is young, perhaps Pliocene or Quaternary, as indicated by the tilted unit B3 sequence on the southwestern flank of the ridge on line 418 (Figure 12). The uplift may be related to the rise of the decollement surface from the base to the top of unit K3; if so, then this rise may begin at the southeastern end of Nuguria Ridge.

At the northeast end of lines 417 and 418, faulting also affects the entire sedimentary section. This faulting, however, appears to be unrelated to the decollement surface beneath the ridge; at least, no clear connection is seen on the seismic data. Additional data are needed both along the trench and northeast of the faults to see if this deformation is part of a seaward-stepped deformation front similar to that observed on lines 410-413.

DISCUSSION

The zone of thrusting along the Kilinailau trench seems to be anomalous, in that the active subduction zone is on the southern side of the arc, and the northern side of the arc is seismically quiet. The regional geology and tectonic models for the Kilinailau Trench region indicate that southward-directed subduction ended by late early Miocene time, volcanism and tectonism slackened or ceased during the middle Miocene, and active northward-directed subduction began in the late Miocene and continues to the present. Within this geologic and tectonic framework, the Kilinailau Trench should have become inactive as a subduction zone by the middle Miocene or earlier and remained inactive to the present; and the North Solomon Island arc should have been locked to and moving with the Pacific plate since the early Miocene.

However, the Kilinailau Trench is tectonically active. Everywhere along the trench, strata at the base of the slope are deformed. Structures are young, and not only deform the youngest sediment in the trench but also have marked bathymetric relief at the sea floor and show few growth features. The young rocks in unit K3, as well as the underlying oceanic rocks in units K1 and K2 are being thrust beneath the slope. The trench does not resemble an inactive subduction zone, in which a relic accretionary wedge is covered by undeformed sediments. In fact, such a geometry, if once present, would now be highly disrupted or obliterated by the ongoing thrusting. The obvious conclusion to be derived from the structural pattern is that the thrust zone is caused by significant and persistent closure between the Pacific plate and the Bougainville-Solomon Islands regions.

Moreover, late Cenozoic tectonic activity caused by this closure is evident along the entire Manus-Kilinailau-KKK trench/fault system. Deformation of young strata occurs at the base of the slope along the eastern part of the Manus Trench (Ryan and Marlow, 1988). Quaternary limestone terraces are uplifted on Mussau Island (Exon and Tiffin, 1984), the Feni and Green islands (Wallace et al, 1983; Bain et al, 1971), Buka Island and Bougainville (Blake and Miezitis, 1967), and Choiseul (Coulson and Vedder, 1986). Late Cenozoic deformation occurs along the Kia-Korigole-Kaipito fault system and its southeastward extension on the northeastern side of the Florida Islands (Kroenke, 1972; Ramsay, 1978; Coulson and Vedder, 1986; Bruns et al, 1986), and may reflect obduction of ocean-plate rocks (Kroenke, 1984). A southwest-dipping zone of low-level seismicity in the Santa Isabel-Florida Islands region may outline subducted, possibly relic, Pacific crust (Cooper, Kroenke, and Resig, 1986). And finally, the folds of the Malaita Anticlinorium and the uplift of Malaita have occurred since the late Miocene (Kroenke, 1972; 1984; Kroenke, Resig and Cooper, 1986).

However, seismicity studies indicate very low levels of earthquake activity along the Manus-Kilinailau-KKK trench/fault system (Johnson, 1979; Hamilton, 1979; Cooper and Taylor, 1985; 1987; Cooper, Kroenke, and Resig, 1986). Virtually all of the seismic activity in the region occurs on the southwestern side of the Solomon Islands. Intense seismicity and a well-developed Benioff zone show that most of the plate convergence between the Australia-India and Pacific plates in the region is taken up by subduction along the New Britain and San Cristobal trenches, and few earthquakes now occur in regions north of the islands. Nevertheless, the Kilinailau trench is presently tectonically active, even though this activity apparently occurs slowly, aseismically, or sporadically; or perhaps seismic events occur at very low magnitudes and are not recorded teleseismically.

Although the Kilinailau Trench deformation clearly indicates closure between the Bougainville-Solomon Islands region and the Pacific plate, the amount of closure is unknown. A minimum figure is 10 to 15 km — the distance the oceanic plate rocks can be traced beneath Nuguria Ridge on the seismic-reflection data. A maximum figure may be about 200 km — the length of the relic ocean plate slab that low-level seismicity suggests lies beneath the arc (Cooper, Kroenke, and Resig, 1986). The best estimate of closure might be made by assuming that the amount of convergence taken up by underthrusting at the Kilinailau Trench is accomodated to the east by a combination of folding in the Malaita Anticlinorium and obduction along Santa Isabel and the Florida Islands. If the amount of possible obduction of the ocean plate plus the amount of shortening in the folds of the Malaita Anticlinorium could be measured, the resultant figure might be the amount of late Cenozoic shortening across the Kilinailau Trench region.

We believe that the Kilinailau Trench deformation began in late Miocene or later time. Post-early Miocene deformation is indicated by uplifted probable Oligocene and early Miocene rocks (unit E2 strata) on Nuguria Ridge. Deformation clearly is contemporaneous with or postdates uplift of the Emirau-Feni Ridge beneath Buka Island and the Green Islands, because unit B2 and B3 strata both onlap and cover deformed unit B1 rocks. Possibly deformation began in the late Miocene when active subduction commenced on the south side of the arc along the New Britain Trench.

As a tectonic mechanism for development of the the Kilinailau Trench thrust zone, we speculate that the deformation is a response to active subduction on the southern side of the Solomon Islands arc. In this model, coupling between the actively subducting Solomon Sea plate and the arc causes a slight northward motion of the arc relative to the Pacific plate. This northward motion reactivates the early Tertiary subduction zone, a zone of weakness, resulting in partial decoupling of the arc from, and in thrusting of the arc over, the Pacific plate. The observed low-level seismic zone through the Santa Isabel-Malaita region is caused by an adjustment of the relic Pacific plate slab to the slowly overriding arc. If this kind of mechanism is in fact operative, then the deformation in the Kilinailau Trench probably began no earlier than the late Miocene, when initiation of subduction on the south side of the arc may have caused major stresses across the arc through partial coupling of the arc and the subducting Solomon Sea plate.

Other mechanisms have been suggested for back-arc thrusting such as gravity tectonics driven by a rising arc or igneous intrusions along the arc. Also, structures similar to those along the Kilinailau Trench could develop as a result of sporadic transform tectonics along the arc. However, the Kilinailau Trench deformation is apparently only part of a much longer zone of deformation which includes the Malaita Anticlinorium. However, the wide zone of folding that forms the Malaita Anticlinorium is unlikely to result from gravity tectonics or from igneous intrusions. Although these mechanisms are unlikely to be the primary cause of the Kilinailau Trench thrusting, they could contribute to the deformation locally.

In a more general sense, this model implies that the observed tectonic activity along the Kilinailau Trench represents internal deformation of a composite plate made up of the Pacific plate and the Solomon Islands arc. We assume that these two elements were sutured when southward subduction stopped in the early Miocene and that they have probably moved in tandem since suturing. In the late Cenozoic, stresses began to be transmitted across the arc with the development of the active subduction zone on the southern side of the arc. These stresses were then accommodated within the composite terrane by northward thrusting along the reactivated early Tertiary subduction zone and by folding within the Malaita Anticlinorium north of the Solomon Islands. The degree of convergence between the island arc and the Pacific plate is probably small, as reflected in the low level of seismicity, and therefore the island-arc system is still primarily coupled to and moving with the Pacific plate. Tectonism along the fold-belt and the Malaita Anticlinorium is a secondary effect caused by the ongoing subduction along the New Britain and San Cristobal Trenches south of the arc system. Substantial convergence along the north side of the arc could only occur if subduction polarity again reversed, and the Manus-Kilinailau-KKK trench-fault system once again became an active subduction zone.

Despite the presence of active deformation, the Manus-Kilinailau-KKK trench-fault system is not now a major plate boundary. Instead, this deformation reflects minor shortening across a sutured terrane boundary within a composite plate. The island-arc system lying between the Manus-Kilinailau-KKK trench-fault system and the New Britain and San Cristobal Trenches constitutes a microplate lying between a highly active and an almost inactive subduction zone, slightly decoupled from, but still moving dominantly with the Pacific plate. The microplate forms a soft margin, squeezed between ocean plate and plateau, riding above both the actively subducting Solomon Sea plate and a remnant subducted slab of the Pacific plate.

CONCLUSIONS

Basement rocks beneath the Kilinailau Trench in the Choiseul-Bougainville-Buka Island regions are layered(?) Cretaceous oceanic basalt of the Ontong Java Plateau. Overlying the basalt are Cretaceous to early Miocene(?) pelagic strata similar to the rocks that cover the Ontong Java Plateau and middle Miocene(?) and younger trench-filling turbidites. Beneath the slope are chaotically bedded Paleogene and early Miocene(?) arc-derived volcanic and sedimentary rocks, plus Miocene(?) and younger accreted rocks and slope-basin turbidites. Trench and slope strata are deformed by folds and thrust faults that effect the youngest strata and have marked bathymetric relief at the sea floor.

The regional tectonic setting of the Kilinailau Trench region is that of an inactive subduction zone, along which subduction ceased in the early Miocene. The middle Miocene was a period of arc-quiescence, followed by a reversal in arc polarity and initiation of subduction on the south side of the arc in the late Miocene. At present, active subduction occurs along the New Britain-San Cristobal trench system. Given this setting, the Kilinailau Trench should be an area with little active tectonism. However, the trench is a site of active back-arc thrusting, with underthrusting of the Pacific plate and overlying strata taking place beneath the slope of the Choiseul and Bougainville regions.

The Kilinailau Trench thrusting reflects closure between the Pacific plate and the island arc of the Bougainville-Solomon Islands region; the amount of closure is unknown. This ongoing closure is perhaps due to partial coupling of the island arc system to the actively subducting Solomon Sea and Australia-India plates at the New Britain-San Cristobal subduction zone. Consequent northward motion of the arc relative to the Pacific plate has reactivated the early Tertiary Kilinailau Trench subduction zone suture, and cause back-arc thrusting. Thus, at present the Bougainville-Solomon Islands arc is moving dominantly with, but is partly detached from and underthrust by, the Pacific plate.

ACKNOWLEDGMENTS

We thank P.E. Hart for aid in processing the seismic data, Susan Vath for drafting support, and T.M. Martin-Funnié for typing support. M.S. Marlow and M.A. Fisher provided reviews which significantly improved the manuscript.

REFERENCES CITED

Andrews, J.E., Packham, G.H., et al, 1975, Initial reports of the Deep Sea Drilling Project, v. 30: Washington, D.C., U.S. Government Printing Office, 753 p.

Bain, J.H.C., Davies, H.L., Hohnen, P.D., Ryburn, R.J., Smith, I.E., Grainger, R., Tingey, R.J., and Moffat, M.R., 1971, Geology of Papua New Guinea: Department of National Development in cooperation with the Geological Survey Division of the Papua New Guinea Department of Lands, Surveys, and Mines, Bureau of Mineral Resources, Geology and Geophysics, scale 1:1,000,000.

Blake D.H., and Miezitis, Y., 1967, Geology of Bougainville and Buka Island, New Guinea: Department of National Development, Bureau of Mineral Resources, Geology and Geophysics, Bulletin 93, PNG 1, Canberra, A.C.T., 56 p.

Bruns, T.R., Cooper, A.K., Mann, D.M., and Vedder, J.G., 1986, Seismic stratigraphy and structure of sedimentary basins in the Solomon Islands region, *in* Vedder, J.G., Pound, K.S., and Boundy, S.Q., eds., Geology and offshore resources of Pacific island arcs—central and western Solomon Islands: Houston, Texas, Circum-Pacific Council for Energy and Mineral Resources Earth Science Series, v. 4, p. 177-214.

Bruns, T.R., Cooper, A.K., and Vedder, J.G., 1986, A multichannel seismic-reflection profile across the Solomon Islands arc, *in* Vedder, J.G., Pound, K.S., and Boundy, S.Q., eds., Geology and offshore resources of Pacific island arcs—central and western Solomon Islands: Houston, Texas, Circum-Pacific Council for Energy and Mineral Resources Earth Science Series, v. 4, p. 215-224.

Chase, T.E., Seekins, B.A., and Lund, K.E., 1986, Submarine topography of the Solomon Islands region, *in* Vedder, J.G., Pound, K.S., and Boundy, S.Q., eds., Geology and offshore resources of Pacific island arcs—central and western Solomon Islands: Houston, Texas, Circum-Pacific Council for Energy and Mineral Resources Earth Science Series, v. 4, p. 21-32.

Coleman, P.J., 1965, Stratigraphical and structural notes on the British Solomon Islands with reference to the first geological map: British Solomon Islands Geological Record (1959-1962), v. 2, p. 17-31.

Coleman, P.J., and Kroenke, L.W., 1981, Subduction without volcanism in the Solomon Islands arc: Geo-Marine Letters, v. 1, p. 129-134.

Coleman, P.J., and Packham, G.H., 1976, The Melanesian Borderlands and India-Pacific Plates' boundary: Earth-Science Reviews, v. 12, p. 197-233.

Colwell, J.B. and Vedder J.G., 1986, Description and interpretation of dredged rocks, Solomon Islands, *in* Vedder, J.G., Pound, K.S., and Boundy, S.Q., eds., Geology and offshore resources of Pacific island arcs—central and western Solomon Islands: Houston, Texas, Circum-Pacific Council for Energy and Mineral Resources Earth Science Series, v. 4, p. 225-236.

Cooper, A.K., Bruns, T.R., and Wood, R.A., 1986, Crustal structure of the Solomon Islands intra-arc basins from sonobuoy seismic studies, *in* Vedder, J.G., Pound, K.S., and Boundy, S.Q., eds., Geology and offshore resources of Pacific island arcs—central and western Solomon Islands: Houston, Texas, Circum-Pacific Council for Energy and Mineral Resources Earth Science Series, v. 4, p. 135-156.

Cooper, P.A., and Taylor, B., 1985, Polarity reversal in the Solomon Islands arc: Nature, v. 314, p. 428-430.

Cooper, P.A., and Taylor, B., 1987a, Seismotectonics of New Guinea: a model for arc reversal following arc-continent collision: Tectonics, v. 6, p. 53-67.

Cooper, P.A., and Taylor, B., 1987b, The spatial distribution of earthquakes, focal mechanisms, and subducted lithosphere in the Solomon Islands, *in* Taylor, B. and Exon, N.F., eds., Marine geology, geophysics, and geochemistry of the Woodlark Basin-Solomon Islands: Houston, Texas, Circum-Pacific Council for Energy and Mineral Resources Earth Science Series, v. 7, p. 67-88.

Cooper, P.A., Kroenke, L.W., and Resig, J.M., 1986, Tectonic implications of seismicity northeast of the Solomon Islands, *in* Vedder, J.G., Pound, K.S., and Boundy, S.Q., eds., Geology and offshore resources of Pacific island arcs—central and western Solomon Islands: Houston, Texas, Circum-Pacific Council for Energy and Mineral Resources Earth Science Series, v. 4, p. 117-122.

Coulson, F.I., and Vedder, J.G., 1986, Geology of the central and western Solomon Islands, *in* Vedder, J.G., Pound, K.S., and Boundy, S.Q., eds., Geology and offshore resources of Pacific island arcs—central and western Solomon Islands: Houston, Texas, Circum-Pacific Council for Energy and Mineral Resources Earth Science Series, v. 4, p. 59-88.

de Broin, C.E., Aubertin, F., and Ravenne, C., 1977, Structure and history of the Solomon-New Ireland region: International Symposium on Geodynamics in Southwest Pacific, Noumea (New Caledonia), 27 August-2 September 1976, Editions Technip, Paris, p. 37-50.

Dix, C.H., 1955, Seismic velocities from surface measurements: Geophysics, v. 20, p. 68-86.

Drummond, K.J., Nishiwaki, C., Corvalan, J., Doutch, H.F., and Craddock, C., 1985, Plate tectonic map of the Houston, Texas, Circum-Pacific region, Pacific basin sheet: Houston, Texas, Circum-Pacific Council for Energy and Mineral Resources, scale 1:17,000,000.

Exon, N.F., and Tiffin, D.L., 1984, Geology and petroleum prospects of offshore New Ireland Basin in northern Papua New Guinea, *in* Watson, S.T., ed., Transactions of the Third Circum-Pacific Energy and Mineral Resources Conference, 22-28 August 1982, Honolulu, Hawaii, p. 623-630.

Exon, N.F., Stewart, W.D., Sandy, M.J., and Tiffin, D.L., 1986, Geology and offshore petroleum prospects of the eastern New Ireland Basin, northeastern Papua New Guinea: BMR Journal of Australian Geology and Geophysics, v. 10, p. 39-51.

Falvey, D.A., and Pritchard, T., 1984, Preliminary paleomagnetic results from northern Papua New Guinea: Evidence for large microplate rotations, *in* Watson, S.T., ed., Transactions of the Third Circum-Pacific Energy and Mineral Resources Conference, 22-28 August 1982, Honolulu, Hawaii, p. 623-630.

Furumoto, A.S., Webb, J.P., Odegard, M.E., and Hussong, D.M., 1976, Seismic studies on the Ontong Java Plateau, 1970: Tectonophysics, v. 34, p. 71-90.

Hamilton, W., 1979, Tectonics of the Indonesian region: U.S. Geological Survey Professional Paper 1078, 338 p.

Hegarty, K.A., Weissel, J.K., and Hayes, D.E., 1983, Convergence at the Caroline-Pacific plate boundary: collision and subduction, *in* Hayes, D.E., The tectonic and geologic evolution of southeast Asian seas and islands: Part 2: Geophysical Monograph 27, American Geophysical Union, Washington D.C., p. 326-348.

Hughes, G.W., and Turner, C.C., 1977, Upraised Pacific Ocean floor, southern Malaita, Solomon Islands: Geological Society of America Bulletin, v. 88, p. 412-424.

Hussong, D.M., Wipperman, L.K., and Kroenke, L.W., 1979, The crustal structure of the Ontong Java and Manihiki oceanic plateaus: Journal of Geophysical Research, v. 84, p. 6003-6010.

Johnson, R.W., 1979, Geotectonics and volcanism in Papua New Guinea: a review of the late Cainozoic: BMR Journal of Australian Geology and Geophysics, v. 4, p. 181-207.

Johnson, T., and Molnar, P., 1972, Focal mechanisms and plate tectonics of the southwest Pacific: Journal of Geophysical Research, v. 77, p. 5000-5032.

Johnson, G.L., Egloff, J., and Hemler, L.G., 1974, Geophysical observations between Hawaii and Australia: Pacific Sciences, v. 28, n. 2, p. 191-205.

Kroenke, L.W., 1972, Geology of the Ontong Java Plateau: Hawaii Institute of Geophysics, Report n. HIG-72-5, University of Hawaii, 119 p.

Kroenke, L.W., 1984, Cenozoic tectonic development of the southwest Pacific: United Nations Economic and Social Commission for Asia and the Pacific Committee for Co-ordination of Joint Prospecting for Mineral Resources in South Pacific Offshore Areas (U.N. ESCAP, CCOP/SOPAC) Technical Bulletin no. 3, 122 p.

Kroenke, L.W., Resig, J.M., and Cooper, P.A., 1986, Tectonic implications of seismicity northeast of the Solomon Islands, *in* Vedder, J.G., Pound, K.S., and Boundy, S.Q., eds., Geology and offshore resources of Pacific island arcs—central and western Solomon Islands: Houston, Texas, Circum-Pacific Council for Energy and Mineral Resources Earth Science Series, v. 4, p. 109-116.

Mann, D.M., 1986, Multichannel seismic operations for CCOP/SOPAC Cruise, Leg 3, Solomon Islands, *in* Vedder, J.G., Pound, K.S., and Boundy, S.Q., eds., Geology and offshore resources of Pacific island arcs—central and western Solomon Islands: Houston, Texas, Circum-Pacific Council for Energy and Mineral Resources Earth Science Series, v. 4, p. 35-40.

Marlow, M.S., Exon, N.F., Ryan, H.F., and Dadisman, S.V., 1988, Offshore structure and stratigraphy of New Ireland basin in northern Papua New Guinea, *in* Marlow, M.S., Dadisman, S.V., and Exon, N.F., Geology and offshore resources of Pacific Island arcs, northern Papua New Guinea: Houston, Texas, Circum-Pacific Council for Energy and Mineral Resources Earth Science Series, v. 9, p. 137-156.

Minster, J.B. and Jordan, T.H., 1978, Present-day plate motions: Journal of Geophysical Research, v. 83, p. 5331-5354.

Pound, K.S., 1986, Correlation of rock units in the central and western Solomon Islands, *in* Vedder, J.G., Pound, K.S., and Boundy, S.Q., eds., Geology and offshore resources of Pacific island arcs—central and western Solomon Islands: Houston, Texas, Circum-Pacific Council for Energy and Mineral Resources Earth Science Series, v. 4, p. 89-98.

Ramsay, W.R.H., 1978, Field, mineralogical and structural observations on some basement rocks, southeast Choiseul, Solomon Islands: Bulletin of the Australian Society of Exploration Geophysicists, v. 9, p. 107-110.

Ramsay, W.R.H., 1982, Crustal strain phenomena in the Solomon Islands, constraints from field evidence, and relationship to the India-Pacific plates boundary: Tectonophysics, v. 87, p. 109-126.

Rickwood, F.K., 1957, Geology of the island of Malaita, *in* Geological reconnaissance of part of the central islands of the British Solomon Islands Protectorate: Colonial Geology and Mineral Resources, v. 10, n. 2, p. 112-145.

Ryan, H.F., and Marlow, M.S., 1988, Multichannel seismic-reflection data collected at the intersection of the Mussau and Manus Trenches, Papua New Guinea, *in* Marlow, M.S., Dadisman, S.V., and Exon, N.F., Geology and offshore resources of Pacific Island arcs, northern Papua New Guinea: Houston, Texas, Circum-Pacific Council for Energy and Mineral Resources Earth Science Series, v. 9, p. 203-210.

Stanton, R.L., 1961, Explanatory notes to accompany a first geological map of Santa Isabel, British Solomon Islands Protectorate: Overseas Geology and Mineral Resources, v. 8, n. 2, p. 127-149.

Taylor, B., 1979, Bismarck Sea: Evolution of a back-arc basin: Geology, v. 7, n. 4, p. 171-174.

Taylor, B., 1984, A geophysical survey of the Woodlark-Solomons region, *in* Exon, N.F., and Taylor, B., eds, Seafloor spreading, ridge subduction, volcanism, and sedimentation in the offshore Woodlark-Solomons region and Tripartite cruise report for *Kana Koki* cruise 82-03-16, leg 4: U.N. ESCAP, CCOP/SOPAC Technical Report n. 34, p. 13-42.

Taylor, B., 1987, A geophysical survey of the Woodlark-Solomons region, *in* Taylor, B. and Exon, N.F., eds., Marine geology, geophysics, and geochemistry of the Woodlark Basin-Solomon Islands: Houston, Texas, Circum-Pacific Council for Energy and Mineral Resources Earth Science Series, v. 7, p. 25-48.

Taylor, B., and Exon, N.F., 1984, An investigation of ridge subduction and related phenomena in the Woodlark-Solomons region-introduction and background, *in* Exon, N.F., and Taylor, B., eds, Seafloor spreading, ridge subduction, volcanism, and sedimentation in the offshore Woodlark-Solomons region and Tripartite cruise report for *Kana Keoki* cruise 82-03-16, leg 4: U.N. ESCAP, CCOP/SOPAC Technical Report n. 34, p. 13-42.

Taylor, B., and Exon, N.F., 1987, An investigation of ridge subduction in the Woodlark-Solomons region: introduction and overview, *in* Taylor, B. and Exon, N.F., eds., Marine geology, geophysics, and geochemistry of the Woodlark Basin-Solomon Islands: Houston, Texas, Circum-Pacific Council for Energy and Mineral Resources Earth Science Series, v. 7, p. 1-24.

Vedder, J.G., Pound, K.S., and Boundy, S.Q., eds., 1986, Geology and offshore resources of Pacific island arcs—central and western Solomon Islands: Houston, Texas, Circum-Pacific Council for Energy and Mineral Resources Earth Science Series, v. 4, 306 p.

Wallace, D.A., Johnson, R.W., Chappell, B.W., Arculus, R.J., Perfit, M.R., and Crick, I.H., 1983, Cainozoic volcanism of the Tabar, Lihir, Tanga, and Feni Islands, Papua New Guinea: geology, whole-rock analyses, and rock-forming mineral compositions: Bureau of Mineral Resources, Australia, Report 243, BMR Microform MF197.

Weissel, J.K., Taylor, B., and Karner, G.D., 1982, The opening of the Woodlark Basin, subduction of the Woodlark spreading system, and the evolution of northern Melanesia since mid-Pliocene time: Tectonophysics, v. 87, p. 253-277.

Winterer, E.L., Reidel, W.R., et al, 1971, Initial reports of the Deep Sea Drilling Project, v. 7: Washington, D.C., U.S. Government Printing Office, 1756 p.

Vedder, J.G., and Bruns, T.R., editors, 1989, Geology and offshore resources of Pacific island arcs—Solomon Islands and Bougainville, Papua New Guinea Regions: Houston, Texas, Circum-Pacific Council for Energy and Mineral Resources Earth Science Series, v. 12.

GEOLOGY OF THE SHORTLAND BASIN REGION, CENTRAL SOLOMONS TROUGH, SOLOMON ISLANDS—REVIEW AND NEW FINDINGS

T.R. Bruns, J.G. Vedder, and A.K. Cooper
U.S. Geological Survey, Menlo Park, California 94025

ABSTRACT

Multichannel seismic-reflection data acquired during 1982 in the Solomon Islands over the Shortland basin region were used to establish a preliminary framework for the seismic stratigraphy, structure, and tectonic history of the basin. Four seismic-stratigraphic units, A through D from oldest to youngest, were identified; these units form distinct sedimentary wedges that can be correlated with geologic units and tectonic events on the surrounding islands. Basin development was largely driven by vertical tectonics on the northeastern and southwestern flanks of the basin. These vertical movements were in turn controlled by tectonic events associated with the subduction and arc-reversal history of the Solomon Islands region.

Additional seismic-reflection data acquired during 1984 resulted in new findings as follows: (1) A basement half graben with as much as 6 km of relief is present in the center of the basin. Unit B strata, of inferred late Oligocene and Miocene age, thicken into and fill the half graben and onlap against the bounding faults. The half graben may have developed as a pull-apart basin or by block rotation in a strike-slip fault or oblique subduction system. Development of the half graben is coincident with and may have been caused by entry of the Ontong Java Plateau into the early Tertiary subduction zone. (2) The maximum sediment thickness in the basin is over 7 km, which is greater than previously reported. Subsidiary depocenters are present beneath Vella Ridge and west of Mono Island, where sediment thicknesses are as much as 3 and 2 km, respectively. (3) We find no obvious connection between Shortland basin and Bougainville basin on the southwestern side of Bougainville because a basement high apparently separates the two basins. The thick sedimentary section in the axial part of Shortland basin could extend northwestward beneath the volcanic rocks of southern Bougainville; however, the basin strata thin toward the island, and the basin may end near or beneath the island. (4) Refraction data confirm that a thick basin fill of interbedded volcaniclastic and volcanic rocks is present beneath the New Georgia wedge region. This observation corrects an erroneous interpretation of thin fill that was based on shallow acoustic basement on seismic-reflection data.

INTRODUCTION

Geographic Setting

The Solomon Islands consist of two uplifted island-arc chains that are separated by New Georgia Sound (Figure 1). New Georgia Sound is in turn underlain by an intra-arc basin, the Central Solomons Trough (Katz, 1980). Herein, we discuss the Shortland basin, which is the westernmost subsidiary basin of the Central Solomons Trough, and the regions surrounding Shortland basin (Figure 2).

Shortland basin is bounded on the west and northwest by Bougainville and the Shortland Islands, on the northeast by Choiseul, on the southeast by Vella Ridge, and on the south by the New Georgia island group (Figure 2). Southeastward, the New Georgia wedge separates Shortland basin from the other subsidiary basins of the Central Solomons Trough. Much of Shortland basin lies beneath 1,300 m of water, and maximum water depths are as great as 1,500 m. Sill depths around the basin are less than 600 m north of Kolombangara and New Georgia, 200 m beneath Bougainville Strait, and 300 m over Vella Ridge. The deepest sill lies in the 1,300 m-deep gap that separates Mono Island and Vella Ridge. South of Mono Island, Vella Ridge, and Vella Lavella, which form the south flank of the basin, the slope drops steeply into the New Britain Trench.

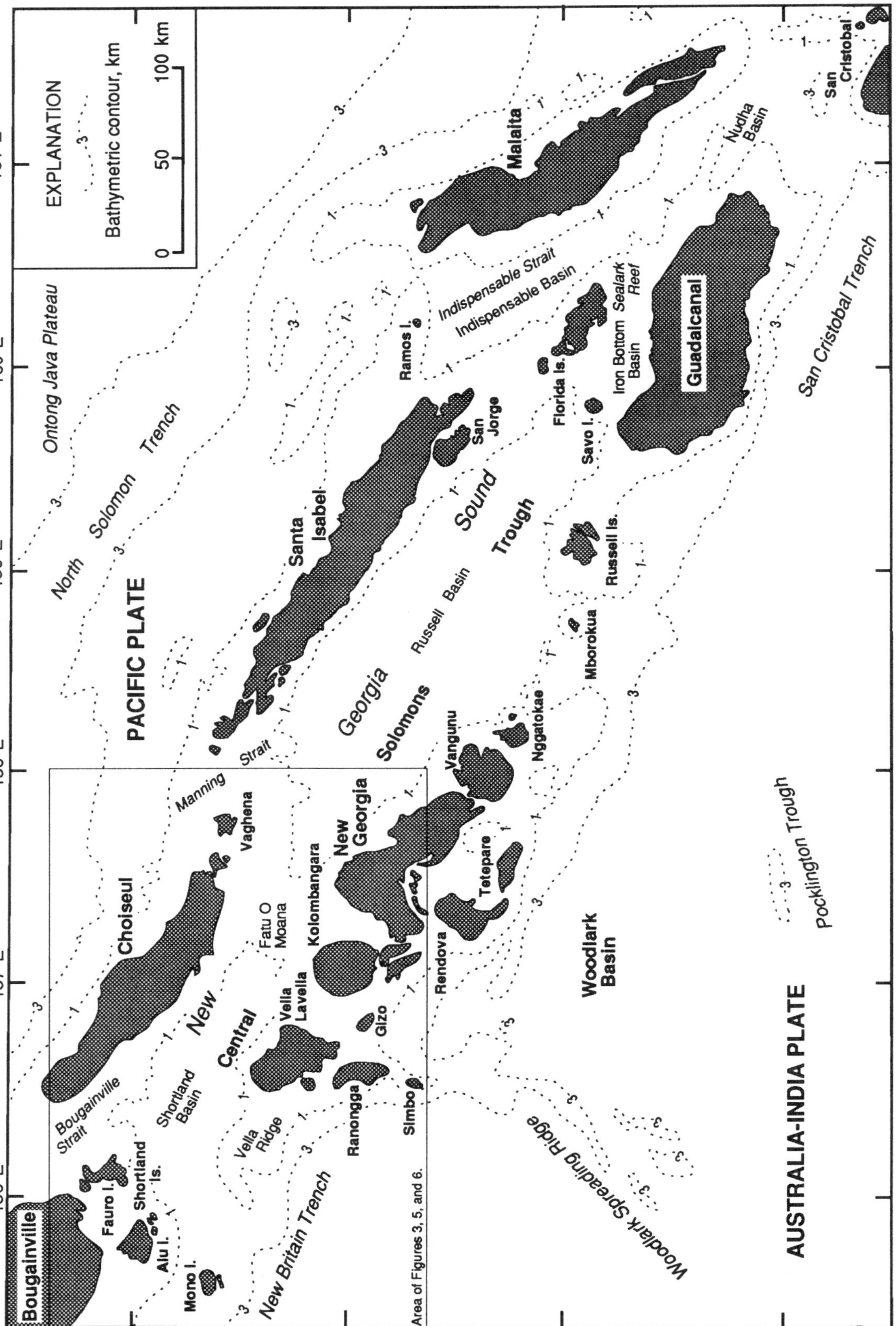

Figure 1. Place names, tectonic features, and bathymetry of the central and western Solomon Islands area.

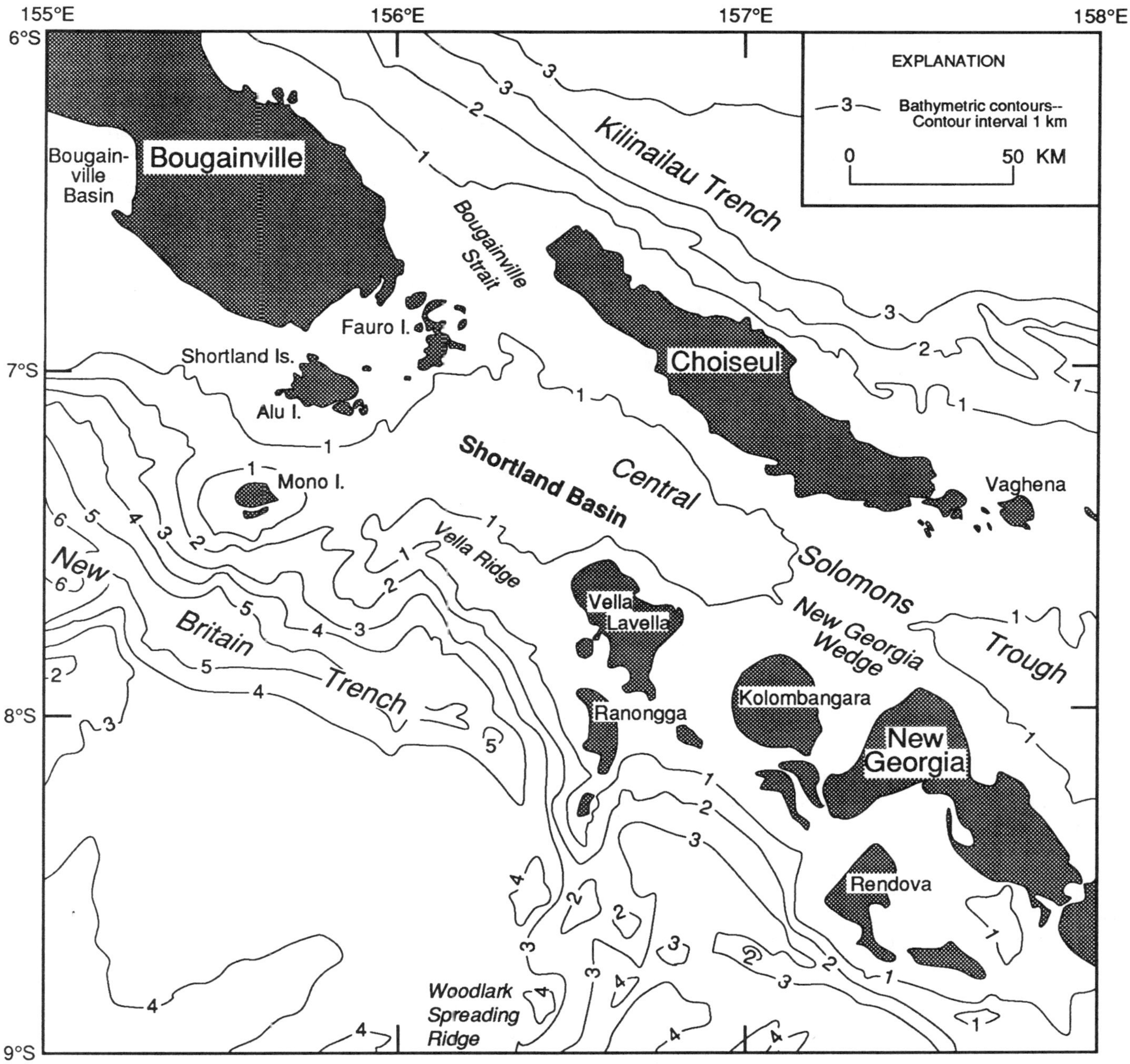

Figure 2. Map of the Shortland basin region. Bathymetry from Chase, Seekins, and Lund (1986) and Chase, Seekins, and Young (this volume).

Previous Work

Marine geological and geophysical surveys in the Solomon Islands region are reviewed in Vedder and Coulson (1986) and Vedder and Bruns (1986; this volume). Major work within the Shortland basin and New Georgia wedge regions included the delineation of the Central Solomons Trough by de Broin, Aubertin, and Ravenne (1977) (part of their Solomon Basin), an interpretation of a seismic profile in Manning Strait by Landmesser (1977), a study of a regional grid of seismic-reflection data across the Central Solomons Trough by Katz (1980), and a study of the petroleum potential of the Central Solomons Trough and adjoining areas by Maung and Coulson (1983) using interpretations of marine geophysical records contained in the files of the Solomon Islands Geological Survey.

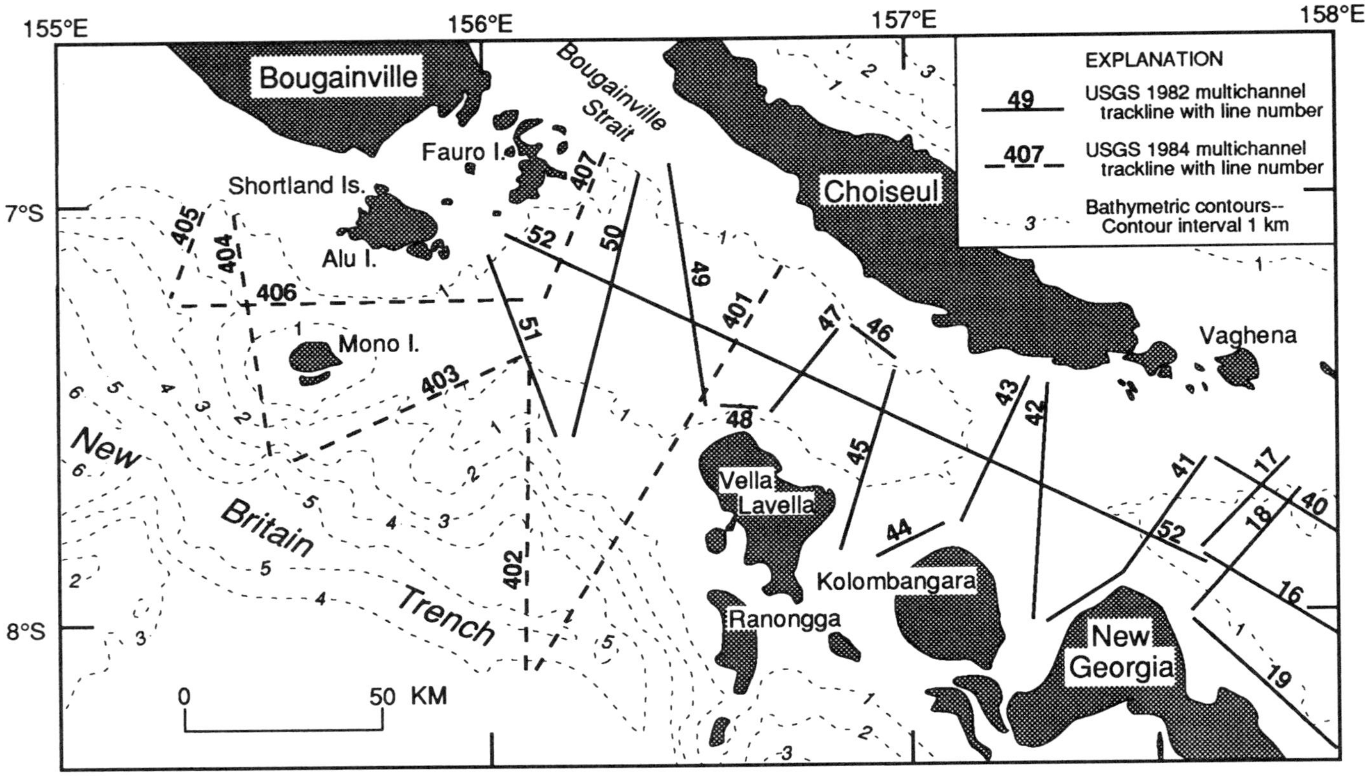

Figure 3. Multichannel seismic-reflection lines acquired from the U.S. Geological Survey research vessel R/V *S.P. Lee* in the Shortland basin region in 1982 and 1984.

In 1982, two cruises acquired data in the Solomon Islands region as part of an Australia-New Zealand-United States Tripartite Agreement coordinated by the United Nations Committee for Co-ordination of Joint Prospecting for Mineral Resources in South Pacific Offshore Areas (CCOP/SOPAC). Scientists aboard the R/V *Kana Keoki* investigated the New Georgia wedge region and the Woodlark Basin south of the Solomon Islands (Taylor and Exon, 1987).

In May, 1982, scientists on board the U.S. Geological Survey research vessel R/V *S.P. Lee* recorded 24-fold seismic-reflection data plus ancillary seismic-refraction, gravity, and magnetic data along about 2,000 km of tracklines over the sedimentary basins in the Solomon Islands region; bottom samples were also acquired (Figures 1 and 3; Vedder, Pound, and Boundy, 1986). These data were used to delineate the seismic stratigraphy, sediment thickness, structure, and depositional history of the basins (Bruns et al, 1986; Bruns, Cooper, and Vedder, 1986; Cooper, Bruns, and Wood, 1986; Cooper, Marlow, and Bruns, 1986).

The 1982 work in the vicinity of Shortland basin showed that sedimentary wedges within the basin could be correlated with the geology and uplift episodes of the surrounding islands and could be used to interpret the development history of the basin. Geological problems that remained after the 1982 survey included delineating the extent of Shortland basin to the west and south, and resolving whether or not this basin connected with Bougainville basin, which underlies the shelf area on the west side of Bougainville (Figure 2; Shaw, 1985; Stewart, Francis, and Pederson, 1987; Hilyard and Rogerson, this volume). Another unsolved problem was the sediment thickness and structure southeast of Shortland basin, where the New Georgia wedge flanks the islands of Kolombangara and New Georgia. A single seismic-refraction line from the 1982 survey suggested the presence of a thick (greater than 5-km) sedimentary sequence (Cooper, Bruns, and Wood, 1986), whereas seismic-reflection data showed a shallow acoustic basement overlain by a thin (2-km or less) sedimentary sequence (Bruns et al, 1986).

In June 1984, additional 24-fold seismic-reflection data and refraction data were acquired by the R/V *S.P. Lee* over Shortland basin and the New Georgia wedge (Figures 2 and 3). The new multichannel data over Shortland basin helped to further delineate the basin extent, especially to the south toward the New Britain Trench and northwest toward Bougainville basin (Figure 2). Refraction data were acquired over the New Georgia wedge to define the sediment thickness and answer unresolved questions about the extent beneath the Central Solomons Trough of volcanic flows and volcaniclastic deposits derived from volcanoes of the New Georgia island group (see Cooper, Cochrane, and Bruns, this volume).

The purpose of this paper is to present new interpretations and reinterpretations based upon data from the combined 1982 and 1984 data sets. These findings include: (1) better definition of a basement half graben in the center of Shortland basin; (2) new maps of structure and sediment thickness in the Shortland basin region; (3) a clearer delineation of the extent of the basin to the south and west; and (4) a reevaluation of the sediment thickness in the region of the New Georgia wedge.

REGIONAL TECTONIC AND GEOLOGIC SETTING

Tectonic Setting

The regional geologic and tectonic setting of the Solomon Islands region is described at length by Kroenke (1984), Coulson and Vedder (1986), Vedder and Coulson (1986), Bruns, Vedder, and Culotta (this volume), Vedder, Bruns, and Cooper (this volume), and Wells (this volume). The following is a brief summary from these papers.

The Solomon Islands are composed of a diverse assemblage of rocks of Mesozoic and Cenozoic age. Basement rocks include oceanic crust of Cretaceous and early Tertiary age and metamorphic equivalents. Overlying this basement are rocks and features that reflect volcanic-arc growth during at least two time periods--late Eocene to early Miocene, and late Miocene and younger. The late Eocene to early Miocene volcanic arc (the North Solomon Island Arc of Kroenke, 1984) lies beneath the northern island chain of the Solomon Islands, including Bougainville, Choiseul, part of Santa Isabel, and Guadalcanal. The late Miocene and younger volcanic arc crosses the early Tertiary arc on Bougainville and underlies the southwestern island chain, including the Shortland Islands, the New Georgia island group, the Russell Islands, Savo Island, and western Guadalcanal. Oligocene through Quaternary volcaniclastic and carbonate rocks cover large parts of the volcanic arc, and lie in the interarc basins such as the Central Solomons Trough.

The regional structure and stratigraphy indicate that four main tectonic episodes affected the Solomon Islands. The basement rocks reflect uplift of Cretaceous and early Tertiary oceanic or oceanic-plateau rocks. During and after uplift, these rocks were markedly deformed, regionally metamorphosed, and eroded, possibly as a result of late Eocene subduction along the northeastern margin of the Solomon Islands.

During late Eocene through early Miocene time, southwestward-directed subduction of the Pacific plate beneath the Solomon Islands created the early Tertiary North Solomon Island Arc (Kroenke, 1984). Growth of the arc was accompanied by an influx of volcaniclastic detritus into the adjacent back-arc basin. These back-arc strata now form the lower units in the Shortland basin and other subbasins of the Central Solomons Trough.

In the middle Miocene, the arc was characterized by volcanic and tectonic quiescence, possibly caused when the thickened Pacific plate crust of the Ontong Java Plateau, which lies immediately north of the Solomon Islands, entered and blocked the subduction zone about 22 to 20 Ma. During the middle Miocene, uplift and faulting occurred on all the islands, and was accompanied by extensive deposition of carbonate rocks.

In the late Miocene, about 10 Ma, subduction shifted to the presently active northward-dipping New Britain and San Cristobal subduction zones on the south side of the Solomon Islands, thus completing a subduction polarity reversal. This subduction shift was accompanied by the development of the late Cenozoic arc that underlies Bougainville and the southern island chain of the Solomon Islands and that closed off the early Tertiary back-arc basin to form the intra-arc basin of the Central Solomons Trough. Deposition into the Central Solomons Trough formed the upper sedimentary wedges of the Shortland basin.

The most recent events to affect the region were the initiation of sea-floor spreading on the Woodlark spreading ridge at about 5 Ma and the subsequent subduction of the newly formed oceanic crust beneath the Solomon Islands. Ridge subduction was accompanied by pronounced tectonic uplift and subsidence along Vella Ridge and in the New Georgia island group, and by near-trench volcanism and markedly low levels of seismicity in the New Georgia island group.

Geology of Islands Surrounding Shortland Basin

Bougainville

Rocks on Bougainville include a succession of volcaniclastic, carbonate, and volcanic rocks initially described by Blake and Miezitis (1967) and recently remapped and redescribed by Hilyard and Rogerson (this volume). The oldest rocks exposed are Eocene(?) and Oligocene(?) basalt, basaltic andesite, and interbedded volcaniclastic rocks (Atamo Volcanics) derived from the early Tertiary arc. These basement rocks are unconformably overlain by shallow-water, late Oligocene to latest early Miocene limestone (Keriaka Limestone), by middle Miocene to Pliocene volcaniclastic debris flows (Toniva Formation and Arawa Conglomerate), and locally by a sequence of middle and late Miocene andesite flows (Aropa Andesite). The late Oligocene to Pliocene rocks are distributed along the southeastern end of Bougainville and probably extend well beneath the adjacent slope, Bougainville Strait, and the Shortland Islands. Igneous rocks of latest Miocene and younger age (Isinai Monzonite, Kupei Complex, and unnamed intrusives) intrude the older rocks. The latest Pliocene and younger volcanic rocks of the Bougainville Group cover most of the northwestern and southeastern parts of the island.

Shortland Islands

The oldest rocks in the Shortland Islands are

volcanic rocks of Oligocene(?) and early Miocene age on Fauro Island; these rocks are similar to those found on nearby Bougainville. Basement rocks on the other islands of the Shortland Islands are mainly formed of intrusive and extrusive rocks of Pliocene(?) and younger age (Coulson and Vedder, 1986; Ridgway and Coulson, 1987).

Pliocene and Pleistocene sedimentary rocks and shallow-marine carbonates occur on Alu and Mono islands and record a transition from hemipelagic deposition of fine volcanic detritus to shallow-water reef construction. On Alu Island, about 300 m of Pliocene siltstone, claystone, and fine-grained sandstone of the Kulitanai Siltstones indicate open-marine shelf environments. These strata are overlain by late Pliocene and Pleistocene calcareous and carbonate rocks and locally derived alluvial deposits. On Mono Island, about 250 m of Pliocene siltstone beds of the Mono Siltstones constitute an outer shelf deposit that is similar in age and lithology to the Kulitanai Siltstones. Overlying and interbedded rocks are reef limestone. As much as 1,400 m of Pliocene(?) to early Pleistocene tuffaceous sandstone beds of the Koria Sandstones are present on the Fauro islands, and probably represent rapid deposition of locally derived volcaniclastic sediment on a shallow shelf.

Choiseul

Choiseul is underlain by an oceanic-basement complex composed of the Choiseul Schists and Voza Lavas. The Choiseul Schists consist largely of metamorphosed Cretaceous(?) and early Tertiary basaltic rocks. The age of the Voza Lavas is uncertain, but interpretations of geological and geochemical data indicate that the Voza Lavas are the protolith for the Choiseul Schists, and therefore are also of Cretaceous(?) and early Tertiary age (Coleman, 1965; Coulson and Vedder, 1986; Pound, 1986; Ridgway and Coulson, 1987).

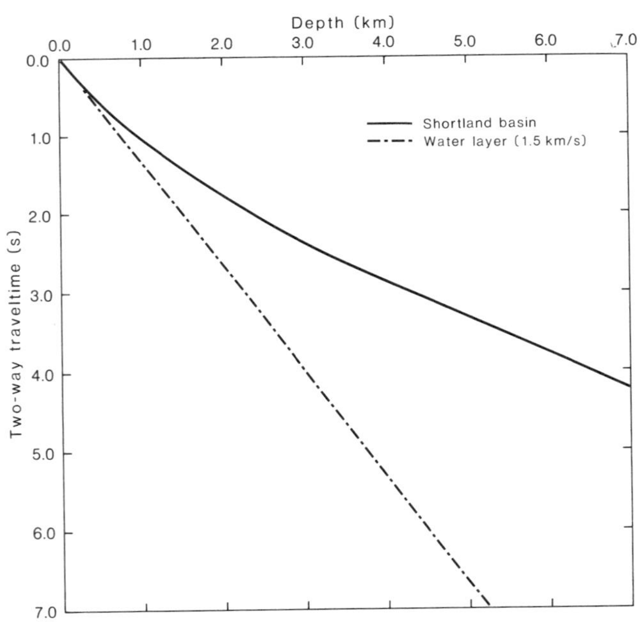

Figure 4. Time-depth conversion curves for the Shortland basin region for converting two-way traveltime on the seismic-reflection records to depth. Water-layer thickness must be added to cumulative sub-seafloor depth to obtain total depth.

Sedimentary rocks directly above the basement schist and basalt are assigned to the late Oligocene and Miocene Mole Formation. The formation is as much as 450 m thick and has a late Oligocene basal conglomerate, the Koloe Breccia Member, which is overlain by Miocene dominantly turbidite-derived siltstone and sandstone and subordinate shallow-water near-shore deposits. The Pliocene Pemba Formation overlies the Mole Formation conformably or with slight discordance. The Pemba Formation has a maximum thickness of 600 m and consists of a lower calcarenite member and an upper calcisiltite member. Both the Mole and Pemba Formations dip and thicken to the west and southwest toward Bougainville Strait and the Central Solomons Trough. Overlying all these units are the Quaternary reef deposits of the Nukiki Limestone Formation, which are now elevated as much as 440 m above sea level at the northwestern end of Choiseul and up to 600 m above sea level at the southeastern end of the island (Coulson and Vedder, 1986; Pound, 1986; Ridgway and Coulson, 1987)

New Georgia Island Group

The New Georgia island group is formed mainly of late Miocene to Holocene calc-alkaline volcanic rocks and minor coeval epiclastic deposits. The volcanic rocks consist of large volumes of olivine basalt, basaltic andesite, and andesite flows. Typically, the volcanic rocks display lateral facies changes from massive flows to volcaniclastic strata. Sedimentary rocks are limited in distribution and are late Pliocene and younger in age. These rocks consist largely of reef platform deposits, although sequences of terrigenous sedimentary rocks locally occur on some of the islands. Uplifted calcareous siltstone deposits on Tetepare are nearly 900 m above the depth at which they were deposited, indicating the magnitude of tectonic uplift throughout the island group (Coulson and Vedder, 1986).

GEOPHYSICAL DATA

Multichannel seismic-reflection data were acquired along seven lines in 1984 in the Shortland basin region (lines 401-407, Figure 3), complementing lines acquired in 1982. These data were recorded using a 1,311 in^3 (21.5 l) airgun array, a Globe Universal Sciences (GUS)[1] seismic-

[1] Any use of trade names is for purposes of identification only and does not imply endorsement by the U.S. Geological Survey, the New Zealand Geological Survey, or the Bureau of Mineral Resources, Geology, and Geophysics, Australia.

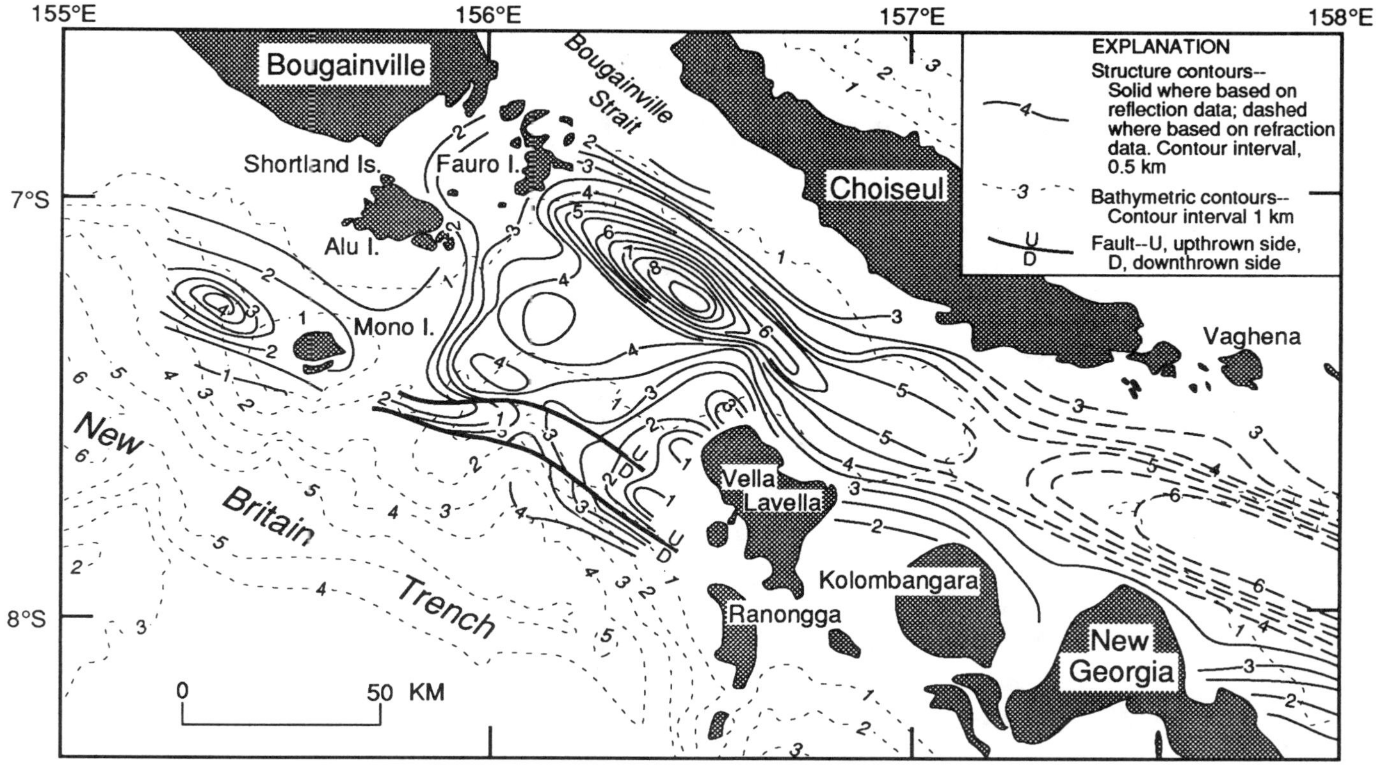

Figure 5. Structure contours at the top of seismic-stratigraphic unit A (acoustic basement). Contours are at the base of the sedimentary fill of the Shortland basin. Solid contours are based on interpreted multichannel seismic-reflection data; dashed contours on seismic-refraction data.

recording system, and a 2,400-m streamer. A shooting geometry with a 50-m shotpoint interval and a 100-m group interval provided 24-fold data. The ship's position was determined by a satellite navigation system that used doppler sonar, speed log, and gyrocompass information to calculate dead-reckoned positions between satellite fixes.

The seismic-reflection data were processed by the U.S. Geological Survey in Menlo Park, California, on a DISCO (trademark of Digicon, Inc.) processing system. The main processing steps included demultiplexing, velocity analysis using coherence estimation and constant-velocity stacks, normal moveout correction, stacking, and filtering. Sections 49 and 50 were specially processed to attenuate the water-bottom multiple. Data for lines 50 and 407 are migrated; the migration velocities were selected from post-stack constant-velocity migrations. Further information about the standard U.S. Geological Survey acquisition and processing methods is given by Mann (1986).

TIME-TO-DEPTH CONVERSION OF SEISMIC DATA

The conversion of seismic-reflection traveltime to depth (Figure 4) is based on seismic-refraction data and on interval velocities measured from seismic-reflection data. Cooper, Bruns, and Wood (1986) used seismic-refraction data to derive a velocity function for Shortland basin of $v=1.42+1.95t$, where v is velocity (km/s) and t is one-way traveltime (s). For reflection times of less than 3 s two-way time, seismic-reflection interval velocities compared favorably with the refraction velocities. Therefore, the linear refraction-velocity function was used to make a regional time-to-depth conversion. Converting the refraction velocity function to a time-depth curve and two-way traveltime yields $D=0.71t+0.24t^2$ for Shortland basin, where D is sub-sea-floor depth (km) and t is two-way sub-seafloor traveltime in seconds for t less than 3 s (Bruns et al, 1986).

For t greater than 3 s, refraction and reflection velocities for sedimentary rocks are poorly determined because most refraction arrivals are from basement rocks (velocities greater than 4.5 to 5 km/s; Cooper, Bruns, and Wood, 1986). Velocities determined from reflection data are poorly determined in the few areas where reflection arrivals from layered rocks below 3 s are present. Therefore, for two-way traveltimes greater than 3 s, a constant interval velocity of 4.5 km/s was used, in agreement with the velocity that Cooper, Bruns, and Wood (1986) used as the dividing line between sedimentary strata and probable igneous and metamorphic basement rocks. This velocity function is different from that Bruns et al (1986), who did

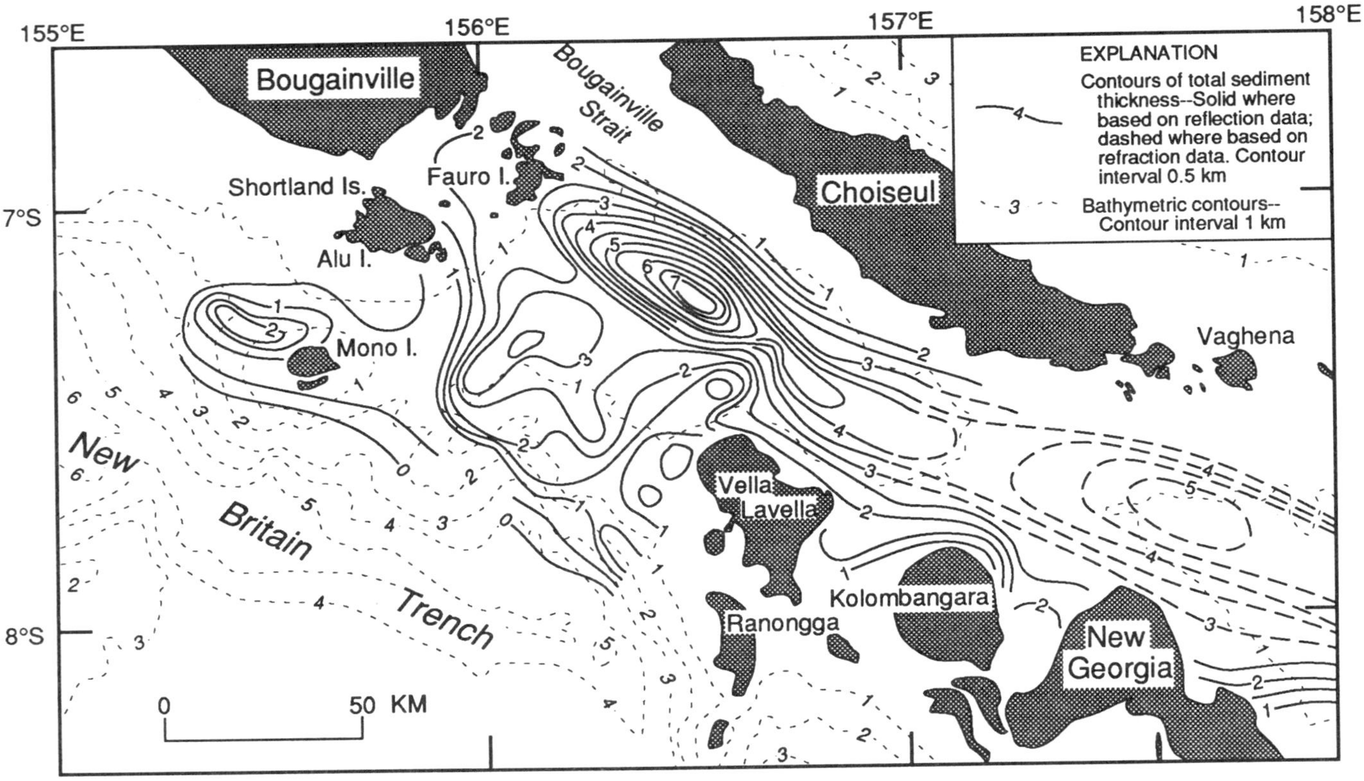

Figure 6. Isopach map of total sediment thickness above seismic-stratigraphic unit A (acoustic basement). Solid contours are based on interpreted multichannel seismic-reflection data; dashed contours on seismic-refraction data.

not use a constant interval velocity for times greater than 3 s. These velocity relationships were used to construct a structure contour map (Figure 5) and a sediment isopach map (Figure 6) from the seismic-reflection data.

SEISMIC STRATIGRAPHY OF SHORTLAND BASIN

Bruns et al (1986) divided the rocks underlying Shortland basin into four seismic-stratigraphic units, designated as A through D from oldest to youngest. The bounding surface of each of these units either is an unconformity or is characterized by an abrupt change in seismic velocity, or both. Two of these units, C and D, were further divided into upper and lower subunits, designated by subscripts "u" and "l" respectively; for example C_l and C_u for subunits C-lower and C-upper. The designation for a seismic-reflection horizon between two units is hyphenated; for example, horizon C-D separates seismic units C and D. The seismic units are shown on the seismic lines of Figures 7-11; a brief description of the units from Bruns et al (1986) follows.

Seismic unit A is acoustic basement and is characterized by discontinuous to choatic seismic reflections along the flanks of the basin and by occasional faint subhorizontal to horizontal reflections in the central part of the basin. On the south side of the basin, the top of unit A lies on a subsurface platform at a depth of about 2 to 3 km (Figures 8 and 9). Along the northern basin margin, unit A dips southward from Choiseul. The maximum depth to the top of unit A in the deepest part of the basin is unclear, because the horizon A-B reflector can only be discontinuously followed there. We interpret the maximum depth to be over 8 km and the maximum sediment thickness to be over 7 km on line 49 (Figures 5, 6, and 8). Seismic-refraction velocities within the unit range from about 4.5 to 5.9 km/s.

Unit B is characterized by laterally continuous parallel reflectors that dip and thicken southward from Choiseul, and we believe that the unit crops out on Choiseul. In the deepest part of the basin, the unit onlaps the south flank of a basement half graben (Figures 7-10). South of the half graben, the unit thins markedly and is truncated in the subsurface by overlying units C and D. Unit B is the thickest of the three basin-fill sequences, reaching a maximum thickness of about 5 km on line 49 (Figure 8). Velocities within the unit from seismic-refraction and seismic-reflection data range from about 3.4 to 4.4 km/s, indicating well-indurated sedimentary rocks.

Both units C and D are characterized by prominent, laterally continuous reflections, although the reflection continuity varies considerably near the basin flanks. Velocities within unit C range from 2.3 to 3.6 km/s, with an abrupt decrease across horizon C-D to velocities of 1.6

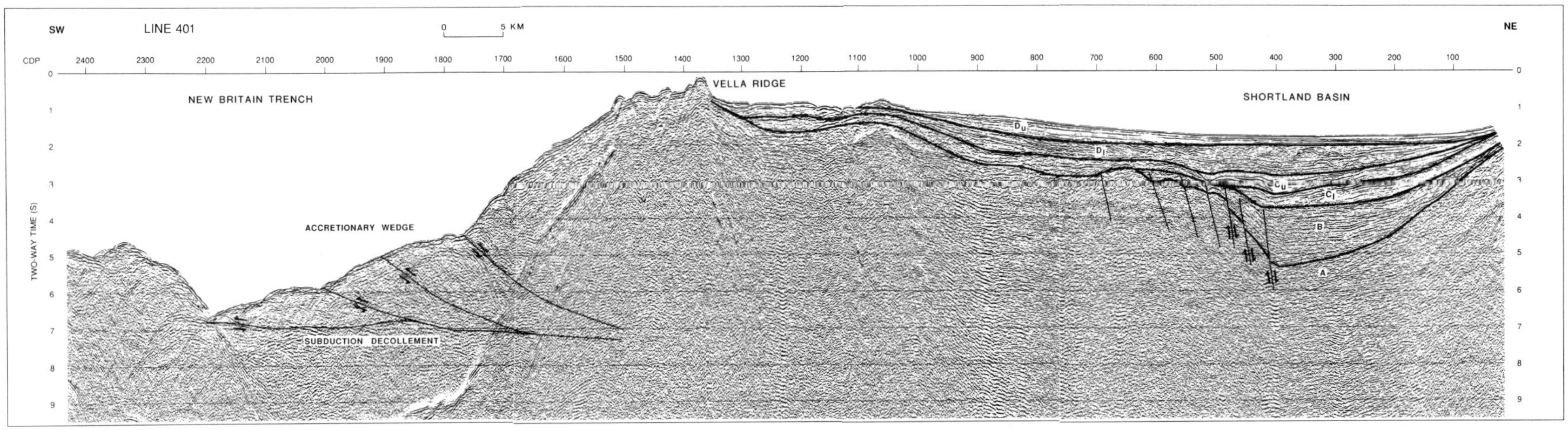

Figure 7. Multichannel seismic-reflection profile 401. This and subsequent profiles show interpreted faults and seismic-stratigraphic units A through D as described in text. CDP, common-depth-point. Vertical exaggeration at the sea floor is about 4:1. Location of seismic profile shown in Figure 3.

to 2.6 km/s in unit D. The velocity decrease across the horizon is generally about 1 km/s.

In the central part of Shortland basin, the unit C subunits form two distinct sediment wedges. Unit C_l dips and thins southward from Choiseul, whereas unit C_u is thickest beneath Vella Ridge and dips and tapers out northward beneath the axis of the basin (Figures 7-10). The contact between the two subunits is a well-defined unconformity in the western part of Shortland basin. However, in the eastern part of the basin and near Bougainville, the wedge shape of the two subunits is absent, both subunits thin, and bedding is concordant across the contact between the subunits.

Unit D is a basin-filling unit that is flat lying and generally undeformed, except near the basin flanks and over Vella Ridge. Seismic reflections from the unit are mainly laterally continuous and of variable amplitude. However, moderately chaotic reflections and a hummocky or mounded appearance characterize strata in the lower part of the unit and near the basin edges, particularly near the Shortland Islands. This reflection character could indicate rapid deposition by debris flows derived from the islands, or the presence of carbonate reefs or reefal debris. Unit D unconformably overlies all older units and thins toward the basin flanks. The unit consists of two subunits, differing mainly in that subunit D_l has a hummocky reflector character and is uplifted on Vella Ridge whereas subunit D_u is everywhere laterally continuous and flat lying.

Geologic Correlation of Seismic Units

Bruns et al (1986) correlated the seismic-stratigraphic units with rock units on Choiseul and other islands surrounding the Shortland basin. Unit A is correlated with the basement rocks of the surrounding islands, and in particular with the Choiseul Schists and the Voza Lavas of Choiseul. The unit could also include the poorly bedded late Oligocene conglomeratic and brecciated rocks at the base of the overlying Mole Formation. Rocks on other islands that correlate with unit A include the Atamo Volcanics and Keriaka Limestone of Bougainville and the basement rocks of the Shortland Islands. The unit may also include volcanic flows and volcaniclastic rocks near the late Cenozoic volcanic centers.

Unit B is correlated with the bedded siltstones and sandstones of the late Oligocene and Miocene Mole Formation, which lies unconformably on the Voza Lavas and Choiseul Schist. Unit C is correlated with the calcarenites and calcisiltites of the Pliocene Pemba Formation of Choiseul and with the Pliocene siltstones of the Shortland Islands, which include the Koria, Alu, and Mono Siltstones. The uppermost part of unit C may have been sampled on Vella Ridge, where dredge samples from approximately the contact between units C and D are as old as earliest Quaternary (Colwell and Vedder, 1986). Finally, unit D is considered to be largely Quaternary.

STRUCTURE AND SEDIMENT THICKNESS

Structure

Shortland basin is surrounded by high-relief islands, and sediments eroded from these islands fill the basin. Although the basin has a complex development history, the resulting basin structure is relatively simple. Strata in the basin dip and thicken basinward from the flanks, and unconformities within, and deformation of, the sedimentary section are dominantly a result of differential uplift of the subsiding basin relative to its margins. The seismic lines shown here (Figures 7-11) illustrate the major features of Shortland basin.

Seismic line 401 (Figure 7) crosses the New Britain Trench, Vella Ridge, and the eastern part of Shortland basin. Based on a bathymetric map by Chase, Seekins, and Young (this volume), the steep slope on the southwestern side of the New Britain Trench (Figure 7, common-depth-points [CDP's] 2200 to 2440) is the flank of a small bathymetric high on the Woodlark Basin ocean plate. At the New Britain Trench, reflectors from oceanic crust can be traced beneath the lower slope for 25 km (Figure 7 at about 7 s beneath CDP's 1700 to 2200) until obscured by the water-bottom multiple. Overlying the oceanic-crust reflectors is an accretionary wedge that developed during subduction of the oceanic crust. Rocks within the wedge are either highly deformed or poorly bedded, as indicated by generally chaotic reflectors. Northeast of CDP 1700, bedded rocks are present beneath the upper slope southwest of Vella Ridge. These rocks taper out at about midslope on line 401 and other seismic lines that cross the slope (see also line 402 in Bruns et al, this volume). The slope strata are continuous with the thick basin strata beneath Vella Ridge (see especially unit C on line 50, Figure 9). Thus, the slope strata were at least in part deposited southward out of Shortland basin prior to the uplift of Vella Ridge.

Seismic lines 401, 49, 50, and 407 (Figures 7-10) illustrate the main structural elements of Shortland basin and show the seismic-stratigraphic wedges that characterize the sedimentary sequence within the basin. At least four episodes of uplift and erosion are reflected in the basin stratigraphy.

First, units A and B are truncated on the southwest side of the basin, where pronounced unconformities at the top of these units extend from the middle of the basin to beneath Vella Ridge (lines 49 and 50, Figures 8 and 9). The unconformities are a regional feature along the entire south island chain between the Shortland Islands and Guadalcanal (Bruns et al, 1986). The unconformity at the top of unit A presumably resulted from uplift and erosion of the southern basin flank during late Eocene and Oligocene subduction, prior to deposition of unit B. The unconformity at the top of unit B may have developed during the late Miocene and early Pliocene initiation of subduction along the New Britain and San Cristobal trenches (Bruns et al, 1986).

A third uplift episode caused the elevation and erosional truncation of units A, B, and C adjacent to

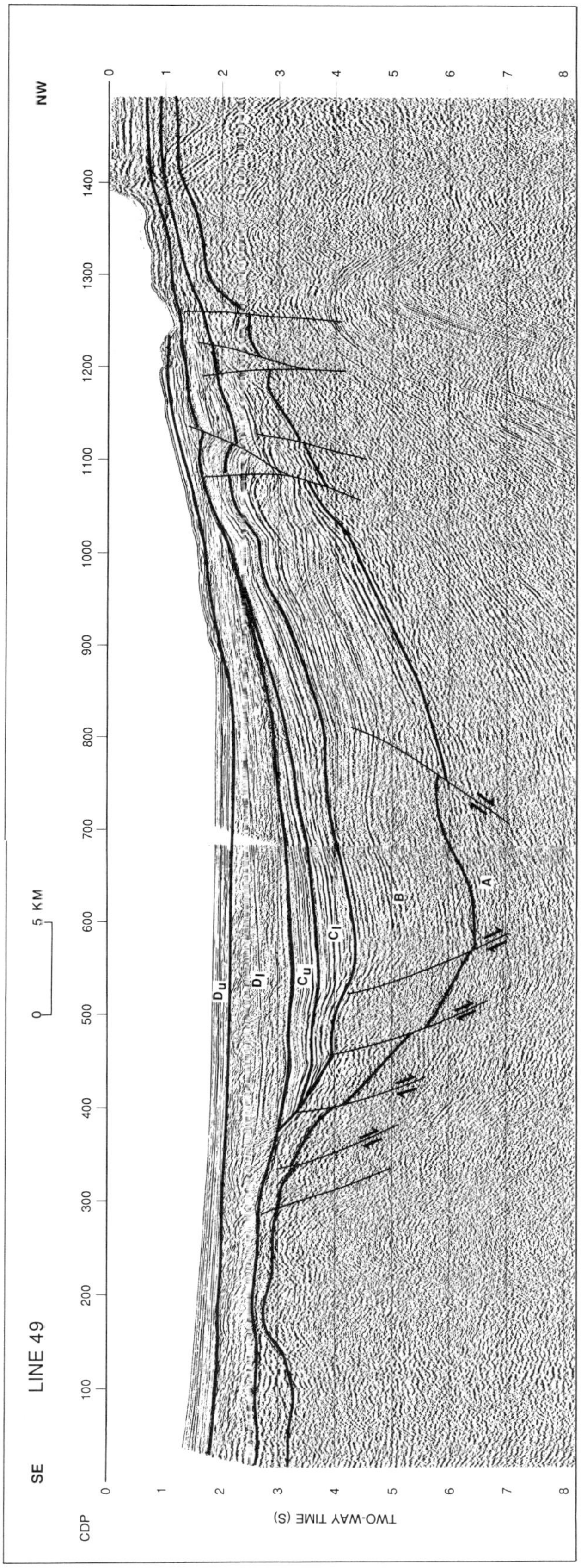

Figure 8. Multichannel seismic-reflection profile 49. Section is processed to attenuate deep-water water-bottom multiple. Profile crosses location of the thickest sedimentary section in Shortland basin. For further explanation, see Figure 7.

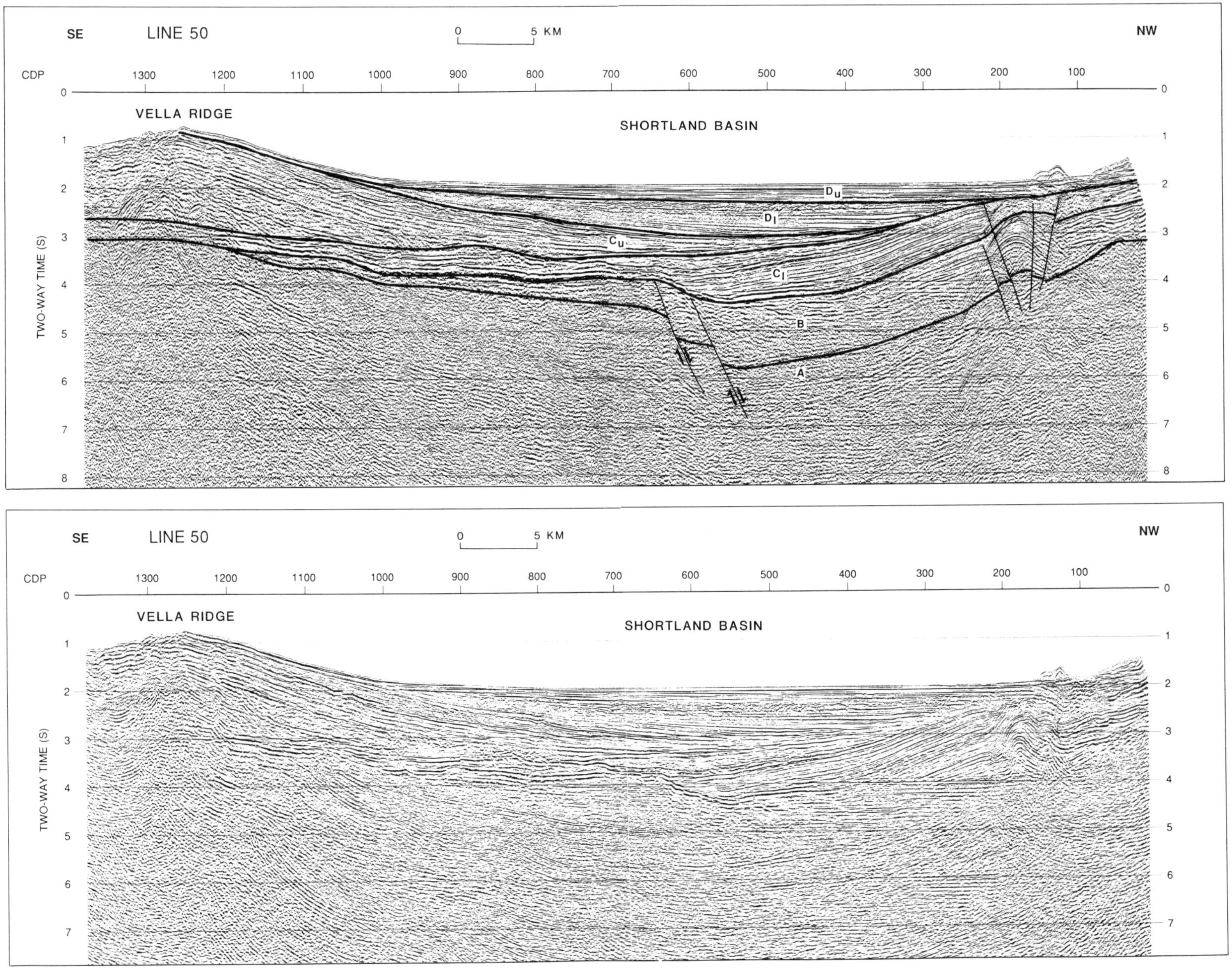

Figure 9. Multichannel seismic-reflection profile 50, unmigrated time section at top, migrated section below. Unmigrated section is processed to attenuate water-bottom multiple. Migrated section resolves diffractions in tightly folded anticline between CDP's 100 and 250. For further explanation, see Figure 7

Choiseul. The uplifted rocks are onlapped by flat-lying unit D strata. Uplift therefore occurred in about the late Pliocene or early Quaternary, coincident with the uplift of Choiseul.

Finally, marked recent uplift has occurred on Vella Ridge. Colwell and Vedder (1986) used analyses of dredged rocks to document a Quaternary history of 1,600 m of uplift followed by 700 m of subsidence of rocks beneath Vella Ridge. When restored to the original depositional depth, these strata were nearly flat lying at the time and depth of deposition. Thus, the present northward dip of units C and D strata is a measure of uplift along the ridge (see line 50, Figure 9). The uplift and subsidence of Vella Ridge was speculatively attributed to subduction of part of the Woodlark spreading ridge and (or) associated transform faults into the New Britain-San Cristobal trench system (Bruns et al, 1986).

Shortland basin rocks are faulted or folded in the center and on the northeastern flank of the basin. In the center of the basin, an extensional basement half graben trends northwestward for at least 80 km between lines 407 and 401 (beneath CDP's 400 to 600, line 401; CDP's 300 to 550, line 49; CDP's 500 to 700, line 50; and CDP's 250 to 400, line 407; Figures 7-10). The structural relief across the faults that bound the south flank of the half graben is over 6 km on line 49 (Figure 8) and decreases westward to 2 km on line 407 (Figure 10) and eastward to about 4 km on line 401 (Figure 7). At the western end of Shortland basin, the half graben trends beneath the Shortland Islands but is not evident on seismic profiles adjacent to Bougainville (see lines 6, 7, and 8 in Bruns et al, 1986). To the east, the half graben turns eastward beneath the western end of line 46 and the northeastern end of line 47 (Figure 3), and trends toward Choiseul.

Strata of unit B thicken southward into Shortland basin and onlap the faulted basement rocks on the south flank of the half graben. The unit B strata are moderately deformed adjacent to the south flank of the half graben, as shown by a gentle fold evident on line 49 in the lower part of the unit B section (Figure 8). The onlap, southward thickening, and early deformation of unit B strata indicate that the half graben must have developed prior to and during early deposition of unit B. The development time for the half graben is therefore about late Oligocene and early Miocene. If this age is correct, then the half graben developed during the time when the Ontong Java Plateau is believed to have entered and blocked the subduction zone north of the Solomon Islands.

Deformed rocks also lie along the northeastern flank of Shortland basin where northwest-trending faults and folds lie in a band at least 30 km long and 5 to 8 km wide on the northern ends of lines 49, 50, and 407 (Figures 8-10). This deformation does not effect unit D (Quaternary) strata and therefore occurred during about the late Pliocene or early Pleistocene, or at about the same time as the late Pliocene uplift of Choiseul. The extent of the deformed zone east and west of the seismic lines 49, 50, and 407 is unknown. The dip of faults within the deformed zone is highly variable (Figure 9), and the structural character suggests features formed as a result of transpression along a strike-slip fault (flower structures of Wilcox, Harding, and Seely, 1973; palm tree structures of Sylvester, 1988).

Line 406 (Figure 11) shows the western end of Shortland basin. The basin strata thin markedly beneath CDP's 1400 to 1700 onto a basement platform (the Mono Island platform on Figure 11) that underlies the Shortland Islands and the area to the west. Along the rest of the line, the sedimentary cover appears to be mainly late Cenozoic (unit D) and is mainly less than 1 km thick. Refraction data also indicate a thin sedimentary cover over basement rocks with no buried low-velocity section beneath the platform (Cooper Cochrane, and Bruns, this volume). Some minor normal faults are evident in the platform rocks. In particular, a scarp is present under about CDP 600; this position is a data gap on the multichannel record (Figure 11), but the scarp is apparent on single-channel seismic-reflection records acquired simultaneously with the multichannel lines.

Sediment Distribution and Basin Extent

The structure contour and sediment thickness maps of Shortland basin (Figures 5 and 6) show that the axis of the basin trends northwestward along the center line of the Central Solomons Trough. The thickest section lies in the basement half graben. The bulk of the sediment in this depocenter lies in unit B and has a maximum thickness of about 5 km. A subsidiary depocenter is present along Vella Ridge where the bulk of the sediment occurs in unit C; this depocenter thins both eastward and westward. However, a small basin west of Mono Island lies along the Vella Ridge trend and contains more than 2 km of sedimentary section in unit D above the acoustic basement. West of this basin, seismic-reflection lines show less than 1 km and commonly less than 0.5 km of sedimentary rocks overlying basement (Figures 6 and 11).

The total sediment thickness given here for Shortland basin is greater than reported by Bruns et al (1986) because of a reinterpretation of the position of the A-B horizon. Better definition of the horizon was obtained by the increased coverage given by the 1984 seismic-reflection data and by reprocessing of the 1982 seismic-reflection lines.

Along the northeast flank of Shortland basin, strata thin onto Choiseul. Nowhere does the thick sequence along the basin axis seem to outcrop on this island. Only a thin sedimentary sequence is present above acoustic basement on the north side of Bougainville Strait (line 403 in Bruns, Vedder, and Culotta, this volume).

At the western end of Shortland basin, the extent and thickness of basin strata are less clear. The thick strata below the basin axis trend beneath the Shortland Islands toward Bougainville. Seismic lines in these areas may not reveal the total sediment thickness, because near-surface volcanic rocks may conceal underlying strata as occurs beneath the New Georgia wedge region southeast of Shortland basin (Figures 1 and 2; Cooper, Cochrane, and

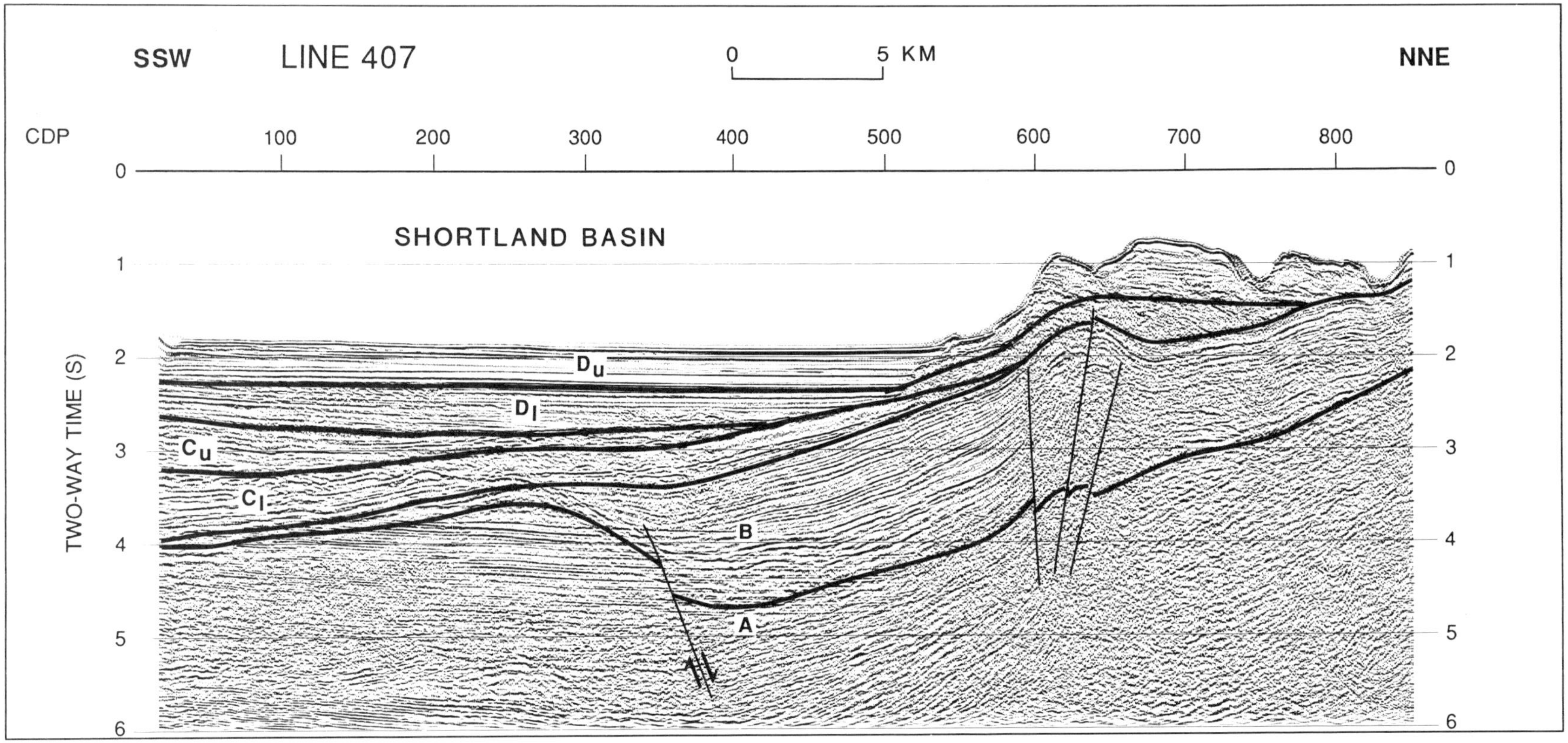

Figure 10. Multichannel seismic-reflection profile 407, migrated time-section. Features in the upper part of the section between CDP's 600 and 800 may be rotated slump blocks off Choiseul. For further explanation, see Figure 7.

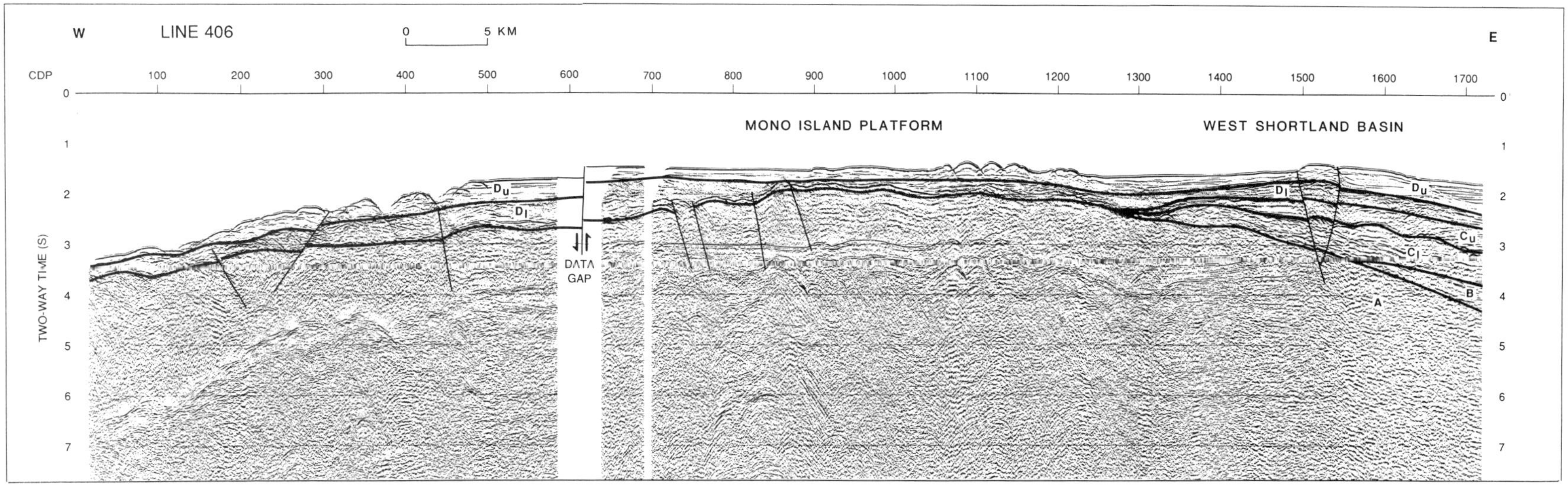

Figure 11. Multichannel seismic-reflection profile 406 across the western part of Shortland basin and the Mono Island platform. For further explanation, see Figure 7.

Bruns, this volume). However, the total sedimentary package thins northwestward toward Bougainville (compare thicknesses on lines 49, 50, and 407). Further, rocks correlated with units C and D of the Shortland basin outcrop on the Shortland Islands. Thus, although part of the basin could be buried beneath the Quaternary volcanic rocks and stratovolcanoes of southwestern Bougainville, the basin strata more likely are thin or uplifted, and Shortland basin may end beneath the Shortland Islands or near Bougainville.

The depositional axis of Shortland basin is on trend with the axis of thick strata in Bougainville basin, which lies along the southwest side of Bougainville (Figure 2; Shaw, 1985; Stewart, Francis, and Pederson, 1987; Hilyard and Rogerson, this volume). If the Shortland basin does end near Bougainville, however, then the two basins are not continuous. A basement ridge or high block (possibly the Aitara High described by Shaw, 1985) along the southwestern side of Bougainville separates Shortland and Bougainville basins there.

On the southwest flank of Shortland basin, strata thin markedly south of Vella Ridge and taper out part way down the slope into the New Britain Trench (Figures 6 and 7). Bruns et al (1986) speculated that during deposition of unit C strata, Shortland basin may have been open southward into the Woodlark Basin and that sediment may have spilled through the Vella Ridge region onto the slope. However, the volume of such material appears to have been small, because only a thin bedded section is discernible beneath the middle and lower parts of the slope. The chaotic character of reflections beneath the slope suggests that most of the slope is underlain by arc-basement rocks, probably rocks of the late Cenozoic arc, or that substantial deformation of strata beneath the slope has occurred as a result of subduction of the Solomon plate beneath the arc.

New Georgia Wedge

The southeastward extent of Shortland basin beneath the New Georgia wedge was interpreted differently by Bruns et al (1986) and Cooper, Bruns, and Wood (1986). The two different interpretations were, respectively, a basin sequence about 2 km thick based on multichannel seismic-reflection data, or a basin sequence about 5 km thick based on seismic-refraction data, a thickness equal to that of the Shortland and Russell basins.

Seismic-refraction data acquired in 1984 convincingly demonstrate that the New Georgia wedge region is underlain by a 5.5-km-thick sequence of interlayered sedimentary and high-velocity rocks, probably volcanic flows and related volcaniclastic debris from the active volcanoes of the New Georgia island group (Cooper, Cochrane, and Bruns, this volume). On seismic-reflection profiles, therefore, the high-velocity rocks form an anomalously shallow acoustic basement, and total basin thickness is obscured. The seismic-stratigraphic units of the Shortland basin, and in particular the acoustic basement, cannot be traced with confidence southeastward from Fatu O Moana, which lies north of Kolombangara (Figure 1). Structure and sediment-thickness contours shown in Figures 5 and 6 for the New Georgia wedge region are based on interpretations of refraction data.

BASIN EVOLUTION

Bruns et al (1986) described a model for the evolution of Shortland basin that is based on the geometry and inferred ages of the sedimentary wedges in the basin. We have somewhat altered that on the basis of interpretations of the 1984 data. The most significant change is the recognition of the half graben that formed during the late Oligocene and Miocene. The character of the half graben was not recognized earlier because of the limited data across the feature. A second significant change is the recognition that unit B strata thin onto Choiseul, rather than being erosionally truncated. As in the previous model, however, the geometry of the basin fill indicates that the evolution of the basin has been primarily driven by vertical uplift and subsidence of the northeastern and southwestern flanks of the basin relative to the subsiding basin depocenter.

Prior to the late Oligocene, rocks that correlate with unit A were uplifted, metamorphosed, and eroded, probably as a consequence of southward directed subduction and related arc growth. The pronounced unconformity at the top of unit A developed during this period.

During the late Oligocene and early Miocene, widespread magmatism and southward subduction of the Pacific plate beneath the Solomon Islands arc caused uplift of the northern basin margin. Sediments shed off the resulting structural high formed unit B in the incipient back-arc Shortland basin (Figure 12A). Also, during the late Oligocene or early Miocene, a basement half graben with relief of 2 to 6 km formed within the back arc. The unit B sediment thickened into and filled the half graben, onlapped the southern flank, and eventually overtopped the half graben and spread southward. The half-graben development is coincident with the arrival of the thick oceanic crust of the Ontong Java Plateau at the subduction zone. The half graben may have developed as a pull-apart basin or by block rotation within the arc as a result of oblique subduction and transpression during the collision and arc-shutdown process. During the middle Miocene, subduction and arc volcanism ceased. The entire basin region subsided, and remnant high areas of the early Cenozoic arc continued to shed into the back-arc basin.

In about the late Miocene, uplift on the south side of the basin led to a shallowing of the basin and to truncation of B (Figure 12B). This uplift may have been caused by compression and volcanic-arc growth associated with the initiation of subduction at the New Britain Trench.

In the early Pliocene, renewed subsidence of the basin allowed deposition of subunit C_1, which probably was derived dominantly from the northern side of the basin (Figure 12C). These sediments covered eroded remnants of the underlying units.

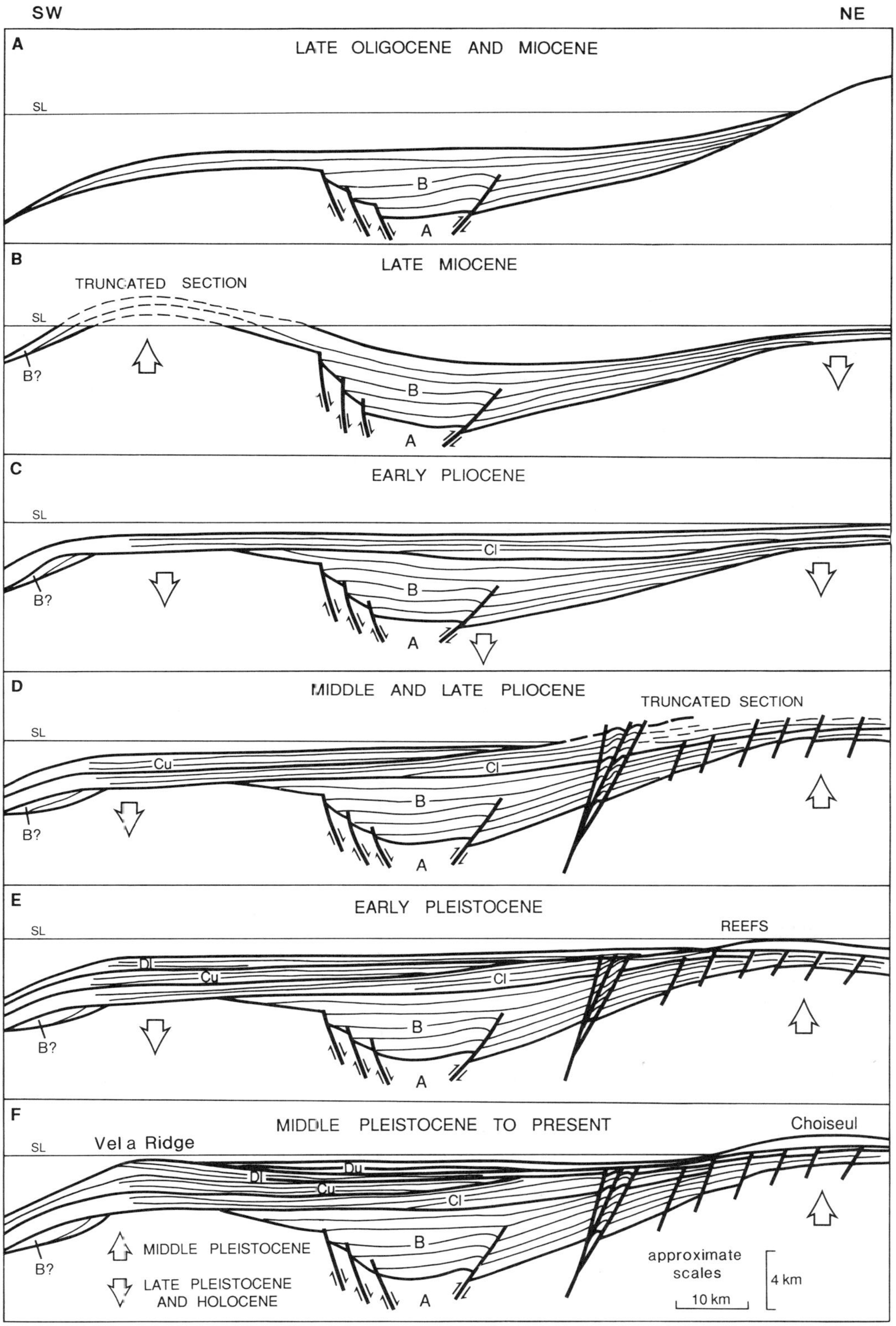

Figure 12. Diagrammatic basin-development history of the central Shortland basin region. Horizontal and vertical scales are approximate. Within limits of seismic control, diagrams are derived from depth conversion and flattening of seismic-reflection horizons. Outside of area of seismic control, diagrams are speculative. Large arrows indicate relative uplift and subsidence during annotated time interval. Seismic-stratigraphic units are as described in text and shown on seismic profiles. Revised from Bruns et al (1986).

In the middle and late Pliocene, renewed uplift of the northern flank of the basin tilted unit B and subunit C_l upward essentially as a packet, and subunit C_l strata were truncated on the north (Figure 12D). Uplift also shifted the subunit C_u depocenter southward relative to the unit B and subunit C_l depocenter to a position beneath the present Vella Ridge. The anticlinal deformation along the northern basin margin probably also occurred during this uplift phase. Shortland basin may have opened southward into Woodlark Basin during this period; however, only a thin sediment cover was deposited onto the slope.

During the early Pleistocene, subsidence and erosion of the northern landmass resulted in the growth of the extensive reef deposits of western Choiseul (Figure 12E). Subunit D_l strata were deposited southward across the strait between the Shortland Islands and Vella Lavella.

Finally, in the late early and middle Pleistocene, renewed uplift began on both sides of the basin (Figure 12F). On the northern side, Quaternary reef deposits were elevated to at least 600 m above sea level on Choiseul, with uplift possibly caused by the development of Pleistocene volcanoes. On the southern side, subunits C_u and D_l on Vella Ridge were elevated above sea level by about 1 Ma (Colwell and Vedder, 1986). Subsequently, during late Pleistocene and Holocene time, these strata again subsided to their present depth of about 700 m. The vertical tectonic movements of Vella Ridge may be caused by subduction of the Woodlark spreading center and associated transform faults into the New Britain-San Cristobal trench system (Bruns et al, 1986).

CONCLUSIONS

Multichannel seismic-reflection data acquired in 1982 were used by Bruns et al (1986) to provisionally define the stratigraphy and structure of the Shortland basin region. Additional data acquired in 1984 resulted in four new findings and reinterpretations:

(1) A basement half graben underlies the center of Shortland basin, and contains the maximum sediment thickness in the basin. Relief across the south flank of the half graben is as much as 6 km but decreases eastward to 4 km near the middle of Choiseul and westward to 2 km adjacent to the Shortland Islands. Basin strata onlap and overtop the south flank of the half graben. The half graben developed in the late Oligocene and early Miocene, coincident with the arrival of the Ontong Java Plateau at the early Tertiary subduction zone. The half graben may have developed as a pull-apart basin or by block rotation in a transform or oblique subduction system.

(2) New maps of structure and sediment thickness show that the depositional sequence in the basin is thicker than previously reported. The maximum sediment thickness is about 7 km in the northwest-trending main basin depocenter in the Central Solomons trough. Subsidiary depocenters are present beneath Vella Ridge and west of Mono Island, where maximum sedimentary-rock thicknesses are as much as 3 and 2 km, respectively.

(3) The regional extent of the Shortland basin to the northwest and southeast is better determined than previously. The thick sedimentary section in the axial part of Shortland basin thins towards the Shortland Islands and Bougainville, and a thick basin fill cannot be traced or inferred beneath the stratovolcanoes and volcanogenic rocks of Bougainville. Thus, Shortland basin could end beneath the Shortland Islands or near Bougainville. A basement high on the southwestern side of Bougainville separates Bougainville basin from Shortland basin. However, a connection between Shortland and Bougainville basins that lies beneath the thick volcanic rocks of southeastern Bougainville cannot be ruled out. On the southwest flank of Shortland basin, sediment moved out of the basin and accumulated on the slope, but these slope strata thin out midway down the slope. Thus, major volumes of sediment probably were not moved southward into the New Britain Trench prior to uplift of Vella Ridge.

(4) Refraction data clearly show that the basin fill beneath the New Georgia wedge region is substantially thicker than was previously interpreted from seismic-reflection data. The new structure and sediment thickness maps reflect this finding and show at least 5 km of interbedded volcanic and volcaniclastic rocks beneath the New Georgia wedge region.

The evolution of Shortland basin was largely driven by vertical tectonics on the northeastern and southwestern flanks of the basin. These vertical movements were in turn driven by tectonic events associated with the subduction and arc-reversal history of the Solomon Islands region. Major uplift of the northern basin margin and early Tertiary faulting within the basin were probably related to early Tertiary subduction and to arrival of the Ontong Java Plateau at the early Tertiary subduction zone. Late Miocene and early Pliocene uplift of the basin margins is coincident with the initiation of subduction along the New Britain-San Cristobal trench system. Uplift possibly was governed by compressional tectonics associated with this subduction. Pliocene and Quaternary uplift of the northern basin margin may have been controlled by late-episode intrusive and volcanic activity on Choiseul. Finally, major recent uplift of the south basin margin may have been caused by subduction beneath the arc of the Woodlark spreading center. Thus, the unique sedimentary wedge geometries in Shortland basin are caused by tectonic events associated with the subduction and arc-reversal history of the Solomon Islands region.

ACKNOWLEDGMENTS

We thank R.C. Culotta, D.M. Mann, and P.E. Hart for processing the seismic data, Susan Vath for drafting support, and T.M. Martin-Funnié for typing support. M.A. Fisher and S.L. Eittreim provided reviews that significantly improved the manuscript.

REFERENCES CITED

Blake, D.H., and Miezitis, Y., 1967, Geology of Bougainville and Buka Island, New Guinea: Department of National Development, Bureau of Mineral Resources, Geology and Geophysics Bulletin 93, PNG 1, Canberra, A.C.T., 56 p.

Bruns, T.R., Cooper, A.K., Mann, D.M., and Vedder, J.G., 1986, Seismic stratigraphy and structure of sedimentary basins in the Solomon Islands region, *in* Vedder, J.G., Pound, K.S., and Boundy, S.Q., eds., Geology and offshore resources of Pacific island arcs—central and western Solomon Islands: Houston, Texas, Circum-Pacific Council for Energy and Mineral Resources Earth Science Series, v. 4, p. 177-214.

Bruns, T.R., Cooper, A.K., and Vedder, J.G., 1986, A multichannel seismic-reflection profile across the Solomon Islands arc, *in* Vedder, J.G., Pound, K.S., and Boundy, S.Q., eds., Geology and offshore resources of Pacific island arcs—central and western Solomon Islands: Houston, Texas, Circum-Pacific Council for Energy and Mineral Resources Earth Science Series, v. 4, p. 215-224.

Chase, T.E., Seekins, B.A., and Lund, K.E., 1986, Submarine topography of the Solomon Islands region, *in* Vedder, J.G., Pound, K.S., and Boundy, S.Q., eds., Geology and offshore resources of Pacific island arcs—central and western Solomon Islands: Houston, Texas, Circum-Pacific Council for Energy and Mineral Resources Earth Science Series, v. 4, p. 21-32.

Coleman, P.J., 1965, Stratigraphical and structural notes on the British Solomon Islands with reference to the first geological map: British Solomon Islands Geological Record (1959-1962), v. 2, p. 17-31.

Colwell, J.B., and Vedder, J.G., 1986, Description and interpretation of dredged rocks, Solomon Islands, *in* Vedder, J.G., Pound, K.S., and Boundy, S.Q., eds., Geology and offshore resources of Pacific island arcs—central and western Solomon Islands: Houston, Texas, Circum-Pacific Council for Energy and Mineral Resources Earth Science Series, v. 4, p. 225-236.

Cooper, A.K., Bruns, T.R., and Wood, R.A., 1986, Crustal structure of the Solomon Islands intra-arc basins from sonobuoy seismic studies, *in* Vedder, J.G., Pound, K.S., and Boundy, S.Q., eds., Geology and offshore resources of Pacific island arcs—central and western Solomon Islands: Houston, Texas, Circum-Pacific Council for Energy and Mineral Resources Earth Science Series, v. 4, p. 135-156.

Cooper, A.K., Marlow, M.S., and Bruns, T.R., 1986, Deep structure of the central and southern Solomon Islands region: implications for tectonic origin, *in* Vedder, J.G., Pound, K.S., and Boundy, S.Q., eds., Geology and offshore resources of Pacific island arcs—central and western Solomon Islands: Houston, Texas, Circum-Pacific Council for Energy and Mineral Resources Earth Science Series, v. 4, p. 157-176.

Coulson, F.I., and Vedder, J.G., 1986, Geology of the central and western Solomon Islands, *in* Vedder, J.G., Pound, K.S., and Boundy, S.Q., eds., Geology and offshore resources of Pacific island arcs—central and western Solomon Islands: Houston, Texas, Circum-Pacific Council for Energy and Mineral Resources Earth Science Series, v. 4, p. 59-88.

de Broin, C.E., Aubertin, F., and Ravenne, C., 1977, Structure and history of the Solomon-New Ireland region: International Symposium on Geodynamics in Southwest Pacific, Noumea (New Caledonia), 27 August-2 September 1976, Editions Technip, Paris, France p. 37-50.

Katz, H.R., 1980, Basin development in the Solomon Islands and their petroleum potential: U.N. ESCAP, CCOP/SOPAC Technical Bulletin, no. 3, p. 59-75.

Kroenke, L.W., 1984, Cenozoic tectonic development of the southwest Pacific: U.N. ESCAP, CCOP/SOPAC, Technical Bulletin, no. 6, 122 p.

Landmesser, C.W., 1977, Evaluation of potential hydrocarbon occurrence in the Solomon Islands: U.N. ESCAP, CCOP/SOPAC Technical Report, v. 1, no. 5, p. 47-53.

Mann, D.M., 1986, Multichannel seismic operations for CCOP/SOPAC Cruise, Leg 3, Solomon Islands, *in* Vedder, J.G., Pound, K.S., and Boundy, S.Q., eds., Geology and offshore resources of Pacific island arcs—central and western Solomon Islands: Houston, Texas, Circum-Pacific Council for Energy and Mineral Resources Earth Science Series, v. 4, p. 35-40.

Maung, T.U., and Coulson, F.I., 1983, Assessment of petroleum potential of the central Solomons Basin: U.N. ESCAP, CCOP/SOPAC Technical Report n. 26, 68 p.

Pound, K.S., 1986, Correlation of rock units in the central and western Solomon Islands, *in* Vedder, J.G., Pound, K.S., and Boundy, S.Q., eds., Geology and offshore resources of Pacific island arcs—central and western Solomon Islands: Houston, Texas, Circum-Pacific Council for Energy and Mineral Resources Earth Science Series, v. 4, p. 89-98.

Ridgway, J., and Coulson, F.I.E., 1987, The geology of Choiseul and the Shortland Islands, Solomon Islands: British Geological Survey Overseas Memoir 8, 134 p.

Shaw, R.D., 1985, Hydrocarbon potential of the Bougainville basin, Papua New Guinea: 14th BMR Symposium, Extended Abstracts—Petroleum Geology of South Pacific Island Countries, Marine Studies Workshop, Canberra, Australia, 22-23 October, p. 19-29.

Stewart, W.D., Francis, G., and Pederson, S.L., 1987, Hydrocarbon potential of Papua New Guinea's Bougainville, southeastern New Ireland basins: Oil and Gas Journal, v. 85, n. 47, p. 83-87.

Sylvester, A.G., 1988, Strike-slip faults: Geological Society of America Bulletin, v. 100, p. 1666-1703.

Taylor, B. and Exon, N.F., 1987, Marine geology, geophysics, and geochemistry of the Woodlark Basin-Solomon Islands: Houston, Texas, Circum-Pacific Council for Energy and Mineral Resources Earth Science Series, v. 7, 363 p.

Vedder, J.G., and Bruns, T.R., 1986, Offshore petroleum potential, Solomon Islands, *in* Vedder, J.G., Pound, K.S., and Boundy, S.Q., eds., Geology and offshore resources of Pacific island arcs—central and western Solomon Islands: Houston, Texas, Circum-Pacific Council for Energy and Mineral Resources Earth Science Series, v. 4, p. 277-294.

Vedder, J.G., and Coulson, F.I., 1986, Regional offshore geology of the Solomon Islands, *in* Vedder, J.G., Pound, K.S., and Boundy, S.Q., eds., Geology and offshore resources of Pacific island arcs—central and western Solomon Islands: Houston, Texas, Circum-Pacific Council for Energy and Mineral Resources Earth Science Series, v. 4, p. 99-108.

Vedder, J.G., Pound, K.S., and Boundy, S.Q., eds., 1986, Geology and offshore resources of Pacific island arcs—central and western Solomon Islands: Houston, Texas, Circum-Pacific Council for Energy and Mineral Resources

Earth Science Series, v. 4, 306 p.
Wilcox, R.E., Harding, T.P., and Seely, D.R., 1973, Basic wrench tectonics: American Association of Petroleum Geologists Bulletin, v. 57, p. 74-96.

Vedder, J.G., and Bruns, T.R., editors, 1989, Geology and offshore resources of Pacific island arcs—Solomon Islands and Bougainville, Papua New Guinea Regions: Houston, Texas, Circum-Pacific Council for Energy and Mineral Resources Earth Science Series, v. 12.

INTERPRETATION OF A MULTICHANNEL SEISMIC-REFLECTION PROFILE NORTHEAST OF THE SOLOMON ISLANDS, FROM THE SOUTHERN FLANK OF THE ONTONG JAVA PLATEAU ACROSS THE MALAITA ANTICLINORIUM TO THE SOLOMON ISLANDS ARC

L.W. Kroenke
Hawaii Institute of Geophysics, University of Hawaii,
2525 Correa Road, Honolulu, Hawaii 96822

INTRODUCTION

The northern margin of the Solomon Islands arc (Pacific Province of Coleman, 1965) abuts the Ontong Java Plateau (OJP) (Figure 1). The margin has been variously interpreted as originally forming either as part of the OJP (Kroenke, 1972; Kroenke, Resig, and Cooper, 1986) or as an integral part of the Solomon Islands arc, unrelated to the OJP (Ramsay, 1978; Turner and Hughes, 1982). The arguments presented over succeeding years by proponents for both hypotheses are cogently summarized and critiqued by Wells (this volume).

During early 1982, the Tripartite cruise of the U. S. Geological Survey vessel, R/V *S.P. Lee* (L7-82-SP) recorded about 2,000 km of 24-fold multichannel seismic-reflection (MCS) data along 52 lines within the Solomon Islands region (Mann, 1986). Seismic line 37 (Figures 2 and 3 on the separate plate which accompanies this volume) completely transects Coleman's Pacific Province, crossing the transition zone between the Solomon Islands arc and the OJP (Figure 1), i.e., the Malaita anticlinorium of Kroenke (1972). Line 37 was originally described by Bruns, Cooper, and Vedder (1986), who noted that the antiforms and synforms of the anticlinorium dominate the structure and that the reflectors in the adjoining North Solomon Trench cannot be traced beneath the adjacent structural high.

Although the origin of the folding and faulting within the Malaita anticlinorium has been, in general, attributed to the collision between the Solomon Islands arc and the OJP, the provenance of the sedimentary section continues to be argued. Kroenke (1972), using single-channel seismic-reflection (SCS) data, was unable to continuously trace OJP reflectors within the North Solomon Trench directly across the adjacent structural highs northeast of Malaita because of the vertical exaggeration of the reflection profiles. Thus he had to correlate the OJP reflectors into the Malaita anticlinorium via a circuitous network of SCS reflection profiles around the less deformed northwestern end of the anticlinorium. In contrast, line 37 permits direct correlation of the reflectors across the adjacent structural highs and is, therefore, reproduced herein. The interpreted time section is shown in Figure 2 and is displayed at a more reduced vertical exaggeration than that shown by Bruns, Cooper, and Vedder (1986) to facilitate along-line correlation. Although basement is highlighted along almost the entire length of the profile, highlighting of fault traces on the profile has been kept to a minimum to avoid obscuring the along-line continuity of reflectors. A migrated version of the profile is shown in Figure 3.

INTERPRETATION OF THE MCS PROFILE

Bruns, Cooper, and Vedder (1986) described the geologic features and structural relationships observable on the line 37 time-section in general terms. The following interpretation and description of structural relationships along the line differ only slightly, and only in detail, from the earlier ones; and the names and locations of major geologic features and structures are the same. The location of line 37 is shown in Figure 1, as are the major physiographic features.

The North Solomon Trench, which is floored with OJP sediments, is located at the northeastern end of line 37. The Santa Isabel-Florida Islands platform is located at the southwestern end of the line. Between the trench and platform lies the Malaita anticlinorium. Folding within the anticlinorium, for the most part, is supratenuous with sediment having moved down-slope off the anticlinal crests into the adjoining synclinal depressions. Deformation, in general, increases from the northeast to the southwest.

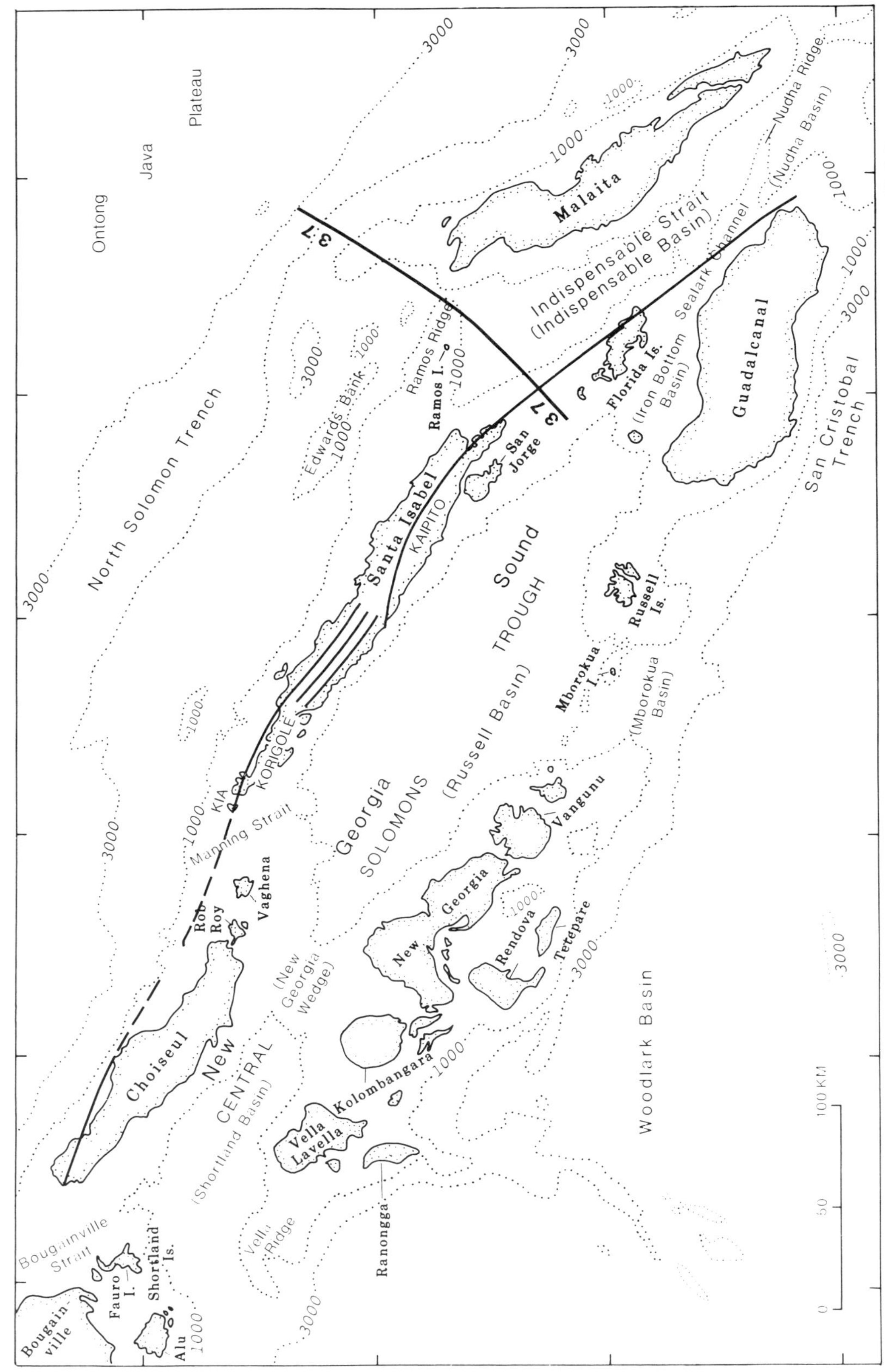

Figure 1. Features of the Solomon Islands arc in the central and western Solomon Islands. The location of line 37 is shown with respect to the arc, the North Solomon Trench, and the Ontong Java Plateau. The Pacific Province of Coleman (1965) lies north of the Kia-Korigole-Kaipito fault.

Despite minor structural offsets of the basement reflector at the northeastern end of line 37, it is apparent on the time section shown in Figure 2 that the sedimentary section overlying the two anticlinal ridges adjacent to the North Solomon Trench (CDP 180-1000) is continuous with the OJP section within the trench. The migrated section shown in Figure 3, although obscured somewhat by over-migrated reflection events in some areas, also demonstrates the continuity of the OJP section with that of the Malaita anticlinorium. Because the second of the two anticlinal ridges (CDP 470-1000) is a direct northwestern extension of Malaita, the island probably constitutes part of the southwestern margin of the OJP.

To the southwest, farther into the anticlinorium, structural deformation increases. On the next anticlinal ridge, between the northwestern extension of Malaita and Ramos Ridge (Figures 2 and 3, CDP 1000-1200), duplex structure appears to be developing. On the northeastern side of Ramos Ridge (CDP 1330-1430), back-thrusting appears to have occurred. At the crest of Ramos Ridge (CDP 1530), igneous intrusions also may be present (shown highlighted in Figure 2 above the dashed basement reflector), making it difficult to trace the basement reflector. The intrusive nature of these reflectors is even more apparent in the migrated section (Figure 3).

Between Ramos Ridge and Indispensible basin, deformation continues to increase. Thrust faulting is common, and the sedimentary section may be fault repeated in several areas. Dikes and sills also may be present, further complicating the structure (CDP 1720-1920). These structures are also more readily discerned on the migrated section (Figure 3).

Beneath the northeastern margin of Indispensible basin, one or more thrust sheets appear to protrude into and deform the ponded basin sediments. These ponded basin sediments were designated as unit I_{CD} by Bruns, Cooper, and Vedder (1986) and Bruns et al (1986) and were interpreted by them to be Pliocene and Quaternary in age. The deformation of the unit, particularly that of the uppermost sediments on the eastern edge of the basin, indicates a relatively recent time for the thrusting.

On the southwestern side of the basin, a thick sequence of low-density sedimentary rocks (Bruns et al, 1986) lies piled at the base of the Santa Isabel-Florida Islands platform. Within the pile, visible on both the time section (Figure 2) and the migrated section (Figure 3), suggestions of rotational fault planes are present, indicative of slumping. In the migrated section (Figure 3), there are also hints of an upward inclination of the basement reflector toward the Santa Isabel-Florida Islands platform and a slight but tantalizing suggestion of a decollement. Additional data acquisition and processing needs to be undertaken, however, before any conclusion can be drawn regarding the nature of this structure.

DISCUSSION

Because reflectors in the North Solomon Trench do not dip beneath the adjacent structural high, it is obvious that neither underthrusting nor subduction has occurred along the present-day trench axis. Instead, igneous basement and the overlying sedimentary section within the trench appear to be continuous with those of the Malaita anticlinorium; the trench axis simply represents the location of a structural hinge with the former locus of subduction lying somewhere between the present-day trench axis and the Kia-Korigole-Kaipito fault system (Figure 1). It is also obvious that Malaita is part of the OJP and, as such, should have moved northward with the plateau on a Pacific plate trajectory.

Musgrave (1985), however, argued that the paleomagnetic poles derived from Malaita paleomagnetic data compare favorably with the Australia polar wander path and that paleolatitudes determined from those data do not fit the paleolatitude curve derived from DSDP site 289 on the OJP (Hammond, Kroenke, and Theyer, 1975). Musgrave's arguments, however, did not appear to take into account any tectonic rotation of Malaita as shown by the increasing structural deformation of the island from west to east and, more importantly, ignore the effect of grain rotation during compaction and deformation subsequent to deposition. Such an effect appears to have markedly decreased the apparent inclination values obtained during the DSDP site 289 paleomagnetic investigation which, in turn, resulted in about half the northward amount of motion for the OJP than would be expected, based on estimates of past Pacific plate motion. These compaction and deformation effects could be expected to be even more severe in the Malaita anticlinorium where supratenuous folding is observed on the reflection profiles.

In view of the foregoing, and based on the strong evidence provided by the line 37 MCS data, it appears that emplacement of igneous basement and deposition of most of the sedimentary section of the Malaita anticlinorium (including all of Malaita and at least the northern half of Santa Isabel) occurred while the anticlinorium was an integral part of the southern margin of the OJP and the Pacific plate..

REFERENCES CITED

Bruns, T.R., Cooper, A.K., and Vedder, J.G., 1986, A multichannel seismic-reflection profile across the Solomon Islands arc, *in* Vedder, J.G., Pound, K.S., and Boundy, S.Q., eds., Geology and offshore resources of Pacific island arcs—central and western Solomon Islands: Houston, Texas Circum-Pacific Council for Energy and Mineral Resources Earth Science Series, v. 4, p. 215-224.

Bruns, T.R., Cooper, A.K., Mann, D.M., and Vedder, J.G., 1986, Seismic stratigraphy and structure of sedimentary basins in the Solomon Islands region, *in* Vedder, J.G., Pound, K.S., and Boundy, S.Q., eds., Geology and offshore resources of the Pacific island arcs—central and western Solomon Islands: Houston, Texas, Circum-Pacific Council for Energy and Mineral Resources Earth Science Series, v. 4, p. 177-214.

Coleman, P.J., 1965, Stratigraphical and structural notes on the British Solomon Islands with reference to the first geological map: British Solomon Islands Geological Record (1959-1962), v. 2, p. 17-31.

Cooper, P.A., Kroenke, L.W., and Resig, 1986, J.M., Tectonic implications of seismicity northeast of the Solomon Islands, *in* Vedder, J.G., Pound, K.S., and Boundy, S.Q., eds., Geology and offshore resources of Pacific island arcs—central and western Solomon Islands: Houston, Texas Circum-Pacific Council for Energy and Mineral Resources Earth Science Series, v. 4, p. 117-122.

Hammond, S.R., Kroenke, L.W., and Theyer, F., 1975, Northward motion of the Ontong-Java Plateau between -110 and -30 m.y.: a paleomagnetic investigation of DSDP Site 289, *in* Andrews, J.E., Packham, G. et al, Initial reports of the Deep Sea Drilling Project, U.S. Government Printing Office, v. 30, p. 415-418.

Kroenke, L.W., 1972, Geology of the Ontong Java Plateau: Hawaii Institute of Geophysics, Report no. HIG-72-5, University of Hawaii, 119 p.

Kroenke, L.W., 1984, Cenozoic tectonic development of the southwest Pacific: U.N. ESCAP, CCOP/SOPAC Technical Bulletin 6, 122 p.

Kroenke, L.W., Resig, J.M., and Cooper, P.A., 1986, Tectonics of the southeastern Solomon Islands: formation of the Malaita anticlinorium, *in* Vedder, J.G., Pound, K.S., and Boundy, S.Q., eds., Geology and offshore resources of Pacific island arcs—central and western Solomon Islands: Houston, Texas, Circum-Pacific Council for Energy and Mineral Resources Earth Science Series, v. 4, p. 109-116.

Mann, D.M., 1986, Multichannel seismic operations for CCOP/SOPAC cruise, Leg 3, Solomon Islands, *in* Vedder, J.G., Pound, K.S., and Boundy, S.Q., eds., Geology and offshore resources of Pacific island arcs—central and western Solomon Islands: Houston, Texas, Circum-Pacific Council for Energy and Mineral Resources Earth Science Series, v. 4, p. 35-40.

Musgrave, R.J., 1985, Paleomagnetism and tectonics of Malaita, Solomon Islands: A cautionary tale for terrane seekers: Third Circum-Pacific Terrane Conference Extended Abstracts, Geological Society of Australia Publication 14, p. 159-163.

Ramsay, W.R.H., 1978, Field, mineralogical and structural observations on some basement rocks, southeast Choiseul, Solomon Islands: Bulletin of the Australian Society of Exploration Geophysicists, v. 9, p. 107-110.

Turner, C.C., and Hughes, G.W., 1982, Distribution and tectonic implications of Cretaceous-Quaternary sedimentary facies in the Solomon Islands: Tectonophysics, v. 87, p. 127-146.

Vedder, J.G., and Bruns, T.R., editors, 1989, Geology and offshore resources of Pacific island arcs—Solomon Islands and Bougainville, Papua New Guinea Regions: Houston, Texas, Circum-Pacific Council for Energy and Mineral Resources Earth Science Series, v. 12.

PETROLOGY AND PHASE CHEMISTRY OF SILICICLASTIC SEDIMENTARY ROCKS DREDGED FROM OFFSHORE SITES IN THE SOLOMON ISLANDS AND NEAR BOUGAINVILLE, PAPUA NEW GUINEA

K.A.W. Crook
Department of Geology, Australian National University, Canberra ACT 2601, Australia
A.R. Niem
Department of Geology, Oregon State University, Corvallis, Oregon 97331-5506, USA

ABSTRACT

Sedimentary rocks dredged from 17 offshore sites in the Solomon Islands and near Bougainville, Papua New Guinea, commonly include mudstone and sandstone. Limestone occurs at nine stations, and rounded metamorphic basement cobbles were sampled at Station 1 west of Guadalcanal. Strata older than mid-Quaternary were sampled only north of Buka Island, where at least 700 m uplift is indicated, at Fatu O Moana, and adjacent to the Russell Islands.

Volcanogenic detritus at several stations is accompanied by detritus of basement provenance comprising metabasites, metamorphic tectonites, and ultramafics. Reworked sedimentary rock fragments are present in a few samples. Phyllosilicate-rich grains of diverse fabric and mineralogy, herein termed papules, which would conventionally be interpreted as altered glassy volcanic rock fragments, constitute more than 5 percent of many samples and appear to be of diverse igneous, metamorphic, and sedimentary origins. Microprobe analyses readily discriminate between arc volcanic- and basement-derived plagioclase, clinopyroxene, and amphibole. Some volcanogenic fragments indicate contemporary volcanism, but detritus from extinct young volcanic piles and from ancient volcanic-sedimentary complexes are not distinguishable.

INTRODUCTION

During Leg 6 of the 1984 R/V *S.P. Lee* Tripartite cruise (L6-84-SP), indurated sedimentary rocks were dredged from ten stations in the Solomon Islands and from one near Bougainville in Papua New Guinea (Figure 1). Two of these stations also yielded rounded clasts which were framework components in unindurated gravels. We here report the results of a study of thirty-eight indurated samples and two gravel samples from these stations. We also include five samples collected by CCOP/SOPAC cruises PNG 79-1 and PNG 79-2 from three sites near Bougainville and ten samples collected by J.G. Vedder from sites close to the Russell Islands (Figure 1)

Sampling sites (Figure 1) are located within the Volcanic and Central Provinces of Coleman (1970) (Figure 2). The Volcanic Province includes the Pliocene and younger Solomon Islands volcanic arc, the adjacent forearc and trench slope on the southwest, and the adjacent flank of the Central Solomons Trough (an intra-arc basin) on the northeast, extending from Guadalcanal to Bougainville (Figure 2). In the Central Province, which adjoins the Volcanic Province to the northeast, Cretaceous-Paleogene metamorphic and igneous basement rocks are overlain by Neogene sedimentary rocks, forming a back-arc massif and the northeastern part of the intra-arc basin (Figure 2).

The geological history of the Solomon Islands, which was reviewed by Coulson and Vedder (1986), is complicated by a reversal of arc polarity within the last 10 Ma (Kroenke, 1984) and by subduction of the presently active Woodlark spreading-ridge system beneath the New Georgia island group (Taylor and Exon, 1987).

Many types of rock are present (Table 1). Mudrock and sandstone are common. In addition to siliciclastic grains, many samples contain a significant biogenic component, dominated by planktic foraminifers. Mud fractions (<30μm) are predominantly clay minerals, accompanied by glass shards in tuffaceous rocks (Table 1). Micritic carbonate is significant or dominant in the mud fractions of 12 samples (Figure 3, Table 2). At Stations 1 and 25, indurated sediments and well-rounded clasts of metamorphic basement rocks were recovered. Limestone was recovered at nine stations (6, 14, 22, 25, 29-32, and

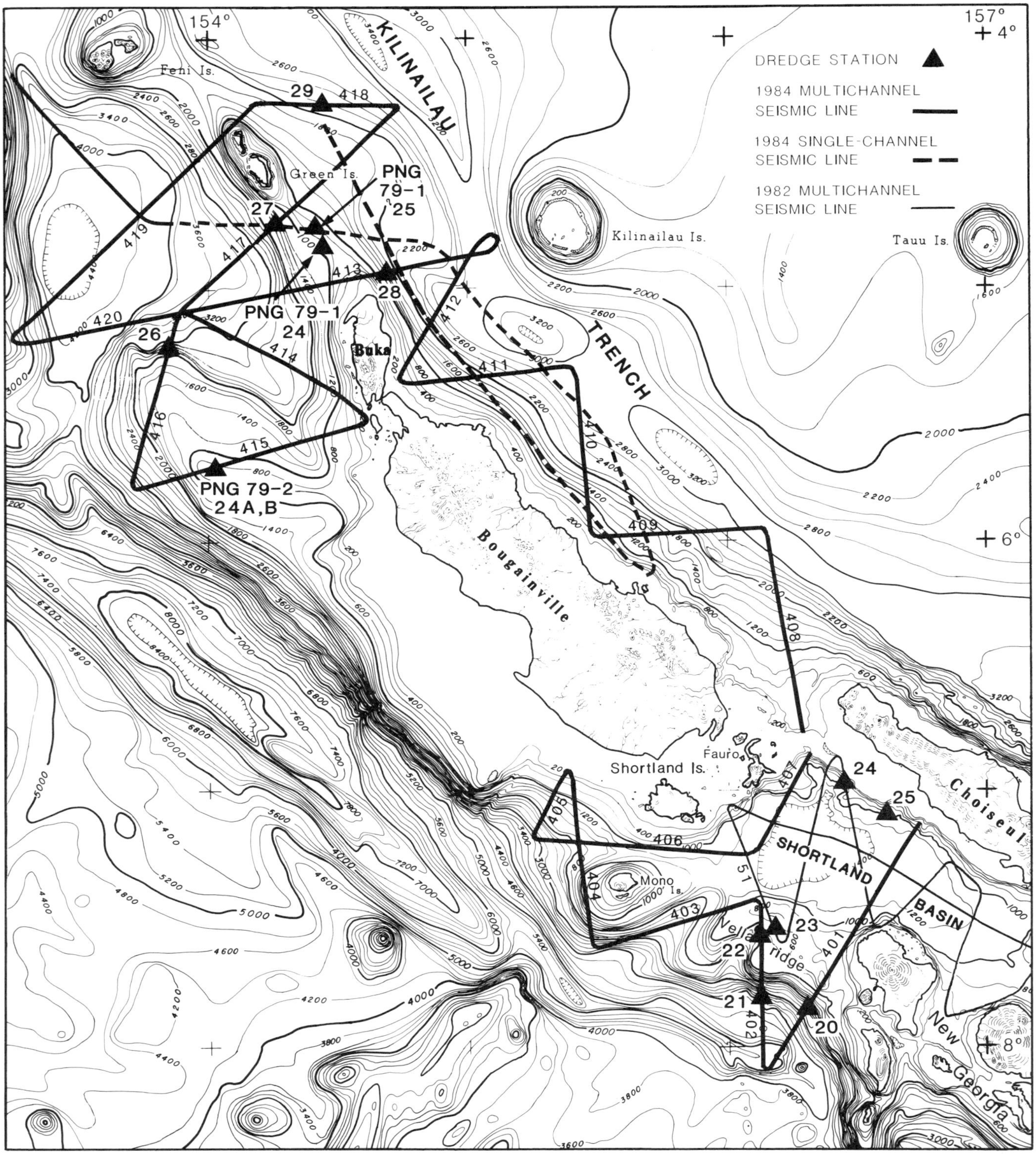

Figure 1. Map showing locations of 1984 *S.P. Lee* (L6-84-SP) dredge stations and other stations that yielded samples

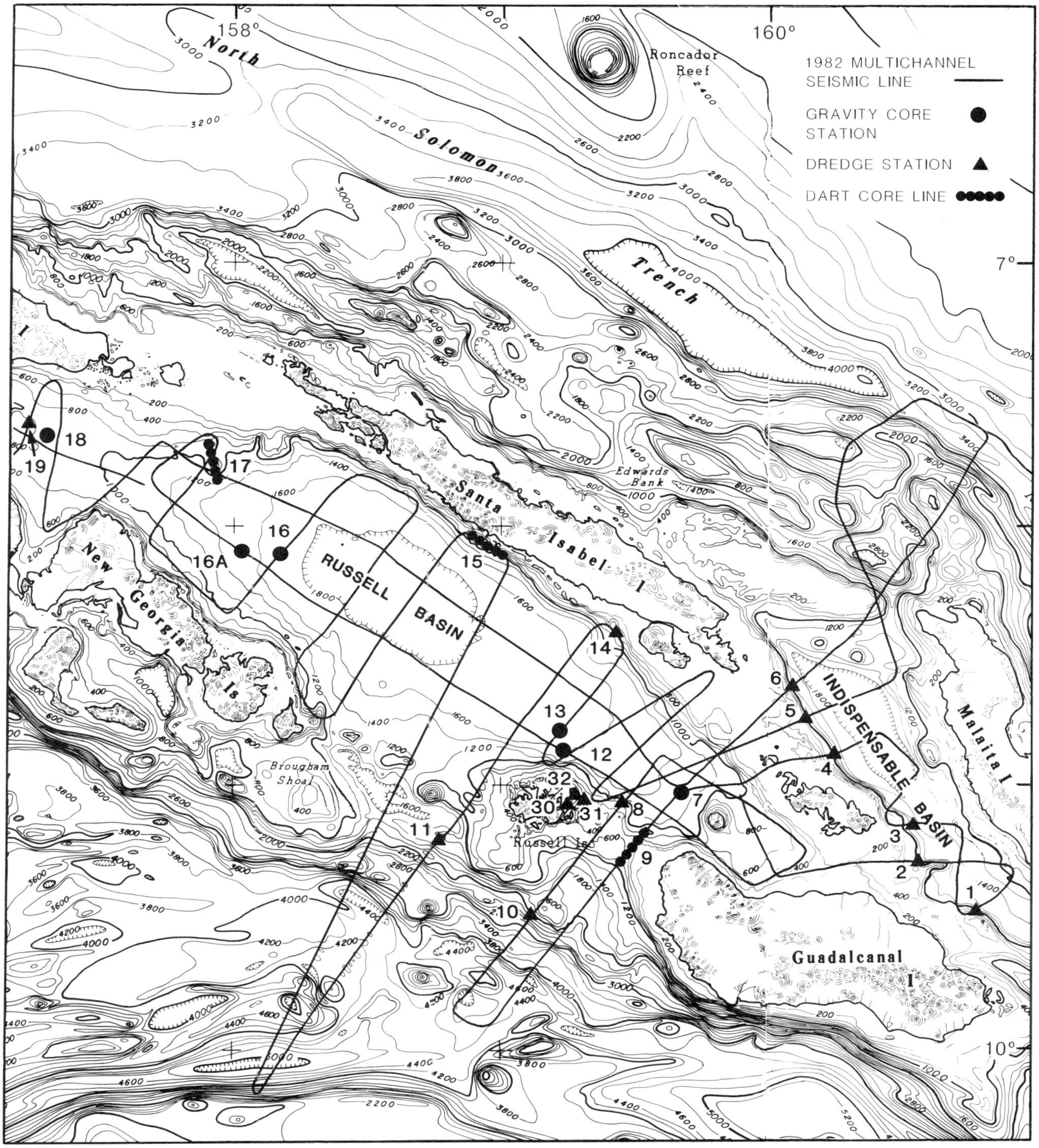

described herein. Bathymetry from Chase, Seekins, and Lund (1986).

Table 1. Details of Samples and Sampling Stations.
(1984 *S.P. Lee* cruise)

Sample No.	Station Depth and Location	Tectonic Setting	Rock Type	Age
1-1-14	700 m 09°27.9′S, 160°45.8′E 09°28.8′S, 160°42.7′E	back-arc massif	petromictic conglomerate	Quaternary (SS)[1]
1-2-1			calcareous foraminiferal litharenitic mudrock	late N22 or N23 (JMR)[2]
2-1-1	750-800 m 09°17.6′S, 160°35.2′E	back-arc massif	calcareous lithic silty mudrock	Quaternary (SS)
2-2-1			dolomitic foraminiferal litharenite	
2-2-2			micritic litharenite	
2-3-1			litharenitic mudrock	late Quaternary (SS)
2-3-2			litharenitic mudrock	late N22 or N23 (JMR)
6-1-1	500-900 m 08°38.4′S, 160°05.1′E	back-arc massif	foraminiferal calc-silty mudrock	late N22 or N23 (JMR)
6-2-1			fossiliferous glauconitic muddy litharenite	Quaternary (SS)
11-1-1	1200-1350 m 09°12.6′S, 158°48.7′E	forearc slope	tuffaceous volcanic lithic silty mudrock	
11-1-3			tuffaceous volcanic lithic silty mudrock	Quaternary
11-1-4			tuffaceous volcanic lithic silty mudrock	
11-1-5			tuffaceous volcanic lithic silty mudrock	
14-1-2	700-500 m 08°25.6′S, 159°27.9′E	back-arc massif	coralgal biolithite/ intra-pel grainstone	
14-2-2			lithic silty mudrock	late N22 or N23 (JMR)
14-2-3			carbonate-cemented volcanic lithic serpa-renite and micrite	
14-3-1			muddy serparenite	
19-1-1	700-500 m 07°37.0′S, 157°15.2′E	intra-arc basin high	foraminiferal volcanic litharenitic mudrock	
19-1-5			foraminiferal volcanic litharenitic mudrock	N21 (JMR)
19-1-7			foraminiferal vlcanic litharenitic mudrock	late Pliocene or Pleistocene (SS)
19-1-9			foraminiferal volcanic litharenitic mudrock	
19-1-10			foraminiferal volcanic litharenitic mudrock	
19-2-2			muddy feldspatho-volcanic litharenite	N22 (JMR) Pleistocene (SS)
21-1-1	3100-3900 m 07°48.7′S, 156°06.9′E	forearc slope	feldspatho-volcanic lithic mudrock	
21-1-2			tuffaceous feldspatho-volcanic lithic mudrock	
21-1-3			feldspatho-volcanic lithic mudrock	
21-1-4			tuffaceous feldspatho-volcanic lithic mudrock	
21-1-5			tuffaceous feldspatho-volcanic lithic mudrock	

Sample No.	Station Depth and Location	Tectonic Setting	Rock Type	Age
22-0-1	850 m	forearc	coralgal biolithite	
22-0-2	07°34.9'S, 156°07.6'E	slope	coral biolithite ?	Quaternary (JMR)
22-0-3			coral biolithite	
22-0-4			coralgal biolithite	
23-1A	620-900 m 07°31.9'S, 156°10.3'E	forearc slope	feldspatho-volcanic lithic foraminiferal biomicrite	late Quaternary (SS)
23-2A			volcanic lithic mudrock	Quaternary (JMR)
25-1	600-850 m	back-arc	coral biolithite	
25-2	07°04.6'S, 156°39.8'E	massif	metadolerite	
25-3			feldspatho-volcanic lithic calcareous mudrock	late N22 or N23 (JMR) late Quaternary (SS)
29-1-1	1500-2000 m	forearc	foraminiferal biomicrite	
29-1-3	04°17.5'S, 154°35.3'E	apron	foraminiferal biomicrite	N21 (JMR); late Pliocene (SS)
30-1-1	190-200 m 09°03.8'S, 159°14.6'E	arc flank of intra-	sparry intraclastic biomicrite	Pliocene or Quaternary (PQ)[3]
30-1-2		arc basin	coralgal biolithite	
30-1-3			sparry intraclastic biomicrite	probably Pleistocene (JMR)
30-2-1			feldspatho-volcanic femic biomicrite	
31-1-1	165-220 09°02.7'S, 159°16.6'E	arc flank of intra-	feldspatho-volcanic femic biomicrite	late early Pliocene (PQ)
31-1-2		arc basin	feldspatho-volcanic femic biomicrite	Pliocene or Pleistocene (JMR)
31-1-3			feldspatho-volcanic femic biomicrite	
31-1-4			feldspatho-volcanic femic biomicrite	
32-1-1	155-190 m 09°02.1'S, 159°16.0'E	arc flank of intra-	coralgal foram biolithite	Pliocene or Quaternary (PQ)
32-1-2		arc basin	coralgal biolithite	
PNG 79-1 24-A	838-900 m 04°48.2'S, 154°25.7'E	arc ridge	foraminiferal femo-feldspathic arenite	Pliocene or Pleistocene probably >1600 m (JMR)[4]
PNG 79-1 24-B			tuffaceous volcanic lithic mudrock	
PNG 79-1 25	875-885 04°42.0'S, 154°22.6'E	arc ridge	tuffaceous mudrock	
PNG 79-2 24A-1	400 m 05°39.1'S, 154°02.1'E	forearc basin	micritic intraclastic biosparite	Pliocene or younger (JMR)
PNG 79-2 24A-2		margin	sparry intraclastic biomicrite	

[1](SS): determined by Samir Shafik (this volume) from nannofossils.

[2](JMR): determined by Johanna M. Resig (this volume) from foraminiferal assemblages.

[3](PQ): determined by Paula Quinterno (this volume) from foraminiferal assemblages.

[4]Depth at which sediments were deposited determined by J.M. Resig from benthic foraminiferal assemblage.

PNG 79-2-24A). Some of these samples are described by Niem (this volume). Many limestone samples contain a minor siliciclastic component (described herein) that, in some cases, forms part of the material infilling primary cavities. Many mudstone and sandstone samples display borings that are partly infilled with mixtures of biogenic and siliciclastic material.

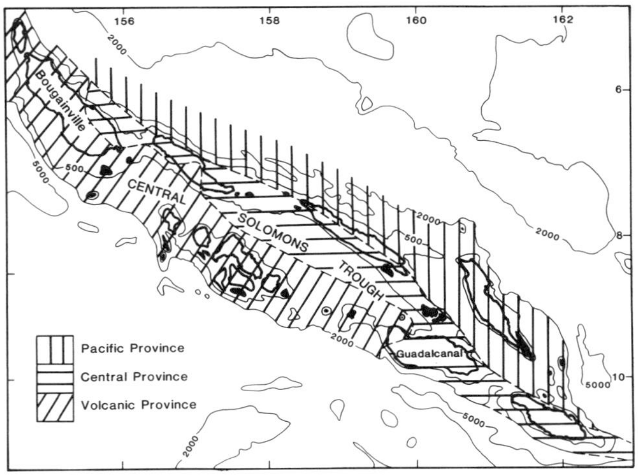

Figure 2. Geological provinces, central and western Solomon Islands (after Coleman, 1970).

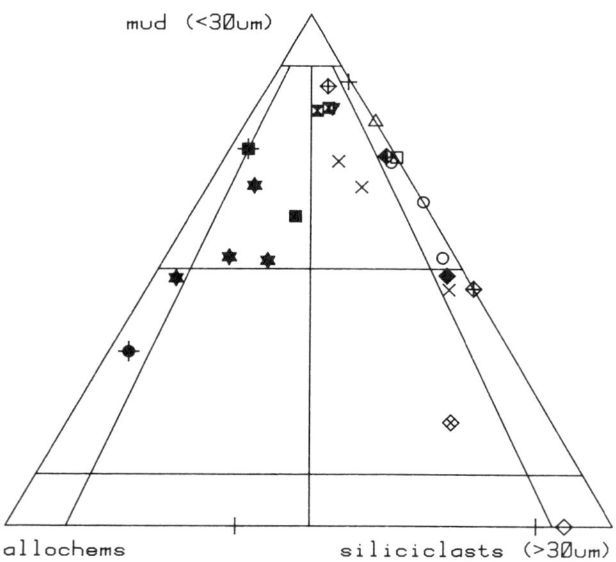

Figure 3. Variation of principal sedimentary components in the sedimentary rocks; allochems are principally biogenic (Table 3). Symbol outlines and transecting lines follow the scheme in Table 3, but black (closed) areas do not. Closed symbols: micrite mud; half-closed symbols: micrite-clay mineral mud; open symbols: clay mineral mud.

The diverse composition in this suite of sediments reflects mixing of three principal components (Figure 3). (1) Siliciclastic mud (<30µm), generally dominated by smectite or smectite-chlorite (Figure 4), is mainly contributed by hemipelagic suspensions; a glass fraction, where present, is wind-distributed volcanic ash and sea-surface pumice grindings. (2) Coarse silt (>30µm), sand, and gravel are contributed by bottom currents, including turbidity currents, and by air-fall of ash or flotation of pumice from volcanic eruptions. (3) The carbonate component includes (a) organic frameworks that grew in place; (b) allochems comprising skeletons and fragments of planktic and less-abundant motile benthic organisms,

Table 2. Percentage of Organic Carbon, and of CO_2 Expressed as Percent $CaCO_3$ (Analyses by AMDEL: Report AC505/85).

Sample No.	Percent Total Organic Carbon (a)	Percent $CaCO_3$(b) (Percent CO_2 x 100/44)
2-1-1B	0.16	38.2
2-2-1B	0.08	32.7
2-2-2B	0.05	33.8(c)
2-3-1B	0.69	22.5
2-3-2B	0.56	21.3
6-1-1	0.68	58.6
6-1-2	0.66	52.3
6-2-1	1.12	24.3
6-3-1	0.44	18.9
11-1-1B	0.03	2.1
11-1-3B	0.04	1.0
11-1-5B	0.04	1.2
14-1-1	0.02	88.4
14-1-2	0.02	88.9
14-2-1	0.13	26.4
14-2-2	0.39	34.5
14-2-3	0.56	45.9(c)
14-3-1	0.90	1.0
19-1-2B	0.05	12.5
19-1-4B	<0.02	13.4
19-1-6B	0.06	21.1
19-1-10B	0.05	25.0
21-1-1B	0.03	0.7
21-1-5B	0.03	0.5
22	<0.02	92.0
23-1A	0.02	70.7
23-2A	0.05	1.9
24	0.68	46.4
25-3B	0.25	50.9
29-1-1	0.04	72.7

(a) Determined by combustion of decarbonated samples in a LECO induction furnace with infra-red detection of CO_2.

(b) Determined by adsorption following acid evolution of CO_2.

(c) Contains dolomite.

including carbonate silt grains (>30μm); and (c) carbonate mud (<30μm) deposited from suspension, in some cases after aggregation into fecal pellets.

A minor siliceous biogenic component, in the form of opaline spicules, is present in some samples. Secondary materials (cements) consist of (1) calcium carbonate, dolomite (samples 2-2-1 and 14-2-3), or phillipsite (samples PNG 79-2-24A and B) in the interstices of rock frameworks; (2) concentrations of phillipsite crystals and pyrite framboids in the body-cavities of some planktic foraminifers; and (3) disseminated pyrite framboids. The carbonate cements are treated in detail by Niem (this volume).

SAMPLING SITES AND SAMPLES RECOVERED

Stations 1, 2, and 6 lie on the eastern flank of the platform that connects Guadalcanal and Santa Isabel (Figure 1), a feature which was exposed during the Pleistocene. At Station 1, 60 percent of the 10 kg of rock recovered consisted of well-rounded cobbles and a small piece of moderately indurated pebble conglomerate. The remainder comprised highly weathered and bored calcareous sandy mudstone.

A full dredge bag (ca. 100 kg) was recovered at Station 2, 95 percent of which comprised cylindroidal and plate-like pieces of calcareous sandy mudstone with abundant claystone fragments. The remaining 5 percent was lithic sandstone.

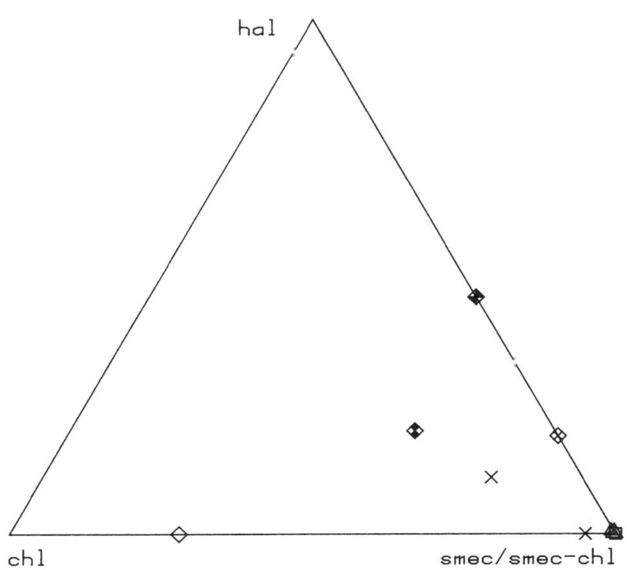

Figure 4. Clay mineralogy of selected samples. hal: halloysite; chl: chlorite; smec/smec-chl: smectite or smectite-chlorite intergrade. Symbols as in Table 3 except that half-closed symbols represent samples with smectite-chlorite intergrade. Open symbols represent samples with smectite.

The 20 kg of material collected at Station 6 was 50 percent soft, light yellowish-gray fossiliferous claystone, 20 percent soft dark to medium grayish-chocolate mudstone (in part oxidized), and 30 percent reef debris stained by iron and manganese oxides and manganese oxides.

Seismic profiles (Figure 5) suggest that the materials sampled at Stations 1, 2, and 6 are young slope sediments (unit I_{CD}, Bruns et al, 1986), except for the reef debris which is derived from a shelf-edge reef up slope (see Niem, this volume).

Station 11 is on a scarp in a shallowly dipping well-bedded sequence (Figure 5) near the top of the inner trench slope southwest of the Russell Islands (Figure 1). Four kg of siltstone and sandstone, partly coated by manganese oxide, was recovered. These rock types are similar to material described by Colwell and Vedder (1986) from Station 3 of the 1982 CCOP/SOPAC Cruise (L7-82-SP).

At Station 14, on the upper inter-arc basin slope close to Santa Isabel (Figure 5), the 6 kg of material recovered was all scree, without fresh surfaces, composed of coralline limestone stained by iron and manganese oxides (70 percent), sandy mudstone (25 percent) and muddy sandstone with serpentinite fragments (5 percent).

Seismic-stratigraphic units C and D (Bruns et al, 1986), which crop out on the flank of Fatu O Moana were sampled at Station 19 (Figure 5), where 90 percent of the 50 kg of material consists of moderately indurated, mottled (bioturbated) yellowish-gray foraminiferal muddy sandstone and mudstone. The remainder of the dredge haul was poorly sorted muddy volcanic sandstone.

Stations 21 to 23 are on successively shallower sites along the inner wall of the New Britain Trench west of Vella Lavella (Figures 1 and 5). The 1 kg of olive-gray sandy mudstone from Station 21 is coated with manganese oxides and is probably scree. Approximately 50 kg of reef-limestone rubble was recovered at Station 22. A ridge underlain by gently-dipping uniform reflectors (unit C_u of Bruns et al, 1986) at Station 23 yielded foraminiferal limestone containing volcanogenic sand and articulated bivalves (70 percent), and sandy mudrock (30 percent). The sandy mudrock resembles material described by Colwell and Vedder (1986) from Station 10 of the 1982 CCOP/SOPAC Cruise (L7-82-SP).

Station 25, for which no seismic record is available, is on the upper slope of the intra-arc basin southwest of Choiseul. The 1 kg of recovered rock consisted of coral limestone fragments from an adjacent shelf-edge reef, a well-rounded metadolerite clast probably river-transported to the shelf edge from basement outcrops during a Pleistocene lowstand of sea level, and some pale-olive bioturbated sandy mudstone probably representing seismic-stratigraphic unit D (Bruns et al, 1986).

At Station 29, 40 kg of sheared buff foraminiferal and nannofossil-rich chalk was recovered from either seismic-stratigraphic unit B1 or B3 (Bruns, Vedder, and Culotta, this volume) on the eastern flank of Nuguria Ridge (Figure 5).

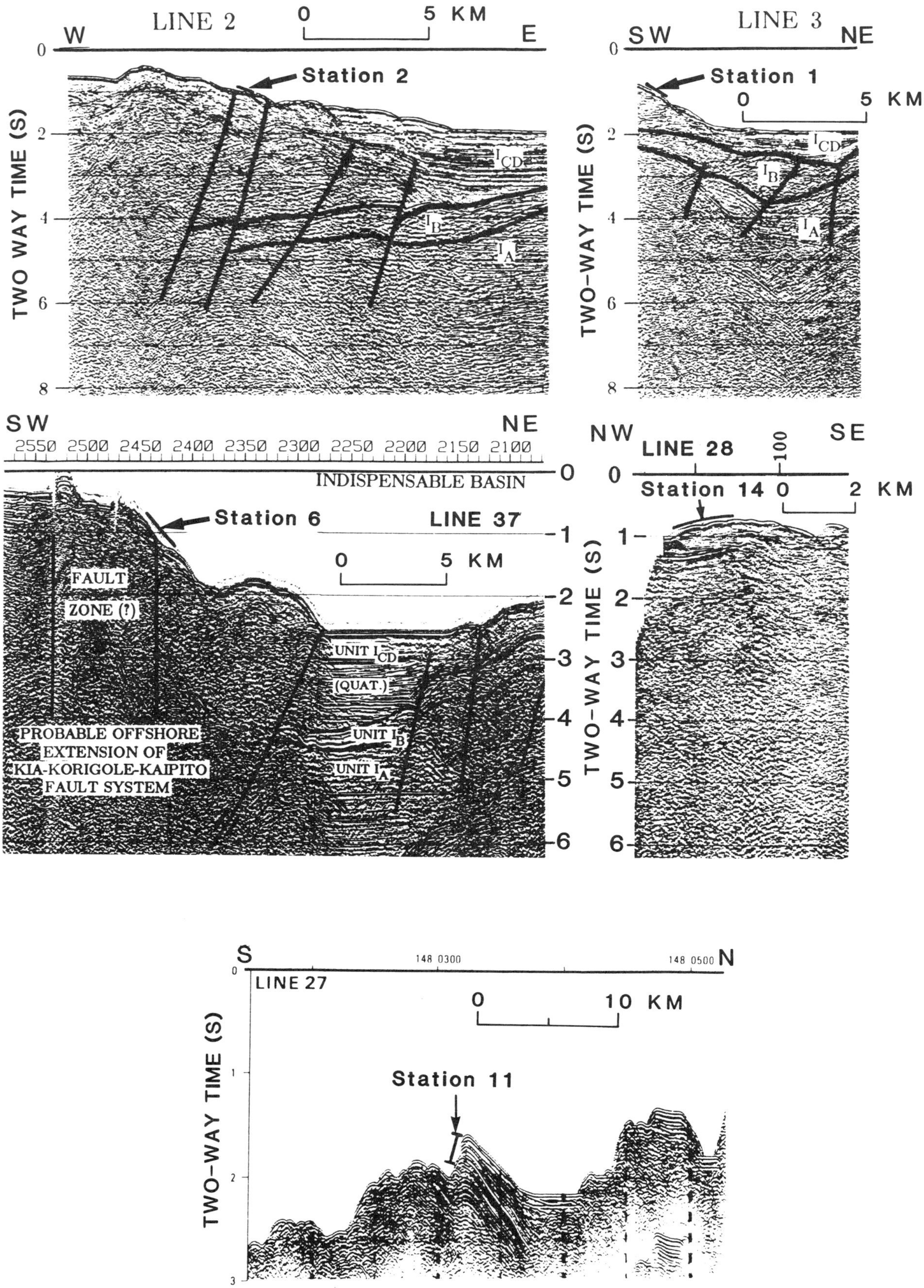
LINE 2
W
E
0 5 KM
Station 2
TWO WAY TIME (S)
I CD
I B
I A
LINE 3
SW
NE
Station 1
0 5 KM
TWO-WAY TIME (S)
SW
NE
2550 2500 2450 2400 2350 2300 2250 2200 2150 2100
INDISPENSABLE BASIN
Station 6
LINE 37
0 5 KM
FAULT
ZONE (?)
UNIT I CD
(QUAT.)
UNIT I B
UNIT I A
PROBABLE OFFSHORE
EXTENSION OF
KIA-KORIGOLE-KAIPITO
FAULT SYSTEM
NW
SE
LINE 28
100
Station 14
0 2 KM
TWO-WAY TIME (S)
S
N
148 0300
148 0500
LINE 27
0 10 KM
Station 11
TWO-WAY TIME (S)

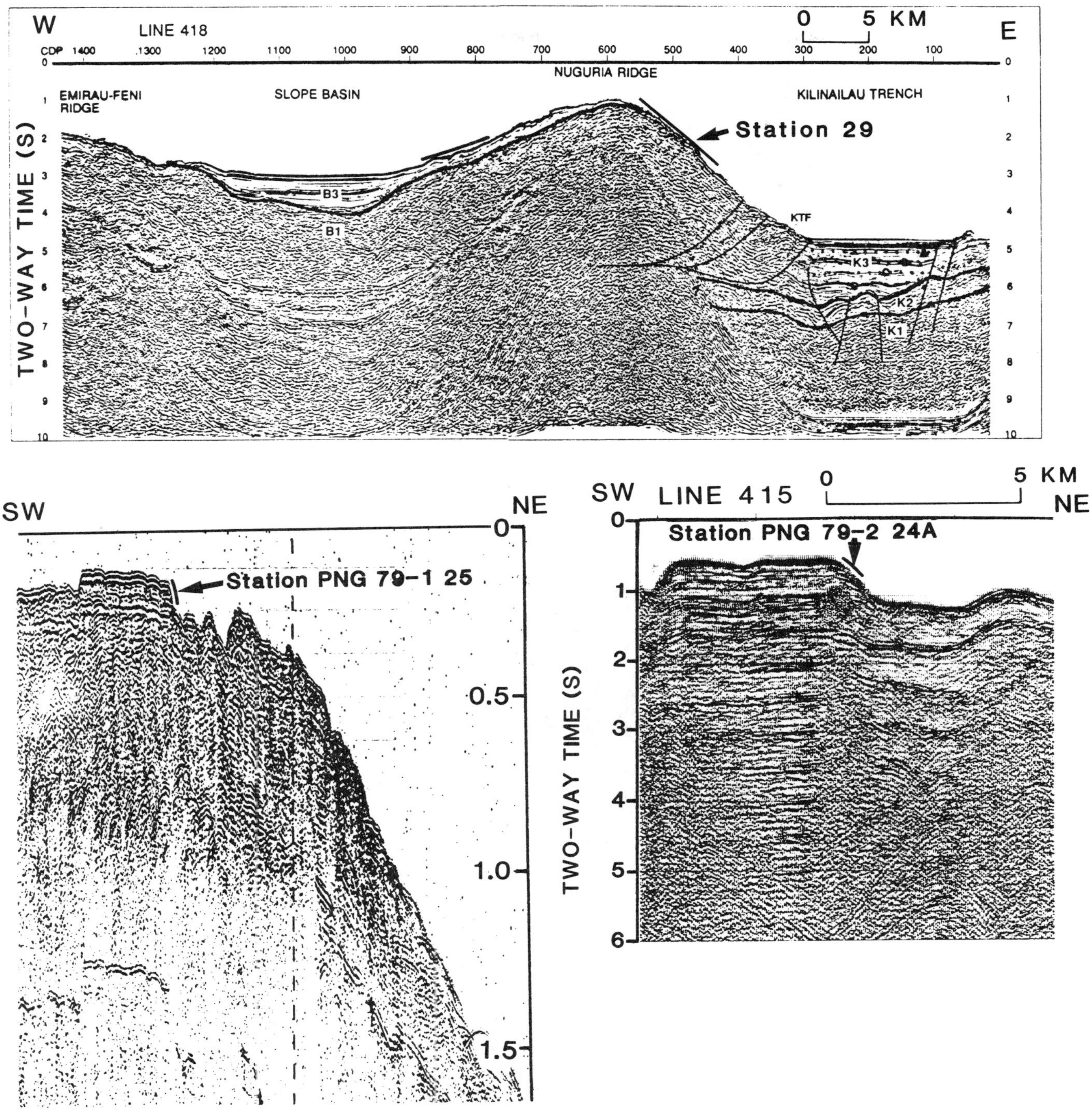

Figure 5. Seismic profiles showing stratigraphic and structural context of 1984 *S.P. Lee* (L6-84-SP) dredge stations. Complete seismic lines shown in following papers: lines 2, 3, 43, and 51 from Bruns et al (1986); line 37 from Bruns, Vedder, and Cooper (1986); line 402 from Bruns and Vedder (this volume); and line 418 from Bruns, Vedder, and Culotta (this volume).

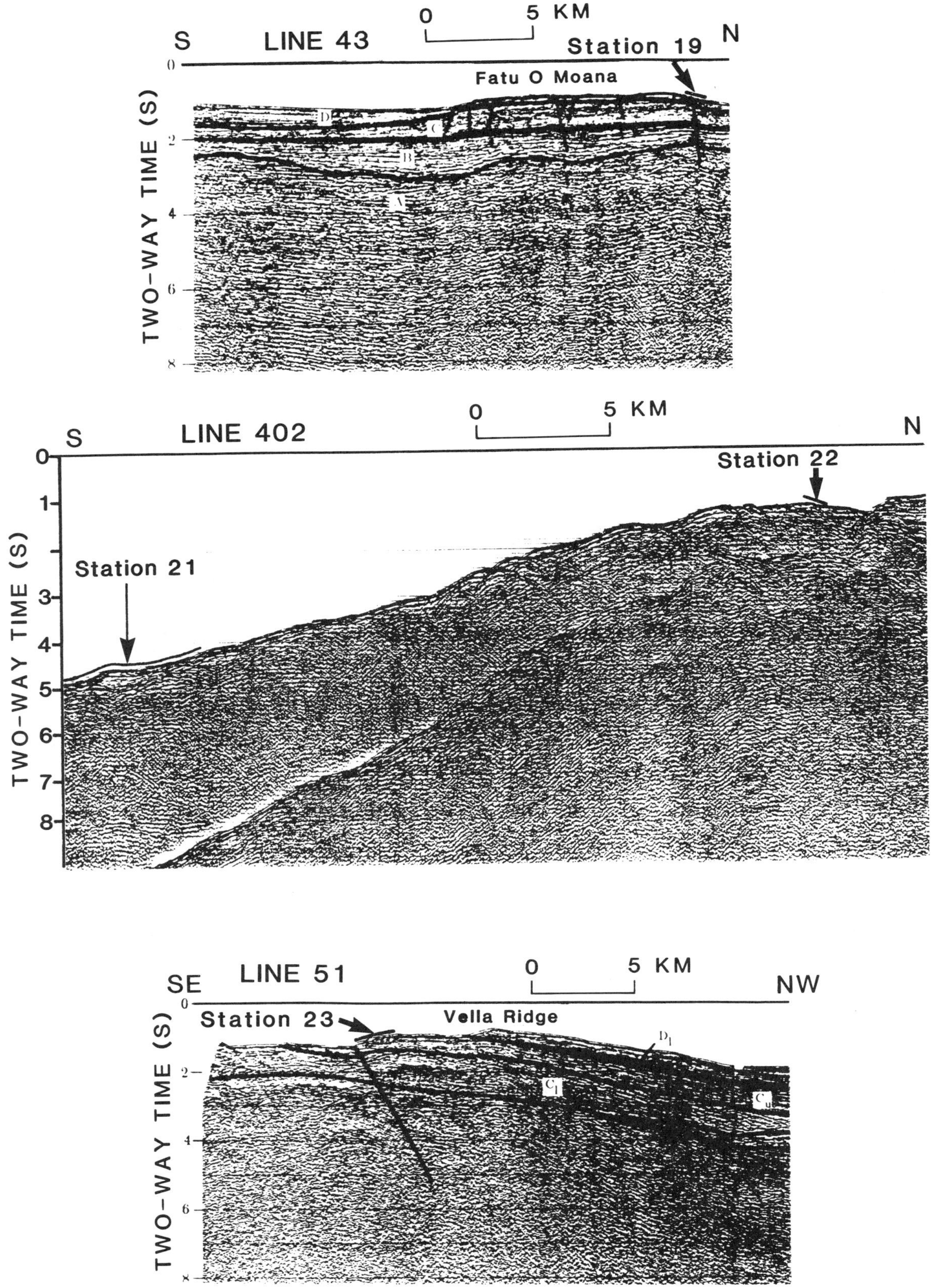

Figure 5. (Cont'd)

Limestone samples from Stations 30, 31, and 32, which were dredged from channels along the northeastern side of the Russell Islands, are discussed by Quinterno (this volume).

Stations PNG 79-1-24 and PNG 79-1-25 on the Emirau-Feni Ridge (Figure 5) yielded tuffaceous mudstone and foraminiferal sandstone that probably came from seismic-stratigraphic unit E4 (Vedder, Bruns, and Cooper, this volume).

Station PNG 79-2-24A on Kaviki Ridge (Figure 5) yielded clastic limestone derived from seismic-stratigraphic units E3 or E4 (Vedder, Bruns, and Cooper, this volume).

TEXTURE AND STRUCTURE OF SAMPLES

All samples of mudrock contain coarse silt that is commonly accompanied by very fine to medium siliciclastic sand. The relative proportions of mud and coarser fractions vary widely, and the mudrock is associated or interlaminated with muddy very fine to fine sandstone, and rarer muddy medium to very coarse sandstone (Table 3). Well-sorted sandstone is present at only two stations (2-2-1, PNG 79-2-24A).

Many samples of mudrock and some pieces of sandstone are texturally inhomogeneous (Table 3). Lamination, where present, is generally disturbed by bioturbation. Irregular to ovoid patches of finer material are present in the mudrock; these patches are probably burrow infills. Some burrows display an ordered internal fabric defined by nested curved laminae; these are striotubules (Brewer, 1964; Fitzpatrick, 1984).

Many samples have been bored. The borings are commonly 0.5-2 mm in diameter but range up to 1 cm. These cavities are generally partly infilled with fine to medium sand, which may be muddy, and silty mud. Similar materials, together with coarse sand, are present as infill in the framework of some limestone samples. Within the borings, the mud may form aggrotubules (Brewer, 1964) composed of fecal pellets. Some of the sand is organized into granotubules and isotubules (Brewer, 1964), which are circular or elliptical in cross section and have a large central cavity. Some tubules are compound structures attached to the walls of borings, but most are simple tubes that show no sign of attachment.

The surfaces of the borings are generally somewhat irregular; a few are coated with a layer of muddy silt <0.2 mm thick. In some samples the host rock contiguous with the borings is stained orange (neoferrans; Brewer, 1964; Fitzpatrick, 1984). Staining is more widespread in a few samples (Table 3) as a consequence of impregnation by iron sesquioxides. The staining presumably is a result of oxidation at the surfaces of outcrop or talus fragments exposed on the sea floor.

Some unusual lithification features are present. Hand specimens from sample 1-2-1 are irregular calcareous, porous, bored manganese-coated crusts that may be part of a hardground. Cylindroidal concretions, 4-25 cm in diameter, 5-40 cm long, with straight circular central cavities 1-3 cm in diameter, predominate at Station 2 (samples 2-1-1 and 2-3-1). These concretions may have formed by precipitation of calcium carbonate around decaying wood fragments. Alternatively, they may be fluid-venting structures similar to those reported by Kulm and Suess (1987) and explored further by Niem (this volume).

FRAMEWORK COMPONENTS

The siliciclastic framework components in the dredged rocks are of quite varied provenance and composition. In general, provenances readily apparent from petrographic examination are volcanogenic material, metabasite, metamorphic tectonite, ultramafic rocks, and sedimentary rocks. The phase chemistry of the framework components is discussed below.

Volcanogenic Materials

Sand and coarse silt grains of volcanic origin are ubiquitous in this suite of sedimentary rocks. Fresh monomineralic mineral grains, most of which are also monocrystalline and euhedral or subhedral consist of: calcic plagioclase, which is commonly zoned; amphibole, which exhibits three schemes of pleochroism—yellow green to greenish brown, yellow brown to dark brown, and straw yellow to bright red brown; pale green clinopyroxene and rare orthopyroxene; and opaque oxides. Biotite occurs at Stations PNG 79-1-24, PNG 79-1-25, and PNG 79-2-24A, and rare grains occur at Station 11; but it is absent in samples from other stations. Andesitic volcanic rock fragmentsare commonly vitrophyric or microlitic. Lathwork and felsitic textures are less common. Most rock fragments are fresh, but the original glass in some fragments is altered. Glass shards are present at Stations 11, 21, PNG 79-1-24, and PNG 79-1-25, together with fresh pumice which is more widespread. Altered pumice, in which the glass has been replaced by phyllosilicates, is common at Station 19 and rare at Station 2. Brown glass chips (hyaloclasts) occur at Station 21 (Table 3). Phases that form monomineralic detrital grains occur as phenocrysts and microphenocrysts in the pumice and other volcanic rock fragments.

Many of these volcanic components are so fresh that they could easily be taken as evidence for contemporaneous volcanicity. Such an inference is clearly warranted where pumice, glass shards, or hyaloclasts are present, but in other cases, such an inference is much less certain. Katz (1986) commented on the prevalence in the Solomon Islands of volcaniclastic material derived from erosion of basement rocks. Although volcanic basement rocks are usually somewhat altered by low-grade metamorphism that resulted in devitrification of glass and growth of secondary minerals, such alteration did not necessarily take place within the edifices of extinct volcanic centers. The deeply entrenched streams that dissect such edifices are capable of supplying substantial

Table 3. Modal Analyses.

Sample no.	symbol	quartz g	quartz gz	quartz gl	plagioclase calcic g	calcic gz	calcic gl	calcic zl	plagioclase sodic g	sodic gz	sodic gl	pyroxene opx g	clinopyroxene g	clinopyroxene gz	clinopyroxene gl	amphibole green or brown g	green or brown gz	green or brown gl	red-brown g	red-brown gl	actinolitic g	actinolitic gz	actinolitic gl	actinolitic zlz	opaques g	opaques gz	opaques gl	Cr spinel g	biotite g	epidote g	epidote gz	epidote gl	epidote zl	epidote zlz	pumpellyite g
1-2-1		0.5		0.3	0.9	0.2										0.9	0.2	0.2																	
f'work	⊠	4.9		0.8	4.5	0.8	0.4		0.8				0.4			4.5	0.4	0.4																	
1-2-1 infill		1.1			3.0											1.5									1.5										
f'work	⧗	4.9			16.0				2.5							6.2	1.2								13.6	1.2									
2-1-1		1.0	0.6		0.2	0.6			0.2							1.0	0.4													0.4					
f'work	◇	4.0	3.7		1.0	1.0			0.3							3.7	3.0													0.7		0.7			
2-2-1 f'work	◇	6.6	3.3	0.3r	6.3		1.7		2.0		2.0	0.3	3.0			3.7					3.7		7.3			1.0				6.0		8.6			
2-2-2	◇	1.6		1.2	2.6		1.6	0.2	0.4		1.0	0.6	1.6			1.4		1.6	1.4	4.6					0.4					1.0	0.4	6.0			
2-3-2		0.8	1.6		0.8					0.6	0.2			0.2		0.4					0.6	2.4	1.8	0.4		0.2				0.4	3.2	0.4	1.4	0.2	
f'work	◇	3.0	4.7	1.3	2.0	1.0	1.0	0.3	0.7	2.7	0.3		1.3	0.7		1.3					3.0	5.7	4.0	0.7		0.3				2.7	13.0	2.3	5.0	1.0	
6-2-1	⟐	0.8	1.4	0.8	0.6		0.4		3.2	0.2	0.8		0.4	0.2		0.2					1.2	1.0	0.6		0.8	0.6		0.4		0.4					
11-1-4	△				0.6	3.2		0.2s						1.8	0.2s	0.2	0.8		0.2																
infill*	▲				36.9		4.0					0.5	11.6		2.7	9.2		0.5	1.3	0.5					3.8		0.5		0.3						
14-2-2	⟐	0.2	0.6		0.2	0.2								0.2			0.2									0.2									
14-2-3 f'work	⟐	1.9			0.4				3.7		13.0		9.3		8.9								0.4												
14-3-1	⟐				2.8	0.3	1.1				0.8		1.1	1.4		0.6												0.3							
19-1-1	×				2.4	0.8	0.8						2.0												0.2	0.2									
19-1-9	×				6.6	1.2	1.0						0.4			4.0	0.2		0.4	0.2					0.2										
19-2-2	×				17.6	0.2	1.0	0.4					6.0			0.2									3.2		0.2								
21-1-2	○				15.0	1.2	2.2						3.4	1.0	1.2	1.6	0.2		0.4						1.0										
21-1-3	○				6.2	0.2	3.2						3.2	0.6	0.2	2.4	0.4	0.8	0.2						0.6										
infill*	●				24.1		1.5						13.4			21.0		1.2	0.5						31.7		0.4								
21-1-4	○				13.0	1.6	4.8						2.0	1.0	0.6	2.4	0.4	0.4	0.2	0.2t					1.0	0.2	0.6								
22-0-4 infill*	✇				50.4		1.2						3.8			22.6			1.6						12.9										
23-1A	□				13.0		0.2						0.2			2.2		0.2																	
23-2A	□				3.8	0.8	1.0						3.8	0.4		4.0	1.0		0.4						1.2	0.4									
25-1 infill*	▼	2.0			6.0		5.5		5.5		7.0					22.4		3.3	0.7		3.5		4.0		1.3					4.2		0.5			5.5
25-3		tr			0.4	2.0	0.2		0.8							1.0	2.8	0.2			tr							tr		tr					
f'work	▽				2.4	8.5	0.7		2.0							9.2	14.6	1.7							0.3										
29-1-1	⊞				0.7	0.6	0.1						0.1			0.1	0.3																		
30-1-1	✡				1.2								0.2			1.2																			
30-1-2 infill*	✶				37.0		3.7						7.4		2.2	25.2			0.7						12.6										
30-2-1	✡				4.7		0.3						1.3		0.2	5.0		0.3							1.0				0.2m						
31-1-1	✡				3.3								1.5		0.2	1.7									0.6										
31-1-3	✡				2.5	0.2	0.3						0.7		0.2	2.5		0.2	0.2	0.3					0.2		0.3								
PNG24-A zone A	+				23.6								3.8			10.6									6.0										
zone B	+				10.0								0.4			4.6	0.2		0.2						1.0										
PNG24-B	+				2.8	0.4	0.2						0.6																0.4						
PNG24A-1	⊕				0.6								0.3			0.5			0.2						0.4				0.21						

	fossils: forams	algae	ostracod	bivalve	gastropod	echinoderm	scaphopod	bryozoa	spicules	matrix: smectite	micrite	cement (CO3)	voids	points	texture	biogenic & other structures
1-2-1	4.2		1.2	0.5		0.7	0.5			69.0		6.1	8.2	572	f sandy mud	fecal pellets, borings, neoferrans, stained
f'work	22.1		11.1	1.2	2.0	3.7	4.5			–	–	–	–	244		
1-2-1 infill	3.0			0.4		1.5				63.3		22.5		267	mud; f sand	strio- & aggrotubules
f'work	25.9			1.2		4.9	1.2			–	–	–	–	81		
2-1-1	0.8		0.2			0.4				72.0		0.2		500	c silty mud	borings with grano- & aggrotubules
f'work	3.3		0.7			0.7				–	–	–	–	300		
2-2-1 f'work	6.3								0.3	–	–	–	–	301	vf sand	few borings, stained
2-2-2	0.6		0.4					0.4			46.4		0.6	500	muddy m sand	few borings
2-3-2	1.8					1.0	0.6			74.0			1.0	500	f sandy mud	bioturbated
f'work	6.3		0.3		0.3	2.0	1.7			–	–	–	–	300		
6-2-1	3.8	2.2		2.2	0.6	1.4	0.4	1.0	0.4b	18.4			9.0	500	muddy f sand	striotubules
11-1-4										80.0h				500	c silty mud	burrows, borings, laminae
infill*										–	–	–	–	371*	mud; f sand	granotubules, isotubules
14-2-2	1.2		0.2		0.2	1.4				86.0			1.2	500	c silty mud	stained
14-2-3 f'work										–	–	–	–	270	m sand	bioturbated c silty mud with m sand blebs
14-3-1										45.7			1.1	363	muddy f sand	
19-1-1	8.4									64.4		0.2c	2.2	500	f-m sandy mud	(some c-vc sand)
19-1-9	9.6		0.2							69.4			1.6	500	f-m sandy mud	(some c-vc sand) borings: mud, f-m sand & iso-
19-2-2	4.2									46.0				500	muddy c-vc sand	irregular mud & sand bodies. /tubules
21-1-2										62.4h				500	vf sandy mud	texture mottled; borings: fecal pellet aggro-
21-1-3	1.4e									70.8				500	vf sandy mud	borings. /tubules & neoferrans
infill*									0.4	–	–	–	–	605*	mud; m sand	simple & compound isotubules
21-1-4	2.0e								0.2	51.2h			1.8	500	m sandy mud	borings
22-0-4 infill*										–	–	–	–	311*	f sand; mud	(boring infill in coral biolithite)
23-1A	20.2		1.0	0.6	0.4	0.2					60.2f			500	m-c sandy mud	patchy distribution of micrite
23-2A										71.8				500	f-m sandy mud	rare borings with m sand infill
25-1 infill*										–	–	–	–	401*	c sand	(infill in interstices of coral biolithite)
25-3	1.2		0.6	2.4		1.4				80.2			0.8	500	vf sandy mud	bioturbated
f'work	14.2		2.4	11.9	1.0	9.2	0.3			–	–	–	–	293		
29-1-1	21.3			0.1		0.1	0.1				67.2		8.5	1000	m sandy mud	
30-1-1	3.5	3.6	0.5	3.3		4.5	1.3b	2.2	0.3	1.0	35.8	13.7	9.9	606	muddy f sand	borings; iso- & granotubules
30-1-2 infill*														135*	vf sand	(infill in borings)
30-2-1	6.7	2.2	0.2n	4.7	0.2	2.2	2.8b	3.4		2.3	40.7		15.7	597	m sandy mud	some borings
31-1-1	8.6	2.3	0.2	2.4	1.5	2.4		5.5	0.2p	1.5	62.2	0.2	3.9	545	m-c sandy mud	borings
31-1-3	8.3	2.8		2.3	1.5	3.5	1.3b	4.8		7.3	33.1	0.7	22.0	601	m-c sandy mud	
PNG24-A zone A	5.6											45.4c	2.8	500	m sand	borings with neosesquioxidans, ferrans; neo-
zone B	27.6											52.4c	2.6	500	foram m sand	/ferrans
PNG24-B	0.4									84.8h		2.4c		500	vf sandy mud	extensive neoferrans
PNG24A-1	13.6	1.2	1.1	2.9	3.1	4.9	0.2	7.0			21.4	30.2	6.4	653	muddy f-m sand	

Sample no.	symbol	volcanic lithic components										papules						metamorphic lithic				sedimentary fragments				serp-entin-ite
		vitric		fels-itic	lathwork		micro-litic		pumice		hyalo-clasts	pale		green		dark		meta-basite	schist	fine-grained		micrite		chert	mud-stone	
		1	1z	1	1	1z	1	1z	1	1z	1	1	1z	1	1z	1	1z	1	1	1	1z	1	1z	1	1	1
1-2-1		0.5	0.2	C.3												0.9	0.7		0.3	1.7	1.6					
f'work	⊠	1.6	0.4	C.8	0.4		0.8									7.8	3.7		1.2	14.8	5.7					
1-2-1 infill																0.4	0.4			1.1						
f'work	■						1.2									3.7	1.2			11.1	3.7					
2-1-1		2.0				0.2			0.2aq							1.4	2.6			7.0	8.6					
f'work	◇	4.7				0.3			0.3a							7.3	9.3			20.3	34.0					
2-2-1 f'work	◇	6.3												0.3		1.0	0.3			17.9	1.0	1.3		0.3		
2-2-2	◇	2.4					0.4							0.4		3.6	1.6			12.0	1.2	1.8		0.2		
2-3-2		1.8														0.4	1.0			2.0	0.8					
f'work	◇	4.3												1.0		3.3	5.0			11.7	3.3					
6-2-1	◈	3.0	0.2									5.4	1.0	3.6		27.0	0.8					2.4	0.6		1.4	
11-1-4	△	0.8	8.8				0.4	1.2	0.2							0.2	0.4									
infill*	▲	8.6		1.6			13.7		3.0							1.1										
14-2-2	⟐											0.8		0.4		2.6	2.0			1.0	0.4	0.6	0.6			
14-2-3 f'work	⟐	11.1		2.2	1.5									1.9		7.4		5.6	1.1					1.1	0.7	30.0
14-3-1	⟐						0.3					19.3	5.2	0.3		3.1		1.4		5.8	1.4				7.7	0.6
19-1-1	×	6.6	0.8		0.4		3.2		6.0a																1.6d	
19-1-9	×	1.0		1.6			1.2	0.2	0.4a	0.2a																
19-2-2	×	9.8			1.6		2.2		6.6a																0.8	
21-1-2	○	1.2			0.6		3.0		4.8		0.8															
21-1-3	○	1.8	1.0		0.6		4.0	0.8	1.6																	
infill*	●	0.5		0.6			4.8																			
21-1-4	○	2.8	1.6	0.8	0.2		2.2		7.8	1.0																
22-0-4 infill*	⊛	3.5		0.3			1.0		2.6																	
23-1A	□	0.4					0.8									0.4										
23-2A	□	2.4			1.2		7.2	0.6																		
25-1 infill*	▼	8.5					1.0		0.5							6.5		10.3						2.0		
25-3		0.2					0.4								0.2	2.4	1.0						0.8			?tr
f'work	▽	0.7					1.0								0.7	8.8	6.1					1.0	2.4			
29-1-1	⊞	0.4					0.1		0.6e																tr d	
30-1-1	✡						0.5															16.9j	1.5k			
30-1-2 infill*	★	5.2a					5.9a																			
30-2-1	✡						1.2a															4.7j			tr	
31-1-1	✡						0.2															2.0j				
31-1-3	✡	0.3a					0.3									0.2						4.0j				
PNG24-A zone A	+						0.6								0.4	0.2						1.0j				
zone B	+						0.2									0.4						0.4j				
PNG24-B	+	1.4a		0.4	tr		1.2	0.4	1.2	3.4																
PNG24A-1	⊕															0.2						5.8j			tr	

a – altered c – phillipsite e – dissolved h – abundant glass fragments k – pellets n – barnacle
b – coral fragments d – volcanic arenite f – micrite + microspar j – diverse intraclasts m – apatite p – calcareous
q – fresh shards in 2nd sample r – 'zl'count s – 'zlz'count t – 'gz'count tr – trace; present, no count
* – based on grain count f'work – count of framework components only

g – monomineralic sand grains gz – monomineralic silt grains gl – sand-sized grains in rock fragments
zlz – silt-sized grains in silt-sized rock fragments 1z – silt-sized rock fragments
zl – silt-sized grains in sand-sized rock fragments 1 – sand-sized rock fragments

Table 4. Chemical Analyses of Cobbles Dredged from 1984 *S.P. Lee* Cruise (L6-84-SP) Station 1.

Major Elements	Sample Numbers												
	1-1-1 (1)	1-1-2 (1)	1-1-3 (1)	1-1-4 (1)	1-1-5 (1)	1-1-6 (1)	1-1-7 (1)	1-1-8 (1)	1-1-9 (1)	1-1-10 (1)	1-1-11 (2)	1-1-12 (1)	1-1-13 (2)
SiO_2	46.74	48.87	48.02	48.17	48.38	48.47	50.00	49.88	47.57	48.47	76.70	47.59	58.22
TiO_2	1.30	1.35	1.29	1.48	1.29	1.31	1.36	1.55	1.47	1.49	0.15	1.49	1.28
Al_2O_3	14.23	13.55	14.29	13.49	13.66	13.86	11.52	13.79	14.68	13.40	10.91	13.99	11.00
Fe_2O_3	3.49	3.66	2.78	3.63	2.80	2.65	5.43	2.57	3.14	4.18	3.90	3.47	4.24
FeO(2)	8.10	7.38	8.06	8.54	8.74	8.26	6.61	7.59	8.48	8.52	0.63	8.58	5.31
MnO	0.18	0.18	0.17	0.19	0.18	0.17	0.18	0.18	0.19	0.17	0.06	0.21	0.21
MgO	7.38	6.92	7.72	6.88	7.14	7.65	5.86	7.19	6.29	6.77	0.70	6.99	4.70
CaO	13.08	11.72	12.44	11.34	12.34	11.43	16.81	12.56	13.28	10.62	10.02	11.65	13.48
Na_2O	3.91	4.15	3.92	3.84	3.44	3.95	0.75	3.50	3.29	3.62	0.23	3.31	0.31
K_2O	0.20	0.21	0.11	0.20	0.12	0.39	0.06	0.08	0.06	0.28	0.08	0.12	0.09
P_2O_5	0.11	0.14	0.12	0.14	0.13	0.12	0.13	0.11	0.14	0.15	0.04	0.15	0.12
S	0.09	0.01	0.01	0.10	0.02	0.04	0.03	0.01	0.02	0.01	--	0.01	--
L.O.I.(2)	2.27	2.18	2.26	2.32	1.79	2.25	1.73	2.07	2.06	2.59	1.34	1.29	1.32
TOTAL	101.05	100.83	101.20	100.22	100.02	100.64	100.47	100.66	100.68	100.20	104.76	98.84	100.28

Analysts: (1) B.W. Chappell, Australian National University; (2) Robert Anisi, Solomon Islands Geological Survey, Honiara

Notes: Samples 11, 13: MnO, MgO, CaO, Na_2O, K_2O by AAS; SiO_2, TiO_2, Al_2O_3, P_2O_5 and total Fe_2O_3 by colorimetry.
Samples 1-10, 12: XRF determinations, except for FeO.
All samples: FeO by titration. XRF determinations of Na_2O are of low precision.
L.O.I.: Lost on ingnition

1-1-1. saussuritized metabasalt (spilite with pillow-interior texture)
1-1-2. fine-grained greenschist (spilite)
1-1-3. fine-grained greenschist
1-1-4. fine-grained greenschist
1-1-5. actinolitic greenschist
1-1-6. sheared saussuritized metabasalt (spilite with pillow-interior texture)
1-1-7. felted greenschist (spilite; metasomatized)
1-1-8. schistose actinolitic metadolerite
1-1-9. fine-grained greenschist (spilite)
1-1-10. fine-grained greenschist
1-1-11. schistose quartz-epidote vein
1-1-12. schistose actinolitic metabasalt
1-1-13. schistose quartz-epidote vein

quantities of older volcanic detritus, including monomineralic grains that are essentially unaltered.

Metabasites

At Stations 1, 14, and 25, fragments of pervasively altered volcanic and holocrystalline mafic rocks are present. Although the primary igneous textures in these grains are preserved, primary minerals are either absent or are represented by relicts of calcic plagioclase and clinopyroxene. Glass is replaced by phyllosilicates, plagioclase is albitized, and pyroxene is replaced by green amphibole and chlorite. Secondary epidote and pumpellyite are present interstitially or in patches that disrupt the primary texture. These secondary minerals may also occur as monomineralic sand grains. The original nature of most of the metabasites is indeterminate. In a few grains, relict textures and primary minerals enable recognition of metaboninite, metadolerite, and ophiolitic metabasalt (spilite). A well-rounded clast of metadolerite was dredged at Station 25 (sample 25-2), and spilite cobbles were recovered at Station 1 (Table 4).

The metabasites were probably derived from a volcanic basement complex, but the timing of their alteration is difficult to establish. The metamorphic assemblage may have formed during hydrothermal alteration while the volcanic rocks were in contact with

Table 5. Chlorite Compositions Determined by Microprobe Analyses at Unocal Research Laboratories, Brea, California.

Major Components	1	2	3
SiO_2	26.48	26.71	26.02
TiO_2	0.05	0.00	0.00
Al_2O_3	20.27	19.39	20.83
FeO	26.18	27.42	26.57
MgO	14.81	14.29	14.42
MnO	0.47	0.14	0.45
CaO	0.05	0.19	0.08
K_2O	0.00	0.00	0.00
Na_2O	0.00	0.00	0.03
Cl	-	0.00	0.03
Total	88.32	88.13	88.41

1: Sample 1-1-2, chlorite bleb in fine-grained greenschist
2: Sample 1-1-6, chlorite with epidote crystals in metabasalt
3: Sample 1-1-10 chlorite in fine-grained greenschist

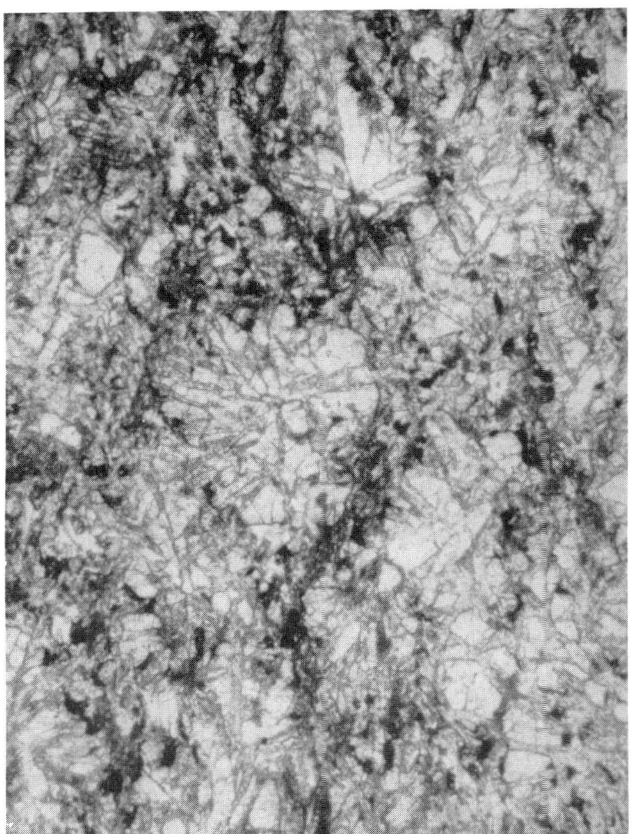

Figure 6. Ophiolitic metabasalt (sample 1-1-6) showing altered feldspar in ophitic relationship to clinopyroxenes, probably from the core of a pillow. Dark bands are spaced shear planes.

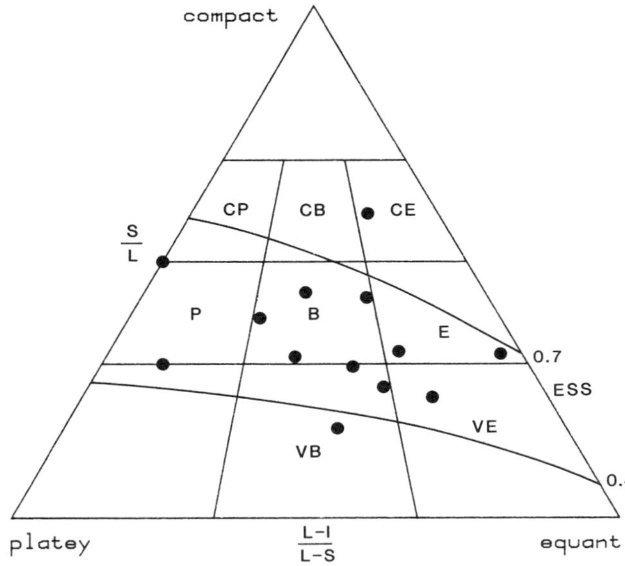

Figure 7. Form indices for cobbles from Station 1. S, I, L: short, intermediate and long axes; ESS: effective settling sphericity; C: compact; P: platey; B: bladed; E: equant; V: very.

seawater soon after their formation, or the metamorphism may be a regional phenomenon that resulted from burial long after volcanic activity ceased. Chlorite compositions (Table 5), when calibrated using the model of Walshe (1986), indicate formation temperatures in the range 239°-266°C. Vein quartz from Station 1 (sample 1-1-11, Table 4) gave $\delta^{18}O$= ±10.5 percent relative to SMOW. The implied formation temperature is 220°C, assuming a $\delta^{18}O$ of 0.00 percent for the fluid phase.

Metamorphic Tectonites

Texturally reconstituted metamorphic rock fragments derived from a metamorphic basement complex are present as sand grains in some samples. The principal rock types are quartz-amphibole schist and schistose to massive actinolitic amphibolite. All samples probably represent metamorphic tectonites, the scale of the fabric in the massive amphibolites being larger than that of the fragments. These rock fragments are accompanied by monomineralic sand grains of quartz (polycrystalline or monocrystalline with undulose extinction), sodic plagioclase, green actinolitic amphibole, and epidote.

These metamorphic rock components closely resemble tectonites retrieved at Station 1, where a suite of cobbles includes fine-grained greenschist, actinolitic schist, and schistose quartz-epidote vein-rock (Table 4). Relict textures resemble those in ophiolitic basalt (Figure 6), but differ from textures typical of MORB (W.E. Cameron, written commun., 1987). Mafic volcanic rocks were evidently the protoliths for these tectonites, which

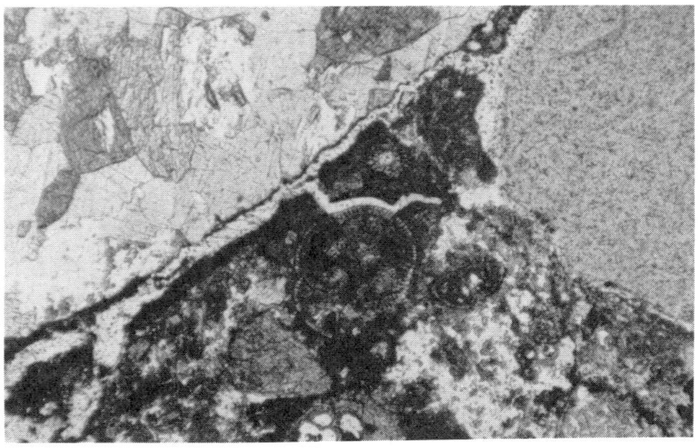

Figure 8. Foraminifer in matrix of conglomerate (sample 1-1-14).

exhibit contorted metamorphic foliations or spaced shear zones. Mineral assemblages indicative of greenschist facies are actinolite-epidote-sphene ± chlorite ± albite ± quartz ± calcite. Relict clinopyroxene is present in some samples.

The cobbles, which have short-axis dimensions of 2.0-5.5 cm, are well rounded. They vary somewhat in form (Figure 7), the mean effective settling sphericity being 0.57 ± 0.11. River or beach abrasion is indicated, followed by incorporation in a weakly lithified marine conglomerate of which sample 1-1-14 is the matrix. In this sample, planktic foraminifers,fresh hornblende gabbro, and metabasite pebbles accompany metamorphic tectonite detritus (Figure 8).

Ultramafics

In some samples from Station 14, serpentinite fragments are a conspicuous component. They are accompanied by grains of red-brown chrome-spinel and rare granitic fragments. Metabasite detritus is also present. The total assemblage indicates that the serpentinite and chrome-spinel were probably derived from the ultramafic parts of an ophiolite complex. The granitic fragments may have been derived from trondjhemites within the ophiolite.

Papules

Many samples contain uniformly pigmented green, yellow, or brown sand grains that lack a distinctive internal fabric when examined in plane-polarized light. Some grains are ovate and well-rounded; others are irregular and angular. Under crossed nicols, these grains display a range of fabrics: some are isotropic, some are aggregate polarizing (argillasepic; Brewer, 1964; Fitzpatrick, 1984), and many show patchy to strongly continuous patterns of birefringence. Interference colors vary from gray to bright orange; higher-order interference colors are occasionally discernible. The characteristics are comparable to those recorded for glaebules and papules by Brewer (1964), and the term papules may conveniently be applied without genetic connotation.

Optical properties suggest that the papules are composed of various phyllosilicates and serpentine minerals. Their crystal-chemical ordering probably varies from incipient to highly regular, and their crystallographic fabrics also vary, from random to a strongly preferred orientation. Pale brown, gray, and colorless papules probably represent small serpentinite fragments. Other papules, strongly pigmented in darkshades of yellow and brown, are probably palagonite or other vitric volcanic rock fragments which have altered to phyllosilicates. Green and yellow-green types may be chlorite or perigenic glauconite. A detailed micromineralogical study would be needed to establish properly the nature and provenance of the various kinds of papules.

Sedimentary Rock Fragments

A few samples contain rare sedimentary rock fragments: mudstone, micrite, recrystallized limestone (sample 1-1-14), chert, volcanic litharenite (sample 19-1-1), and muddy very fine feldspathic arenite (sample 29-1-1). The chert is associated with metamorphic detritus and may have come from chert bodies within a metamorphic basement complex. The various arenites evidently represent older volcaniclastic rocks and provide definitive evidence for incorporation of reworked volcanogenic materials into this suite of sedimentary rocks. The micrite fragments are probably intrabasinal, but the mudstone fragments may be extrabasinal. Reworked microfossils are reported by Resig (this volume) and Shafik (this volume) from Stations 2, 6, and 19. At Station 23, reworked foraminifers infilled with smectite mud are present in biomicrite (sample 23-1A).

FRAMEWORK COMPOSITIONS AND SPATIAL VARIABILITY

Comprehensive modes, showing abundances of components discussed above, have been determined for many samples (Table 3). Most of these determinations are based on standard point counting, in which grains <30μm are treated as matrix so that coarse silt grains (>30μm) are counted with sand grains in calculating the values of framework parameters (Table 6). Coarse silt is abundant in these muddy sediments, and the silt composition differs appreciably from the sand fraction (e.g., sample 2-3-2 framework count, Table 3). Silt-size components and monomineralic grains in rock fragments have been shown separately on Table 3 to facilitate comparison of the data with sandstone modes determined using Gazzi-Dickinson counting conventions (Zuffa, 1980).

Table 6. Abbreviations for Siliciclastic Framework Parameters and Indices.

Abbreviation	Definition	Figure
Q:	quartz as monomineralic sand and silt-size grains (g and gz components, Table 3)	Figure 9
F:	plagioclase as monomineralic sand and silt-size grains (g and gz components, Table 3)	Figure 9
R:	all other siliciclasts except green papules and mudstone clasts	Figure 9
q:	all quartz	Figure 10
f:	all plagioclase	Figure 10
l:	aphanitic volcanic and metamorphic lithic components, pale and dark papules, serpentinite, and extrabasinal sedimentary rock fragments	Figure 10
fmg:	all pyroxene, amphibole, opaques, chrome spinel, biotite, epidote, and pumpellyite	Figure 10
met:	metamorphic components: quartz, sodic plagioclase, actinolitic amphibole, epidote, pumpellyite, and metamorphic rock fragments	Figure 11
sp:	serpentinite, pale papules, and chrome-spinel	Figure 11
vmin:	volcanic mineral grains: all calcic plagioclase, pyroxene, non-actinolitic amphibole, opaques, and biotite	Figure 11
lv:	volcanic aphanitic lithic components and dark papules	Figure 12
vxl:	vitric volcanic lithic components, dark papules, pumice, and hyaloclasts	Figure 12
mic:	microlitic volcanic lithic components	Figure 12
fl:	felsitic and lathwork volcanic lithic components	Figure 12
g(m):	monomineralic sand and silt-size grains (g and gz components, Table 3) comprising calcic plagioclase, pyroxene, non-actinolitic amphibole, opaques, and biotite	Figures 13, 14
pxt:	sand-size grains of calcic plagioclase, pyroxene, non-actinolitic amphibole, opaques, and biotite occurring in rock fragments (gl components, Table 3)	Figures 13, 14
vlith:	"lv" plus silt-size grains in rock fragments (zl and zlz components, Table 3) comprising calcic plagioclase, pyroxene, non-actinolitic amphibole, opaques, and biotite	Figures 13, 14
l(a):	altered volcanic lithic components, pumice, and dark papules	Figure 14
l(f):	all fresh components of "vlith"; i.e. vlith minus l(a) components	Figure 14
MPI:	metamorphic provenance index: (met + sp)/(vmin + lv)	Figure 11
VPI:	volcanic provenance index: (g(m) + pxt)/vlith	Figures 13, 14
CMI:	comminution maturity index: g(m)/pxt	Figure 13
VAI:	volcanic alteration index: l(a)/(l(f) + g(m) + pxt)	Figure 14

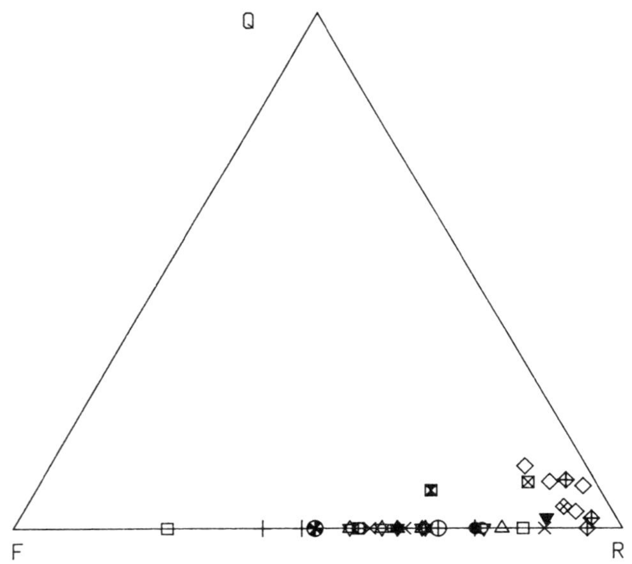

Figure 9. Silicilastic framework compositions; Q, F, and R parameters defined in Table 6; symbols as in Table 3.

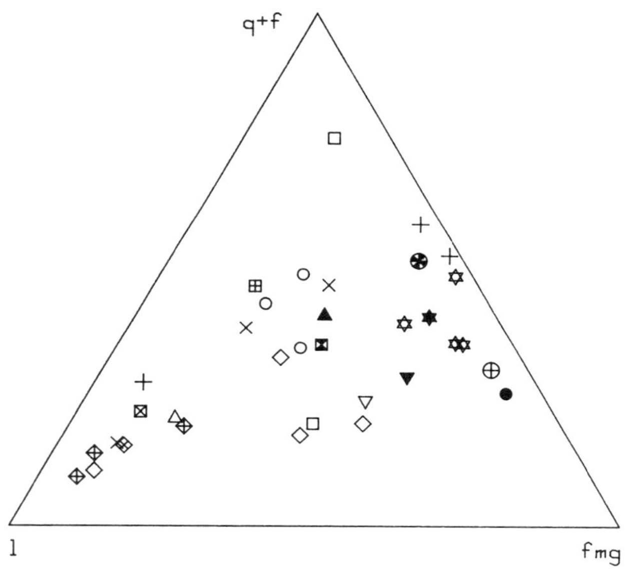

Figure 10. Silicilastic framework compositions, showing abundances of quartz plus feldspar (q+f), lithic (l), and ferromagnesian and related heavy mineral (fmg) components. See Table 6 for parameter definitions; symbols as in Table 3.

In cases where grains within borings or coral frameworks were few and dispersed, modes were determined by grain counting. All siliciclastic grains in the area chosen were counted. Each monomineralic grain was assigned to a sand-size grade on the basis of its long-axis dimension. Rock fragments containing phenocrysts were treated as composites of smaller lithic-hosted mineral grains and lithic components, with counts for each being

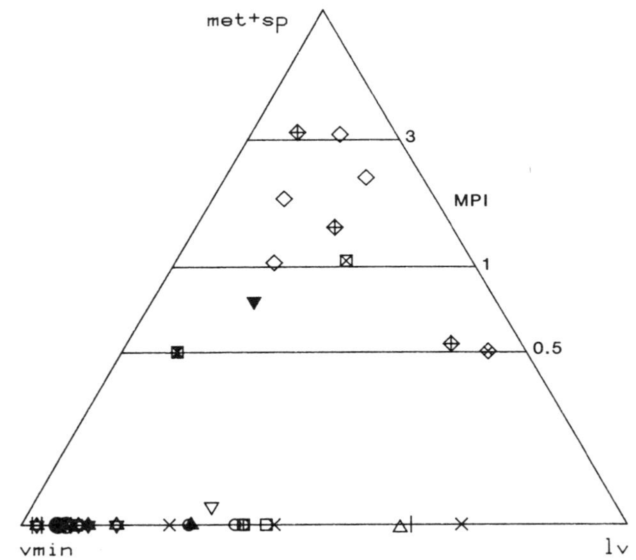

Figure 11. Relative abundances of basement (met + sp) and young volcanic (vmin, lv) components. Parameters and indices (MPI) defined in Table 6; symbols as in Table 3.

assigned to appropriate size grades. Final approximations to point-count modes were obtained by multiplying observed abundances by a geometric series of factors, ranging from x1 for very fine sand to x16 for very coarse sand, followed by summation of the weighted abundances for each component.

The siliciclastic frameworks of these rocks are quartz-poor volcanic or volcanic-metamorphic litharenite and feldspatho-litharenite (Folk, 1980) (Figure 9). Many samples contain significant amounts of ferromagnesian (femic) and related heavy minerals (fmg, Figure 10), but most are not placers. Similar abundances are reported by Crook and Taylor (1987) for samples from other localities in this region. The infills of borings are enriched in fmg components relative to their host-rock frameworks (Table 3; closed and partly closed symbols in Figure 10). These are probably biogenic placers that resulted from grain selection by organisms occupying the borings.

Important first-order variations in provenance are evident in Figure 11. In the Solomon Islands Central Province (Coleman, 1970), metamorphic detritus is present at all dredge stations (1, 2, 6, 14, and 25). Metamorphic detritus predominates at Stations 1 and 2 and is accompanied by ultramafic detritus at Station 14 and, in trace amounts, at Station 25. The metamorphic detritus was derived from basement sources on Guadalcanal (Mbirao Volcanics and Mbirao Metabasics described by Hackman, 1980), Santa Isabel and San Jorge (Sigana Volcanics, San Jorge Ultramafite summarized by Coulson and Vedder, 1986), and Choiseul (Voza Lavas, Choiseul Schists summarized by Coulson and Vedder, 1986). Crook (*in* Geology Department, University of Sydney, 1956) described serparenites similar to sample 14-3-1 from

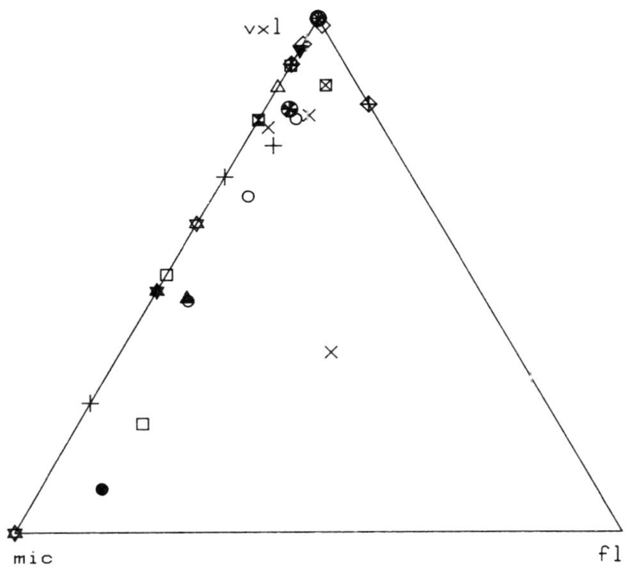

Figure 12. Provenance differences in volcanogenic materials indicated by proportions of different lithic textural components. vxl: vitric and other related components; mic: microlitic; fl: felsitic and lathwork. Detailed definitions in Table 6; symbols as in Table 3.

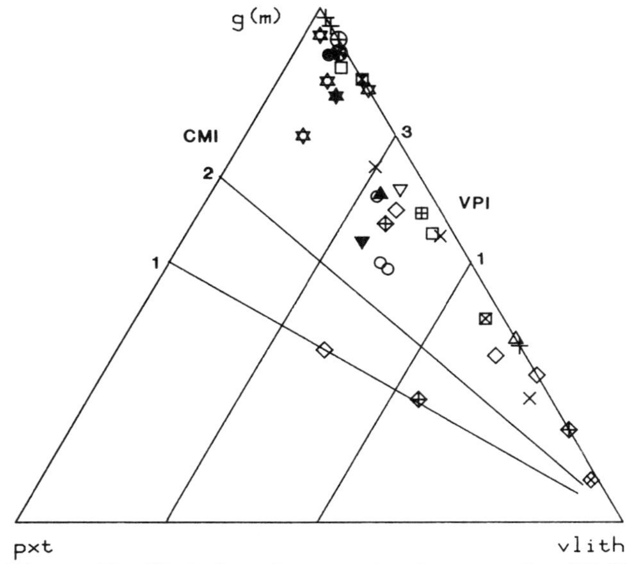

Figure 13. Variations in comminution maturity (CMI) and volcanic provenance (VPI) in volcanogenic materials. g(m): mineral grains; pxt: phenocrysts; vlith: lithics. Detailed definitions in Table 6; symbols as in Table 3.

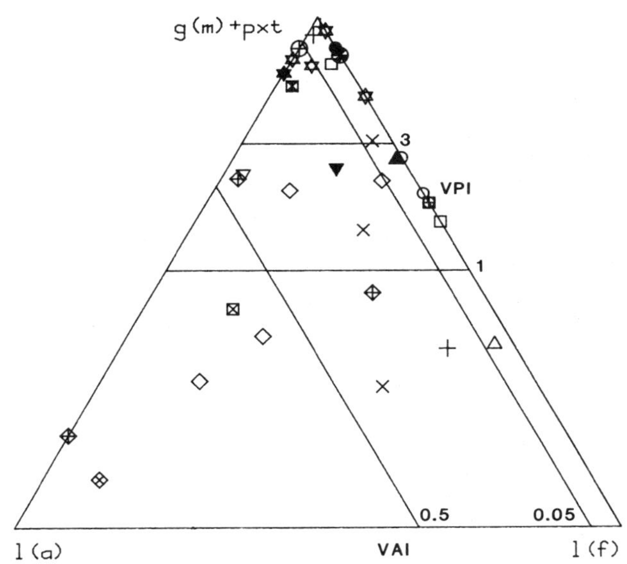

Figure 14. Alteration (VAI) and provenance (VPI) of volcanogenic materials. g(m)+pxt: all mineral grains; l(a): altered lithics; l(f): fresh lithics. Detailed definitions in Table 6; symbols as in Table 3.

San Jorge adjacent to Station 14. Some indication of variations in volcanic provenance is provided by changes in relative abundances of young volcanic (vmin and lv) components (Table 6 and Figure 11).

Interpretation of the volcanogenic fraction requires consideration of three aspects: variations due to changes in provenance, maturation of the detritus by comminution, and maturation as a result of weathering or other kinds of alteration. Crook and Taylor (1987) considered these aspects in their interpretation of a suite of Solomon Islands offshore samples, most of which contained abundant fresh volcanogenic detritus. Although appreciable altered volcanic material was present in seven of the 31 samples analyzed, they did not consider the ways in which alteration might perturb the provenance and maturity indices used in their interpretations.

Differences in provenance are shown by variations in two sets of parameters. First, the proportions and types of the four principal volcanic lithic textural components (vitric, microlitic, felsitic, and lathwork) may vary (Figure 12). Thus, microlitic components are notably abundant at intra-arc basin (Stations 19, 30, and 31) and forearc slope (Stations 11, 21-23, and PNG 79-1-24) localities in the Solomon Islands Volcanic Province of Coleman (1970).

Second, the relative proportions of phenocrysts, microphenocrysts, and aphanitic groundmass ("lithic") components in primary volcanic rocks will vary, and the variation between volcanic edifices will probably be larger than that within the volcanic rocks of a single edifice. During the formation of volcaniclastic units, comminution of the volcanic rocks will release monomineralic sand and silt grains. Thus, an index of comminution maturity (CMI, Table 6 and Figure 13) can be used to assess the degree of comminution of fresh volcaniclastic rocks (Crook and Taylor, 1987). Furthermore, the volcanic provenance index (VPI, Table 6 and Figures 13, 14), when applied to fresh volcaniclastic rocks, provides a measure of the relative abundances of phenocrysts and other textural components at the volcanic

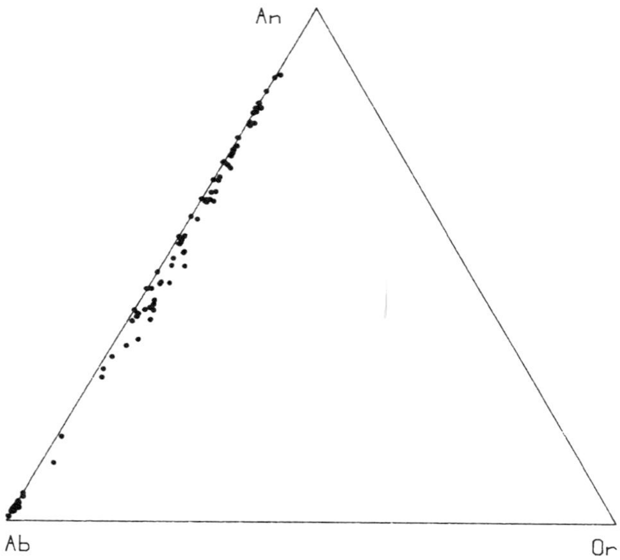

Figure 15. Plagioclase compositions.

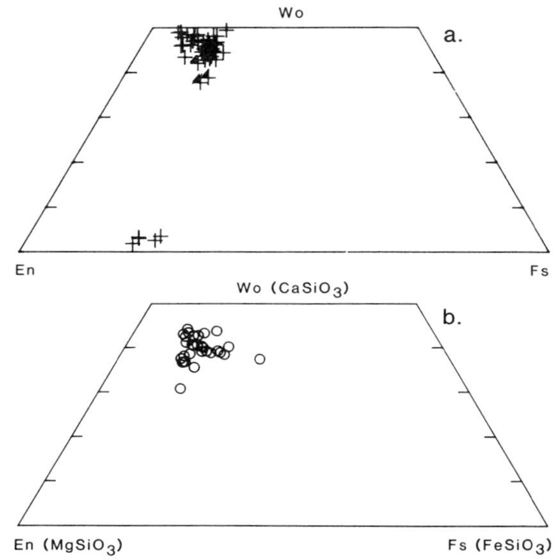

Figure 16. Pyroxene compositions; a: pyroxenes from island-arc volcanic rocks (crosses) and metabasite (partly closed crosses); b: pyroxenes from metabasite.

source. Both of these indices will be inaccurate if the volcaniclastic rocks have been weathered or altered in other ways, because alteration will preferentially destroy aphanitic materials, particularly glass, and will simultaneously increase the proportions of monomineralic sand and silt grains. To compensate for the effects of alteration, a volcanic alteration index (VAI, Table 6 and Figure 14) can be employed.

These three indices are shown on Figures 13 and 14; VAI is low (<0.05) in samples from forearc stations and near Bougainville, implying that values of the other indices are unlikely to be perturbed. In these samples, VPI shows no systematic variation. Crook and Taylor (1987) showed elsewhere that CMI is grainsize dependent, ranging from 1.0 ± 0.5 for coarse sand through 7 ± 3 for fine sand to infinity for coarse silt. Although several of the unaltered samples have CMI values in this "normal" range, those from Stations 23, 29-31, and PNG 79-1-24, and infills from Stations 21 and 22, have high CMIs, implying substantial comminution, perhaps by explosive volcanism. Some altered samples (VAI >0.2) from back-arc Stations 2, 6 and 14 have "normal" CMI-values despite their evident alteration. The initial comminution maturity of this detritus may have been unusually low.

The extent to which volcaniclastic materials in this suite have been reworked is difficult to establish. Pumice and finely comminuted glass shards at forearc Stations 2, 11, 21, PNG 79-1-24 and PNG 79-1-25 attest to contemporary volcanicity, as does the presence of pumice in infills at Stations 22 and 25 and secondary porosity after pumice fragments at Station 29. If the alteration of pumice fragments at Stations 2 and 19 is post-depositional, contemporary volcanicity is again implied. However, these interpretations are complicated by the presence of reworked volcaniclastic rock fragments at Stations 19 and 29, occurrences of reworked microfossils at Stations 2, 6, 19, and 23, and perhaps by occurrences of mudstone fragments at Stations 2, 6, 14, 19, 30, and PNG 79-2-24A. The possibility of substantial reworking of volcaniclastic material cannot be ruled out at many stations, particularly those where VAI is high.

PHASE CHEMISTRY

Techniques

Chemical analyses of selected minerals were carried out on carbon-coated polished thin sections using energy-dispersive methods on the TPD electron microprobe at the Research School of Earth Sciences, Australian National University (ANU). Analytical conditions were: accelerating voltage 14kV, beam current 3nA, beam diameter 2μm, counting period 80 s at a total count rate of approximately 400 cps. As many as 11 elements were determined simultaneously. Data were computer reduced at the time of analysis (Ware, 1973). Data on instrumental and matrix corrections and on the precision and accuracy of this probe are given in Reed and Ware (1973, 1975). Analytical data, which are available from K.A.W. Crook on request, are archived at the ANU Microprobe Laboratory under numbers 300349-300433, 300750-300857, 301135-301260, 304192-304418, and 304531-304554.

The phases analyzed are representative of the detrital components in samples of the siliciclastic rocks and limestone, in which they occur as monomineralic grains and components in rock fragments; or they are major components of igneous and metamorphic cobbles retrieved at Stations 1 and 25, occurring as primary

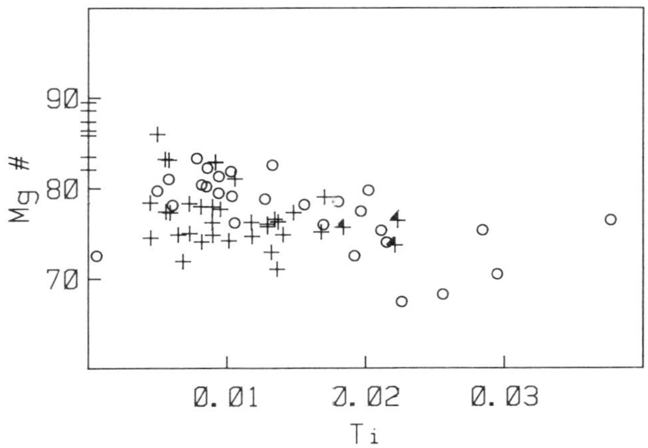

Figure 17. Magnesium numbers—100*Mg/(Mg + Fe)—versus Ti contents of clinopyroxenes; crosses: arc-volcanic clinopyroxenes; circles and partly closed crosses: clinopyroxenes in metabasite or from metabasite detritus.

pyrogenic components, secondary (metamorphic) components, or components in veins.

Feldspars

All of the feldspars are plagioclase. Compositions (Figure 15) reflect the two major provenances represented in this suite of sedimentary rocks and clasts. The albite and sodic oligoclase represent detritus of metamorphic provenance, including grains in metabasites and metamorphic tectonites. The plagioclases, which range in composition from calcic oligoclase to bytownite, are of volcanic provenance representing mainly island-arc volcanic detritus. Two analyses in this group are of relict plagioclase (An_{49}, An_{59}) in a metadolerite (sample 25-2). Overall, the compositional range is the same as that recorded by Crook and Taylor (1987, Figure 18) from elsewhere in the Solomon Islands offshore region.

Pyroxenes

Orthopyroxene is uncommon. Compositions range from En_{70} to En_{77} (Figure 16). All examples are accompanied by clinopyroxenes and are derived from island-arc volcanic rocks.

Clinopyroxene is both widespread and common. Petrographic criteria enabled easy discrimination between clinopyroxenes of volcanic (island arc) provenance (Figure 16a) and those occurring as relict phases in metabasites (Figure 16b). Only three clinopyroxene grains were misidentified as to provenance (partly closed symbols in Figures. 16a, 17, and 18), being ascribed to a volcanic provenance, whereas their phase chemistry shows them to be of metabasite provenance.

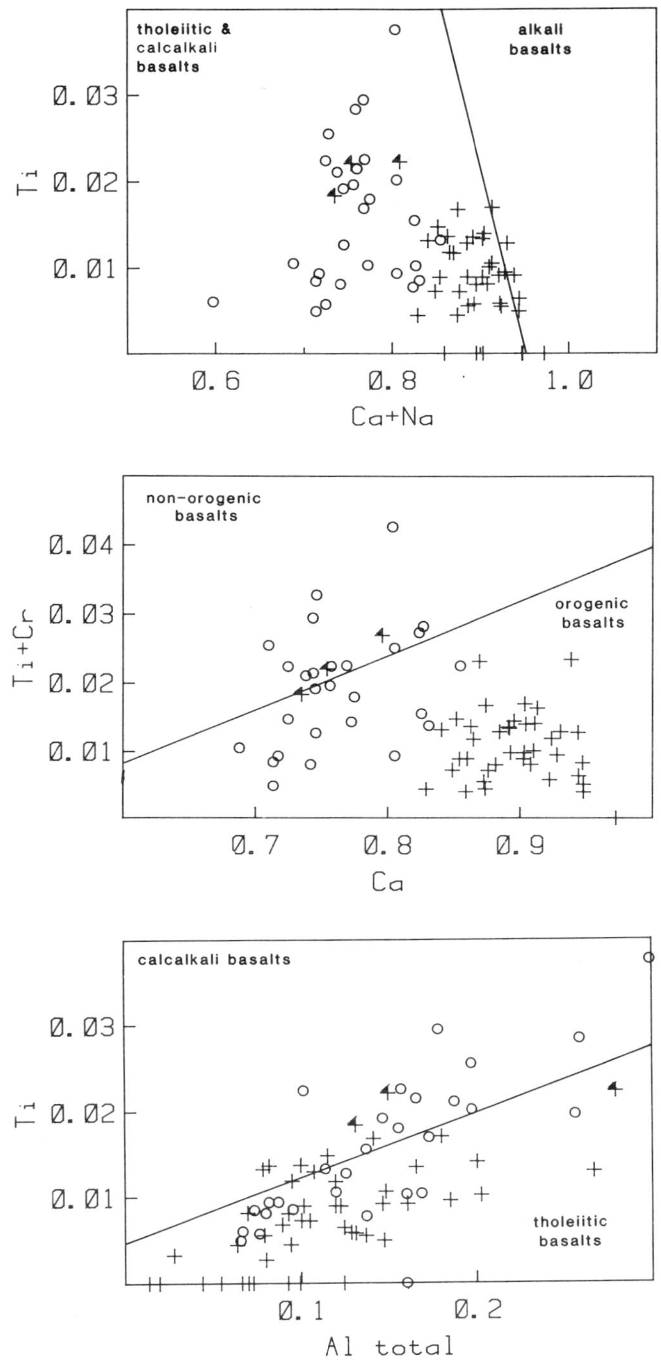

Figure 18. Petrogenetic affinities of clinopyroxenes using discriminatory criteria for basaltic clinopyroxenes of Leterrier et al (1982). Symbols as in Figure 17.

Most of the arc volcanic clinopyroxenes have Mg numbers less than 80. Those with Mg numbers greater than 80 are not confined to samples collected at forearc stations, in contrast to the geographic distribution noted by Crook and Taylor (1987).

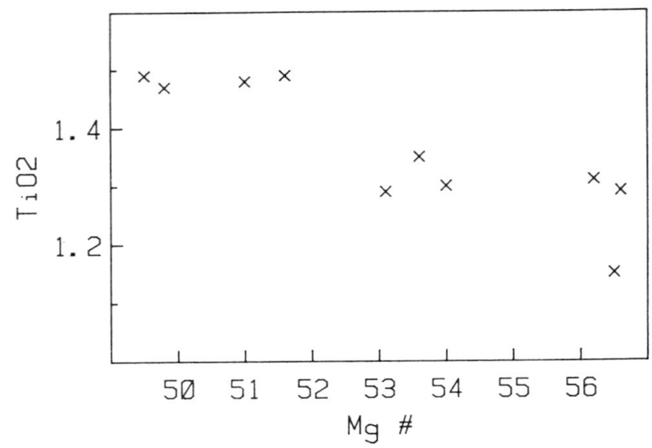

Figure 19. TiO_2 versus Mg number for selected analyses from the suite of metabasites and related metamorphic tectonites (Table 1), representative of ophiolitic metabasalt.

The petrogenetic affinities of the clinopyroxenes have been further examined using criteria for discriminating among basaltic clinopyroxenes proposed by Leterrier et al (1982). The arc volcanic clinopyroxenes (crosses, Figure 18), which were probably derived from rocks ranging in composition from basalt to high-silica andesite, show a similar range of composition to those recorded by Crook and Taylor (1987, Figure 21), except that Ti content ranges slightly higher for grains with low total Al than was observed in that study.

The metabasite clinopyroxenes are distinctive, particularly on the plot of Ti+Cr versus Ca (Figure 18). On this plot, according to Leterrier et al (1982), the field of nonorogenic basalts contains 81 percent of the pyroxenes from this tectonic setting used in the preparation of the discrimination diagram. However, only 47 percent of the metabasite clinopyroxenes fall in this field, several of the analyses showing unusually low Ti+Cr relative to Ca. Most of the plotted analyses (circles, Figure 18) represent clinopyroxenes in the suite of metabasite and related metamorphic tectonite cobbles from Station 1 (Table 4) and from the metadolerite clast (sample 25-2).

The petrogenetic affinities of the suite from Station 1 have been examined using whole-rock chemical analyses (Table 1). TiO_2 contents in this suite show a systematic decrease with increasing Mg number (Figure 19), a pattern characteristic of tholeiitic basalt. W.E. Cameron (written commun., 1987) has noted that the textural characteristics of this suite are typical of spilite and ophiolitic metabasalt. Although ophiolites are commonly regarded as ancient oceanic lithosphere, i.e. representative of the non-orogenic basalt suite, many ophiolites occur in arc-related tectonic settings even though this may not have been the setting in which they originally formed. Thus, while the affinities of the tholeiitic (metabasite) clinopyroxenes lie more with the non-orogenic suite of

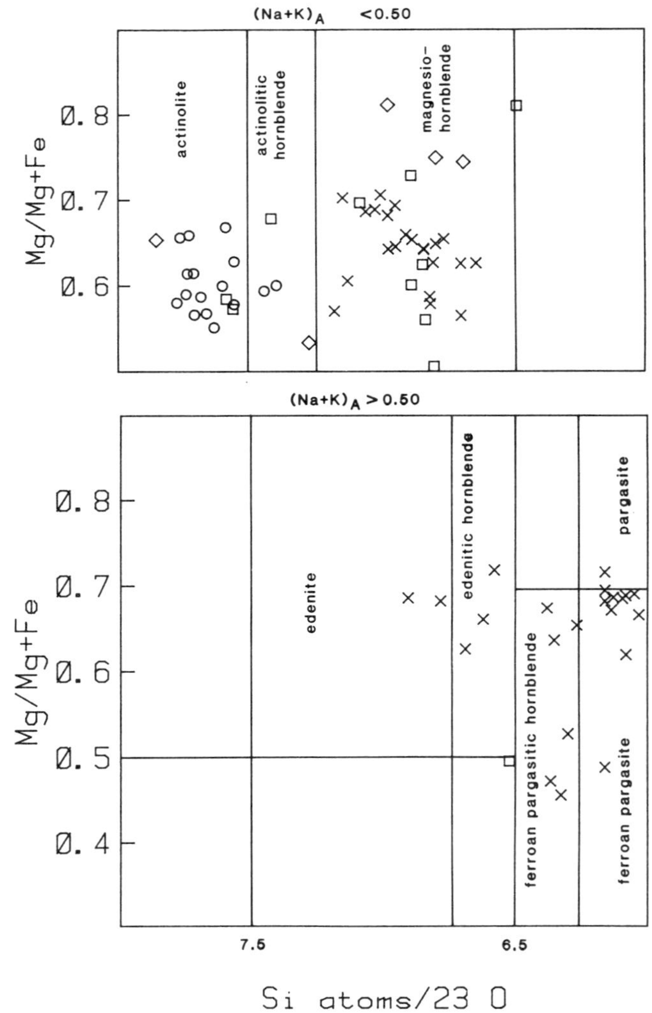

Figure 20. Amphibole compositions and nomenclature after Leake (1978); circles: metamorphic amphiboles (samples 1-1-12, 1-1-13, and 25-2); diamonds: amphiboles from Guadalcanal basement rocks; squares: other amphiboles identified as probably metamorphic; crosses: amphiboles of arc-volcanic provenance.

Leterrier et al (1982), the differences noted above are likely to be real, reflecting the somewhat enigmatic and uncertain tectonic affinities of ophiolitic metabasalt.

Amphiboles

The amphibole analyses (Figure 20) include analyses of grains from samples 5347 (schistose amphibolite, Mbirao Volcanics) and 5831 (amphibolitic gabbro, Marau Ultrabasics) from the basement complex in eastern Guadalcanal. These analyses were included as a check on the provenance of the cobbles collected at Station 1. However, the compositions of the two groups of samples differ appreciably.

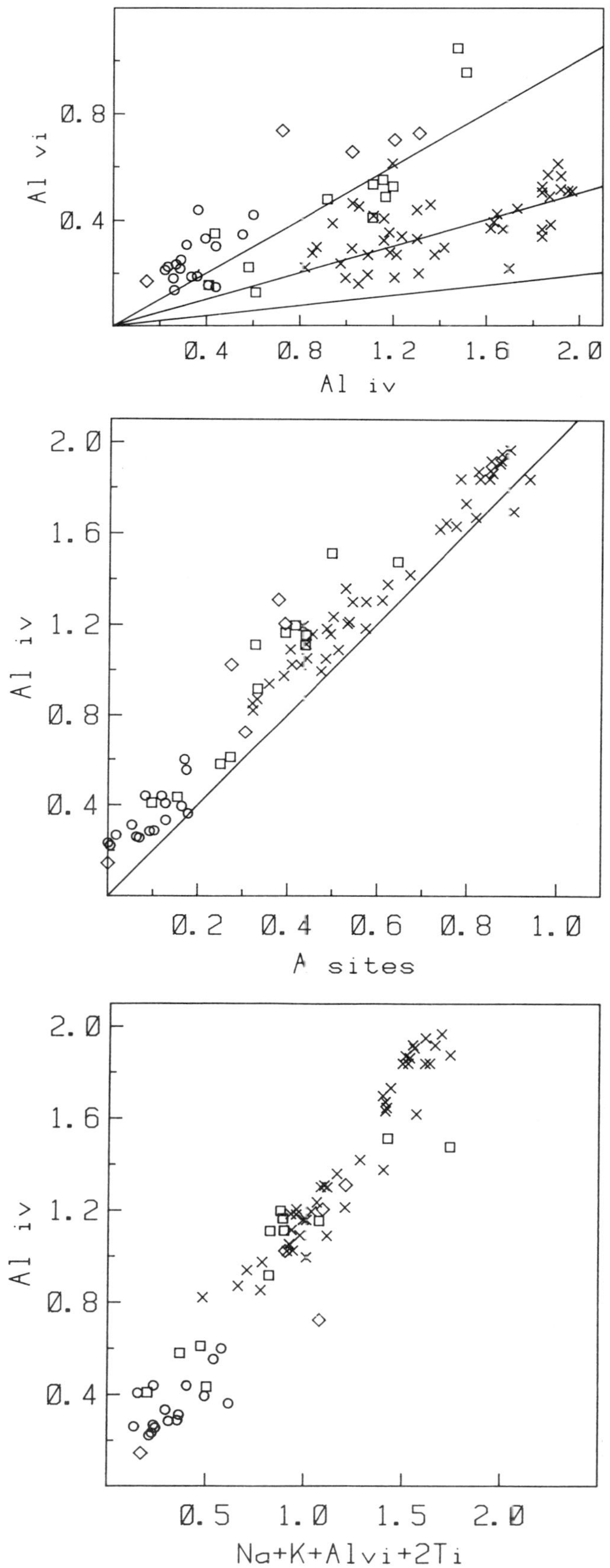

The compositional range in Figure 20 is similar to that reported by Crook and Taylor (1987, Figure 23). However, within this compositional range, amphiboles of known or probable metamorphic provenance show considerable variation in composition, many being magnesio-hornblende. The distinctiveness of the metamorphic amphiboles from those of island-arc volcanic provenance is clearly shown in the partitioning of aluminum between tetrahedral and octahedral sites. Octahedral aluminum (Al^{vi}) is generally twice as abundant in metamorphic amphiboles as in volcanic amphiboles (Figure 21A).

The compositional variations shown in Figure 21 result from Al-tschermakite ($Al^{iv}Al^{vi} \rightleftharpoons Mg^{vi}Si^{iv}$) and edenite ($Al^{iv}(Na+K)^{A} \rightleftharpoons Total^{A}Si^{iv}$) substitutions. The complete substitution scheme:

$$Al^{iv} \rightleftharpoons (Na+K)^{A} + Al^{vi} + 2Ti$$

is confirmed by the correlation shown on Figure 21C. These compositional variations are similar to those reported by Crook and Taylor (1987, Figure 24) from elsewhere in the Solomon Islands and by Crook (1987) from sites in the western Solomon Sea.

DISCUSSION

Most of the sequences sampled are apparently mid-Quaternary and younger (Table 1). However, late Pliocene and Pleistocene strata were sampled north of Buka Island (Stations 29, and PNG 79-1-24), and at Fatu O Moana (Station 19), confirming descriptions by Crook and Taylor (1987). Limestone samples close to the Russell Islands (Stations 30 and 31), which are assigned a Pliocene or Pleistocene age (Table 1) may be part of the Pleistocene limestone which encircles the volcanic core of the islands (Coulson and Vedder, 1986).

Bathymetric estimates based on benthic foraminiferal assemblages were made for all samples by Johanna Resig (Table 1), except for sample 23-2 which lacks benthics. The depths at which sediments were deposited are the same as those from which samples were collected for all stations except PNG 79-1-24,where at least 700 m uplift of the arc ridge north of Buka Island is indicated since the sampled sediment accumulated in the Pliocene or Pleistocene.

Detritus derived from the igneous and metamorphic basement rocks of the Central Province is present at all stations within that province. The basement rock

Figure 21. Structural chemistry of amphiboles; A: distribution of aluminum between tetrahedral and octahedral sites; B: occupancy of alkali ion sites vs. aluminum in tetrahedral sites; C: plot of combined edenite and Al+Ti tschermakite substitution in amphiboles. Symbols as in Figure 20.

fragments give a good indication of the range of rock types present in the source areas, except for the sedimentary rocks (refer to Coulson and Vedder, 1986, Figure 4), and the mineralogy and fabrics of the basement rock fragments are distinctive. Clinopyroxene and amphibole from basement sources are readily discriminated by microprobe analysis. These observations suggest that basement massifs within ancient island-arc terranes, which are now incorporated in fold belts, will readily be detected and characterized petrologically from the study of associated sedimentary sequences.

In some samples, there is evidence for recycling of volcaniclastic material, and such redeposition cannot be ruled out for many of the other samples. Consequently, some attempt has been made to develop criteria to help decide whether volcaniclastic units indicate contemporary volcanicity, the dissection of young but extinct volcanic piles, or the erosion of ancient volcanic-sedimentary complexes. Some criteria, such as the presence of glass shards, hyaloclasts, and pumice fragments as indicators of contemporary volcanicity, seem satisfactory; but the value and significance of indices such as those of comminution maturity, volcanic provenance, and volcanic alteration will require a much larger and geographically extensive data base for their assessment.

The grains termed papules in this study warrant some comment because they constitute more than 5 percent of many samples. Papules are a commonly encountered grain type in volcaniclastic rocks. Most workers (ourselves included) have treated papules as altered glassy volcanic fragments, but the present study suggests that their origins may be more diverse. Detailed micro-mineralogical and fabric studies on these papules may produce valuable data on their pre-and post-depositional alteration and provenance.

ACKNOWLEDGMENTS

We acknowledge the assistance of shipboard colleagues. Patrick Coleman (University of Western Australia) and Colin Midobatu (Solomon Islands Government) assisted with aspects of Solomon Islands geology. Bruce Chappell and Nick Ware (Australian National University) and Robert Anisi (Solomon Islands Government) assisted with chemical analyses. Warrington Cameron (Australian National University) provided valuable guidance in the interpretation of the metabasites, to which Brian Harrold (Australian National University) and Brian Smith (Unocal Research Laboratory) contributed data on the temperatures of formation of minerals. Johanna Resig (Hawaii Institute of Geophysics), Samir Shafik (Bureau of Mineral Resources, Geology, and Geophysics, Australia) and Paula Quinterno (U.S. Geological Survey) provided paleontological determinations.

REFERENCES CITED

Brewer, R., 1964, Fabric and mineral analysis of soil: San Francisco, John Wiley and Sons, 470 p.

Bruns, T.R., Cooper, A.K., Mann, D.M., and Vedder, J.G., 1986, Seismic stratigraphy and structure of sedimentary basins in the Solomon Islands region, *in* Vedder, J.G., Pound, K.S., and Boundy, S.Q., eds., Geology and offshore resources of Pacifgic island arcs—central and western Solomon Islands: Houston, Texas, Circum-Pacific Council for Energy and Mineral Resources Earth Science Series, v. 4, p. 177-214.

Bruns, T.R., Cooper, A.K., and Vedder, J.G., 1986, Seismic stratigraphy and structure of sedimentary basins in the Solomon Islands region, *in* Vedder, J.G., Pound, K.S., and Boundy, S.Q., eds., Geology and offshore resources of Pacific island arcs—central and western Solomon Islands: Houston, Texas, Circum-Pacific Council for Energy and Mineral Resources Earth Science Series, v. 4, p. 215-223.

Chase, T.E., Seekins, B.A., and Lund, K.E., 1986, Submarine topography of the Solomon Islands region, *in* Vedder, J.G., Pound, K.S., and Boundy, S.Q., eds., Geology and offshore resources of Pacific island arcs—central and western Solomon Islands: Houston, Texas, Circum-Pacific Council for Energy and Mineral Resources Earth Science Series, v. 4, p. 21-31.

Coleman, P.J., 1970, Geology of the Solomon and New Hebrides Islands as part of the Melanesian re-entrant, Southwest Pacific: Pacific Science, v. 24, p. 289-314.

Colwell, J.B., and Vedder, J.G., 1986, Description and interpretation of dredged rocks, Solomon Islands, Houston, Texas: Circum-Pacific Council for Energy and Mineral Resources Earth Science Series, v. 4, p. 225-236.

Coulson, F.I. and Vedder, J.G., 1986, Geology of the central and western Solomon Islands, Houston, Texas: Circum-Pacific Council for Energy and Mineral Resources Earth Science Series, v. 4, p. 59-88.

Crook, K.A.W., 1987, Petrology and mineral chemistry of sedimentary rocks from the western Solomon Sea: Geo-Marine Letters, v. 6, p. 203-209.

Crook, K.A.W., and Taylor, G.R., 1987, Sedimentary petrology and mineral chemistry, Woodlark Basin-Solomon Islands offshore region, *in* Taylor, B., and Exon, N.F., eds., Marine geology, geophysics, and geochemistry of the Woodlark Basin-Solomon Islands: Houston, Texas, Circum-Pacific Council for Energy and Mineral Resources Earth Science Series, v. 7, p. 295-346.

Fitzpatrick, E.A., 1984, Micromorphology of soils: New York, Chapman and Hall, 433 p.

Folk, R.L., 1980, Petrology of sedimentary rocks: Austin, Texas, Hemphill Publishing Co., 182 p.

Geology Department, University of Sydney, 1956, Geological reconnaissance of parts of the central islands of the British Solomon Islands Protectorate: Overseas Geology and Mineral Resources, v. 6, p. 267-306.

Hackman, B.D., 1980, The geology of Guadalcanal, Solomon Islands: Institute of Geological Sciences Overseas Memoir v. 6, 115 p.

Katz, H.R., 1986, Stratigraphy of the SW Pacific, *in* Cronan, D.S., ed., Sedimentation and mineral deposits in the southwestern Pacific Ocean: London, Academic Press, p. 5-82.

Kulm, L.D., and Suess, E., 1987, Fluid venting structures on the Oregon continental shelf: EOS, v. 68, p. 1487.

Kroenke, L. W., 1984, Cenozoic tectonic development of the Southwest Pacific: UN ESCAP, CCOP/SOPAC Technical Bulletin no. 6, 122 p.

Leake, B.E., 1978, Nomenclature of amphiboles: Canadian Mineralogist, v. 16, p. 501-520.

Leterrier, J., Maurey, R.C., Thonon, P., Girard, D., and Marchal, M., 1982, Clinopyroxene composition as a method of identification of the magmatic affinities of paleo-volcanic series: Earth and Planetary Science Letters, v. 59, p. 139-154.

Reed, S.J.B., and Ware, N.G., 1973, Quantitative electron microprobe analysis using a lithium-drifted silicon detector: X-ray Spectrometry, v. 2, p. 69-74.

Reed, S.J.B., and Ware, N.G., 1975, Quantitative electron microprobe analysis of silicates using energy-dispersive x-ray spectrometry: Journal of Petrology, v. 16, p. 499-519.

Taylor, B., and Exon, N.F., 1987, An investigation of ridge subduction in the Woodlark-Solomons region: introduction and overview, *in* Taylor, B., and Exon, N.F., eds., Marine geology, geophysics, and geochemistry of the Woodlark Basin-Solomon Islands: Houston, Texas, Circum-Pacific Council for Energy and Mineral Resources Earth Science Series, v. 7, p. 1-24.

Walshe, J.L., 1986, A six-component chlorite solid solution model and the conditions of chlorite formation in hydrothermal and geothermal systems: Economic Geology, v. 81, p. 681-703.

Ware, N.G., 1973, Annotated computer programs for electron probe microanalysis using a lithium-drifted silicon detector: Report, Research School of Earth Sciences, Australian National University, Canberra.

Zuffa, G.G., 1980, Hybrid arenites: their composition and classification: Journal of Sedimentary Petrology, v. 50, p. 21-30.

Vedder, J.G., and Bruns, T.R., editors, 1989, Geology and offshore resources of Pacific island arcs—Solomon Islands and Bougainville, Papua New Guinea Regions: Houston, Texas, Circum-Pacific Council for Energy and Mineral Resources Earth Science Series, v. 12.

CORAL AND LIMESTONE, CARBONATE CEMENTS IN SILICICLASTIC ROCKS, AND CHALK FROM DREDGE SAMPLES, SOLOMON ISLANDS AND BOUGAINVILLE, PAPUA NEW GUINEA

A.R. Niem
Department of Geology, Oregon State University, Corvallis, Oregon 97331-5506, USA

ABSTRACT

A mixture of intensively bored and cemented Quaternary coralgal limestone (boundstone) and fresh corals was dredged from four stations in the Solomon Islands intra-arc basins and forearc area. Paleoecology and geographic positions suggest that the limestones were initially formed in shallow, warm, wave-agitated forereef and forereef rubble depositional environments. Limestone blocks and coral edifices, undercut by boring organisms, were toppled by storm waves and funneled to slope depths by rock slides and other gravity-driven processes. Solitary ahermatypic corals and serpulid worms subsequently colonized the manganese-stained reef limestone rubble. A multichannel seismic-reflection record indicates that some reefs, which were rubble sources, have subsided tectonically since the Pleistocene.

Submarine aragonite and high-Mg peloidal cements, as well as multiple borings by clionid sponges and other endolithic marine organisms developed during growth and after the demise of the forereefs. Wave-abraded carbonate-skeletal sand and silt partly infill borings and reef-framework crevices. Reservoir-quality porosities and permeabilities of the coralgal limestones and fresh corals range from 37.1 percent to 49.7 percent and 1,743 md to 6,444 md, respectively.

Low-Mg blocky, sparry calcite cement and neomorphic textures in a subrounded coral cobble suggest freshwater diagenesis in the phreatic zone prior to rounding and transportation to slope depths. The cobble was probably derived from an uplifted Pleistocene reef-limestone terrace on the nearby island of Choiseul.

Stable carbon and oxygen isotopes and geochemistry of the biotics and cements in limestones and a late Pliocene chalk (from northwest of Bougainville) fall within the ranges of common shallow-marine limestones, aragonite, and high- and low-Mg calcite cements, chalks, and biotics. However, very depleted values of $\delta^{13}C$ (-29.4‰ to -46.7‰ PDB) and significantly elevated values of $\delta^{18}O$ (2.4‰ to 6.2‰ PDB) from a Quaternary calcite-cemented silty mudstone (a carbonate chimney) and dolomite-cemented litharenites suggest that these cements were probably derived from oxidation of methane. Precipitation of these microcrystalline cements probably took place around vents of escaping pore fluids in near-surface siliciclastic sediments in a manner much like that proposed for the origin of carbonate chimneys and cemented silts on the modern Oregon-Washington accretionary margin. The carbonate chimney and dolomite-cemented sandstone occur along a major fault zone bordering Indispensable basin. This fault zone may have provided avenues for escaping pore fluids. However, microcrystalline dolomite and calcite cements in siliciclastic samples dredged from other stations indicate that this process is widespread throughout the Solomon Islands.

INTRODUCTION

Coralgal reef-limestone and coral fragments were recovered from four dredge sites (Stations 6, 14, 22, and 25; Figures 1A and 1B) in the Solomon Islands-Bougainville, Papua New Guinea area during the 1984 Tripartite cruise of the R/V *S.P. Lee* (L6-84-SP). At Station 6 between Santa Isabel and the Florida Islands, several manganese-stained reef-limestone and coral-rubble blocks, up to 0.5 m across, and minor soft pteropod-bearing mudstone were dredged from depths between 500 and 900 m off a steep forereef escarpment (Figures 2 and 3A). The fault-controlled escarpment (Bruns et al, 1986) is capped by a drowned Pleistocene reef on the outer edge of a broad platform that rims the southwestern flank of Indispensable Basin southeast of Santa Isabel (Figure 1A).

Dredge Station 14 is located on the steep intra-arc slope between the southeastern end of Santa Isabel and the

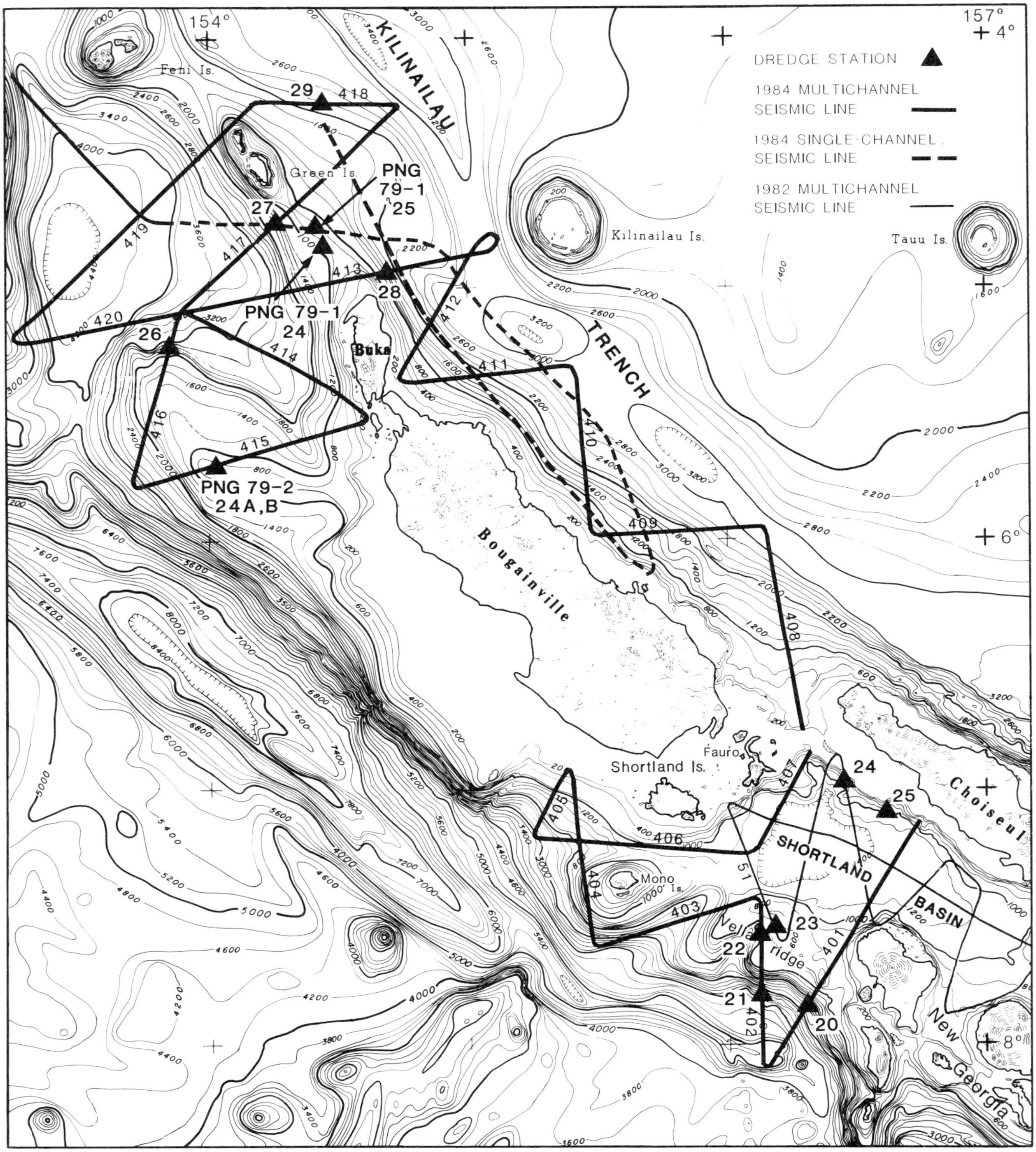

Figures 1A and 1B. Location maps of dredge samples, Solomon Islands and Bougainville, Papua New Guinea, 1982 cruise and Leg 6, 1984 cruise of R/V *S.P. Lee*. Bathymetry from Chase, Seekins, and Lund (1986).

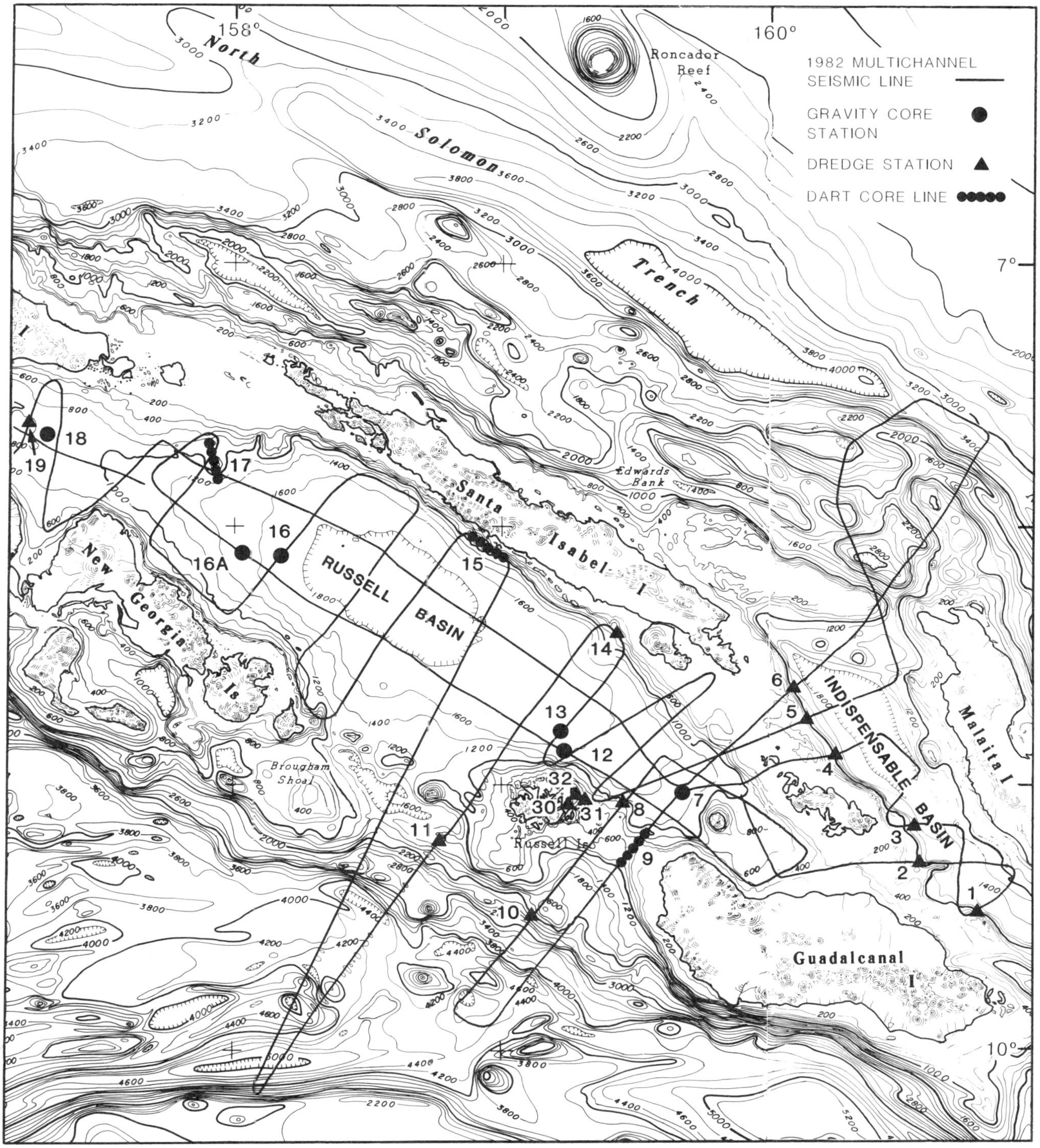
158°
160°
7°
10°
North
Solomon
Trench
Roncador Reef
1982 MULTICHANNEL SEISMIC LINE
GRAVITY CORE STATION
DREDGE STATION
DART CORE LINE
Edwards Bank
Santa
Isabel
RUSSELL BASIN
New Georgia Is
INDISPENSABLE BASIN
Malaita I
Brougham Shoal
Russell Is
Guadalcanal I

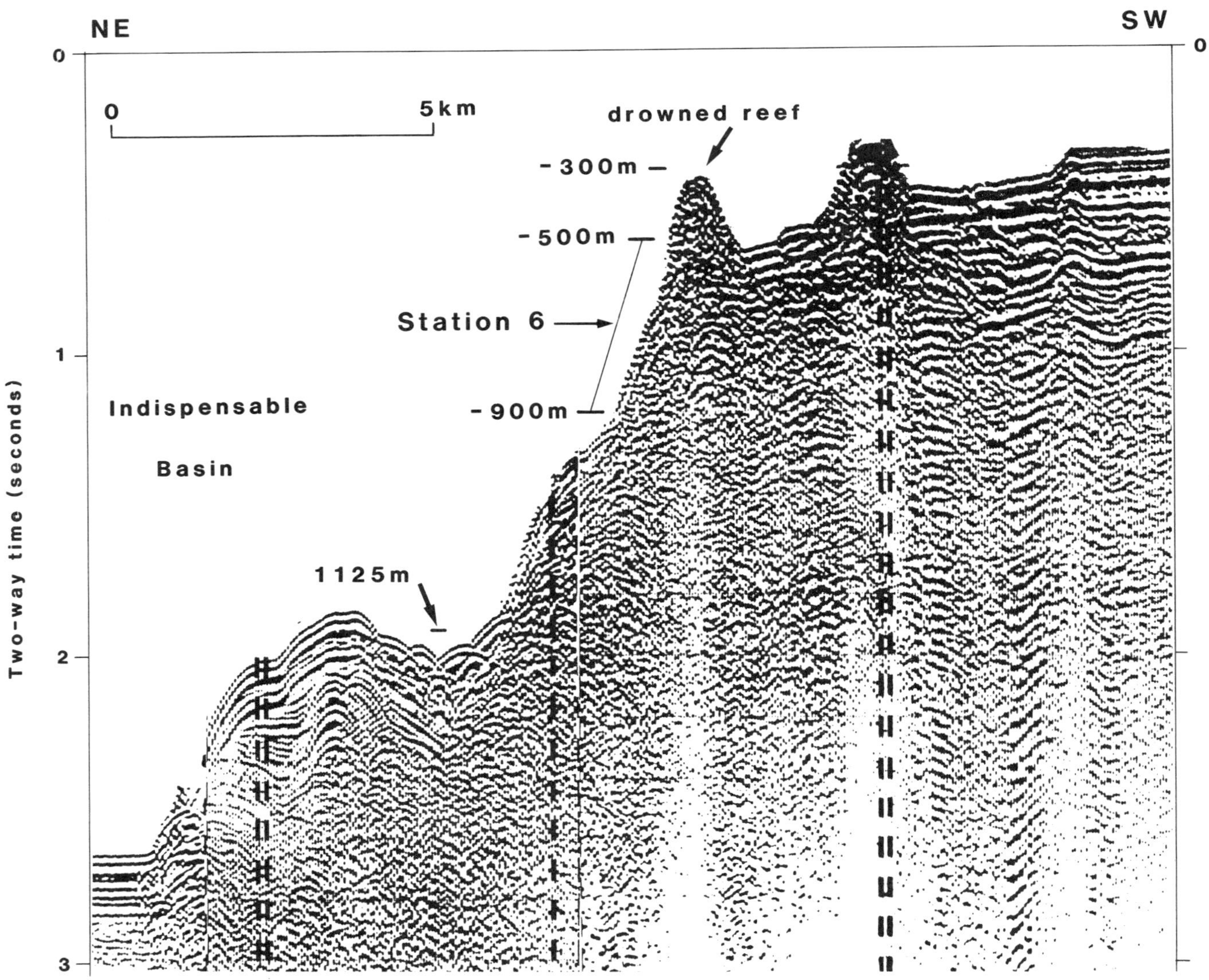

Figure 2. Section of multichannel seismic-reflection record from 1982 R/V *S.P. Lee* cruise illustrating the position of Station 6, southwest slope of Indispensable basin, Solomon Islands. Dredged Pleistocene limestone samples are coralgal forereef rubble collected at depths between 500 and 900 m on a steep escarpment below a drowned reef at 300 m.

Figure 3A (facing page). Typical block of coralgal boundstone dredged at Station 6. Note abundant elongate and semicircular clionid sponge borings and small barnacle in boring (B and arrow) on porous outer part of slabbed sample (49.7 percent porosity, 6,444 md permeability) compared to dense, carbonate-cemented unbored interior (37.1 percent porosity, 4,553 md permeability). White letter (M) marks a bivalve scar. Note younger solitary ahermatypic coral (A) with well-developed septae attached to manganese-stained surface of reef limestone. **B.** Surface view of colonial coral *Galaxea facicularis*, illustrating widely separated hexagonal corallites with well-preserved septae. Forereef talus dredged off Santa Isabel at Station 14. Note irregular cavernous surface, some of which is covered by dark ferromanganese stain (photographically lightened) that has been colonized by a variety of delicately ornamented calcareous serpulid worm tubes, bryozoans, and an oyster spat. **C.** Transverse slab of fresh unaltered scleractinian coral *Physogyra lichtensteini* with thin shelled bivalve in boring (B). Note abundant open corallite pores. Dredge Station 22, northwest of Vella Lavella. **D.** Porous *Porites* sp. colonial coral (P) encrusting a white nodule of coralline algae with concentric layering, sample 22-6. Note open chambers of high-spired cerithid gastropod (arrow) encrusted by coralline algae.

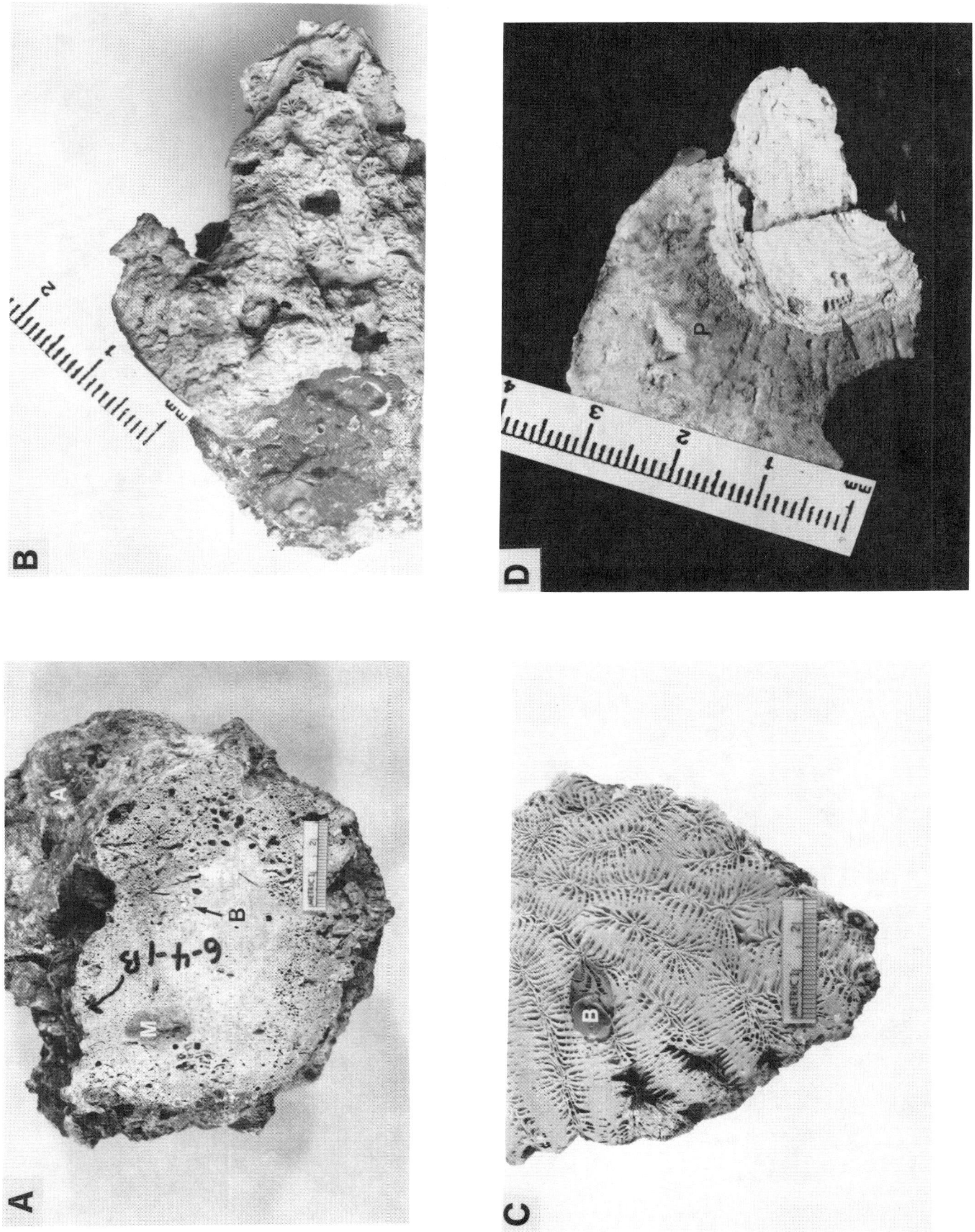

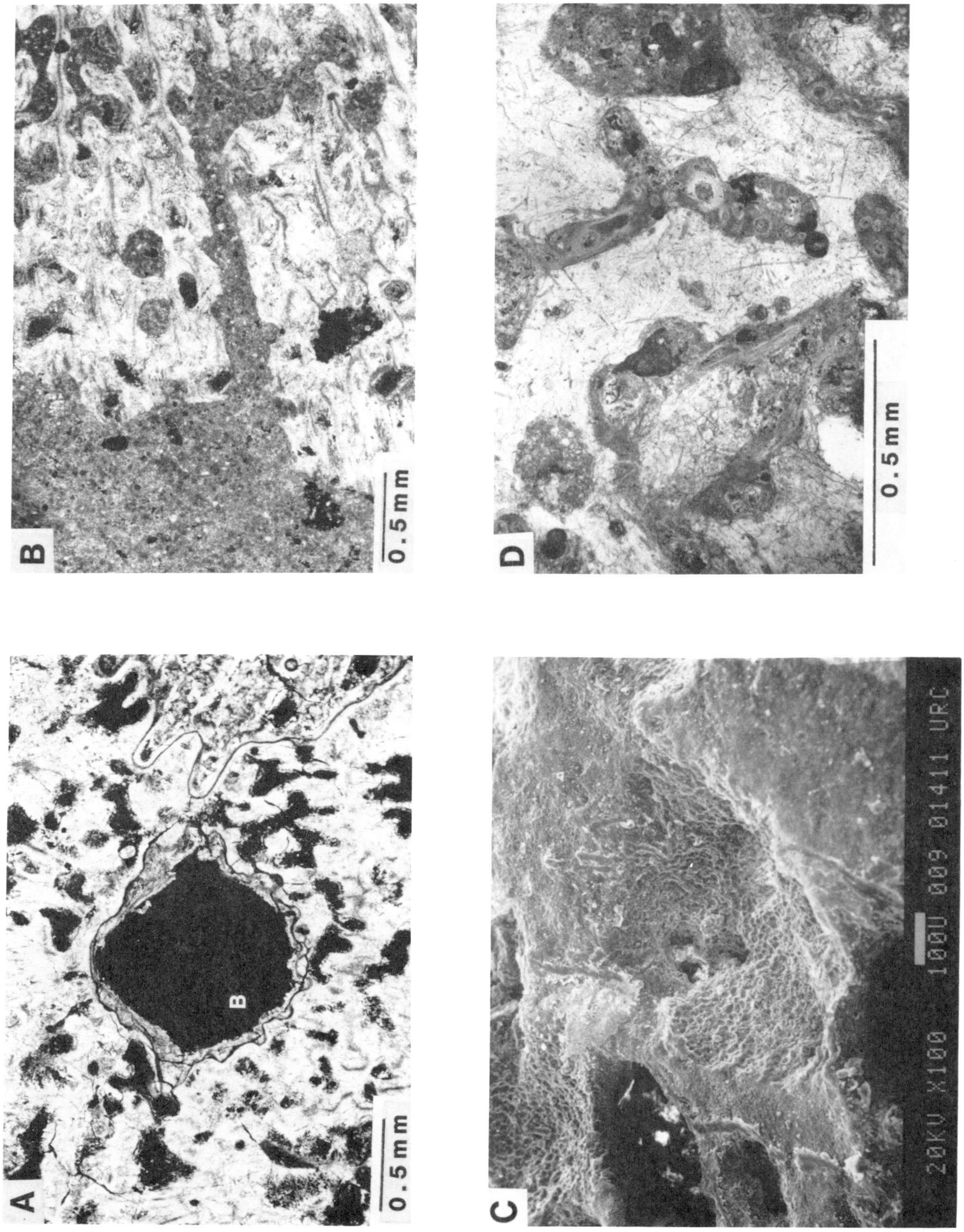
A
B
0.5 mm
B
0.5 mm
C
20KV X100 100U 009 01411 URC
D
0.5 mm

Russell basin of the Central Solomons Trough (Vedder and Coulson, 1986) (Figure 1A). Pleistocene and modern reefs occur on the outer shelf immediately northwest of this station (Solomon Islands Department of Geological Surveys, 1969). Based upon interpretation of Uniboom seismic records, Colwell and Tiffin (1986) postulated that reef rubble is strewn across the sea floor in the vicinity of Station 14. Fragments of ferromanganese-stained corals (Figure 3B), coralgal limestone, and small blocks of bedded very fine-grained lithic sandstone and silty claystone were recovered from this site from depths between 600 m and 750 m.

Dredge Station 22 lies along an irregular escarpment of a submerged platform (Vella Ridge of Bruns et al, 1986) which extends northwest of Vella Lavella on the southwestern margin of the Shortland basin (Figure 1B). Dredging yielded approximately 50 kg of manganese-stained, fresh, unaltered (Figure 3C) to diagenetically well-cemented and intensely bored reef and coral rubble. Coral and coralgal boundstone clasts range in size from angular cobbles to boulders.

Four subrounded cobble-size fragments of coral were recovered together with rounded metadolerite cobbles and a block of bioturbated feldspatho-volcaniclastic mudstone at Station 25 off Choiseul on a steep slope that extends southwestward into the Shortland basin (Figure 1B).

Similar coralgal limestones and a rhodolith clast were dredged around the Russell Islands and the southeastern end of Santa Isabel on an earlier R/V *S.P. Lee* cruise (L7-82-SP) (Colwell and Vedder, 1986) (Figure 1A, Stations 1/2, 1/4, and 1/7); on an auxiliary cruise in 1984 (Crook and Niem, this volume; Quinterno, this volume) (Figure 1B, Stations 30, 31, and 32); and on an earlier CCOP/SOPAC cruise near Bougainville (Crook and Niem, this volume).

LITHOLOGY, TEXTURE, AND STRUCTURES

Reef limestones dredged from the four stations are dominantly altered fragments of intensively bored, massive colonial scleractinian corals that are encrusted by and encrust coralline algae (coralgal boundstone) (Figure 3D), vermetid gastropods, bryozoans, and gypsinid foraminifers. There is abundant pore-filling peloidal submarine cement and internal carbonate sediment filling cavities and borings (e.g., Stations 6 and 22). Several generations of borings, cementation, encrustation, and erosion are present in many samples. For example, some earlier borings, filled with white carbonate skeletal sand and silt and/or peloidal micrite cement, are truncated by later open and filled borings. Sediment filling, cementation, boring, and encrusting is so extensive in some samples that the original skeletal microstructure of the scleractinian coral is nearly obliterated (Figure 3A). Other samples from the same stations are fragments of fresh coral with few or no borings or submarine cements (Figure 3C).

A few coralgal samples contain abundant open borings in the outer margins, whereas the intensively bored interior is densely cemented with peloidal high-Mg calcite (Figures 3A, 4A, and 4B). Clionid sponge galleries with their distinctive scalloped microsculpture (Bromley, 1978; Warme and McHuron, 1978) are the most abundant type of boring; barnacle borings are subordinate (e.g., Figures 3A, 4A, 4B, and 4C). According to Warme (1977), other boring organisms include sipunculids, polychaete worms, vermetid gastropods, *Lithophaga*, pholads, and echinoids. Floral microborings formed by endolithic algae and fungi are also abundant (Figure 4D). A few borings contain multiple nested shells of bivalves. Unusual, very small (<100 μ in diameter), circular to elongate tubes of multilayered high-Mg calcite are present in clionid sponge borings (Figure 4D), and some cut across the earlier micrite-peloidal cements. The tubes are locally lined with submicroscopic bladed crystals of calcite. The origin of these enigmatic multilayered small tubes is not clear; they may be skeletal tubes of minute polychaete worms (Majewske, 1974).

The outer bored margin of several other large fragments of corals (such as *Porites*) is densely cemented or filled with carbonate skeletal sand and silt. This dense margin surrounds an unbored interior that has abundant small, open coral mesopores. The carbonate sand and silt that fills the outer borings is tightly cemented by nearly contemporaneous peloidal high-Mg submarine cement which may have acted as an early diagenetic seal that protected the porous inner part of the coral mass from further cementation.

Figure 4A (facing page). Photomicrograph of cross section of a barnacle boring (B) that creates a large void in scleractinian coral *Porites lutea*. Tiny needles of early aragonite submarine cement partly line abundant open intraskeletal pores and borings (dark) in coral; porous outer part of sample 6-4-1 in Figure 3A; crossed nicols, 40 X. **B**. Photomicrograph of intensively bored coral *Porites lutea* (light) showing complete and incomplete infilling of clionid sponge borings (note microscalloped walls) by peloidal high-Mg calcite cement and angular silt-size sponge chips of aragonite (white). Dense interior of sample 6-4-1 in Figure 3A, plane-polarized light, 40 X. C. Scanning electron photomicrograph of bowl-shaped clionid sponge boring with typical microscalloped surface. Boring in coral *Galaxea facicularis* in sample 14-1-1 in Figure 3B. White bar is 100 μ long. **D**. Photomicrograph (at high magnification) of bored coral *Porites* sp. in sample 22-6, showing tiny rod-shaped endolithic algal and fungal borings in aragonite coral (white). Larger clionid sponge borings with characteristic microcuspate boundaries are filled with minute, multi-layered calcareous worm(?) tubes, darker peloidal-microcrystalline high-Mg calcite cement, and scattered silt-size sponge chips (white). Plane-polarized light, 100 X.

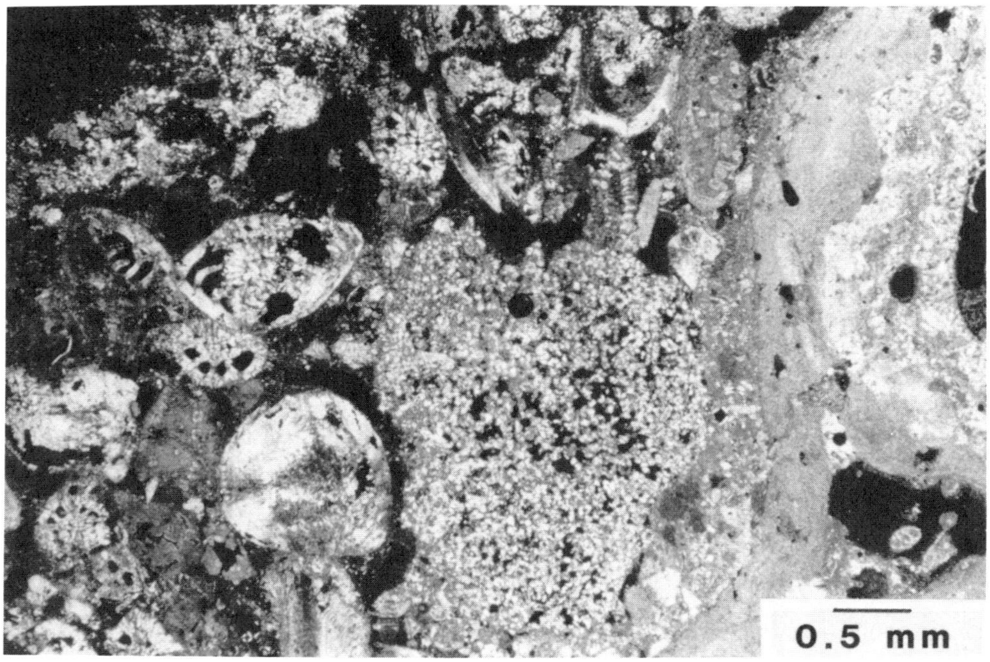

Figure 5. Cavity in coralgal limestone filled with grainstone of benthic foraminifers including shallow-water miliolids and amphisteginids, scattered bits of molluscan shells, and a carbonate lithoclast. Wall of cavity (right side of photo) is formed by encrusting coralline algae. Note high interparticle and intraskeletal (foraminifer) porosity (dark areas). Sample 22-3, crossed nicols, 20 X.

Internal carbonate sediment filling borings and coral cavities consists of silt-size sponge chips of aragonite coral in a peloidal to microcrystalline high-Mg calcite cement (Figure 4B). Some carbonate internal sediment is a biotic grainstone composed of benthic shallow-marine foraminifers (including miliolids and amphisteginids) (Figure 5), rare echinoderm spines, ostracods and bryozoans, fragments of coralline algae and coral, tunicate spicules (ascidians) (Shafik, this volume) or alcyonarian spicules (Resig, this volume), fragments of gastropods (including juvenile *Voluta* sp. and *Lima* sp.; T. DeVries, oral commun., 1985) and pelecypods, and *Halimeda* plates. Many biotics are tightly bound by a peloidal to micritic high-Mg calcite cement. In a few borings, a geopetal fabric of silt-size sponge chips and resedimented peloids is preserved.

A minor amount of dark-gray hemipelagic planktonic foraminifer- and pteropod-bearing terrigenous silt and clay is incorporated in some pores and open cavities of the coralgal boundstone samples (see Crook and Niem, this volume). This fine terrigenous sediment is a result of contamination during the dredging operation.

The outer surface of many coralgal limestone blocks is stained dark-gray to black by manganese or iron sesquioxides. Encrusting or nesting on the manganese-stained surfaces, particularly in crevices, cavities, and other cryptic environments are calcareous serpulid worm tubes (Figure 3B), foraminifer-covered worm tubes, cheilostome bryozoans, clionid sponges, the gastropod *Pyramidella*, ostreid and chamid pelecypods (T. DeVries, oral commun., 1985), and ahermatypic corals. Some of these encrusters are, in turn, coated with ferromanganese stain, whereas others clearly post-date the stain, indicating several generations of encrustation.

DEPOSITIONAL ENVIRONMENT

Most limestone samples were dredged from a forereef-rubble facies and deeper resedimented slope facies (Figure 2). Paleoecology and geographic position indicate that the fresh corals and coralgal limestones originally formed in a clear, warm, shallow-marine, strongly wave-agitated oxygenated environment, perhaps in the forereef or coral-head biozone. Such environments surround many of the islands and shallow platforms in the Solomon Islands chain (Solomon Islands Department of Geological Surveys, 1969).

Hermatypic corals identified by Mitchell Colgan (University of California, Santa Cruz) in the dredge samples include massive fragments of *Porites* sp., *Porites lutea* (Stations 22 and 6), *Pachyseris rugosa* (Station 6), *Galaxea facicularis* (Station 14), *Physogyra lichtensteini*, *Chypastrea* sp., *Acropora (Isopora) cuneata* (Station 22), and *Diploastrea heliopora* (Station 25). Although most of these coral species are not restricted to particular reef facies, this assemblage is most characteristic of the forereef, i.e. <80 m deep (M. Colgan, written commun., 1986). For example, Colgan reports that *Pachyseris rugosa* is presently found at depths >15 m near Guam.

In addition, although known from depths as great as 156 m, the abundance of encrusting coralline algae suggests initial formation of the coralgal limestone in less than 20 to 30 m of clear, warm water (Johnson, 1961; M. Colgan, oral commun., 1985). These algae, together with encrusting foraminifers, such as gypsinids, and benthic foraminifers, such as *Amphistegina* and miliolids in sediment-filled cavities and borings, imply fairly shallow (<40 m) turbulent water similar to that on the active reef flat or forereef (Resig, this volume; G. Chaproniere and R.C. Tjalsma, oral commun., 1986). Broken and wave-abraded skeletal-carbonate sand and silt (fragments of foraminifers, mollusks, echinoid spines, tunicate spicules, *Halimeda*, and coralline algae) occur as internal sediment fill only in reef cavities, crevices, and borings that were apparently protected from strong wave turbulence. Carbonate silt and micrite in the cavities and borings were probably produced *in situ* mainly by endolithic borers and grazers (e.g., parrot fish) in and on the reef (e.g., clionid sponges for the abundant silt-size chips of aragonite).

The Solomon Island samples display multiple generations of coral framework growth and encrustation by gypsinid foraminifers and coralline algae. Borings and crevices are filled with submarine peloidal cements and wave-abraded skeletal carbonate sands. These features are common in modern forereefs of the Caribbean (Warme, 1977) and over much of the reef facies of the Great Barrier Reef of Australia (Marshall, 1986). Clionid sponge borings (some open, others filled with sediment and/or cement), which are prolific in the Solomon Islands coralgal dredge samples (Figures 3A and 4A-D), are also the most abundant type of boring on the modern forereef of northern Jamaica at depths >70 m (Warme, 1977). Similarly, on modern Bermuda reefs, endolithic borers cannot establish themselves in areas of highest wave turbulence and rapid sediment influx and, therefore, tend to be common in the forereef environment (Bromley, 1978). Contemporaneous submarine peloidal-micritic cements in the bored coralgal limestone samples from the Solomon Islands are also most abundant on the windward reef-flat and forereef environments of the Holocene Great Barrier Reef (Marshall and Davies, 1981; Marshall, 1986), in modern Belize reefs (James et al, 1976), and in living Jamaican reefs (Land and Moore, 1980).

The mixture of fresh, uncemented coral masses and diagenetically more altered coralgal limestone samples from Stations 6, 14, and 22 is similar in texture and general faunal composition to an exposed forereef coral-head facies and coralgal reef-rubble facies of a Pleistocene barrier reef in Florida described by Lighty (1985). The Florida coral-head facies also consists of a variety of fresh, massive and branching scleractinian corals that are extensively encrusted and bored. Abraded sand-size biotic fragments of coral, coralline algae, *Halimeda*, and mollusks fill cavities and crevices in this facies. Many of the Solomon Island samples are from the downslope forereef-rubble facies which typically consists of angular gravel-size coral talus. The talus is so extensively bored, encrusted with coralline algae and encrusting benthic foraminifers (gypsinids), and infilled with submarine peloidal cement that the original skeletal microstructure of the corals is nearly impossible to determine.

Also, the geographic position of the Solomon Islands dredge sites on steep slopes below submerged forereefs and reef flats on the southeast-facing windward sides strongly supports the interpretation of the deposit as forereef rubble and associated reef rubble on a forereef slope that formed by rock falls, slides, and debris flows off the forereef (Enos and Moore, 1983). Some samples (e.g., Station 6) appear to have been dredged directly from drowned-reef and forereef-rubble aprons or talus (Figure 2). For example, the drowned reef at Station 6 is in >350 m of water.

Most dredged forereef samples at other stations (e.g., 14 and 25), however, show evidence of further redeposition into deeper water as angular to rounded gravel to boulder talus downslope from the steep forereef and forereef-rubble zone. These samples have been transported down slope through repeated slumping, rock falls, rock avalanches, and debris flows (Cook and Mullins, 1983). From Uniboom seismic records and dredging, Colwell and Tiffin (1986) interpreted the mixture of coarse terrigenous sediment and coralgal debris at the base of the slope to be formed by submarine slumps and slides and by turbidity current processes in the Central Solomons Trough.

Field studies of modern Caribbean and southwest Pacific reefs show that intensive boring and other bioerosion (e.g., browsing parrot fish) of the base and substrate of coral colonies weaken and undermine coralgal forereef edifices (Warme, 1977). Severe tropical storm waves commonly topple overhanging bored limestone masses of dead coral heads and dislodge living unbored coral heads and branches downslope into the forereef-rubble facies. Such dislodgement downslope results in a mixture of unbored fresh coral heads and extensively diagenetically cemented and bored coralgal limestone similar to that dredged at Stations 6 and 22. In these deeper water forereef-slope rubble environments, manganese increasingly coats exposed surfaces that are not buried in sediment; and cryptic ahermatypic corals, bryozoans, and serpulid worms attach themselves to displaced coralgal blocks.

As illustrated in Figure 3B, many dredged coral and coralgal limestone samples are coated with a dark manganese stain over much of the exposed surface and encrusted by stained and unstained ahermatypic corals and serpulid worm tubes. A few solitary ahermatypic corals attached themselves to the upper surface of overturned and displaced heads of colonial hermatypic corals. These displaced heads also show partial manganese staining of exposed upper surfaces and incomplete burial and filling of pores on the lower surface by brown terrigenous and hemipelagic muds, suggesting resedimentation. Ahermatypic corals of the family Carophyllidae have a wide habitat range of 0 to 2,000 m or more. A few large dredged clasts of stained coral masses and coralgal limestone (e.g., Station 22) display fresh, unstained fractured surfaces, suggesting that they may have been broken from drowned or submerged reef outcrops; but more likely, these samples were broken during dredging from larger olistostromal reef blocks.

At Station 25, cobble-size fragments of *Diploastrea heliopora*, which contain freshwater low-Mg sparry calcite cement and neomorphic textures (see petrology and isotope sections), were recovered together with rounded cobbles of metadolerite (Crook and Niem, this volume). The rounding of these clasts, the freshwater calcite cement, and the neomorphic textures suggest recycling from emergent or uplifted Pleistocene reefs and Tertiary metavolcanics on nearby Choiseul (Station 25, Figure 1B). The cobbles of both rock types were rounded by fluvial and/or wave processes. Eventually, these clasts were funneled downslope via slides or other gravity-driven processes (debris flows or high-density turbidites described by Middleton and Hampton, 1973) or were rafted by tree roots to the deep-marine depositional site.

PALEOBATHYMETRY AND TECTONICS

Based upon petrography of dredged coralgal limestone and recognition of submerged reefs on seismic records from an earlier (1982) Tripartite cruise, Colwell and Vedder (1986) reported that substantial intra-arc subsidence has occurred locally in the Solomon Islands since the Pleistocene. Moreover, Resig (1986) concluded from benthic foraminiferal paleoecology that 300 to 550 m of subsidence occurred on the shelf platform that links Guadalcanal, the Russell Islands, and the New Georgia island group (Figure 1A).

During the 1984 R/V *S.P. Lee* cruise, forereef rubble was dredged from the escarpment between 500 m to 900 m below the drowned Pleistocene(?) reef at Station 6 on the southwestern rim of Indispensable basin (Figure 2) and reaffirms the earlier conclusions concerning subsidence. The submerged reef near Station 6 is presently at >350 m, and talus samples indicate that during growth of these coral heads and coralgal reef limestones in the forereef, water depths were <20 to 80 m (see depositional environment section). Even if the reefs at Station 6 were formed at lowest sea-level stand during the maximum Pleistocene glaciation, some subsidence of the Santa Isabel-Florida Islands-Guadalcanal platform is required. The multichannel seismic record suggests a younger (Holocene?) reef in shallower depths to the southwest on the platform (Figure 2).

Colwell and Vedder (1986) suggested that solution tubes in Quaternary calcite-cemented corals and dolomite-cemented calcarenite and rounded pebbles of crystal tuff dredged from the submerged platform northwest of Vella Lavella during the 1982 R/V *S.P. Lee* cruise indicate local subaerial exposure during part of the Quaternary. No evidence of subaerial exposure (vadose or phreatic cement or dissolution features) was found in any coralgal limestone samples dredged downslope from this platform during the 1984 R/V *S.P. Lee* cruise (Station 22 near Colwell and Vedder's sites 1/10 and 1/11, Figure 1B). Perhaps parts of the platform were never emergent or the cemented forereef talus was formed prior to emergence.

AGE

Most of the scleractinian corals that constitute the dredged limestone samples are long-ranging species that are not age diagnostic (M. Colgan, 1986, written commun.). For example, the colonial corals *Porites* sp. and *Porites lutea* (Stations 6 and 22) range from Eocene to Holocene. *Galaxea facicularis* (Station 14) ranges from Miocene to Holocene, and *Diploastrea heliopora* (Station 25) has a known range of Cretaceous to Holocene.

Associated benthic foraminifers and coccoliths are, however, diagnostic of a Pleistocene to Holocene age. Resig (this volume) suggests that the carbonate-cemented skeletal sand that infills cavities in the coralgal limestones at Station 22 are probably Quaternary, based upon the presence of the foraminifers *Calcarena hispida* and *Baculogypsina sphaerulata*. Rare nannofossils in the sediment-filled cavities and borings indicate a Quaternary (Station 6) to probably Quaternary age (Station 25) (Shafik, this volume).

Uplifted Pleistocene reef terraces and modern fringing, patch, and barrier reefs built on a Pleistocene reef foundation are common surrounding all the Solomon Islands and on adjacent shallow-water platforms (Hackman, 1980; Coulson and Vedder, 1986; Pound, 1986; Vedder and Coulson, 1986; Solomon Islands Department of Geological Surveys, 1969). These uplifted reefs are probably the principal sources for the resedimented forereef limestone and coral rubble samples dredged at slope depths nearby (see depositional environment section). The fresh appearance of some coral samples (e.g., sample 22-2; Figure 3C), which are characterized by only minor aragonite cement, manganese staining, and boring, suggest they may be Holocene in age. In addition, the fresh colonial coral *Physogyra lichtensteini* from Station 22 has been recognized only in the Holocene (M. Colgan, 1986, oral commun.).

In contrast, the other more diagenetically altered, intensively bored, and well-cemented manganese-stained reef samples may be Pleistocene. Thus, there may be a mixture of Pleistocene and Holocene material at some dredge sites (e.g., Stations 6 and 22). However, submarine cementation and intensity of borings are quite variable over even modern reef tracts such as the Great Barrier Reef of Australia (Marshall, 1986; Kiene, 1985). A mixture of intensively bored and densely submarine-cemented limestone with compact coralgal framework and unbored and uncemented open-branching coral genera (such as *Physogyra lichtensteini* and *Pachyseris rugosa* of Stations 22 and 6) are fairly common within the same Holocene reef (J. Marshall and W. Kiene, 1986, oral commun.).

Our attempt to radiocarbon-date fresh coral samples from two sites (Station 22 and 6) was unsuccessful (A. Hogg, written commun., 1987). The count rates are within

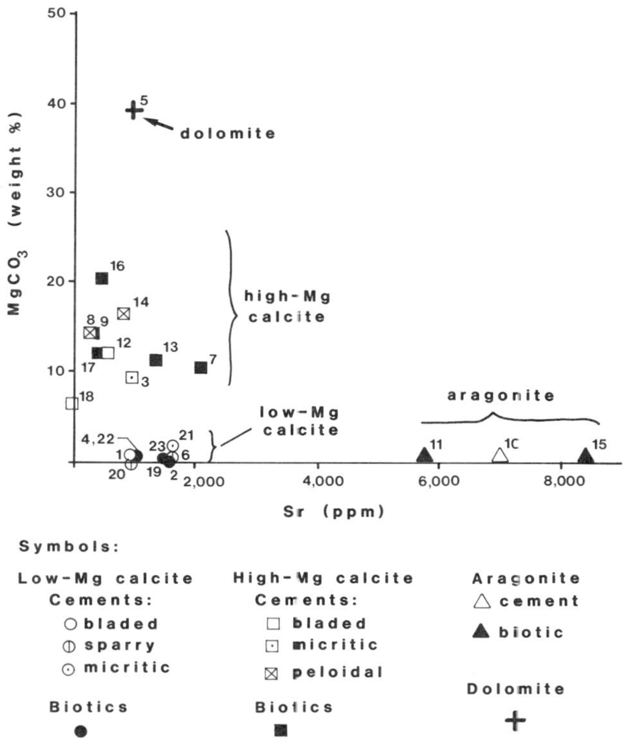

Figure 6. Binary graph of $MgCO_3$ versus Sr, showing range of compositions of aragonite, high-Mg calcite, low-Mg calcite, and dolomite in cements and biotics in coralgal limestone, corals, litharenites, mudrocks, and foraminiferal chalk from 1984 R/V *S.P. Lee* cruise. Numbers refer to samples and dredge sites listed in Table 1.

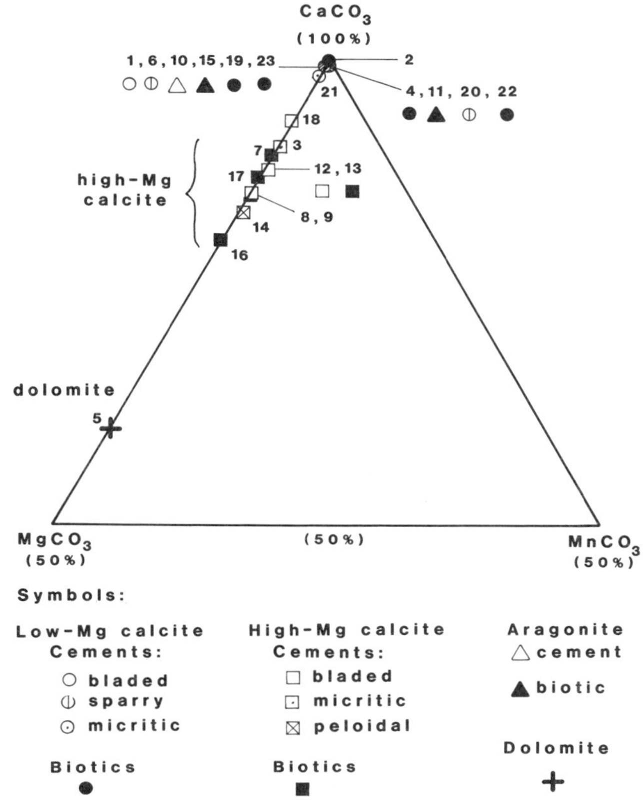

Figure 7. Ternary diagram of $CaCO_3$-$MgCO_3$-$MnCO_3$, showing compositional variation of cements and biotics of the dredged corals, coralgal boundstones, chalks, and siliciclastic sediments. Numbers refer to samples and dredge sites listed in Table 1.

two sigma of background and, therefore, undatable. Colwell and Vedder (1986) reported that a fresh aragonitic coral fragment dredged from west of the Russell Islands during the 1982 R/V *S.P. Lee* cruise yielded a radiocarbon age of >31,000 years B.P. In summary, a Pleistocene age appears likely for many of these coralgal reef limestones and coral fragments although some fragments may be as young as Holocene.

PETROLOGY

Backscatter electron microscopy and microprobe analysis for Ca, Mg, Fe, Mn, and Sr cations (Table 1) verified and quantified carbonate compositions that previously had been determined qualitatively by thin-section study and staining methods. Microprobe analysis methods and standards are discussed in Appendix I. The materials studied include biotic constituents and cements in the Quaternary coralgal boundstone and coral fragments (Stations 6, 22, and 25), late Pliocene foraminiferal chalk (Station 29), and Quaternary carbonate-cemented foraminiferal litharenite and calcareous lithic silty mudstone (carbonate chimney, Station 2).

Graphs of composition ($MgCO_3$ versus Sr and $CaCO_3$-$MgCO_3$-$MnCO_3$, Figures 6 and 7) show that four carbonate minerals, 1) aragonite (Sr >5,700 ppm), 2) high-Mg calcite (>9 weight percent $MgCO_3$), 3) low-Mg calcite (Sr <2,000 ppm and <2 weight percent $MgCO_3$), and 4) dolomite (Table 1) are present as biotic constituents and cements in the coralgal limestone and carbonate-cemented siliciclastics.

Biotic Constituents

Chemical compositions of the biotic constituents (Figures 6 and 7) compare favorably with the compositional ranges of similar fossil groups described by Bathurst (1975) and Scholle (1978).

Aragonite occurs as needle-shaped sclerodermites in the fresh unaltered trabeculae of the widespread scleractinian corals (97.8 weight percent $CaCO_3$; Sr = 5,750 ppm;

Table 1. Composition (weight percent) of Carbonates in Solomon Islands Samples from Microprobe Analysis.

Analysis Designation	$CaCO_3$	$MgCO_3$	$MnCO_3$	$FeCO_3$	Sr (ppm)
1	199.76	0.71	0.00	0.00	1,015
2	100.04	0.00	0.00	0.36	1,600
3	88.65	9.04	0.07	0.37	1,015
4	101.73	0.32	0.15	0.00	1,025
5	58.86	39.06	0.00	0.46	1,015
6	100.23	0.61	0.10	0.00	1,655
7	90.76	10.51	0.00	0.00	2,110
8	85.91	14.32	0.02	0.00	250
9	83.87	14.24	0.17	0.00	340
10	98.84	0.28	0.00	0.00	7,020
11	97.82	0.40	0.05	0.00	5,750
12	92.93	12.22	0.12	0.00	590
13	85.75	11.71	0.00	0.00	1,350
14	85.98	16.76	0.10	0.00	845
15*	67.63	0.29	0.00	0.00	8,455
16	83.27	20.19	0.00	0.00	500
17	83.95	12.35	0.00	0.00	340
18	94.30	6.57	0.00	0.78	0
19	104.80	0.38	0.00	0.00	1,520
20	103.45	0.00	0.12	0.00	1,025
21	99.79	1.92	0.00	0.00	1,655
22	93.92	0.77	0.14	0.00	1,025
23+	80.55	0.75	0.02	0.00	1,600

* Spot nearly destroyed by beam; similar result with second attempt.
+ Very small biotic fragment.

Analysis Designation	Station Sample Number and Description
1	2-1-1, hollow dumbbell-shaped concretion (carbonate chimney), calcite cement growing into empty *Globigerina* chamber
2	2-1-1, hollow dumbbell-shaped concretion (carbonate chimney), *Globigerina* test wall
3	2-1-1, hollow dumbbell-shaped concretion (carbonate chimney), micritic calcite cement next to glass shard
4	2-2-1, B, dolomite-cemented litharenite, *Globigerina* test wall
5	2-2-1, B, dolomite-cemented litharenite, dolomite peloidal cement possibly pseudomorphed after coccoliths
6	2-2-1, B, dolomite-cemented litharenite, blocky sparry calcite cement that fills a chamber of a broken foraminifer test
7	6-4-1, coralgal reef block, encrusting gypsinid foraminifer
8	6-4-1, coralgal reef block, high-Mg calcite exterior of peloid (within chamber of a gypsinid foraminifer)
9	6-4-1, coralgal reef block, microscopic biotic(?) nucleus of accretionary peloid (within chamber of a gypsinid foraminifer)
10	22-2, fresh scleractinian coral (*Physogyra lichtensteini*), aragonite needles growing into void in coral skeleton
11	22-2, fresh scleractinian coral (*Physogyra lichtensteini*), coral skeleton
12	22-3, coralgal limestone, bladed high-Mg calcite cement that infills miliolid foraminifer test
13	22-3, coralgal limestone, miliolid foraminifer test wall
14	22-3, coralgal limestone, microcrystalline peloidal cement

Analysis Designation	
15	22-3, coralgal limestone, ascidian(?) spicule
16	22-3, coralgal limestone, coralline (red) algae
17	22-3, coralgal limestone, minute calcareous worm(?) tube
18	22-3, coralgal limestone, submicroscopic bladed calcite crystals (cement) filling open space in interior of minute calcareous worm(?) tube
19	25-1A, coral cobble, neomorphic sparry dirty calcite that comprises the septae of coral *Diploastrea heliopora*
20	25-1A, coral cobble, blocky equant sparry calcite cement filling pore space between septae of coral *Diploastrea heliopora*
21	25-1A, coral cobble, micrite fill in corallite void (*Diploastrea heliopora*)
22	29-1, late Pliocene chalk, globorotalid foraminifer test wall
23	29-1, late Pliocene chalk, coccolith in chamber of foraminifer test

sample 22-2). Aragonite also comprises the white silt-size scalloped fragments of coral produced by boring clionid sponges and the scattered microscopic star-shaped tunicate (ascidian) or alcyonarian spicules (Sr = 8,455 ppm; Figures 6, 8A, and 8B). Staining with Feigal solution shows that aragonite also forms sand-size fragments of pelecypod and gastropod shells and *Halimeda* in the skeletal grainstone that infills larger borings and crevices in the coralgal limestone samples.

High-Mg calcite (i.e., >9 weight percent $MgCO_3$) comprises abundant skeletal biotic constituents in the coralgal boundstone. These biotic fragments include miliolid foraminifer tests (11.7 weight percent $MgCO_3$; Figure 8A), encrusting benthic gypsinid foraminifers (10.5 weight percent $MgCO_3$), sponge chips (14.2 weight percent $MgCO_3$; Figure 8B), coralline algae (20.2 weight percent $MgCO_3$), and microscopic calcareous polychaete worm(?) tubes (12.4 weight percent $MgCO_3$; Table 1; points 7, 9, 13, 16, and 17 on Figures 6 and 7).

Low-Mg calcite (<2 weight percent $MgCO_3$) forms biotics such as planktonic foraminifers (e.g., globorotalids, 0.8 weight percent $MgCO_3$, Figure 8C; and *Globigerina* sp., 0.0 to 0.3 weight percent $MgCO_3$; Figures 9A and 11) in the carbonate-cemented siliciclastics, and coccoliths (0.8 weight percent $MgCO_3$) in the late Pliocene chalk (Figure 8D).

Cements and Diagenesis

The dredged coralgal boundstones and coral masses display two histories of diagenesis: 1) those that are relatively unaltered with minor cementation, and 2) those that have a complex paragenetic sequence of carbonate cementation and some incipient recrystallization that has obliterated much of the original depositional texture.

Aragonite

An early submarine aragonite micrite cement, although minor, is common to most limestone samples, particularly the relatively unaltered corals in which it is the only cement (e.g., sample 22-2). Acicular aragonite has grown syntaxially in mesopores within the aragonitic skeleton of the scleractinian corals as tiny elongate needles (<4μ) which partially to completely occlude the pores (Figure 4A). The aragonite cement is 98.8 weight percent $CaCO_3$ and Sr = 7,020 ppm (Figure 6 and Table 1). Coarsely acicular aragonite is a common early submarine cement in modern coral reefs throughout the southwest Pacific and Caribbean (Harris, Kendall, and Lerche, 1985; Marshall, 1986). Rare early aragonite needle cement has also formed on biotic fragments in the volcaniclastic foraminiferal packstone from Station 23 (Figure 9C) and has been followed by blocky equant low-Mg sparry calcite cement pore fill. In a few samples, a second generation of aragonite-mesh cement (Marshall, 1986) consisting of randomly oriented needle-shaped crystals, up to 40 μ long, has nucleated on earlier formed silt-size peloidal micritic submarine cement and fills pores in the coralgal boundstone.

High-Magnesium Calcite

The most common cement in the coralgal boundstones (e.g., samples 6-4-1c, 14-1-2, 22-3, 22-6) is a microcrystalline to cryptocrystalline peloidal high-Mg calcite which appears powdery white (sucrosic) in hand specimen. The brown to nearly opaque peloids are oval to subspherical aggregates of cryptocrystalline (<4 μ) high-Mg calcite and range in diameter from 20 to 60 μ. The peloids are tightly cemented by a clear microcrystalline high-Mg calcite (4 to 30 μ) or occur as uncemented aggregates. A few peloids have accreted around a nucleus of high-Mg sponge chips and aragonite ascidian spicules (Figure 8B); many, however, appear to lack a nucleus. The amount of $MgCO_3$ in the peloids ranges from 14.3 weight percent to 16.8 weight percent (Figures 8A and 8B; points 8 and 14 on Figure 6). A similar to slightly higher amount of $MgCO_3$ (i.e., 16.3 percent to 18.4 percent) has been reported in high-Mg peloidal calcite cements in reefs off Belize and Panama (James et al, 1976; Macintyre, 1977). The peloids fill most borings (Figure 4B),

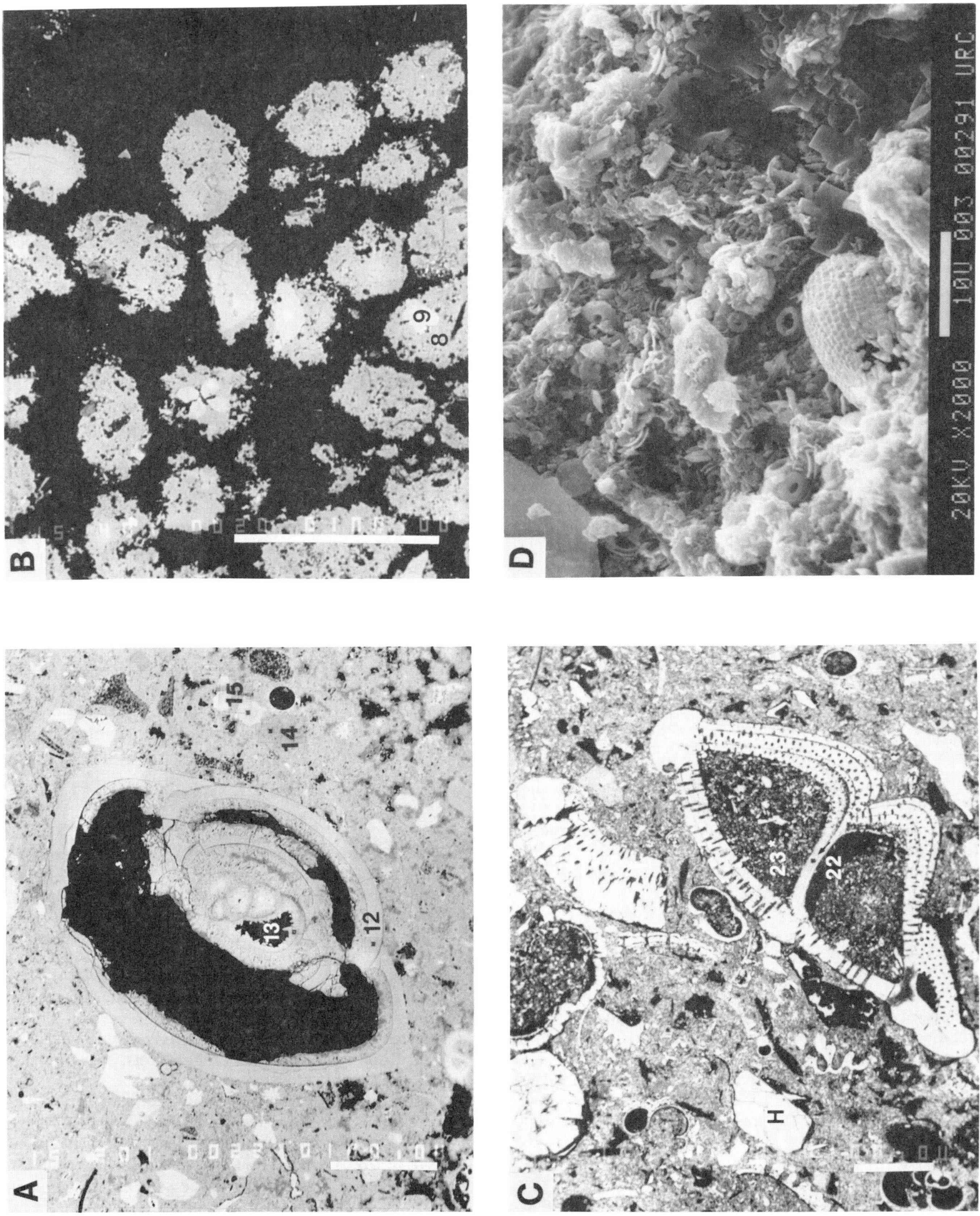
A
B
C
D
20KU X2000 10U 003 00291 URC

interparticle pore spaces between sand- and silt-size biotics (Figure 8A), natural coral-growth crevices and cavities, and growth chambers of various biota such as gypsinid foraminifers. In contrast, the minor aragonite-needle cement typically fills only vacated coral skeletal meso-pores.

Peloidal high-Mg calcite is common as an early submarine cement in borings and intraskeletal pores in the modern Great Barrier Reef of Australia (Marshall, 1986; Marshall and Davies, 1981) and in Holocene Caribbean and Florida reefs (Macintyre, 1977, 1984; Lighty, 1985; Land and Moore, 1980). A two-stage origin is suggested by Macintyre (1985) for these peloidal cements: 1) precipitation of spherical peloids followed by 2) precipitation of clear microcrystalline high-Mg calcite cement that bonds the peloids together.

Most peloidal cements fill and therefore post-date the borings and coral cavities in the Solomon Islands samples. However, some unfilled late-stage borings by pholads, barnacles, and clionid sponges cut across the earlier cement-filled borings; and a few coralline algae and corals overlie peloidal cement-filled borings, suggesting that the cements were formed during and after coral growth and activity by the most recent borers on the reef.

High-Mg calcite also occurs as microcrystalline (<5 µ long) bladed crystals that have grown as isopachous to randomly oriented crystals in empty chambers of miliolid foraminifer tests (point 13 on Figure 8A), worm tubes, and restricted intraskeletal pores of other biotic fragments. The amount of $MgCO_3$ in the high-Mg bladed calcite cement in the reef limestone from Station 22 is slightly higher than in the high-Mg biotic substrate on which this cement nucleated. The cement is 12.2 weight percent $MgCO_3$ compared to 11.7 weight percent $MgCO_3$ in the biotic constituent (points 13 and 12, Figure 6 and Table 1). The bladed high-Mg calcite cements differ slightly in composition from the more abundant peloidal high-Mg calcite cements (common in borings in the Solomon Islands coralgal limestones) which generally have a higher $MgCO_3$ content (e.g., 14.3 to 16.8 weight percent $MgCO_3$, Figure 6, Table 1). Microcrystalline high-Mg calcite cement also fills interparticle pores in the hollow lithic silty mudstone concretion (carbonate chimney, Figures 10 and 11) from Station 2 (9.0 weight percent $MgCO_3$; Table 1, n. 3).

Dissolution, Low-Magnesium Calcite, and Neomorphism

Most coralgal boundstone and coral samples show no evidence of emergence. Absent are subaerially formed dissolution features and cements, such as vugs and channelways, widespread blocky or equant sparry low-Mg calcite cements, and neomorphic textures that typically develop in the phreatic zone, as well as vadose-whisker, meniscus, pisolitic and microstalactitic cements (Harris, Kendall, and Lerche, 1985; Scholle, 1978). Only early submarine peloidal high-Mg calcite and aragonite-needle cements occur.

However, the rounded cobble of the coral *Diploastrea heliopora* (25-1C) shows extensive petrographic evidence of subaerial fresh-water cementation and neomorphism. Minor low-Mg micrite cement (1.9 weight percent $MgCO_3$) and biotic silt and sand partially fill intraskeletal voids between septae and are superseded by clear void-

---

Figure 8A (facing page). Backscatter electron photomicrograph of high-Mg calcite miliolid foraminifer in a matrix of porous microcrystalline to pelletal high-Mg calcite cement and scattered aragonite sponge chips (irregular, scalloped white fragments) and star-shaped tunicate (ascidian) or alcyonarian spicules. Outline of peloidal cement and microcrystalline high-Mg calcite over-growth and intercement porosity (dark area) well-displayed in lower right. Note thin-bladed isopachous crystals of high-Mg calcite forming rim cement that partly infills intraskeletal chambers of the miliolid foraminifer (dark area is intraskeletal void). Numbers refer to microprobe chemical compositions plotted on Figures 6 and 7 and presented in Table 1. Sample 22-3. Vertical bar at left is 100 µ long. **B.** Backscatter electron photomicrograph of porous high-Mg peloids (medium gray) nucleated on white high-Mg calcite sponge chips and aragonite star-shaped ascidian spicules. Microprobe analysis of point 8 is: 85.9 weight percent $CaCO_3$, 14.3 weight percent $MgCO_3$, 0.02 weight percent $MnCO_3$, 86.33 mole percent Ca, and 250 ppm Sr. Point 9 is 83.9 weight percent $CaCO_3$, 14.2 weight percent $MgCO_3$, 0.17 weight percent $MnCO_3$, and 340 ppm Sr. Sample 6-4-1. Vertical bar at left is 100 µ long. **C.** Backscatter electron photomicrograph of globorotalid and *Globigerina* sp. with rare sand-size hornblende (point H) in matrix of microcrystalline coccoliths and micrite cement. Opaque areas include tiny rod-shaped intraskeletal pores within foraminifer test, megapore in foraminifer chamber, and submicroscopic interparticle voids within the microcrystalline calcite-coccolith matrix. Late Pliocene foraminiferal chalk, Station 29. Numbers refer to microprobe analyses: 22 = low-Mg calcite of foraminifer test; 23 = low-Mg calcite of coccolith; H = hornblende. See ternary plots in Figures 6 and 7 and Table 1. Large globorotalid foraminifer at center right is probably *Globorotalid menardii* (D'Orbigny) var. *miocenica* (Palmer) as illustrated and described by Postuma (1971) (R.C. Tjalsma, 1985, oral commun.). Bar is 100 µ long. **D.** Scanning electron photomicrograph of late Pliocene chalk composed of disk and star-shaped nannofossils with scattered larger planktonic foraminifers and blocky microcrystalline calcite cement. Disk-shaped coccoliths consist of low-Mg calcite as determined by microprobe analysis (point 23 on Figure 8C). Station 29. Bar is 10 µ long.

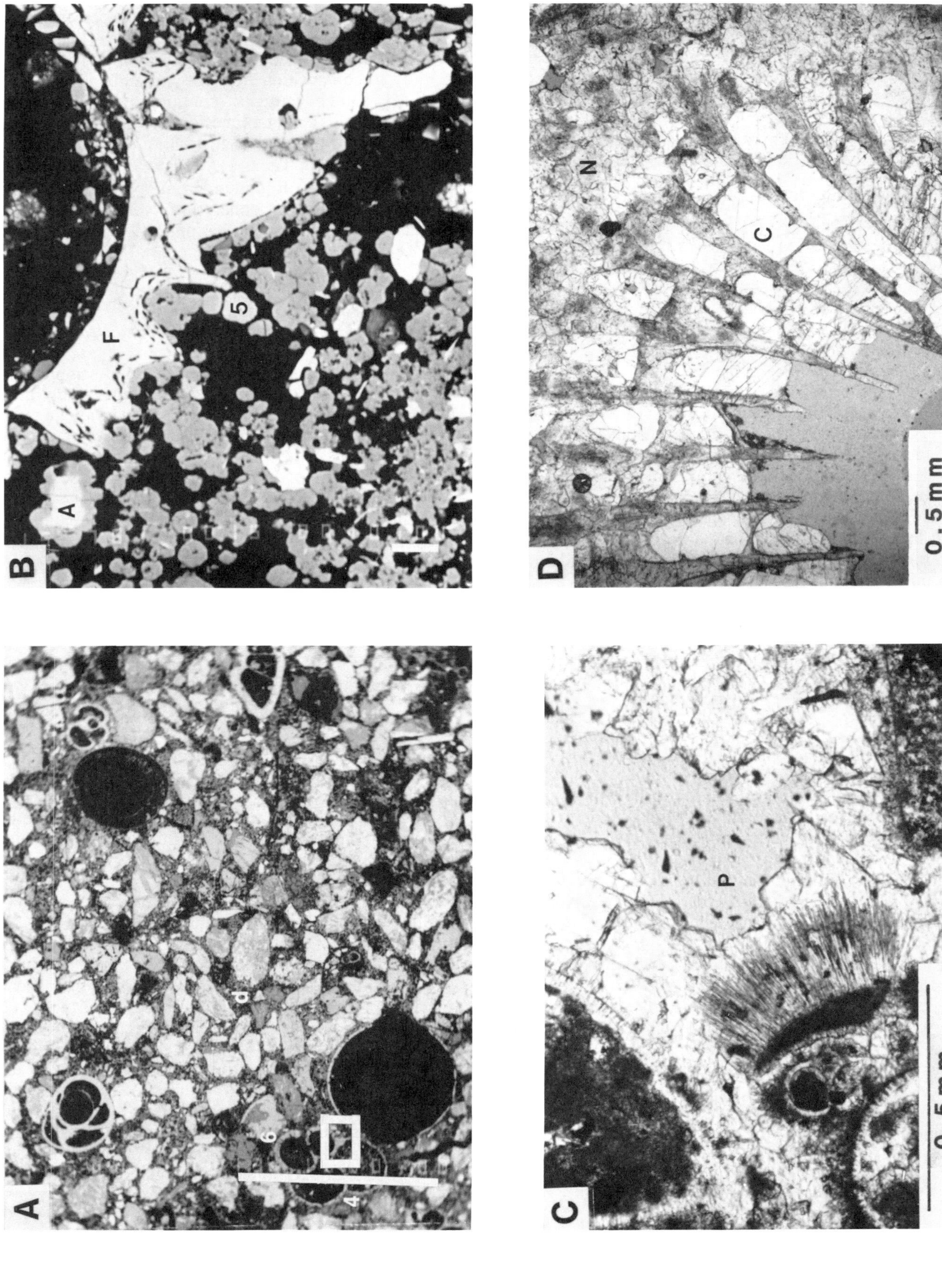
A
B
C
D
F
A
5
6
4
d
N
C
P
0.5mm
0.5mm

Figure 10. Slabbed lower half of hollow dumbbell-shaped carbonate chimney (lithic silty mudstone) from Station 2. Left: top view; right: longitudinal slabbed section, showing long, cylindrical, open central tube coated with manganese oxide and a younger unstained attached serpulid worm tube (S) and solitary ahermatypic coral (C). These features, the carbon and oxygen isotopes, and the intensively bored outer surface (top view) suggest that the open chimney was subsequently exposed to deep-sea boring and colonizing organisms at a former vent of escaping methane-enriched pore fluids.

filling, blocky sparry low-Mg calcite (Figure 9D and point 20 on Figure 6). Rare early submarine acicular aragonite cement, which developed syntaxially on some septae, is also superseded by the clear low-Mg calcite cement. In addition, the septae and coral walls consist of a mosaic of large (up to 0.5 mm in diameter) dirty coarse pseudospar or anhedral low-Mg calcite crystals (0.4 weight percent $MgCO_3$) (Figure 9D; Table 1, n. 19). Some dirty pseudospar crystals in the coral wall and septae are in optical continuity with the clear crystals of low-Mg sparry cement in the adjacent interseptal voids. The pseudospar formed during neomorphic transformation of aragonite to low-Mg calcite (Folk, 1965).

Low-Mg calcite also occurs as rare sparry calcite that fills foraminifer tests (0.6 weight percent $MgCO_3$) in the dolomite-cemented lithic arenite from Station 2 (Figure 9A), and submicroscopic bladed crystals of an intermediate-Mg calcite (6.6 weight percent $MgCO_3$) line the interior cavity of minute calcareous worm tubes in coralgal limestone from Station 22. The very small size of these crystals (<2 μ) precludes a precise chemical analysis by microprobe because the spot size is so large relative to the crystal size. These bladed crystals of calcite may be transitional in composition between low- and high-Mg calcite or may have been high-Mg calcite that lost some Mg through pore-water leaching similar to that described by

---

Figure 9A (facing page). Backscatter electron photomicrograph of microcrystalline dolomite cement in planktonic foraminifer-bearing litharenite. Medium-gray finely crystalline dolomite (d) infills primary interparticle porosity. Numbers refer to microprobe chemical analyses of test of *Globigerina* sp. (point 4), and lighter gray blocky, sparry low-Mg calcite pore fill of foraminifer chamber (point 6). Outlined area is Figure 9B. Sample 2-2-1. Bar is 1,000 μ long. **B**. Backscatter electron photomicrograph of edge of foraminifer test in dolomitic foraminiferal litharenite outlined in lower left of Figure 9A. Dolomite is a medium-gray interparticle microcrystalline cement. Note peloidal to spherical form of dolomite (medium gray); some appears to be a replacement of disc-shaped coccoliths(?) with central hole (dark area), whereas dolomite peloids accreted around microscopic detrital high-Mg calcite or aragonite biotic fragments (A) and on larger fragment of *Globigerina* sp. test (F). Microprobe analysis of point 5 is dolomite: 58.9 weight percent $CaCO_3$, 39.1 weight percent $MgCO_3$, and 0.5 weight percent $FeCO_3$ (see Table 1 and Figure 7). Bar is 10 μ long. **C**. Rare, early diagenetic acicular aragonite needles of submarine cement syntaxially grown on carbonate biotic substrate superseded by blocky low-Mg calcite cement that partly fills pore (P) of volcaniclastic foraminiferal packstone. Note darker gray micrite infilling some foraminifer tests. Sample 23-18C. Plane-polarized light, 100 X. **D**. Photomicrograph of septae and wall of the coral *Diploastrea heliopora* recrystallized to dirty, blocky, sparry neomorphic low-Mg calcite (N); clear sparry low-Mg calcite cement (C) partially infills intraskeletal voids between septae. Station 25. Plane-polarized light, 20 X.

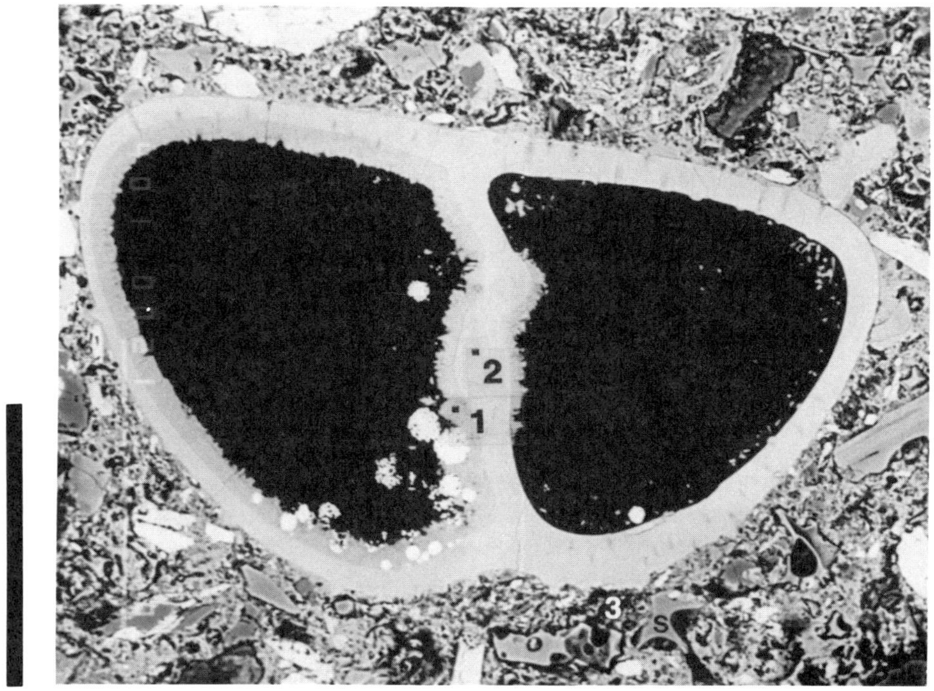

Figure 11. Backscatter electron photomicrograph of part of the carbonate chimney (Figure 10), which consists of high-Mg calcite-cemented lithic silty mudstone. Note low-Mg calcite planktonic foraminifer test (point 2) with fringe of microscopic isopachous low-Mg calcite cement (point 1 and Table 1) grown into interior chamber (void) of test (dark). Pyrite framboids (white circular aggregate objects within chamber) post-date the cement. Rhyolitic glass shard (s) (76 percent SiO_2, microprobe analysis) is set in micritic cement (point 3). Numbers refer to microprobe chemical compositions plotted on Figures 6 and 7 and presented in Table 1. Bar is 100 μ long.

Saller (1986) for high-Mg calcite cements in Eocene reef limestones of Enewetak Atoll.

Dolomite

In a litharenite sample dredged at Station 2, dolomite occurs as a widespread microcrystalline to peloidal cement that infills interparticle pore space between siliciclastic framework grains and low-Mg calcite planktonic foraminifers (Figures 9A and 9B). The peloidal form of the dolomite suggests that it may have replaced or is a pseudomorph of coccoliths or submicroscopic high-Mg calcite peloids. The dolomite may also have been directly precipitated from pore water in this form. The dolomite is 58.9 weight percent $MgCO_3$ and 39.1 weight percent $CaCO_3$ (Table 1, point 5 on Figure 7). X-ray diffraction analysis shows the dolomite to be a well-crystallized, well-ordered calcium dolomite (Rivette, 1986, oral commun.). None of the other samples studied for this report contain dolomite. Microcrystalline dolomite cements have been reported in methane-derived carbonate chimneys, crusts, and slabs on the outer continental shelf and slope of Oregon (Schroeder, Kulm, and Muehlberg, 1987; Ritger, Carson, and Suess, 1987). Like the Solomon Islands samples, some carbonate chimneys and crusts from offshore Oregon are composed of sand- and silt-size siliciclastic framework grains cemented by microcrystalline dolomite.

Micrite

A micritic to locally sparry blocky calcite void-filling cement occurs in a foraminiferal feldspatho-volcaniclastic packstone recovered at Station 23 (Figure 9C). Micrite also weakly cements the late Pliocene nannofossil-globigerinid chalk recovered at Station 29 northwest of Bougainville (Figures 8C and 8D).

A few aragonitic fossils, such as echinoid spines, in the foraminiferal packstone from Station 23 have undergone dissolution, perhaps as a result of being transported downslope below the aragonite compensation depth. Blocky calcite and micritic cement precipitated in the adjacent interparticle pores outline the molds of these former aragonitic biotics.

Most of the coralgal samples were dredged from below the aragonite saturation depth for the Pacific Ocean (≈300 m, according to Scholle, Arthur, and Ekdale, 1983). The lack of aragonite dissolution and lack of conversion of high-Mg calcite to dolomite suggests: 1) these dredged samples were not in deep water long enough for substantial dissolution; 2) marine waters in the Solomon Islands may

Table 2. Stable Carbon and Oxygen Isotopic Composition (per mil relative to PDB) of Dredge Samples.

Plotting Code	Sample No.	$\delta^{13}C$	$\delta^{18}O$	Hand Specimen Description
A	2-3-1	-0.4	-1.0	Foraminifer-rich hemipelagic mud
B	6-4-1	1.3	1.1	Serpulid worm tubes attached to Mn-stained coralgal reef limestone rubble
C	6-4-1B	1.4	-1.1	Reef limestone rubble - scleractinian coral, *Porites lutea* - porous exterior
D	6-4-1C	3.2	0.2	Reef limestone rubble - scleractinian coral, *Porites lutea* - dense peloidal cemented interior
E	6-5-1	0.4	-1.6	Fresh coral, *Pachyseris rugosa*
F	12-0-90	1.3	-1.4	Coccolith silty clay with a few foraminifers; core
G	13-75	0.9	-0.8	Foraminiferal turbidite sand; core
H	16-0-90	0.8	-2.0	Foraminifer-coccolith-rich silty clay; core
I	22-4A	1.5	-1.5	Carbonate sandy mud in coral cavity
J	29-1	0.8	-0.6	Late Pliocene nannofossil chalk
K	16-206	1.1	<0.1	Foraminiferal sand; core
L	22-2	-0.7	-3.9	Fresh scleractinian coral, *Physogyra lichtensteini*
M	22-4B	1.7	0.1	Coralline algae encrusting scleractinian coral, *Acropora (Isopora) cuneata*
N	22-5	1.4	2.0	Solitary ahermatypic coral of Family Caryophyllidae
O	22-5B	-2.6	-4.4	Fresh colonial coral, *Porites* sp.
P	22-7	1.4	-1.1	Microcrystalline carbonate peloidal cement (micrite) filling cavity or boring in scleractinian coral *Porites* sp.
Q	23-18C	-46.7	2.5	Sparry calcite-cemented *Globigerina* packstone
R	25-1A	-4.2	-5.4	Neomorphosed scleractinian coral *Diploastrea heliopora*
S	2-1-1	-31.5	2.4	Microcrystalline calcite-cemented lithic silty mudstone (hollow dumbbell-shaped carbonate chimney)
T	2-2-1,B	-29.4	6.2	Dolomite-cemented foraminiferal litharenite

not be substantially undersaturated with respect to aragonite; or 3) the coralgal dredge samples were out of contact with open sea water (i.e., buried by terrigenous sediment or protected by manganese stain (Saller, written commun., 1988).

All the biotics and carbonate cements (aragonite, high- and low-Mg calcite, dolomite) contain very low percentages of Mn and Fe (range 0.00 to 0.17 weight percent $MnCO_3$ and 0.00 to 0.78 weight percent $FeCO_3$, respectively; Table 1; Figure 7). Such low contents of Mn and Fe indicate that these are fairly pure calcites, aragonites, and dolomites which precipitated in oxidized waters. Slight increases in trace amounts of Mn, Sr, and Fe can cause aragonite and high-Mg biotic and other carbonate constituents to appear lighter gray (white) in back-scatter electron photomicrographs (Figures 8A and 8B).

STABLE CARBON AND OXYGEN ISOTOPES

Oxygen and carbon isotopic analyses were performed on 20 samples of coralgal reef limestones, corals, foraminifer-coccolith oozes, muds, and foraminifer turbidite sands from cores, microcrystalline peloidal cement in reef limestones, serpulid worm tubes, and coralline algae from nine stations (Table 2). Preparation methods are described in Appendix I. Results of the analyses were plotted on $\delta^{18}O$ (PDB) versus $\delta^{13}C$ (PDB) binary graphs and compared with carbon and oxygen isotopic compositional fields outlined by Hudson (1977) and Milliman (1974) for reef limestones, corals, deep-water limestones, biogenic oozes (chalks), shallow-water cements, and methane-derived cements and dolomites (Figures 12 and 13). These isotopic fields have been modified to include isotopic compositional data for biotics and cements for limestones and carbonate sediments from the Pacific and Caribbean from Saller (1986), Gonzalez and Lohmann (1985), and Gross and Tracey (1965).

Three dredge samples of fresh unaltered to weakly aragonite-cemented scleractinian colonial corals *Porites*

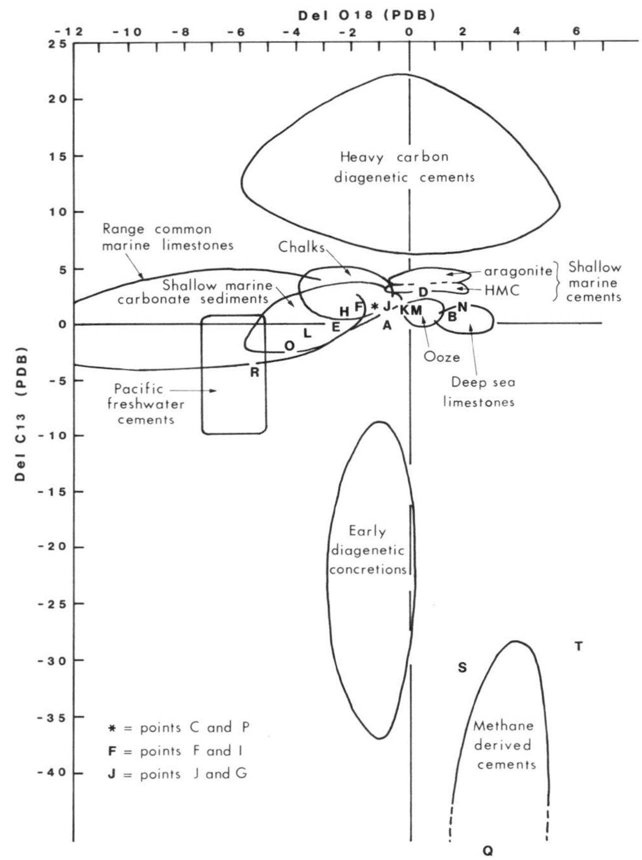

Figure 12. Range of values of $\delta^{13}C$ and $\delta^{18}O$ of marine limestones, chalks, shallow-marine cements, modern Pacific carbonate sediments, Pacific freshwater cements and concretions compared to stable carbon and oxygen isotopic values for Solomon Islands and Papua New Guinea coralgal limestone, biotics, and carbonate cements (see Table 2 for key to letters). Compositional fields after Hudson (1977) and modified with isotopic compositional data from Saller (1986, Pacific freshwater cements), Gonzalez and Lohmann (1985, shallow-marine cements), and Gross and Tracey (1965, unaltered modern Pacific shallow-water carbonate sediments).

lutea, Physogyra lichtensteini (Figure 3C), and *Pachyseris rugosa* from Stations 22 and 6 (points O, L, and E on Figures 12 and 13) are depleted in $\delta^{18}O$ relative to sea water (SMOW). All three samples plot within the compositional ranges of carbon and oxygen isotopic ratios recognized by Milliman (1974) and Gonzalez and Lohmann (1985) for modern hermatypic scleractinian corals (Figure 13) and within the field for common shallow-marine limestones determined by Hudson (1977) and unaltered modern Pacific shallow-marine carbonate sediments of Gross and Tracey (1965) (Figure 12).

The diagenetically altered scleractinian coral *Diploastrea heliopora* from Station 25 (Figure 9D) also plots within the limits of carbon and oxygen isotopic ratios for hermatypic scleractinian corals. However, this altered coral is more depleted in $\delta^{13}C$ than the three fresh unaltered corals dredged from Stations 6 and 22 and plots in the field for Pacific freshwater cements delineated by Saller (1986) (point R on Figure 12). This $\delta^{13}C$ signature supports the petrographic evidence that this rounded recycled coral cobble had previously undergone neomorphic alteration and subaerial freshwater cementation in the phreatic zone (see geochemistry section).

Both the solitary ahermatypic coral (22-5) and the serpulid worm tube that are attached to the dark manganese oxide coated colonial coral limestone have values that plot in the isotopic fields characteristic of deep-marine limestones (i.e., positive $\delta^{18}O$ values) (Points N and B, Figures 12 and 13). These positive values probably reflect the fact that these two marine organisms grew in colder, deeper marine waters after the colonial coral limestone blocks rolled or slid downslope. Manganese oxides commonly coat or stain displaced coral limestone blocks and pebbles at depth in the southwest Pacific where sedimentation rates are low (M. Colgan, 1986, oral commun.).

The dense, bored interior of the coralline algae-encrusted colonial coral *Porites lutea* (sample 6-4-1C, Figure 3A, point D on Figures 12 and 13), is enriched in $\delta^{13}C$ and contains positive $\delta^{18}O$ values in contrast to the negative or depleted $\delta^{18}O$ values of uncemented aragonitic hermatypic corals (points E and L), including fresh samples of the same genus (point O). The values from this material plot in the shallow-water limestone and shallow cement field (Figures 13 and 12, respectively), possibly reflecting the abundance of high-Mg calcite peloidal submarine cement (>50 percent) in the intensely bored interior of the coral. The temperature at which inorganic submarine peloidal high-Mg calcite cement was precipitated, assuming equilibrium with SMOW (i.e., $\delta^{18}O$ = O) is calculated to be 16°C (using the calcite equation of Craig, 1965). Gonzalez and Lohmann (1985) found that early submarine carbonate cements better reflect the expected sea-water temperature in Holocene reef limestone than do the skeletal biotics, which tend to form out of equilibrium with sea water.

Similarly, the high-Mg calcite red coralline algae contains slightly positive or enriched $\delta^{18}O$ and $\delta^{13}C$ iso-

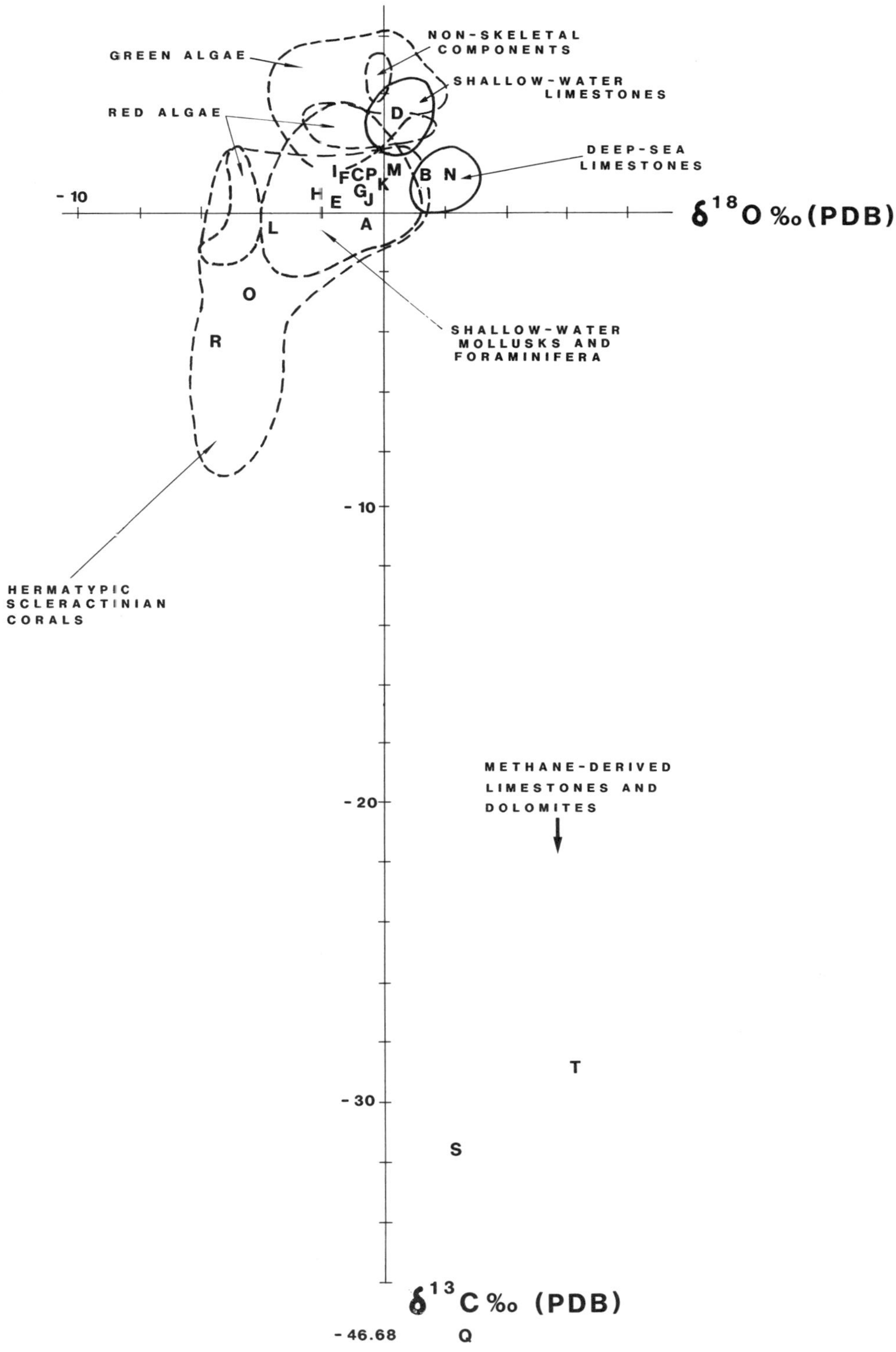

Figure 13. Stable carbon and oxygen isotopic values for Solomon Islands and Papua New Guinea dredged coralgal limestones, biotics, and carbonate cements (see Table 2 for key to letters) compared with values of $\delta^{13}C$ and $\delta^{18}O$ for different groups of shallow-water carbonate skeletal organisms, limestones, and dolomites. Compositional fields after Milliman (1974) and modified with isotopic compositional data from Gonzalez and Lohmann (1985, corals, green algae, and red algae).

topic ratios and does not plot within the expected values for coralline algae (point M on Figure 13). Keith and Weber (1965) also found that red coralline algae show marked deviations from predicted oxygen isotopic equilibrium.

The oxygen and carbon isotopic compositions of white biotic-rich mud that fills cavities and borings in the coralgal reef limestone (Points I and P, samples 22-4A and 22-7) plot within or very close to the expected $\delta^{18}O$ and $\delta^{13}C$ isotopic ranges of common marine limestones and chalks (Figure 12) and of green calcareous algae, shallow-water foraminifers, and mollusks (Figure 13). These isotopic compositions reflect the composition of the skeletal biotics that comprise the carbonate sandy mud. These fine normal-marine carbonate clasts probably were washed in from reef and backreef environments where there are abundant mud-size fragments of green calcareous algae, sand-size benthic foraminifers, the green algae *Halimeda*, and comminuted molluscan fragments.

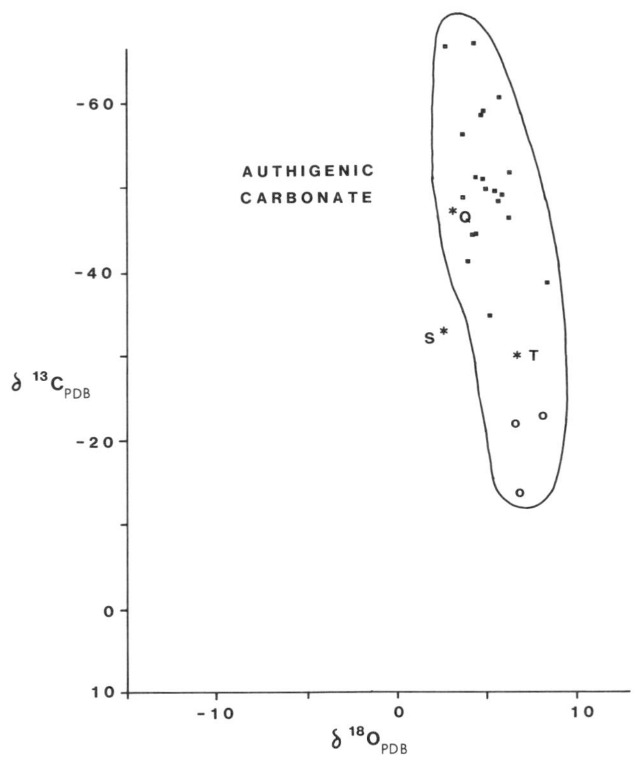

Figure 14. Range of stable carbon and oxygen isotopic compositions (per mil relative to PDB) of authigenic methane-derived carbonates and carbonate chimneys from the accretionary wedge of the Oregon and Washington continental margin (modified from Ritger, Carson, and Suess, 1987 and Schroeder, Kulm, and Muehlberg, 1987). Solomon Islands samples (*) include Q = 23-1-C (foraminiferal packstone); T = 2-2-1 (dolomite-cemented foraminiferal litharenite); and S = 2-1-1 (carbonate chimney).

Pelagic and hemipelagic sediments cored at Stations 13 and 16 and dredged at Station 29 plot in the fields of deep-sea oozes, chalks, unaltered modern Pacific shallow-marine carbonate sediments, and common marine limestones on Figure 12. These samples include *Globigerina*-coccolith-pteropod muds (point H; Station 16, 0-90 cm), foraminifer sand turbidite (point K; Station 16, 206 cm; point G, Station 13, 75 cm), and coccolith-foraminifer chalk (point J; sample 29-1). All of these samples and a foraminifer-rich hemipelagic mud dredged at Station 2 (point A; Figures 12 and 13) have values that plot within the area occupied by foraminifer and shallow-water mollusks (Figure 13). The same samples are slightly depleted in $\delta^{18}O$ and indicate sea-water temperatures of 16° to 21°C (using the calcite equation of Craig, 1965, and assuming equilibrium with SMOW).

The carbonate chimney (lithic silty mudstone) (point S; sample 2-1-1), dolomite-cemented foraminiferal litharenite (point T; sample 2-2-1), and the low-Mg calcite-cemented volcaniclastic packstone (point Q, sample 23-1C) are very depleted in $\delta^{13}C$ (i.e., -29.4‰ to -46.7‰ PDB) compared to all the shallow-water coralgal limestones, biotics, and deep-sea foraminifer-coccolith oozes (chalk) and hemipelagic mud samples (Figures 12 and 13). Furthermore, they are slightly enriched in $\delta^{18}O$ (i.e., 2.4‰ to 6.2‰ PDB) (Table 2). The 2.4‰ value of $\delta^{18}O$ is to be expected from a sea water source at this temperature and latitude; but the 6.2‰ value of $\delta^{18}O$ is out of isotopic equilibrium, possibly reflecting a previously $\delta^{18}O$-enriched pore-water source, perhaps from greater burial depth (E. Suess, oral commun., 1988).

$\delta^{18}O$ and $\delta^{13}C$ values of these dredged samples plot within or near the isotopic compositional field of carbonate cements formed by oxidation of biogenically derived methane (Figure 12). The stable isotope composition of these three samples also lies within or near the range of moderately to extremely depleted light carbon and slightly enriched oxygen isotopic values (per mil relative to PDB) of lithologically similar methane-derived authigenic carbonate concretions, crusts, and carbonate chimneys reported by Ritger, Carson, and Suess (1987) and by Schroeder, Kulm, and Muehlberg (1987) from the accretionary outer continental shelf and slope of Oregon and Washington (Figure 14). The microcrystalline calcite, aragonite, and dolomite authigenic cements in the Holocene crusts, concretions, and chimneys from the Oregon-Washington continental margin apparently precipitated on and within the upper few cm or m of terrigenous and hemipelagic siliciclastic sediments. The cements appear to be produced by CO_2-enriched pore waters from oxidation of biogenic methane that are being expelled through vents or chimneys on and just below the sea floor (Ritger, Carson, and Suess, 1987; Schroeder, Kulm, and Muehlberg, 1987; Kulm, Suess, and Snavely, 1988). The methane is thought to move upward from a shallow zone of biologic methanogenesis through the sediment column by burial-induced compaction or more likely by tectonic compression of the accretionary wedge sediments (Ritger, Carson, and Suess, 1987).

The carbonate chimney from Station 2 (Figure 10) is similar in shape to hollow cylindrical chimneys illustrated by Schroeder, Kulm, and Muehlberg (1987) and Kulm, Suess, and Snavely (1988). Carbonate cements in the Solomon Islands concretionary mudstone, sandstone, and chimney, as well as in the Oregon-Washington dredge samples apparently coexist and are intergrown with authigenic pyrite framboids (Figure 11), suggesting nearby coexisting sulfate-reducing conditions. As hypothesized by Ritger, Carson, and Suess (1987), during the course of migration through the uppermost sediment, biogenic methane is oxidized by microbial activity and the resulting carbonate is precipitated with Ca^{++} and Mg^{++} as microcrystalline interstitial carbonate cement near the base of the zone of sulfate reduction (Suess and Whiticar, 1988). The stable carbon and oxygen isotopic signatures of the Quaternary carbonate chimney, dolomitic foraminiferal litharenite, and foraminiferal packstone from Stations 1, 2, and 23 probably formed by similar biochemical processes related to fluid venting.

The origin of the methane is not clear. Biogenic methane is normally more highly fractionated (i.e., $\delta^{13}C$ = -75‰ to -90‰ PDB, Claypool, Presly, and Kaplan, 1973) than indicated by the $\delta^{13}C$ measured in the Solomon Islands samples. The Solomon Islands samples have $\delta^{13}C$ values of -29‰ to -47‰ that overlap the range of $\delta^{13}C$ values commonly attributed to thermogenic methane (i.e., $\delta^{13}C$ = -30‰ to -50‰, Whiticar, Suess, and Wehner, 1985). However, Ritger, Carson, and Suess (1987), who found a similar range of moderately depleted light $\delta^{13}C$ percentages in the Oregon-Washington material, suggested that the transfer of the isotopic composition of the methane carbon to the carbonate mineral is complex. This range of values may reflect increasing depletion of the biogenic methane pool during oxidation in a closed system of the upper surface sediments as shown by Suess and Whiticar (1988). In such a case, the resulting carbon isotopic composition of the bicarbonate is only moderately depleted, i.e., -40‰ to -60‰ PDB rather than -75‰ to -90‰. Mixing of isotopically heavier non-methane carbon during upward pore-fluid migration is another possibility. It is clear, however, that a source for biologically formed methane is present in the near-surface sediments of the Central Solomons Trough and, furthermore, that the growth of chimney structures indicates fluid venting.

Biogenic methane, formed by fermentative decomposition of organic matter by microorganisms during shallow burial of hemipelagic muds, was detected in gravity cores at several stations in the Russell basin in New Georgia Sound (Kvenvolden and Niem, this volume). Some of these gravity-core stations are associated with acoustically transparent "diapir-like" anomalies that truncate strong horizontal sediment reflectors on high-resolution seismic records. These anomalies may be caused by biogenic gas escaping from depth to the near-surface sediments (Kvenvolden and Niem, this volume; Colwell and Tiffin, 1986). It is noteworthy that Stations 1 and 2, where the hollow carbonate chimneys, porous carbonate hardground or crust, and dolomitic litharenite were dredged, are located along a steep escarpment. The escarpment is the trace of a fault zone that delineates the southwestern boundary of Indispensable basin (Figure 1A; Vedder and Coulson, 1986; Bruns and others, 1986; Wells, this volume). This fault zone may have created fractures that permit upward escape of methane-bearing pore fluids. However, similar microcrystalline calcite and dolomite cements are widespread in Quaternary siliciclastic sediments dredged in the Solomon Islands. For example, Colwell and Vedder (1986) reported microcrystalline dolomite cement in calcarenite at Station 1/11 on the platform west of Vella Lavella (Figure 1B), and Crook and Niem (this volume) note its occurrence in volcanic lithic arenite at Station 14 south of Santa Isabel (Figure 1A). Therefore, such methane-derived early carbonate cementation apparently occurs throughout the Solomon Islands intra-arc basins, back-arc massifs, and forearc basins where methane-bearing pore-waters are venting through the zone of sulfate reduction in near-surface sediments.

POROSITY AND PERMEABILITY

Five dredge samples were submitted for quantitative analysis of porosity and permeability (Table 3). One-inch diameter drilled plugs (dry) were analyzed by K. Northrup of Unocal Science and Technology Division (US&T) research labs, using a Temco permeameter. The lithologies include dolomite-cemented foraminiferal litharenite from Station 2, densely cemented and porous sections of coralgal limestone (Station 6, Figure 3A), a fresh *Porites* coral (Station 22), and a late Pliocene coccolith chalk (Station 29, Figures 8C and 8D). Porosities and permeabilities range from 34.7 percent to 49.7 percent and 5.66 md to 6,444 md (Table 3).

The highest porosity and permeability (49.7 percent and 6,444 md) are in the intensively bored periphery of the dredged coralgal reef limestone (Figures 3A and 4A). Abundant open meso- and megapores (porosity terminology of Choquette and Pray, 1970) formed mainly by interconnected galleries of spherical to ellipsoidal (3 to 5 mm diameter) and tubular clionid sponge, sipunculid, and polychaete worm borings (Warme, 1975, 1977) largely account for these high measurements. In addition, there are a few larger open pholad and barnacle borings (Figure 4A) and tiny elongate micropores formed by algal and fungal borings. Large mega-voids are present as irregular channelways and reef crevices in the coral growth framework. Intraskeletal mesopores and open corallite calyxes are widespread in the coral *Porites lutea* that forms the bulk of the sample, and minor interparticle and intraparticle porosity occurs within the biotic skeletal sands (i.e., in foraminifer chambers [Figure 5], vermetid gastropod whorls, and calcareous worm tubes).

Table 3. Porosity and Permeability of Solomon Islands Dredge Samples.

Sample designation and lithology	Porosity (percent)	Permeability (md)	Bulk Density (g/cc)	Plug Integrity
2-2-1 Dolomite-cemented foraminiferal litharenite	34.7	13.80	2.02	2
22-1B Fresh coral, *Porites* sp., possibly *P. lutea*	46.3	1743	1.59	2
6-4-1 Coralgal limestone, largely *Porites lutea* porous, bored exterior of block	49.7	6444	1.48	3
6-4-1 Coralgal limestone, largely *Porites lutea*, dense, well-cemented interior of block	37.1	4553	1.84	2
29-1 Late Pliocene foraminifer-nannofossil chalk	41.2	5.66	1.19	3

Plug Integrity Code: 1 = very good to excellent; 2 = good to fair, some minor flaws; 3 = poor, major flaws; 4 = very poor, broken or out of round

The dense well-cemented interior of the coralgal reef limestone (Figure 3A) displays 25 percent to 29 percent reduction of measured porosity and permeability relative to the periphery (reduced to 37.1 percent and 4,553 md). The reduction is attributed mainly to fewer open borings and more extensive submarine aragonite-needle cement-filling coral skeletal mesopores and to widespread high-Mg calcite peloidal cement filling and occluding many of the older interior borings and the interparticle and intraparticle pores in the carbonate biotic-sediment fill (e.g., Figures 4B, 4D, and 8A).

The measurements for the dense well-cemented interior of the coralgal reef limestone are probably more indicative of expected values of porosity and permeability for buried submarine-cemented Pleistocene coralgal forereefs. The higher values of porosity and permeability for the intensively bored periphery of the reef rubble are local and attributable to surficial secondary, eogenetic and syndepositional boring processes and probably do not represent the reservoir potential of these forereef limestones in the subsurface.

A porosity evolution scheme, which is deduced from cross-cutting relationships of pore types and the paragenetic sequence of submarine calcite and aragonite cements in thin sections of the coralgal limestone samples, is envisioned as follows:

Stage 1 - formation of primary growth-framework depositional porosity (irregular channelways, vugs, and crevices, Figure 3C) by growing coralgal forereef; some borings (e.g., barnacles, pholads, lithophagids, vermetid gastropods, fungi, and algae); and minor reduction of porosity by growth of microscopic aragonite isopachous cement in pores in the open framework of corals. Some reduction of primary porosity and size of borings by filling with washed-in wave-abraded carbonate biotic sand and mud (e.g., fragments of foraminifers, mollusks, and spicules) that has only minor interparticle porosity (Figure 5). Some early eogenetic and depositional high-Mg peloidal submarine calcite cement contemporaneously precipitated in these biotic sands and open borings by sea water flushing in and out of the forereef system followed by minor aragonite cementation of some peloid-filled voids.

Stage 2 - demise of coralgal and associated reef organisms and formation of intraparticle porosity (i.e., skeletal micro- and mesopores) in dead corals, foraminifers (Figure 8A), coralline algae, gypsinid foraminifers, and calcareous worm tubes; continued precipitation of early eogenetic submarine cements (microscopic aragonite needles and bladed and peloidal high-Mg calcite) from seawater tidally pumped through the porous reef system; extensive cementation of loose carbonate biotic sands and silt-size peloids in borings and reef crevices by second-stage high-Mg microcystalline calcite cement; and subsequent reduction of most primary and eogenetic (e.g., borings) porosity.

Stage 3 - extensive large galleries of surficial borings by clionid sponges, polychaete worms, and sipunculids; fewer by pholads and vermetid gastropods; and tiny micritic borings by algae and fungi on the exposed surfaces of dead corals and coralgal masses (e.g., Figure 3A). Much of the surficial boring took place both during life and after death of the reef, as is common on the Great Barrier Reef (Marshall and Davies, 1981; Kiene, 1985) and Bermuda (Warme, 1975), and after downslope displacement as forereef talus to greater depths (Lighty, 1985). A few live clionid sponges on and within reef talus blocks were dredged at Stations 6 and 22. Some dissolution of the cements in borings and aragonite biotics also created minor secondary porosity in the reef rubble and in some carbonate-cemented siliciclastics. All these stages are mutually overlapping to some extent.

A fresh and weakly cemented porous unbored coral (*Porites lutea*), dredged from Station 22 displays higher primary porosity (46.3 percent) than the dense interior of the well-cemented bored coralgal reef limestone (37.1 percent) composed mainly of the same coral genus from Station 6 (Figure 3A). However, the fresh coral from Station 22 has significantly lower permeability (1,744 md compared to 4,553 md). Porosity results mainly from abundant intraskeletal mesopores and corallite calyxes and less commonly from minor open borings. Approximately 50 percent of the mesopores are partially filled with peloidal high-Mg micritic cement. Apparently, the open intraskeletal mesoporosity of this coral sample is not as well interconnected, and, therefore, the effective porosity and permeability are lower than in the more intensively bored coralgal limestone from Station 6. In any case, the three plugs from the two samples (22-1B and 6-4-1, Table 3) illustrate the range of high reservoir-quality porosities (37.1 percent to 49.7 percent) and permeabilities (1,743 md to 6,444 md) that could be expected in bored, fresh, and diagenetically submarine-cemented Pleistocene coralgal forereef tract and forereef rubble in the Solomon Islands.

However, greatly reduced reservoir porosities and permeabilities may be present in more deeply buried and older reef limestones in which the limestone has been exposed to extensive neomorphism and low-Mg sparry calcite cementation in the mesogenetic zone (Choquette and Pray, 1970) and/or subaerial uplift in the phreatic zone in the telogenetic zone (Harris, Kendall, and Lerche, 1985). The coral cobble (*Diploastrea heliopora*) from Station 25 is an example of this freshwater and neomorphic effect (Figure 9D).

The soft late Pliocene foraminiferal-coccolith chalk (sample 29-1) was dredged from a slope basin on Nuguria Ridge, northwest of Bougainville (see Bruns, Vedder, and Culotta, this volume). This chalk has high interparticle and intraskeletal porosity (41.2 percent) but unusually low permeability (5.66 md). The low permeability is probably due, in part, to collapse of the small chalky plug during testing in the permeameter (Table 3). Most sand-size mesoporosity (estimated 15 percent) in the chalk is intraparticle in isolated open chambers of *Globigerina* and globorotalid tests (Figure 8C). The greatest proportion of porosity (26 percent) is, however, microporosity. The microporosity is in the form of intraparticle pores in coccolith and foraminifer tests and intercrystalline pores between microscopic high-Mg calcite crystals and between coccoliths (Figure 8D). The lack of good interconnection between these micropores may also account for the low permeability.

Similarly, the dolomite-cemented foraminiferal litharenite dredged at Station 2 (sample 2-2-1) (Figure 9A) displays high porosity (34.7 percent) and low permeability (13.80 md). In this case, too, the microporosity consists largely of unconnected or isolated empty chambers in planktonic foraminifer tests (chiefly globigerinids and globorotalids; Figure 9A) and microporosity in intercrystalline areas within the microcrystalline dolomite cement (Figure 9B). Some micropore porosity within foraminifer tests has been reduced by infilling with low-Mg sparry blocky calcite and/or microcrystalline dolomite cement. Although porosity is high in this sandstone, permeability is low because of the lack of interconnecting pores in the submicrocrystalline dolomite cement which infills most of the available interparticle spaces between sand-sized siliciclastic framework grains (Figure 9A). Micrite and aragonite-needle cement followed by widespread sparry calcite cement has significantly reduced inter- and intraparticle porosity and presumably also the permeability in the foraminiferal packstone from Station 23 (Figure 9C). Early diagenetic, submarine-carbonate cementation is a widespread and common type of cementation in the Solomon Islands (Colwell and Vedder, 1986; Crook and Niem, this volume). This type of cementation has, along with compaction, significantly reduced porosity in the majority of the dredged Quaternary sandstone, mudstone, and coralgal limestone samples.

ACKNOWLEDGMENTS

Foraminifers and coccoliths in reef limestone samples were examined by J.M. Resig and S. Shafik. Foraminifers in thin sections were studied by R.C. Tjalsma of Unocal Science and Technology Division (US&T) and G.C.H. Chaproniere of the Australian Bureau of Mineral Resources, Geology, and Geophysics (BMR), who provided additional information on age and paleoecology. M. Colgan, University of California, Santa Cruz; T. DeVries, Oregon State University; and E. J. Moore, U.S. Geological Survey (retired) identified corals and mollusks. A.H. Saller (US&T), J.F. Marshall (BMR), and W.E. Kiene (Australian National University) helped in understanding the significance of the borings and the chemistry of the carbonate cements. T. Elliott (US&T) and A. Chivas (Australian National University) provided valuable assistance in interpreting the isotopic data. Financial support for making thin sections, microprobe, SEM, and isotopic analysis, and travel to Australian National University for A.R. Niem was provided by the U.S. Geological Survey, Unocal Science and Technology Division, and Oregon State University. K.L. Cameron and K.A.W. Crook arranged logistics in Australia.

APPENDIX I—TECHNIQUES

Thin Section, Staining, and Porosity and Permeability Measurements

Thirty (30) coral and coralgal boundstone samples were slabbed, impregnated with blue epoxy, and thin sectioned. With the assistance of J.B. Colwell (Australian Bureau of Mineral Resources, Geology, and Geophysics), carbonate thin sections were stained for aragonite, high-Mg and low-Mg calcite, and dolomite using Feigal solution, titan yellow, EDTA, and alizarin red-S, according to the techniques of Friedman (1959) and Choquette and Trusell (1977). A Siemens X-ray diffractometer at Unocal Science and Technology Division (US&T) research laboratories in Brea, California, was used to investigate further the mineralogy of carbonate cements in the calcareous lithic silty mudstone (hollow carbonate chimney) and foraminiferal litharenite recovered at Station 2. Measurements of porosity and permeability were made by K. Northrup (US&T) on 3-cm diameter plugs drilled in two reef samples from Stations 6 and 22 and on dolomite-cemented foraminiferal litharenite and chalk samples, using a Temco permeameter (Table 3).

Microprobe Analysis

Several thin sections of the dredge samples were polished and then coated with approximately 150 Å of carbon for microprobe analysis of cation compositions (Ca, Mg, Fe, Mn, and Sr) of carbonate biotics and cements at US&T. The carbon-coated sections were studied on a television monitor and photographed, using a backscatter electron (BSE) image mode on a JEOL 733 microprobe. Spot chemical analyses were conducted on individual carbonate grains and crystals selected on the backscatter image. These quantitative cation analyses of calcite and aragonite in coral and algal samples and of calcite and dolomite cements in foraminiferal litharenite samples and carbonate chimneys were performed by the microprobe's 4-wave-length dispersive spectrometer (WDS) and a 1-wave-length Tracor Northern energy-dispersive spectrometer (EDS) operated by M. Bell and L. Bisaha of US&T.

For the five elements of interest in this analysis, an accelerating voltage of 5 keV would be appropriate for magnesium and strontium, but that is inadequate for detection of calcium, manganese, and iron. A minimum of 10 keV and preferably 20 keV is required for detection of iron. Therefore, an accelerating voltage of 13 keV was chosen as a compromise voltage. It is recognized that this somewhat degrades the detection limit for strontium. In order to minimize sample destruction that results from high current and/or a focused beam, a beam current of 5 nA with a beam diameter of 10 μ was used for a 20-s count time. Wavelength-dispersive spectrometers (WDS), which are more sensitive, were used for detection of strontium and manganese. Energy-dispersive spectrometers (EDS) measured calcium, manganese and iron. WDS and EDS were run simultaneously.

All peak-fitting, deconvolution, overlap, and matrix corrections were accomplished using a Tracor Northern TN 2000 MECCA data collection and software analysis package which includes ZAF and Bence-Albee reduction algorithms. A suite of carbonate reference minerals (calcite, dolomite, and siderite) of known quantitative chemical composition were obtained from the Smithsonian Institution for use as standards, and strontium titanate ($SrTiO_3$) and spessartine garnet were used for strontium and manganese calibration.

SEM Analysis

Several coralgal reef limestone, carbonate-cemented foraminiferal litharenite and mudstone, and chalk samples underwent scanning electron microscopic (SEM) analysis at US&T. Five- to thirty-gram splits of the dredge samples were first sputter coated with 100 to 200 Å of gold/palladium alloy in a Polaron 5000 sputter coater and then analyzed and photographed on an AmRay 1000-Å scanning electron microscope operated by J. Jaco of US&T. The AmRay SEM was fitted with a KEVEX EDS X-ray system to do semi-quantitative chemical analysis of all minerals and cements in the dredged samples. Typical operating parameters were 20 kv accelerating voltage using a spot size of less than 100 Å.

Stable Carbon and Oxygen Isotopic Analysis

Sixteen (16) coralgal limestones, corals, and carbonate cements in siliciclastic sandstones and carbonate chimneys were prepared for stable oxygen and carbon isotopic analysis by D. Van der Linde under the direction of T. Elliott, both of US&T, by placing 10 to 15 mg of powdered sample in a drying oven overnight at 90°C. The powdered samples were reacted with anhydrous phosphoric acid at 25°C, following the general method of McCrea (1950). Isotopic ratios of the evolved CO_2 were measured using a Finnigan MAT 251-EM mass spectrometer, with a working standard calibrated against NBS standards 18, 19, 20. Estimated precisions are ± 0.2 per mil for oxygen and ± 0.1 per mil for carbon.

REFERENCES CITED

Bathurst, R.G.C., 1975, Carbonate sediments and their diagenesis: Second enlarged edition, Developments in Sedimentology 12, New York, Elsevier Scientific Publishing Co., 658 p.

Bromley, R.G., 1978, Bioerosion of Bermuda reef: Palaeogeography, Palaeoclimatology, and Palaeoecology, v. 23, p. 169-197.

Bruns, T.R., Cooper, A.K., Mann, D.M., and Vedder, J.G., 1986, Seismic stratigraphy and structure of sedimentary basins in the Solomon Islands region, *in* Vedder, J.G., Pound, K.S., and Boundy, S.Q., eds., Geology and offshore resources of Pacific island arcs — central and western Solomon Islands: Houston, Texas, Circum-Pacific Council for Energy and Mineral Resources Earth Science Series, v. 4, p. 177-214.

Chase, T.E., Seekins, B.A., and Lund, 1986, Submarine topography of the Solomon Islands region, in Vedder, J.G., Pound, K.S., and Boundy, S.Q., eds., Geology and offshore resources of Pacific island arcs—central and western Solomon Islands: Houston, Texas, Circum-Pacific Council for Energy and Mineral Resources Earth Science

Series, n. 4, p. 21-31.

Choquette, P.W., and Pray, L.C., 1970, Geological nomenclature and classification of porosity in sedimentary carbonates: American Association of Petroleum Geologists Bulletin, v. 54, p. 207-250.

Choquette, P.W., and Trusell, F.C., 1977, A procedure for making the titan yellow stain for Mg-calcite permanent: Journal of Sedimentary Petrology, v. 47, p. 639-641.

Claypool, G.E., Presly, B.J., and Kaplan, I.R., 1973, Gas analyses in sediments from Legs 10, 11, 13, 14, 15, 18, and 19, *in* Initial Reports of the Deep Sea Drilling Project: Washington, D.C., U.S. Government Printing Office, v. 19, p. 879-884.

Colwell, J.B., and Tiffin, D.L., 1986, Recent depositional patterns in the Central Solomons Trough of the Solomon Islands, *in* Vedder, J.G., Pound, K.S., and Boundy, S.Q., eds., Geology and offshore resources of Pacific island arcs — central and western Solomon Islands: Houston, Texas, Circum-Pacific Council for Energy and Mineral Resources Earth Science Series, v. 4, p. 243-254.

Colwell, J.B., and Vedder, J.G., 1986, Description and interpretation of dredged rocks, Solomon Islands, *in* Vedder, J.G., Pound, K.S., and Boundy, S.Q., eds., Geology and offshore resources of Pacific island arcs — central and western Solomon Islands: Houston, Texas, Circum-Pacific Council for Energy and Mineral Resources Earth Science Series, v. 4, p. 225-236.

Cook, H.E., and Mullins, H.T., 1983, Basin margin environment, *in* Scholle, P.A., Bebout, D.G., and Moore, C.H., eds., Carbonate depositional environments: American Association of Petroleum Geologists Memoir 33, p. 539-617.

Coulson, F.K., and Vedder, J.G., 1986, Geology of the central and western Solomon Islands, *in* Vedder, J.G., Pound, K.S., and Boundy, S.Q., eds., Geology and offshore resources of Pacific island arcs — central and western Solomon Islands: Houston, Texas, Circum-Pacific Council for Energy and Mineral Resources Earth Science Series, v. 4, p. 59-87.

Craig, H., 1965, The measurement of oxygen isotope paleotemperatures, *in* Tongiorgi, E., ed., Stable isotopes in oceanographic studies and paleotemperatures: Consiglio Nazionale delle Richerche, Laboratoirio de Geologic Nucleare, Pisa, p. 161-182.

Enos, Paul, and Moore, C.H., 1983, Fore-reef slope environment, *in* Scholle, P.A., Bebout, D.G., and Moore, C.H., eds., Carbonate depositional environments: American Association of Petroleum Geologists Memoir 33, p. 507-537.

Folk, R.L., 1965, Some aspects of recrystallization in ancient limestones, *in* Pray, L.C., and Murray, R.C., eds., Dolomitization and limestone diagenesis: a symposium: Society of Economic Paleontologists and Mineralogists Special Publication 13, p. 14-48.

Friedman, G.M., 1959, Identification of carbonate minerals by staining methods: Journal of Sedimentary Petrology, v. 29, n. 1, p. 87-97.

Gonzalez, L.A., and Lohmann, K.C., 1985, Carbon and oxygen isotopic composition of Holocene reefal carbonates: Geology, v. 13, p. 811-814.

Gross, M.G., and Tracey, J.I., Jr., 1965, Oxygen and carbon isotopic composition of limestones and dolomites, Bikini and Eniwetok atolls: Science, v. 151, p. 1082-1084.

Hackman, B.D., 1980, The geology of Guadalcanal, Solomon Islands: London, Overseas Memoirs of the Institute of Geologic Science, n. 6, 115 p.

Harris, P.M., Kendall, C.G., St. C., and Lerche, Ian, 1985, Carbonate cementation - a brief review, *in* Schneidermann, Nahum and Harris, P.M., eds., Carbonate cements: Society of Economic Paleontologists and Mineralogists Special Publication N. 36, p. 79-95.

Hudson, J.D., 1977, Stable isotopes and limestone lithifications: Journal of Geological Society London, v. 133, p. 637-660.

James, N.P., Ginsburg, R.N., Marszalet, D.S., and Choquette, P.W., 1976, Facies and fabric specificity of early subsea cements in shallow Belize (British Honduras) reefs: Journal of Sedimentary Petrology, v. 46, p. 523-544.

Johnson, J.H., 1961, Limestone-building algae and algal limestones: Colorado School of Mines, Boulder, Colorado, 297 p.

Keith, M.L., and Weber, J.N., 1965, Systematic relationships between carbon and oxygen isotopes in carbonates deposited by modern corals and algae: Science, v. 150, p. 498-501.

Kiene, W.E., 1985, Biological destruction of experimental coral substrates at Lizard Island, Great Barrier Reef, Australia: Proceedings of the Fifth International Coral Reef Congress, Tahiti, v. 5, p. 339-344.

Kulm, L.D., Suess, Erwin, and Snavely, P.D., Jr., 1988, Fluid venting structures on the northern Oregon continental shelf: NOAA Symposium Series for Undersea Research, v. 6, n. 2, p. 20-45.

Land, L.S., and Moore, C.H., 1980, Lithification, micritization, and syndepositional diagenesis of biolithites on the Jamaican island slope: Journal of Sedimentary Petrology, v. 50, p. 357-370.

Lighty, Robin, 1985, Preservation of internal reef porosity and diagenetic sealing of submerged early Holocene barrier reef, southeast Florida shelf, *in* Schneidermann, N., and Harris, P.M., eds., Carbonate cements: Society of Economic Paleontologists and Mineralogists Special Publication N. 36, p. 123-151.

Macintyre, I.G., 1977, Distribution of submarine cement in a modern Caribbean fringing reef, Galeta Point, Panama: Journal of Sedimentary Petrology, v. 47, p. 503-516.

Macintyre, I.G., 1984, Extensive submarine lithification in a cave in the Belize Barrier Reef platform: Journal of Sedimentary Petrology, v. 54, p. 11-25.

Macintyre, I.G., 1985, Submarine cements—the peloidal question, *in* Schneidermann, N., and Harris, P.M., eds., Carbonate cements: Society of Economic Paleontologists and Mineralogists Special Publication N. 36, p. 109-116.

Majewske, O.P., 1974, Recognition of invertebrate fossil fragments in rocks and thin sections: Leiden, Netherlands, E.J., Brill, 101 p.

Marshall, J.F., 1986, Regional distribution of submarine cements within an epicontinental reef system, central Barrier Reef, Australia, *in* Schroeder, J.H., and Purser, B.H., eds., Reef diagenesis: Berlin-Heidelberg, Springer-Verlag, p. 8-26.

Marshall, J.F., and Davies, P.J., 1981, Submarine lithification on windward reef slopes: Capricorn-Bunker Group, southern Great Barrier Reef: Journal of Sedimentary Petrology, v. 51, p. 953-960.

McCrea, J.M., 1950, On the isotopic chemistry of carbonates and a paleotemperature scale: Journal of Chem. Phys., v. 18, p. 849-857.

Middleton, G.V., and Hampton, M.A., 1973, Mechanics of

flow and deposition, *in* Middleton, G.V., and Bouma, A.H., eds., Turbidites and deep-water sedimentation: Pacific Section Society of Economic Paleontologists and Mineralogists, Short Course Notes, p. 1-38.

Milliman, J.D., 1974, Marine carbonates: Springer-Verlag, Berlin, 375 p.

Postuma, J.A., 1971, Manual of planktonic Foraminifera: New York, Elsevier Publishing Co., 420 p.

Pound, K.S., 1986, Correlation of rock units in the central and western Solomon Islands, *in* Vedder, J.G., Pound, K.S., and Boundy, S.Q., eds., Geology and offshore resources of Pacific island arcs — central and western Solomon Islands: Houston, Texas, Circum-Pacific Council for Energy and Mineral Resources Earth Science Series, v. 4, p. 89-97.

Resig, J.M., 1986, Foraminiferal stratigraphy and paleobathymetry of dredged rock, R/V *S.P. Lee*, Solomon Islands, *in* Vedder, J.G., Pound, K.S., and Boundy, S.Q., eds., Geology and offshore resources of Pacific island arcs — central and western Solomon Islands: Houston, Texas, Circum-Pacific Council for Energy and Mineral Resources Earth Science Series, v. 4, p. 255-260.

Ritger, S., Carson, B., and Suess, E., 1987, Methane-derived authigenic carbonates formed by subduction-induced pore-water expulsion along the Oregon/Washington margin: Geological Society of America Bulletin, v. 98, p. 147-156.

Saller, A.H., 1984, Petrologic and geochemical constraints on the origin of subsurface dolomite, Enewetak Atoll: an example of dolomitization by normal seawater: Geology, v. 12, n. 4, p. 217-220.

Saller, A.H., 1986, Radiaxial calcite in lower Miocene strata, subsurface Enewetak Atoll: Journal of Sedimentary Petrology, v. 56, no. 6, p. 743-762.

Scholle, P.A., 1978, A color illustrated guide to carbonate rock constituents, textures, cements, and porosities: American Association of Petroleum Geologists Memoir 27, 241 p.

Scholle, P.A., Arthur, M.A., and Ekdale, A.A., 1983, Pelagic environment, *in* Scholle, P.A., Bebout, D.G., and Moore, C.H., eds., Carbonate depositional environments: American Association of Petroleum Geologists Memoir 33, p. 619-691.

Schroeder, N.A.M., Kulm, L.D., and Muehlberg, G.E., 1987, Carbonate chimneys on the outer continental shelf; evidence for fluid venting on the Oregon margin: Oregon Geology, v. 49, n. 8, p. 91-96.

Solomon Islands Department of Geological Surveys, 1969, Geological map of the British Solomon Islands: Department of Geological Surveys, Solomon Islands, scale 1:1,000,000.

Suess, E., and Whiticar, M.J., 1988, Methane-derived CO_2 in pore fluids expelled from the Oregon subduction zone, *in* Okada, H., and Cadet, J.P., eds., KAIKO Symposium: Geology, geochemistry and biology of the subduction zone: Paleogeography, Paleoclimatology, Paleoecology, Special Issue.

Vedder, J.G., and Coulson, F.I., 1986, Regional offshore geology of the Solomon Islands, *in* Vedder, J.G., Pound, K.S., and Boundy, S.Q., eds., Geology and offshore resources of Pacific island arcs — central and western Solomon Islands: Houston, Texas, Circum-Pacific Council for Energy and Mineral Resources Earth Science Series, v. 4, p. 99-108.

Warme, J.E., 1975, Borings as trace fossils and the processes of marine bioerosion, *in* Frey, R.W., ed., The study of trace fossils: a synthesis of principles, problems, and procedures in ichnology: New York, Springer-Verlag, p. 181-227.

Warme, J.E., 1977, Carbonate borers—their role in reef ecology and preservation: American Association of Petroleum Geologists Studies in Geology N. 4, p. 261-279.

Warme, J.E., and McHuron, E.J., 1978, Marine borers: Trace fossils and geologic significance, *in* Trace fossil concepts: Society of Economic Paleontologists and Mineralogists Short Course N. 5, p. 67-118.

Whiticar, M.J., Suess, E., and Wehner, H., 1985, Thermogenic hydrocarbons in surface sediments of the Bransfield Strait, Antarctic Peninsula: Nature, v. 314, p. 87-90.

Vedder, J.G., and Bruns, T.R., editors, 1989, Geology and offshore resources of Pacific island arcs—Solomon Islands and Bougainville, Papua New Guinea Regions: Houston, Texas, Circum-Pacific Council for Energy and Mineral Resources Earth Science Series, v. 12.

SEDIMENTOLOGY AND CLAY MINERALOGY OF QUATERNARY SEDIMENTS FROM NEW GEORGIA SOUND, SOLOMON ISLANDS

J.B. Colwell
Bureau of Mineral Resources, Geology, and Geophysics
GPO Box 378, Canberra 2601, Australia

A.R. Niem, W.A. Niem
Department of Geology, Oregon State University
Corvallis, Oregon 97331-5506

ABSTRACT

During 1984, the U.S. Geological Survey R/V *S.P. Lee* took six gravity cores and nineteen short dart cores in the New Georgia Sound area of the Solomon Islands. The cores were taken in order to provide samples for hydrocarbon gas analysis, to identify slump sediments, and to determine bedrock compositions. All of the cores are entirely Quaternary in age and lie above both the present-day aragonite and calcite compensation depths. None of the cores penetrated bedrock.

The sediments comprising the cores are typically hemipelagic and consist of a mixture of biogenic (mainly planktonic foraminifera, pteropods, and calcareous nannofossils) and terrigenous (largely volcanogenic) components. Calcium carbonate contents typically range from 10 to 40 percent, and total organic carbon contents range from 0.2 to 0.5 percent. Most of the sediments are bioturbated and consist of slightly sandy silty clays and clayey silts interbedded with sandy turbidite layers. Relatively sand-rich sediments occur in areas on the flanks of the sound, probably partly in response to current winnowing.

Smectite, chloritic intergrades, chlorite, kaolinite-halloysite and minor illite dominate the clay fraction of the cores. Three major clay-mineral suites can be recognized. Suite 1 is dominated by abundant well-crystallized chlorite and poorly crystallized smectite and is largely confined to the eastern end of New Georgia Sound. Suite 2, which occurs mainly in the central part of the sound, is characterized by chloritic intergrades, subordinate poorly crystallized kaolinite-halloysite, and minor to moderate amounts of chlorite. By contrast, suite 3 is dominated by smectite with subordinate to minor amounts of kaolinite-halloysite and chlorite, and this suite appears, from its position and mineralogy, to be a mixture of the other two clay-mineral suites.

The distribution of the clay-mineral suites reflects provenance. Source for the clay minerals and sand- and silt-size terrigenous material include Late Cretaceous(?) to early Tertiary metamorphosed and unmetamorphosed basalts and andesites, and late Tertiary and Quaternary, predominantly calc-alkaline, basaltic andesites and andesites.

INTRODUCTION

As is the case throughout much of the southwest Pacific, there are only limited data pertaining to the modern sediments occurring adjacent to the Solomon Islands. Studies undertaken by Valencia (1973), Colwell (1986), Crook and Taylor (1987), and Exon, Colwell, and Bolton (1987), showed that the region is dominated by hemipelagic sedimentation that is the product of both the biogenic input from the water column and terrigenous (largely volcanogenic) input from the land.

In this paper, we describe the sedimentology and clay mineralogy of six gravity cores and nineteen short dart cores taken in 1984 by the U.S. Geological Survey R/V *S.P. Lee* in the New Georgia Sound area of the Solomon Islands. The cores, which range in length from 25-296 cm, are entirely Quaternary in age (Shafik, this volume; G. Chaproniere, oral commun.). This work was done as part of the Tripartite II program involving scientists from the United States, Australia and New Zealand.

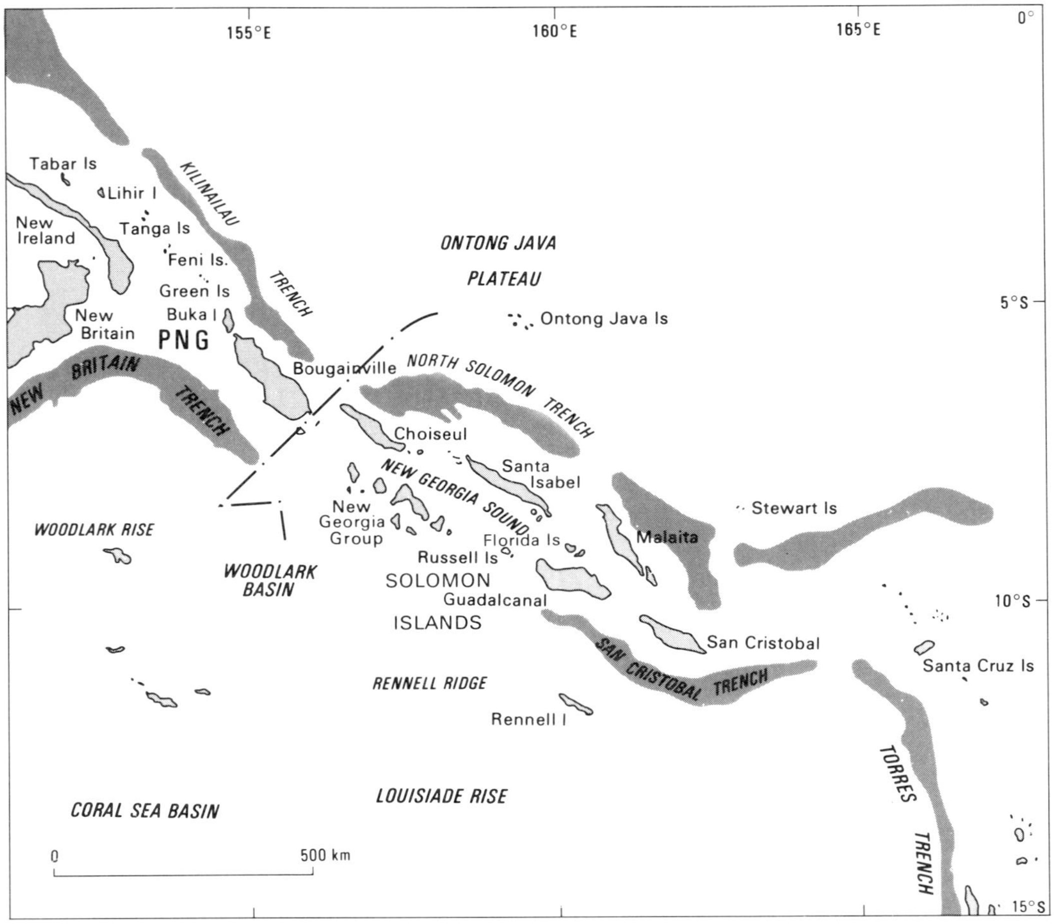

Figure 1. Place names in the Solomon Islands region. Area includes eastern Papua New Guinea and northern Vanuatu.

Regional Setting

The Solomon Islands (Figure 1) form part of the Melanesian Borderlands of Coleman and Packham (1976) and consist of a twin chain of islands bounded on the north by the Ontong Java Plateau and on the south by the Woodlark and Coral Sea Basins. Coleman (1965, 1970) divided the islands into three geologic provinces: the Pacific Province, consisting of Malaita and the northeastern part of Santa Isabel; the Central Province, consisting of San Cristobal, Guadalcanal, the Florida Islands, southwestern Santa Isabel, and Choiseul; and the Volcanic Province, consisting of the Russell Islands, Savo, the New Georgia island group, Shortland Islands, and northwestern Guadalcanal. Although problems are inherent in this simple subdivision (see Coulson and Vedder, 1986; Pound, 1986), in general, the Pacific Province is thought to be either wholly an overthrust and folded portion of the southwestern edge of the oceanic Ontong Java Plateau (Kroenke, 1972, 1984) or in part, an uplifted belt of forearc rocks (Coulson and Vedder, 1986). The Central Province, on the other hand, is composed of an intensely faulted, partly metamorphosed core of pre-late Eocene(?) mafic lavas and associated intrusions and ultramafic bodies overlain by a thick Paleogene and Neogene sedimentary sequence and younger lavas. The Volcanic Province mainly consists of late Miocene and younger basalts and andesites associated with northeastward subduction of the Australia-India plate beneath the Pacific plate. New Georgia Sound (The Slot), which is the site of the present study, lies in an intra-arc position between the relatively young volcanic rocks of the New Georgia island group and the Russell Islands to the south and the mainly older volcanic, metamorphic, intrusive, and sedimentary rocks of Guadalcanal, the Florida Islands, Santa Isabel, and Choiseul to the east and north (Figure 1). Water depths are generally less than 1,800 m (Chase, Seekins, and Young, this volume).

SITE SELECTION AND ANALYTICAL TECHNIQUES

Coring was undertaken for a number of reasons, namely to obtain sediment samples for hydrocarbon gas analysis (mainly Stations 12, 16, 16A, and 18), to identify slump sediments extending northwest from Savo Island at the eastern end of New Georgia Sound (Stations 7 and 13) and in the case of the dart cores, to obtain samples of bedrock (Stations 9, 15, and 17). All sites (Figure 2) were selected on the basis of high-resolution (Uniboom) seismic-reflection data recorded on an earlier cruise by the *S.P. Lee* in 1982 (Vedder, Pound, and Boundy, 1986; Colwell and Tiffin, 1986; Figure 3). Unfortunately, none of the dart cores recovered bedrock, cores at Stations 7 and 13 failed to reach the main body of slump sediments, and the cores sited over areas of apparent gassy sediments contain only small quantities of hydrocarbon gases (Kvenvolden and Niem, this volume). These negative results are apparently due to the failure of the coring devices to penetrate below a thin cover of Quaternary sediment, which apparently blankets older strata even in areas of steep slopes and high current activity. Maximum depth of penetration of the cores was 2.95 m.

Standard analytical techniques (e.g., Carver, 1971; Colwell and Vedder, 1986a) were applied to the cores. The cores were split, photographed, logged, and subsampled onboard ship. Representative samples were distributed to the U.S. Geological Survey, the Bureau of Mineral Resources, Geology and Geophysics (BMR), Australia, and the Solomon Islands Geological Survey. One longitudinal half of each core was frozen for storage at the U.S. Geological Survey's core archives in Menlo Park, California.

Petrographic and grain-size analyses were carried out at BMR and Oregon State University. Total organic carbon (TOC) and CO_2 chemical analyses were undertaken at the Australian Mineral Development Laboratories using acid evolution and a LECO induction furnace. Clay mineralogy was determined by X-ray diffraction at Union Oil Company of California's (UNOCAL) research laboratories in Brea, California. Scanning electron microscopy (SEM) and transmission electron microscopy (TEM) work was performed at UNOCAL's research laboratories and Oregon State University, respectively. Paleontological examinations were made by G. Chaproniere (foraminifera) and S. Shafik (nannofossils), both of BMR. Gas analyses, which were made on parts of the cores using a Carle Model 311 gas chromatograph soon after the cruise, are reported elsewhere (Kvenvolden and Niem, this volume). Full details of the techniques and the identification criteria used in the clay-mineral analysis and SEM and TEM work are given in Appendices I and II.

DESCRIPTION OF THE SEDIMENTS

As noted by Colwell (1986), Exon, Colwell, and Bolton (1987), and this study, most of the Quaternary sediments in the deeper parts of New Georgia Sound are bioturbated, slightly sandy, hemipelagic silty clays and clayey silts interbedded with sandy turbidite layers. The sediments are composed of a mixture of biogenic components (mostly planktonic foraminifera, pteropods, and calcareous nannofossils), silt- and sand-size terrigenous detritus (mainly plagioclase feldspar, clinopyroxene, lithic grains, opaque minerals, and hornblende), and clay minerals. From the virtual absence of glass shards, there appears to be very little direct input of pyroclastic material. Calcium carbonate contents typically range from 10 to 40 percent and total organic carbon values from 0.2 to 0.5 percent. A brief description of the main core groupings based upon lithology is given below.

Stations 7, 12, and 13 (Figures 4, 5, and 6)

Gravity cores, ranging in length from 125 to 295 cm, were taken at Stations 7, 12, and 13 at the eastern end of New Georgia Sound (Figure 2) in water depths of 1,290 to 1,682 m. The cores typically consist of mottled (burrowed), pale-olive (10Y6/2) to light olive-grey (5Y5/2), slightly sandy, calcareous to slightly calcareous, clayey silts and silty clays interbedded with some sandy or silty turbidite layers. The upper 10 cm of the cores is typically oxidized to dark yellowish-brown. The biogenic fraction of the sediments is composed of calcareous nannofossils, planktonic foraminifera, and trace amounts of pteropods, whereas the terrigenous fraction consists mainly of volcanogenic material (plagioclase feldspar, clinopyroxene, lithic grains, hornblende, and opaque minerals), clay minerals, and trace amounts of biotite and quartz.

Several turbidite layers are present in the cores. They are dark-grey to black, graded with sharp basal contacts, and range in thickness from 1 to 4 cm. In some cases (e.g., Station 7, 78-90 cm), the turbidites occur as a series of thin layers interbedded with the dominant mud lithology. Generally the turbidites consist of very fine-grained, feldspathic sands and silts containing weathered lithic fragments (mainly lithic and crystal tuffs), minor clinopyroxene, biotite, hornblende, opaque minerals, and biogenic carbonate. Calcium carbonate contents are typically less than 10 percent. Glass shards are absent. In one sample (Station 13, 134-138 cm), the turbidite layer consists of a heavy-mineral lag deposit of fine-grained magnetite and pale-green augite with trace amounts of other components (mainly plagioclase, lithic grains, and planktonic foraminifera).

Station 9

Nine short (<50 cm long) dart cores were attempted at this station in a transect ranging in water depth from 893 to 369 m across the Russell Island-Guadalcanal platform (Figure 2). Although eight of the attempts recovered sediment, none of the cores reached the target of bedrock. All of the cores are late Quaternary in age (Shafik, this volume).

The cores range from greyish-olive, hemipelagic silty clays to olive-grey, largely feldspathic sands. Typically the sand fraction consists of plagioclase feldspar, planktonic foraminifera, pteropods, hornblende, lithic grains (including some metamorphics), and opaque minerals. By contrast, the mud fraction is composed of silt-size feldspar, pyroxene, and lithics together with clay minerals (mainly smectite and chlorite) and calcareous nannofossils. Mud values decrease from approximately 75 percent in Core 9A (893 m water depth) to 10 percent in Core 9H (399 m water depth). Traces of biotite and quartz occur in a few samples. All of the cores, except those in less than 600 m of water, possess a thin (1-5 cm thick) surface oxidation layer.

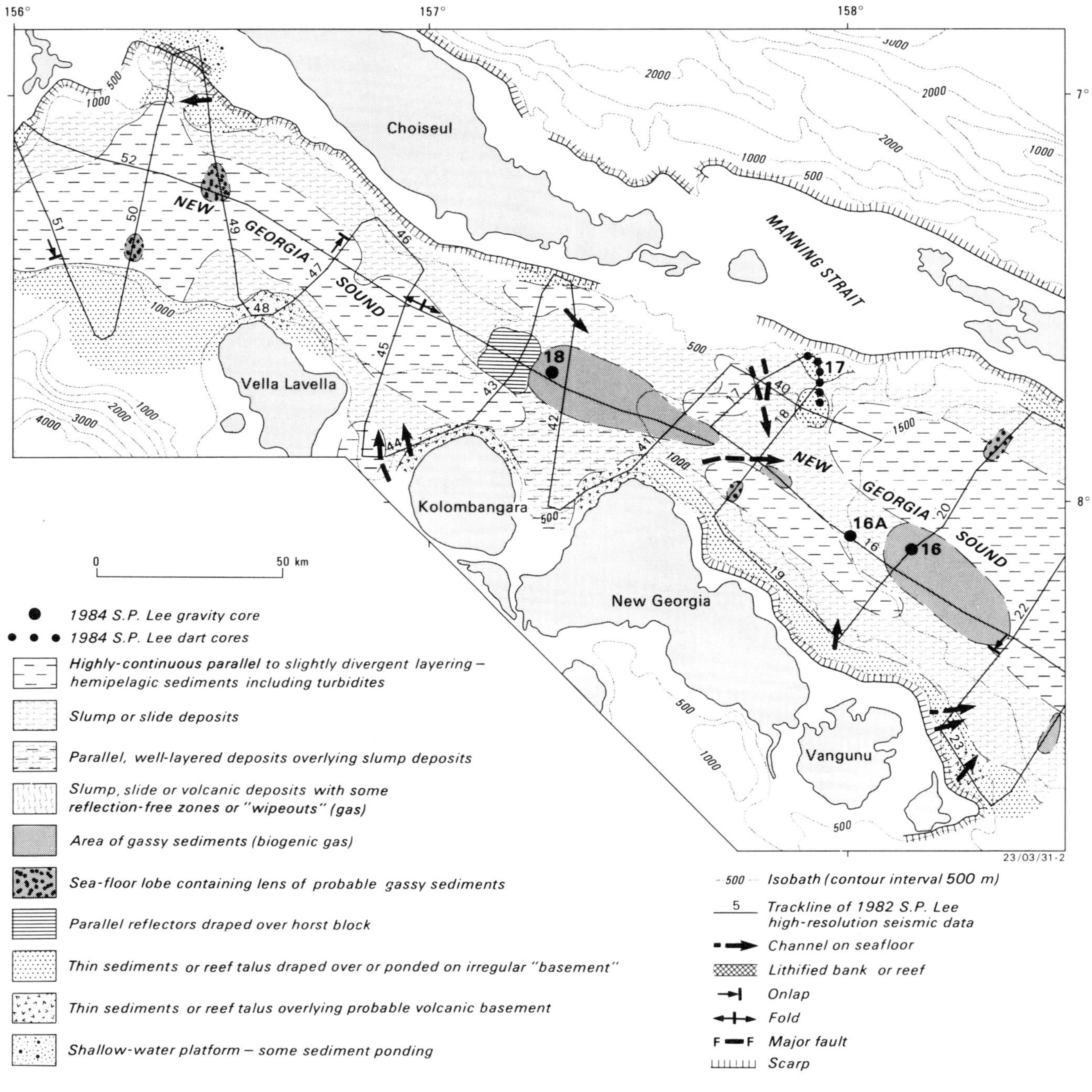

Figure 2. Pattern of modern sedimentation in New Georgia Sound showing the sites of 1984 *S.P.Lee* gravity and dart cores. Modified from Colwell and Tiffin (1986).

Station 16 (Figure 7)

A 219 cm-long gravity core was recovered from Station 16 in the central part of New Georgia Sound in 1,672 m of water. This core was one of three (12, 16 and 18) sited over what appeared to be gaseous sediments on high-resolution seismic-reflection records (Figures 2, 3). Although the core was taken to provide samples for gas analysis, only low concentrations of mainly methane gas were recovered (Kvenvolden and Niem, this volume).

Typically, the core is made up of mottled (burrowed), yellowish-grey (5Y7/2) to light olive-grey (5Y5/2), very slightly sandy calcareous silty clays (Figure 8) interbedded with carbonate- or volcanic-rich, sandy turbidite layers at 70-85 cm and 190-206 cm. The upper 12 cm of the core is oxidized. Between 120 and 130 cm, a series of thin (<1 cm), dark olive-grey, feldspathic, volcanic-rich clayey silt layers occur interbedded with the clay. These layers, which have sharp basal contacts and gradational upper contacts, are interpreted as probable distal turbidites. They contain only minor amounts of calcareous debris.

Two sandy turbidite layers (Figure 7) display well-developed graded bedding. The upper layer at 70-85 cm ranges from medium grained at its base to fine grained at its top and consists of a slightly muddy calcareous (planktonic foraminifera and pteropods) sand. Terrigenous material is mainly plagioclase feldspar and volcanic rock fragments with minor amounts of pyroxene and hornblende.

The lower turbidite layer at 190-206 cm consists of a dark olive-grey to black, coarse-grained volcanic-rich sand grading into very slightly muddy fine-grained sand of similar composition at the top. The basal contact is sharp, and the upper contact is gradational. The non-carbonate fraction consists of opaque minerals, altered glass, and highly altered volcanic rock fragments with lesser amounts of plagioclase feldspar and a trace of pyroxene. In many cases, the rock fragments are almost completely replaced by secondary clay minerals and iron oxides. Approximately 20 percent of the sand consists of calcium carbonate, mainly planktonic foraminifera and pteropod fragments.

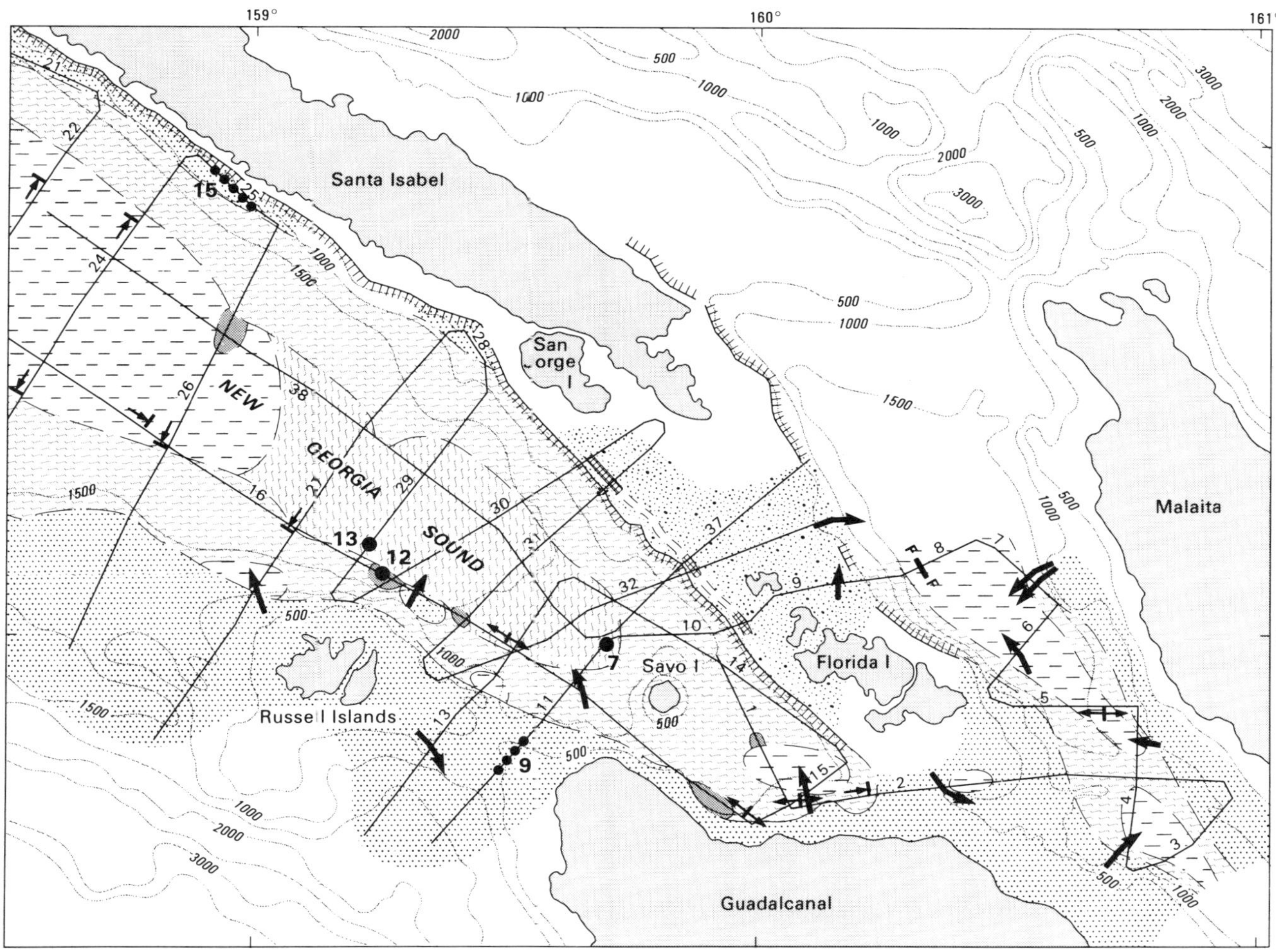

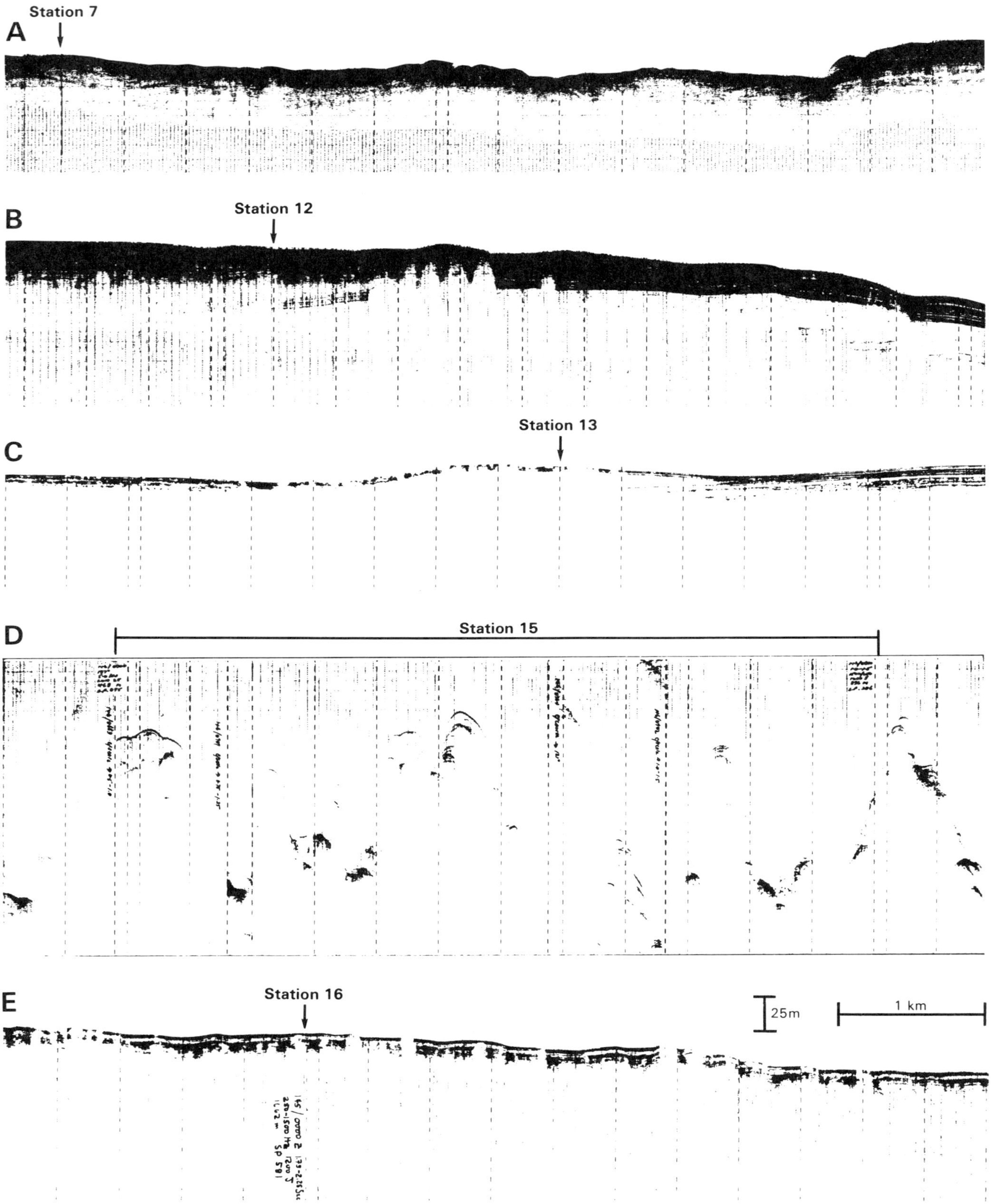

Figure 3. Sections of high-resolution (Uniboom) seismic-reflection records recorded by the *S.P. Lee* in 1982 showing the sites of core stations.

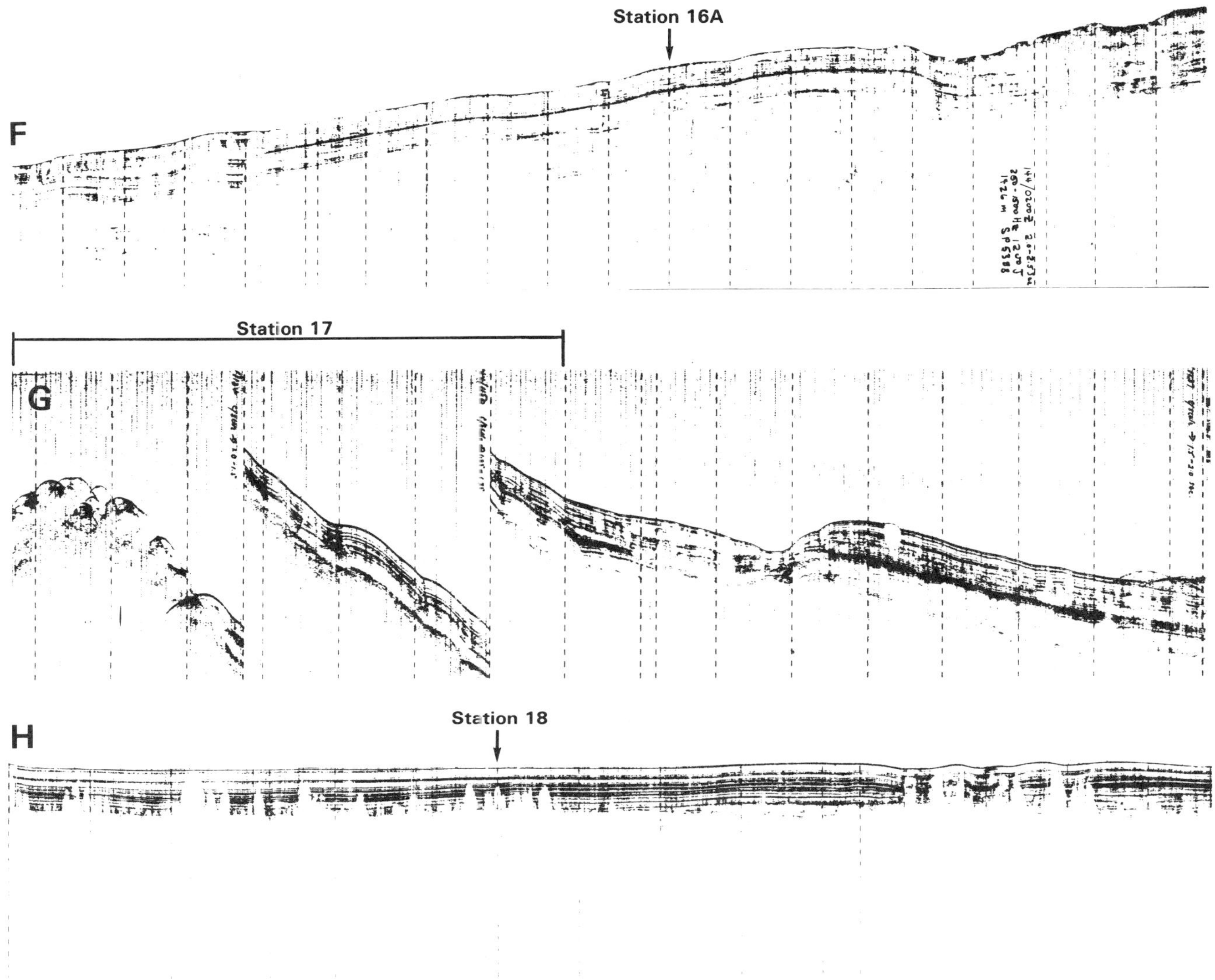

Stations 15 and 17

Eleven short dart cores (maximum length 49 cm) were recovered at Stations 15 and 17 in 895 to 458 m of water south and southwest of Santa Isabel (Figure 2). All of the cores are late Quaternary in age. Typically, these cores consist of greyish-olive (10Y4/2) to olive-grey (5Y3/2), foraminifera-rich sandy muds and muddy sands. The sediments are hemipelagic and are made up of a mixture of lithic fragments (including intermediate volcanics and volcaniclastic sandstones), terrigenous clays, biogenic carbonate (principally planktonic foraminifera, in some cases, infilled with pyrite), volcanic detritus (plagioclase feldspar, hornblende, pyroxene, and opaque minerals), sponge spicules, and wood fragments. Carbonate contents are generally between 15 and 20 percent. All of the cores reflect the dilution of biogenic carbonate by terrigenous material from Santa Isabel, and at Station 17, probably the effects of current winnowing in the Manning Strait area between Choiseul and Santa Isabel (Figure 2). Fresh glass shards occur in minute amounts in Core 15E.

Stations 16A and 18 (Figures 9 and 10)

Gravity cores were taken at Stations 16A and 18 in the western part of New Georgia Sound in 1,444 and 913 m of water, respectively (Figure 2). The cores, which have a similar lithology, range in length from 264 to 266 cm. They consist of slightly mottled (burrowed), yellowish-grey (5Y7/2) to light olive-grey (5Y5/2), slightly sandy calcareous silty clays. The sediments are a mixture of biogenic and terrigenous components including calcareous nannofossils, planktonic foraminifera, clays, pteropods, and minor amounts of volcanogenic sand and silt (feldspar, pyroxene, altered glass, lithic grains, hornblende, and opaque minerals). Calcium carbonate values typically

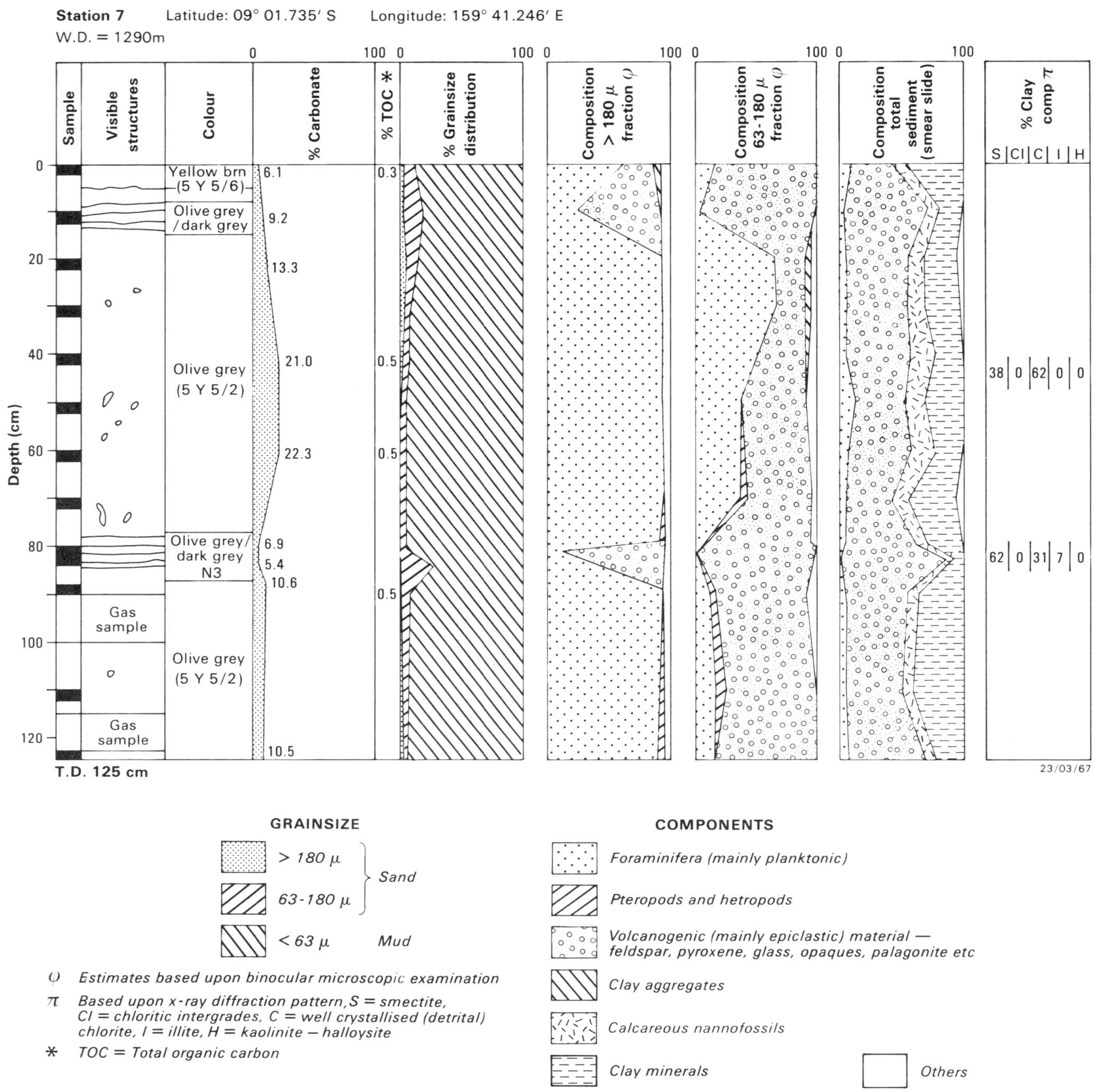

Figure 4. Log of gravity core taken at Station 7.

range from 35 to 50 percent. Total organic carbon values vary from 0.2 to 0.7 percent. Core 16A has a 6-cm thick surface oxidation layer, whereas Core 18 has a 3-cm thick layer. Both cores have relatively low gas contents (Kvenvolden and Niem, this volume). A thin volcanic-rich sandy turbidite layer similar to those in other cores occurs between 110 and 112 cm in Core 18.

CLAY MINERALOGY

Smectite, chloritic intergrades, chlorite, kaolinite-halloysite, and minor illite dominate the clay fraction of cores recovered from New Georgia Sound (Table 1). Figures 11A through 11D show the relative abundance of each clay-mineral type from the tops of the cores across the

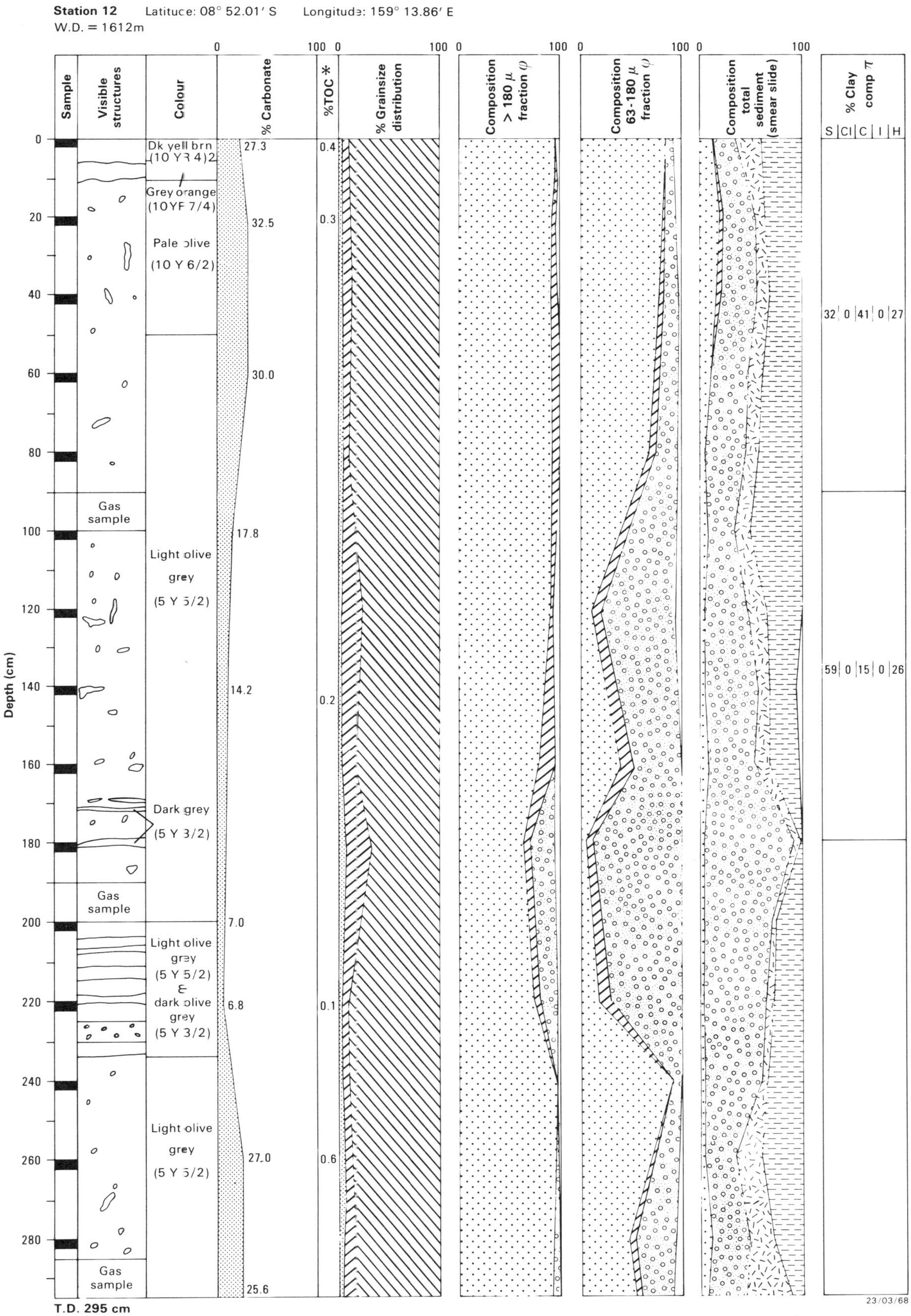

Figure 5. Log of gravity core taken at Station 12 (explanation the same as for Figure 4).

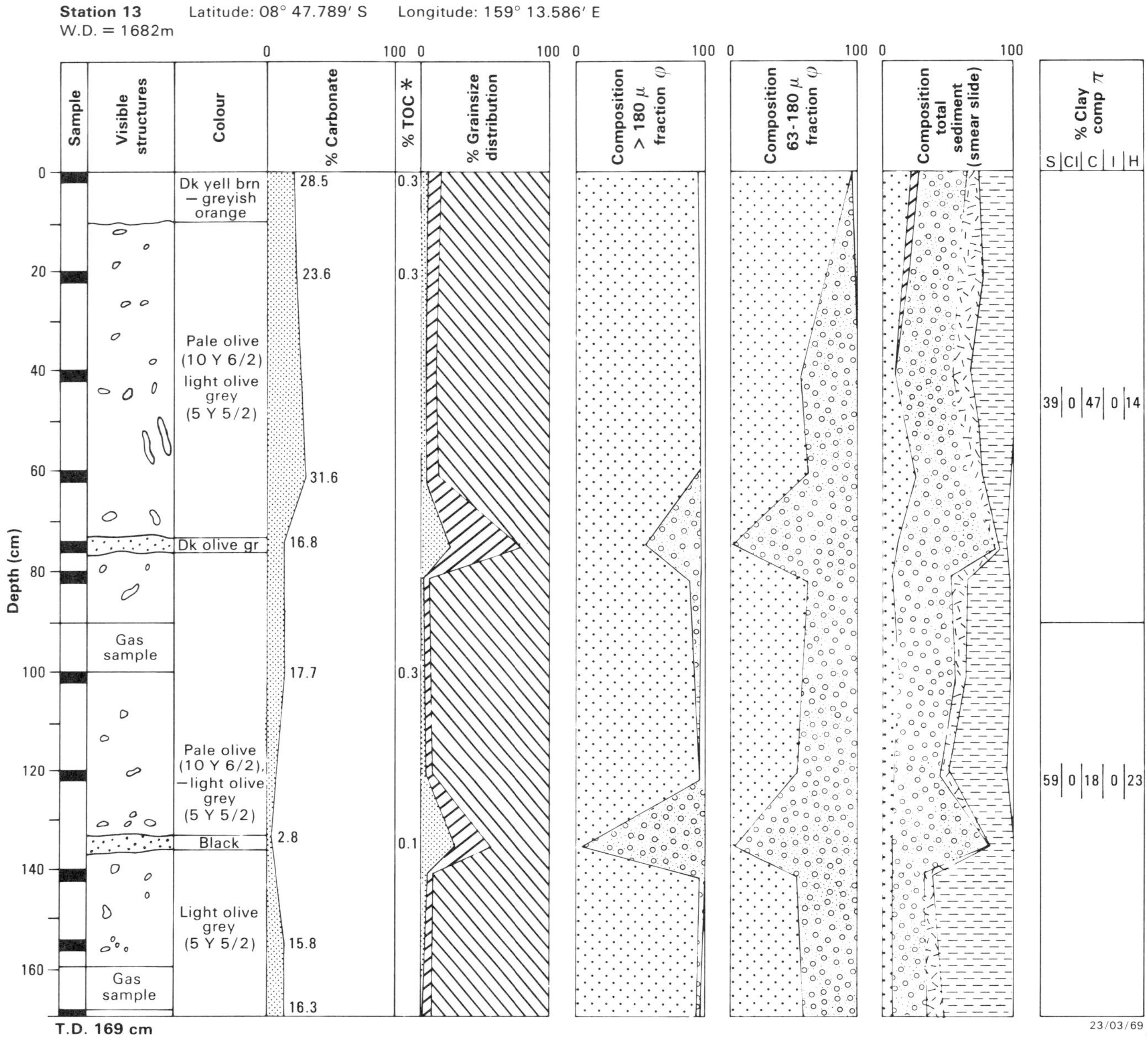

Figure 6. Log of gravity core taken at Station 13 (explanation the same as for Figure 4).

basin. In addition, X-ray diffraction analysis of the fine silt fraction (2 to 15 mm) in many samples shows abundant calcite (fragments of microfossils) and terrigenous plagioclase, pyroxene, amphibole, and quartz (Table 1). Further discussion of clay mineral identification and determination of their relative abundance appears in Appendix II

The clay composition of sediments recovered from New Georgia Sound can be broadly separated into three suites (Table 1).

Suite 1: Well-crystallized chlorite and poorly crystallized smectite dominate the clay fraction at Stations 7 and 9 in the eastern part of New Georgia Sound (Figures 2, 11A, 11B, and 12). Chloritic intergrades occur only in Cores 9E and 9H (Table 1).

Suite 2: In contrast, chloritic intergrades and subordinate poorly crystallized kaolinite-halloysite are the principal clays from Stations 15, 16, 16A, 17, and 18 in the central part of New Georgia Sound (Figures 2, 11C, 11D, and 13). Minor to moderate amounts of chlorite are also present (Figure 11A; Table 1).

Suite 3: Smectite with subordinate to minor amounts of kaolinite-halloysite and chlorite (Table 1) constitute the clays at Stations 12 and 13 north of the Russell Islands (between stations at the eastern end and central part of New Georgia Sound (Figure 2). This suite

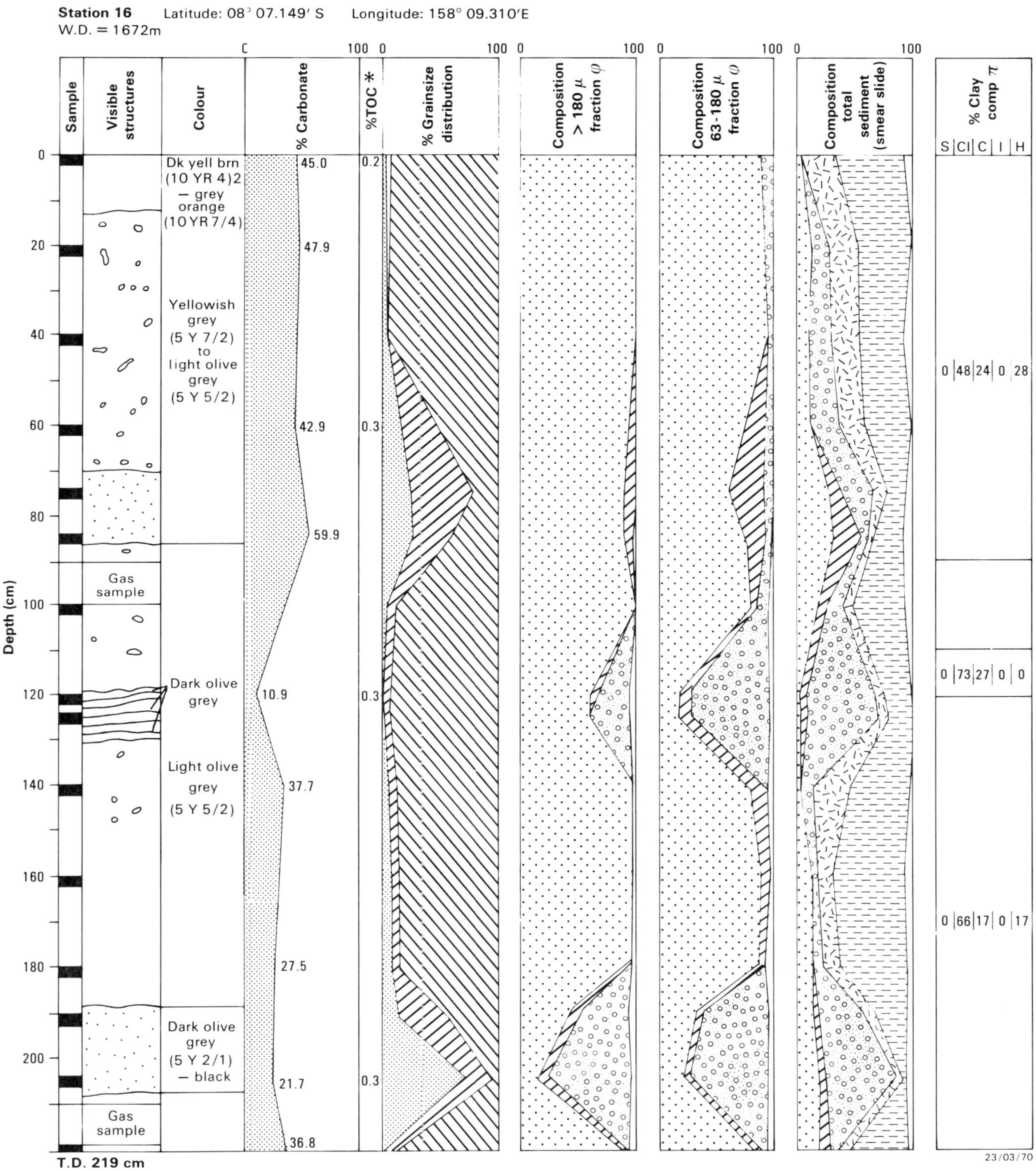

Figure 7. Log of gravity core taken at Station 16 (explanation the same as for Figure 4).

Figure 8. SEM photograph of silty clay in Core 16 (206 cm) showing prominent coccoliths and aggregates of clay minerals. Scale bar = 10 µm.

appears to be a mixture of the other two clay mineral suites, although it lacks chloritic intergrades.

Trace to minor amounts of illite occur in some samples in suites 1, 2, and 3 throughout the basin (Table 1).

DISCUSSION AND PROVENANCE

The grain size and composition of the sediments in New Georgia Sound are controlled by a number of factors. These include the nature of terrigenous input from the islands, the biological productivity of the water column, water depth, and the shape of the basin beneath the sound.

Terrigenous input to New Georgia Sound consists mainly of clay minerals with lesser amounts of sand and silt-size debris, mainly volcanic. Clay minerals in near-surface marine sediments rarely undergo major compositional changes (other than cation exchange and neoformation of smectite, chlorite, and glauconite) and generally reflect the onland detrital clay mineralogy in soils (Dixon, 1977; Eberl, 1984). This appears to be largely the case for the New Georgia Sound core samples. With the exception of the well-crystallized chlorite, the three clay mineral suites are typical of those produced in weathering of acid tropical soils developed on oceanic volcanic islands in a warm humid climate (Siefferman and Millot, 1969; Eswaran and DeConinck, 1971; Bleeker and Parfitt, 1974; Eswaran, 1979; Muller, Reidmuller, and Schwaighofer, 1984; Velde, 1985). The Solomon Islands, which lie between 6°30'S latitude and 10°S latitude, have a warm average temperature of 20°C (range 22° to 32°C) and high rainfall (114 to 635 mm), using Guadalcanal as an example (Hackman, 1980). The resulting soils include laterites (oxisols) and podzols (ulfisols) (Grover, 1958; Hackman, 1980).

As shown in Figure 11(A) and Table 1, the abundance of well-crystallized chlorite increases systematically southeastward from small proportions at Stations 15, 16, 16A, 17, and 18 in the northwest to moderate amounts at Stations 12 and 13, to dominance of the chlorite clay-mineral suite at Stations 7 and 9. This southeastward increase suggests that the source of well-crystallized chlorite is on Guadalcanal and that the chlorite-rich suite becomes diluted by other clay minerals (such as kaolinite-halloysite and chloritic intergrades) to the northwest. Differential flocculation in sea water of various clay mineral types (Eberl, 1984) may also explain some of these regional variations.

Multiple sources are likely for the clay minerals. The well-crystallized detrital chlorite that dominates suite 1 (Stations 7 and 9) may have been derived via rivers and longshore drift from the chlorite-bearing greenschist-amphibolite basement rocks exposed on east-central Guadalcanal (e.g., chlorite schist in the Mbirao Metabasics described by Hackman, 1980), possibly recycled through the widespread Neogene siliciclastic-volcaniclastic strata on north-central and western Guadalcanal (Figure 12). Crook and Niem (this volume) also recognize chemically similar well-crystallized Mg-rich chlorite in greenschist cobbles in Quaternary conglomerate dredged at Station 1 and in on-shore samples of Mbirao Lava (greenschist). Alterna-

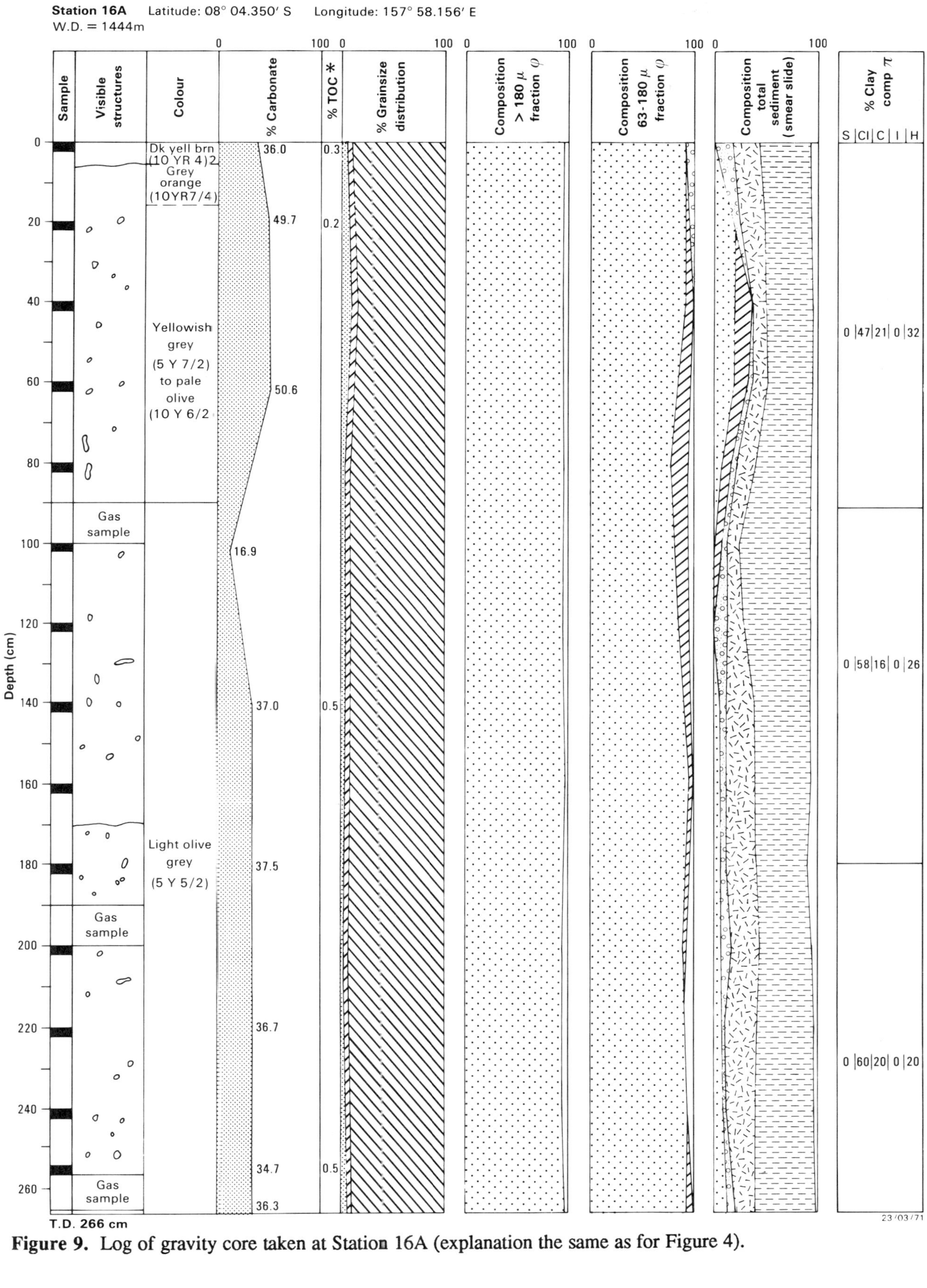

Figure 9. Log of gravity core taken at Station 16A (explanation the same as for Figure 4).

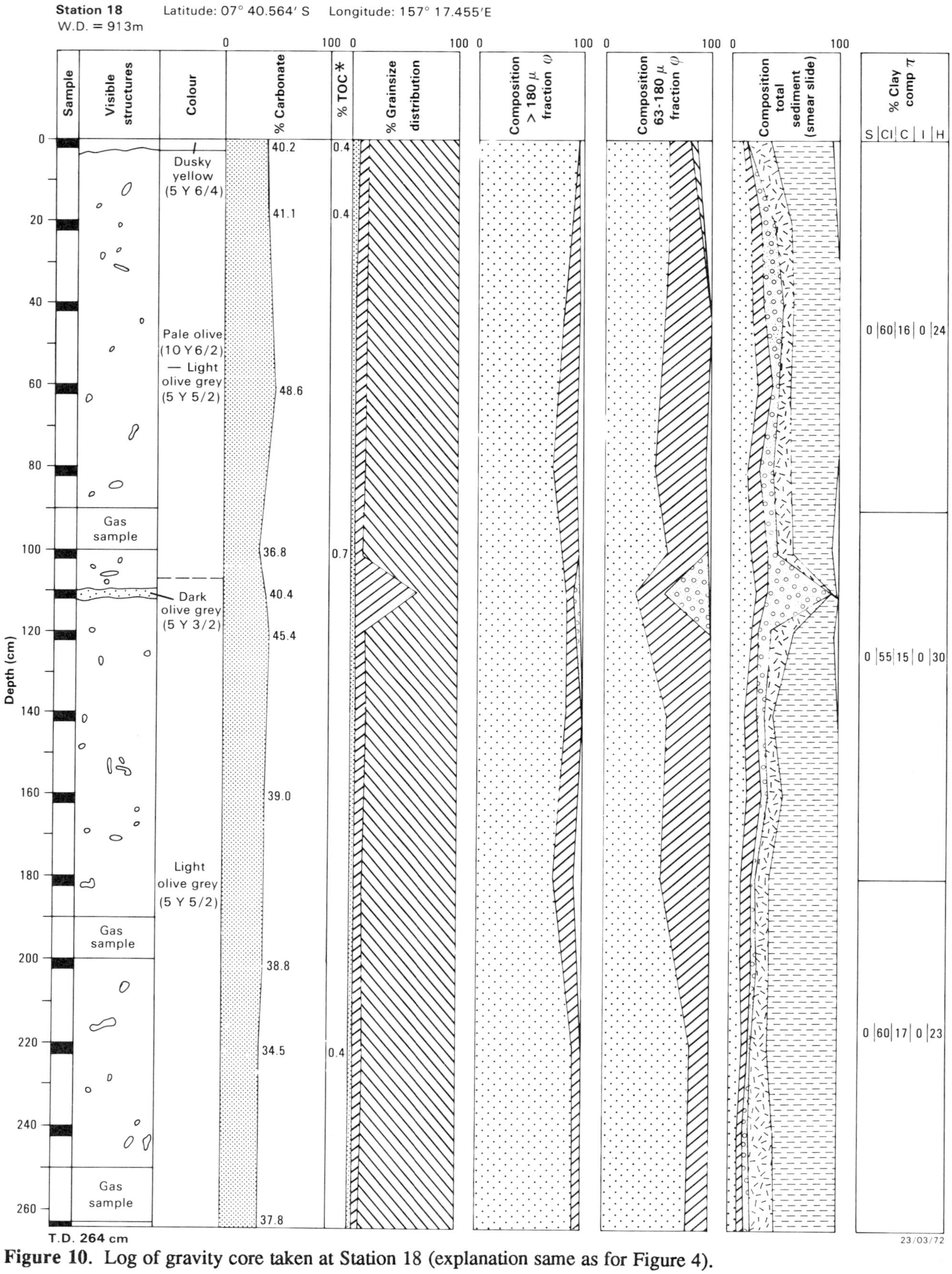

Figure 10. Log of gravity core taken at Station 18 (explanation same as for Figure 4).

Table 1. Clay Mineralogy (<2 μm Fraction) of Solomon Islands Dart and Gravity Core Samples.

Sample no.	Smectite %*	Chl. Int. %	Chlorite %	Illite %	Kaol.-Halloy. %	Other minerals (2-15 μm silt fraction)
SUITE 1						
7T(44-52 cm)	38	--	62	tr	--	Ca2,Q,P,Py,Hb
7B(78-83 cm)	62	--	31	7	--	P2,Ca,Q,Hb,Px
SA(29-35 cm)	38	--	56	6	--	Ca2,Hb,P,Q,Px
9B	39	poss	57	4	--	Ca2,Hb,P,Q,Px
9C	40	poss	50	tr	--	Ca2,Hb,P,Q,Px
9D	38	poss	62	tr	--	Ca2,Hb,P,Q,Px1
9E	--	45	55	tr	--	Ca1,Hb,P,Q,Px1
9H	--	35	58	7	--	Ca2,Hb,P,Q,Px1
SUITE 3						
12T(0-90 cm)	32	--	41	--	27	Ca3,Hb,P,Q,Px
12B(90-180 cm)	59	--	15	tr	26	Ca3,Hb,P,Px
13T(0-90 cm)	39	--	47	tr	14	Ca3,Hb,P,Q
13B(100-155 cm)	59	--	18	tr	23	Ca2,Hb,P1,Q,P
SUITE 2						
15A	--	50	10	--	40	Ca,Px2,Hb
15B	--	43	11	--	46	Ca,Px2
15C	vm.tr?	34	13	--	53	Ca,Px2
15D	--	37	16	--	47	Ca,Px1
15F	--	52	14	--	34	Ca2,Px2
15G	--	50	14	--	36	Ca2,Px2
16-1(0-90 cm)	--	48	24	--	28	Ca2,Hb,P
16-2(110-120 cm)	--	73	27	tr	tr	Ca2,Hb,P,Px
16-3(100-206 cm)	--	66	17	tr	17	Ca3,Hb,P,Q
16A-1(0-90 cm)	--	47	21	--	32	Ca3,Q,Hb
16A-2(100-180 cm)	--	58	16	tr	26	Ca3,Px2
16A-3(200-260 cm)	--	60	20	tr	20	Ca3,Hb
17A(29-30 cm)	--	57	14	--	29	Ca3,Px,Hb,P
17D(29-30 cm)	--	55	15	--	30	Ca3,Px2,Hb,P
18A(0-90 cm)	--	60	16	--	24	Ca3,Px2,Hb,P
18B(100-180 cm	--	55	15	--	30	Ca3,Px2,Hb,P
18C(200-240 cm)	--	60	17	--	23	Ca3,Px,Hb,P

%: Percent smectite, chloritic intergrade, chlorite, illite, and kaolinite-halloysite based on semiquantitative measurements of relative peak heights. tr = trace, poss = possibly present, vm.tr = very minor trace

Q = quartz, P = plagioclase, Px = pyroxene, Hb = hornblende, Ca = calcite. Numbers following mineral key refer to peak intensity: 1 = minor, 2 = moderate, 3 = strong, no number = trace.

tively, according to Hackman (1980), hydrothermal chlorite altered from clinopyroxenes is present in unmetamorphosed Mbirao pillow basalts and in younger Tertiary gabbro and diorite intrusives (e.g., Poha Diorite) as well as in mineralized zones (e.g., Gold Ridge) on central and western Guadalcanal. Similar mineralized Cretaceous to early Tertiary volcanic, greenschist, and plutonic rocks on the other islands bordering New Georgia Sound, such as Santa Isabel, the Florida Islands, and Choiseul (Coulson and Vedder, 1986; Pound, 1986; Figure 12), may be additional but minor sources for detrital chlorite in all three clay mineral suites.

The abundant smectite and minor chloritic intergrades in the samples from Stations 7 and 9 (suite 1) and from Stations 12 and 13 (suite 3, Table 1) are probably derived from weathering of intermediate volcanic mafic minerals (e.g., pyroxene, amphibole, or calcic plagioclase) and volcanic glass in lavas, breccias, and pyroclastic debris. Smectite commonly forms from basaltic-andesitic lavas and silicic tuffaceous parent materials in poorly drained soils (i.e., stagnant conditions) in both tropical and temperate climates (Drever, 1971; Colman, 1982; Glasmann, 1982; Eberl, 1984; Glasmann and Simonson, 1985; R.A. Eggleton, C. Foudoulis, and D. Varkevisser, written

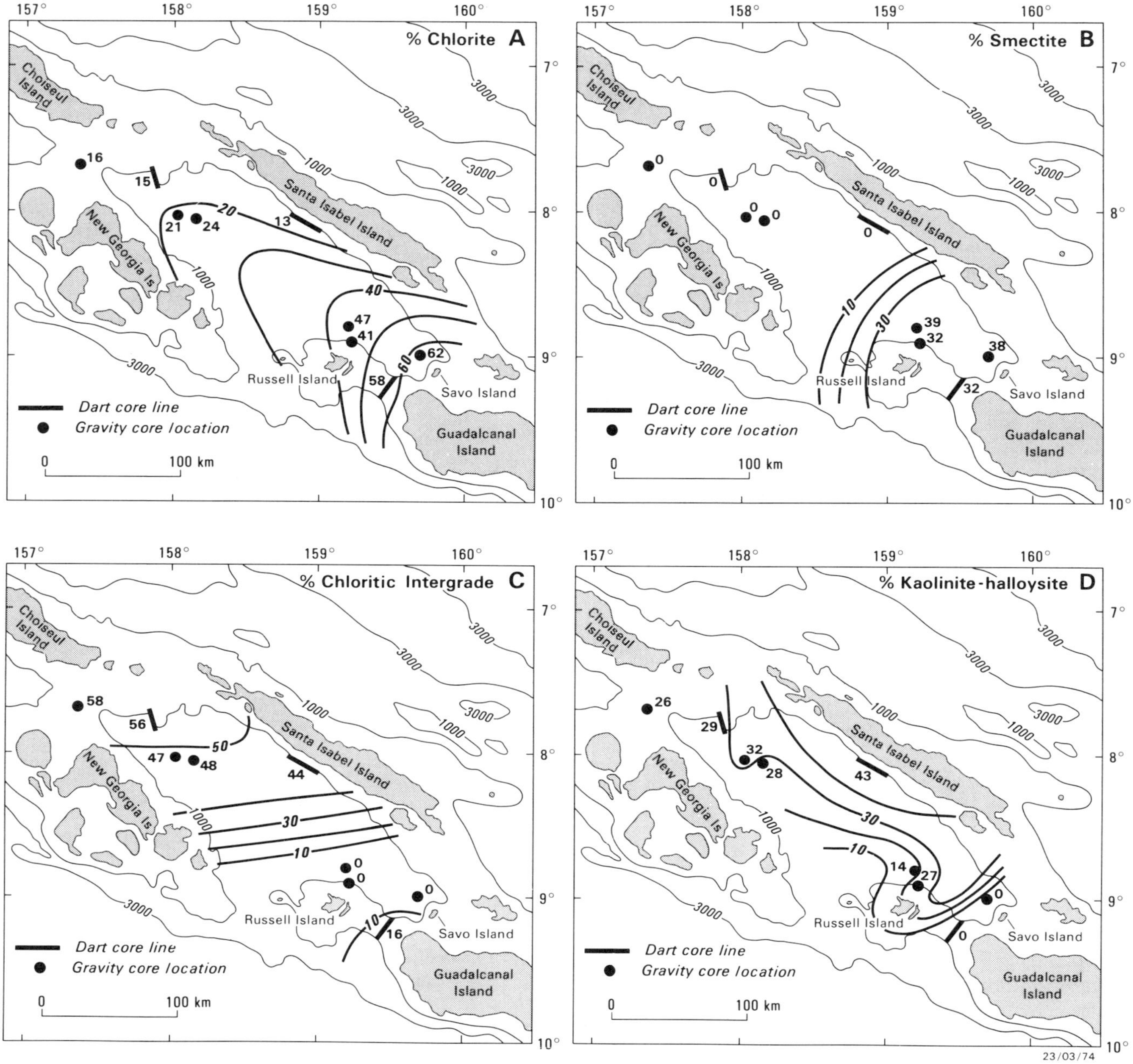

Figure 11. Geographic variations in percent: (A) chlorite, (B) smectite, (C) chlorite intergrades, and (D) kaolinite-halloysite in surface and near-surface sediments from New Georgia Sound. All of the sediments are late Pleistocene-Holocene in age (G. Chaproniere, oral commun.). Percentages shown at the dart core lines are an average of values along the line.

commun.). The Quaternary Gallego Volcanics of northwestern Guadalcanal and Savo Island, Neogene lavas of western Guadalcanal, and the Pliocene basaltic andesite breccias and basaltic lavas of the Russell Islands are likely sources (Figure 12). Smectite can also form diagenetically by alteration of volcanic glass in fine ashes in submarine environments (Velde, 1985). However, clay-altered pseudomorphs of glass shards and fresh shards are rare in smear slides of the coarser fraction in these cores. The bulk of the smectite clay is associated with detrital sand- and silt-size volcanic rock and mineral fragments and, therefore, is likely to have been derived from deep soils and weathered intermediate to mafic volcanic parent materials washed off nearby volcanic islands. The minor amounts of detrital illite which occur in clay suites 1 and 3 may have come from hydrothermally altered feldspar in the volcanic rocks on Guadalcanal and the Russell Islands and/or may be associated with well crystallized silt- to fine sand-size

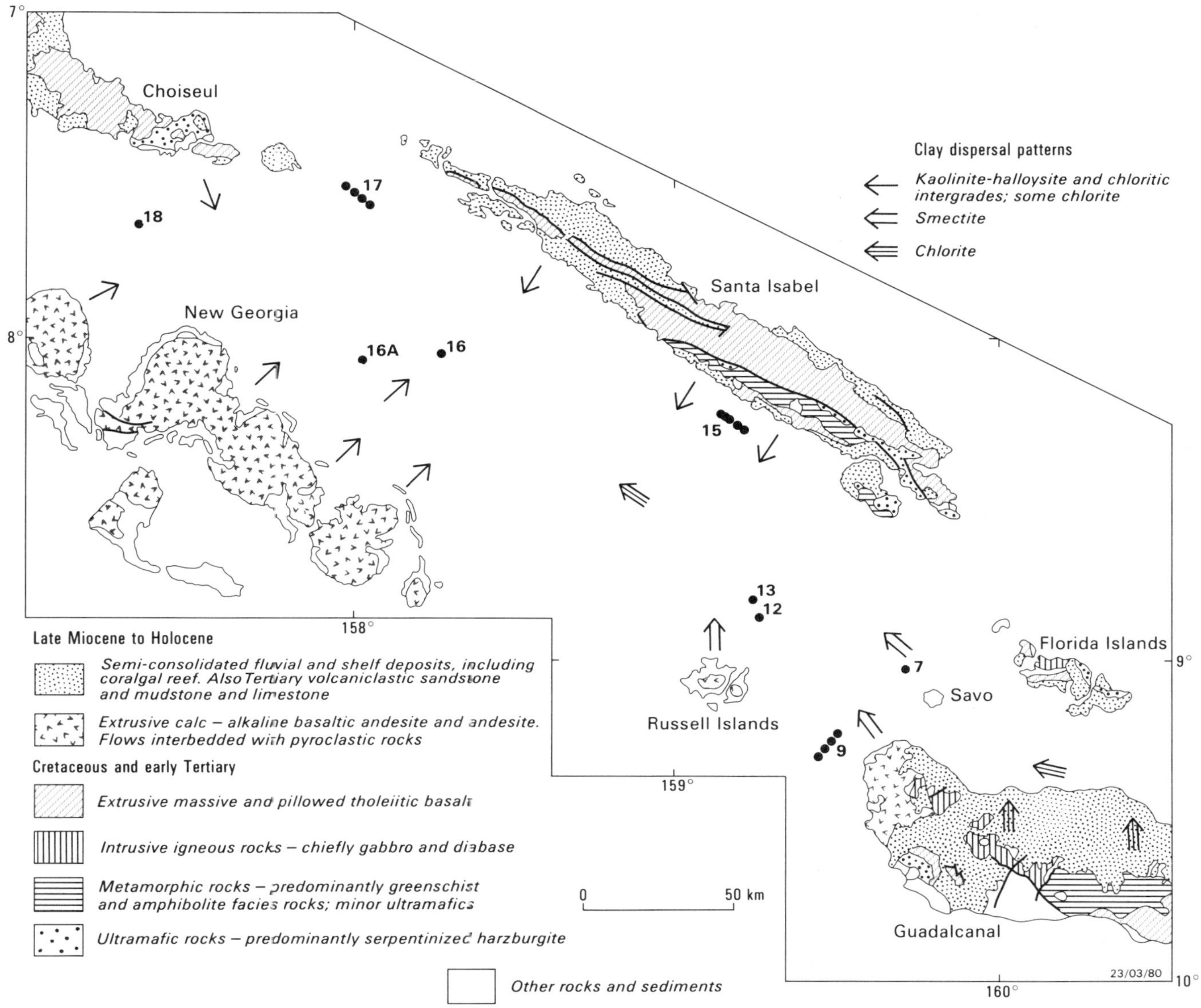

Figure 12. Generalized geologic map of the islands bordering central and eastern New Georgia Sound showing interpreted clay dispersal patterns.

biotite flakes noted in grain mounts of cores from Stations 7 and 9.

In suite 2 (Stations 15, 16, 16A, 17, and 18), the mixture of kaolinite, halloysite, and chloritic intergrades suggests derivation primarily from well-drained acidic tropical soils formed on late Tertiary to Quaternary basaltic to andesitic lavas of the New Georgia island group (Stanton and Bell, 1969) and Cretaceous to early Tertiary basaltic, metamorphic, and ultramafic rocks of Santa Isabel and the southeastern part of Choiseul (Figure 12; Coulson and Vedder, 1986). A mixture of thin, broken pseudohexagonal to polygonal plates of kaolinite and rod-shaped to tubular 7Å (dehydrated halloysite occurs in suites 2 and 3 (Figure 14). Kaolinite-group clays (7Å halloysite and kaolinite) and chloritic intergrades typically form in tropical and humid temperate climates from further leaching of smectite-altered andesitic and basaltic parent materials (e.g., plagioclase and augite) (Craig and Loughnan, 1964; Drever, 1971; White and Sarcia, 1978; Glasmann, 1982; Muller, Reidmuller, and Schwaighofer, 1984; Glasmann and Simonson, 1985). Smectite can be transformed to halloysite and kaolinite in tropical soil microenvironments with more extensive leaching and wetting and drying (seasonal dessication) (Borchardt, 1977). Halloysite can form in soils on the wetter parts (windward sides) of volcanic islands where there is more extensive leaching, whereas smectite forms on the same parent material on dry leeward slopes (Eberl, 1984). Halloysite is

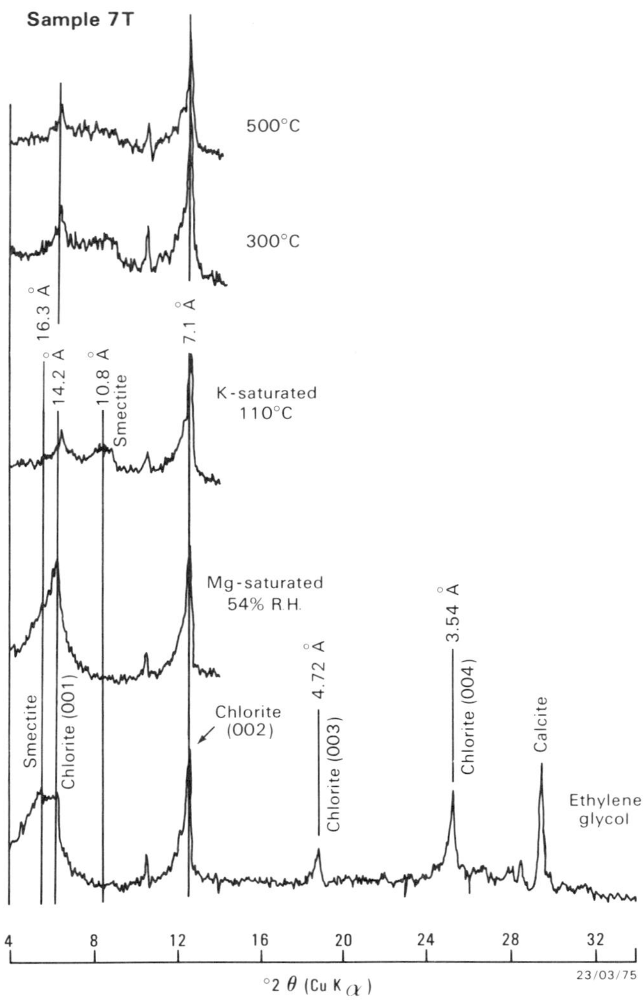

Figure 13. Typical X-ray diffraction pattern of smectite and well-crystallized chlorite (sharp peak) in the eastern part of New Georgia Sound.

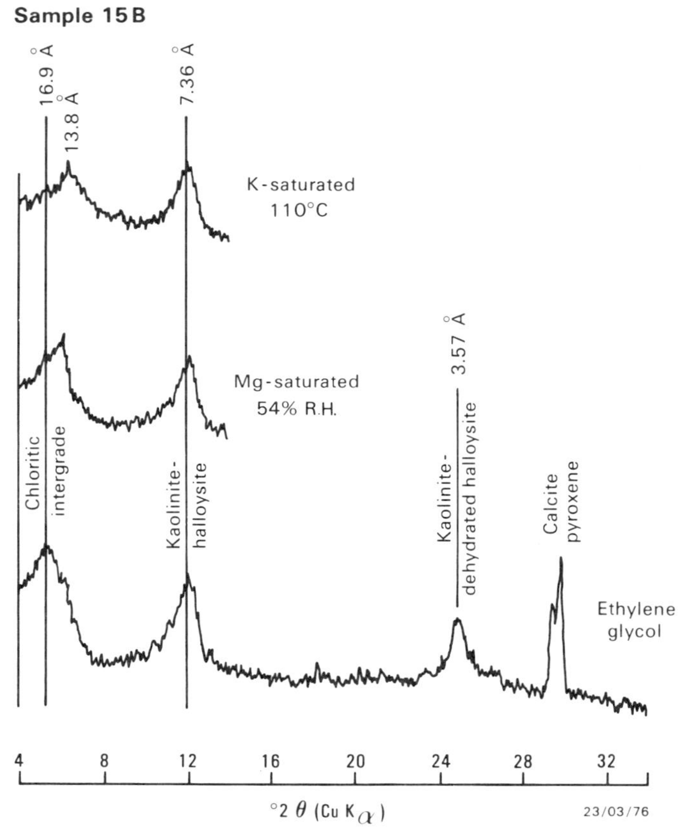

Figure 14. Typical X-ray diffraction pattern of chloritic intergrades and kaolinite-halloysite in the western part of New Georgia Sound.

more common than kaolinite in weathered volcanic materials in these temperate and tropical soil microenvironments (Bleeker and Parfitt, 1974; Glasmann, 1982) but may undergo mineral transformation to more stable kaolinite in time (Fieldes, 1955; Dixon, 1977). Kaolinite and halloysite are most abundant at Station 15 south of Santa Isabel (Figure 11D). Although the adjacent onland geology is similar to that on parts of Guadalcanal, Cretaceous and early Tertiary volcanics and metamorphics lie on the relatively wet western side of Santa Isabel, probably enhancing the degradation of smectite to kaolinite-group minerals.

Another potential, but minor, source for kaolinite may be local kaolinitized and hydrothermally mineralized Tertiary lavas intruded by gabbroic and dioritic plutons on the islands, such as on Gold Ridge and in the Gallego Volcanics on Guadalcanal (Hackman, 1980; Figure 12). Exon, Colwell, and Bolton (1987) also suggest that some kaolinite clay could be transported to the Solomon Islands offshore region as wind-borne dust derived from the Australian continent.

The abundant chloritic intergrades in suite 2 may reflect further weathering of the more mafic volcanic smectite clays in soils initially formed under stagnant wet conditions (Glasmann and Simonson, 1985). Some chloritic intergrades could be a result of degradation of well-crystallized chlorite as reported in Italy by Sartori, Riffaldi, and Levi-Minzi (1979); others may reflect the uptake of Mg cations in smectite in marine pore waters. Chlorite has been reported forming from smectite in the marine environment (Velde, 1985). Hence, all these clay minerals would be expected in the fine-grained hemipelagic sediments of New Georgia Sound.

Most of the sand- and silt-size terrigenous material in the New Georgia Sound sediments is volcanogenic. This composition reflects the widespread occurrence of volcanics and associated intrusive rocks of various types and ages on the islands bordering the study area. These rocks range from Late Cretaceous(?) to early Tertiary metamorphosed and unmetamorphosed basalts and andesites on Guadalcanal, the Florida Islands, Santa Isabel, and Choiseul to late Tertiary and Quaternary predominantly calc-alkaline basaltic andesite and andesite on Guadalcanal, the New Georgia island group, Savo, and Choiseul (Coulson and Vedder, 1986). Specific sources probably include the Pliocene to Quaternary Gallego Volcanics on Guadalcanal, Savo, and the correlative volcanic rocks on the northwestern end of the New Georgia island group for much of the hornblende, the basaltic andesites and andesites throughout the New Georgia island group for much of the plagioclase feldspar and green clinopyroxene, and diorites such as the Poha and Lungga Diorites of northern Guadalcanal for much of the quartz and biotite recorded at Stations 7 and 9 at the eastern end of New Georgia Sound. The absence of fresh glass shards (except in minute amounts in Core 15E) and recognizable ash layers in the cores (as opposed to volcanic-rich turbidite layers) suggests that there has been minimal direct input of material to the New Georgia Sound from ash falls during the late Quaternary. Most of the non-biogenic material in the cores is epiclastic, reflecting vigorous weathering and erosion of the islands as a result of marked tectonic uplift during the Quaternary (for example, Colwell and Vedder, 1986b) as well as the hot and wet climate. The importance of rivers in the supply of terrigenous material to the offshore areas was well documented when at Station 15 the *S.P. Lee* passed over a large plume of turbid flood water extending several kilometers offshore from the mouth of the Korighole River.

The relatively high content of calcareous plankton (mostly foraminifera, pteropods, and nannofossils) in the sediments is shown by calcium carbonate contents which typically range from 10 to 40 percent. All of the cores are sited above the aragonite and calcite compensation depths, which, at present, lie throughout the region at approximately 1,800 m and 4,700-4,800 m, respectively, according to Exon, Colwell, and Bolton (1987). A slight decrease in average carbonate content with depth in the core occurs in several cases (e.g., Station 12; Figure 5). This change in carbonate content may be due to either a decrease in terrigenous input through time (possibly related to waning island uplift or to a slight change in climate), increased productivity rates of the phytoplankton and zooplankton in the upper water column, or alternatively, to the effects of carbonate-dissolution cycles such as those recorded by Saito, Burckle, and Hays (1975) in cores from the equatorial Pacific.

The abundance of bioturbation and the yellowish-brown oxidized upper part of all cores suggest a well-developed benthic infauna and oxygenated bottom waters in New Georgia Sound. These conditions may explain, in part, the relatively low total organic carbon values noted in the samples. Reducing conditions transform the oxidized accumulating yellowish-brown mud to greenish-grey after a few centimeters of burial.

All of the sandy layers in the cores recovered in water deeper than 500 m are turbidites. This finding is consistent with the view of Colwell and Tiffin (1986) that mass movements and density currents play a major role in the transport of both terrigenous and biogenic sediments into the deeper parts of New Georgia Sound away from its relatively steep margins. The sandy sediments recovered in relatively shallow water at Stations 9, 15 and 17 reflect their proximity to land (close to sources of terrigenous material and in areas of current action that may have removed the mud).

SUMMARY AND CONCLUSIONS

Quaternary sediments cored in New Georgia Sound are hemipelagic. They consist of a mixture of biogenic (mainly planktonic foraminifera, pteropods, and calcareous nannofossils) and terrigenous (largely volcanogenic) components. All of the sampled sediments lie above the present-day aragonite and calcite compensation depths.

Regional variations in the clay mineralogy of surface sediments through New Georgia Sound reflect provenance. Such variations are far less pronounced in the sand- and silt-size terrigenous fractions which tend to be dominated by detritus (mainly plagioclase feldspar and pyroxene) derived from the weathering of widespread basaltic and andesitic volcanic rocks on the islands. Very little material appears to be supplied via direct ash falls.

Turbidity currents (and possibly other mass movements) are important conveyors of sand and silt into the deeper parts of New Georgia Sound.

ACKNOWLEDGMENTS

We are grateful to the technical staff of Oregon State University, BMR, and UNOCAL Research Laboratory for their assistance with the analyses. Particular thanks are extended to J. Reed Glasmann and James Wood of UNOCAL and Tony Eggleton of the Geology Department of the Australian National University for assistance with the clay-mineral preparation and identification. Alfred Soeldner of Oregon State University operated the transmission electron microscope; R.W. Jaco of UNOCAL ran the scanning electron microscope. One of us (A.R.N.) thanks Oregon State University for a travel grant that enabled him to work on the manuscript while a Visiting Fellow in the Department of Geology, Australian National University, Canberra. The head of that Department, Ken Campbell, is thanked for his hospitality. N. Exon and J. Reed Glasmann reviewed the manuscript. Figures were drawn by B. Pashley of the BMR Drawing Office. Colwell publishes with permission of the Director, Bureau of Mineral Resources, Canberra.

APPENDIX I
LABORATORY PROCEDURES USED IN CLAY-MINERAL ANALYSIS AND SEM AND TEM WORK

Each of the samples used in the clay-mineral analysis and SEM and TEM work represents several centimeters of the upper part of the dart core and/or the top half, middle part, or bottom half of the longer gravity

Figure 15. TEM micrograph of <2 µm clay fraction, Station 12, showing some rod-shaped to tubular halloysite, very thin, broken pseudohexagonal to polygonal plates of kaolinite and chlorite, and a fragment of a coccolith test (c). Smectite appears as very thin, fluffy particles without well-defined borders.

cores. The intervals sampled are listed in Table 1 and are shown diagrammatically on Figures 4-7, 9 and 10.

The preparation of samples for analysis involved dispersion of 1 to 2 grams of mud sample in dilute $NaPO_3$, centrifugation into clay (<2 µm) and fine silt (2 to 15 µm) fractions, and Mg- and K-saturation of the clay separates by washing three times in 1N $MgCl_2$ and distilled water and in 1N KCl and distilled water, respectively. Oriented clay films were mounted on glass slides, following the paste method of Theisen and Harward (1962).

A computerized Philips/Norelco diffractometer equipped with Ni-beta filtered Cu K-alpha radiation was used to analyze both the K- and Mg-saturated clay fractions and silt separates. Mg-saturated clay separates were run ethylene glycol-saturated and at 54 percent relative humidity. K-saturated clay samples were scanned under dry air conditions after heating at 110°C, 300°C, and 500°C each for two hours, following the method of Glasmann and Simonson (1985).

Three samples (Stations 15, C and D and 16, 206 cm) were mounted on metal stubs and covered with gold/palladium alloy in a Polaron 5000 sputter coater for scanning electron-microscope analysis (SEM). An AMRAY 1000A SEM fitted with a KEVEX EDX system was used at UNOCAL research laboratories in Brea, California. Further characterization of the morphology of the Mg-saturated clay fraction by transmission electron microscopy (TEM) was performed on 7 samples at Oregon State University on a Philips TEM 300 operated at 100 KV and 20 µamp and equipped with a liquid nitrogen decontamination system. Clay samples were prepared by dropping dilute clay suspensions onto Formvar-coated copper grids.

APPENDIX II
CLAY MINERAL IDENTIFICATIONS AND MEASUREMENTS OF RELATIVE ABUNDANCES

Estimates of relative clay abundances were made by measuring d(001) peak heights of different clay minerals, according to a method modified after Griffin (1971). These data, although based on limited sampling and only semiquantitative, were plotted on a map of New Georgia Sound in order to illustrate the general distribution of clay mineral types of approximately the same age (late

Pleistocene to Holocene) (Figures 11A through 11D). Data from the middle and bottom of gravity cores (Table 1) were not used in construction of these maps because of the relatively wide variation of Pleistocene ages of these sediments as indicated by foraminiferal assemblages (G. Chaproniere, 1986, oral commun.). Chloritic intergrade refers to a system intermediate between chlorite and vermiculite or smectite (Carstea, Harward, and Knox, 1970). In New Georgia Sound, the chloritic intergrade appears to be a system between chlorite and smectite. Both smectite clay and chlorite intergrades expand to a similar range of d(001) basal spacing of 16.4 to 17.3 Å with ethylene glycol treatment (Figures 13 and 14). Smectite clay collapses to 10 to 11 Å with K-saturation and dry air treatment at 100°C and swells to 13.5 to 15.2 Å with Mg-saturation in 54 percent relative humidity (Figure 13). Chloritic intergrade, unlike smectite, however, shows only limited or minor collapse (i.e., to 13.2 to 15.2 Å) upon K-saturation and drying at 100°C (Figure 13), 300°C, and 500°C and minor shift of peak position with Mg-saturation at 54 percent relative humidity. Glasmann and Simonson (1985) noted a similar shift of diffraction peaks for chloritic intergrades in soils from the Oregon Coast Range (USA).

Poorly crystallized or disordered kaolinite-halloysite is recognized by broad asymmetrical d(001) and d(002) peaks ranging from 7.18 to 7.49 Å and 3.56 to 3.59 Å, respectively (Figure 14). These peak ranges include mixtures of both the clay mineral kaolinite and the clay mineral dehydrated (7 Å) halloysite (metahalloysite). Dehydrated halloysite has broad basal reflections at 7.4 Å, 4.43 Å, and 3.60 Å (Brindley, 1981). The basal reflections in the Solomon Islands clay samples most commonly occur at 7.36 Å and 3.59Å, very close to 7 Å dehydrated halloysite rather than well crystallized kaolinite, which commonly has sharp reflections at 7.15-7.20 Å and 3.57 Å (Brindley, 1981). These samples, however, lack the characteristic major basal reflection at 4.43 Å (approximately 19.92° 2θ) that is typical of halloysite (Brindley, 1981). Brindley, Santos, and Santos (1963) showed that nearly equal mixtures of kaolinite and dehydrated 7 Å halloysite can give a disordered kaolinite diffraction pattern which lacks this definitive peak. In addition, very thin sheets and small crystals of kaolinite can form asymmetrical basal d(001) reflections at higher Å (i.e., >7.20 Å) and lower 2θ values than normally expected for the clay mineral kaolinite (Brindley and Brown, 1980). TEM photographs of Solomon Island samples with this diffraction pattern reveal both the tubular or rod-shaped morphology typical of halloysite and broken pseudohexagonal to polygonal plates characteristic of the clay mineral kaolinite (Sudo et al, 1981; Brindley and Brown, 1980) (Figure 15).

Well-crystallized chlorite is characterized by sharp symmetrical peaks at 14.2 Å (001), 7.10 Å (002), 4.72 Å (003), and 3.57 Å (Figure 13). Identification of chlorite was confirmed through SEM analysis by the small pseudohexagonal platy morphology and the distinctive Fe, Mg, Al, and Si energy-dispersive spectrum (EDS) (Welton, 1984). Minor illite is characterized by a non-expandable peak at 10 Å.

REFERENCES CITED

Bleeker, P., and Parfitt, R.L., 1974, Volcanic ash and its clay mineralogy at Cape Hoskins, New Britain, Papua New Guinea: Geoderma, v. 11, p. 123-135.

Borchardt, G.A., 1977, Montmorillonite and other smectite minerals, *in* Dixon, J.B., Weed, S.B., Kittrick, J. A., Milford M.H., and White, J.L., eds., Minerals in soil environments: Soil Science Society of America, Madison, Wisconsin, p. 293-330.

Brindley, G.W., 1981, X-ray identification (with ancillary techniques) of clay minerals, *in* Longstaffe, F.J., ed., Short course in clays and the resource geologist: Mineralogical Association of Canada, p. 22-38.

Brindley, G.W., and Brown, G., 1980, Crystal structures of clay minerals and their X-ray identification: London, Mineralogical Society, 495 p.

Brindley, G.W., Santos, P. de Souza, and Santos, H. de Souza, 1963, Mineralogical studies of kaolinite-halloysite clays: Part 1, Identification problems: American Mineralogist, v. 48, p. 897-910.

Carstea, D.D., Harward, M.E., and Knox, E.G., 1970, Formation and stability of hydroxy-Mg interlayers in phyllosilicates: Clays and Clay Minerals, v. 18, p. 213-222.

Carver, R.E., 1971, Procedures in sedimentary petrology: New York, Wiley, 653 p.

Coleman, P.J., 1965, Stratigraphic and structural notes on the British Solomon Islands with reference to the first geological map: British Solomon Islands Geological Record (1959-1962), v. 2, p. 17-31.

Coleman, P.J., 1970, Geology of the Solomon and New Hebrides Islands, as part of the Melanesian re-entrant, Southwest Pacific: Pacific Science, v. 24, p. 289-314.

Coleman, P.J., and Packham, G.H., 1976, The Melanesian Borderlands and India-Pacific Plates' boundary: Earth Science Reviews, v. 12, p. 197-233.

Colman, S.M., 1982, Chemical weathering of basalts and andesites: evidence from weathering rinds: U.S. Geological Survey Professional Paper 1246.

Colwell, J.B., 1986, Preliminary descriptions of gravity cores from New Georgia Sound, Solomon Islands, *in* Vedder, J.G., Pound, K.S., and Boundy, S.Q., eds., Geology and offshore resources of Pacific island arcs—central and western Solomon Islands: Houston, Texas, Circum-Pacific Council for Energy and Mineral Resources Earth Science Series, v. 4, p. 237-241.

Colwell, J.B., and Tiffin, D.L., 1986, Recent depositional patterns in the Central Solomons Trough of the Solomon Islands, *in* Vedder, J.G., Pound, K.S., and Boundy, S.Q., eds., Geology and offshore resources of Pacific island arcs—central western Solomon Islands: Houston, Texas, Circum-Pacific Council for Energy and Mineral Resources Earth Science Series, v. 4, p. 243-254.

Colwell, J.B., and Vedder, J.G., 1986a, Sampling methods, Solomon Islands, *in* Vedder, J.G., Pound, K.S., and Boundy, S.Q., eds., Geology and offshore resources of Pacific island arcs—central and western Solomon Islands:

Houston, Texas, Circum-Pacific Council for Energy and Mineral Resources Earth Science Series, v. 4, p. 49-56.

Colwell, J.B. and Vedder, J.G., 1986b, Description and interpretation of dredged rocks, Solomon Islands, *in* Vedder, J.G., Pound, K.S., and Boundy, S.Q., eds., Geology and offshore resources of Pacific island arcs—central and western Solomon Islands: Houston, Texas, Circum-Pacific Council for Energy and Mineral Resources Earth Science Series, v. 4, p. 225-236.

Coulson, F.I., and Vedder, J.G., 1986, Geology of the central and western Solomon Islands, *in* Vedder, J.G., Pound K.S., and Boundy, S.Q., eds., Geology and offshore resources of Pacific island arcs—central and western Solomon Islands: Houston, Texas, Circum-Pacific Council for Energy and Mineral Resources Earth Science Series, v. 4, p. 59-87.

Craig, D.C., and Loughnan, F.C., 1964, Chemical and mineralogical transformations accompanying the weathering of basic volcanic rocks from New South Wales: Australian Journal of Soil Research, v. 2, p. 218-234.

Crook, K.A.W. and Taylor, G.R., 1987, Sedimentary petrology and mineral chemistry, Woodlark Basin-Solomon Islands offshore region, *in* Exon, N. and Taylor B., eds., Marine geology, geophysics, and geochemistry of the Woodlark Basin-Solomon Islands: Houston, Texas, Circum-Pacific Council for Energy and Mineral Resources Earth Science Series, v. 7, p. 295-346.

Dixon, J.B., 1977, Kaolinite and serpentine group minerals, *in* Dixon, J.B., Weed, S.B., Kittrick, J.A., Milford, M.H., and White, J.L., eds., Minerals in soil environments: Soil Science Society of America, Madison, Wisconsin, p. 357-403.

Drever, J.I., 1971, Chemical weathering in a subtropical igneous terrain, Rio Ameca, Mexico: Journal of Sedimentary Petrology, v. 41, n. 4, p. 951-961.

Eberl, D.D., 1984, Clay mineral formation and transformation in rocks and soils: Philosophical Transactions of the Royal Society of London, Series A, v. 311, p. 241-257.

Eswaran, H., 1979, The alteration of plagioclases and augites under differing pedo-environmental conditions: Journal of Soil Science, v. 30, p. 547-555.

Eswaran, H., and DeConinck, F., 1971, Clay mineral formations and transformations in basaltic soils in tropical environments: Pedologie, v. 21, p. 181-210.

Exon, N.F., Colwell, J.B., and Bolton, B.R., 1987, Sedimentology of Quaternary sediment cores from the Solomon Islands offshore region, *in* Exon, N. and Taylor, B., eds., Marine geology, geophysics, and geochemistry of the Woodlark Basin-Solomon Islands: Houston, Texas, Circum-Pacific Council for Energy and Mineral Resources Earth Science Series, v. 7, p. 241-263.

Fieldes, M., 1955, Clay mineralogy of New Zealand soils. Part II: Alophane and related mineral colloids: New Zealand Journal of Science of Technology, v. 37, p. 336-350.

Glasmann, J.R., 1982, Alteration of andesite in wet, unstable soils of Oregon's western Cascades: Clays and Clay Minerals, v. 30, n. 4, p. 253-263.

Glasmann, J.R., and Simonson, G.H., 1985, Alteration of basalt in soils of western Oregon: Soil Science Society of America Journal, v. 49, n., 1, p, 262-273.

Griffin, G.M., 1971, Interpretation of X-ray diffraction data, *in* Carver, R.E., ed., Procedures in sedimentary petrology: New York, Wiley-Interscience, p. 541-569.

Grover, J.C., 1958, The Solomon Islands—Geological exploration and research, 1953-56: Geological Survey of the British Solomon Islands Memoir n. 2.

Hackman, B.D., 1980, The geology of Guadalcanal, Solomon Islands: Overseas Memoir 6, Institute of Geological Science, London, 115 p.

Kroenke, L.W., 1972, Geology of the Ontong Java Plateau: Hawaii Institute of Geophysics, Report n. HIG-72.5, University of Hawaii, 119 p.

Kroenke, L.W., 1984, Solomon Islands: San Cristobal to Bougainville and Buka, *in* Kroenke, L.W., ed., Cenozoic tectonic development of the southwest Pacific: U.N. ESCAP CCOP/SOPAC Technical Bulletin, n. 6, ch. 4, 22 p.

Muller, H.W., Riedmuller, G., and Schwaighofer, B., 1984, Weathering products of andesitic rocks from Sulawesi, Indonesia: Clay Minerals, v. 19, p. 21-28.

Pound, K.S., 1986, Correlation of rock units in the central and western Solomon Islands, *in* Vedder, J.G., Pound, K.S., and Boundy, S.Q., eds., Geology and offshore resources of Pacific island arcs—central and western Solomon Islands: Circum-Pacific Council for Energy and Mineral Resources Earth Science Series, v. 4, p. 89-97.

Saito, T., Burckle, L.H., and Hays, J.D., 1975, Late Miocene to Pleistocene biostratigraphy of equatorial Pacific sediments: Micropaleontology Special Publication, n. 1, p. 226-244.

Sartori, F., Riffaldi, R., and Levi-Minzi, R., 1979, Occurrence of chloritic intergrades in the Recent sediments of Arno River [Italy]: Clay Minerals, v. 14, p. 47-65.

Siefferman, G., and Millot, G., 1969, Equatorial and tropical weathering of recent basalts from Cameroon: allophane, halloysite, metahalloysite, kaolinite, and gibbsite, *in* Heller, L., ed., Proceedings of the International Clay Conference, Tokyo, Japan, 1969: Israel University Press, Jerusalem, p. 417-430.

Stanton, R.L., and Bell, J.D., 1969, Volcanic and associated rocks of the New Georgia Group, British Solomon Islands Protectorate: Overseas Geology and Mineral Resources, v. 10, n. 2, p. 113-145.

Sudo, T., Shimoda, S., Yotsumoto, H., and Aita, S., 1981, Electron micrographs of clay minerals: Developments in Sedimentology 31, New York, Elsevier Scientific Publishing Company, 203 p.

Theisen, A.A., and Harward, M.E., 1962, A paste method of preparation of slides for clay mineral identification by X-ray diffraction: Soil Science Society of America, Proceedings, v. 26, p. 90-91.

Valencia, M.J., 1973, Calcium carbonate and grain-size analysis of surface sediments, western equatorial Pacific: Pacific Science, v. 27, p. 290-303.

Vedder, J.G., Pound, K.S., and Boundy, S.Q., eds., 1986, Geology and offshore resources of Pacific island arcs—central and western Solomon Islands: Houston, Texas, Circum-Pacific Council for Energy and Mineral Resources Earth Science Series, v. 4, 306 p.

Velde, B., 1985, Clay Minerals—a physico-chemical explanation of their occurrence: Developments in Sedimentology 40, New York, Elsevier Scientific

Publishing Company, 427 p.

Welton, J.E., 1984, SEM petrology atlas: American Association of Petroleum Geologists, Methods in Exploration Series, 237 p.

White, R.W., and Sarcia, C., 1978, Natural and artificial weathering of basalt, northwestern United States: Bureau de Recherches Geologiques et Minieres Bulletin, Series 2, n. 1, p. 1-29.

Vedder, J.G., and Bruns, T.R., editors, 1989, Geology and offshore resources of Pacific island arcs—Solomon Islands and Bougainville, Papua New Guinea Regions: Houston, Texas, Circum-Pacific Council for Energy and Mineral Resources Earth Science Series, v. 12.

THE MICROPALEONTOLOGY OF SEDIMENTARY UNITS FROM THE SOLOMON ISLANDS

G.W. Hughes
Robertson Research International
Ty'n-y-Coed, Llanrhos, Llandudno, Gwynedd, U.K.

INTRODUCTION

The sedimentary successions exposed on the individual islands of the Solomon Islands are highly varied, both in lithological type and age. Since the initial reconnaissance mapping of 1950, age determination of the fossiliferous sedimentary rock has been performed using micropaleontology. Because concepts of biostratigraphic zonation have changed considerably during the last thirty years or so, the purpose of this paper to place the age determinations of varied assignment within a unified scheme of international acceptance. These revisions, it is hoped, will provide a consistent basis for future onshore, and possibly offshore, correlations.

The establishment of a series of zones from the Paleocene to Holocene by Blow (1979) enabled global assignment of deep-marine planktonic foraminiferal-bearing sediments to strictly defined zones. Paleogene zones range from P1 to P22 and extend from the Paleocene to Oligocene, and the Neogene, Zones N4 to N23, extends from the basal Miocene to Pleistocene. Mesozoic planktonic foraminifera are best assigned to a scheme established by Robaszynski and Caron (1979) and Robaszynski et al (1984). Shallow marine successions rarely contain planktonic foraminifera, and within sediments of such environments, larger benthonic foraminifera commonly occur and may be assigned an age through the system of the Letter Stage. The most commonly used scheme is that of Adams (1970), parts of which have subsequently been revised (Adams, 1983).

Inferred paleobathymetry based upon the composition of foraminiferal assemblages is given in terms of inner neritic (0-100 m), outer neritic (100-200 m), and bathyal (200 m and deeper).

Each island, or island group will be dealt with individually, and the ages will be examined within the context of the existing stratigraphic nomenclature. The most comprehensive illustration of these units and their ages is that presented by Pound (1986, Figure 1), and should be consulted; the islands within the Eastern Outer Islands group are not, however, included within Pound's contribution. The islands will be described in southeastward order, commencing with the Shortland Islands and ending with the Eastern Outer Islands (Figure 1).

The sources of the micropaleontological data are varied, but mostly derived from the published works of Coleman (1963, 1965, 1966, 1968), Coleman and McTavish (1964), McTavish (1966, 1968), and Coleman, McGowran, and Ramsay (1978) together with determinations made by the author in 1970-1981 during his assignment with the Geological Survey of the Solomon Islands as geologist/micropaleontologist (Hughes, Procter, and Hackman, 1975; Hughes, 1977, 1981a-f, 1982; and Hughes, unpublished). Published and unpublished details of samples collected for micropaleontological analysis from all of the islands and much of the unpublished faunal data from New Georgia and Santa Isabel are held in the files of the Geological Survey of the Solomon Islands.

It is beyond the constraints of this paper to provide tables displaying all of the foraminiferal species recovered from any single formation. Instead, the writer has restricted the named species to those which have direct age or paleobathymetric value. The information is presented in brief style so that the conclusions, and reasons for their achievement, can be easily read and understood by non-micropaleontologists.

THE SHORTLAND ISLANDS

Summary of the Geology

The Shortland Islands consist of numerous small islands, the largest of which are, from northeast to southwest, Fauro, Alu and Mono (Figure 1).

On Fauro, the Masamasa Volcanics, of probable Miocene age, are overlain by the Koria Sandstone, of

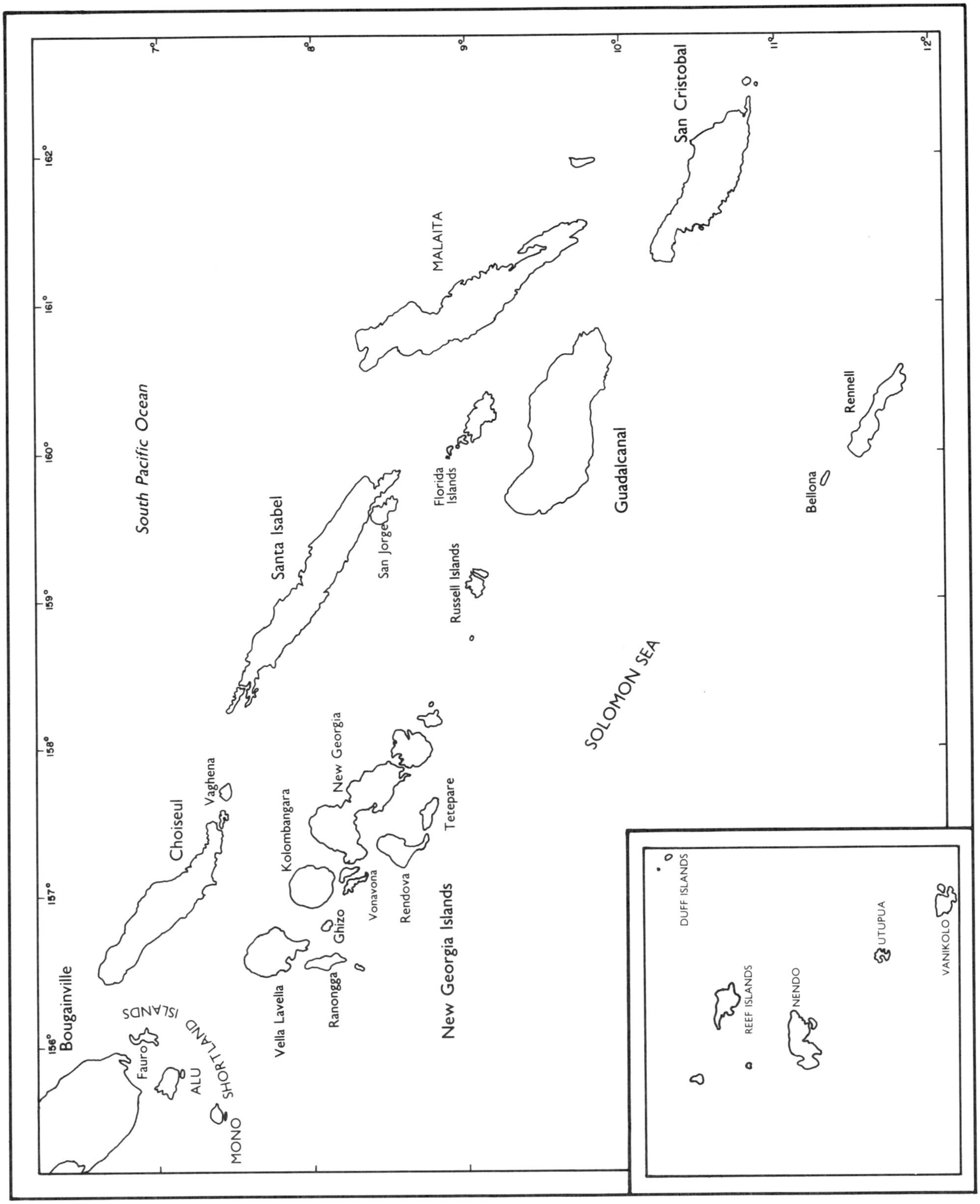

Figure 1. Map showing islands mentioned in the text.

probable Pliocene age. The Togha Pyroclastics overlie the Koria Sandstone and are of Pliocene-Pleistocene age and represent a late volcanic phase which is associated with the Oema Andesite and probably with the Fauro Dacite.

On Alu, the local basement is composed of the Alu Basalts, together with the intrusive Hisiai Complex, of probable Miocene age, and is overlain by the Kulitanai Siltstones, of early Pliocene age. These units are overlain by the late Pliocene to Pleistocene Lofang Limestones and the Pleistocene Toanapina Conglomerate.

Basalts form a local basement on Mono Island, and are considered to be of Miocene age. These basalt flows are overlain by the Kughala Limestone, of late Miocene age, and by the Mono Siltstone, of early and late Pliocene age. The Kolehe Limestone is a facies variation of the Mono Siltstone, and is also of Pliocene age. The Bare Reef Limestones are of possible Pliocene to Pleistocene age and overlie the Mono Siltstones. Pleistocene Soanatalu Silts and Totoko Reef Limestones are the youngest sediments in the succession.

Detailed information is provided in Hughes (1981f, 1982), Ridgway and Coulson (in press), Turner and Hughes (1982), and Turner and Ridgway (1982).

Island: Fauro

Stratigraphic Unit: Koria Sandstone.
Age: Pliocene(?).
Evidence: Regional stratigraphy.
Paleoenvironment: Inner neritic.

Island: Alu (Shortland Island)

Stratigraphic Unit: Kulitanai Siltstones.
Age: Early Pliocene (Zone N19/20).
Evidence: *Globorotalia crassaformis, Globigerinoides fistulosus, Sphaeroidinella dehiscens.*
Paleoenvironment: An upper bathyal environment is suggested by the presence of the following benthonic foraminifera, in addition to the deep-marine planktonic forms: *Ceratobulimina pacifica, Chilostomella ovoidea, Dentalina leguminiformis,* and *Hoeglundina elegans.*

Stratigraphic Unit: Lofang Limestones.
Age: Late Pliocene (Zone N21).
Evidence: *Globigerinoides fistulosus.*

Age: Pleistocene.
Evidence: *Calcarina spengleri.*
Paleoenvironment: An inner neritic to outer neritic, partly reefal depositional environment is inferred.

Stratigraphic Unit: Toanapina Conglomerate.
Age: Pleistocene.
Evidence: Stratigraphic position above the Lofang Limestones.
Paleoenvironment: Shallow inner neritic.

Island: Mono

Stratigraphic Unit: Kughala Limestone.
Age: Late Miocene (Zone N18).
Evidence: *Orbulina universa, Globorotalia pseudopima,* absence of Pliocene restricted forms, and stratigraphic position beneath lower Pliocene Mono Siltstones
Paleoenvironment: An inner neritic environment is based upon the poor character of the planktonic assemblage and the presence of *Amphistegina* species.

Stratigraphic Unit: Mono Siltstone.
Age: Early Pliocene (Zone N19/20).
Evidence: *Sphaeriodinella dehiscens, Pulleniatina obliquiloculata* (dextrally coiled).

Age: Late Pliocene (Zone N21).
Evidence: *Globorotalia tosaensis.*
Paleoenvironment: An upper bathyal environment is based upon the character of the planktonic assemblage together with the presence of *Ceratobulimina pacifica.*

Stratigraphic Unit: Kolehe Limestones.
Age: Pliocene (Zone N19/20-N21).
Evidence: *Globigerinoides fistulosus, Globorotalia multicamerata.*
Paleoenvironment: Outer neritic.

Stratigraphic Unit: Bare Reef Limestones.
Age: Pliocene (?) to Pleistocene.
Evidence: *Cycloclypeus carpenteri.*
Paleoenvironment: Inner neritic, reefal.

Stratigraphic Unit: Soanatalu Silts.
Age: Pleistocene.
Evidence: *Calcarina spengleri.*
Paleoenvironment: Inner neritic.

Stratigraphic Unit: Totoko Reef Limestones.
Age: Pleistocene.
Evidence: Stratigraphic position above the Soanatalu Silts.
Paleoenvironment: Inner neritic, reefal.

THE NEW GEORGIA ISLAND GROUP

Summary of the Geology

The New Georgia island group consists of individual and coalesced volcanic centers, together with a minor proportion of sedimentary rocks of mostly late Miocene to Pleistocene age. Seemingly anomalous early Miocene strata have been recently identified on Ghizo and possibly on Kolombangara (Figure 1), but their relationship with the surrounding succession is not clear.

On Kohinggo Island, the Pliocene and Pleistocene Kohinggo Formation represents an unusual clastic unit which forms the core of an extensive shallow-marine Pleistocene reefal limestone sequence consisting of the

Vizolo Arenites and the Raised Reef Limestones. Tetepare Island consists of a volcanic basement of probable Pliocene and Pleistocene age, which is overlain by the Pleistocene sediments of the Tetepare Formation.

On Vella Lavella, the coalesced volcanic centers form a basement upon which the Pliocene to Pleistocene sediments of the Oula Formation were deposited.

Kolombangara Island is composed of the Kolombangara Volcanic Formation which contains rare lenses of limestone.

Ghizo Island contains a complex succession in which the Mari Limestone and Kaindo Arenite Members of the Kongga Formation are closely associated with volcanic rocks. The generally accepted late Miocene to Pliocene age for this formation has recently been discovered to possibly extend into the early Miocene.

Ranongga Island contains a volcanic basement upon which rest the Vori and Lea Members of the Pliocene to Pleistocene Tuara Sedimentary Formation.

Rendova Island consists mostly of volcanic rocks, but these are locally overlain by the Pliocene to Pleistocene sediments of the Tetepare Formation.

New Georgia itself contains rare sedimentary rocks of Pleistocene age, which were obtained from water-well bores drilled in the Noro region. The microfauna in these rocks closely resembles that of the Vizola Arenites of Kohinggo but is not described in this study.

Little is published on the regional geology of the New Georgia island group, but regional maps prefixed NG, published by the Solomon Islands Geological Survey, contain a brief summary of the geology of individual islands. Aspects of the Tetepare Formation are considered by Hughes, Varol, and Dunkley (in press). General descriptions of the geology were given by Hughes (1982), Coulson and Vedder (1986), and Pound (1986).

Island: Kohinggo

Stratigraphic Unit: Kohinggo Formation.
Age: Late Pliocene to early Pleistocene (Zone N21-N22).
Evidence: *Globorotalia tosaensis*.
Paleoenvironment: The presence of the deep marine forms *Globorotalia tosaensis*, *Globorotalia menardii*, and *Sphaeroidinella dehiscens* indicates water depths in excess of 100 m. The scarcity of benthonic foraminifera suggests adverse bottom conditions.

Stratigraphic Unit: Vizolo Arenites Formation.
Age: Pleistocene.
Evidence: *Baculogypsina sphaerulata* and *Calcarina spengleri*.
Paleoenvironment: The presence of *Baculogypsina sphaerulata*, *Calcarina spengleri*, *Heterostegina suborbicularis*, *Peneroplis planatus*, *Peneroplis pertusus*, *Alveolinella quoyi*, *Operculinella venosa*, *Operculina complanata*, *Marginopora vertebralis*, *Sorites marginalis*, *and Cycloclypeus* sp. suggests deposition within a back-reef setting.

Stratigraphic Unit: Raised Reef Limestones.
Age: Pleistocene.
Evidence: Regional stratigraphy and stratigraphic position above the Kohinggo Formation, which is late Pliocene to early Pleistocene in age.
Paleoenvironment: The combined presence of species of *Cycloclypeus* and *Alveolinella*, together with the rare planktonic foraminifera *Orbulina universa* and *Globorotalia tumida*, suggests the outer margin of a carbonate shelf.

Island: Tetepare

Stratigraphic Unit: Tetepare Formation.
Age: Pleistocene (Zone N22-N23).
Evidence: *Globorotalia truncatulinoides*.
Paleoenvironment: A deep-marine, bathyal, depositional environment is indicated by the presence of *Melonis pompilioides*, *Ehrenbergina pacifica*, *Ceratobulimina pacifica*, *Uvigerina hispida*, and *Laticarinina pauperata*.

Island: Vella Lavella

Stratigraphic Unit: Oula Formation.
Age: Pleistocene (Zone N22).
Evidence: *Globorotalia truncatulinoides*, *Globorotalia tosaensis*, *Neogloboquadrina acostaensis*.

Age: Pleistocene.
Evidence: *Calcarina spengleri*, *Calcarina hispida*.

Age: Late Pliocene (Zone N21).
Evidence: *Globigerinoides fistulosus*, *Globorotalia tosaensis*, *Sphaeroidinella dehiscens*.
Paleoenvironment: A deep-marine depositional environment is inferred for the Oula Formation, well within the bathyal regime. Deep-marine benthonic species include *Stilostomella abyssorum*, *Planulina ungeriana*, *Planulina marialana*, *Uvigerina asperula*, *Dentalina subsoluta*, and numerous *Nodosaria* species. These species, together with a rich and diverse planktonic forminiferal assemblage in which keeled globorotalids are abundant, form a deep-marine autochthonous assemblage in which allochthonous, larger benthonic foraminifera are present. The allochthonous species include *Amphistegina adiata*, *Amphistegina papillosa*, *Operculina* sp., *Elphidium* sp., and miliolids and are considered to have been transported from a shallow-marine, probably reef-associated environment into the deep-marine depositional setting.

Island: Kolombangara

Stratigraphic Unit: Kolombangara Volcanic Formation.
Age: Late Pliocene (Zone N21).
Evidence: *Globorotalia tosaensis, Globigerinoides fistulosus, Pulleniatina obliquiloculata* (sinistrally coiled).
Paleoenvironment: The planktonic assemblage was recovered from a pelagic limestone bed within volcanic arenites.

Age: Miocene(?) to early Pliocene.
Evidence: *Lepidocyclina* sp.
Paleoenvironment: The isolated sample containing this genus may denote the presence of pre-volcanic sedimentary rocks. If *in situ*, the sample represents a shallow-marine environment.

Stratigraphic Unit: Reefal Limestones.
Age: Pleistocene.
Evidence: *Calcarina spengleri*.
Paleoenvironment: A shallow-neritic environment is inferred to have included a fringing reef with associated lagoonal sediments.

Island: Ghizo

Stratigraphic Unit: Konggu Formation, Mari Limestone Member and Kaindo Arenite Member.
Age: Early Miocene.
Evidence: *Lepidocyclina* sp. cf. *japonica* (upper[?] Te). *Globigerina tripartita, Globoquadrina praedehiscens* (Zone N4).
Paleoenvironment: The larger benthonic foraminifera have been transported from their shallow-marine environment into a bathyal environment containing a rich planktonic foraminiferal assemblage and deep-marine benthonic forms which include *Gyroidina soldani altiformis*.
Note: This age determination resulted from recent re-examination and intensive sample disaggregation; the stratigraphic relationship of the single sample containing this assemblage is not known.

Age: Late Miocene (Zone N17-N18).
Evidence: *Globorotalia tumida* group and *Sphaeroidinellopsis seminulina*, in the absence of *Sphaeroidinella* spp. and *Pulleniatina primalis*.
Paleoenvironment: An upper bathyal environment is indicated. Deep marine benthonic foraminifera include *Planulina wuellerstorfi, Ehrenbergina bicornis, Uvigerina crassicostata, Oridorsalis umbonata*, and costate nodosarids.

Age: Early Pliocene (Zone N19/20).
Evidence: *Sphaeroidinella dehiscens immatura, Pulleniatina obliquiloculata, Globoquadrina altispira*.
Paleoenvironment: (As in late Miocene).

Island: Ranongga

Stratigraphic Unit: Vori Member of theTuara Sedimentary Formation.
Age: Late Pliocene (Zone N21).
Evidence: *Globorotalia tosaensis, Globigerinoides fistulosus, Pulleniatina obliquiloculata* (dextrally coiled), *Sphaeroidinella dehiscens*.
Paleoenvironment: An upper bathyal environment is inferred, based upon the presence of a rich planktonic foraminiferal assemblage associated with deep-water benthonic forms which include *Oridorsalis umbonatus, Melonis affinis, Globocassidulina subglobosa*, costate nodosarids, *Pullenia bulloides*, and *Pyrgo denticulata*.

Stratigraphic Unit: Lea Member of the Tuara Sedimentary Formation.
Age: Early Pleistocene (Zone N22).
Evidence: *Globorotalia tosaensis, Globorotalia truncatulinoides*, rare *Globigerinoides fistulosus* (basal N22).

Age: Late Pleistocene (Zone N23).
Evidence: *Globorotalia truncatulinoides*. Absence of *Globorotalia tosaensis*.
Paleoenvironment: An upper bathyal environment is inferred, based upon an assemblage similar to the one recovered from the Vori Member but with an increased proportion of shallow marine forms, including *Elphidium crispum*.

Island: Rendova

Stratigraphic Unit: Tetepare Formation.
Age: Late Pliocene (Zone N21).
Evidence: *Pulleniatina obliquiloculata* (dextrally coiled), *Globorotalia tosaensis*.

Age: Early Pleistocene (Zone N22).
Evidence: *Globorotalia tosaensis, Globorotalia truncatulinoides, Pulleniatina finalis*.

Age: Late Pleistocene (Zone N23).
Evidence: *Globorotalia truncatulinoides*. Absence of *Globorotalia tosaensis, Pulleniatina finalis*.
Paleoenvironment: The Tetepare Formation was deposited in an upper bathyal environment, based upon the abundance of deep-marine planktonic foraminifera, together with the presence of deep benthonic forms which include *Planulina wuellerstorfi, Osangularia culter, Oridorsalis umbonatus, Karreriella bradyi*, and *Hoeglundina elegans*.

CHOISEUL

Choiseul consists of an Eocene volcanic and metamorphic basement, upon which lie the Koloe Breccias and the fine-grained clastic sediments of the Mole Formation of Oligocene to early Pliocene age. Locally, the Mole Formation contains the basal Vuasa Limestone Member of late Oligocene to early Miocene age.

Upper Pliocene Pemba Beds overlie the Mole Formation and contain the Sui Calcarenite and Mbani Calcisiltite Members. The Vaghena Formation occurs only on Vaghena Island off the southeastern tip of Choiseul. The Maetambe Volcanics and Nukiki Limestones represent the Pleistocene deposits, which overlie both the Mole and Pemba formations.

Details of the geology of Choiseul are described by Ridgway and Coulson (in press), as well as in Hughes (1982), Turner and Hughes (1982), Coulson and Vedder (1986), and Pound (1986).

Stratigraphic Unit: Vuasa Limestone Member of the Mole Formation.
Age: Late Oligocene (lower Te).
Evidence: *Lepidocyclina (Eulepidina) ephippioides, Heterostegina borneensis.*

Age: Early Miocene (upper Te).
Evidence: *Miogypsinoides thecidaeformis, Miogypsina dehartii, Spiroclypeus higginsi*, and *Spiroclypeus leupoldi.*

Age: Early to middle Miocene (upper Te-lower Tf).
Evidence: *Lepidocyclina sumatrensis, Lepidocyclina irregularis.*
Paleoenvironment: A shallow-marine, reef-associated environment is inferred.

Stratigraphic Unit: Mole Formation.
Age: Late Miocene (N17-N18).
Evidence: *Orbulina universa, Sphaeroidinellopsis seminulina, Globorotalia tumida plesiotumida.*

Age: Early Pliocene (N19/20).
Evidence: *Sphaeroidinella dehiscens, Globigerina venezuelana.*
Paleoenvironment: The clastic sediments of the Mole Formation were deposited in a deep-marine environment within the upper bathyal zone, i.e., deeper than 200 m. Deep-marine benthonic foraminiferal indicators include *Dentalina legumen, Glandulina aequalis, Globocassidulina subglobosa, Globulina gibba, Karreriella novangliae, Lenticulina* sp., and *Melonis pompilioides.*

Stratigraphic Unit: Sui Calcarenite Member of the Pemba Formation.
Age: Early Pliocene (Zone N19/20).
Evidence: *Sphaeroidinella dehiscens immatura, Globoquadrina altispira altispira, Globigerinoides fistulosus, Globorotalia crassaformis.*

Stratigraphic Unit: Mbani Calcisiltite Member of the Pemba Formation.
Age: Late Pliocene (Zone N21).
Evidence: *Globorotalia tosaensis, Globorotalia dutertrei, Globigerinoides fistulosus, Globorotalia crassaformis.*
Paleoenvironment: An upper bathyal depositional environment is concluded for the Pemba Formation, based upon the high planktonic to benthonic ratio and the following deep-marine benthonic foraminifera: *Dentalina legumen, Globocassidulina subglobosa, Ceratobulimina pacifica, Gyroidina acuta, Hoeglundina elegans, Karreriella* sp., *Melonis pompilioides, Nodosaria filiformis, Nodosaria radicula glanduliniformis, Pyrgo denticulata*, and *Stilostomella abyssorum.*

Stratigraphic Unit: Vaghena Formation.
Age: Early Pliocene.
Evidence: *Globoquadrina venezuelana, Sphaeroidinella dehiscens immatura.*
Paleoenvironment: An upper bathyal depositional environment is suggested by the presence of the following benthonic foraminifera: *Chilostomella oolina, Dentalina aciculata, Dentalina abyssorum, Ehrenbergina trigona, Glandulina laevigata, Globocassidulina subglobosa, Hoeglundina elegans, Globulina gibba*, and *Saracenaria latifrons.*

Stratigraphic Unit: Nukiki Limestone Formation.
Age: Pleistocene.
Evidence: *Baculogypsina sphaerulata, Calcarina spengleri.*
Paleoenvironment: A shallow-marine environment associated with a fringing reef is inferred.

THE FLORIDA ISLANDS

A basement complex consisting of Vatilay Gabbro, Kasika Metabasics and Nggela Ultramafics is overlain by the Ghumba and Siota Beds of Oligocene to Pliocene age and early Miocene age, respectively. The Florida Limestone is of late Miocene age and possibly ranges into the Pleistocene.

The results of recent geologic mapping have not yet been published. The available published data are included in Hughes (1982), Coulson and Vedder (1986), Pound (1986), Neef (1979) and Turner and Hughes (1982).

Stratigraphic Unit: Siota Beds.
Age: Early Miocene (upper Te).
Evidence: *Eulepidina ephippioides, Lepidocyclina martini.*
Paleoenvironment: An upper bathyal setting is inferred, and the allochthonous large foraminifera are present within clastic strata which contain early Miocene nannofossils.

Stratigraphic Unit: Ghumba Beds.
Age: Late Oligocene (Zone P22).
Evidence: *Globigerinita dissimilis, Globigerina ciperoensis, Globigerina tripartita.*

Age: Middle Miocene (Zone N9).
Evidence: *Orbulina universa, Globigerinoides sicanus.*

Age: Late Miocene to early Pliocene (Zone N17-N19/20).
Evidence: *Pulleniatina primalis.*
Paleoenvironment: An outer neritic to upper bathyal environment of deposition is concluded.

Stratigraphic Unit: Florida Limestone.
Age: Late Miocene (Zone N17-N18).
Evidence: *Globigerina venezuelana, Sphaeroidinellopsis seminulina,* and stratigraphic position.
Paleoenvironment: The presence of a rich planktonic foraminiferal assemblage combined with the presence of uvigerinids, *Melonis affinis*, and *Lenticulina* sp. suggests a depositional environment within the outer neritic to possibly upper bathyal regime.

SANTA ISABEL

The island of Santa Isabel consists of a Cretaceous basement complex, which includes the Vitoria Microgabbro, the San Jorge Kolomola Ultramafite, and the Sigana Volcanics, together with metabasic and metasedimentary rocks. The oldest unmetamorphosed sedimentary rocks belong to the Tanakau Group and range in age from late Paleocene to Pliocene.

The results of recent detailed mapping are not yet available. The publications of most value to date include Stanton (1961), Hughes (1982), Turner and Hughes (1982), Coulson and Vedder (1986), and Pound (1986). Generic and specific names in parenthesis are those assigned by Blow (1979) and in current usage.

Stratigraphic Unit: Tanakau Group.
Age: Late Paleocene (Zone P3-P4).
Evidence: *?Subbotina (Globigerina) pseudobulloides, ?Subbotina (Globigerina) inconstans, Truncorotaloides (Globigerina) angulata/conicotruncana, Truncorotaloides (Globorotalia) acuta, Globorotalia* cf. *chapmani*.

Age: Late Paleocene (Zone P5-P6).
Evidence: *Truncorotaloides (Globigerina) velascoensis, Truncorotaloides (Globorotalia) aequus (aequa), Truncorotaloides (Globorotalia)* cf. *acutispira, Globorotalia* cf. *imitata, Globorotalia laevigata.*

Age: Late Eocene to early Oligocene.
Evidence: *Chiloguembelina* spp., *Pseudohastigerina* spp., *Globoquadrina* spp.

Age: Late Oligocene to early Miocene (lower Te-upper Te).
Evidence: *Lepidocyclina* spp., *Heterostegina* cf. *borneensis, Spiroclypeus* spp.

Age: Middle Miocene.
Evidence: *Orbulina universa, Sphaeroidinellopsis* sp., and stratigraphic position.
Age: Late Miocene (Zone N18).
Evidence: *Pulleniatina primalis, Sphaeroidinellopsis seminulina, Globorotalia tumida.*

Age: Pliocene.
Evidence: *Sphaeroidinella dehiscens*; also see Varol (this volume).
Paleoenvironment: The rich planktonic foraminiferal content of limestone samples examined from the Tanakau Group suggests very deep marine conditions. Rare deep-marine benthonic species include *Planulina wuellerstorfi, Pleurostomella alternans, Pullenia bulloides, Gyroidina soldani, Globocassidulina subglobosa, Saracenaria italica, Nodosaria vertebralis,* and *Uvigerina* spp.

GUADALCANAL

The island of Guadalcanal contains one of the most varied stratigraphic successions within the Solomons Islands. This variation resulted from lateral changes of lithological facies across the island from west to east. The basement complex includes the Mbirao Metabasics, Mbirao Lavas, Ghausava, Suta, and Marau Ultramafics, Poha Diorite, Guadalcanal Gabbro, and Suta Volcanics.

The Tetekanji Limestones occur within this basement complex. The Kavo Greywacke Beds form the earliest clastic strata to overlie the basement, together with limestone of the Mbetilonga Limestone. These units are, in turn, overlain locally by the Toni Formation, Lungga Beds, and Honiara Beds of clastic character, and elsewhere by the reefal Mbonehe Limestone.

The geology of central and eastern Guadalcanal was described in detail by Hackman (1980), and western Guadalcanal by Hughes (1977). Regional descriptions include Hughes (1982), Turner and Hughes (1982), Coulson and Vedder (1986), and Pound (1986).

Stratigraphic Unit: Tetekanji Limestones
Age: Cretaceous(?).
Evidence: Regional stratigraphic relationships.
Paleoenvironment: Bathyal(?).

Stratigraphic Unit: Kavo Greywacke Beds.
Age: Early(?) to middle Miocene.
Evidence: *Orbulina* sp. in some samples.
Paleoenvironment: An outer neritic to bathyal environment is inferred.

Stratigraphic Unit: Mbetilonga Limestone of the Mbetilonga Group.
Age: Early Miocene (upper Te).
Evidence: *Miogypsina* spp. and *Miogypsinoides* spp., *Spiroclypeus higginsi, Spiroclypeus leupoldi, Miogypsina indonesiensis, Miogypsina polymorphina, Miogypsina thecidaeformis, Miogypsinoides dehaarti, Lepidocyclina japonica, Lepidocyclina sumatrensis, Heterostegina borneensis.*
Paleoenvironment: A shallow inner neritic, reefal environment is inferred.

Stratigraphic Unit: Mbonehe Limestone.
Age: Early Miocene (upper Te).
Evidence: *Lepidocyclina martini, Lepidocyclina ferreroi, Cycloclypeus* cf. *indopacificus Cycloclypeus martini.*
Paleoenvironment: A shallow inner neritic, reefal environment is inferred.

Stratigraphic Unit: Charikangge Grit of the Toni Formation.
Age: Late Miocene (Zone N17-N18).
Evidence: *Pulleniatina primalis, Sphaeroidinellopsis* sp. in the absence of *Sphaeroidinella* spp.
Paleoenvironment: Deposition in an outer neritic environment is implied, into which there was a common influx of reworked early and middle Miocene larger foraminifera.

Stratigraphic Unit: Lungga Beds.
Age: Early Pliocene (Zone N19/20).
Evidence: *Sphaeroidinella dehiscens immatura, Globorotalia plesiotumida.*

Age: Late Pliocene (Zone N21).
Evidence: *Globorotalia tosaensis.*
Paleoenvironment: An upper bathyal environment is inferred based upon the presence of deep-marine benthonic foraminifera, including *Favocassidulina favus, Melonis pompilioides,* costate nodosarids, spinose uvigerines, and *Laticarinina pauperata.*

Stratigraphic Unit: Honiara Beds.
Age: Pleistocene.
Evidence: *Calcarina spengleri, Baculogypsina sphaerulata.*
Paleoenvironment: A shallow inner neritic, lagoonal environment is inferred.

MALAITA

The island of Malaita has comparatively simple stratigraphy, consisting of the Cretaceous basaltic basement of the Fo'onda Clastics and Fiu Lavas of the Alite Volcanics; these are overlain successively by the Kwara'ae Mudstones and the 'Are'are Limestones and their lateral equivalents. The Hauhui Conglomerates, together with the Tomba Silts and Rokera Limestones occur in the uppermost part of the sequence and are of Pleistocene age.

The detailed geology of southern Malaita was described by Hughes and Turner (1976, 1977), as well as by Hughes (1982), Turner and Hughes (1982), Coulson and Vedder (1986), and Pound (1986).

Stratigraphic Unit: Kwara'ae Mudstones of the Malaita Sediments.
Age: Late Cretaceous and/or older.
Evidence: Stratigraphic position beneath the 'Are'are Limestones.

Stratigraphic Unit: 'Are'are Limestones (Lower Cretaceous to Pleistocene) of the Malaita Sediments.
Age: Early Cretaceous (upper Albian).
Evidence: *Planomalina buxtorfi, Rotalipora ticinensis.*

Age: Late Eocene (Zones P15-P16).
Evidence: *Globigerinatheka semiinvoluta,* with *Globigerinatheka index, Globorotalia cerroazulensis,* and *Pseudohastigerina micra.*

Age: Early Oligocene (Zones P18-P20).
Evidence: *Globigerina ampliapertura,* in the absence of Eocene forms.

Age: Late Oligocene (Zone P21-P22).
Evidence: *Globorotalia opima nana, Globorotalia kugleri,* in the absence of *Globigerinoides* spp.

Age: Early Miocene (Zones N4-N6).
Evidence: *Globigerinoides trilobus, Globiberinoices obliquus, Globigerinita dissimilis,* and *Globigerinita unicava.*

Age: Middle Miocene (Zones N9).
Evidence: *Orbulina universa, Praeorbulina transitoria.*

Age: Middle Miocene (Zones N10-N15).
Evidence: *Globorotalia lenguaensis.*

Age: Late Miocene (Zones N17-N18).
Evidence: *Pulleniatina obliquiloculata,* in the absence of *Globorotalia tumida* and *Sphaeroidinella dehiscens.*

Age: Pliocene (Zones N19/20-N21).
Evidence: *Globigerinoides fistulosus, Globorotalia puncticulata,* and *Globorotalia tosaensis.*

Age: Pleistocene (Zones N22-N23).
Evidence: *Globorotalia truncatulinoides, Baculogypsina sphaerulata* (larger foraminifera).
Paleoenvironment: No benthonic foraminifera have been recorded from the Cretaceous to Miocene limestone succession, but the planktonic foraminiferal assemblage contains a sufficient proportion of deep-marine forms to suggest very deep marine conditions of deposition. The close association between the oldest limestones and

chert beds, together with their stratigraphic position above the radiolarian ooze known as the Kwara'ae Mudstones, suggest a depth of 4,000 m, using the estimated calcite compensation depth.

The Pliocene limestones contain a similarly deep-marine planktonic assemblage, together with deep-marine benthonic foraminifera, which include *Favocassidulina favus*, *Ceratobulimina pacifica*, and *Stilostomella abyssorum*, and which indicate water depths between 500 m and 2,000 m.

A Pleistocene assemblage contains deep-marine *Nodosaria* spp. together with allochthonous, or shallow-marine derived, *Baculogypsina sphaerulata*, *Cellanthus craticulatus*, and *Amphistegina radiata*. Common keeled globorotalids in this sample, together with *Pulleniatina* spp. indicate water depths in excess of 200 m.

Stratigraphic Unit: Hauhui Conglomerates.
Age: Pleistocene.
Evidence: *Calcarina spengleri*.
Paleoenvironment: A shallow-marine, high-energy environment is suggested by the coarse clastic lithology and the presence of an exclusively benthonic foraminiferal assemblage containing *Calcarina spengleri*, *Spiroloculina* sp. *Operculinella venosa*, and *Operculina complanata*.

SAN CRISTOBAL

The Warahito Lavas form the basement on this island, together with ultramafic rocks and rare occurrences of the pelagic Hautarau Limestones. These units are overlain by the Paleogene Ravo Limestones and reefal limestones of the Cristobal Group. Pliocene sediments overlie these limestone units, although their spatial relationship is not fully understood owing to incomplete mapping.

Details of the geology of western San Cristobal were described by Jeffrey (1975a, 1975b), and general descriptions were given in Hughes (1982), Coulson and Vedder (1986), and Pound (1986).

Stratigraphic Unit: Ravo Limestones.
Age: Paleocene to Eocene.
Evidence: *Globorotalia* spp. "of Paleocene-Eocene affinity"
Paleoenvironment: An upper bathyal environment is inferred.

Stratigraphic Unit: Cristobal Group.
Age: Early Miocene (upper Te).
Evidence: *Lepidocyclina martini*.
Paleoenvironment: An inner(?) neritic environment is inferred.

Age: Pliocene.
Evidence: *Pulleniatina obliquiloculata* (dextrally coiled), *Globorotalia crassaformis*, *Globigerinoides fistulosus*.
Paleoenvironment: An outer(?) neritic environment is inferred.

EASTERN OUTER ISLANDS

Of the numerous islands that constitute this volcanic group, sedimentary rocks have been mapped only on Nendo where they are closely associated with the early Miocene Mengalu Volcanics, and the overlying Malue, Notapu, Tepiai, and Kanenggo Beds and Luemaonda Reefal Limestones of Miocene to Pliocene age.

Detailed descriptions of the geology of Nendo and other islands in the group were provided by Hughes, Craig, and Dennis (1981).

Island: Nendo (Santa Cruz)

Stratigraphic Unit: Mengalu Volcanics.
Age: Early Miocene (upper Te).
Evidence: *Eulepidina ephippioides*, *Spiroclypeus margaritatus*, *Heterostegina borneensis*, *Lepidocyclina* sp. cf. *sumatrensis* and *Miogypsina* sp. cf. *thecidaeformis*. These forms constitute an allochthonous reefal assemblage within the submarine volcaniclastics.

Age: Late Oligocene or early Miocene (26±2 Ma).
Evidence: Potassium-argon radiometric dating.
Paleoenvironment: Submarine flanks of a volcano. Intermittent activity permitted colonization by larger foraminifera, which were transported from a shallow-marine environment to one downslope.

Stratigraphic Unit: Nolua Volcanics.
Age: Early Miocene (upper Te).
Evidence: *Eulepidina ephippioides*.
Paleoenvironment: These volcaniclastics also represent a marine site of deposition and contain both reworked and autochthonous foraminifera. The foraminiferal assemblage is sparser than that recovered from the Mengalu Volcanics.

Stratigraphic Unit: Malue Beds.
Age: Miocene to Pliocene (upper Tf-Tg/h).
Evidence: Poor.
Paleoenvironment: The presence of *Operculina complanata*, *Cycloclypeus* sp. cf. *indopacificus* and *Cycloclypeus* sp. cf. *carpenteri*, together with the common alga *Halimeda* sp., suggests a back-reef environment.

Stratigraphic Unit: Notapu Beds.
Age: Miocene(?) to Pliocene.
Evidence: Regional stratigraphy.
Paleoenvironment: The common occurrence of miliolid foraminifera suggests a moderately shallow marine environment, possibly associated with a reef-flat setting.

Stratigraphic Unit: Luemaonda Reefal Limestones.
Age: Middle Miocene to Pliocene.
Evidence: The presence of *Sphaeroidinella dehiscens* within parts of this stratigraphic unit indicates a Pliocene to Pleistocene age. The presence of *Lepidocyclina* cf. *radiata* elsewhere confirms a Miocene to Pliocene age.

Stratigraphic Unit: Tepiai Beds.
Age: Late Miocene to Pliocene.
Evidence: Regional stratigraphy.
Paleoenvironment: The presence of rare *Cycloclypeus* sp. cf. *carpenteri*, together with the alga *Halimeda* sp. suggests a reefal environment.

Stratigraphic Unit: Kanenggo Beds.
Age: Late Miocene to Pliocene.
Evidence: Regional stratigraphic evidence.
Paleoenvironment: Because turbid conditions precluded the establishment of a microfauna, the environment of deposition is uncertain.

SUMMARY

The data presented here serve as an abstracted and condensed presentation of the ages and paleobathymetry of stratigraphic units along most of the length of the Solomon Islands. It is not within the scope of this paper to review the geodynamic history which would explain the varying stratigraphic sections that are exposed on each of the islands as well as throughout the Solomon Islands region. The list of references cited contains the data sources from which more detailed microfaunal assemblage information may be extracted.

ACKNOWLEDGMENTS

I fully acknowledge the use of published works by Coleman (1963, 1965, 1966, 1978), Coleman and McTavish (1964), McTavish (1966, 1968), Van Deventer and Postuma (1973), and Neef (1979), upon which much of the present work is based. Samples examined by the author would not have been available for study without the assistance of the staff of the Geological Survey of the Solomon Islands. Thanks are here paid to my colleague Dennis Clowser of Robertson Research International, whose careful checking of the typescript has considerably improved the consistency and style of the contribution, and also to Stefanie Taylor for her careful typing. I also wish to acknowledge the assistance of Robertson Research International for the use of their drafting facilities, and also for their permission to publish this paper.

REFERENCES CITED

Adams, C.G., 1970, A reconsideration of the East Indian letter classification of the Tertiary: Bulletin of the British Museum (Natural History), v. 19, n. 3, p. 87-137.

Adams, C.G., 1983, Neogene larger foraminifera, evolutionary and geological events in the context of datum planes, *in* Tsuchi, ed., Final report no. IGCP Project 114, p. 47-67.

Blow, W.H., 1979, The Cainozoic Foraminiferida, v.I. and II: Leiden, E.J. Brill, 1413 p.

Coleman, P.J., 1963. Tertiary larger foraminifera of the British Solomon Islands, southwest Pacific: Micropaleontology, v. 9, no. 1, p. 1-38.

Coleman, P.J., 1965, Assemblages of Tertiary larger foraminifera in the Solomon Islands and New Hebrides archipelago: Annual Report of the New Hebrides Geological Survey, p. 48-51.

Coleman, P.J., 1966, Upper Cretaceous (Senonian) bathyal pelagic sediments with Globotruncana from the Solomon Islands: Journal Geological Society of Australia, v. 13, n. 2, p. 439-447.

Coleman, P.J., 1978, Reflections on outer Melanesian Tertiary larger foraminifera, *in* The Crespin Volume: Bureau of Mineral Resources, Australia, Bulletin 192, p. 31-36.

Coleman, P.J., McGowran, B., and Ramsay, W.R.H., 1978, New, early Tertiary, ages for basal pelagites, northeast Santa Isabel, Solomon Islands (central southwest flank, Ontong Java Plateau): Bulletin Australian Society Exploration Geophysics, v. 9, no. 3, p. 110-114.

Coleman, P.J., and McTavish, R.A., 1964, Association of larger and planktonic foraminifera in single samples from middle Miocene sediments, Guadalcanal, Solomon Islands, southwest Pacific: Journal of the Royal Society of Western Australia, v. 47, pt. 1, p. 13-24.

Coulson, F.I. and Vedder, J.G., 1986, Geology of the central and western Solomon Islands, *in* Vedder, J.G., Pound, J.S., and Boundy, S.Q., eds., Geology and offshore resources of Pacific Island Arcs—central and western Solomon Islands: Houston, Texas, Circum-Pacific Council for Energy and Mineral Resources Earth Science Series, v. 4, p. 59-88.

Hackman, B.D. 1980, The geology of Guadalcanal, Solomon Islands: Institute of Geological Sciences, Overseas Memoir 6, 115 p.

Hughes, G.W., 1977, The geology and foraminiferal micropaleontology of the Lungga and Itina areas, western Guadalcanal, Solomon Islands: Unpubl. doctoral dissertation, University College of Wales, Aberystwyth, United Kingdom.

Hughes, G.W., 1981a, Catalogue of age determination of Solomon Islands rocks: Occasional Paper, Geological Survey, Honiara, Guadalcanal, Solomon Islands, unpaginated.

Hughes, G.W., 1981b, The geology of the Oaka and Siruka Bay area, Choiseul; an explanation of the 1:50,000 geological map sheet CH 10: British Technical Cooperation Western Solomons Mapping Project, Rept. n. 10, 92 p.

Hughes, G.W., 1981c, The geology of the Panggoe area, Choiseul; an explanation of the 1:50,000 geological map sheet CH 7: British Technical Cooperation Western Solomons Mapping Project, Rept. n. 5, 55 p.

Hughes, G.W., 1981d, The geology of the Ririo area, Choiseul; an explanation of the 1:50,000 geological map sheet CH 4: British Technical Cooperation Western Solomons Mapping Project, Rept. n. 5, 55 p.
Hughes, G.W., 1981e, The geology of Vaghena Island, Choiseul; an explanation of the 1:50,000 geological map sheet CH 12: British Technical Cooperation Western Solomons Mapping Project, Rept, n. 12, 59 p.
Hughes, G.W., 1981f, The micropaleontology of the Shortland Islands and Choiseul: Report 17, Western Solomons Mapping Project, Honiara, 42 p.
Hughes, G.W., 1982, Stratigraphic correlation between sedimentary basins of the ESCAP region: v. VIII, Solomon Islands, *in* ESCAP Atlas of Stratigraphy, III, New York, United Nations, Mineral Resources Development Series, n. 48, p. 115-130.
Hughes, G.W., Craig, P.M. and Dennis, R.A., 1981, Geology of the Eastern Outer Islands: Solomon Islands Geological Survey Division Bulletin 4, 108 p. 6 maps.
Hughes, G.W., Procter, W.D., and Hackman, B.D., 1975, North 'Are'are, Malaita geological map sheet ML 12: Geological Survey Division, Solomon Islands Ministry of Natural Resources, scale 1:50,000.
Hughes, G.W. and Turner, C.C., 1976, Geology of southern Malaita: Solomon Islands Geological Survey Bulletin 2, 80 p.
Hughes, G.W., and Turner, C.C., 1977, Upraised Pacific Ocean floor, southern Malaita, Solomon Islands: Geological Society of America Bulletin, v. 88, p. 412-424.
Hughes, G.W., Varol, O., and Dunkley, P.N., (in press), Pleistocene uplift of Tetepare Island, Solomon Group, as implied by foraminiferal and calcareous nannofossil evidence: Royal Society of New Zealand.
Jeffrey, D.H., 1975a, Arosi, San Cristobal geological map sheet SC1: Geological Survey Division, Solomon Islands Ministry of Natural Resources, scale 1:50,000.
Jeffrey, D.H., 1975b, Arosi - West Bauro, San Cristobal geological map sheet SC2: Geological Survey Division, Solomon Islands Ministry of Natural Resources, scale 1:50,000.
McTavish, R.A., 1966, Planktonic foraminifera from the Malaita Group, British Solomon Islands: Micropaleontology v. 12, n. 1, p. 1-36.
McTavish, R.A., 1968, Tertiary faunas of smaller foraminifera from the British Solomon Islands: Geological Records British Solomon Islands, 3, p. 66-71.
Neef, G., 1979, Cenozoic stratigraphy of Small Nggela Island, Solomon Islands—early Miocene deposition in a forearc basin followed by Pliocene patch reef deposition: New Zealand Journal of Geology and Geophysics, v. 22, n. 1, p. 53-70.
Pound, K.S., 1986, Correlation of rock units in the central and western Solomon Islands, *in* Vedder, J.G., Pound, K.S. and Boundy, S.Q., eds., Geology and offshore resources of Pacific Island Arcs—central and western Solomon Islands: Houston, Texas, Circum-Pacific Council for Energy and Mineral Resources Earth Science Series, v. 4, p. 89-97.
Ridgway, J. and Coulson, F.I., eds., (in press), The geology of Choiseul and the Shortland Islands Solomon Islands: Geological Survey Division, Solomon Islands Ministry of Natural Resources, Bulletin 16.
Robaszynski, F., and Caron, M., 1979, Atlas de Foraminifères planctoniques du Crétacé moyen. Parts 1-2: Cahiers Micropaléontol., v. 1-2, p. 1-185 and 1-181.
Robaszynski, F., Caron, M., Gonzalez Donoso, J.M., and Wonders, A.A.H., 1984, Atlas of Late Cretaceous globotruncanids: Revue de Micropaléontologie, v. 26, n. 34, p. 145-305.
Stanton, R.L., 1961, Explanatory notes to accompany a first geological map of Santa Ysabel, British Soloman Islands Protectorate: Overseas Geology and Mineral Resources, v. 8, n. 2, p. 127-149.
Turner, C.C., and Hughes, G.W., 1982, Distribution and tectonic implications of Cretaceous-Quaternary sedimentary facies in Solomon Islands: Tectonophysics, v. 87, p. 127-146.
Turner, C.C. and Ridgway, J., 1982, Tholeiitic, calc-alkaline and alkaline(?) igneous rocks of the Shortland Islands, Solomon Islands: Tectonophysics, v. 87, p. 335-354.
Van Deventer, J., and Postuma, J.A., 1973, Early Cenomanian to Pliocene deep-marine sediments from north Malaita, Solomon Islands: Journal of the Geological Society of Australia, v. 20, pt. 2, p. 145-152.

Vedder, J.G., and Bruns, T.R., editors, 1989, Geology and offshore resources of Pacific island arcs—Solomon Islands and Bougainville, Papua New Guinea Regions: Houston, Texas, Circum-Pacific Council for Energy and Mineral Resources Earth Science Series, v. 12.

CALCAREOUS NANNOFOSSIL STUDY OF THE CENTRAL AND WESTERN SOLOMON ISLANDS

O. Varol
Robertson Research International, Ty'n-y-Coed, Llanrhos, Llandudno, Gwynedd, U.K.

ABSTRACT

A total of 126 samples were examined for their calcareous nannofossil content to determine the age of selected sedimentary formations which outcrop in the central and western Solomon Islands. Calcareous nannofossil Zones NN18 to NN21, of late Pliocene to late Pleistocene age, are assigned to the Tetepare Formation which outcrops on Tetepare and Rendova islands. The Tuara Sedimentary Formation on Ranongga Island is placed in calcareous nannofossil Zones NN16 to NN20, of late Pliocene to late Pleistocene age. Calcareous nannofossil Zones NN1/2 to NN12, of early Miocene to early Pliocene age, are assigned to the Konggu Formation which outcrops on Gizo Island. Sediments ranging in age from Late Cretaceous (*Eiffellithus eximius* Zone) to late Pleistocene (Zone NN20) were identified from the Tanakau Group, which is distributed on Santa Isabel, San Jorge, Sulei, and Karuo islands. A new genus, *Hughesius*, and eight new species, *Hughesius gizoensis*, *Helicosphaera jakubowskii*, *H. magnifica*, *H. girgisii*, *H. mullerae*, *Scyphosphaera praeglobulata*, *S. pseudorecurvata* and *Anacanthoica solomonica* are described. Eleven new combinations are also introduced; these are: *Geminilithella subtilis*, *G. lordii*, *G. petaliformis*, *Calcidiscus tropicus*, *Pontosphaera ovata*, *P. sparsiforatus*, *P. trenus*, *P. vigintiforatus*, *Reticulofenestra producta*, *R. stavensis*, and *R. tenuistriatus*.

INTRODUCTION

The regional geology and stratigraphic correlations of the Solomon Islands were summarized by Hughes (1982). Among the islands discussed by Hughes (1982), Santa Isabel and the New Georgia island group were some of the least studied.

According to Hughes (1982), little is known about the facies relationships and ages of the strata that overlie the volcanic basement complex (Sigana Volcanics) of Santa Isabel. In the New Georgia island group, sedimentary rocks are limited in distribution along the frontal part of the active arc where limestone and arenite beds were dated by Hughes (1982) as Pliocene and Pleistocene based on foraminiferal content. According to Dunkley (1983), Pleistocene coral reefs are present at Ranongga Island. On the uplifted islands of Tetepare and Rendova, sequences of deep-water sediments of Pleistocene age are present (Coulson and Vedder, 1986; Hughes, Varol, and Dunkley, in press). The calcareous nannofossil and foraminiferal contents of sediments on Tetepare Island were studied by Hughes, Varol, and Dunkley (in press) to estimate the rate of Pleistocene uplift of this island.

The aim of the present study is to date selected sedimentary formations in the central and western Solomon Islands using calcareous nannofossils (Figure 1). Emphasis is placed on the New Georgia island group and Santa Isabel. The zonal and age assignment (Figure 2) is given for each sample together with the ranges of stratigraphically important species (Figure 3).

The sample numbers used in this study are catalogued in the Geological Survey, Ministry of Lands and Energy, Honiara, Solomon Islands. Splits of these samples are also stored in Honiara.

STRATIGRAPHY

A biostratigraphic study on samples from the Tetepare Formation, Konggu Formation, Tuara Sedimentary Formation, and Tanakau Group has been carried out using calcareous nannofossils. The results of this study are discussed below with reference to the individual formations.

The absence of certain zones and their assignable age is possibly the result of a condensed sequence, and/or wide

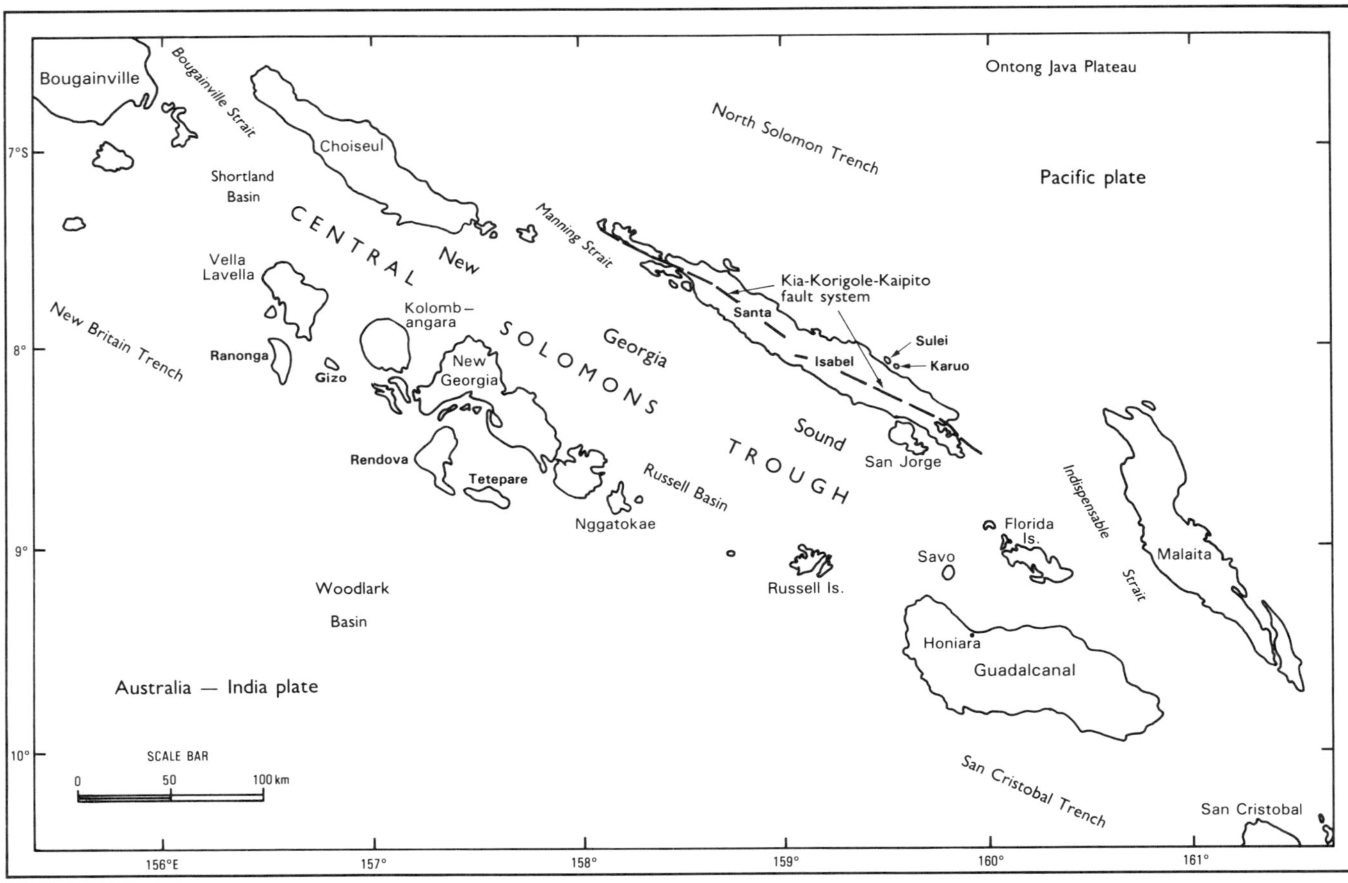

Figure 1. The central and western Solomon Islands.

sampling intervals, and/or faulting, and/or unconformities. Because of the limited availability of geological information, the phenomena responsible for the absent zones are difficult to determine.

Tetepare Formation

This formation has been studied using samples from the islands of Tetepare and Rendova of the New Georgia island group.

Tetepare Island. Twenty-seven samples were examined for their calcareous nannofossil content (see Hughes, Varol, and Dunkley, in press). Samples from TP118 to TP13 (18 samples) are assigned to Zone NN19, samples from TP100 to TP12 (4 samples) are assigned to Zone NN20, and samples from TP97 to TP94 (5 samples) are assigned to Zone NN21. All identified zones are of Pleistocene age.

Rendova Island. Fifteen samples were examined for their calcareous nannofossil content. Zone NN18, of late Pliocene age, is assigned to sample RE317. Zone NN19, of early Pleistocene age, is assigned to samples RE52, RE38, RE318, RE39, RE68, and RE61, whereas Zone NN20, of late Pleistocene age, is assigned to samples RE62, RE37, and RE256.

Konggu Formation

The Konggu Formation outcrops on Gizo Island in the New Georgia island group. Two members, namely the Kaindo Arenite and Mari Limestone Members, were sampled for their calcareous nannofossil content.

Three samples were studied from the Kaindo Arenite Member. Calcareous nannofossil Zone NN10, of late Miocene age, is assigned to samples GZ4 and GZ5. Sample GZ8 is barren of calcareous nannofossils.

Ten samples were studied from the Mari Limestone Member. Sample GZ41 is dated as early Miocene, Subzone NN1/2d, and is the oldest zonal assignment for the sedimentary rocks in the New Georgia island group. Zone NN10, of late Miocene age, is assigned to samples GZ50 and GZ51, whereas Zone NN11a, also of late Miocene age, is assigned to sample GZ38. Zone NN12a, of early Pliocene age, is assigned to samples GZ39 and

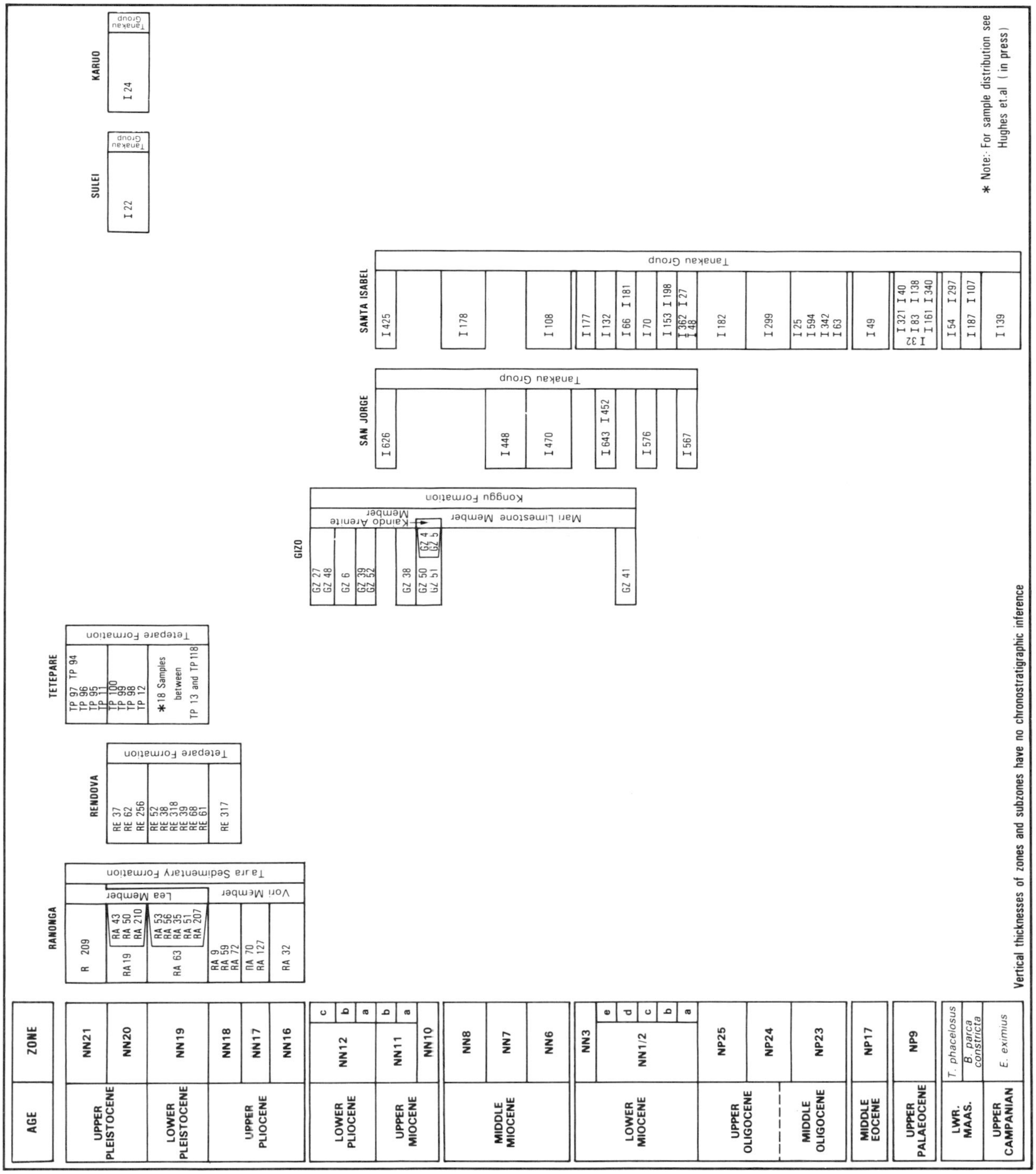

Figure 2. Zonal and age assignment of stratigraphic units for studied samples from the central and western Solomon Islands.

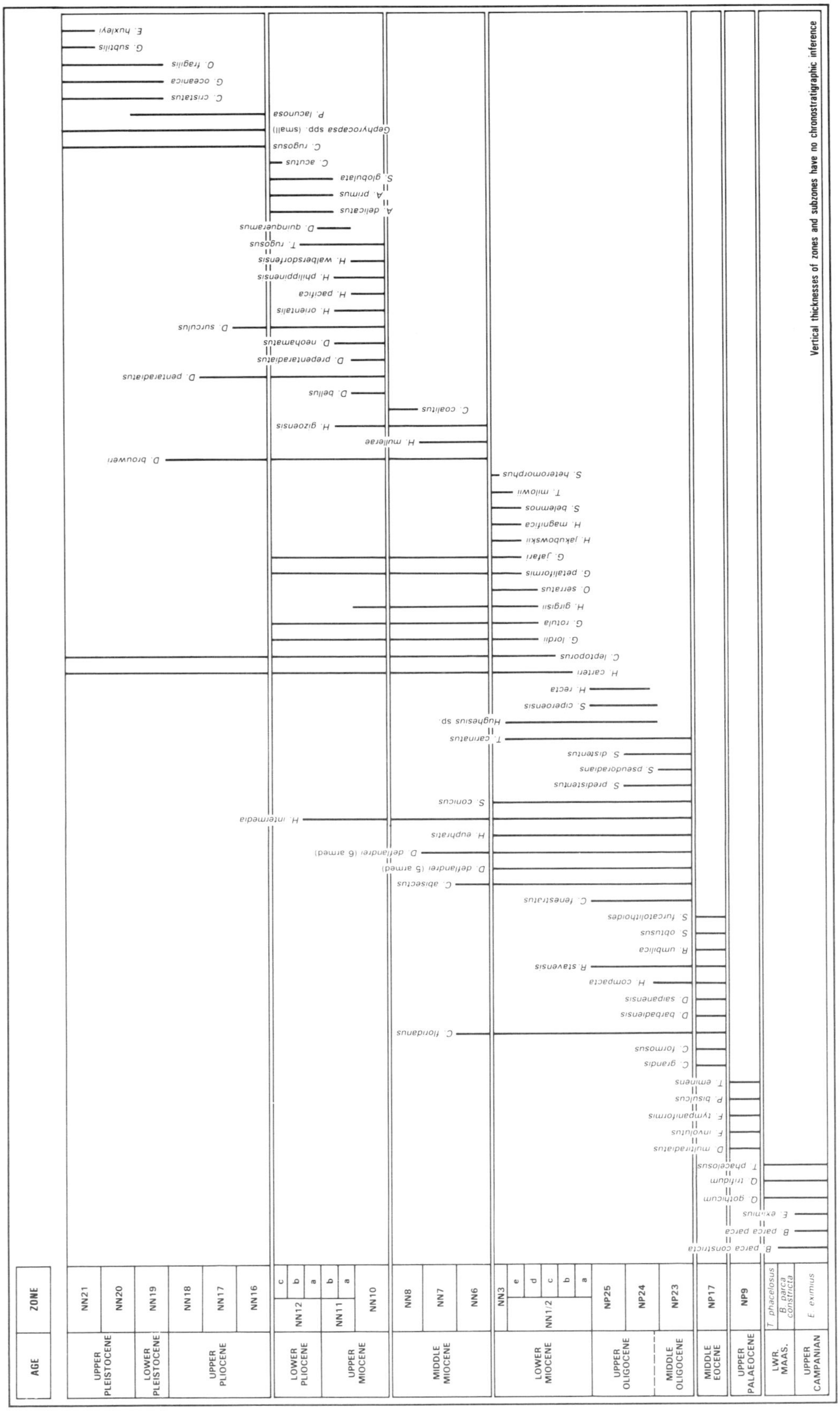

Figure 3. Ranges of stratigraphically useful calcareous nannofossil species found in the central and western Solomon Islands.

GZ52. Sample GZ6 is placed in Subzone NN12b, also of early Pliocene age, and Subzone N12c is assigned to samples GZ27 and GZ48. Sample GZ40 is barren of calcareous nannofossils.

Tuara Sedimentary Formation

The nannofloral content of the Vori and Lea Members of the Tuara Sedimentary Formation was studied on the basis of samples from Ranongga Island in the New Georgia island group.

A total of 12 samples from the Lea Member were examined. The calcareous nannofossil assemblages in samples RA35, RA51, RA53, RA207, and RA56 are assigned to Zone NN19, of early Pleistocene age. Samples RA43, RA50, and RA210, however, are assigned to Zone NN20, of late Pleistocene age. Sample RA209 is also of late Pleistocene age, within Zone NN21. Samples RA54, RA214, and RA52 are barren of calcareous nannofossils.

Eight samples were studied from the Vori Member. Sample RA32 is assigned to Zone NN16, samples RA70 and RA127 are assigned to Zone NN17, and samples RA9, RA59, and RA72 are assigned to Zone NN18. All these zones are of late Pliocene age. Zone NN19, of early Pleistocene age, and Zone NN20, of late Pleistocene age, are represented in samples RA63 and RA19, respectively.

Tanakau Group

The calcareous nannofossil content of samples from the Tanakau Group of Santa Isabel, San Jorge, Sulei, and Karuo islands was examined in order to supplement the sparse biostratigraphic data from the Santa Isabel area.

San Jorge Island. Nine samples were examined for their calcareous nannofossil content. Samples I567 and I576 are assigned to calcareous nannofossil Subzones NN1/2a and NN1/2c of early Miocene age. Samples I643 and I452 are placed in Subzone NN1/2e, also of early Miocene age. Sample I470 is assigned to Zone NN6, of middle Miocene age; and sample I448 contains an assemblage representing Zone NN7, also of middle Miocene age. Sample I626 is assigned to Subzone NN11b, of late Miocene age. Samples I430 and I566 are barren of calcareous nannofossils.

Santa Isabel Island. A total of 41 samples from this island were studied. The *Eiffellithus eximius* Zone of the upper Campanian Stage is identified in sample I139. Samples I107 and I187 are assigned to the *Broinsonia parca constricta* Zone, and samples I54 and I297 are assigned to the *Tranolithus phacelosus* Zone, both of which are in the lower Maastrichtian Stage. Samples I32, I40, I83, I138, I161, I321, and I340 are identified as belonging to Zone NP9, of late Paleocene age. Sample I49 is the only sample assignable to Zone NP17, of middle Eocene age. Samples I25, I594, I342 ,and I63 are assigned to Zone NP23, of middle Oligocene age, whereas samples I299 and I182 are assigned to Zones NP24 and NP25, respectively, both of late Oligocene age. Zone NN1/2, of early Miocene age, is assigned as follows: Subzone NN1/2a is identified in samples I27, I48, and I362, Subzones NN1/2b is in samples I153 and I198, Subzone NN1/2c in sample I70, Subzone NN1/2d in samples I66, I181, and Subzone NN1/2e is in sample I132. Sample I177 is assigned to Zone NN3, of early Miocene age. Zones NN6 and NN8, of middle Miocene age, are assigned to samples I108 and I178, respectively. Sample I425 is assigned to Subzone NN11b, of late Miocene age. Samples I184, I136, I188, I211, I26, I625, I411, I344, and I318 are barren of calcareous nannofossils.

Sulei Island. Only a single sample, I22, was examined; it is assigned to Zone NN20, of late Pleistocene age.

Karuo Island. Sample I24, the only sample examined from this island, is assigned to Zone NN20, of late Pleistocene age.

CALCAREOUS NANNOFOSSIL BIOSTRATIGRAPHY

Eiffellithus eximius, *Broinsonia parca constricta*, and *Tranolithus phacelosus* Zones were identified in the upper Campanian to lower Maastrichtian interval. For the Cenozoic, the calcareous nannofossil results are described mainly with reference to the standard Tertiary and Quaternary zonation scheme of Martini (1971). However, the absence of some of Martini's marker species, possibly owing to environmental and/or preservational factors, has necessitated the selection of alternative species to define some of the zonal boundaries. This modification of the zonation scheme of Martini (1971) is discussed in the remarks section for each zone where applicable. Some of the zones used by Martini have been subdivided into subzones which are described under the appropriate zone headings. These alternative markers are based on worldwide experience and have been selected after an extensive survey of the literature.

The zonation scheme of Okada and Bukry (1980) also has been used worldwide by many nannopaleontologists and was found to be applicable to the samples studied from the Solomon Islands.

Eiffellithus eximius Zone

Stage: Upper Campanian.

Definition: Interval from the first occurrence of *Quadrum trifidum* to the last occurrence of *Eiffellithus eximius*.

Remarks: A poorly preserved, low-diversity calcareous nannofossil assemblage was obtained. Species recovered include: *Quadrum trifidum*, *Quadrum gothicum*, *Broinsonia parca parca*, *Broinsonia parca constricta*,

Calculites obscurus, Arkhangelskiella cymbiformis, Micula staurophora, and *Watznaueria barnesae.* The top of this zone can also be defined by the last occurrence of *Broinsonia parca parca.*
Distribution: Santa Isabel Island (I139).

Broinsonia parca constricta Zone

Stage: Lower Maastrichtian.
Definition: Interval from the last occurrence of *Eiffellithus eximius* to the last occurrence of *Broinsonia parca constricta.*
Remarks: The samples examined within this zone yielded poorly preserved, low-diversity calcareous nannofossil assemblages. Species recovered include: *Quadrum trifidum, Quadrum gothicum, Watznaueria barnesae, Broinsonia parca constricta, Arkhangelskiella cymbiformis,* and *Retecapsa crenulata.*
Distribution: Santa Isabel Island (I107, I187).

Tranolithus phacelosus Zone

Stage: Lower Maastrichtian.
Definition: Interval from the last occurrence of *Broinsonia parca constricta* to the last occurrence of *Tranolithus phacelosus.*
Remarks: Poorly preserved low-diversity nannofossil assemblages were obtained from the samples within this zone. Species recorded include: *Micula staurophora, Watznaueria barnesae, Quadrum trifidum, Quadrum gothicum,* and *Tranolithus phacelosus.*
Distribution: Santa Isabel Island (I54, I297).

Discoaster multiradiatus Zone (NP9)

Age: Late Paleocene.
Definition: Interval from the first occurrence of *Discoaster multiradiatus* to the first occurrence of *Tribrachiatus bramlettei.*
Remarks: Poorly preserved, low-diversity calcareous nannofossil assemblages were obtained from the samples within this zone. *Fasciculithus tympaniformis* is the most abundant species and occurs together with subordinate *Discoaster multiradiatus, Fasciculithus involutus, Fasciculithus bobii, Discoaster mohleri,* and *Prinsius bisulcus.* The last occurrence of *F. tympaniformis* could be used for the approximate upper limit of this zone.
Distribution: Santa Isabel Island (I32, I40, I83, I138, I161, I321, I340).

Discoaster saipanensis Zone (NP17)

Age: Middle Eocene.
Definition: Interval from the last occurrence of *Chiasmolithus solitus* to the last occurrence of *Chiasmolithus grandis* and/or *Sphenolithus obtusus.*
Remarks: A moderately to poorly preserved and low-diversity calcareous nannofossil assemblage was observed in the one sample assignable to this zone. *Reticulofenestra umbilica* and *Sphenolithus moriformis* are common, whereas *Discoaster barbadiensis, Discoaster saipanensis, Chiasmolithus grandis, Sphenolithus obtusus, Sphenolithus furcatolithoides, Reticulofenestra stavensis* (distinguished from *R. bisecta* by being larger than 10µ), *Cyclicargolithus floridanus, Helicosphaera compacta,* and *Coccolithus formosus* are rare. The upper limit of the zone is defined by the last occurrence of *Chiasmolithus grandis* and/or *Sphenolithus obtusus* because of the absence of *Chiasmolithus oamaruensis.* This definition is at variance with Martini's (1971) definition of Zone NP17. The base of this zone can also be approximated by the last occurrence of *Sphenolithus spiniger* and/or *Helicosphaera seminulum.*
Distribution: Santa Isabel Island (I49).

Sphenolithus predistentus Zone (NP23)

Age: Middle Oligocene.
Definition: Interval from the last occurrence of *Reticulofenestra umbilica* to the first occurrence of *Sphenolithus ciperoensis.*
Remarks: Poorly preserved, moderately high diversity calcareous nannofossil assemblages were recovered from the samples within this zone. *Sphenolithus moriformis* and *Cyclicargolithus floridanus* are abundant, whereas *Cyclicargolithus abisectus* (larger than 10µ), *Reticulofenestra stavensis, Clausicoccus fenestratus, Coronocyclus nitescens, Sphenolithus conicus, Sphenolithus distentus,* and *Sphenolithus predistentus* are rare in these samples.
Distribution: Santa Isabel Island (I25, I63, I342, I594).

Sphenolithus distentus Zone (NP24)

Age: Middle to late Oligocene.
Definition: Interval from the first occurrence of *Sphenolithus ciperoensis* to the last occurrence *Sphenolithus distentus.*
Remarks: A moderately well preserved and highly diverse calcareous nannofossil assemblage was obtained from the one sample within this zone. *Cyclicargolithus floridanus, Sphenolithus moriformis,* and

Reticulofenestra dictyoda are clearly the most abundant species. Other species observed include: *Sphenolithus predistentus*, *Sphenolithus distentus*, *Sphenolithus ciperoensis*, *Reticulofenestra stavensis*, and *Clausicoccus fenestratus*. The occurrence of *Sphenolithus pseudoradians* indicates the presence of the lowermost part of Zone NP24. The absence of *Helicosphaera recta* also supports this zonal assignment since *H. recta* usually first appears in the lower part of Zone NP24 just above the evolutionary appearance of *Sphenolithus ciperoensis*. The last occurrence of *Helicosphaera compacta* can be used to subdivide this zone into two subzones.

Distribution: Santa Isabel Island (I299).

Sphenolithus ciperoensis Zone (NP25)

Age: Late Oligocene.

Definition: Interval from the last occurrence of *Sphenolithus distentus* to the last occurrence of *Helicosphaera recta* and/or *Sphenolithus ciperoensis*.

Remarks: A moderately well preserved and diverse calcareous nannofossil assemblage was obtained from the one sample within this zone. *Cyclicargolithus floridanus* together with *Sphenolithus moriformis* dominates the assemblage. Rarely occurring species in this sample include: *Clausicoccus fenestratus*, *Discoaster deflandrei* (5 and 6 armed), *Reticulofenestra stavensis*, *Cyclicargolithus abisectus*, *Sphenolithus ciperoensis*, *Sphenolithus conicus*, and *Helicosphaera recta*.

Distribution: Santa Isabel Island (I182).

Triquetrorhabdulus carinatus Zone/
Discoaster druggii Zone (NN1/NN2)

Age: Early Miocene.

Definition: Interval from the last occurrence of *Helicosphaera recta* and/or *Sphenolithus ciperoensis* to the last occurrence of *Triquetrorhabdulus carinatus*.

Remarks: According to Martini and Muller (1986), it is difficult to distinguish Zone NN1 from Zone NN2 due to the scarcity or absence of the marker species *Discoaster druggii* employed by Martini (1971). Moreover, *D. druggii* is difficult to identify with certainty in particularly poorly preserved assemblages (Varol, 1983; Perch-Nielsen, 1985). Nevertheless, it is possible to subdivide this interval into five subzones using the evolutionary appearance of *Helicosphaera carteri*, *Calcidiscus leptoporus*, *Geminilithella lordii* and/or *G. rotula*, and *Geminilithella jafari* and/or *G. petaliformis*. However, their exact position in relation to the evolutionary appearance of *Discoaster druggii* cannot be determined with certainty.

Sphenolithus conicus Subzone (NN1/2a)

Age: Early Miocene.

Definition: Interval from the last occurrence of *Helicosphaera recta* and/or *Sphenolithus ciperoensis* to the first occurrence of *Helicosphaera carteri*.

Remarks: Poorly preserved, low-diversity calcareous nannofossil assemblages were observed in the samples within this interval. Species recovered include: *Sphenolithus conicus*, *Sphenolithus moriformis*, *Cyclicargolithus abisectus*, *Cyclicargolithus floridanus*, *Discoaster deflandrei* (5 and 6 armed), and *Hughesius* sp.

Distribution: Santa Isabel Island (I48, I27, I362). San Jorge Island (I567).

Helicosphaera carteri Subzone (NN1/2b)

Age: Early Miocene.

Definition: Interval from the first occurrence of *Helicosphaera carteri* to the first occurrence of *Calcidiscus leptoporus*.

Remarks: Moderately well preserved and highly diverse calcareous nannofossil assemblages were obtained from the samples within this interval. Species recovered include: *Helicosphaera carteri*, *Sphenolithus conicus*, *Cyclicargolithus abisectus*, *Cyclicargolithus floridanus*, *Discoaster deflandrei* (5 and 6 armed), *Hughesius* sp., *Coronocyclus nitescens*, and *Clausicoccus obrutus*. The presence of extremely rare specimens of *Reticulofenestra stavensis* is attributed to reworking rather than being considered a locally extended range as it is not recorded in the underlying subzone.

Distribution: Santa Isabel Island (I153, I198).

Calcidiscus leptoporus Subzone (NN1/2c)

Age: Early Miocene.

Definition: Interval from the first occurrence of *Calcidiscus leptoporus* to the first occurrence of *Geminilithella lordii* and/or *Geminilithella rotula*.

Remarks: Moderately well preserved and highly diverse calcareous nannofossil assemblages were recovered from the samples within this subzone. *Sphenolithus moriformis* and *Cyclicargolithus floridanus* dominate the assemblages. *Hughesius* sp., *Helicosphaera euphratis*, *Calcidiscus leptoporus*, *Helicosphaera carteri*, *Sphenolithus conicus*, and *Discoaster deflandre* (5 and 6 armed) were rarely recorded. The first occurrence of *Reticulofenestra pseudoumbilica* was observed in this subzone.

Distribution: Santa Isabel Island (I70). San Jorge Island (I576).

Geminilithella lordii Subzone (NN1/2d)

Age: Early Miocene.
Definition: Interval from the first occurrence of *Geminilithella lordii* and/or *G. rotula* to the first occurrence of *Geminilithella jafari* and/or *G. petaliformis*.
Remarks: Moderately well preserved and highly diverse calcareous nannofossil assemblages were obtained from the samples within this subzone. *Sphenolithus moriformis* and *Cyclicargolithus floridanus* dominate the assemblages. Rarely occurring species include: *Helicosphaera carteri*, *Helicosphaera euphratis*, *Coronocyclus nitescens*, *Sphenolithus conicus*, *Hughesius* sp., *Discoaster deflandrei* (5 and 6 armed), *Geminilithella lordii*, and *G. rotula*. *Orthorhabdus serratus* and *H. girgisii* have their first occurrence in this subzone.
Distribution: Santa Isabel Island (I66, I181). Gizo Island (GZ41).

Geminilithella jafari Subzone (NN1/2e)

Age: Early Miocene.
Definition: Interval from the first occurrence of *Geminilithella jafari* and/or *G. petaliformis* to the last occurrence of *Triquetrorhabdulus carinatus*.
Remarks: A well-preserved, highly diverse calcareous nannofossil assemblage was obtained from the samples within this subzone. *Cyclicargolithus floridans* and *Sphenolithus moriformis* are the most abundant species. Other species recovered include: *Cyclicargolithus abisectus*, *Helicosphaera carteri*, *Helicosphaera euphratis*, *Helicosphaera obliqua*, *Calcidiscus leptoporus*, *Discoaster deflandrei* (5 and 6 armed), *Triquetrorhabdulus milowi*, *Sphenolithus belemnos*, *Orthorhabdus serratus*, and *Sphenolithus conicus*. *Hughesius* sp. displays its last occurrence in this zone, whereas *Triquetrorhabdulus milowi*, *Helicosphaera magnifica*, and *H. jakubowskii* have their first occurrence. *Sphenolithus belemnos* also has its first occurrence at the base of this subzone, but some doubtful forms were observed well below, as low as Zone NP25. *Coronocyclus nitescens* is not found above this subzone and suggests possible ecological restriction of its distribution. The occurrence of extremely rare specimens of *Helicosphaera recta* is attributed to reworking.
Distribution: Santa Isabel Island (I132). San Jorge Island (I452, I643).

Sphenolithus belemnos Zone (NN3)

Age: Early Miocene.
Definition: Interval from the last occurrence of *Triquetrorhabdulus carinatus* to the last occurrence of *Sphenolithus belemnos*.
Remarks: A poorly to moderately preserved and diverse calcareous nannofossil assemblage was obtained from the sample within this zone. Species recovered include: *Sphenolithus heteromorphus*, *Helicosphaera carteri*, *Triquetrorhabdulus milowi*, *Calcidiscus leptoporus*, *Discoaster deflandrei* (5 and 6 armed), *Cyclicargolithus floridanus*, *Sphenolithus conicus*, and *Cyclicargolithus abisectus*. The presence of *Sphenolithus heteromorphus* is suggestive of the upper part of Zone NN3. The first occurrence of *Helicosphaera crouchii*, *H. waltrans*, and *H. vedderi* were observed in this zone but are extremely rare. The only forms definitely assignable to *Discoaster druggii* are present in this zone; however, forms closely resembling it were observed as low as Subzone NN1/2d.
Distribution: Santa Isabel Island (I177).

Discoaster exilis Zone (NN6)

Age: Middle Miocene.
Definition: Interval from the last occurrence of *Sphenolithus heteromorphus* to the last occurrence of *Cyclicargolithus floridanus*.
Remarks: The upper limit of this zone is defined by the last occurrence of *Cyclicargolithus floridanus* following the usage of Bukry (1975), Chi (1979) and Varol (1983). Because *Discoaster kugleri*, used by Martini (1971) to define the upper boundary of this zone, is not recorded, the definition of this zone differs from Martini's. Poorly preserved low-diversity calcareous nannofossil assemblages were recovered. Species include: *Cyclicargolithus floridanus*, *Calcidiscus leptoporus*, *Discoaster brouweri*, *D. deflandrei* (6 armed), *Helicosphaera carteri*, *Sphenolithus moriformis*, *H. mullerae*, and *H. californiana*.
Distribution: Santa Isabel Island (I108). San Jorge Island (I470).

Discoaster kugleri Zone (NN7)

Age: Middle Miocene.
Definition: Interval from the last occurrence of *Cyclicargolithus floridanus* to the first occurrence of *Catinaster coalitus*.
Remarks: The lower limit of this zone is at variance with Martini (1971) as mentioned for the *Discoaster exilis* Zone. A poorly preserved and moderately high diversity calcareous nannofossil assemblage was obtained from the single sample that represents this zone. Species include: *Discoaster intercalaris*, *D. deflandrei* (6 armed), *Calcidiscus leptoporus*, *Discoaster brouweri*, *Helicosphaera carteri*, *Helicosphaera mullerae*, and *Geminilithella lordii*. The last occurrence of *Helicosphaera mullerae* approximates the upper limit of this zone.
Distribution: San Jorge Island (I448).

Catinaster coalitus Zone (NN8)

Age: Middle Miocene.
Definition: Interval from first occurrence of *Catinaster coalitus* to the first occurrence of *Discoaster hamatus*.
Remarks: A poorly preserved and moderately low diversity calcareous nannofossil assemblage was recovered from the single sample that represents this zone. *Coccolithus pelagicus*, *Reticulofenestra pseudoumbilica*, and *Sphenolithus abies* are common, whereas *Helicosphaera carteri*, *Calcidiscus leptoporus*, and *Catinaster coalitus* are rare. The common occurrence of *Hughesius gizoensis* was noted in this sample.
Distribution: Santa Isabel Island (I178).

Discoaster calcaris Zone (NN10)

Age: Late Miocene.
Definition: Interval from the last occurrence of *Discoaster hamatus* to the first occurrence of *Discoaster quinqueramus*.
Remarks: Well-preserved, highly diverse calcareous nannofossil assemblages were obtained from the samples within this zone. *Reticulofenestra minuta*, *R. pseudoumbilica*, and *Sphenolithus abies* are the most abundant species. Other species include: *Discoaster pentaradiatus*, *Discoaster intercalaris*, *Discoaster variabilis*, *Discoaster challengeri*, *Helicosphaera intermedia (=H. rhomba)*, *Helicosphaera carteri*, *Calcidiscus leptoporus*, and *Geminilithella lordii*. The following species appear to have their first occurrences in this zone: *Helicosphaera pacifica*, *Discoaster neohamatus*, *Tetralithus symenoidesa*, *Discoaster prepentaradiatus*, *Discoaster bellus*, *Discoaster surculus*, *Helicosphaera orientalis*, *H. philippinensis*, and *Discoaster pentaradiatus*. The recorded ranges of some of these species, however, are influenced by sampling interval and/or ecological and preservational conditions.

The last occurrences of *Helicosphaera pacifica*, *H. walbersdorfensis*, *Discoaster bellus*, and *Discoaster prepentaradiatus* are in this zone. The absence of *Catinaster coalitus* in this zone indicates that possibly only the upper part of Zone NN10 is present.
Distribution: Gizo Island (GZ4, GZ5, GZ50, GZ51).

Discoaster quinqueramus Zone (NN11)

Age: Late Miocene.
Definition: Interval from first to last occurrence of *Discoaster quinqueramus*.
Remarks: This zone has been subdivided into two subzones using the evolutionary appearance of *Amaurolithus delicatus* and/or *Amaurolithus primus* following Okada and Bukry (1980).

Discoaster berggrenii Subzone (NN11a)

Age: Late Miocene.
Definition: Interval from the first occurrence of *Discoaster quinqueramus* to the first occurrence of *Amaurolithus delicatus* and/or *Amaurolithus primus*.
Remarks: A well-preserved, highly diverse calcareous nannofossil assemblage was recorded from the single sample that represents this subzone. *Reticulofenestra minuta* and *Sphenolithus abies* dominate the assemblage. The diversity and abundance of the genus *Scyphosphaera* increase slightly in this subzone, and *S. globulata* makes its first evolutionary appearance. *Helicosphaera orientalis*, *H. philippinensis*, *Hughesius gizoensis*, *Tetralithus symenoidesa*, and *Discoaster neohamatus* display their last occurrences in this subzone. Other species in this sample include: *Discoaster quinqueramus*, *Discoaster pentaradiatus*, *Discoaster intercalaris*, *Helicosphaera intermedia*, *Triquetrorhabdulus rugosus*, *Geminilithella lordii*, *Scyphosphaera intermedia*, *Scyphosphaera globulosa*, *Scyphosphaera pulcherrima*, and *Scyphosphaera campanula*.
Distribution: Gizo Island (GZ38).

Amaurolithus primus Subzone (NN11b)

Age: Late Miocene.
Definition: Interval from the first occurrence of *Amaurolithus delicatus* and/or *Amaurolithus primus* to the last occurrence of *Discoaster quinqueramus*.
Remarks: Well-preserved, highly diverse calcareous nannofossil assemblages are present in the samples that represent this subzone. *Sphenolithus abies* and *Reticulofenestra pseudoumbilica* are the most abundant species. Species which are rare in these samples include: *Amaurolithus primus*, *Amaurolithus delicatus*, *Triquetrorhabdulus rugosus*, *Discoaster quinqueramus*, *Helicosphaera intermedia*, *Scyphosphaera globulata*, *Scyphosphaera apsteinii*, *Scyphosphaera deflandrei*, *S. pseudorecurvata*, *S. quasitubifera*, *S. praeglobulata*, and *S. brevisa*. The diversity of the genus *Scyphosphaera* is greater than in the underlying subzone.
Distribution: Santa Isabel Island (I425). San Jorge Island (I626).

Amaurolithus tricorniculatus Zone (NN12)

Age: Early Pliocene.
Definition: Interval from the last occurrence of *Discoaster quinqueramus* to the first occurrence of *Ceratolithus rugosus* and/or last occurrence of *Ceratolithus acutus*.
Remarks: A three-fold subdivision of Zone NN12 is proposed using the last occurrence of *Triquetrorhabdulus rugosus* and/or *Helicosphaera intermedia* and the first occurrence of *Ceratolithus acutus*.

Triquetrorhabdulus rugosus Subzone (NN12a)

Age: Early Pliocene.
Definition: Interval from the last occurrence of *Discoaster quinqueramus* to the last occurrence of *Triquetrorhabdulus rugosus* and/or *Helicosphaera intermedia*.
Remarks: The samples from this subzone yielded moderately well preserved and highly diverse calcareous nannofossil assemblages which are dominated by *Sphenolithus abies*, *Reticulofenestra minuta*, and *R. pseudoumbilica*. Other species include: *Triquetrohabdulus rugosus*, *Helicosphaera intermedia*, *Helicosphaera carteri*, *Discoaster pentaradiatus*, *Discoaster brouweri*, *Discoaster challengeri*, *Amaurolithus delicatus*, and *Discoaster surculus*.
Distribution: Gizo Island (GZ39, GZ52).

Discoaster intercalaris Subzone (NN12b)

Age: Early Pliocene.
Definition: Interval from the last occurrence of *Triquetrorhabdulus rugosus* and/or *Helicosphaera intermedia* to the first occurrence of *Ceratolithus acutus*.
Remarks: A moderately to poorly preserved and diverse calcareous nannofossil assemblage was obtained from the single sample that represents this subzone. *Sphenolithus abies* and *Reticulofenestra minuta* are the most abundant species and occur together with rare specimens of *Discoaster pentaradiatus*, *Amaurolithus delicatus*, *Discoaster surculus*, *Geminilithella jafari*, *Geminilithella lordii*, *Geminilithella petaliformis*, *Discoaster intercalaris*, *Discoaster brouweri*, and *Scyphosphaera amphora*.
Distribution: Gizo Island (GZ6).

Ceratolithus acutus Subzone (NN12c)

Age: Early Pliocene.
Definition: Interval from the first occurrence of *Ceratolithus acutus* to the first occurrence of *Ceratolithus rugosus* and/or last occurrence of *Ceratolithus acutus*.
Remarks: Moderately to poorly preserved and diverse nannofossil assemblages were recovered from the samples that represent this subzone. *Reticulofenestra minuta* and *Sphenolithus abies* are the most abundant species. The increase in abundance of *Helicosphaera carteri* and *Calcidiscus leptoporus* is noteworthy. Other species include: *Discoaster intercalaris*, *Discoaster brouweri*, *Discoaster pentaradiatus*, *Ceratolithus acutus*, *Geminilithella lordii*, *Geminilithella jafari*, *Geminilithella petaliformis*, and *Hayaster perplexus*.
Distribution: Gizo Island (GZ27, GZ48).

Discoaster surculus Zone (NN16)

Age: Late Pliocene.
Definition: Interval from the last occurrence of *Sphenolithus abies* to the first occurrence of *Discoaster surculus*.
Remarks: The last occurrence of *Sphenolithus* abies is used to mark the base of this zone instead of "large" *Reticulofenestra pseudoumbilica*. This definition is at variance with that of Martini (1971) because the exact size of his "large" *R. pseudoumbilica* was not given. If a minimum size limit of about 7-8μ is assumed for the species *R. pseudoumbilica*, specimens definitely are present well above the last occurrence of *Sphenolithus abies*, within Zone NN16. Smaller individuals (between 5 and 7μ) of *R. pseudoumbilica* are here included in the taxon *R. tenuistriatus*, which displays a last occurrence that seems to be diachronous over a wide area and is present as high as Zone NN18.

A moderately well preserved and highly diverse calcareous nannofossil assemblage was observed in the sample that represents this zone. *Pseudoemiliania lacunosa* and *Helicosphaera carteri* are common, whereas *Discoaster brouweri* (including 3 armed), *Discoaster surculus*, *Discoaster pentaradiatus*, *Ceratolithus rugosus*, small *Gephyrocapsa* spp. (smaller than 3μ), and *Reticulofenestra pseudoumbilica* are rare.
Distribution: Ranongga Island (RA32).

Discoaster pentaradiatus Zone (NN17)

Age: Late Pliocene.
Definition: Interval from the last occurrence of *Discoaster surculus* to the last occurrence of *Discoaster pentaradiatus*.
Remarks: Well-preserved, highly diverse calcareous nannofossil assemblages were recorded from the samples that represent this zone. *Pseudoemiliania lacunosa* is the most common species together with *Helicosphaera carteri*. Other species include: *Ceratolithus rugosus*, *Discoaster pentaradiatus*, *Discoaster brouweri*, *Discoaster challengeri*, and *Calcidiscus leptoporus*.
Distribution: Ranongga Island (RA70, RA127).

Discoaster brouweri Zone (NN18)

Age: Late Pliocene.
Definition: Interval from the last occurrence of *Discoaster pentaradiatus* to the last occurrence of *Discoaster brouweri* and/or the first occurrence of *Gephyrocapsa oceanica*.
Remarks: The samples examined yielded well-preserved and highly diverse calcareous nannofossil assemblages. *Pseudoemiliania lacunosa* and *Helicosphaera carteri* are the most common species and occur together with rare

specimens of *Discoaster brouweri*, *Ceratolithus cristatus*, small *Gephyrocapsa* spp., *Pontosphaera japonica*, and *Caldiscus leptoporus*. An interval between the extinction of *Discoaster brouweri* and the first occurrence of *Gephyrocapsa oceanica* may be present in the studied area; however, the sampling gap and some reworking make this supposition difficult to prove.

Distribution: Ranongga Island (RA9, RA59, RA72). Rendova Island (RE317).

Pseudoemiliania lacunosa Zone (NN19)

Age: Early Pleistocene.

Definition: Interval from the last occurrence of *Discoaster brouweri* and/or the first occurrence of *Gephyrocapsa oceanica* to the last occurrence of *Pseudoemiliania lacunosa*.

Remarks: This zone was subdivided into four subzones by Gartner (1977) using the following events listed in order: last occurrence of *Calcidiscus tropicus (=C. macintyrei)*, last occurrence of *Helicosphaera sellii*, and the first acme of small *Gephyrocapsa* spp. In the studied area, the last occurrences of *C. tropicus* and *H. sellii* are not useful, as discussed by Hughes, Varol, and Dunkley (in press). The extended ranges of *H. sellii* and *C. tropicus* are possibly the result of extensive reworking rather than differences in species concepts because these events seem to be easily distinguished in other areas. The acme of small *Gephyrocapsa* spp. is readily recognized. A quantitative study over this biostratigraphic interval could be useful to determine whether true extinctions of *Calcidiscus tropicus* and *Helicosphaera sellii* can be separated from isolated and extremely rare higher occurrences.

Well-preserved and highly diverse calcareous nannofossil assemblages were obtained from the samples that represent this zone. *Pseudoemiliania lacunosa*, small *Gephyrocapsa* spp. (smaller than 3µ), and *Gephyrocapsa oceanica* (larger than 3µ) are the most abundant species. Other rarely to commonly occurring species include: *Helicosphaera carteri*, *Calcidiscus leptoporus*, *Ceratolithus cristatus*, *Umbilicosphaera sibogae*, *Scyphosphaera* spp., *Syracosphaera* spp., *Oolithus fragilis*, *Florisphaera profunda*, and large *Gephyrocapsa oceanica* (larger than 6µ).

Distribution: Ranongga Island (RA53, RA63, RA56, RA35, RA51, RA207). Rendova Island (RE52, RE38, RE318, RE39, RE68, RE61). Tetepare Island (TP118, TP117, TP116, TP112, TP111, TP110, TP109, TP15, TP108, TP107, TP106, TP105, TP14, TP104, TP103, TP102, TP101, TP13).

Gephyrocapsa oceanica Zone (NN20)

Age: Late Pleistocene.

Definition: Interval from the last occurrence of *Pseudoemiliania lacunosa* to the first occurrence of *Emiliania huxleyi* and/or *Geminilithella subtilis*.

Remarks: The samples examined yielded well-preserved and highly diverse calcareous nannofossil assemblages which are dominated by *Gephyrocapsa oceanica*. Other species include: rare to common small *Gephyrocapsa* spp., *Oolithus fragilis*, *Florisphaera profunda*, *Calcidiscus leptoporus*, *Umbilicosphaera sibogae*, *Helicosphaera carteri*, and *Scyphosphaera* spp.

Distribution: Tetepare Island (TP100, TP99, TP98, TP12). Rendova Island (RE62, RE37, RE256). Ranongga Island (RA19, RA50, RA43, RA210). Sulei Island (I22). Karuo Island (I24).

Emiliania huxleyi Zone (NN21)

Age: Late Pleistocene.

Definition: Interval above the first occurrence of *Emiliania huxleyi* and/or *Geminilithella subtilis*.

Remarks: Well-preserved, highly diverse calcareous nannofossil assemblages which are dominated by *Gephyrocapsa oceanica* and *Emiliania huxleyi* were recorded from the samples that represent this zone. *Geminilithella subtilis*, *Helicosphaera carteri*, *Hayaster perplexus*, *Oolithus fragilis*, *Calcidiscus leptoporus*, and *Umbilicosphaera sibogae* are common, whereas *Florisphaera profunda*, *Syracosphaera pulchra*, *Scapholithus fossilis*, and *Scyphosphaera* spp. are rare in these samples.

Distribution: Ranongga Island (RA209). Tetepare Island (TP97, TP96, TP95, TP11, TP94).

CONCLUSIONS

The following biostratigraphic results are based upon a study of calcareous nannofossils in samples from the central and western Solomon Islands.

A late Pliocene to late Pleistocene age (Zones NN18-NN21) is assigned to the Tetepare Formation which outcrops on Tetepare and Rendova islands.

A late Pliocene to late Pleistocene age (Zones NN16 to NN21) is assigned to the Tuara Sedimentary Formation from Ranongga Island. The Lea Member is restricted to Zones NN19 to NN21 of the Pleistocene, whereas the Vori Member ranges from Zone NN16 to Zone NN20, of late Pliocene to Pleistocene age.

An early Miocene to early Pliocene age (Subzones NN1/2d to NN12c) is assigned to the Konggu Formation and its Mari Limestone Member. The Kaindo Arenite Member is restricted to a late Miocene age (Zone NN10).

An exceptionally long age range of Late Cretaceous (Campanian Stage) to Pleistocene (*Eiffellithus eximius* Zone to Zone NN20) is assigned to the Tanakau Group which occurs on San Jorge, Santa Isabel, Sulei and Karuo islands.

ACKNOWLEDGMENTS

I am grateful to the management of Robertson Research International for permission to publish this paper. I express my gratitude to Dr. G. W. Hughes, who introduced me to the geology of the Solomon Islands, provided the geological background information, furnished the samples, and reviewed the manuscript. I sincerely thank Mr. M. Jakubowski and Dr. M. H. Girgis for their constructive review of the manuscript and useful suggestions.

This study would not have been successful without the help of the Photographic, Laboratory, Editing, Drafting, and SEM Departments of Robertson Research. I also thank Ms. D. Starr for typing the manuscript.

APPENDIX

TAXONOMIC NOTES

Anacanthoica solomonica Varol, n. sp.
Plate 1, Figure 13

Acanthoica sp. Sachs and Skinner, 1973, Plate 4, Figures 19-21.

Diagnosis: Species of *Anacanthoica* with an almost elliptical coccosphere and small cone-shaped coccoliths.

Holotype: Plate 1, Figure 13.

Type level and locality: Upper Miocene, Solomon Islands.

Dimensions of holotype: Width of base in individual coccolith, 1.6μ. Maximum height of individual coccolith, 3.1μ.

Description: This small calyptrolith has a basal plate of various shapes which is usually convex distally and concave proximally. The distal process tapers strongly toward the distal end. Each individual coccolith is formed by identical crystals, and therefore it is a holococcolith. *A. solomonica* is the only fossil form of *Anacanthoica* so far known.

Remarks: *A. solomonica* is similar to recent forms of *A. aculeata* and *A. cidaris* but differs by the elliptical shape of its coccosphere and angular outline of the basal plate of its coccolith.

Occurrence: *A. solomonica* is recorded only in the upper Miocene sequence of the Solomon Islands. Sachs and Skinner (1973) reported this species in upper Pliocene to lower Pleistocene sequences of the Gulf Coast area. *A. solomonica* is also present in the middle Miocene sequence of Malta and the upper Miocene sequence of southern Turkey.

Calcidiscus tropicus (Kamptner) Varol, n. comb.
Plate 1, Figure 11

Coccolithus tropicus Kamptner, 1955, p. 32, Plate 6, Figures 79a, b.

Cyclococcolithus macintyrei Bukry and Bramlette, 1969, pp. 132-133, Plate 1, Figures 1-3.

Cyclococcolithus tropicus (Kamptner) Gartner, Chen, and Stanton, 1983, p. 46.

Remarks: *C. tropicus* is distinguished from *C. leptoporus* by being larger than 10μ in diameter, a criterion which is most useful for routine light microscope examination. The last occurrence of *C. tropicus* was employed as a subzonal marker within Zone NN19 (Gartner, 1977). In this study, however, *C. tropicus* occurs as high as Zone NN21, but this is possibly due to reworking. For a more extensive synonomy see Gartner, Chen, and Stanton (1983).

Discoaster Tan, 1927 emend. Tan 1931

Type species: *Discoaster pentaradiatus* Tan, 1927.

Synonyms: *Hemi-discoaster* Tan, 1927
Eu-discoaster Tan, 1927
Helio-discoaster Tan, 1927

Remarks: Tan (1927) caused some confusion in the way he introduced *Discoaster, Hemi-discoaster, Eu-discoaster*, and *Helio-discoaster*. First, *Discoaster* was used as a genus in binary combination of species without description, and therefore it was invalid according to I.C.B.N. article 32 until it was validated in 1931 by Tan. Second, *Hemi-discoaster, Eu-discoaster*, and *Helio-discoaster* were also invalid in spite of their description as genera, according to I.C.B.N. article 33.4, because they were misplaced in their rank. It is inferred that Tan (1927) placed these taxa above the rank of *Discoaster*, which was treated as a genus in binary combination of different species names. In conclusion, all the genera originally introduced by Tan (1927) were invalid and he validated only *Discoaster* in 1931; therefore *Hemi-discoaster, Eu-discoaster*, and *Helio-discoaster* are invalid and must be rejected.

Geminilithella lordii (Varol) Varol, n. comb.
Plate 4, Figures 16-18

Umbilicosphaera lordii Varol, 1982, pp. 248, 251, Figure 4(5).

Remarks: This strictly circular species has equal-sized, single-cycled shields which are non-birefringent under polarized light. *G. lordii* is distinguished from *G. rotula* by lacking a birefringent wall around the central area under cross-polarized light.

Geminilithella petaliformis (Varol) Varol n. comb.
Plate 4, Figure 25

Umbilicosphaera petaliformis Varol, 1982, p. 251, Figure 4(7).

Remarks: This small form is difficult to distinguish from *G. jafarii* under the light microscope, but *G. petaliformis* differs from *G. jafarii* by the presence of inflated shields and the petal-like elements of the shields.

Geminilithella subtilis (Muller) Varol, n. comb.
Plate 1, Figures 1a, 5, Plate 4, Figures 26-27

Cyclococcolithus subtilis Muller, 1976, pp. 50-51, Plate 1, Figures 5-6.

Remarks: This species differs from the other members of the genus by being elliptical. It is similar, however, to other members of the genus in that under cross-polarized light its equal-sized single-cycled shields are non-birefringent, and the wall around the central area is birefringent.

Helicosphaera jakubowskii Varol, n. sp.
Plate 2, Figure 5; Plate 3, Figures 1-3

Diagnosis: A small *Helicosphaera* with a granulated central area and a flange that extends well beyond the limit of the shields.

Derivation of name: In honor of Mr. M. Jakubowski, Nannopaleontologist, Robertson Research International, U.K.

Holotype: Plate 2, Figure 5.

Type level and locality: Lower Miocene, Solomon Islands.

Dimensions of holotype: Length, 5.5µ.

Description: This small species has a granulated central area which is birefringent under cross-polarized light. The shields and well-extended flange beyond the limit of the shields are non-birefringent, and the central area covered by the blanket (extension of the wall on the distal surface of the distal shield) is birefringent in cross-polarized light.

Remarks: *H. jakubowskii* is distinguished from *H. magnifica* by its granulated central area and the less extended flange along its length. *H. jakubowskii* is distinguished from *H. granulata* by its optical character and smaller size. In *H. granulata*, the blanket covers the whole distal surface of the specimen which is totally birefringent under cross-polarized light.

Occurrence: This species is present in Subzones NN1/2e to Zone NN3 in the Solomon Islands and occurs in Zone NN1/2e in Cyprus and southern Turkey.

Helicosphaera girgisii Varol, n. sp.
Plate 3, Figures 11-13

Diagnosis: A small *Helicosphaera* with a closed central area and a flange which does not extend beyond the limit of the shields.

Derivation of name: In honor of Dr. M.H. Girgis, Nannopaleontologist, Robertson Research International, U.K.

Holotype: Plate 3, Figure 11.

Type level and locality: Upper Miocene, Solomon Islands.

Dimensions of holotype: Length, 5.8µ. Width, 3.3µ.

Description: This small elongated *Helicosphaera* has a closed central area which is birefringent under polarized light. Both distal shield and flange (external end of distal shield), which do not extend beyond the limit of shields, are not birefringent under cross-polarized light.

Remarks: This species is very similar to *H. walbersdorfensis*; however, its central area is closed, whereas the latter has an oblique bar as an optical continuation of the shields and two oblique openings in the central area. *H. girgisii* is distinguished from *H. magnifica* by the absence of a large well-extended flange.

Occurrence: In the Solomon Islands, *H. girgisii* ranges from Subzone NN1/2d to Zone NN10; it also occurs in Subzones NN1/2d and NN1/2e in Cyprus and southern Turkey and in Zones NN5 to NN8 in the Caribbean area.

Helicosphaera magnifica Varol, n. sp.
Plate 3, Figures 4-9

Diagnosis: Medium-sized *Helicosphaera* with a flange that extends well beyond the limit of the shields and covers almost the whole length of the species.

Holotype: Plate 3, Figures 6-7.

Type level and locality: Lower Miocene, Solomon Islands.

Dimensions of holotype: Length of shields, 6.1µ. Width of shields, 3.3µ. Length of flange, 8µ. Width of flange, 4.7µ.

Description: This medium-sized *Helicosphaera* has a closed central area which is birefringent under cross-polarized light. The distal shield and the large flange, which extends well beyond the limit of the shields and covers its whole length, are non-birefringent to weakly birefringent due to overgrowth of secondary calcite.

Remarks: *H. magnifica* is distinguished from *H. girgisii* by its large extending flange. *H. jakubowskii* is distinguished from *H. magnifica* by having a granulated central area and a shorter flange along the long axis of form.

Occurrence: This species is found only in Subzones NN1/2e to Zone NN3 in the Solomon Islands but also occurs in Subzone NN1/2e in Cyprus.

Helicosphaera mullerae Varol, n. sp.
Plate 2, Figure 7; Plate 3, Figures 16-21

Helicosphaera walbersdorfensis Muller, 1974b, *partim*, Plate 4, Figures 35-37, *non* Plate 2, Figure 15.

Helicosphaera walbersdorfensis Muller - Theodoridis, 1984, Plate 13, Figure 5; Plate 14, Figure 3; Plate 19, Figures 9-11; Plate 26, Figure 4.

Diagnosis: A small species of Helicosphaera with a flange extending well beyond the limit of shields and an oblique bar which is an optical continuation of the shield.

Derivation of name: In honor of Dr. C. Muller, Nannopaleontologist, Geologisch-Paläontologisches Inst. der Universität, Frankfurt am Main., F.R.G.

Holotype: Plate 2, Figure 7.

Type level and locality: Middle Miocene, Solomon Islands.

Dimensions of holotype: Length, 6.1µ. Width, 3.4µ.

Description: This small form has a blanket which is confined to the central area and is birefringent under cross-polarized light, whereas the distal shield and flange, which extend well beyond the limit of the shields, are non-birefringent under cross-polarized light. The central area has an oblique bar in optical continuity with the shield. In strongly etched assemblages, the bar is almost lost, but its remnants can always be seen. In

PLATE 1

Figures 1a, 5	*Geminilithella subtilis* (Muller) Varol, n. comb. Proximal views, upper Pleistocene
Figures 1b, 2a, 3	*Emiliania huxleyi* Kamptner Distal views, upper Pleistocene
Figures 2b, 4a, 7, 8	*Florisphaera profunda* Okada and Honjo Plan views, upper Pleistocene
Figure 4b	*Oolithus fragilis* (Lohmann) Martini and Muller Proximal view, Pleistocene
Figure 6	*Calcidiscus leptoporus* (Murray and Blackman) Loeblich and Tappan Distal view, Pleistocene
Figures 9-10	*Scyphosphaera pseudorecurvata* Varol, n. sp. Oblique distal views, upper Miocene, Holotype (Figure 9)
Figure 11	*Calcidiscus tropicus* (Kamptner) Varol, n. comb. Distal view, lower Pleistocene
Figures 12, 15	*Scyphosphaera conica* Kamptner Side views, upper Miocene
Figure 13	*Acanthoica solomonica* Varol n. sp. View of coccosphere, upper Miocene, Holotype
Figure 14	*Scyphosphaera globulosa* Kamptner Oblique distal view, upper Miocene

a
b
1
a
b
2
3
b
a
4
5
6
7
8
9
10
11
12
13
14
15

PLATE 2

Figures 1-4	*Helicosphaera walbersdorfensis* Muller 1, 2 Proximal views, upper Miocene 3, 4 Distal views, upper Miocene
Figure 5	*Helicosphaera jakubowskii* Varol, n. sp. Proximal view, lower Miocene, Holotype
Figures 6, 11	*Helicospharea carteri* (Wallich) Kamptner 6 View of coccosphere, upper Miocene 17 Distal view, upper Miocene
Figure 7	*Helicosphaera mullerae* Varol, n. sp. Proximal view, middle Miocene, Holotype
Figure 8	*Geminilithella jafari* (Muller) Backman Distal view, lower Miocene
Figure 9	*Scapholithus fossilis* Deflandre Plan view, Pleistocene
Figure 10	*Helicosphaera intermedia* Martini Proximal view, upper Miocene
Figure 12	*Pontosphaera scutellum* Kamptner Proximal view, upper Miocene
Figure 13	*Pontosphaera sparsiforatus* (Kamptner) Varol, n. comb. Proximal view, upper Miocene
Figures 14-15	*Scyphosphaera praeglobulata* Varol, n. sp. Oblique distal view, upper Miocene, Holotype (Figure 15)
Figure 16	*Scyphospharea pulcherrima* Deflandre Oblique distal view, upper Miocene
Figure 17	*Amaurolithus delicatus* Gartner and Bukry Plan view, upper Miocene
Figure 18	*Amaurolithus primus* (Bukry and Percival) Gartner and Bukry Plan view, upper Miocene
Figure 19	*Florisphaera profunda* Okada and Honjo Plan view of broken coccosphere, upper Pleistocene

1
2
3
4
5
6
7
8
9
10
11
12
13
14
15
16
17
18
19

PLATE 3

Figures 1-3	*Helicosphaera jakubowskii* Varol, n. sp.
Figures 4-9	*Helicosphaera magnifica* Varol, n. sp. Holotype (Figures 6-7)
Figure 10	*Helicosphaera* cf. *magnifica* Varol, n. sp.
Figures 11-13	*Helicosphaera girgisii* Varol, n. sp. Holotype (Figure 11)
Figures 14-15	*Helicosphaera californiana* Bukry
Figures 16-21	*Helicosphaera mullerae* Varol, n. sp.
Figures 22-25	*Helicosphaera walbersdorfensis* Muller
Figures 26-28	*Helicosphaera philippinensis* Muller
Figure 29	*Helicosphaera pacifica* Muller and Bronnimann
Figure 30	*Helicosphaera intermedia* Martini
Figures 31-32	*Helicosphaera valtrans* Theodoridis
Figures 33-34	*Helicosphaera vedderi* Bukry
Figure 35	*Helicosphaera crouchii* Bukry

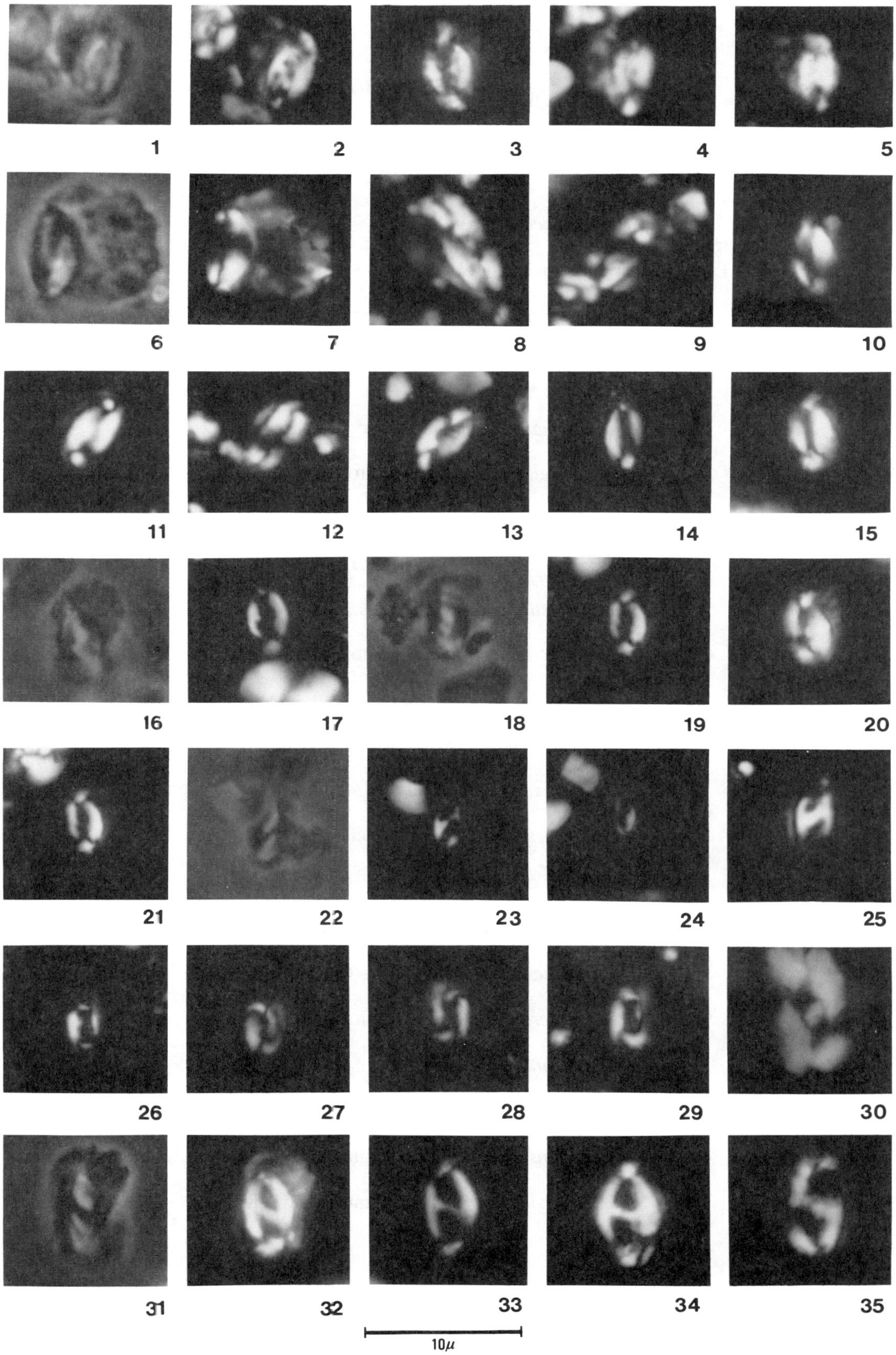
1
2
3
4
5
6
7
8
9
10
11
12
13
14
15
16
17
18
19
20
21
22
23
24
25
26
27
28
29
30
31
32
33
34
35
10μ

PLATE 4

Figures 1-2	*Scyphosphaera quasitubifera* Varol
Figures 3-4	*Scyphosphaera praeglobulata* Varol, n. sp.
Figures 5-6	*Scyphosphaera brevisa* Varol
Figures 7-8	*Scyphosphaera pseudorecurvata* Varol, n. sp.
Figures 9-13	*Hughesius gizoensis* Varol, n. gen., n. sp. Holotype (Figures 9-10)
Figure 14	*Hughesius* sp.
Figure 15	*Triquetrorhabdulus milowii* Bukry
Figures 16-18	*Geminilithella lordii* (Varol) Varol, n. comb.
Figures 19-20	*Geminilithella rotula* (Kamptner) Backman
Figure 21	*Florisphaera profunda* Okada and Honjo
Figures 22-24	*Geminilithella jafari* (Muller) Backman
Figure 25	*Geminilithella petaliformis* (Varol) Varol, n. comb.
Figures 26-27	*Geminilithella subtilis* (Muller) Varol, n. comb.
Figure 28	*Cyclicargolithus floridanus* (Roth and Hay) Bukry
Figure 29	*Reticulofenestra tenuistriatus* (Kamptner) Varol, n. comb.
Figure 30	*Reticulofenestra minutula* (Gartner) Haq and Berggren
Figure 31	*Reticulofenestra minuta* Roth
Figures 32-33	*Sphenolithus belemnos* Bramlette and Wilcoxon
Figure 34	*Reticulofenestra pseudoumbilica* (Gartner) Gartner
Figure 35	*Pontosphaera ovata* (Levin and Joerger) Varol, n. comb.
Figure 36	*Sphenolithus conicus* Bukry
Figure 37a, 38a	*Sphenolithus heteromorphus* Deflandre, 1953
Figure 37b, 38b	*Sphenolithus moriformis* (Bronnimann and Stradner) Bramlette and Wilcoxon
Figures 39-41	*Triquetrorhabdulus rugosus* Bramlette and Wilcoxon

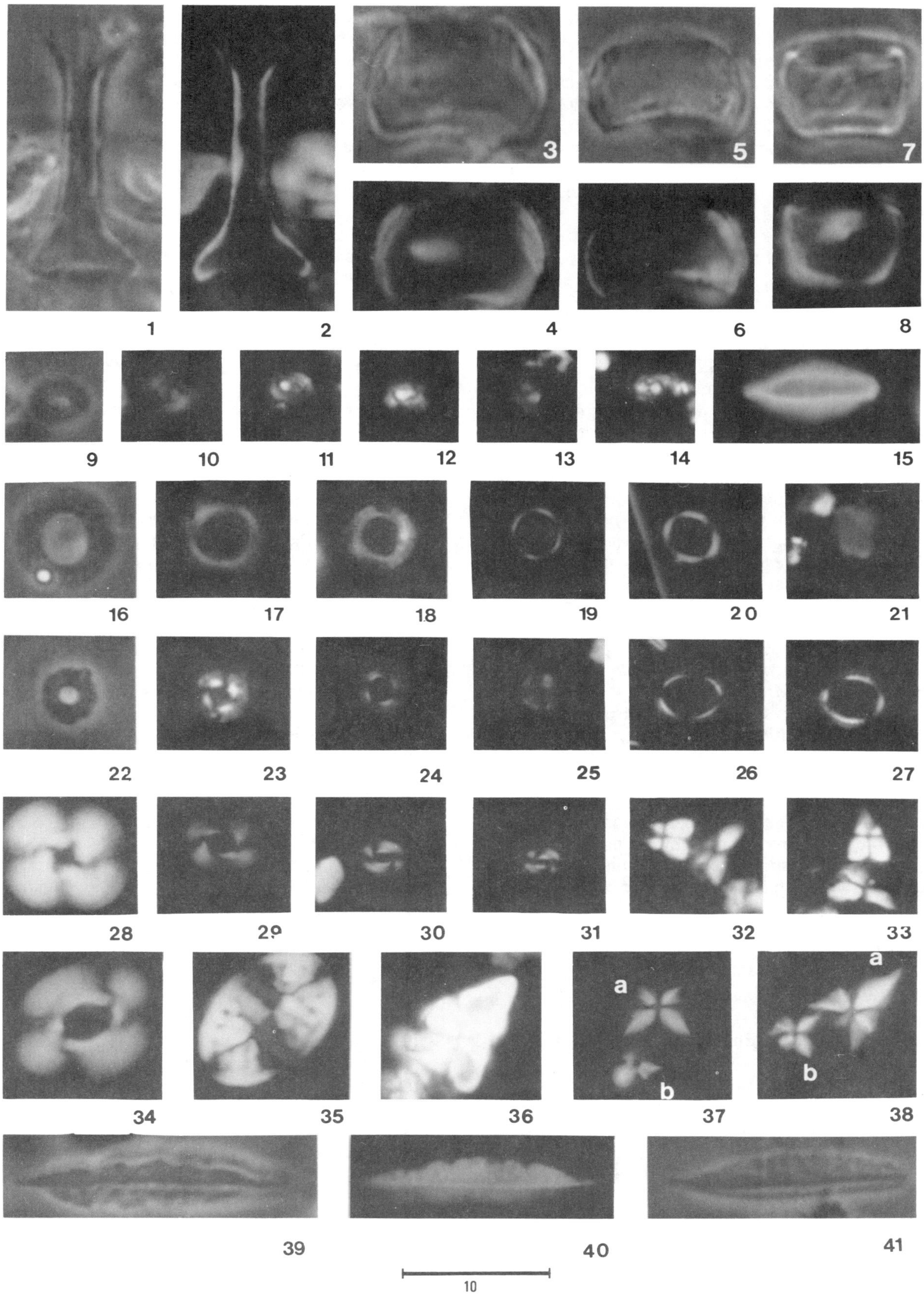
1 2 3 4 5 6 7 8
9 10 11 12 13 14 15
16 17 18 19 20 21
22 23 24 25 26 27
28 29 30 31 32 33
34 35 36 37 38
a b
39 40 41
10

this respect, this new species is distinguished from *H. californiana*, which has a slit in the central area.

Remarks: *H. californiana* is very similar to *H. mullerae*, but it is more elongate and lacks an oblique bar in the central area. *H. californiana* has a worldwide total range from the upper part of Zone NN3 to Zone NN6, whereas *H. mullerae* ranges from the upper part of Zone NN5 to Zone NN7.

Occurrence: *H. mullerae* is recorded only in Zones NN6 to NN7 in the present study.

Helicosphaera philippinensis Muller
Plate 3, Figures 26-28

Helicosphaera philippinensis Muller, 1981, p. 429, Plate 1, Figures 7-9, 11, 10(?), 12(?).

Helicosphaera orientalis Black - Theodoridis, 1984, p. 118, Plate 13, Figure 4; Plate 17, Figure 11; Plate 18, Figures 2-5; Plate 25, Figure 6.

Remarks: The forms included within *H. philippinensis* have a rectangular outline and an inverse bar in optical continuity with the shield. In this respect, it is difficult to discern whether or not Plate 1, Figures 10 and 12 of Muller (1981) belong to this species since these figures give the impression that the bar is not in optical continuity with the shield. *H. philippinensis* is distinguished from *H. orientalis* by the presence of inverse slits in the central area; *H. orientalis* has a closed central area in well-preserved specimens or sutures along the long axis of forms in poorly preserved specimens.

It is important to separate these two species as they have different stratigraphic ranges. *H. orientalis* is restricted to upper Miocene sequences, whereas *H. philippinensis* ranges stratigraphically from the middle to upper Miocene worldwide. *H. pacifica*, another rectangular species, is distinguished from *H. orientalis* and *H. philippinensis* by having a perforated plate in the central area. All three of these species have similar optical properties. The distal shield is non-birefringent, whereas the wall (formed by elements parallel to long axis, possibly connecting proximal and distal shields) is birefringent, and the blanket is absent or confined to the central area.

Occurrence: *H. philippinensis* is recorded only in Zones NN10 to NN11 in the present study.

Helicosphaera vedderi Bukry, 1981
Plate 3, Figures 33-34

Helicosphaera vedderi Bukry, 1981, p. 463, Plate 6, Figures 8-17.

Helicosphaera seminulum seminulum Bramlette and Sullivan - Clochiatti, 1971, Plate 17, Figures 3-4.

Remarks: This rare species is usually about 7-10μ in length and has an oblique bar in the central area as a continuation of the shield. Under cross-polarized light, the wall is birefringent, whereas the shield and the slightly protruding flange are non-birefringent. Both the distal and proximal side of *H. vedderi* were well illustrated by Clochiatti (1971) under the name *H. seminulum seminulum* from foraminiferal Zone N8 (nannofossil Zones NN5 part - NN4 part) of Hodna, Algeria. The blanket is present and confined to the central area. The flange is extended slightly beyond the limit of the shield. In *H. waltrans*, the flange extends well beyond the shields, and the opening in the central area is usually triangular in shape. This species is similar to *H. crouchii*, in which a central opening dominates the central area, and the bar is almost aligned to a short axis and slightly offset in the middle. *H. vedderi*, *H. waltrans*, and *H. crouchii* all have similar optical properties and similar ranges (between uppermost NN3 and lower NN5 Zones worldwide). *H. mediterranea* is distinguished from these species by having a completely birefringent shield in cross-polarized light because a blanket covers the whole distal surface of the distal shield. This species has a worldwide range from Subzone N1/2e to the lower part of Zone NN4.

Helicosphaera walbersdorfensis Muller
Plate 2, Figures 1-4; Plate 3, Figures 22-25

Helicosphaera walbersdorfensis Muller, 1974b, *partim*, pp. 392-393, Plate 2, Figure 15; Plate 14(?), Figures 45-46; *non* Plate 4, Figures 35-37.

Helicosphaera minuta Muller, 1981, pp. 428-429, Plate 1, Figures 1-3; Plate 1, Figures 4-6.

Helicosphaera stalis Theodoridis, 1984, pp. 127-128, Plate 13, Figure 6; Plate 20, Figures 10-12; Plate 21, Figures 1-12; Plate 26, Figures 5-6.

Remarks: The holotype illustrated by Muller (1974b; Plate 2, Figure 15) is a distal view of a small *Helicosphaera* with an oblique bar and flange which does not extend beyond the limit of the shields and is without a blanket. In these respects, the specimens presented in this study in Plate 2, Figures 3-4 are identical to the holotype. The forms illustrated by Theodoridis (1984) under a light microscope as *H. stalis* are identical to Plate 3, Figure 23. (A number of specimens from the Caribbean area were observed under both the light and electron microscopes; these specimens are identical to those illustrated on Plate 2, Figures 3-4 and Plate 3, Figure 23.) However, the paratypes illustrated by Muller (1974b; Plate 4, Figures 35-37) have flanges which extend well beyond the shields and thus differ from the holotype.

This species is distinguished from *H. vedderi* by its smaller size (3.5-5.5μ) and more elongated shape. The flange never extends beyond the limit of the shield, and is lacking a blanket. The two species also have different stratigraphic ranges. *H. walbersdorfensis* has diachronous first and last occurrences and a total range from the upper part of Zone NN5 to Zone NN11, of middle to late Miocene age worldwide. In the Solomon Islands, this species is recorded only in upper Miocene sequences; however, it occurs in the upper part of Zone NN5 in Egypt and the Caribbean area and has a first occurrence in the upper part of Zone NN7 in Malta. It has a last occurrence at about Zone NN10 in China and within Zone NN11 in West Africa. Its range is variable in southeast Asia.

Hughesius Varol, n. gen.

Type species: *Hughesius gizoensis* Varol, n. gen., n. sp.

Diagnosis: Small elliptical placoliths with two equal-sized single-cycled shields and central area occupied by a variable number of plates.

Derivation of name: In honor of Dr. G. W. Hughes, Micropaleontologist, Robertson Research International, U.K.

Description: These small elliptical forms have 20 to 30 elements in their proximal and distal shields which are equal in size. The members of this genus show no birefringence under cross-polarized light.

Remarks: *Hughesius* is distinguished from the similarly constructed *Clausicoccus* by having equal-sized single-cycled shields which are non-birefringent under cross-polarized light; in *Clausicoccus*, a wall and proximal, double-cycled, shield are birefringent under cross-polarized light.

Hughesius gizoensis Varol, n. gen., n. sp.
Plate 4, Figures 9-13

Gen. *et* sp. indet. Varol, 1985, p. 153, Plate 1, Figure 12.

Diagnosis: Small, elliptical placolith with single-cycled, equal-sized shields and two equal-sized plates in the central area divided along the short axis of the placolith.

Holotype: Plate 4, Figures 9-10.

Type level and locality: Upper Miocene, Gizo Island, Solomon Islands.

Dimensions of holotype: Maximum length, 4μ.

Description: This small form has about 15 to 25 radially arranged elements in its shields. The central area is occupied by two almost equal-sized plates which are separated along the short axis of the placolith. Under cross-polarized light, the species show no birefringence.

Remarks: This specimen is distinguished from *Hughesius* sp., whose stratigraphic range is restricted to upper Oligocene to lower Miocene, by the presence of two equal-sized plates in the central area instead of four.

Occurrence: *H. gizoensis* is present from Zone NN6 to Zone NN11 in various parts of world including Papua New Guinea, southern Turkey, China, southeast Asia, and the Caribbean area. This species also occurs in Gizo and Santa Isabel islands in the Solomon Islands.

Hughesius sp.
Plate 4, Figure 14

Remarks: This small elliptical placolith is distinguished from *H. gizoensis* by having four plates in the central area. *Hughesius* sp. is also easily distinguished from *C. obrutus* by being non-birefringent under cross-polarized light; *C. obrutus* has a proximal shield and wall around the central area which is birefringent under cross-polarized light.

Occurrence: *Hughesius* sp. is present in sequences from the upper Oligocene (Zone NP24) to the lower Miocene (Subzone NN1/2e) in the Solomon Islands, including Gizo and Santa Isabel islands, southern Turkey, China, southeast Asia, Cyprus, and the Middle East.

Reticulofenestra stavensis (Levin and Joerger)
Varol n. comb.

Coccolithus stavensis Levin and Joerger, 1967, p. 165, Plate 1, Figures 7a-d.

Remarks: *R. stavensis* is distinguished from *R. bisecta* by its large size (greater than 10μ). The last occurrence of this species is at the Oligocene/Miocene boundary in the studied area.

Reticulofenestra producta (Kamptner) Varol, n. comb.

Ellipsoplacolithus productus Kamptner, 1963, p. 172, Plate 8, Figures 42,44. *Dictyococcites productus* (Kamptner) Backman, 1980, pp. 49-50, Plate 4, Figures 1a, 2a, 6, 7.

Remarks: Species of *Reticulofenestra* with a closed central area are distinguished in this study using various criteria including size, which is consistent with the respective holotypes. *R. producta* has a placolith length less than 3μ, whereas *R. hesslandii* has a placolith length ranging between 3 and 6μ. Placolith lengths between 6 and 10μ are referred to *R. bisecta.*

Reticulofenestra tenuistriatus (Kamptner) Varol, n. comb.
Plate 4, Figure 29

Coccolithus tenuistriatus Kamptner, 1963, p. 160, Plate 2, Figures 14-15; text Figure 16.

Remarks: Species of *Reticulofenestra* with an open central area are distinguished in this study using various criteria which are consistent with the respective holotypes, including size. Forms that have placolith maximum length less than 3μ are placed in *R. minuta,* whereas those that have placolith maximum length ranging between 3 and 5μ are referred to as *R. minutula. R. tenuistriatus* has a maximum length between 5 and 7μ, whereas *R. pseudoumbilica* has a maximum length of more than 7μ. There is a distinct extinction level for *R. pseudoumbilica* in the lower part of Zone NN16 between the last occurrences of *Sphenolithus abies* and *Discoaster tamalis.*

Scyphosphaera pseudorecurvata Varol, n. sp.
Plate 1, Figures 9-10; Plate 4, Figures 7-8

Diagnosis: Lophodolith has a subrectangular outline in side view; its height is shorter than its width, and the apical end curves strongly inward.

Holotype: Plate 1, Figure 9.

Type level and locality: Upper Miocene, Gizo Island, Solomon Islands.

Dimensions of holotype: Maximum height, 4.5μ. Maximum width, 9μ. Width of base, 6.6μ. Width of apical end, 6.4μ.

Description: The wall is slightly concave-convex above the inflated basal plate. The apical end is strongly curved inward and gives a subrectangular appearance to the species. The wall forms an "edge" at the base, and it has only longitudinal ornamentation.

Remarks: This species is easily distinguished from similar forms such as *S. brevisa* by its subrectangular outline, inflated basal plate, and strongly inward-curved apical end; *S. brevisa* has a strongly arched basal plate and a simple apical end.

Scyphosphaera praeglobulata Varol, n. sp.
Plate 2, Figures 14-15, Plate 4, Figures 3-4

Diagnosis: Lophodolith has a strongly arched basal plate and strongly concave-convex wall. Its width of apical opening is much smaller than its width of base.

Holotype: Plate 2, Figure 15.

Type level and locality: Upper Miocene, Gizo Island, Solomon Islands.

Dimensions of holotype: Maximum height, 6.5µ. Maximum width, 11.5µ. Width of base, 10µ. Width of apical end, 6.2µ.

Description: The basal plate is strongly arched as in *S. brevisa* [Figure 3(7) of Varol, 1984 may not belong to this species; therefore, compare holotype Figure 3(5)]. However, the wall in *S. praeglobulata* is strongly concave-convex, and the ratio of width of base to apical end is much greater; in *S. brevisa*, the width of the base is approximately equal to the width of the apical end. This species is known only in upper Miocene sequences, whereas *S. brevisa* ranges stratigraphically from the middle Miocene to upper Miocene. *S. praeglobulata* is easily distinguished from *S. globulosa* and *S. globulata* by having a width which is much greater than its height; however, it is similar to *S. globulata* in having a much smaller diameter at the apical end than at its base; the diameter of the apical end is almost equal to the diameter of the base in *S. globulosa*. *S. praeglobulata* has a much more strongly arched basal plate in comparison to *S. globulosa* and *S. globulata*.

Occurrence: *S. praeglobulata* is present in the upper Miocene sequences of Gizo and Santa Isabel islands of the Solomon Islands and in southern Turkey and Egypt.

Pontosphaera vigintiforatus (Kamptner)
Varol, n. comb.

Discolithus vigintiforatus Kamptner, 1948, p. 5, Plate 1, Figures 8a, b.

Remarks: This species has two complete cycles of perforations in its basal plates. In this study, species of *Pontosphaera* in the Miocene to Holocene sequences are distinguished in routine light microscope examination based on the number of complete cycles of perforations together with other criteria which are consistent with their respective holotypes.

Pontosphaera trenus (Kamptner) Varol, n. comb.

Discolithus trenus Kamptner, 1967, p. 138, text Figure 15.

Remarks: *P. trenus* is distinguished from *P. sparsiforatus* by being smaller (less than 10µ). This species has one complete single cycle of perforations in its basal plate. In this respect, *P. trenus* is easily separated from *P. vigintiforatus* and *P. multipora* which have two and three complete cycles of perforations in their basal plates, respectively.

Pontosphaera sparsiforatus (Kamptner) Varol, n. comb.
Plate 2, Figure 13

Discolithus sparsiforatus Kamptner, 1948, p. 5, Plate 2, figs. 13a, b.

Remarks: This species is larger than 10µ and has only one complete cycle of perforations in its basal plate; it was found to be stratigraphically confined to the lowermost Pliocene to middle Miocene. The forms with a single complete cycle of perforations, which are confined to the center of the basal plate, are known only in the upper Miocene Zone NN10. *P. sparsiforatus* is easily distinguished from *P. pacifica* and from *P. callosa* which have four and five complete cycles of perforations in their basal plates, respectively.

Pontosphaera ovata (Levin and Joerger) Varol, n. comb.
Plate 4, Figure 35

Discolithina ovata Levin and Joerger, 1967, p. 167, Plate 2, Figures 6a-d.

Pontosphaera segmenta (Bukry and Percival) Varol, p. 133, Plate 1, Figures 26-28.

Remarks: The number of elements in the basal plates is highly variable; in the present material, the number varies between 12 and 24. *P. ovata* has a rim, but in poorly preserved assemblages, the rim is absent. The species is possibly a poorly preserved morphotype of one of the *Pontosphaera* species, but it is difficult to assign to any other available species name. This species generally is limited to upper Oligocene and lower Miocene sequences. *P. scutellum* is easily distinguished from *P. ovata* by being large (>10µ) and possessing a smooth non-perforated basal plate. Some specimens of *P. scutellum* have randomly arrayed rare perforations in the basal plate, whereas *P. ovata* has distinct elements aligned along the short axis of the lophodolith.

MESOZOIC SPECIES CONSIDERED

Arkhangelskiella cymbiformis Vekshina, 1959
Broinsonia parca constricta Hattner et al, 1980
Broinsonia parca parca (Stradner) Bukry, 1969
Cribrosphaerella ehrenbergii (Arkhangelsky) Deflandre, 1952
Eiffellithus eximius (Stover) Perch-Nielsen, 1968
Eiffellithus turriseiffelii (Deflandre) Reinhardt, 1965
Micula staurophora (Gardet) Stradner, 1963
Calculites obscurus (Deflandre) Prins and Sissingh *in* Sissingh, 1977
Prediscosphaera cretacea (Arkhangelsky) Gartner, 1968
Quadrum gothicum (Deflandre) Prins and Perch-Nielsen *in* Manivit et al, 1977
Quadrum trifidum (Stradner) Prins and Perch-Nielsen *in* Manivit et al, 1977

Retecapsa crenulata (Bramlette and Martini) Grun and Allemann, 1975
Tranolithus phacelosus Stover, 1966
Watznaueria barnesae (Black) Perch-Nielsen, 1968

CENOZOIC SPECIES CONSIDERED

Anacanthoica solomonica Varol n. sp.
Amaurolithus delicatus Gartner and Bukry, 1975
Amaurolithus primus (Bukry and Percival) Gartner and Bukry, 1975
Braarudosphaera bigelowii (Gran and Braarud) Deflandre, 1947
Calcidiscus leptoporus (Murray and Blackman) Loeblich and Tappan, 1978
Calcidiscus premacintyrei Theodoridis, 1984
Calcidiscus tropicus (Kamptner) Varol, n. comb.
Calciosolenia murrayi Gran, *in* Murray and Hjort, 1912
Catinaster coalitus Martini and Bramlette, 1963
Ceratolithus acutus Gartner and Bukry, 1974
Ceratolithus cristatus Kamptner, 1950
Ceratolithus rugosus Bukry and Bramlette, 1968
Chiasmolithus grandis (Bramlette and Sullivan) Radomski, 1968
Clausicoccus fenestratus (Deflandre and Fert) Prins, 1979
Clausicoccus obrutus (Perch-Nielsen) Prins, 1979
Coccolithus cavus Hay and Mohler, 1967
Coccolithus formosus (Kamptner) Wise, 1973
Coccolithus pelagicus (Wallich) Schiller, 1930
Coccolithus subpertusus (Hay and Mohler) Heck and Prins, 1987
Coronocyclus nitescens (Kamptner) Bramlette and Wilcoxon, 1967
Coronosphaera mediterranea (Lohmann) Gaarder *in* Gaarder and Heimdal, 1977
Cyclicargolithus abisectus (Muller) Wise, 1973
Cyclicargolithus floridanus (Roth and Hay) Bukry, 1971
Discoaster adamanteus Bramlette and Wilcoxon, 1967
Discoaster barbadiensis Tan, 1927
Discoaster bellus Bukry and Percival, 1971
Discoaster brouweri Tan, 1927
Discoaster calcaris Gartner, 1967
Discoaster challengeri Bramlette and Riedel, 1954
Discoaster deflandrei Bramlette and Riedel, 1954
Discoaster druggii Bramlette and Wilcoxon, 1967
Discoaster intercalaris Bukry, 1971
Discoaster kugleri Martini and Bramlette, 1963
Discoaster mohleri Bukry and Percival, 1971
Discoaster multiradiatus Bramlette and Riedel, 1954
Discoaster neohamatus Bukry and Bramlette, 1969
Discoaster pentaradiatus Tan, 1927
Discoaster prepentaradiatus Bukry and Percival, 1971
Discoaster quinqueramus Gartner, 1969
Discoaster saipanensis Bramlette and Riedel, 1954
Discoaster surculus Martini and Bramlette, 1963
Discoaster variabilis Martini and Bramlette, 1963
Emiliania huxleyi (Lohmann) Hay and Mohler *in* Hay et al, 1967
Fasciculithus bobii Perch-Nielsen, 1971a
Fasciculithus hayi Haq, 1971
Fasciculithus involutus Bramlette and Sullivan, 1961
Fasciculithus tympaniformis Hay and Mohler *in* Hay et al, 1967
Florisphaera profunda Okada and Honjo, 1973
Geminilithella jafari (Muller) Backman, 1980
Geminilithella lordi (Varol) Varol, n. comb.
Geminilithella petaliformis (Varol) Varol, n. comb.
Geminilithella rotula (Muller) Backman, 1980
Geminilithella subtilis (Muller) Varol, n. comb.
Gephyrocapsa oceanica Kamptner, 1943
Gephyrocapsa protohuxleyi McIntyre, 1970
Hayaster perplexus (Bramlette and Riedel) Bukry, 1973
Helicosphaera ampliaperta Bramlette and Wilcoxon, 1967
Helicosphaera californiana Bukry, 1981
Helicosphaera carteri (Wallich) Kamptner, 1954
Helicosphaera compacta Bramlette and Wilcoxon, 1967
Helicosphaera crouchii Bukry, 1981
Helicosphaera euphratis Haq, 1966
Helicosphaera girgisii Varol, n. sp.
Helicosphaera intermedia Martini, 1965
Helicosphaera magnifica Varol, n. sp.
Helicosphaera mediterranea Muller, 1981
Helicosphaera mullerae Varol, n. sp.
Helicosphaera jakubowski Varol, n. sp.
Helicosphaera obliqua Bramlette and Wilcoxon, 1967
Helicosphaera orientalis Black, 1971
Helicosphaera philippinensis Muller, 1981
Helicosphaera pacifica Muller and Bronnimann, 1974
Helicosphaera perch-nielsenae Haq, 1971
Helicosphaera recta Haq, 1966
Helicospahera rhomba Bukry, 1971
Helicosphaera sellii Bukry and Bramlette, 1969
Helicosphaera vedderi Bukry, 1981
Helicosphaera walbersdorfensis Muller, 1974b
Heliolithus kleinpellii Sullivan, 1964
Hellodosphaera dalmatica (Kamptner) Okada and McIntyre, 1977
Homozygosphaera schilleri (Kamptner) Okada and McIntyre, 1977
Homozygosphaera wetsteinii (Kamptner) Halldal and Markali, 1955
Hughesius gizoensis Varol, n. gen., n. sp.
Lithostromation perdurum Deflandre, 1942b
Micrascidites vulgaris Deflandre and Deflandre-Rigaud, 1956
Neosphaera coccolithomorpha Lecal-Schlauder, 1950
Oolithus fragilis (Lohmann) Martini and Muller, 1972
Orthorhabdus serratus Bramlette and Wilcoxon, 1967
Pontosphaera callosa (Martini) Varol, 1982
Pontosphaera japonica (Takayama) Nishida, 1971
Pontosphaera messinae Bartolini, 1970
Pontosphaera multipora (Kamptner) Roth, 1970
Pontosphaera ovata (Levin and Joerger) Varol, n. comb.
Pontosphaera pacifica Burns, 1973
Pontosphaera scutellum Kamptner, 1952
Pontosphaera sparsiforatus (Kamptner) Varol, n. comb.
Pontosphaera syracusana Lohmann, 1902

Pontosphaera trenus (Kamptner) Varol, n. comb.
Pontosphaera vigintiforatus (Kamptner) Varol, n. comb.
Pyrocyclus hermosus Roth and Hay *in* Hay et al, 1967
Pyrocyclus inversus Hay and Towe, 1962
Prinsius bisulcus (Stradner) Hay and Mohler, 1967
Pseudoemiliania lacunosa (Kamptner) Gartner, 1969
Reticulofenestra bisecta (Hay, Mohler and Wade) Roth, 1970
Reticulofenestra dictyoda (Deflandre) Stradner *in* Stradner and Edwards, 1968
Reticulofenestra hesslandii (Haq) Roth, 1970
Reticulofenestra minuta Roth, 1970
Reticulofenestra minutula (Gartner) Haq and Berggren, 1978
Reticulofenestra producta (Kamptner) Varol, n. comb.
Reticulofenestra pseudoumbilica (Gartner) Gartner, 1969
Reticulofenestra stavensis (Levin and Joerger) Varol, n. comb.
Reticulofenestra tenuistriatus (Kamptner) Varol, n. comb.
Reticulofenestra umbilica (Levin) Martini and Ritzkowski, 1968
Rhabdosphaera clavigera Murray and Blackman, 1898
Rhabdosphaera stylifera Lohmann, 1902
Scapholithus fossilis Deflandre, 1954
Scyphosphaera amphora Deflandre, 1942a
Scyphosphaera ampla Kamptner, 1955
Scyphosphaera apsteinii Lohmann, 1902
Scyphosphaera brevisa Varol, 1982
Scyphosphaera campanula Deflandre, 1942a
Scyphosphaera catharrellus Kamptner, 1955
Scyphosphaera cohenii Boudreaux and Hay, 1969
Scyphosphaera conica Kamptner, 1955
Scyphosphaera cylindrica Kamptner, 1955
Scyphosphaera deflandrei Muller, 1974a
Scyphosphaera globulata Bukry and Percival, 1971
Scyphosphaera globulosa Kamptner, 1955
Scyphosphaera intermedia Deflandre, 1942a
Scyphosphaera porosa Kamptner, 1967
Scyphosphaera praeglobulata Varol, n. sp.
Scyphosphaera priformis Kamptner, 1955
Scyphosphaera pseudorecurvata Varol, n. sp.
Scyphosphaera pulcherrima Deflandre, 1942a
Scyphosphaera quasitubifera Varol, 1982
Scyphosphaera recta (Deflandre) Kamptner, 1955
Scyphosphaera recurvata Deflandre, 1942a
Sphenolithus abies Deflandre, 1954
Sphenolithus belemnos Bramlette and Wilcoxon, 1967
Sphenolithus ciperoensis Bramlette and Wilcoxon, 1967
Sphenolithus conicus Bukry, 1971
Sphenolithus distentus (Martini) Bramlette and Wilcoxon, 1967
Sphenolithus furcatolithoides Locker, 1967
Sphenolithus heteromorphus Deflandre, 1953
Sphenolithus moriformis (Bronnimann and Stradner) Bramlette and Wilcoxon, 1967
Sphenolithus obtusus Bukry, 1971
Sphenolithus predistentus Bramlette and Wilcoxon, 1967
Sphenolithus pseudoradians Bramlette and Wilcoxon, 1967
Sphenolithus radians Deflandre, 1952
Syracosphaera histrica Kamptner, 1941
Syracosphaera pulchra Lohmann, 1902
Tetralithoides symeonidesii Theodoridis, 1984
Thoracosphaera heimii (Lohmann) Kamptner, 1941
Toweius eminens (Bramlette and Sullivan) Perch-Nielsen, 1971a
Toweius pertusus (Sullivan) Romein, 1979
Triquetrorhabdulus carinatus Martini, 1965
Triquetrorhabdulus dorsalis Varol, 1982
Triquetrorhabdulus milowii Bukry, 1971
Triquetrorhabdulus rugosus Bramlette and Wilcoxon, 1967
Umbellosphaera irregularis Paasche *in* Markali and Paasche, 1955
Umbilicosphaera sibogae (Weber-van Bosse) Gaarder, 1970

SELECTED REFERENCES

Arkhangelskii, A.D., 1912, Verkhnemelovyya otlozheniya vostoka evropeiskoi Rossii (Upper Cretaceous deposits of east European Russia): Material Geologie Rossie, v. 25, 631 p.

Backman, J., 1980, Miocene-Pliocene nannofossils and sedimentation rates in the Hatton-Rockall basin, NE Atlantic Ocean: Stockholm Contributions in Geology, v. 36, n. 1, p. 1-91.

Bartolini, C., 1970, Coccoliths from sediments of the western Mediterranean: Micropaleontology, v. 16, n. 2, p. 129-154.

Black, M., 1964, Cretaceous and Tertiary coccoliths from Atlantic seamounts: Palaeontology, v. 7, p. 306-316.

Black, M., 1967, New names for some coccolith taxa: Geological Society of London, Proceedings, n. 1640, p. 139-145.

Black, M., 1971, The systematics of coccoliths in relation to the palaeontological record, *in* Funnell, B.M., and Riedel, W.R., eds., The micropalaeontology of oceans: Cambridge University Press, p. 611-624.

Boudreaux, J.E., and Hay, W.W., 1969, Calcareous nannoplankton and biostratigraphy of the late Pliocene-Pleistocene-Recent sediments in the Submarex cores: Revista Española de Micropaleontologia, v. 1, n. 3, p. 249-292.

Braarud, T., Bramlette, M.N., Deflandre, G., Kamptner, E., Loeblich, A.R. Jr., Martini, E. and Tappan, H., 1964, *Rhabdosphaera* Haeckel, 1894, (Coccolithophorida): Proposed validation under the plenary powers and designation of a lectotype for *Coccolithus oceanicus* Schwarz, 1894. Z.N.(S.) 1658: Bulletin of Zoological Nomenclature, v. 21, p. 397-400.

Bramlette, M.N., and Martini, E., 1964, The great change in calcareous nannoplankton fossils between the Maestrichtian and Danian: Micropaleontology, v. 10, p. 291-322.

Bramlette, M.N., and Riedel, W.R., 1954, Stratigraphic value of discoasters and some other microfossils related to Recent coccolithophores: Journal of Paleontology, v. 28, p. 385-403.

Bramlette, M.N., and Sullivan, F.R., 1961, Coccolithophorids and related nannoplankton of the early Tertiary in California: Micropaleontology, v. 7, p. 129-174.

Bramlette, M.N., and Wilcoxon, J.A., 1967, Middle Tertiary calcareous nannoplankton of the Cipero Section, Trinidad, W.I.: Tulane Studies in Geology, v. 5, p. 93-131.

Brönnimann, P., and Stradner, H., 1960, Die Foraminiferen und Discoasteridenzonen von Kuba und ihre interkontinentale Korrelation: Erdöl-Zeitschrift, v. 76, p. 364-369.

Bukry, D., 1969, Upper Cretaceous coccoliths from Texas and Europe: University of Kansas Paleontological Contributions, Paper 51 (Protista 2), 79 p.

Bukry, D., 1971, Cenozoic calcareous nannofossils from the Pacific Ocean: Transactions of the San Diego Society of Natural History, v. 16, p. 303-327.

Bukry, D., 1973, Phytoplankton stratigraphy, Deep Sea Drilling Project Leg 20, western Pacific Ocean, *in* Heezen, B.C., MacGregor, I.D., and shipboard scientific party, Initial Reports of the Deep Sea Drilling Project, v. 20, p. 307-317.

Bukry, D., 1975, Coccolith and silicoflagellate stratigraphy, northwestern Pacific Ocean, Deep Sea Drilling Project Leg 32, *in* Larson, R.L., Moberley, R., and shipboard scientific party, Initial Reports of the Deep Sea Drilling Project, v. 32, p. 677-701.

Bukry, D., 1981, Pacific coast coccolith stratigraphy between Point Conception and Cabo Corrientes, Deep Sea Drilling Project, Leg 63, *in* Yeats, R.S., Haq, B.U., and shipboard scientific party, Initial Reports of the Deep Sea Drilling Project, v. 63, p. 445-471.

Bukry, D., and Bramlette, M.N., 1968, Stratigraphic significance of two genera of Tertiary calcareous nannofossils: Tulane Studies in Geology, v. 6, p. 149-155.

Bukry, D., and Bramlette, M.N., 1969, Some new and stratigraphically useful calcareous nannofossils of the Cenozoic: Tulane Studies in Geology, v. 7, p. 131-142.

Bukry, D., and Percival, S.F., 1971, New Tertiary calcareous nannofossils: Tulane Studies in Geology, v. 8, p. 123-146.

Burns, D.A., 1973, Structural analysis of flanged coccoliths in sediments from the southwest Pacific Ocean: Revista Española de Micropaleontologia, v. 5, n. 1, p. 147-160.

Chi, W.R., 1979, Calcareous nannoplankton biostratigraphy of the Nantou area, central Taiwan: Petroleum Geology of Taiwan, v. 16, p. 131-165.

Clochiatti, M., 1971, Contribution a l'étude du nannoplancton calcaire du neogene d'Afrique du nord: Muséum National d'Histoire Naturelle, Memoires Sér. C, v. 23, 135 p.

Coulson, F.I., and Vedder, J.G., 1986, Geology of the central and western Solomon Islands, *in* Vedder, J.G., Pound, K.S., and Boundy, S.Q., eds., Geology and offshore resources of Pacific Island arcs—central and western Solomon Islands: Houston, Texas, Circum-Pacific Council for Energy and Mineral Resources, Earth Science Series, v. 4, p. 59-87.

Deflandre, G., 1942a, Coccolithophorides fossiles d'Oranie; genres *Scyphosphaera* Lohmann et *Thorosphaera* Ostenfeld: Société Histoire Naturelle de Toulouse, Bulletin, v. 77, p. 125-137.

Deflandre, G., 1942b, Possibilités morphogénétiques comparées du calcaire et de la silice, à propos d'un nouveau type de microfossile calcaire de structure complexe, *Lithostromation perdurum* n. gen. n. sp.: Académie des Sciences de Paris, Comptes Rendus Hebdomadaires des Séances, v. 214, n. 23, p. 917-919.

Deflandre, G., 1947, *Braarudosphaera* gen. nov., type d'une famille nouvelle de Coccolithophoridés actuels à éléments composites: Académie des Sciences de Paris, Comptes Rendus Hebdomadaires des Séances, v. 225, p. 439-441.

Deflandre, G., 1953, Hétérogénéité intrinséque et pluralité des éléments dans les coccolithes actuels et fossiles: Académie des Sciences de Paris, Comptes Rendus Hebdomadaires des Séances, v. 237, p. 1785-1787.

Deflandre, G. and Fert, C., 1954, Observations sur les Coccolithophoridés actuels et fossiles en microscopie ordinaire et électronique: Annales de Paléontologie, v. 40, p. 115-176.

Deflandre, G., and Deflandre-Rigaud, M., 1956, *Micrascidites* manip. nov., sclérites de Didemnidés (Ascidies, Tuniciers) fossiles du Lutétien du Bassin Parisien et du Balcombien d'Astralie: Société Géologique de France, Comptes Rendus Sommaire des Séances, n. 3-4, p. 47-49.

Dunkley, P.N., 1983, Volcanism and the evolution of the ensimatic Solomon Islands arc, *in* Shimozuru, D., andYokoyama, I., eds., Arc volcanism: Physics and Tectonics: Tokyo, Terrapub, p. 225-241.

Gaarder, K.R., 1970, Three new taxa of Coccolithinae: Nytt Magasin for Botanikk, v. 17, p. 113-126.

Gaarder, K.R., and Heimdal, B.R., 1977, A revision of the genus *Syracosphaera* Lohmann (Coccolithineae): "Meteor" Forschungsergebnisse, v. D/24, p. 54-71.

Gardet, M., 1955, Contribution a l'étude des coccolithes des terrains Néogènes de l'Algérie: Algeria, Service de la Carte Geologique, ser. 2, Bulletin 5, p. 477-550.

Gartner, S., Jr., 1967, Calcareous nannofossils from Neogene of Trinidad, Jamaica, and Gulf of Mexico: University of Kansas, Paleontological Contributions, v. 29, p. 1-7.

Gartner, S., Jr., 1968, Coccoliths and related calcareous nannofossils from Upper Cretaceous deposits of Texas and Arkansas: University of Kansas, Paleontological Contributions, Paper 48 (Protista 1), p. 1-56.

Gartner, S., Jr., 1969, Correlation of Neogene planktonic foraminifera and calcareous nannofossil zones: Transactions, Gulf Coast Association of Geological Societies, v. 19, p. 585-599.

Gartner, S., Jr., 1977, Calcareous nannofossil biostratigraphy and revised zonation of the Pleistocene: Marine Micropaleontology, v. 2, n. 1, p. 1-25.

Gartner, S., Jr., and Bukry, D., 1974, *Ceratolithus acutus* Gartner and Bukry n. sp. and *Ceratolithus amplificus* Bukry and Percival, nomenclatural classification: Tulane Studies in Geology, v. 11, p. 115-118.

Gartner, S., Jr., Chen, M.P., and Stanton, R.J., 1983, Late Neogene nannofossils, biostratigraphy, and paleoceanography of the northeastern Gulf of Mexico and adjacent areas: Marine Micropaleontology, v. 8, p. 17-50.

Gran, H.H., 1912, Pelagic plant life, *in* Murray, J., and Hjort, J., eds., The depths of the oceans: London, Macmillan, p. 307-386.

Gran, H.H., and Braarud, T., 1935, A quantitative study of the phytoplankton in the Bay of Fundy and the Gulf of Maine (including observations on hydrography, chemistry and turbidity): Journal of the Biological Board of Canada, v. 1, p. 279-467.

Grasse, P.P., 1952, Traité de zoologie, Anatomie, systematique, biologie, Tome 1, fasc. 1: Masson, Paris, 1071 p.

Grun , W., and Allemann, F., 1975, The Lower Cretaceous of Caravaca (Spain); Berriasian calcareous nannoplankton of the Miravetes Section (Subbetie Zone, Province of Murcia): Eclogae Geologicae Helveticae, v. 68, n. 1, p. 147-211.

Haeckel, E., 1894, Systematische Phylogenie der Protisten und Pflanzen: Berlin, Reimer, 400 p.

Halldal, P., and Markali, J., 1955, Electron microscope studies on coccolithophorids from the Norwegian Sea, the Gulf Stream, and the Mediterranean: Norske Videnskaps - Akademi i Oslo, Matematisk-Naturvidenskapelig Klasse, v. 1, p. 1-27.

Haq, B.U., 1966, Electron microscope studies on some upper Eocene calcareous nannoplankton from Syria: Stockholm Contributions in Geology, v. 15, p. 23-37.

Haq, B.U., 1971, Paleogene calcareous nannoflora Parts I-IV: Stockholm Contributions in Geology, v. 25, p. 1-158.

Haq, B.U., and Berggren, W.A., 1978, Late Neogene calcareous plankton biochronology of the Rio Grande Rise (South Atlantic Ocean): Journal of Paleontology, v. 52, n. 6, p. 1167-1194.

Hattner, T.G., Wind, F.H., and Wise, S.W., Jr., 1980, Santonian-Campanian boundary: comparison of nearshore-offshore calcareous nannofossil assemblages: Cahiers de Micropaleontologie, v. 1980, n. 3, p. 9-26.

Hay, W.W., 1977, Calcareous nannofossils, *in* Ramsay, A.T.S., ed., Oceanic micropalaeontology: New York, Academic Press, p. 1055-1200.

Hay, W.W., and Mohler, H.P., 1967, Calcareous nannoplankton from early Tertiary rocks at Pont Labau, France, and Paleocene-Eocene correlations: Journal of Paleontology, v. 41, p. 1505-1541.

Hay, W.W., Mohler, H.P., Roth, P.H., Schmidt, R.R., and Boudreaux, J.E., 1967, Calcareous nannoplankton zonation of the Cenozoic of the Gulf Coast and Caribbean-Antillean area, and transoceanic correlation: Transactions of the Gulf Coast Association of Geological Societies, v. 17, p. 428-480.

Hay, W.W., Mohler, H.P., and Wade, M.E., 1966, Calcareous nannofossils from Nal'chik (northwest Caucasus): Eclogae Geologicae Helveticae, v. 59, p. 379-399.

Hay, W.W., and Towe, K.M., 1962, Electron microscope examination of some coccoliths from Donzacq (France): Eclogae Geologicae Helveticae, v. 55, p. 497-517.

Hughes, G.W., 1982, Stratigraphic correlation between sedimentary basins of the ESCAP region, v. 8, Solomon Islands, ESCAP Atlas of Stratigraphy III: United Nations Mineral Resources Development Series, n. 48, p. 115-130.

Hughes, G.W., Varol, O., and Dunkley, P.N., Pleistocene uplift of Tetepare Island, Solomon Group as implied by foraminiferal and calcareous nannofossil evidence: Royal Society of New Zealand (in press).

Huxley, T.H., 1868, On some organisms living at great depths in the North Atlantic Ocean: Quarterly Journal of Microscopical Science, Ser. 2, n. 8, p. 203-212.

Kamptner, E., 1927, Beitrag zur Kenntnis adriatischer Coccolithophoriden: Archiv für Protistenkunde, v. 58, p. 173-184.

Kamptner, E., 1937, Neue und bemerkenswerte Coccolithineen aus dem Mittelmeer: Archiv für Protistenkunde, v. 89, p. 279-316.

Kamptner, E., 1941, Die Coccolithineen dur Südwestküste von Istrien: Naturhistorisches Museum in Wien, Annalen, v. 51, p. 54-149.

Kamptner, E., 1943, Zur Revision der Coccolithineen-Spezies *Pontosphaera huxleyi* Lohm: Akademie der Wissenschaften in Wien, Mathematische-Naturwissenschaftliche Klasse, Anzeiger, v. 80, p. 43-49.

Kamptner, E., 1948, Coccolithen aus dem Torton des inneralpinen Wiener Beckens: Osterreichische Akademie der Wissenschaften, Mathematsche-Naturwissenschaftliche Klasse, Sitzungsberichte, Part 1, v. 157, p. 1-16.

Kamptner, E., 1950, Über den submikroskopischen Aufbau der Coccolithen: Österreichische Akademie der Wissenschaften, Mathematische-Naturwissenschaftliche Klasse, Anzeiger, v. 87, p. 152-158.

Kamptner, E., 1952, Das mikroskopische Studium des Skelettes der Coccolithineen (Kalkflagellaten): Mikroskopie, v. 7, p. 375-386.

Kamptner, E., 1954, Untersuchungen über den Feinbau der Coccolithen: Archiv für Protistenkunde, v. 100, p. 1-90.

Kamptner, E., 1955, Fossile Coccolithineen-Skelettreste aus Insulinde: Koninklijke Nederlandse Akademie van Wetenschappen Afdeeling Naturkunde, Verhandelingen, v. 50, n. 2, p. 1-105.

Kamptner, E., 1963, Coccolithineen-Skelettreste aus Tiefseeablagerungen des Pazifischen Ozeans: Naturhistorisches Museum in Wien, Annalen, v. 66, p. 139-204.

Kamptner, E., 1967, Kalkflagellaten-Skelettreste aus Tiefseeschlamm des Südatlantischen Ozeans: Naturhistorisches Museum in Wien, Annalen, v. 71, p. 117-198.

Lecal-Schlauder, J., 1950, Notes préliminaires sur les coccolithophoridés d'Afrique du Nord: Sociéte d'Histoire Naturelle de Afrique Nord, Bulletin, v. 40, 160-167.

Levin, H.L., 1965, Coccolithophoridae and related microfossils from the Yazoo Formation (Eocene) of Mississippi: Journal of Paleontology, v. 39, p. 265-272.

Levin, H.L., and Joerger, A.P., 1967, Calcareous nannoplankton from the Tertiary of Alabama: Micropaleontology, v. 13, p. 163-182.

Locker, S., 1967, Neue Coccolithophoriden (Flagellata) aus dem Alttertiär Norddeutschlands: Geologie (Berlin), v. 16, p. 361-364.

Loeblich, A.R., Jr., and Tappan, H., 1963, Type fixation and validation of certain calcareous nannoplankton genera: Proceedings of the Biological Society of Washington, v. 76, p. 191-196.

Loeblich, A.R., Jr., and Tappan, H., 1966, Annotated index and bibliography of the calcareous nannoplankton: Phycologia, v. 5, n. 2 and 3, p. 81-216.

Loeblich, A.R., Jr., and Tappan, H., 1978, The coccolithophorid genus *Calcidicus* Kamptner and its synonyms: Journal of Paleontology, v. 52, n. 6, p. 1390-1392.

Lohmann, W., 1902, Die Coccolithophoridae, eine Monographie der Coccolithen bildenden Flagellaten, zugleich ein Beitrag zur Kenntnis des Mittelmeerauftriebs: Archiv für Protistenkunde, v. 1, p. 89-165.

Lohmann, H., 1912, Untersuchungen über das Pflanzen und Tierleben der Hochsee, Zugleich ein Bericht über die biologischen Arbeiten auf der Fahrt der "Deutschland" von Bremerhaven nach Buenos Aires in der Zeit vom 7. Mai bis 7 September 1911: Veröffentlichungen des Instituts für Meereskunde und des Geographischen Instituts an der Universität, Berlin, N.F., Geographisth-Naturwisschaften Reihe, v. 1, p. 1-92, 2 tables.

Lohmann, H., 1919, Die Bevölkerung des Ozeans mit Plankton nach den Ergebnissen der Zentrifugenfänge während der Ausreise der "Deutschland" 1911: Archiv für Biontologie (Ges. Naturf. Freunde Berlin), v. 4, n. 3, p. 1-617. This in 2 parts: pt. 1, p. 1-470; pt. 2, p. 471-617.

McIntyre, A., 1970, *Gephyrocapsa protohuxleyi* sp. n., a possible phyletic link and index fossil for the Pleistocene: Deep Sea Research, v. 17, p. 187-190.

Manivit, H., Perch-Nielsen, K., Prins, B., and Verbeek, J.W., 1977, Middle Cretaceous calcareous nannofossil biostratigraphy: Koninklijke Nederlandse Akademie van Wetenschappen, Proceedings, v. B80, n. 3, p. 169-181.

Markali, J., and Paasche, E., 1955, On two species of *Umbellosphaera*, a new marine coccolithophorid genus: Nytt Magasin for Botanikk, v. 4, p. 95-100.

Martini, E., 1965, Mid-Tertiary calcareous nannoplankton from Pacific deep-sea cores, *in* Whittard, W.F., and Gradshaw, R.B., eds., Submarine geology and geophysics, Proceedings of the 17th Symposium of the Colston Research Society of London: Butterworths, p. 393-411.

Martini, E., 1971, Standard Tertiary and Quaternary calcareous nannoplankton zonation: Planktonic Conference, Proceedings, v. 2, p. 739-777

Martini, E., and Bramlette, M.N., 1963, Calcareous nannoplankton from the experimental Mohole drilling: Journal of Paleontology, v. 37, p. 845-856.

Martini, E., and Muller, C., 1972, Nannoplankton aus dem nördlichen Arabischen Meer. "Meteor" Forschungsergebnisse, v. C/10, p. 63-74.

Martini, E., and Muller, C., 1986, Current Tertiary and Quaternary calcareous nannoplankton stratigraphy and correlations: Newsletters on Stratigraphy, v. 16 , n. 2, p. 99-112.

Martini, E., and Ritzkowski, S., 1968, Die Grenze Eozän/Oligozän in der typus-region des Unteroligozäns (Helmstedt, Egeln, Latdorf), *in* Colloque sur l'Eocène, Paris, mai 1968: France, Bureau de Recherches Géologiques et Minières, Mémoir, n. 69, p. 233-237.

Muller, C., 1970, Nannoplankton-Zonen der Unteren Meeresmolasse Bayerns: Geologica Bavarica, v. 63, p. 107-118.

Muller, C., 1974a, Calcareous nannoplankton, Leg 25 (western Indian Ocean), *in* Simpson, E.S.W., Schlich, R., and shipboard scientific party, Initial Reports of the Deep Sea Drilling Project, v. 25, p. 579-633.

Muller, C., 1974b, Nannoplankton aus dem Mittel-Miozän von Walbersdorf (Burgenland): Senckenbergiana Lethaea, v. 55, p. 389-405.

Muller, C., 1976, Nannoplankton-Gemeinschaften aus dem Jung-Auartar des Golfes von Aden und des Roten Meeres: Geologische Jahrbuch, v. D17, p. 33-37.

Muller, C., 1981, Beschreibung neuer *Helicosphaera*-Arten aus dem Miozän und Revision biostratigraphischer Reichweiten einiger neogener Nannoplankton-Arten: Senckenbergiana Lethaea, v. 61, n. 3/6, p. 427-435.

Muller, C., and Bronnimann, P., 1974, Eine neue Art der Gattung *Helicosphaera* Kamptner aus dem Pazifischen Ozean: Eclogae Geologicae Helveticae, v. 67, p. 661-662.

Murray, G., and Blackman, V.H., 1898, On the nature of the coccospheres and rhabdosphaeres: Philosophical Transactions of the Royal Society of London, v. 190B, p. 427-441.

Murray, J., and Hjort, J., 1912, eds., The depths of the oceans: London, Macmillan.

Nishida, S., 1971, Nannofossils from Japan; IV, Calcareous nannoplankton fossils from the Tonohama Group, Shikoku, southwest Japan: Transactions and Proceedings of the Palaeontological Society of Japan, new ser., n. 83, p. 143-161.

Okada, H., and Bukry, D., 1980, Supplementary modification and introduction of code numbers to the low-latitude coccolith biostratigraphic zonation (Bukry, 1973; 1975): Marine Micropaleontology, v. 5, n. 3, p. 321-325.

Okada, H., and Honjo, S., 1973, The distribution of oceanic coccolithophorids in the Pacific: Deep Sea Research, v. 20, p. 355-374.

Okada, H., and McIntyre, A., 1977, Modern coccolithophores of the Pacific and North Atlantic Oceans: Micropaleontology, v. 23, p. 1-55.

Perch-Nielsen, K., 1968, Der Feinbau und die Klassifikation der Coccolithen aus dem Maastrichtien von Dänemark: Kongelige Danske Videnskabernes Selskab, Biologiscke Skrifter, v. 16, n. 1, p. 1-96.

Perch-Nielsen, K., 1969, Die Coccolithen einiger dänischer Maastrichtien-und Danienlokalitäten. Dansk Geologisk Forening, Meddelelser, v. 19, p. 51-66.

Perch-Nielsen, K., 1971a, Einige neue Coccolithen aus dem Paläozän der Bucht von Biskaya: Bulletin of the Geological Society of Denmark, v. 20, Part 4, p. 347-361.

Perch-Nielsen, K. 1971b, Neue Coccolithen aus dem Paläozän von Dänemark, der Bucht von Biskaya und dem Eozän der Labrador See: Bulletin of the Geological Society of Denmark, v. 21, n. 1, p. 51-66.

Perch-Nielsen, K., 1971c, Elektronenmikroskopische Untersuchungen an Coccolithen und verwandten Formen aus dem Eozän von Dänemark: Kongelige Danske Videnskabernes Selskab, Biologiske Skrifter, v. 18, n. 3, p. 1-73.

Perch-Nielsen, K., 1985, Cenozoic calcareous nannofossils, *in* Bolli, H.M., Saunders, J.B., and Perch-Nielsen, K., eds., Plankton stratigraphy: Cambridge University Press, Cambridge Earth Science Series, p. 427-554.

Prins, B., 1979, Notes on nannology; 1, *Clausicoccus*, a new genus of fossil coccolithophorids: International Nannoplankton Association Newsletter, v. 1, n. 1, p. N2-N4.

Radomski, A., 1968, Calcareous nannoplankton zones in Palaeogene of the western Polish Carpathians: Polskie Towarzystwo Geologiczne, Rocznik, v. 38, p. 454-605.

Raffi, I., and Rio, D., 1979, Calcareous nannofossil biostratigraphy of Deep Sea Drilling Project Site 132 - Leg 13 (Tyrrhenian Sea - Western Mediterranean): Rivista Italiana di Paleontologia e Stratigrafia, v. 85, p. 127-172.

Reinhardt, P., 1965, Neue Familien für fossile Kalkflagellaten (Coccolithophoriden, Coccolithineen): Deutsche Akademie der Wissenschaften zu Berlin, Monatsberichte, v. 7, p. 30-40.

Romein, A.J.T., 1979, Lineages in early Paleogene calcareous nannoplankton. Utrecht Micropalaeontological Bulletins, v. 22, p. 1-231.

Roth, P.H., 1970, Oligocene calcareous nannoplankton biostratigraphy: Eclogae Geologicae Helveticae, v. 63, p. 799-881.

Sachs, J.B., and Skinner, H.C., 1973, Calcareous nannofossils and the late Pliocene-early Pleistocene biostratigraphy, Louisiana Continental Shelf: Tulane Studies in Geology, v. 10, n. 3, p. 113-162.

Schiller, J., 1930, Coccolithineae, *in* Rabenhorst, L., ed., Kryptogamen-Flora von Deutschland, Österreich und der Schweiz: Akademie der Verlagsgeschaften, Leipzig, v. 10, pt. 2, p. 89-267.

Schwarz, E.H.L., 1894, Coccoliths: Annals and Magazine of Natural History, ser. 6, v. 14, p. 341-346.

Sissingh, W., 1977, Biostratigraphy of Cretaceous calcareous nannoplankton: Geologie en Mijnbouw, v. 56, p. 37-65.

Stover, L.E., 1966, Cretaceous coccoliths and associated nannofossils from France and the Netherlands, Micropaleontology, v. 12, p. 133-167.

Stradner, H., 1963, New contribution to Mesozoic stratigraphy by means of nannofossils: Proceedings of the sixth world Petrologic Congress (Frankfurt am Main, 1963), sect. 1, paper 4, 16 p.

Stradner, H., and Edwards, A.R., 1968, Electron microscopic studies on upper Eocene coccoliths from Oamaru Diatomite, New Zealand: Austria, Geologische Bundesanstalt, Jahrbuch (Wien), spec. v. 13, p. 1-66.

Sullivan, F.R., 1964, Lower Tertiary nannoplankton from the California Coast Ranges. I. Paleocene: University of California Publications in Geological Sciences, v. 44, p. 163-227.

Sullivan, F.R., 1965, Lower Tertiary nannoplankton from the California Coast Ranges. II. Eocene: University of California Publications in Geological Sciences, v. 53, p. 1-74.

Takayama, T., 1967, First report on nannoplankton of the upper Tertiary and Quaternary of the southern Kwanto Region, Japan: Austria, Geologische Bundesanstalt, Jahrbuch (Wien), v. 110, p. 169-198.

Tan, Sin Hok, 1927, *Discoasteridae incertae sedis*: Koninklijke Akademie van Wetenschappen te Amsterdam, Proceedings of the Section of Sciences, v. 30, p. 411-419.

Tan, Sin Hok, 1931, *Discoasteridae, Coccolithinae,* and *Radiolaria*: Leidsche Geologische Mededeelingen, v. 5, p. 92-114.

Theodoridis, S., 1984, Calcareous nannofossil biozonation of the Miocene and revision of the helicoliths and discoasters: Utrecht Micropaleontological Bulletins, v. 32, p. 1-271.

Varol, O., 1982, Calcareous nannofossils from the Antalya Basin, Turkey: Neues Jahrbuch für Geologie und Paläontologie, Monatshefte, v. 4, p. 244-256.

Varol, O., 1983, Proposed calcareous nannofossil zonation scheme for the Miocene to Holocene of South East Asia: Bulletin of the Geological Society of Malaysia, v. 16, p. 37-46.

Varol, O., 1984, New Neogene calcareous nannofossil taxa from Malta and southern Turkey: Neues Jahrbuch für Geologie und Paläontologie, Monatschefte, v. 6, p. 375-384.

Varol, O., 1985, Miocene calcareous nannofossils from the Mut Basin, southern Turkey: Journal of Micropaleontology, v. 4, n. 2, p. 127-139.

Vekshina, V.N., 1959, Kokkolitofirdy maastrikhskikh otlozhenii Zapadno-Sibirskoi nizmennosti (coccolithophoridae of the Maastrichtian deposits of the west Siberian lowland), *in* Khalfin, L.L., ed., Materialy po paleontologii i stratigrafii zapadnoi Sibiri: Sibirskii Nauchno-Issledovatelskii Institut Geologii, Geofiziki i Mineralnogo Syrya, Trudy, v. 2, p. 56-81.

Wallich, G.C., 1877, Observations on the coccosphere: Annals and Magazine of Natural History, ser. 4, v. 16, p. 322-339.

Weber-Van Bosse, A., 1901, Etudes sur les Algues de l'Archipel Malaisien: Ann. Jard. Bot. Buitenzorg, v. 17 ser. 2.2, p. 126-141.

Wise, S.W., 1973, Calcareous nannofossils from cores recovered during Leg 18, Deep Sea Drilling Project: Biostratigraphy and observations on diagenesis, *in* Kulm, L.D., von Huene, R., and shipboard scientific party, Initial Reports of the Deep Sea Drilling Project, v. 18, p. 569-615.

Vedder, J.G., and Bruns, T.R., editors, 1989, Geology and offshore resources of Pacific island arcs—Solomon Islands and Bougainville, Papua New Guinea Regions: Houston, Texas, Circum-Pacific Council for Energy and Mineral Resources Earth Science Series, v. 12.

CALCAREOUS NANNOFOSSILS FROM THE SOLOMON ISLANDS—BOUGAINVILLE REGION, R/V *S.P. LEE* CRUISE L6-84-SP

S. Shafik
Bureau of Mineral Resources, Geology and Geophysics
GPO Box 378, Canberra 2601, Australia

INTRODUCTION

Sampling was successful at 24 stations out of the 31 stations occupied. Bedrock sampling was tried by means of chain-bag dredges and dart corers, whereas soft-sediment sampling was attempted using a gravity corer.

Dredging was mainly along the eastern margin of the Guadalcanal-Santa Isabel platform, west of Vella Lavella, and north of Buka (Figure 1). Most of the samples containing calcareous nannofossils are Quaternary in age. The exceptions are some fragments of muddy sandstone (types 1 and 2) recovered at Station 19, and calcilutite at Station 29, which are assigned a late Pliocene age. Material recovered by dart coring is assigned mainly a late Pleistocene to Holocene age, indicating a widespread late Quaternary veneer covering the bedrock, even where the slope is steep.

Gravity coring resulted in the recovery of late Quaternary sediments from the Russell Basin.

RESULTS

Station 1

Water depth at Station 1 is about 700 m. Two rock types were recovered; the predominant rock lithology is a metavolcanic conglomerate which is almost barren of coccoliths. Only rare and poorly preserved small *Gephyrocapsa* spp. and a single specimen of *Umbilicosphaera sibogae* were identified. The other lithology is an indurated calcareous sandstone which yielded small *Gephyrocapsa* spp., *Calcidiscus leptoporus*, poorly preserved *Reticulofenestra pseudoumbilica*, and a very poorly preserved coccolith mimicking *Cyclicargolithus abisectus*. Although the nannofossils are extremely rare in the preparations studied from this station, suggesting that the taxa are *in situ*, contamination could not be ruled out. The age of the sample is likely to be Quaternary on the presence of *Gephyrocapsa* and the absence of *Discoaster brouweri*.

Station 2

Three rock types were recovered from water depths between 750 and 800 m at Station 2. A coccolith preparation from toroid-like sandy mudstone (rock type 1 of this station) yielded common specimens of poorly preserved *Gephyrocapsa oceanica*, suggesting a Quaternary age. The second rock type consisting of plate-shaped pieces of sandy mudstone with abundant claystone fragments proved to be barren of calcareous nannofossils. Nannofossils were encountered in three preparations from the third rock type, a poorly indurated volcaniclastic sandstone, and the identified taxa are admixtures of Quaternary and older forms. Although preservation is generally moderate and the fossils are few to common, there is some evidence of rapid burial. *Emiliania huxleyi* is present and *Gephyrocapsa* includes forms alien to *G. oceanica* (such as *G. omega*) and *G. caribbeanica* types. Also present are *Helicosphaera wallichii*, *H. carteri*, *Ceratolithus cristatus*, *Calcidiscus leptoporus*, *Scapholithus fossilis* and questionable small *Pseudoemiliania lacunosa*. Reworked forms include *Coccolithus pelagicus*, *Cyclicargolithus abisectus*, *Discoaster brouweri*, *D. pentaradiatus*, *D. druggii*, *Reticulofenestra* sp., and *Sphenolithus* sp. The occurrence of *Emiliania huxleyi* and *Gephyrocapsa* spp. indicates a late Quaternary age (late Pleistocene to Holocene).

Station 6

Four sedimentary rock types were obtained at Station 6 where water depths range between 500 and 900 m. The predominant rock type is claystone that yielded abundant well-preserved coccoliths. Small *Gephyrocapsa*,

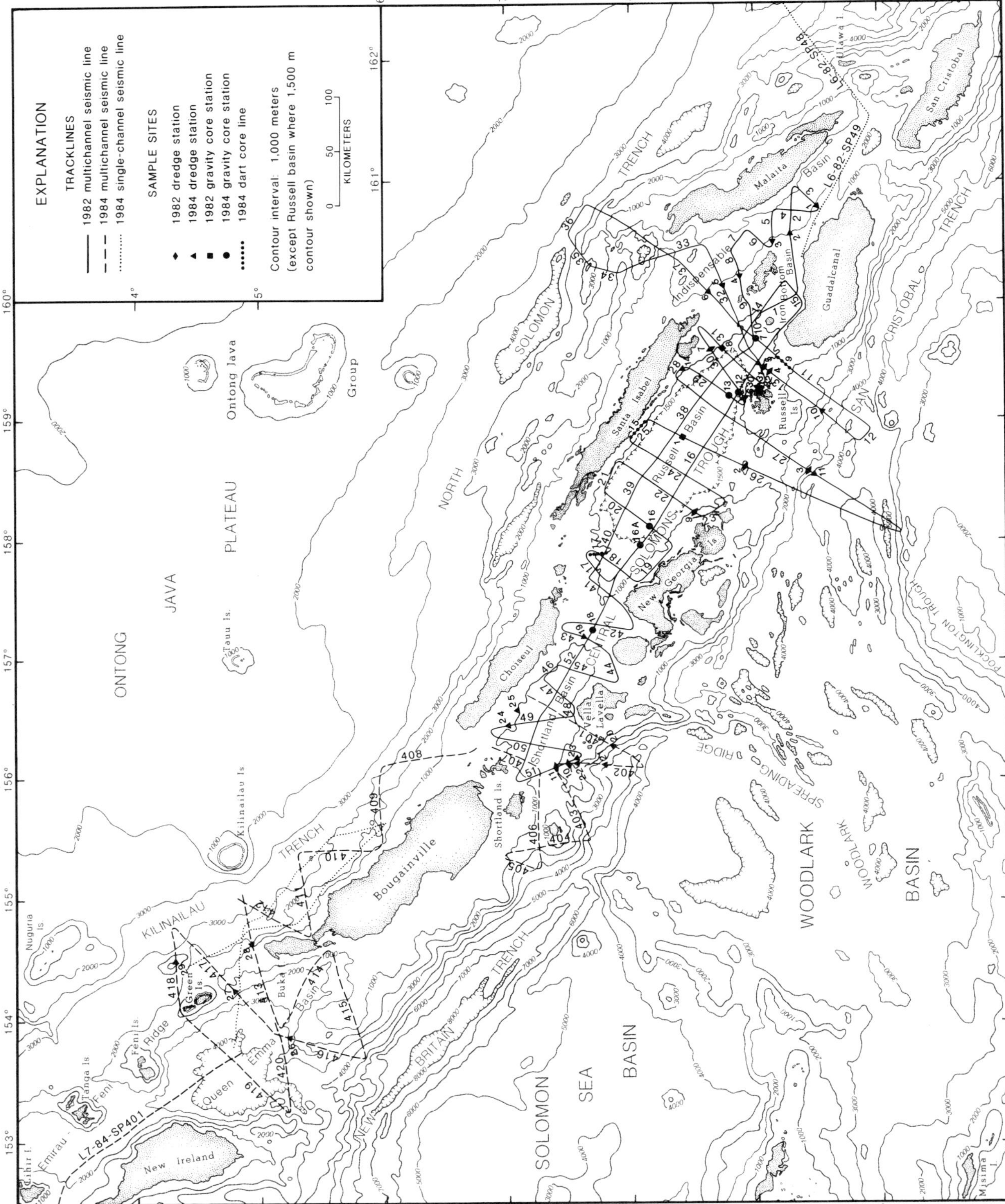

Figure 1. Gravity core, dart core, and dredge stations occupied during R/V *S.P. Lee* cruise L6-84-SP.

including *G. aperta*, *Calcidiscus leptoporus*, questionable *Emiliania huxleyi*, *Helicosphaera wallichii*, *Gephyrocapsa omega*, *Umbilicosphaera sibogae*, a form similar to *Pontosphaera plana*, and *Thoracosphaera* sp. were identified as being indigenous. A late Quaternary age is suggested. Reworked taxa include *Cyclicargolithus abisectus*, *C. floridanus*, and *Sphenolithus abies*. Another preparation from the same mudstone yielded a similar indigenous assemblage, in which *Gephyrocapsa omega* also is present. Reworked forms are, however, more diversified and include *Coccolithus pelagicus*, *Cyclicargolithus abisectus*, *Discoaster druggi*/*D. kuglerii* group, *D. variabilis*, *Sphenolithus abies*/*S. neoabies* group, and *S. miriformis*, suggesting redeposition from upper Tertiary sources.

Rock type 2 is a mudstone that contains glass shards and extremely rare coccoliths. The taxa identified are questionable *Emiliania huxleyi*, small *Calcidiscus leptoporus*, *Helicosphaera carteri*, *Gephyrocapsa oceanica*, small *Gephyrocapsa* spp., and *Cyclicargolithus abisectus; C. abisectus* is thought to be reworked. The age is suggested as Quaternary, and the rarity of the fossils is thought to be a result of dilution by the volcanic components of the mudstone.

Rock type 3 was described on board ship as surface oxidation layer of rock type 2. It yielded a few nannofossil forms, including abundant small *Gephyrocapsa* spp., *Gephyrocapsa oceanica*, small *Helicosphaera carteri*, *Calcidiscus leptroporus*, questionable *C. macintyrei*, and *Umbilicosphaera sibogae*, which collectively suggest a Quaternary age. In addition, *Coccolithus pelagicus*, *Cyclicargolithus abisectus* and *Discoaster pentaradiatus* were identified, but these species are thought to be reworked.

A preparation from the reef debris (rock type 4) contains rare nannofossils. These are *Gephyrocapsa oceanica*/*G. omega* group, very small *Gephyrocapsa* spp., *Calcidiscus leptoporus*, questionable *Emiliania huxleyi*, *Helicosphaera wallachii*, *Umbilicosphaera sibogae*, *Scapholitus fossilis*, and *Cyclicargolithus abisectus*. The age is Quaternary, with *C. abisectus* being reworked. The presence of ascidian spicules is taken to indicate shallow-water deposition.

Station 7

At Station 7, a gravity core recovered 125 cm of hemipelagic mud including thin volcanic-rich layers, from a water depth of 1,290 m. A preparation from the bottom of the core (corecatcher) yielded abundant coccoliths including *Emiliania huxleyi*, small *Gephyrocapsa* spp., *Gephyrocapsa oceanica*, *Helicosphaera carteri*, *Calcidiscus* sp. cf. *C. leptoporus*, forms mimicking *Pontosphaera plana*, plus *Umbilicosphaera sibogae* and *Scapholithus fossilis*. The age is late Pleistocene to Holocene. Reworked taxa are extremely rare. Evidence for shallow-water deposition includes the presence of ascidian spicules.

Station 9

Eight dart cores recovered sediments at Station 9. Core L6-84-SP-9A consists of 45 cm of hemipelagic mud from a water depth of 893 m. A preparation from the base of the core yielded abundant coccoliths, including *Calcidiscus leptoporus*, *Ceratolithus cristatus*, *C. telesmus*, rare *Emiliania huxleyi*, *Gephyrocapsa oceanica*/*G. omega* group, small *Gephyrocapsa* spp., *Umbilicosphaera sibogae*, and *Scapholithus fossilis*. The age is regarded as late Pleistocene to Holocene.

Bottom samples from cores L6-84-SP-9B and L6-84-SP-9C (which recovered 25 cm and 38 cm of hemipelagic mud from water depths of 718 m and 685 m, respectively) contain coccolith assemblages similar to that recorded for L6-84-SP-9A, and the ages of both are late Pleistocene to Holocene.

Core L6-84-SP-9D recovered 38 cm of sandy hemipelagic mud that is rich in volcanic components. The taxa identified from the based of this core include *Calcidiscus leptoporus*, *Emiliania huxleyi*, *Gephyrocapsa oceanica*, and small *Gephyrocapsa* spp., indicating a late Pleistocene to Holocene age. *Reticulofenestra pseudoumbilica* identified in the same preparation is regarded as being reworked. The occurrence of ascidian spicules is taken to indicate deposition in shallow water.

Core L6-84-SP-9E recovered 43 cm of hemipelagic muddy sand and sandy mud from a water depth of 582 m. Abundant small *Gephyrocapsa* spp., common *Emiliania huxleyi*, *Ceratolithus cristatus*, *Calcidiscus leptoporus*, *Gephyrocapsa oceanica*/*G. omega*, and *Helicosphaera wallichii* were identified in a preparation from the base of this core. The age is regarded to be late Pleistocene to Holocene. Deposition in shallow water is indicated by the occurrence of ascidian spicules.

Core L6-84-SP-9F recovered 26 cm of volcanic-rich muddy sand from a water depth of 525 m. In a preparation from the base of this core, coccoliths are not abundant, probably as a result of dilution by the volcanic components. The presence of *Emiliania huxleyi*, *Gephyrocapsa oceanica*, small *Gephyrocapsa* spp., *Calcidiscus leptoporus*, and *Umbilicosphaera sibogae* suggests a late Pleistocene to Holocene age.

Core L6-84-SP-9G recovered 23 cm of muddy sand from a water depth of 420 m. *Calcidiscus leptoporus*, a form similar to *C. macintyrei*, *Emiliania huxleyi*, *Gephyrocapsa caribbeanica*, *G. oceanica*, *G. omega*, small *Gephyrocapsa* spp., and *Scapholithus fossilis* were identified from a preparation from the base of this core. A late Pleistocene to Holocene age is indicated.

Core L6-84-SP-9H recovered 52 cm of muddy sand and sandy mud from a water depth of 399 m. In a preparation from the base of this core, several coccolith species were identified; volcanic components abound. The taxa include *Emiliania huxleyi*, *Helicosphaera carteri*, *Crenalithus doronicoides*, questionable *Calcidiscus macintyrei*, *Umbilicosphaera sibogae*, *Gephyrocapsa omega*, and *G. oceanica*. The age is considered to be late Pleistocene to Holocene.

Station 11

Blocks of partly manganese-coated siltstone and fine-grained sandstone were recovered at Station 11 where the water depth ranges between 1,200 and 1,350 m. Five preparations from five of these blocks were examined. Coccoliths are extremely rare, mainly small *Gephyrocapsa* spp. *Calcidiscus* sp. and *Gephyrocapsa oceanica* were also identified. The age is considered to be Quaternary on the assumption that the taxa identified are *in situ* and not due to contamination.

Station 12

At Station 12, a gravity core recovered 295 cm of bioturbated mud with thin beds of volcaniclastic mud from a water depth of 1,612 m. A coccolith-rich preparation from the 294 cm-level contained *Calcidiscus leptoporus, Emiliania huxleyi, Gephyrocapsa oceanica, G. omega,* small *Gephyrocapsa* spp., *Helicosphaera carteri*, and *Scapholithus fossilis*, which collectively indicate a late Pleistocene to Holocene age. *Thoracosphaera* sp. and some ascidian spicules are also present.

Station 13

At Station 13, a gravity core recovered 169 cm of bioturbated sandy mud with sandy turbidite layers from a water depth of 1,682 m. The coccolith assemblage identified from the corecatcher includes questionable *Ceratolithus simplex, Calcidiscus leptoporus, Emiliania huxleyi*, rare *Gephyrocapsa omega*, small *Gephyrocapsa* sp., *Helicosphaera carteri*, a form similar to *Ponosphaera plana, Scapholithus fossilis, Syraccosphaera* sp., and *Umbilicosphaera sibogae* suggesting a late Pleistocene to Holocene age. In the nannofossil preparation examined, volcanic components are present, and the foraminiferal tests are sediment filled.

Station 14

Three types of lithologies were dredged at Station 14 where water depth ranges between 500 and 700 m. The samples obtained are all scree (displaced downslope). The predominant lithology, which consists of coralline limestone, was not examined. The other two lithologies, which are basically mudstones, are almost barren of coccoliths. In one preparation, only traces of small *Gephyrocapsa* spp. are believed to be *in situ* and not due to contamination from younger sediments. Other preparations were found to be barren of coccoliths.

Station 15

Seven dart cores were successfully attempted at Station 15. Core L6-84-SP-15A recovered 30 cm of greyish-olive sandy mud from a water depth of 778 m. In a preparation from the base of this core, coccoliths are rare, probably due to dilution by volcanic components. Ascidian spicules and other siliceous microfossils also are present. Taxa identified include *Calcidiscus leptoporus, Emiliania huxleyi*, small *Gephyrocapsa* spp., *Helicosphaera carteri, Umbilicosphaera sibogae*, and *Scapholithus fossilis*; *Thoracosphaera* sp. also is present. The age is considered to be late Pleistocene to Holocene.

Core L6-84-SP-15B recovered 39 cm of sandy mud from a water depth of 614 m. The base of this core yielded *Ceratolithus cristatus, C. telesmus, Emiliania huxleyi*, small *Gephyrocapsa* spp., *Gephyrocapsa oceanica, Helicosphaera carteri*, a form similar to *Pontosphaera plana, Syracosphaera globulata*, and *Cyclicargolithus abisectus*. The age is regarded to be late Pleistocene to Holocene, with *C. abisectus* being reworked.

Core L6-84-SP-15C recovered 32 cm of sandy mud from a water depth of 456 m. In a preparation from the base of the core, coccoliths are not very abundant because of the copious occurrence of volcanic particles, mineral grains, ascidian spicules, siliceous microfossils, and foraminiferal tests. Taxa identified include *Calcidiscus leptoporus, Emiliania huxleyi, Gephyrocapsa oceanica/G. omega* group, small *Gephyrocapsa* spp., and *Scapholithus fossilis*. The age is considered to be late Pleistocene to Holocene.

Core L6-84-SP-15D recovered 37 cm of sandy mud from a water depth of 458 m. Coccoliths are greatly reduced in abundance because of the occurrence of various volcanic components, ascidian spicules, and foraminiferal tests. Taxa identified from the base of this core include *Calcidiscus leptoporus,Emiliania huxleyi, Gephyrocapsa oceanica, Helicosphaera* sp. and *Umbilicosphaera sibogae*. The age is regarded to be late Pleistocene to Holocene.

Core L6-84-SP-15E recovered 25 cm of fine-grained sandy mud from a water depth of 895 m. Taxa identified from the base of this core include *Ceratolithus cristatus, C. telesmus, Calcidiscus leptoporus, Emiliania huxleyi, Gephyrocapsa oceanica*, small *Gephyrocapsa* spp., and *Helicosphaera carteri*. Ascidian spicules also are present in the same preparation. The age is considered to be late Pleistocene to Holocene.

Core L6-85-SP-15F recovered 33 cm of fine-grained sandy mud from a water depth of 752 m. Coccoliths are not abundant because of dilution by various volcanic components. Taxa identified from a preparation from the base of this core include *Calcidiscus leptoporus, Emiliania huxleyi* and *Gephyrocapsa oceanica*. Ascidian spicules and *Thoracosphaera* sp. are present in the same preparation. The age is regarded to be late Pleistocene to Holocene.

Core L6-85-SP-15G recovered 33 cm of fine-grained sandy mud from a water depth of 752 m. Coccoliths are not abundant because of dilution by various volcanic components. Taxa identified in a preparation from the base of the core include *Calcidiscus lleptoporus*, *Emiliania huxleyi*, *Gephyrocapsa oceanica*, small *Gephyrocapsa* spp., and a form similar to *Pontosphaera plana*. The age is considered to be late Pleistocene to Holocene.

Stations 16 and 16A

Two gravity cores were successfully retrieved at Stations 16 and 16A: Core L6-84-SP-16 recovered 219 cm of bioturbated mud containing volcanic-rich sandy turbidite layers from a water depth of 1,672 m, and core L6-84-SP-16A recovered 266 cm of bioturbated hemipelagic muds from a water depth of 1,444 m.

A preparation from the corecatcher of Core L6-84-SP-16 yielded *Calcidiscus leptoporus*, *Emiliania huxleyi*, *Gephyrocapsa oceanica*, *G. omega*, small *Gephyrocapsa* spp., *Helicosphaera carteri*, *H. wallichii*, a form similar to *Pontosphaera plana*, *Rhabdosphaera stylifera*, *Scapholithus fossilis*, and *Thoracosphaera* sp. The age is regarded as late Pleistocene to Holocene.

The corecatcher of Core L6-84-SP-16A yielded a rich coccolith assemblage. Taxa identified include *Calcidiscus leptoporus*, *Emiliania huxleyi*, *Gephyrocapsa caribbeanica*, *G. oceanica*, small *Gephyrocapsa* spp., *Helicopshaera carteri*, *Scapholithus fossilis*, *Syracosphaera* sp. cf. *S. globulata*, *Sphenolithus abies/S. neoabies* and *Umbilicosphaera sibogae*. *Thoracosphaera* sp. is also present. The age is regarded as late Pleistocene to Holocene, with the sphenoliths being reworked.

Station 17

Four dart cores were retrieved at Station 17. Core L6-84-SP-17A recovered 31 cm of muddy sand from a water depth of 735 m. A preparation from the base of this core revealed the occurrence of *Calcidiscus leptoporus*, *Ceratolithus* sp., *Emiliania huxleyi*, *Gephyrocapsa oceanica*, small *Gephyrocapsa* spp., *Helicosphaera carteri*, questionable *H. sellii*, and *Thoracosphaera* sp. The age is considered to be late Pleistocene to Holocene.

Core L6-84-SP-17B recovered 34 cm of muddy sand from a water depth of 701 m. The base of this core contains *Ceratolithus cristatus*, *Calcidiscus leptoporus*, *Emiliania huxleyi*, small *Gephyrocapsa* spp., and ascidian spicules. The age is late Pleistocene to Holocene.

Core L6-84-SP-17C recovered 36 cm of muddy sand from a water depth of 764 m. Foraminiferal tests (partially sediment filled), siliceous microfossils, and coccoliths are present in a preparation from the base of this core. The coccolith taxa include *Calcidiscus leptoporus*, *Ceratolithus cristatus*, *Crenalithus doronicoides*, *Emiliania huxleyi*, *Gephyrocapsa ericsonia*, *G. oceanica*, small *Gephyrocapsa* spp., *Helicosphaera carteri*, and *Umbilicosphaera sibogae*. *Thoracosphaera* sp. is also present. The age is regarded as late Pleistocene to Holocene.

Core L6-84-SP-17D recovered 48 cm of muddy sand from a water depth of 37 m. The base of this core yielded *Calcidiscus leptoporus*, *Emiliania huxleyi*, *Gephyrocapsa oceanica/G. omega*, small *Gephyrocapsa* spp., *Helicosphaera wallichii*, and *Thoracosphaera* sp. In addition, foraminiferal tests (partially filled) and siliceous microfossils are present. The age of this core is regarded as late Pleistocene to Holocene.

Station 18

At Station 18, a gravity core recovered 264 cm of bioturbated mud with a thin sandy turbidite layer from a water depth of 913 cm. The coccolith assemblage from the corecatcher includes *Calcidiscus leptoporus*, *Emiliania huxleyi*, *Gephyrocapsa oceanica*, *G. omega*, small *Gephyrocapsa* spp., *Helicosphaera carteri*, a form similar to *Pontosphaera plana*, *Umbellosphaera irregularis*, *Umbilicosphaera sibogae* and *Scapholithus fossilis*. *Thoracosphaera* sp. is also present. The age is considered to be late Pleistocene to Holocene.

Station 19

Two rock types were obtained at Station 19, where water depth ranges between 700 and 750 m. Rock type 1 is a moderately indurated, burrowed fine-grained muddy sandstone. Ten preparations were examined from ten rock fragments of this type lithology, and all yielded poorly preserved nannofossils. The age of the rock fragments ranges from late Pliocene to Pleistocene. Reworked older taxa are common. The Pleistocene age is given to those samples which lack *Calcidiscus macintyrei* and discoasters, particularly *D. brouweria* and *D. triradiatus*, but contain the association *Gephyrocapsa* spp., *Calcidiscus leptoporus*, and *Helicosphaera carteri*. The late Pliocene age is based on the presence of *Discoaster brouweri* in association with *Calcidiscus leptoporus*, *C. macintyrei*, *Ceratolithus* spp., *Helicosphaera wallichii*, and *Pseudoemiliania lacunosa*, in the absence of species of *Gephyrocapsa*. It is uncertain whether taxa such as *Sphenolithus abies/S. neoabies*, *Reticulofenestra pseudoumbilica*, and *Coccolithus pelagicus*, in preparations assigned to the late Pliocene, are reworked or not.

Rock type 2 is volcanic-rich muddy sandstone which contains rare, poorly-preserved nannofossils. In one preparation, *Calcidiscus leptoporus*, *Coccolithus pelagicus*, *Discoaster brouweri*, and *Helicosphaera carteri* were identified, and a late Pliocene age is indicated. In another preparation, *Discoaster brouweri* is absent, and a few specimens of small *Gephyrocapsa* spp., *Calcidiscus leptoporus*, *Helicosphaera carteri*, *Reticulofenestra pseudoumbilica*, and ?*Pseudoemiliania lacunosa* are present. The age of the latter is regarded as Pleistocene, with *R. pseudoumbilica* being reworked from older sediments.

Station 21

Several rock fragments, probably all scree, were obtained at Station 21 where water depths range between 3,100 and 3,900 m. Six of these rock fragments were examined for coccoliths but were be barren.

Station 22

Reefal limestone rubble was recovered at Station 22 from a water depth of 850 m. Except for extremely rare, very small, unidentifiable placoliths, all preparations examined from this station are nonfossiliferous.

Station 23

Two rock types were obtained at Station 23 where water depth ranges between 620 and 900 m. The predominant lithology is feldspathic sandstone. Virtually no coccoliths are present in a preparation from this sandstone, except a very few specimens of ?*Emiliania huxleyi*, *Umbilicosphaera sibogae*, and small nondescript placoliths. In another preparation from this rock type, coccoliths are similarly rare but more diversified. Among the taxa identified are *Calcidiscus leptoporus*, *Crenalithus doronicoides*, questionable *Emiliania huxleyi*, *Gephyrocapsa caribbeanica*, *G. oceanica*, *G. omega*, small *Gephyrocapsa* spp., *Helicosphaera carteri*, and *Umbilicosphaera sibogae*. The age is late Quaternary based on the assumption that these fossils are *in situ* and not the result of contamination.

The other rock type, which is bioturbated volcaniclastic muddy sandstone, is barren of calcareous nannofossils.

Station 24

One sediment type, an olive-grey hemipelagic mud, was recovered at Station 24 where water depth is about 1,200 m. Coccoliths are abundant and include *Calcidiscus leptoporus*, *Crenalithus doronicoides*, *Emiliania huxleyi*, *Gephyrocapsa oceanica*, *G. omega*, small *Gephyrocapsa* spp., *Helicosphaera carteri*, *H. wallichii*, *Rhabdosphaera stylifera*, *Scapholithus fossilis*, *Umbellosphaera irregularis*, and *Umbilicosphaera sibogae*. This assemblage is considered to be late Pleistocene to Holocene. Deposition in shallow water is suggested by the presence of ascidian spicules.

Station 25

Three rock types were recovered at Station 25 where water depth ranges between 600 and 850 m. Two of the three rock types were examined. In a preparation from a fragment of coral limestone (rock type 1 of this station), coccoliths are extremely rare, being represented by a few specimens of poorly preserved *Calcidiscus leptoporus* and small nondescript placoliths. The age of this coral limestone is uncertain but probably is Quaternary.

The other rock type examined from this station is a bioturbated muddy sandstone/sandy mudstone. It contains abundant coccoliths. Taxa identified include *Calcidiscus leptoporus*, *Crenalithus doronicoides*, *Emiliania huxleyi*, *Gephyrocapsa caribbeanica*, *G. oceanica*, *G. omega*, small *Gephyrocapsa* spp., *Helicosphaera carteri*, *H. wallichii*, *Scapholithus fossilis*, and *Umbilicosphaera sibogae*. These species indicate a late Pleistocene to Holocene age.

Station 29

A richly fossiliferous calcilutite was recovered at Station 29 where water depth ranges between 1,500 and 2,000 m. Three preparations from three different fragments were examined, and all yielded abundant coccoliths. The assemblages in the three preparations are similar, except for the notable occurrence of *Discoaster decorus* in one only. Main elements include *Calcidiscus leptoporus*, *G. macintyrei*, *Ceratolithus rugosus*, *Coccolithus pelagicus*, *Crenalithus doronicoides*, *Discoaster brouweri*, *D. surculus*, *Hayaster perplexus*, *Helicosphaera carteri*, *H. sellii*, *H. wallichii*, *Oolithothus fragilis*, *Pseudoemiliania lacunosa* (or a very similar form), *Reticulofenestra* sp. (related to *R. pseudombilica*), and *Umbilicosphaera sibogae*. Two preparations are assigned a late Pliocene age, and the third, with *Discoaster decorus*, is considered slightly older.

Vedder, J.G., and Bruns, T.R., editors, 1989, Geology and offshore resources of Pacific island arcs—Solomon Islands and Bougainville, Papua New Guinea Regions: Houston, Texas, Circum-Pacific Council for Energy and Mineral Resources Earth Science Series, v. 12.

TERTIARY AND QUATERNARY MICROFOSSILS FROM THE RUSSELL ISLANDS

P. J. Quinterno
U. S. Geological Survey, Menlo Park, CA 94025, USA

ABSTRACT

Arenaceous limestone, dredged from water depths between 220 and 165 m off the Russell Islands, contains two different assemblages of planktonic foraminifers with markedly different states of preservation. The poorly preserved assemblage indicates a possible age of late early Pliocene (upper part of N 19/20); the better preserved assemblage contains *Globorotalia truncatulinoides* and indicates an early Pleistocene (N22) to Holocene age.

One piece of limestone from the same dredge haul encloses a fragment of fine-grained tuffaceous calcarenite that contains early Tertiary planktonic foraminifers and calcareous nannofossils ranging from mid-Eocene through Oligocene age. This is the oldest reported date from sedimentary rocks in this island group.

INTRODUCTION

The Russell Islands lie between New Georgia and Guadalcanal on the southwestern side of the Solomon Islands arc (Figure 1) and are remnants of an emergent volcano consisting of late Miocene(?) and Pliocene(?) basaltic andesite breccia overlain by basaltic lavas. Uplifted siltstone and reef limestone of Pleistocene age surround the volcanic core (Coulson and Vedder, 1986).

Most samples of sedimentary rocks from the 1982 R/V *S. P. Lee* cruise (L7-82-SP) to the Solomon Islands yielded Quaternary ages; the remainder were assigned only a general Pliocene through Quaternary age (Resig, 1986). During the 1984 R/V *S. P. Lee* cruise (L6-84-SP), sedimentary rocks of late Pliocene age (N21) were dredged at 2 sites; the remainder were Quaternary (Resig, this volume).

Soon after the 1984 cruise, Brian Bailey, H. W. Chaney, and J. G.Vedder visited the Russell Islands, where they obtained offshore samples which included fossiliferous sandy limestone, biolithite, and biomicrite (Crook and Niem, this volume). A rectangular, wire-basket shell dredge with dimensions of 100x60x40 cm was used to collect the samples; it was dragged under alternating low power and current drift through the deep channels northeast of Yandina. The largest haul was at site 31 where about 45 kg of limestone, including blocks as large as 50x25x20 cm, filled the dredge. Selected samples from three dredge sites were analyzed for foraminifers and calcareous nannofossils. Approximate locations of these sites are shown on Figure 2, and the estimated coordinates and water depths are listed on Table 1.

METHODS

Because the samples were well-indurated, the foraminifers were separated from the matrix by placing pieces of rock in a porcelain mortar and striking the rock fragments with a porcelain pestle until the fragments were reduced in size to about 0.5 to 1 cm. After each blow, the finest particles were removed from the mortar, so that any foraminifers present would not be crushed during successive blows. The sample was washed over a 0.062 mm sieve, transferred to filter paper and air dried. All planktonic and benthic foraminifers that withstood the crushing were picked and sorted on cardboard slides. Planktonic foraminiferal zonal designations follow the Neogene zonal scheme of Blow (1969), and species ranges are from Kennett and Srinivasan (1983).

Calcareous nannofossils were studied in smear slides using a light microscope at magnifications of 500-1000x. The calcareous nannoplankton zonal scheme follows Okada and Bukry (1980).

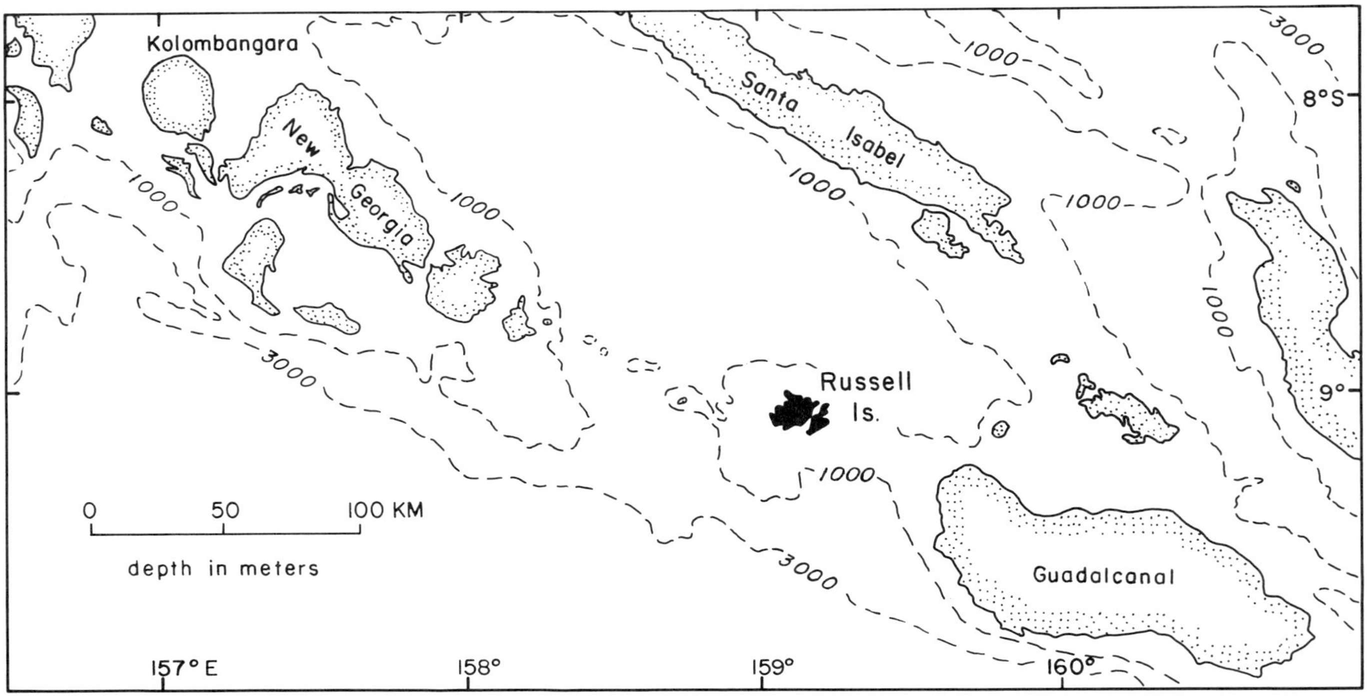

Figure 1. Map of the Russell Islands and adjoining areas.

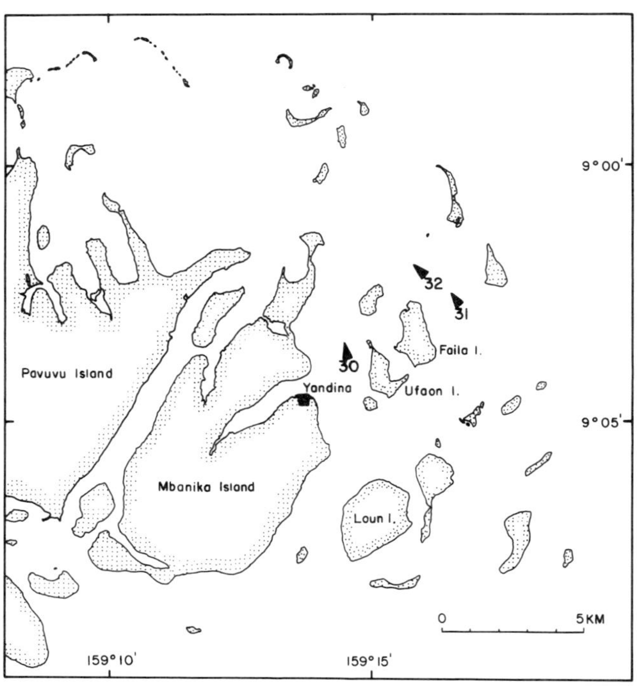

Figure 2. Dredge sites in the northeastern Russell Islands. 1984

RESULTS

Planktonic Foraminifers

Foraminifers are rare and not age diagnostic in 30-1-1 (biolithite). They are more abundant in 32-1-1 (biomicrite), but lack good biostratigraphic markers; therefore, only a general age range of Pliocene to Holocene can be assigned.

Planktonic and benthic foraminifers are common in sample 31-1-1 (biolithite); planktonic species are listed in Table 2. The rock contains two different assemblages with markedly different states of preservation. Most of the foraminifers are filled; a few are quite well-preserved and appear to be younger than the majority. The rock is well-cemented but has many cavities and a rough surface partly

Table 1. Approximate dredge-site positions and water depths, Russell Islands, 1984.

Station	Latitude (S)	Longitude (E)	Depths (meters)
30	9°03.8′	159°14.6′	200-190
31	9°02.7′	159°16.6′	220-185
32	9°02.1′	159°16.0′	190-155

Table 2. Planktonic foraminifers in samples 31-1-1 and 31-1-4.

Dentoglobigerina sp. aff. *D. altispira*
Globigerina bulloides
G. falconensis
G. sp. aff. *G. praecalida*
Globigerinita glutinata
Globigerinoides conglobatus
G. elongatus
G. immaturus
G. obliquus
G. ruber
G. sacculifer
G. triloba
Globorotalia crassaformis
G. sp. aff. *G. fimbriata*
G. sp. aff. *G. inconstans*
G. sp. aff. *G. lenguaensis*
G. menardii
G. obesa
G. sp. aff. *G. scitula*
G. truncatulinoides
G. tumida G. spp.
Hastigerina siphonifera
Orbulina universa
Paragloborotalia acostaensis
P. humerosa
Pulleniatina obliquiloculata
Sphaeroidinellopsis seminulina

covered by encrusting organisms. Microscopic examination of the surface and a cut slab reveals that the foraminifers within the rock are filled, but that some specimens on the surface or in the cavities are hollow and quite well-preserved; these are probably Holocene.

The planktonic foraminifers within the rock indicate an age of late early Pliocene (upper part of N19/20). This conclusion is based on the presence of several specimens of *Globorotalia crassaformis* (N19/20 to present); the absence of *Globorotalia tosaensis* (N21-N22) and *G. truncatulinoides* (N22 to Holocene) except as noted below; and abundant dextral coiling *Pulleniatina obliquiloculata*. Bronnimann and Resig (1971) and Resig (1986) used the coiling changes of *Pulleniatina* to determine the biostratigraphy of samples from the western tropical Pacific. Saito (1976) correlated coiling changes of *Pulleniatina* from Pliocene through Quaternary with the paleomagnetic stratigraphy in 10 low-latitude deep-sea cores. He found that from about 3.6 Ma to about 2.5 Ma (latest early Pliocene, Zone N19/20, to late Pliocene, Zone N21), *Pulleniatina* is almost entirely dextrally coiled. During the remainder of the Pliocene, sinistral forms are equal to or dominate over dextral forms. Coiling directions fluctuate between sinistral, dextral, and random during the early Quaternary; from about 750 ka to the present, the coiling mode is almost 100 percent dextral. It is doubtful that the foraminifers in 31-1-1 belong to this latest dextral coiling event, because *Sphaeroidinellopsis seminulina*, *Globigerinoides obliquus*, and *Dentoglobigerina altispira* are present and indicate an age no younger than earliest Pleistocene. *Pulleniatina primalis* (N17B-N20) is absent; if it were present, the argument for the late early Pliocene age would be stronger.

Some specimens of *Globorotalia crassaformis* in 31-1-1 are poorly preserved and probably belong to the early Pliocene assemblage; some are well preserved and probably a result of recent contamination. Only one specimen of *Globorotalia truncatulinoides* was found in 31-1-1. Although the evolutionary appearance of this species approximates the Pliocene-Pleistocene boundary (Kennett and Srinivasan, 1983), the specimen is much better preserved than the majority of the foraminifers in this sample. I conclude that it is Holocene and was probably lodged in one of the cavities of the rock; it is, therefore, not indicative of the age of the rock.

Sample 31-1-4 is sandy limestone that contains a pebble-size fragment and several smaller fragments of weakly indurated fine-grained tuffaceous calcarenite. A light-gray foraminiferal mud fills some of the cavities around the tuffaceous fragment. Foraminifers obtained from the tuffaceous material do not permit a conclusive age determination. The majority range from Pliocene to Holocene, but several specimens resembling the early Tertiary species, *Globorotalia inconstans* are present. The mixed ages may be a result of reworking during deposition, or may be a result of mixing the less indurated (younger?) gray mud with the more consolidated tuffaceous material when sampling for foraminifers.

Calcareous Nannoplankton

Calcareous nannoplankton are rare and poorly preserved in all the samples; they are not age diagnostic in 30-1-1, 31-1-1, and 32-1-1. However, the fragment of tuffaceous calcarenite in dredge sample 31-1-4 contains a sparse flora that indicates a mid-Eocene through Oligocene age. Species present are *Dictyococcites scrippsae*, *Markalius inversus*, *Sphenolithus* sp. aff. *S. distentus*, and *Zygrhablithus bijugatus*. If the sphenolith is, in fact, *Sphenolithus distentus*, then the age would be mid-late Oligocene (CP18- 19a).

A similar, but sparser, flora is present in the limestone that encloses the tuffaceous fragment in 31-1-4, but the presence of these nannofossils may be due to contamination. There are small cavities scattered throughout the limestone that are filled with tuffaceous material and gray mud; some of this material may have been mixed with the limestone during nannofossil sampling.

CONCLUSIONS

This report documents the occurrence of microfossils older than any that previously have been reported from the Russell Islands. Planktonic foraminifers in one sample (31-1-1) indicate an age of late early Pliocene (upper part of N19/20). However, this age determination is based, in part, on the absence of a species (*Globorotalia tosaensis*) and, therefore, is somewhat questionable. Planktonic foraminifers and calcareous nannofossils from an apparent clastic fragment in another sample (31-1-4) indicate an early Tertiary (mid-Eocene through Oligocene) age. Previously, only rocks of late Miocene(?), Pliocene(?), and Quaternary age were reported. Seismic-reflection profiles in the vicinity show that unit A (acoustic basement) and unit B (probable late Oligocene[?] and Miocene age) approach the seafloor surface (Bruns et al, 1986, Figure 15, line 30; Figure 16, line 32). Moreover, crossings of the ridge that connects Guadalcanal and the Russell Islands indicate the presence of presumed volcanic and basement rocks at or near the surface (Vedder and Bruns, 1986). Therefore, it is reasonable to believe that early Tertiary rocks formed seafloor outcrops near the site of dredge sample 31-1-4 and were the source of the early Tertiary fossils in the clastic fragments within the limestone.

Hughes (this volume) reports other localities where older than expected rocks are exposed in the southwestern chain of the Solomon Islands. He found early Miocene microfossils in sedimentary rocks on Ghizo and possibly Kolombangara of the New Georgia island group, where only late Miocene to Pleistocene rocks previously had been reported.

ACKNOWLEDGMENTS

I thank George Chaproniere (Bureau of Mineral Resources, Australia) and Johanna Resig (University of Hawaii) for their assistance with foraminiferal taxonomy and their constructive reviews of this paper; and J. G. Vedder (U.S. Geological Survey) and Sean Stone (U.S. Geological Survey) for helpful discussions.

REFERENCES CITED

Blow, W. H., 1969, Late middle Eocene to Recent planktonic foraminiferal biostratigraphy, *in* Bronnimann, P., and Renz, H. H., eds., First International Conference on Planktonic Microfossils, Geneva, 1967, Proceedings, v. 1, p. 199-421.

Bronnimann, Paul, and Resig, J. M., 1971, A Neogene Globigerinacean biochronologic time-scale of the southwestern Pacific, *in* Winterer, E.L., and others, Initial Reports of the Deep Sea Drilling Project, v. 7: Washington, D. C., U. S. Government Printing Office, p. 1235-1469.

Bruns, T. R., Cooper, A. K., Mann, D. M., and Vedder, J. G., 1986, Seismic stratigraphy and structure of sedimentary basins in the Solomon Islands region, *in* Vedder, J. G., Pound, K. S., and Boundy, S. Q., eds., Geology and offshore resources of Pacific island arcs—central and western Solomon Islands: Houston, Texas, Circum-Pacific Council for Energy and Mineral Resources Earth Science Series, v. 4, p. 177-214.

Coulson, F. I., and Vedder, J. G., 1986, Geology of the central and western Solomon Islands, *in* Vedder, J. G., Pound, K. S., and Boundy, S.Q., eds., Geology and offshore resources of Pacific island arcs—central and western Solomon Islands: Houston, Texas, Circum-Pacific Council for Energy and Mineral Resources Earth Science Series, v. 4, p. 59-87.

Kennett, J. P., and Srinivasan, M. S., 1983, Neogene planktonic Foraminifera, a phylogenetic atlas: Pennsylvania, Hutchinson Ross Publishing Company, 265 p.

Okada, Hisatake, and Bukry, D., 1980, Supplementary modification and introduction of code numbers to the low-latitude coccolith biostratigraphic zonation (Bukry, 1973; 1975): Marine Micropaleontology, v. 5, p. 321-325.

Resig, J. M., 1986, Foraminiferal stratigraphy and paleobathymetry of dredged rock, R/V *S. P. Lee* cruise, Solomon Islands, *in* Vedder, J. G., Pound, K. S., and Boundy, S. Q., eds., Geology and offshore resources of Pacific island arcs—central and western Solomon Islands: Houston, Texas, Circum-Pacific Council for Energy and Mineral Resources Earth Science Series, v. 4, p. 255-260.

Saito, Tsunemasa, 1976, Geologic significance of coiling direction in the planktonic foraminifera *Pulleniatina*: Geology, v. 4, p. 305-309.

Vedder, J. G., and Bruns, T. R., 1986, Offshore petroleum potential, Solomon Islands, *in* Vedder, J. G., Pound, K. S., and Boundy, S. Q., eds., Geology and Offshore Resources of Pacific island arcs—central and western Solomon Islands: Houston, Texas, Circum-Pacific Council for Energy and Mineral Resources Earth Science Series, v. 4, p. 277-291.

Vedder, J.G., and Bruns, T.R., editors, 1989, Geology and offshore resources of Pacific island arcs—Solomon Islands and Bougainville, Papua New Guinea Regions: Houston, Texas, Circum-Pacific Council for Energy and Mineral Resources Earth Science Series, v. 12.

FORAMINIFERAL AGE AND PALEOBATHYMETRY OF DREDGE AND DART-CORE SAMPLES, 1984 R/V *S.P. LEE* CRUISE, SOLOMON ISLANDS AND EASTERN PAPUA NEW GUINEA[1]

J.M. Resig
Hawaii Institute of Geophysics, University of Hawaii,
2525 Correa Road, Honolulu, Hawaii 96822, USA

INTRODUCTION

Planktonic and benthic foraminifera in 21 samples of dredged rock as well as in samples from 19 cores taken in the Solomon Islands on cruise L6-84-SP of the R/V *S.P. Lee* were examined for evidence of age and water depth at deposition. The consolidated samples were partly disaggregated through crushing and boiling in water. All samples were sieved through a 62 μm screen. The sand fraction was spread on a tray and the foraminifera identified. Total samples were searched for planktonic index species, which were recorded. The benthic species recorded occurred in one or two trays of sample; and although they are representative, they do not compose the total population.

Age determinations employ the planktonic zonation of Blow (1969) together with the coiling changes of *Pulleniatina* (Saito, 1976) that were found applicable to the western equatorial Pacific (Brönnimann and Resig, 1971). Paleodepths are estimated from published (Hughes, 1977; Burke, 1981) and unpublished (Resig) data on the bathymetry of Holocene benthic foraminifera from the Solomon Islands area. Depth ranges of some of these benthic species are given in Resig (1986).

RESULTS

A compendium of data that indicates the presence of planktonic and benthic foraminiferal species in respect to the sample location and depth is shown in Table 1 and summarized below.

Age (Zone)

The samples partition into three groups according to age; absence of foraminifera in a fourth group precludes age determinations. Station and sample sites are shown on Figure 1.

1. Late Pliocene (N. 21): Station 29 (samples 1-3), Station 19 (samples 1-7A). Based on the presence of *Globorotalia tosaensis* in the absence of *G. truncatulinoides*. *Pulleniatina* coils right in sample 29-1-3, which also contains *Globoquadrina altispira*, indicating an age of about 2.8-3.1 Ma, near the base of Zone N. 21; *Pulleniatina* coils left in sample 19-1-7A, indicating the top of Zone N. 21.

2. Early Quaternary (N. 22): Station 19 (volcanic sandstone). Based on the concurrence of *Globorotalia tosaensis* and *Globorotalia truncatulinoides* and the presence of *Globigerinoides fistulosus*, *Globoquadrina pseudofoliata*, and a 3:1 ratio of left-to-right coiling *Pulleniatina*.

3. Undetermined age (foraminifera absent): Station 11 (samples 1-2A, 1-3A, 1-5A), Station 21 (samples 1-2A, 1-4A, dredge wire), Station 26.

4. Late Quaternary (late N. 22 or N. 23): all other samples. Based on the presence of red *Globigerina rubescens*, *Globorotalia truncatulinoides*, right-coiled *Pulleniatina* and occasional *Globigerina calida* and *Globigerinella adamsi*. Station 22 sample is a calcareous sandstone dated as probably Quaternary on the presence of the benthic foraminifera *Calcarina hispida* and *Baculogypsina sphaerulata*, which have not been reported in older strata.

Reworked Pliocene or late Miocene *Globoquadrina altispira* was noted in Station 2 (sample 3-1A).

[1]Hawaii Institute of Geophysics Contribution No. 2044

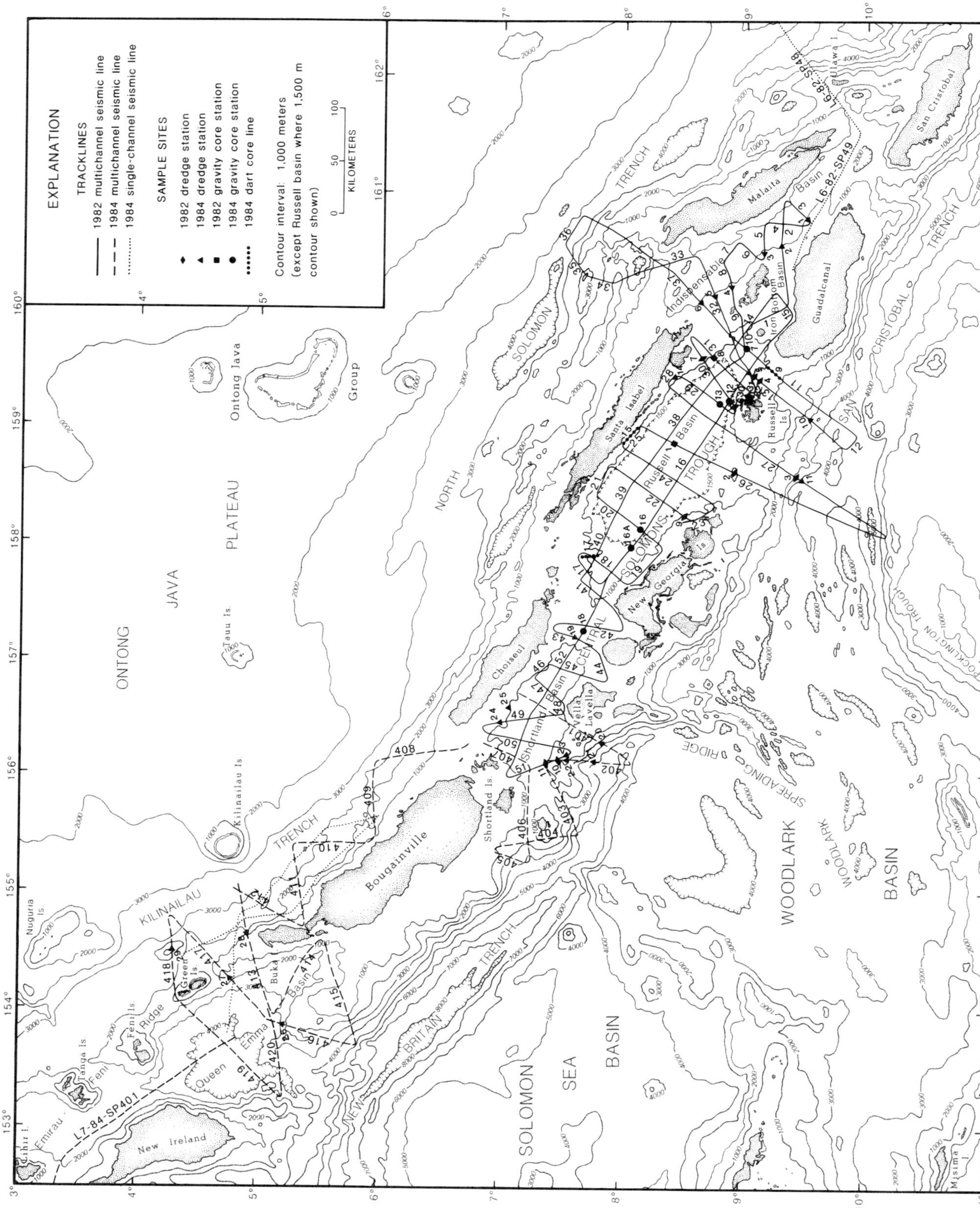

Figure 1. Station and sample sites for *S.P. Lee* cruise L6-84-SP.

Table 1. Some Diagnostic Foraminifera, R/V *S.P. Lee*, L6-84-SP.
(Note: Incomplete records. Not all species listed. Samples not searched for each benthic species listed.)

PLANKTONIC SPECIES (QUATERNARY – PLIOCENE)

STATION	LAT (S)	LONG (E)	WATER DEPTH (m)	SAMPLE	*Globigerina rubescens* (red)	*Pulleniatina obliquiloculata* (rt.)	*Globigerina calida*	*Globigerinella adamsi*	*Globoratalia truncatulinoides*	*Globoratalia tosaensis*	*Pulleniatina obliquiloculata* (lft.)	*Globigerinoides fistulosus*	*Globoquadrina pseudofoliata*	*Globoquadrina altispira*
1	9°27.93′	160°45.78′	700	1-1-2	x	x								
2	9°17.61′	160°35.24′	750-800	2-3-1a	x	x								x
				2-3-2a		x	x		x					
6	8°38.44′	160°05.12′	500-900	6-1-1		x			x					
				6-1-2		x			x					
				6-2-1		x			x					
9A	9°12.11′	159°32.37′	893	0 cm, 45 cm	x	x								
9B	9°12.83′	159°32.15′	718	0 cm, 25 cm	x	x								
9C	9°12.98′	159°32.16′	685	0 cm, 38 cm	x	x								
9D	9°12.73′	159°31.67′	642	0 cm, 38 cm	x	x								
9E	9°12.73′	159°31.56′	582	0 cm, 42 cm	x	x			x					
9F	9°12.76′	159°31.46′	525	0 cm, 26 cm	x	x								
9G	9°12.72′	159°31.29′	420	0 cm, 23 cm	x	x			x					
9H	9°12.68′	159°31.23′	399	0 cm, 52 cm	x	x			x					
11	9°12.55′	158°48.66′	1200-1350	11-1-1a	x	x								
				11-1-2A										
				11-1-3A										
				11-1-5A										
14	8°25.60′	159°27.93′	500-700	14-2-1		x								
				14-2-2		x								
15A	8°07.29′	159°02.35′	778	0 cm, 30 cm	x	x								
15B	8°06.97′	159°02.26′	614	0 cm, 30 cm	x	x								
15C	8°06.66′	159°01.89′	456	0 cm	x	x								
15D	8°06.8′	159°01.15′	458	0 cm, 37 cm	x	x			x					
15E	8°06.4′	158°58.1′	895	0 cm, 25 cm	x	x			x					
15F	8°05.55′	158°56.7′	752	0 cm, 29 cm	x	x								
15G	8°04.5′	158°54.9′	752	0+30 cm	x	x	x		x					
17A	7°47.71′	157°55.81′	735	0 cm, 31 cm	x	x		x	x					
17B	7°46.55′	157°54.58′	701	0 cm, 34 cm	x	x			x					
17C	7°46.21′	157°54.91′	764	0 cm, 36 cm	x	x			x					
17D	7°45.48′	157°54.93′	737	48 cm	x	x			x					
19	7°36.98′	157°15.18′	700-750	volc. ss.		x			x	x	x	x	x	
				19-1-7A						x	x			
21	7°48.65′	156°06.93′	3100-3900	21-1-2A	no foraminifera									
				21-1-4A	no foraminifera									
				wire	no foraminifera									
22	7°34.88′	156°07.63′	850	ls.										
23	7°31.91′	156°10.31′	620-900	23-2B		x			x					
24	7°01.42′	156°24.62′	1200	sdy. md.	x	x								
25	7°04.81′	156°39.75′	600-650	25-3		x			x					
26	5°13.33′	153°52.70′	?	wire	no foraminifera									
29	4°17.54′	154°35.29′	1500-2000	29-1-3		x				x				x

BENTHIC SPECIES — SUBLITTORAL <300 m

SAMPLE	*Amphistegina lessonii*	*Baculogypsina sphaerulata*	*Calcarina hispida + spengleri*	*Cymbaloporetta bradyi*	*Elphidium macellum*	Miliolid species	*Anomalinella rostrata*	*Amphistegina papillosa*	*Rectobolivina raphana*	*Bolivina hantkeniana*	*Bolivinita quadrilatera*	*Cancris auriculus*	*Planorbulinella larvata*	*Heterolepa praecincta*	*Bolivina schwageri*	*Anomalina glabrata*
(Depth Range)	< 40	< 40	< 40	< 40	40	40	1-150	1-260	10-260	12-72	16-1570	30-150	36-300	40-300	63-260	63-320
1-1-2						x										
2-3-1a										x						
2-3-2a	x				x							x			x	
6-1-1								x	x							
6-1-2	approx. same as sample 6-1-1															
6-2-1	x				x		x		x	x	x			x		
0 cm, 45 cm (9A)																
0 cm, 25 cm (9B)																
0 cm, 38 cm (9C)																
0 cm, 38 cm (9D)																
0 cm, 42 cm (9E)																
0 cm, 26 cm (9F)																
0 cm, 23 cm (9G)				x						x						x
0 cm, 52 cm (9H)	approx. same as sample 9G															
11-1-1a	benthic foraminifera few and fragmented															
11-1-2A	few foramniferal fragments only															
11-1-3A	no foraminifera															
11-1-5A	few foraminiferal fragments only															
14-2-1			x			x										
14-2-2																
0 cm, 30 cm (15A)											x		x			
0 cm, 30 cm (15B)											x					
0 cm (15C)				x		x			x					x		x
0 cm, 37 cm (15D)																
0 cm, 25 cm (15E)											x					x
0 cm, 29 cm (15F)																
0+30 cm (15G)																
0 cm, 31 cm (17A)											x					
0 cm, 34 cm (17B)																
0 cm, 36 cm (17C)																
48 cm (17D)																
volc. ss.																
19-1-7A																
21-1-2A																
21-1-4A																
wire (21)																
ls.	x	x	x	x	x	x										
23-2B	no benthic foraminifera															
sdy. md.											x					
25-3																
wire (26)																
29-1-3																

MIDDLE SUBLITTORAL-UPPER BATHYAL 60 - 1600 m

SAMPLE	*Bolivina pacifica*	*Cibicidoides pseudoungerianus*	*Bolivina striatula*	*Trifarina bradyi*	*Cassidulina neocarinata*	*Rectobolivina columellaris*	*Uvigerina proboscidea vadescens*	*Bolivina glutinata*	*Hoeglundina hyalina*	*Sphaeroidina bulloides*
(Depth Range)	63-666	63-790	63-1225	63-1570	63-3451	100-1225	100-1225	100-1570	100-1570	100-1570
1-1-2		x								x
2-3-1a					x					
2-3-2a		x								x
6-1-1	x				x					
6-1-2										
6-2-1		x								
0 cm, 45 cm (9A)		x		x		x				x
0 cm, 25 cm (9B)	approx. same as sample 9A									
0 cm, 38 cm (9C)	approx. same as sample 9A									
0 cm, 38 cm (9D)	approx. same as sample 9A									
0 cm, 42 cm (9E)						x	x			
0 cm, 26 cm (9F)	approx. same as sample 9E									
0 cm, 23 cm (9G)		x			x	x	x		x	
0 cm, 52 cm (9H)										
11-1-1a										
11-1-2A										
11-1-3A										
11-1-5A										
14-2-1		x			x				x	x
14-2-2	approx. same as sample 14-2-1									
0 cm, 30 cm (15A)	approx. same as sample 15C									
0 cm, 30 cm (15B)	approx. same as sample 15C									
0 cm (15C)		x			x			x	x	
0 cm, 37 cm (15D)	approx. same as sample 15C									
0 cm, 25 cm (15E)		x			x					x
0 cm, 29 cm (15F)	approx. same as sample 15E									
0+30 cm (15G)	approx. same as sample 15E									
0 cm, 31 cm (17A)		x	x	x				x	x	
0 cm, 34 cm (17B)	approx. same as sample 17A									
0 cm, 36 cm (17C)	approx. same as sample 17A									
48 cm (17D)	approx. same as sample 17A									
volc. ss.		x								x
19-1-7A		x								
21-1-2A										
21-1-4A										
wire (21)										
ls.										
23-2B										
sdy. md.		x		x	x					
25-3		x							x	
wire (26)										
29-1-3										

UPPER BATHYAL 300 - 1600 m

SAMPLE	*Melonis novozealandicus*	*Textularia saggitula fistulosa*	*Melonis multicameratus*	*Bulimina marginata*	*Trifarina pauperata*	*Bolivina robusta*	*Cassidulina* sp. (*crassa* auct.)	*Rectobolivina dimorpha pacifica*
(Depth Range)	260-368	260-368	344-989	344-1216	344-1320	344-1570	344-1570	344-1570
1-1-2				x		x		
2-3-1a						x		x
2-3-2a			x				x	
6-1-1						x		
6-1-2								
6-2-1				x				x
0 cm, 45 cm (9A)	x		x		x	x		
0 cm, 25 cm (9B)								
0 cm, 38 cm (9C)								
0 cm, 38 cm (9D)								
0 cm, 42 cm (9E)	x		x			x		
0 cm, 26 cm (9F)								
0 cm, 23 cm (9G)						x	x	
0 cm, 52 cm (9H)								
11-1-1a								
11-1-2A								
11-1-3A								
11-1-5A								
14-2-1	x					x	x	
14-2-2								
0 cm, 30 cm (15A)						x		
0 cm, 30 cm (15B)								
0 cm (15C)	x	x				x		x
0 cm, 37 cm (15D)								
0 cm, 25 cm (15E)						x		x
0 cm, 29 cm (15F)								
0+30 cm (15G)								
0 cm, 31 cm (17A)			x			x		x
0 cm, 34 cm (17B)								
0 cm, 36 cm (17C)								
48 cm (17D)								
volc. ss.						x		
19-1-7A								
21-1-2A								
21-1-4A								
wire (21)								
ls.								
23-2B								
sdy. md.						x		
25-3			x					
wire (26)								
29-1-3								

UPPER BATHYAL-LOWER BATHYAL OR ABYSSAL 300 - >4000 m; LOWER BATHYAL-ABYSSAL 1600 - >4000 m

SAMPLE	*Cassidulina subglobosa*	*Rotorbinella lobatula*	*Bulimina aculeata*	*Pullenia bulloides*	*Cassidulina pacifica*	*Rectobolivina columellaris* str.	*Seabrookia pelucida*	*Bolivina subreticulata*	*Anomalina colligera*	*Bolivina karreriana*	*Bolivina albatrossi*	*Bulimina rostrata*	*Nuttallides rugosus*	*Uvigerina bradyana*	*Astrononion schwageri*	*Melonis affinis*	*Oridorsalis umbonatus*	*Ceratobulimina pacifica*	*Uvigerina asperula*	*Robertinoides* sp.	*Ehrenbergina trigona*	*Cibicidoides mundula*
(Depth Range)	260-4344	290-3419	344-2712	344-4332	360-865	360-989	360-1225	360-1225	510-620	500-790	500-1570	510-2660	500-3419	500-3419	500-4332	500-4332	500-4332	593-1890	610-2839	790-865	1611-2686	1747-4245
1-1-2				x																		
2-3-1a					x					x							x					
2-3-2a					x																	
6-1-1		x	x	x			x	x		x	x								x			
6-1-2																						
6-2-1										x		x										
0 cm, 45 cm (9A)	x	x	x	x							x							x				
0 cm, 25 cm (9B)																						
0 cm, 38 cm (9C)																						
0 cm, 38 cm (9D)																						
0 cm, 42 cm (9E)			x		x				x	x	x											
0 cm, 26 cm (9F)																						
0 cm, 23 cm (9G)			x			x				x	x		x	x					x			
0 cm, 52 cm (9H)																						
11-1-1a																						
11-1-2A																						
11-1-3A																						
11-1-5A																						
14-2-1		x	x	x						x	x	x							x			
14-2-2									•													
0 cm, 30 cm (15A)																						
0 cm, 30 cm (15B)		x	x		x																	
0 cm (15C)		x								x		x	x				x					
0 cm, 37 cm (15D)																						
0 cm, 25 cm (15E)			x		x							x			x			x		x		
0 cm, 29 cm (15F)																						
0+30 cm (15G)																						
0 cm, 31 cm (17A)		x							x			x	x					x				
0 cm, 34 cm (17B)																						
0 cm, 36 cm (17C)																						
48 cm (17D)																						
volc. ss.	x			x													x					
19-1-7A				x										x				x				
21-1-2A																						
21-1-4A																						
wire (21)																						
ls.																						
23-2B																						
sdy. md.		x	x									x										
25-3	x		x		x												x					
wire (26)																						
29-1-3	x			x											x	x	x				x	x

Water Depth

The primary benthic foraminiferal assemblages of the samples were all indicative of the water depths in which the samples were taken, with the exception of the species inhabiting less than 40 m water depth contained in the calcareous, skeletal sandstone of Station 22 (850 m, between Vella Lavella and Mono Island) This rock, which is probably Quaternary in age, showed no apparent physical evidence of having been broken from outcrop.

Samples from Stations 1 and 2 northeast of Guadalcanal, Station 6 north of the Florida Islands, and Stations 14 and 15 southwest of Santa Isabel contain specimens of shallow-water species, including *Amphistegina, Elphidium*, and various miliolids that were transported downslope and mixed with the indigenous mid-depth assemblages. The volcanic sandstone from Station 23 did not include enough benthic foraminifera species to interpret water depth.

DISCUSSION AND CONCLUSIONS

The planktonic foraminifera and nannofossil age determinations (Shafik, this volume) agree that samples from Station 29 (1-3) and Station 19 (1-7A) are late Pliocene in age, the majority of the samples are late Quaternary in age, and some samples contain reworked microfossils. The Pliocene samples occur in two different settings: the rock from Station 29 was dredged at depths between 1,500 and 2,000 m on the Pacific slope of the island chain northeast of the Green Islands, whereas rocks from Station 19 were dredged in New Georgia Sound at depths of 700 to 750 m between Choiseul and the New Georgia island group. In addition to the Pliocene rocks, an early Quaternary volcanic sandstone was dredged at Station 19.

The prevalent benthic foraminiferal assemblages indicate that deposition of all but one sample occurred in about the depths the samples were taken. The only anomalous sample is a shallow-water skeletal sandstone of Quaternary age from between Vella Lavella and Mono Island. No evidence has been presented that this sample was *in situ* and that it was emplaced through tectonic subsidence at the recovery site. Examination of rocks dredged just north of Station 22 on a previous R/V *S.P. Lee* cruise (Resig, 1986, Stations 10 and 11) showed rapid uplift on the order of 600 m in 0.5 m.y. during the early Quaternary. This tectonic activity could have dislodged shallow-water deposits and may have sustained their downslope transport. Direct evidence of downslope transport occurs in the vicinity of Guadalcanal and Santa Isabel where several samples bear tests of shallow-water foraminifera mixed with the indigenous deeper-water assemblages. Many of the samples also contain alcyonarian spicules mixed with deep-water foraminiferal assemblages. Alcyonarian octocorals are common in the reef environment, and their spindle-shaped spicules are readily transported downslope. It is concluded that on the basis of foraminifera, the samples collected by dredging and dart coring during the 1984 R/V *S.P. Lee* cruise show no appreciable vertical tectonic displacement at the sample sites since deposition in late Pliocene and Quaternary time. Downslope transport of biogenous grains is evident in 20 percent of the samples. A single rock sample may have been transported downslope from shallow to deep water.

ACKNOWLEDGMENTS

Loren W. Kroenke and Scot Izuka of the Hawaii Institute of Geophysics provided useful commentary and review of this report.

REFERENCES CITED

Blow, W.H., 1969, Late middle Eocene to Recent planktonic foraminiferal biostratigraphy, *in* Bronnimann, P., and Renz, H.H., eds., Proceedings of the First International Conference on Planktonic Microfossils: Leiden, E.J. Brill, p. 199-421.

Brönnimann, P., and Resig, J.M., 1971, A Neogene globigerinacean biochronologic time-scale of the southwestern Pacific, *in* Winterer, E.L., et al, Initial reports of the Deep Sea Drilling Project: Washington, D.C., U.S. Government Printing Office, v. 7, p. 1235-1469.

Burke, S.C., 1981, Recent benthic foraminifera of the Ontong Java Plateau: Journal of Foraminiferal Research, v. 11, p. 1-19.

Hughes, G .W., 1977, Recent foraminifera from the Honiara Bay area, Solomon Islands: Journal of Foraminiferal Research, v. 7, p. 45-57.

Resig, J.M., 1986, Foraminiferal stratigraphy and paleobathymetry of dredged rock, R/V *S.P. Lee* cruise, Solomon Islands, *in* Vedder, J.G., Pound, K.S., and Boundy, S.Q., eds., Geology and offshore resources of Pacific island arcs—central and western Solomon Islands: Houston, Texas, Circum-Pacific Council for Energy and Mineral Resources Earth Science Series, v. 4, p. 255-260.

Saito, T., 1976, Geological significance of coiling direction in the planktonic foraminifera *Pulleniatina*: Geology, v. 4, p. 305-309.

Vedder, J.G., and Bruns, T.R., editors, 1989, Geology and offshore resources of Pacific island arcs—Solomon Islands and Bougainville, Papua New Guinea Regions: Houston, Texas, Circum-Pacific Council for Energy and Mineral Resources Earth Science Series, v. 12.

HYDROCARBON GASES IN SEDIMENTS OF THE SOLOMON ISLANDS AREA

K. A. Kvenvolden
U.S. Geological Survey, Menlo Park, California 94025, USA

A. R. Niem
Oregon State University, Corvallis, Oregon 97331-5506, USA

ABSTRACT

Small concentrations of hydrocarbon gases were found in 15 sediment samples from six stations in the area of the Central Solomons Trough. Methane was the most abundant hydrocarbon gas ranging in concentration from 190 to 2,600 nanoliters/liter of wet sediment (nL/L). Ethane, measured in only three samples, ranged from 8 to 62 nL/L, whereas propane, measured in six samples, ranged from 4 to 79 nL/L. Unsaturated hydrocarbon gases were present in most samples: ethene (8 to 250 nL/L) and propene (10 to 81 nL/L). Butanes (isobutane, 10 nL/L; n-butane, 20 nL/L) as well as a small amount of heavier hydrocarbons were detected in one sample. With the exception of this one sample, the remaining samples contain a background of hydrocarbon gases which appear to be ubiquitous in oceanic sediment. This background probably results from low-level, microbial activity taking place in the sediment. The single, anomalous sample hints at the presence of thermogenic hydrocarbons; however, the relatively low concentrations and the occurrence of the sample stratigraphically between samples having background values reduce the significance of this anomaly with regard to possible hydrocarbon seepage from deeply buried petroleum deposits.

INTRODUCTION

Low concentrations of hydrocarbon gases, methane (C1), ethane (C2), propane (C3), isobutane (i-C4), and n-butane (n-C4) as well as the unsaturated hydrocarbon gases ethene (C2:1) and propene (C3:1) are common in oceanic sediment worldwide. C1 is most abundant with amounts usually less than about 10,000 nanoliters/liter of wet sediment (nL/L). The other hydrocarbon gases generally occur in amounts of at least two to three orders of magnitude lower. These hydrocarbons are dissolved in the interstitial water and probably result from low-level microbial processes taking place within the sediment (Claypool and Kvenvolden, 1983).

Superimposed in this background of hydrocarbons are occasional anomalous occurrences of hydrocarbon gases resulting from high levels of microbial activities and from the active seepage of thermogenically-derived hydrocarbons from deeply buried petroleum-type deposits. These anomalies may lead to gas-charged sediment and to the identification of possible energy resources within an area. (For example, see Kvenvolden et al, 1979).

On the U.S. Geological Survey R/V *S.P. Lee* cruise L6-84-SP, a brief survey was undertaken to determine the hydrocarbon gases in sediment from the Solomon Islands area of the Pacific Ocean. Gravity cores at six stations (Figure 1) in the Central Solomons Trough (Colwell, Vedder et al, 1984) provided 15 sediment samples for gas analyses (Table 1). Water depths at these stations ranges from 1,118 to 1,682 m, and the sediment type recovered was hemipelagic mud with minor amounts of sand. Except for Station 7, these cores were taken over or near the areas interpreted by Colwell and Tiffin (1986) to contain shallow-depth, gas-produced seismic features. Examples of these seismic anomalies are shown in Figure 2.

The procedures for sampling involved gravity coring to subbottom depths less than 3 m. Ten-cm long segments of sediment core (8-cm diameter) were removed for gas extraction and analysis (Kvenvolden and Redden, 1980). Each segment was placed in a metal can, previously prepared with septa to provide for the removal

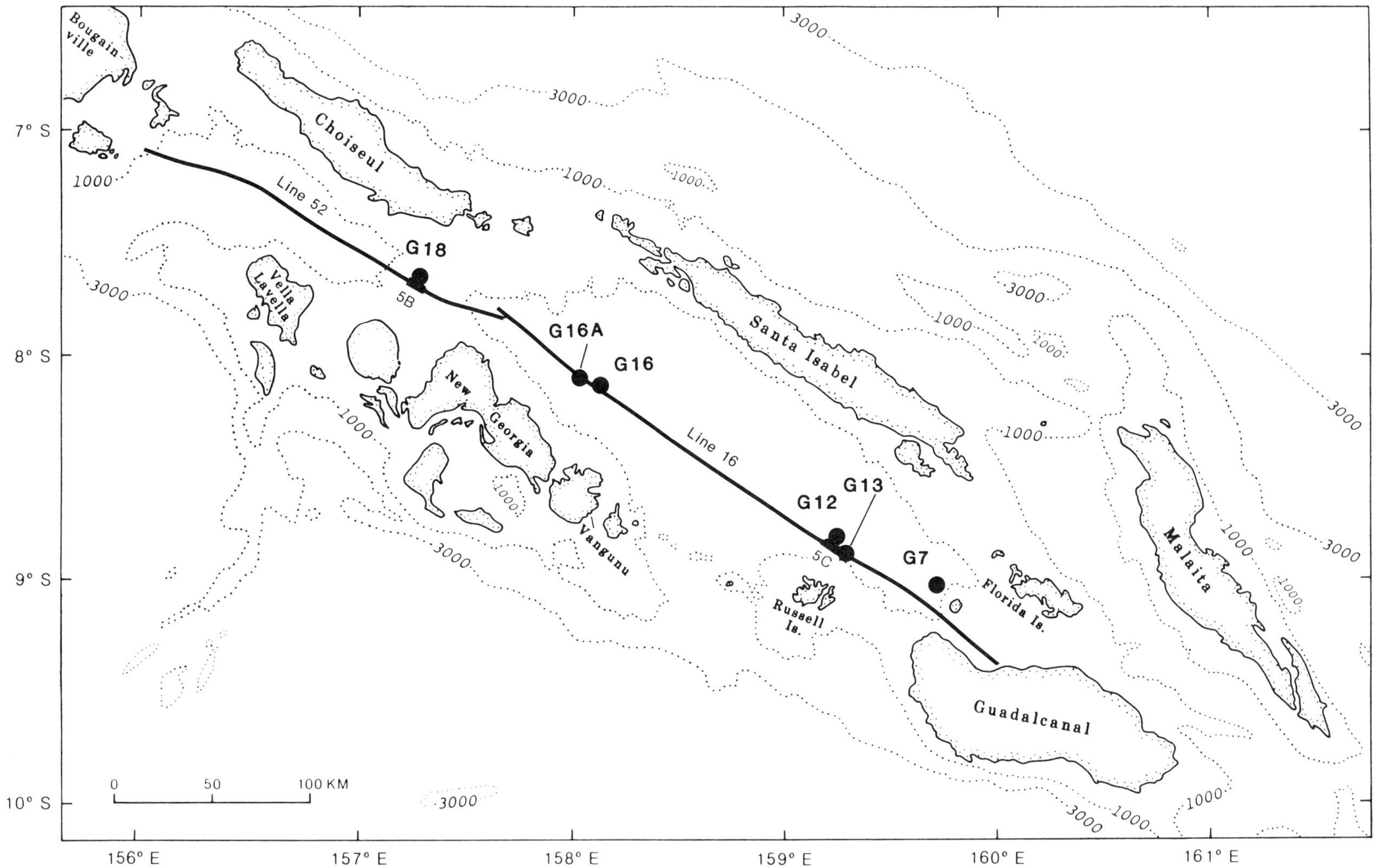

Figure 1. Map of the Central Solomons Trough showing sites where samples were obtained for gas analyses and areas (5B and 5C) inferred from seismic records to contain shallow accumulations of gas. Bathymetric contours are in meters.

Table 1. Concentrations of hydrocarbon gases (nL/L of wet sediment) in sediment from the Central Solomons Trough.

Core number and interval (cm)	Water depth (m)	Latitude	Longitude	C1	C2	C2:1	C3	C3:1	C1 / C2+C3	C2 / C2:1
				—nanoliters per liter wet sediment—						
G7 90-100	1290	09° 01.7′ S	159° 41.2′ E	990	nd	nd	nd	10	nc	nc
G7 115-125				890	nd	23	nd	19	nc	nc
G12 90-100	1612	08° 52.0′ S	159° 13.9′ E	360	nd	15	nd	10	nc	nc
G12 190-299				1100	nd	8	nd	10	nc	nc
G12 284-294				490	nd	23	4	72	120	nc
G13 90-100	1682	08° 47.8′ S	159° 13.6′ E	190	nd	nd	nd	10	nc	nc
G13 159-169				240	nd	23	8	19	30	nc
G16 90-100	1672	08° 07.2′ S	158° 09.3′ E	690	nd	15	nd	nd	nc	nc
G16 209-219				420	nd	23	nd	10	nc	nc
G16A 90-100	1444	08° 04.4′ S	157° 58.2′ E	770	nd	15	nd	10	nc	nc
G16A 190-200				2500	62	250	79	72	18	0.25
G16A 246-256				460	nd	15	nd	nd	nc	nc
G18 90-100	1118	07° 40.6′ S	157° 17.5′ E	930	nd	15	4	14	240	nc
G18 190-200				2600	8	23	8	14	160	0.35
G18 254-264				1400	16	60	20	81	40	0.26

nd = not detected; nc = not calculated.

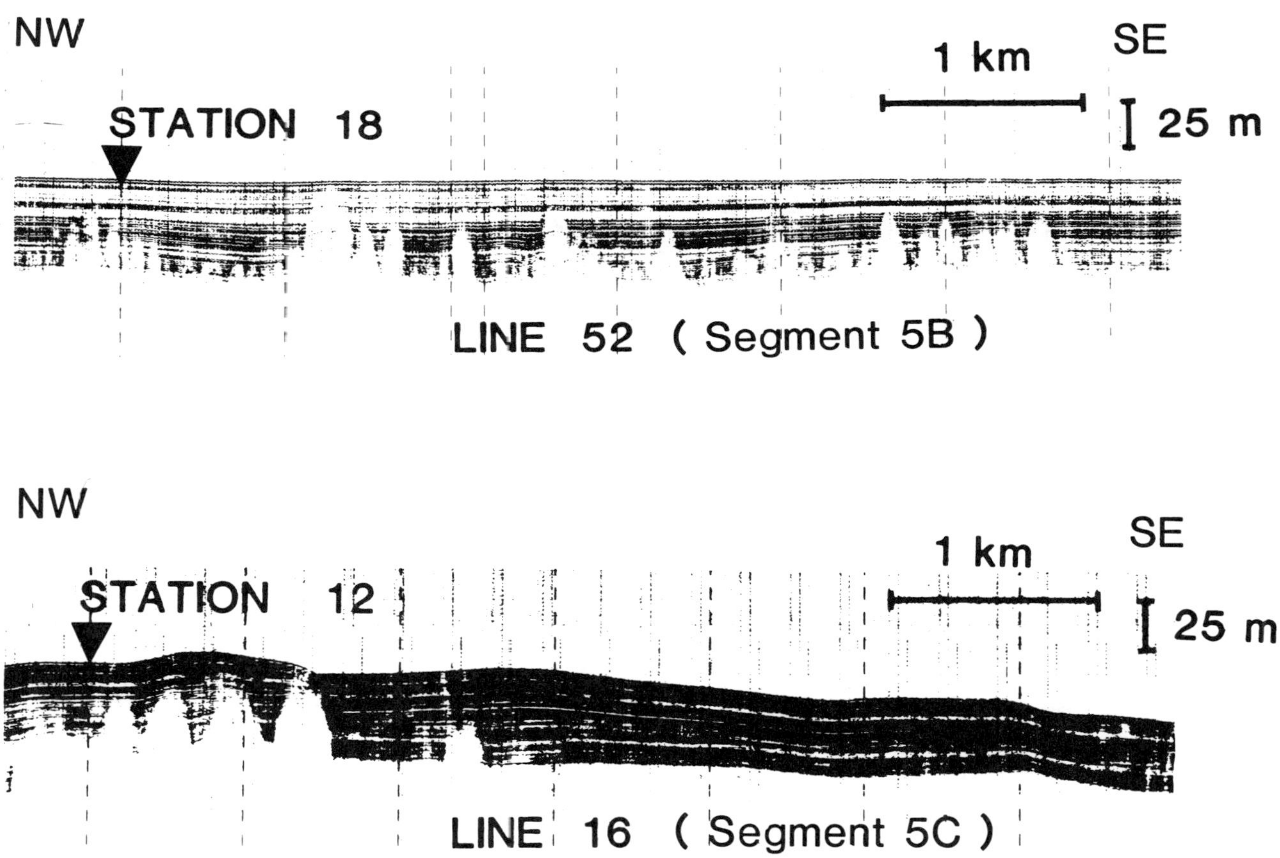

Figure 2. Examples of Uniboom seismic records of acoustic anomalies, located in Figure 1, and believed to be caused by gas in the near-surface sediment.

of gas after extraction. To each can was added sufficient degassed water to establish a 100-cc headspace, after which the can was sealed. The headspace was purged with helium (100 cc/min for 10 min), and the can was then shaken for 10 min to extract gas from sediment into the helium-filled headspace. One cc of this headspace mixture was analyzed by gas chromatography. The following partition coefficients were used to correct for gas solubilities during extraction: C1 = 0.8; C2, C3, i-C4 = 0.7; C2:1 and C3:1 = 0.6. Concentrations are reported as nanoliters of gas per liter of wet sediment (nL/L).

RESULTS AND DISCUSSION

Small concentrations of hydrocarbon gases were found in each of the 15 samples analyzed (Table 1). C1 was most abundant, ranging in concentration from 190 to 2,600 nL/L. C2, detected in only three samples, ranged from 8 to 62 nL/L, whereas C3, measured in six samples, ranged from 4 to 79 nL/L. One sample contained i-C4 (10 nL/L) and n-C4 (20 nL/L) as well as an unquantified amount of C5+. Unsaturated hydrocarbons were present in most samples: C2:1 (8 to 250 nL/L) and C3:1 (10 to 81 nL/L).

Figure 1 shows the sampling stations. Stations 7, 16, and 16A were located away from seismic anomalies, and Stations 12, 13, and 18 were located on seismic anomalies. Figure 2 shows Stations 12 and 18 located on Uniboom records exhibiting features attributed to shallow gas; the records correspond to segments 5C and 5B on lines 16 and 52, respectively (Figure 1). The concentrations of hydrocarbon gases obtained from samples associated with seismic anomalies were not significantly different from results obtained at stations away from the anomalies. Only one sample at Station 16A (190-200 cm) has a gas composition which is unusual. C1, at a concentration of 2,500 nL/L, is accompanied by heavier hydrocarbons, each one of which is at least ten times greater in abundance than the same compound in other samples. The higher amounts of C2, C3, and C4 suggest the possibility of thermogenic gas. More significance would be attached to this observation if the underlying sample at this station also had unusually high concentrations of gas; however, the deeper sample at this station had only background levels of hydrocarbon gases.

Our study has demonstrated that low concentrations of hydrocarbon gases are present in sediments of the Solomon Islands area. These gases probably result from low-level microbial activity within the sediment and represent the background of hydrocarbon gases of this area. In no instance did the results of our gas analysis correlate with the gas-related features on marine seismic records. Although one sample had some characteristics associated with thermogenic gas, the evidence is insufficient to suggest a possible petroleum-related source.

REFERENCES CITED

Claypool, G.E., and Kvenvolden, K.A., 1983, Methane and other hydrocarbon gases in marine sediment: Annual Review of Earth and Planetary Science, v. 11, p. 299-327.

Colwell, J.B., and Tiffin, D.L., 1986, Recent depositional patterns in the Central Solomons Trough of the Solomon Islands, *in* Vedder, J.G, Pound, K.S., and Boundy, S.Q., eds., Geology and offshore resources of Pacific island arcs—Central and western Solomon Islands: Houston, Texas, Circum-Pacific Council for Energy and Mineral Resources Earth Science Series, v. 4, p. 243-254.

Colwell, J.B., Vedder, J.G., et al, 1984, Initial report on 1984 R/V *S.P. Lee* Cruise: UN CCOP/SOPAC Cruise Report No. 91, 22 p.

Kvenvolden, K.A., and Redden, G.D., 1980, Hydrocarbon gases in sediment of the shelf, slope, and basin of the Bering Sea: Geochimica et Cosmochimica Acta, v. 44, p. 1145-1150.

Kvenvolden, K.A., Weliky, K., Nelson, C.H., and DesMarais, D.J., 1979, Submarine seep of carbon dioxide in Norton Sound, Alaska: Science, v. 205, p. 1264-1266.

Vedder, J.G., and Bruns, T.R., editors, 1989, Geology and offshore resources of Pacific island arcs—Solomon Islands and Bougainville, Papua New Guinea Regions: Houston, Texas, Circum-Pacific Council for Energy and Mineral Resources Earth Science Series, v. 12.

GEOLOGIC SETTING AND PETROLEUM PROSPECTS OF BASIN SEQUENCES, OFFSHORE SOLOMON ISLANDS AND EASTERN PAPUA NEW GUINEA

J.G. Vedder, T.R. Bruns
U.S. Geological Survey, Menlo Park, California 94025, USA

ABSTRACT

Late Cenozoic intra-arc basins in the New Ireland-Guadalcanal segment of the Melanesian arc system developed in response to complex interaction along the boundary between the Pacific and Australia-India plates. Among the associated tectonic events were arc-polarity reversal and subduction of the Woodlark spreading-ridge system. Back-arc depositional aprons derived from the ancestral arc were transformed into intra-arc depositional basins by uplift along the newly active arc after reversal. Sequences of strata, chiefly volcaniclastic and as much as 6 to 8 km thick, that accumulated in these basins may have the ingredients necessary for the generation and entrapment of hydrocarbons.

The Central Solomons Trough, which underlies New Georgia Sound, is a composite intra-arc basin that contains faulted and folded late Oligocene(?) and younger strata as much as 7.0 km thick. Deformed Oligocene and younger strata about 8.0 km thick underlie the deep-water basin west of Buka Island between Bougainville and New Ireland. A down-faulted basin beneath Indispensable Strait locally contains as much as 4.5 km of Miocene(?) and younger strata, and an offshore extension of the Mbokokimbo basin on eastern Guadalcanal may include 6.0 km or more of late Miocene and younger strata. At least 2.0 km of slightly deformed late Oligocene(?) and younger strata underlie parts of both Manning and Bougainville straits.

Offshore areas that may be favorable targets for future petroleum exploration in the central and western Solomon Islands and in eastern Papua New Guinea include New Georgia Sound, Indispensable Strait, Manning Strait, Bougainville Strait, the shallow-water platform southwest of Bougainville, and the open-water expanse between New Ireland and Buka Island. The only exploratory well in the entire region was drilled in 1975 on the island platform southwest of Bougainville where 1,682 m of strata, chiefly Miocene and Pliocene volcaniclastic beds, were penetrated in Oceanic Exploration Co., L'Etoile No. 1 well.

Potential reservoir rocks as well as structural and stratigraphic traps may occur in the Central Solomons Trough and the adjoining straits. Also, the areas southwest of Bougainville and west of Buka Island contain underlying structures and sedimentary sequences that may incorporate reservoir rocks and traps. Source rocks, however, have not been positively identified, either onshore or offshore, and the thermal history of the basins is uncertain.

INTRODUCTION

The stratigraphy and petroleum potential of offshore areas in the central and western Solomon Islands were discussed in preliminary fashion by Bruns et al (1986) and by Vedder and Bruns (1986). They presented a summary of previous work, together with the results of the 1982 R/V *S.P. Lee* cruise. The purpose of this paper is to review, update, and supplement the earlier reports by using new data collected during the 1984 R/V *S.P. Lee* cruise, particularly in the Bougainville-New Ireland region of eastern Papua New Guinea (Figure 1).

As in the previous reports (Bruns et al, 1986; Vedder and Bruns, 1986), much of the focus is on the Central Solomons Trough of Katz (1980)—a composite intra-arc structural and depositional basin that extends southeastward from Bougainville beneath New Georgia Sound to Guadalcanal (Figure 2). Emphasis is placed on the Central Solomons Trough because it contains sedimentary sequences and structures that could be tested for hydrocarbon accumulations with currently available technology. The subsidiary basins that form this composite intra-arc trough are Shortland basin, Russell basin, and Iron Bottom basin. Features contiguous to the Central Solomons Trough that may be targets for future petroleum exploration include Indispensable basin west of

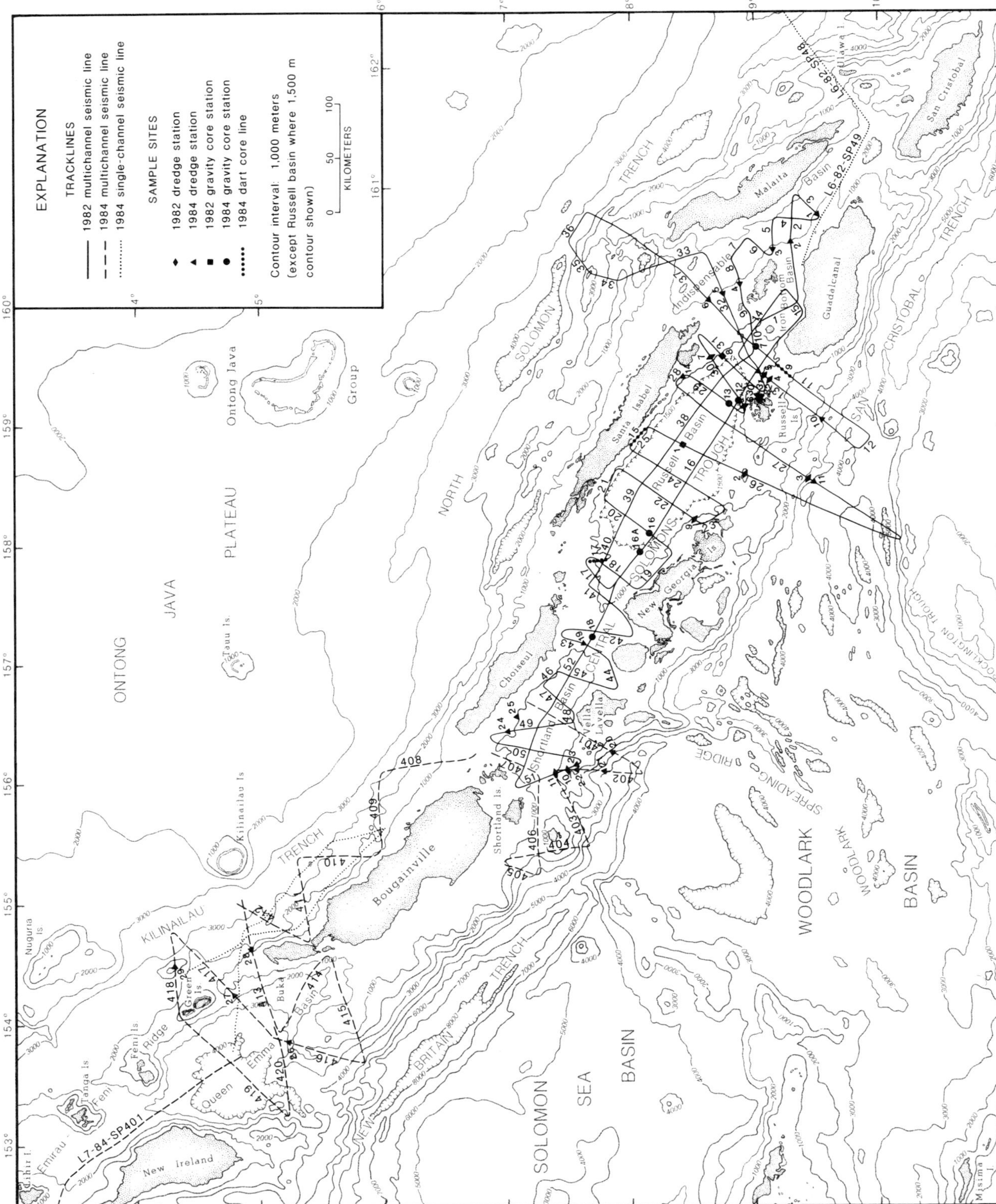

Figure 1. Ship tracklines and sample sites, R/V *S.P. Lee* cruises L7-82-SP and L6-84-SP in the central and western Solomon Islands and eastern Papua New Guinea. Supplemental tracklines from other cruise legs designated L6-82-SP and L7-84-SP.

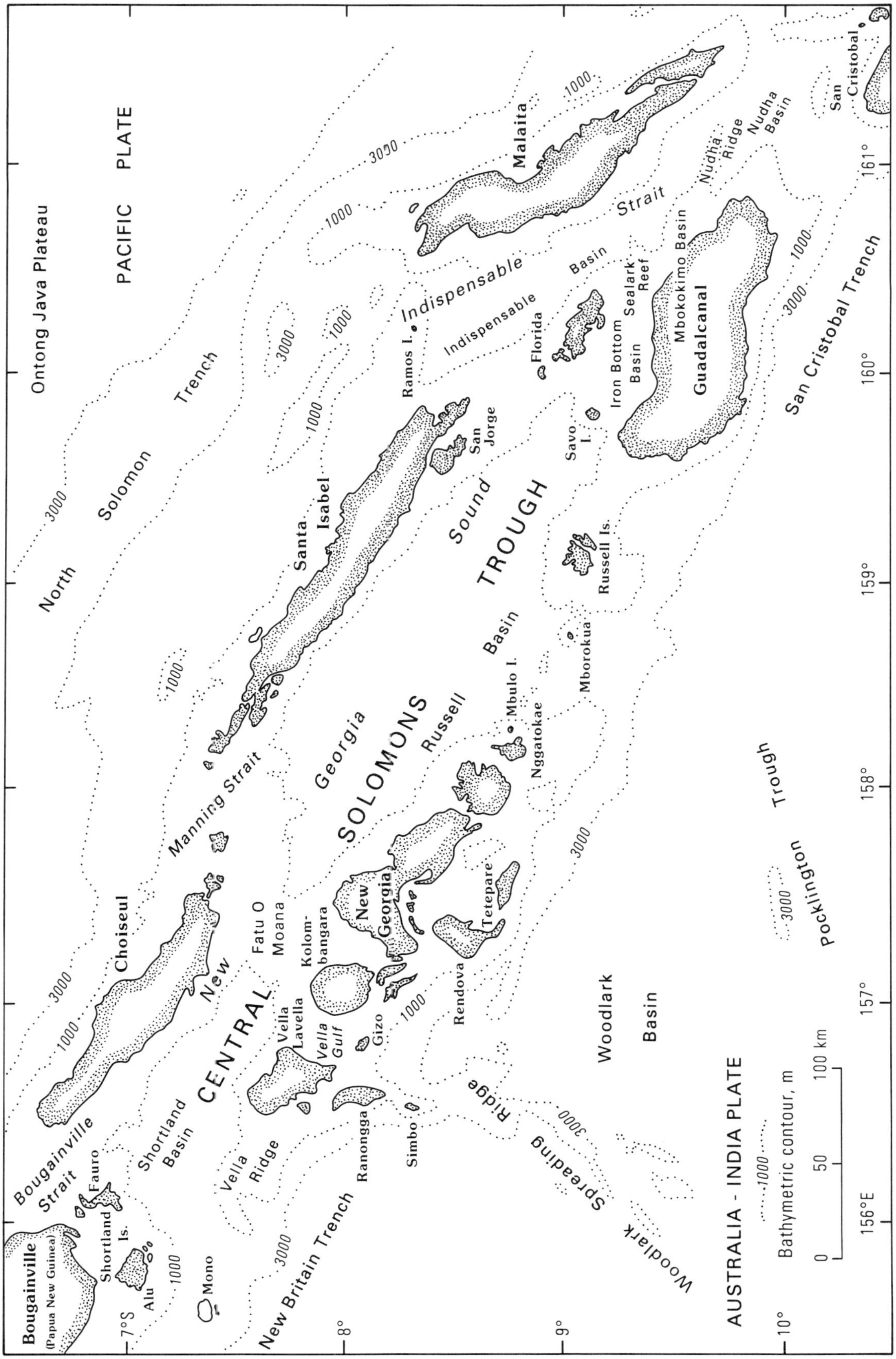

Figure 2. Place names and tectonic features in the central and western Solomon Islands.

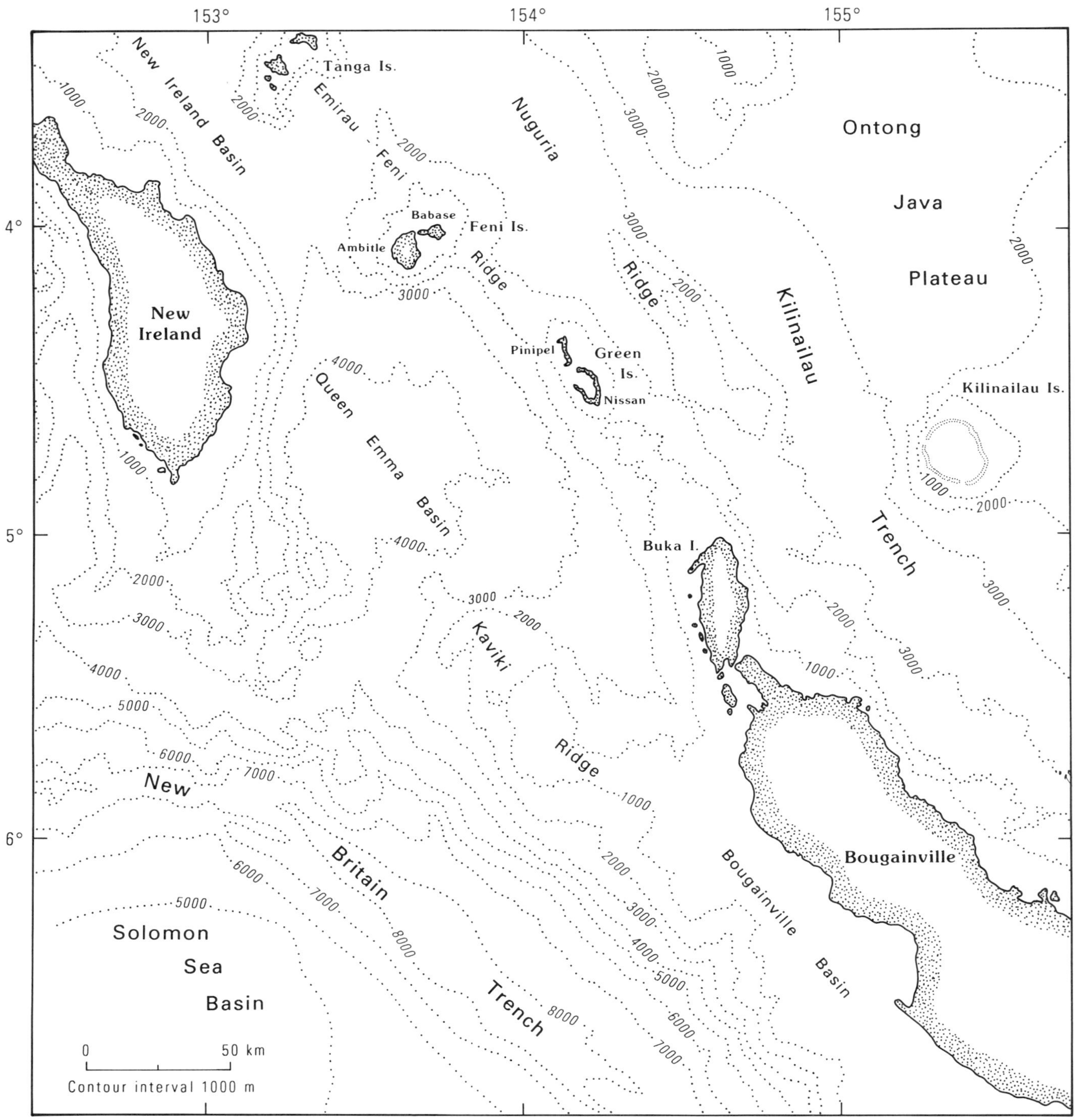

Figure 3. Place names and tectonic features in part of eastern Papua New Guinea.

Malaita, Nudha basin east of Guadalcanal, and the subsea part of Mbokokimbo basin northeast of Guadalcanal.

Other basins are given less attention because water depths are excessive, structures are little known, or stratigraphic sequences are thin. Queen Emma basin, which lies between Bougainville and New Ireland (Figure 3), contains an exceptionally thick sedimentary sequence (Vedder, Bruns, and Cooper, this volume), but water depths generally are more than 2,000 m. Partly filled deep-water basins of unknown structural and depositional character occur east of Malaita and north of San Cristobal (Figure 1). Folded strata north of Santa Isabel and west of northern

Malaita (Katz, 1982, Figure 3; Coleman and Kroenke, 1981, Figure 2; Kroenke, Resig, and Cooper, 1986, Figure 3) probably consist mainly of thin sequences of largely pelagic sediments. Mborokua basin west of the Russell Islands is small and incompletely investigated (Bruns et al, 1986). Although highly deformed sedimentary rocks form parts of the inner wall of the Kilinailau Trench, and slightly deformed strata fill the trench (Bruns, Vedder, and Culotta, this volume), most of these sequences are in water deeper than 1,500 m.

Most of the island platforms, subsea ridges, and trench slopes are not considered to have significant petroleum potential because they generally are covered by thin veneers of sedimentary rocks that overlie igneous and metamorphic basement rocks. Exceptions are the thicker sedimentary sequences that occur locally beneath Manning and Bougainville straits and Vella Ridge (Bruns et al, 1986). Another exception is the thick sedimentary sequence in Bougainville basin, which underlies the island platform southwest of Bougainville (Shaw, 1985; Stewart, Francis, and Pederson, 1987).

We have arranged this report so that brief accounts of previous offshore work and geologic setting of the basins are followed by discussions of the stratigraphy and structure of individual depocenters and selected platforms and ridges. Succeeding sections deal with positive and negative conditions that affect the petroleum resource potential in areas with the best prospects, followed by the conclusions derived therefrom. Because the Central Solomons Trough is more than twice as large as the other basins, this composite basin is described first. Under most headings, geologic features are taken up in geographic order from northwest to southeast.

PREVIOUS OFFSHORE WORK

A broad network of reconnaissance geophysical tracklines covers much of the offshore Solomon Islands-eastern Papua New Guinea region. This reconnaissance work began in 1969 and resulted in publicly available oil company and institutional surveys that are listed in Table 1. The only known detailed surveys were made by a consortium of companies in the Bougainville basin area between 1970 and 1974 (Shaw, 1985). Publications that deal with either regional or local assessments of offshore petroleum potential are scant.

Grover (1955) implied that stratigraphic and structural conditions suitable for entrapment of petroleum might be present in the inter-island areas, and he was the first to note the possibility of offshore accumulations of hydrocarbons in the Solomon Islands. Landmesser (1977) concluded that the Manning Strait area between Choiseul and Santa Isabel offered a favorable subsea prospect. He based this conclusion upon a preliminary evaluation of island geology and offshore seismic-reflection profiles. Katz (1980) indicated that structural and stratigraphic traps might be present in Miocene reef deposits along basin margins in the Central Solomons Trough and that such deposits could contain petroleum, despite reported low porosities and permeabilities and lack of evidence for source rocks. Hughes (1982) delineated late Cenozoic basins and noted that Neogene organic-rich sediments were deposited on the flanks of Choiseul, Santa Isabel, and Guadalcanal. He also mentioned that extensive pelagic limestone deposits on and around Malaita, eastern Santa Isabel, and Ulawa have hydrocarbon potential. Maung and Coulson (1983) assessed the offshore petroleum potential of the central and western Solomon Islands by using inferences from island geology together with interpretations of marine geophysical records—chiefly single-channel and multichannel seismic-reflection profiles. They concluded that the number of anticlinal structures and reefal buildups is sufficient to warrant further investigation. Shaw (1985) and Stewart, Francis, and Pederson (1987), using marine seismic-reflection data and drill-hole information from the exploratory well in the Bougainville basin, inferred that a variety of attractive targets, including both clastic and carbonate reservoirs, are present in the basin. They convincingly demonstrated that the unproductive well was not a valid test of the petroleum potential of the area.

GEOLOGIC SETTING AND NOMENCLATURE

Central Solomons Trough

The name Central Solomons Trough is used herein for the composite intra-arc structural and depositional basin that extends nearly 450 km from the Shortland Islands into the channel between the Florida island group and Guadalcanal (Figure 2). This definition is virtually the same as that of Katz (1980). The Central Solomons Trough is only partly equivalent to the Solomon Basin, which was described by de Broin, Aubertin, and Ravenne (1977) as a single regional feature that stretches nearly 900 km from southeastern New Ireland through Bougainville to the area north of Guadalcanal. The Central Solomons Basin as used by Maung and Coulson (1983) is more extensive than Katz's Central Solomons Trough and includes the semi-enclosed sea that stretches from Bougainville all the way to eastern San Cristobal. The Central Solomons Basin described by Hughes (1982) is the same as Katz's (1980) Central Solomons Trough. Some workers have used the term basin interchangeably for a morphologic entity and for a structural-depositional depression. The individual structural-depositional sub-basins that make up the composite basin (Shortland basin, Russell basin, and Iron Bottom basin) are considered here to be geologic features rather than bathymetric depressions. The morphologic expression of the Central Solomons Trough, as well as that of its sub-basins, is summarized in Table 2.

In the context of its regional geologic setting, the Central Solomons Trough is one of several intra-arc basins that lie along the complex boundary between the Australia-India and Pacific plates in the segment of the Melanesian

Table 1. Summary of Exploration for Petroleum Resources in the Central and Western Solomon Islands and the Bougainville Area, Papua New Guinea, 1969-1984.

DATE	SURVEYED BY	SURVEY LOCATION	LENGTH (KM)	DESCRIPTION
October-November 1969	Magellan Petroleum (Australia), Ltd.	Shortland Islands shelf, Bougainville Strait, Manning Strait, Indispensable Strait, region between Malaita and Guadalcanal	769 (discontinuous traverses)	Seismic survey (airgun and sparker).
1969-1970	Bataafse Internationale Petroleum Maatschappij N.V.	Shortland Islands, Choiseul, Santa Isabel, Guadalcanal, Malaita		Field examination of exposed sedimentary rocks.
1970	Compagnie General de Geophysique *for* Australian Bureau of Mineral Resources	Queen Emma basin, Emirau-Feni Ridge, Kilinailau Trench	?	Single-channel seismic reflection.
1970-1974	Shell Development (Australia), Pty. Ltd., Amoco Australia Exploration Company, Australian Gulf Oil Company, Hematite Petroleum	Bougainville basin and vicinity	1800 (within the prospecting license area)	Seismic-reflection surveys.
May 1971	Shell International Petroleum	Shortland Islands, New Georgia Sound, Malaita, Indispensable Strait, San Cristobal, and Rennell Island	2764	Multichannel seismic reflection. Magnetometer. Gravimeter.
October 1971	Western Geophysical and Teledyne Exploration International, Inc. *for* Southern Pacific Petroleum	Manning Strait	141	12-fold digital seismic reflection.
October-November 1971	Teledyne Exploration Company	Shortland Islands, Bougainville Strait, Choiseul, Santa Isabel, Florida Islands, north coast Guadalcanal, Malaita, Ulawa	2458	Seismic survey (sparker).
June 1972	Mobil Oil Corporation	North coast San Cristobal, Malaita, north coast Guadalcanal, Russell Islands area, Manning Strait, north coast Choiseul, Bougainville Strait, south coast New Georgia	4447	Multichannel seismic reflection. Sonobuoy seismic refraction.
1972	Australian Gulf Oil Company	Shortland Islands shelf, Bougainville Strait, north coast Choiseul, New Georgia Sound, south coast Santa Isabel, region between Malaita, Guadalcanal, and San Cristobal	2480	Single-channel and multichannel seismic reflection. Sonobuoy refraction profile. Magnetometer. Gravimeter.

DATE	SURVEYED BY	SURVEY LOCATION	LENGTH (KM)	DESCRIPTION
1973	Western Geophysical Company	Shortland Islands shelf, Bougainville Strait, north coast Choiseul, Manning Strait, Russell basin, Iron Bottom basin, Sealark Channel, and area east of Guadalcanal	1000 (approx- imate)	Multichannel seismic reflection. Other(?).
1975	AUSTRADEC III Centre National d'Exploit- ation des Océans, Office de la Re- cherche Scientifique et Technique Outre- Mer, Institut Français du Pétrole Comité d'Etudes Pétrolières Marines	Central Solomon Islands, Woodlark Basin, New Ireland forearc.	5000	Multichannel seismic reflection (12- and 24-fold).
December 1978	Western Geophysical *for* Pacific Energy and Minerals, Ltd.	Area between Guadalcanal, Savo Island and the Florida Islands	100	Multichannel seismic reflection.
1979	CCOP/SOPAC	New Georgia Sound	4064 (trackline spacing 15-25 km)	Single-channel seismic reflection.
1981	CCOP/SOPAC	New Ireland forearc (including northern Queen Emma basin)	?	Single-channel seismic reflection.
May-June 1982	CCOP/SOPAC R/V *S.P. Lee* (L7-82-SP)	Shortland Islands, New Georgia Sound, Indispensable Strait, Iron Bottom Sound, Malaita fold belt, New Georgia-Guadalcanal forearc area	5500 (trackline spacing 10-30 km)	Multichannel seismic reflection (24-fold). High-resolution Uni- boom seismic reflection. 3.5 kHz high-resolution seismic reflection. 12 kHz bathymetric profiles. Wide-angle seismic reflection/- refraction sonobuoy profiles. Magnetometer. Gravimeter.
1982	U.S. Geological Survey	Guadalcanal, Florida Islands, Malaita		Field examination and laboratory analysis of outcrop samples and cores.
June 1984	CCOP/SOPAC R/V *S.P. Lee* (L6-84-SP)	Central and western Solomon Islands, eastern Papua New Guinea	2000	Multichannel seismic reflection (24-fold). High-resolution Uni- boom seismic reflection. 3.5 kHz high-resolution seismic reflection. 12 kHz bathymetric profiles. Wide-angle seismic reflection/- refraction sonobuoy profiles. Magnetometer. Gravimeter.

Table 2. Bathymetric Morphology and Dimensions of the Central Solomons Trough and Its Sub-basins.

Name	Bathymetric expression	Bounding features	General configuration	Axial trend	Maximum length (km)	Maximum width (km)	Water depth axial part (m)	Other
Central Solomons Trough (composite)	Narrow trough	Shortland Islands (NW) SW island chain NE island chain Sealark Reef (SE)	Elongate, flat-floored, steep-sided, narrow-ended	N60°W	450	100	1000-1800	Disrupted by Savo volcano near SE end
Shortland basin	Asymmetric depression	Shortland Islands (NW) Vella Ridge (SW) Choiseul (NE) Fatu O Moana (SE)	Rhomboid, flat-floored, steep-sided on NW and NE, seafloor slopes gradually upward to SE	N60°W	140	60	1000-1500	Breached by submarine canyon at W corner
Russell basin	Boat-shaped depression	Fatu O Moana (NW) New Georgia-Russell Islands-Guadalcanal ridge (SW) Santa Isabel (NE) Savo Island (SE)	Elongate-elliptical, flat-floored, steep-sided on SW and NE	N60°W	300	100	1000-1800	
Iron Bottom basin	Double-troughed depression	Savo Island (NW) Guadacanal (SW) Florida Islands (NE) Sealark Reef (NE)	Rhomboid, steep-sided on SW and NE, shoals and narrows to SE	N65°W	45	30	300-1000	Bifurcates and deepens on the flanks of Savo volcano

arc system that extends from Papua New Guinea into Vanuatu. In the vicinity of the Solomon Islands, interaction between these plates since early Tertiary time apparently resulted in two episodes of subduction and arc magmatism that included reversal of arc polarity (Kroenke, 1984; Wells, this volume). There is no evidence that any continental rocks have been involved in these events.

Using the tectonic reconstructions of Coleman and Kroenke (1981), Kroenke (1984), and Wells (this volume), the sequence of events leading to basin formation seems to have been: 1) pre-Oligocene southward-directed subduction of the Pacific plate beneath the Solomons Islands region, resulting in deformation and local metamorphism of oceanic rocks; 2) Oligocene to early Miocene widespread magmatism and emplacement of large volumes of island-arc tholeiitic volcanic rocks; 3) early Miocene collision with the leading edge of the Ontong Java Plateau, causing slowing and eventual cessation of subduction; 4) middle Miocene quiescence, during which subduction was accommodated elsewhere; and 5) late Miocene to Holocene northward-directed subduction of the Australia-India plate beneath the Solomons Islands region, resulting in a reversal of arc polarity and a second phase of volcanism together with uplift, extension, and formation of a set of intra-arc basins. These late Cenozoic events probably intensified after the inception of Pliocene to Holocene subduction of the Woodlark spreading-ridge system (Exon and Taylor, 1984; Taylor and Exon, 1987). According to Dunkley (1983), active subduction of the Woodlark Basin caused eruption of unusual magmas in the New Georgia island group, rapid uplift of the arc accompanied by local development of tensional features, obliteration of the trench, high heat flow along the clogged segment of the trench, and reduced seismicity.

Rapid sedimentation in the Central Solomons Trough probably began in the late Oligocene and continues to the present. Until the late Miocene reversal in arc polarity, the basin sediments accumulated mainly as back-arc volcaniclastic aprons that coalesced southwest of the Oligocene volcanic centers (Choiseul, Santa Isabel), possibly as subordinate forearc sheets of sediment that collected in now-dismembered basins along the northeastern flank of the arc (Guadalcanal, Florida island group). Sedimentation rates slowed in the early and middle Miocene allowing local development of carbonate sequences. Arc-polarity reversal near the end of Miocene time resulted in back-arc deposition of volcaniclastic sediment wedges that were derived from the newly formed volcanic chain southwest of the old arc. These wedges probably spread northward into the incipient intra-arc basin. They subsequently enlarged in volume as volcanism and uplift increased along the newly formed arc axis in the Pliocene. As noted by Katz (1980), the late Oligocene to Pliocene strata in the basin probably are not composed entirely of primary volcaniclastic detritus but contain substantial amounts of material derived from uplifted older, locally metamorphosed, arc rocks as well. Katz's opinion is supported by the findings of Colwell, Niem, and Niem (this volume) and Crook and Niem (this volume). During episodes of volcanic quiescence, reef

Table 3. Bathymetric Morphology and Dimensions of Basins Apart from the Central Solomons Trough.

Name	Bathymetric expression	Bounding features	General configuration	Axial trend	Maximum length (km)	Maximum width (km)	Water Depth axial part (m)	Other
Queen Emma basin	Broad asymmetric trough	New Ireland (NW) Kaviki Ridge (SW) Emirau-Feni Ridge (NE) Bougainville (SE)	Elliptical below 3500 m, triangular above 2500 m, flanks steepen above 3000 m, narrows and shoals to SE	N30°W	200	125	3000-4800	Partly breached on SW by re-entrant from New Britain Trench
Bougainville basin[1]	Island platform	Kaviki Ridge (NW) New Britain Trench slope (SW) Bougainville (NE) Shortland Islands (SE)	Elongate, irregular, underlies reef platform and forereef slope	N45°W	>150	50	<200	Extends on-shore at Bougainville and Alu Island
Mborokua basin	Ridge-crest depression	Nggatokae and Mbulo (NW) Brougham Shoal (SW) Mborokua (NE) Russel Islands platform (SE)	Elongate, irregular, narrow, slightly curved	N55°W	50	25	1500-1700	Perched on forearc ridge
Indispensable basin	Narrow asymmetric trough	Ramos Ridge (NW) Santa Isabel-Florida Islands platform and and Nudha Ridge (SW) Malaita and Alite Reef (NE)	Elongate, convergent to SE, flat-floored, floor locally slopes slopes SW, steep-sided on SW	N35°W (NW) N50°W (SE)	>200	55	1400-1900	Apparently merges with basins north of San Cristobal
Nudha basin	Narrow symmetric trough	Rua Sura Island (SW) Guadalcanal (SW) Nudha Ridge (NE)	Elongate, steep-sided, deepens and opens SE	N60°W	>70	15	500-1500	Merges with SE end of Indispensable basin

[1]Former bathymetric basin now filled and blanketed by coralline-algal reef.

limestone and related shelf and slope carbonate beds formed around dormant emergent craters and insular basement highs.

Extension, accompanied by differential subsidence and faulting during the Pliocene and Pleistocene, resulted in a deepening of the basin, probably in large part as a result of subduction of the Woodlark spreading-ridge system. At the same time, rapid influx of sediment caused by uplift along the basin margins created depositional sequences as much as 3.5 km thick at places in the trough (Bruns et al, 1986). Vertical tectonism prevailed along the southwestern edge of the central and western Solomon Islands in the Quaternary. The maximum estimated rates range from at least 2.0 km/m.y. of downwarp in the Central Solomons Trough (Katz, 1980) to as much as 3.7 km/m.y. of uplift in the Rendova-Tetepare area of the New Georgia island group (Hughes, Varol, and Dunkley, 1984). Smaller amounts of Quaternary vertical movement were reported between Vella Ridge and Guadalcanal, where local uplift of 0.5 km to 2.0 km and local subsidence of as much as 1.4 km occurred (Chivas et al, 1982; Bruns et al, 1986; Colwell and Vedder, 1986; Resig, 1986; Bruns, Vedder, and Cooper, this volume). At Choiseul on the northeastern side of the Central Solomons Trough, at least 0.6 km of vertical tectonism elevated and tilted Pleistocene fringing reefs (Strange, 1981).

Dunkley (1983) attributed uplift along the southwestern flank of the New Georgia island group to compressive stresses developed in the overriding plate, in response to subduction of buoyant young oceanic lithosphere and increased coupling between the plates. The estimated amounts of uplift attest to the exceptionally rapid rates of tectonism in the central and western Solomon Islands and contributed to the more than 9 km of structural relief between the basement floor of Russell basin and the highest peak on Guadalcanal.

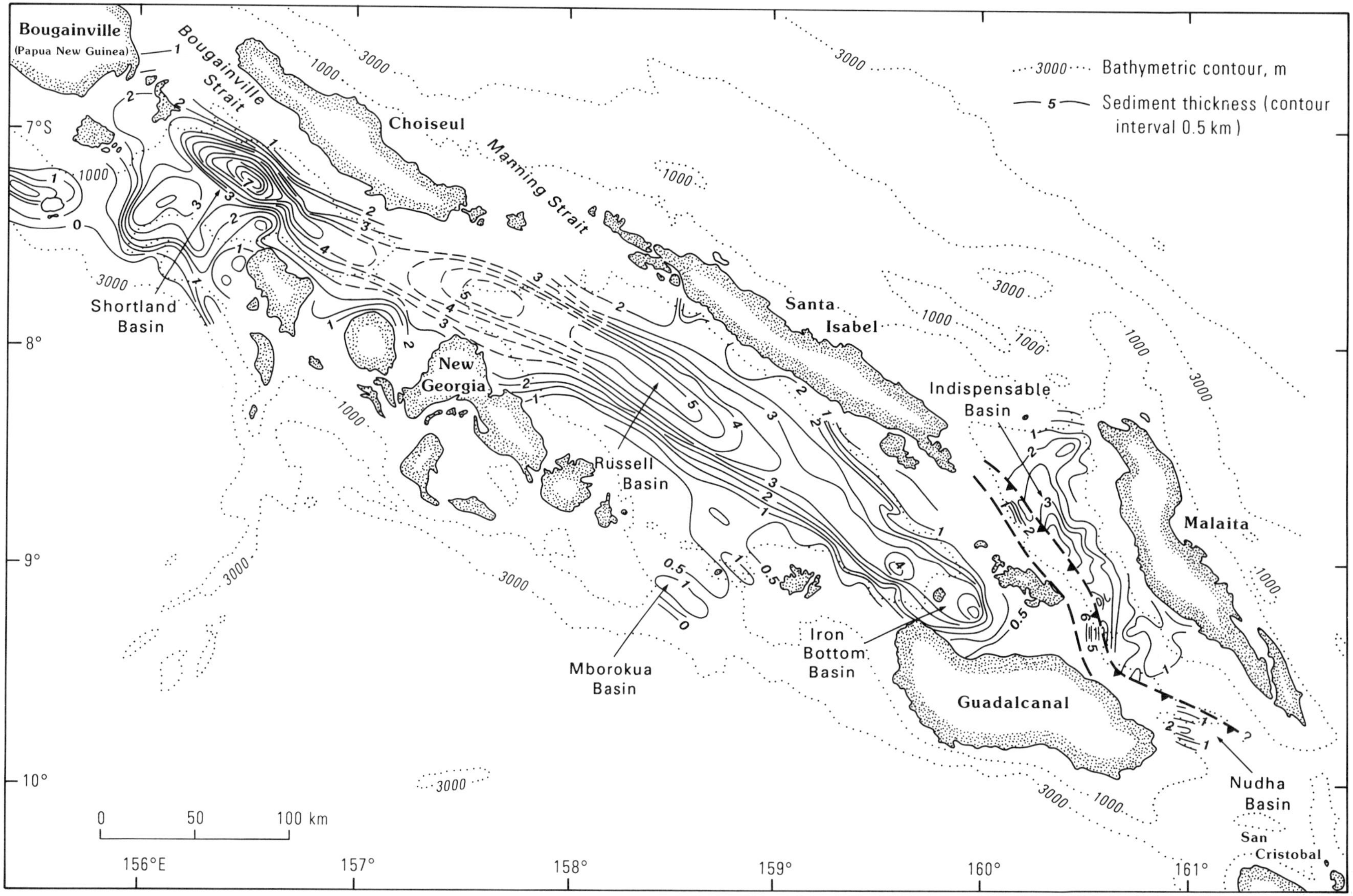

Figure 4. Contours showing sediment thickness, Central Solomons Trough and adjoining basins in the central and western Solomon Islands. Solid contours—thicknesses based upon seismic-reflection data interpreted by Bruns et al (1986) and Bruns, Vedder, and Cooper (this volume; dashed contours—thicknesses based upon seismic-refraction data interpreted by Cooper, Cochrane, and Bruns (this volume).

Other Large Composite Basins

Aside from the Central Solomons Trough, other large, morphologically well-defined basins (Table 3) that are part of the Solomon Islands arc system include Queen Emma basin, which underlies the deep-water area between New Ireland and Bougainville (Figures 2 and 3), and Indispensable basin, which lies between Malaita and the Florida Islands. Smaller basins that occur in different arc settings are described under subsequent headings.

Queen Emma basin is a subsided structural-depositional trough that formed initially in a forearc setting along the early Cenozoic New Ireland-Bougainville segment of the Melanesian arc system (Vedder, Bruns, and Cooper, this volume). After the late Cenozoic change in arc polarity, the basin continued to evolve in a similar but reversed setting. Queen Emma basin is separated from the New Ireland Basin of Exon and Tiffin (1984) by a buried basement ridge over which the stratigraphic units thin markedly, and from the Bougainville basin of Shaw (1985) by the southeastern part of Kaviki Ridge. Uplift along the southern part of Emirau-Feni Ridge may have amounted to as much as 0.7 km after deposition of Pliocene or Pleistocene foraminiferal arenite (Resig *in* Crook and Niem, this volume).

Because Indispensable basin lies in a structural depression along an uplifted, possibly obducted piece of the Ontong Java Plateau (Coleman and Kroenke, 1981; Kroenke, 1984; Kroenke, this volume), this basin and its subsidiary, Nudha basin, are in a geologic environment that contrasts with the intra-arc setting of the Central Solomons Trough. Indispensable basin is believed to have resulted from a downwarp along the leading edge of the plateau during its collision with the early Cenozoic arc (Bruns et al, 1986), and thus is part of Kroenke's (1984) Malaita anticlinorium.

Bougainville basin, which underlies the platform southwest of Bougainville (Shaw, 1985; Stewart, Francis, and Pederson, 1987) is discussed in a subsequent section. Smaller basins, including the offshore part of Mbokokimbo basin and Mborokua basin, both of which are little known (Bruns et al, 1986), also are discussed below.

BASIN SEQUENCES AND STRUCTURES IN THE CENTRAL SOLOMONS TROUGH

Shortland Basin

The northwestern part of New Georgia Sound between the Shortland Islands, Choiseul, and Vella Lavella is underlain by the Shortland basin (Table 2). This structural and depositional feature constitutes one of the three sub-basins that make up the Central Solomons Trough (Figure 2).

The sedimentary section is thickest in the central part of the Shortland basin, where the sequence is as much as 7 km thick (Figure 4). Seismic-reflection profiles show submarine unconformities at intervals throughout the basin section (Bruns et al, 1986; Bruns, Vedder, and Cooper, this volume). These onlapping and downlapping unconformities resulted from late Miocene and later episodes of volcanism and uplift on the basin flanks combined with the deposition of sedimentary wedges that thin seaward. Three lenticular sedimentary bodies are delimited by the most conspicuous of these unconformities (Figure 5). The sedimentary body (unit B), which at places is as much as 3 km thick, probably is composed chiefly of late Oligocene and Miocene strata whose source was an uplifted part of the early Cenozoic arc to the north and northeast, where Oligocene and older mafic volcanic rocks and schistose basement make up a large part of the ancestral arc. The upper two bodies (units D and C), each of which is as much as 3 km thick, were derived mainly from Pliocene and Quaternary volcanic centers to the south and west and presumably consist largely of coeval volcaniclastic strata, together with subordinate intercalated flows.

Faults and folds trend northwest and west in the Shortland basin. North- and northeast-trending faults similar to those on adjacent islands may be present, but they are not apparent on the widely spaced seismic-reflection profiles. Folds in the central part of the basin deform only the Miocene(?) and older parts of the section (units A and B), whereas along the northeastern flank, folding affects all but the youngest Quaternary strata (Figure 5, lines 49 and 407). The upper two lenticular sedimentary bodies were gently arched over Vella Ridge on the southern margin of the basin (Figure 5, line 50) in response to middle Quaternary uplift along the entire southwest flank of the ridge (Bruns et al, 1986).

New Georgia Wedge

A lenticular mass of relatively high velocity material was revealed at shallow depth (0 to about 4 km) by sonobuoy refraction data along the northeast margin of the New Georgia island group (Cooper, Bruns, and Wood, 1986). This thick body, called the New Georgia wedge, has no apparent bathymetric expression (Figure 6). Midway between Choiseul and Kolombangara, a horst-like structure, Fatu O Moana (Taylor and Exon, 1983), is superimposed on the northwestern end of the wedge (Figure 7). To the southeast, the wedge extends into Russell basin and thins southeastward along the southwestern flank of the basin. The New Georgia wedge is a recognizable subsurface feature from Vella Gulf to the vicinity of Mbulo Island and extends about halfway across the Central Solomons Trough. Pliocene and Pleistocene effusive rocks and volcaniclastic strata generated by the New Georgia volcanoes probably are the main constituents of the New Georgia wedge. These rocks correlate with units D and C in the Shortland and Russell basins. However, the lower part of the wedge may be as old as late Miocene. Even though interpretations of seismic-refraction data indicate that the wedge is more than 4.0 km thick, the

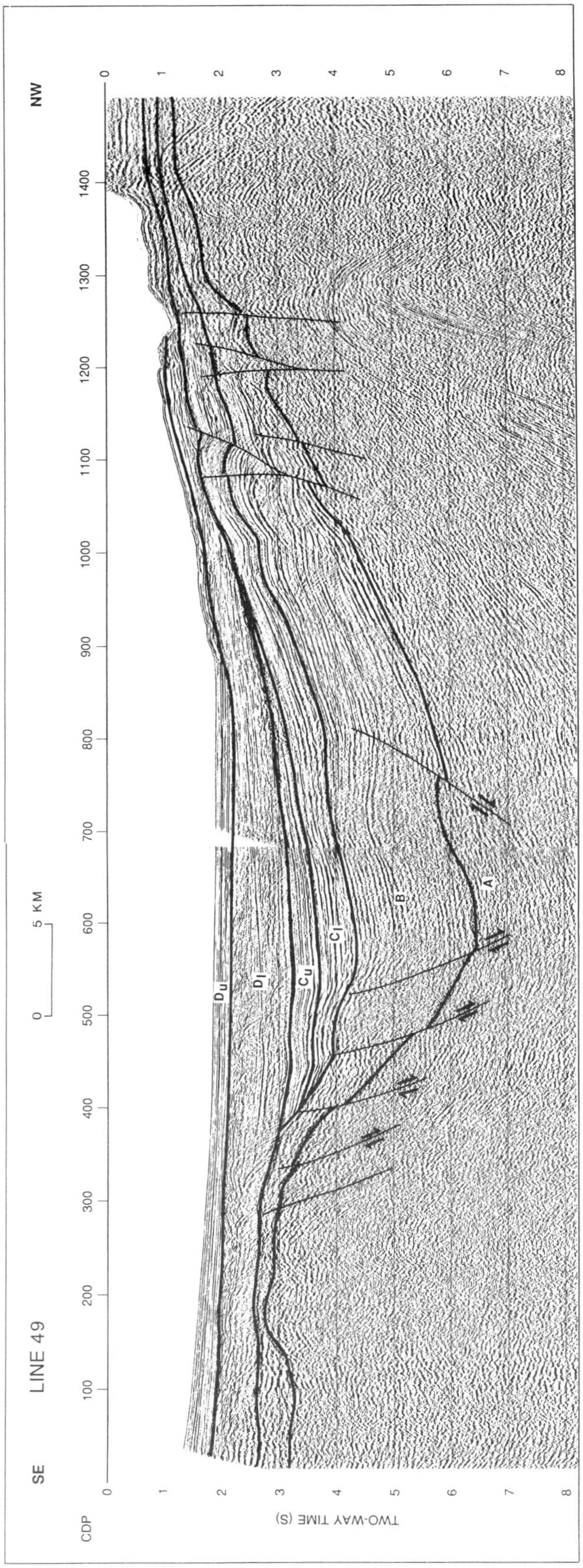
LINE 49
SE
NW
CDP
TWO-WAY TIME (S)
0 5 KM
Du
Dl
Cu
Cl
B
A

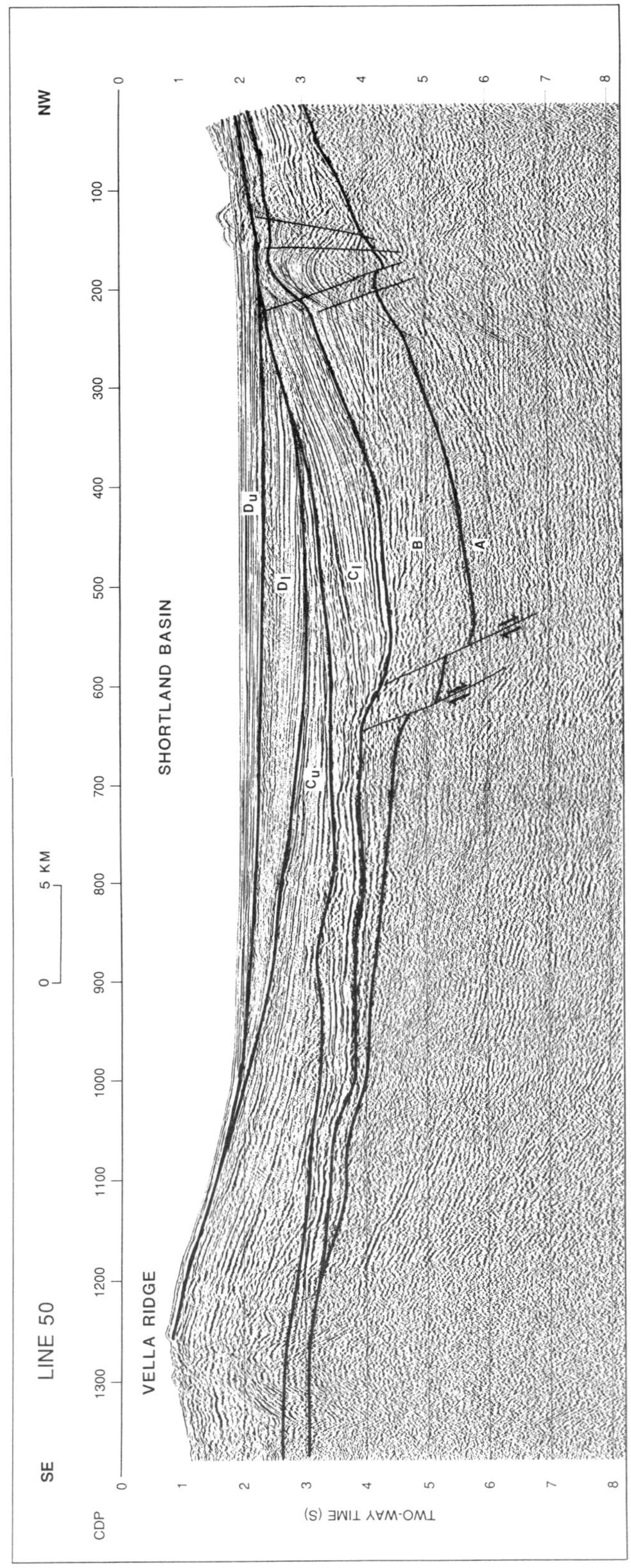
LINE 50
SE
NW
CDP
TWO-WAY TIME (S)
0 5 KM
SHORTLAND BASIN
VELLA RIDGE
Du
Dl
Cu
Cl
B
A

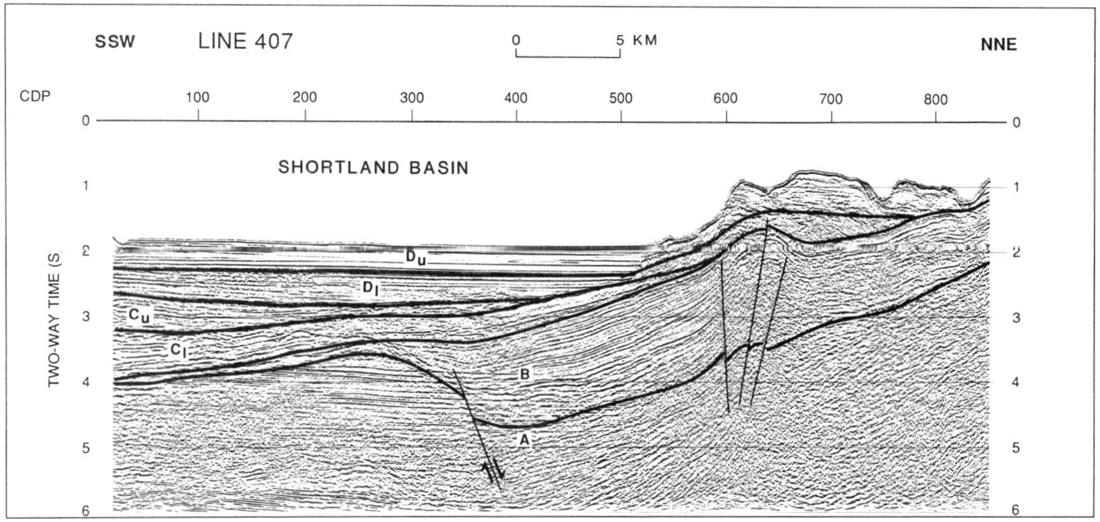

Figure 5. Interpreted multichannel seismic-reflection profiles, Shortland basin, Central Solomons Trough. Sediment wedges are bounded by unconformities. Unit A is acoustic basement; unit B, late Oligocene(?) and Miocene strata; and units C and D, Pliocene and younger strata. Subscripts u and l indicate upper and lower. Line 407 is time-migrated.

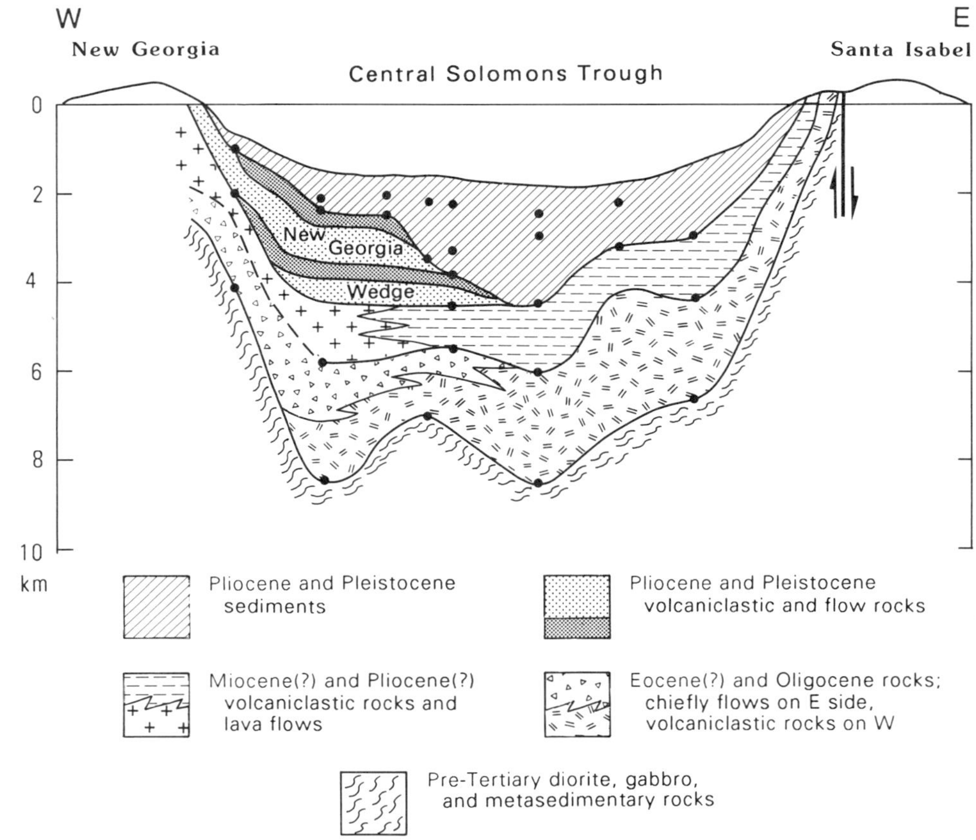

Figure 6. Schematic cross section of the Central Solomons Trough showing the New Georgia wedge. Line of section is W to E between central New Georgia and central Santa Isabel. Section based upon island geology, refraction velocities, seismic-reflector geometry, and gravity models. Large dots represent velocity discontinuities derived from sonobuoy refraction profiles and compiled on Figure 7 of Cooper, Bruns, and Wood (1986).

maximum thickness is unknown, because seismic reflections were not returned from below the high-velocity wedge. The late Miocene age of the lower part is inferred from the corresponding age of the earliest volcanism in the New Georgia island group. Northeast-trending faults cut the wedge and the overlying Quaternary strata; the throw on some of these faults is as much as 500 m.

Russell Basin

Seismic-reflection profiles show as much as 5.0 km of sedimentary section above acoustic basement in the Russell basin, the sub-basin of the Central Solomons Trough that adjoins the Shortland basin on the southeast. This thick sedimentary sequence includes three well-defined unconformities that correlate with those in the Shortland basin (Bruns et al, 1986). These unconformities divide the Russell basin sequence into three stratigraphic units (B, C, and D, Figure 8), in which volcaniclastic mudstone and sandstone are believed to be the dominant rock types. Although carbonate deposits derived from fringing reefs may be interlayered with the volcaniclastic strata, such carbonate lenses, if present, are thin, non-persistent, and not discernible on seismic-reflection profiles. Sources for all of these sedimentary rocks lay mainly to the north, south, and east. Sediment movement probably included transport along the axis of the trough.

Faults are sporadic within the basin. However, normal faults are common in areas adjacent to the main islands. These nearshore normal faults generally trend parallel to the prevailing bathymetric slopes. An exception to this trend is the inferred north-northeast orientation of faults that transect the Russell Islands and the slope to the north. Low-amplitude, northwest-trending folds and faults south of San Jorge deform strata as young as Holocene (Figure 8, line 30); the folds persist for about 20 km and trend N50°W to N65°W. Reverse offset is apparent on at least one fault that cuts Pliocene and older strata near the center of the basin (Figure 8, line 16). In the southeastern part of Russell basin, folds of unknown trend deform all but the youngest strata (upper part, unit D) in the Quaternary sequence (Figure 8, line 16).

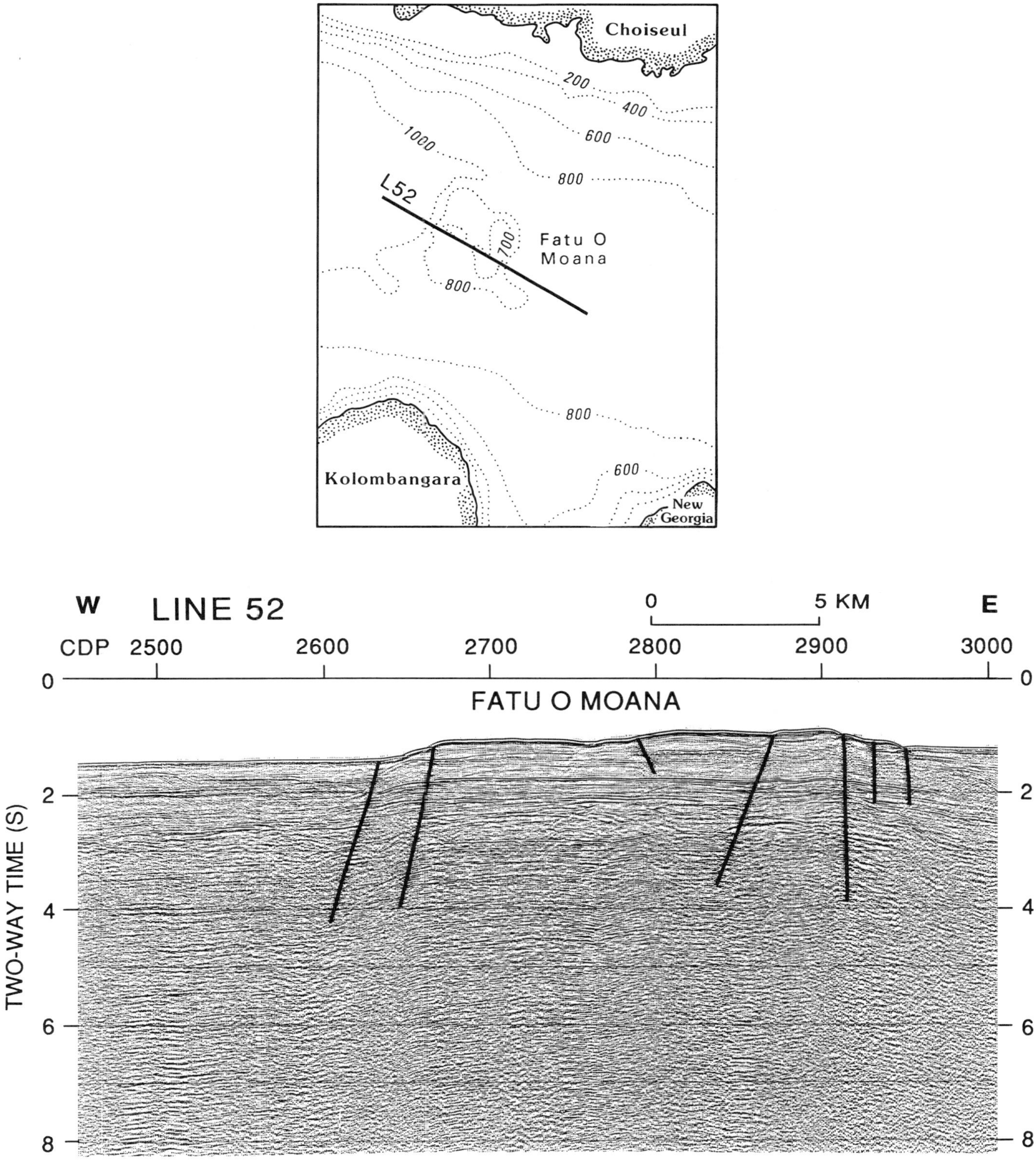

Figure 7. Generalized bathymetry and multichannel seismic-reflection profile across Fatu O Moana, Central Solomons Trough. Seismic-stratigraphic units correlate with those shown in Figure 5.

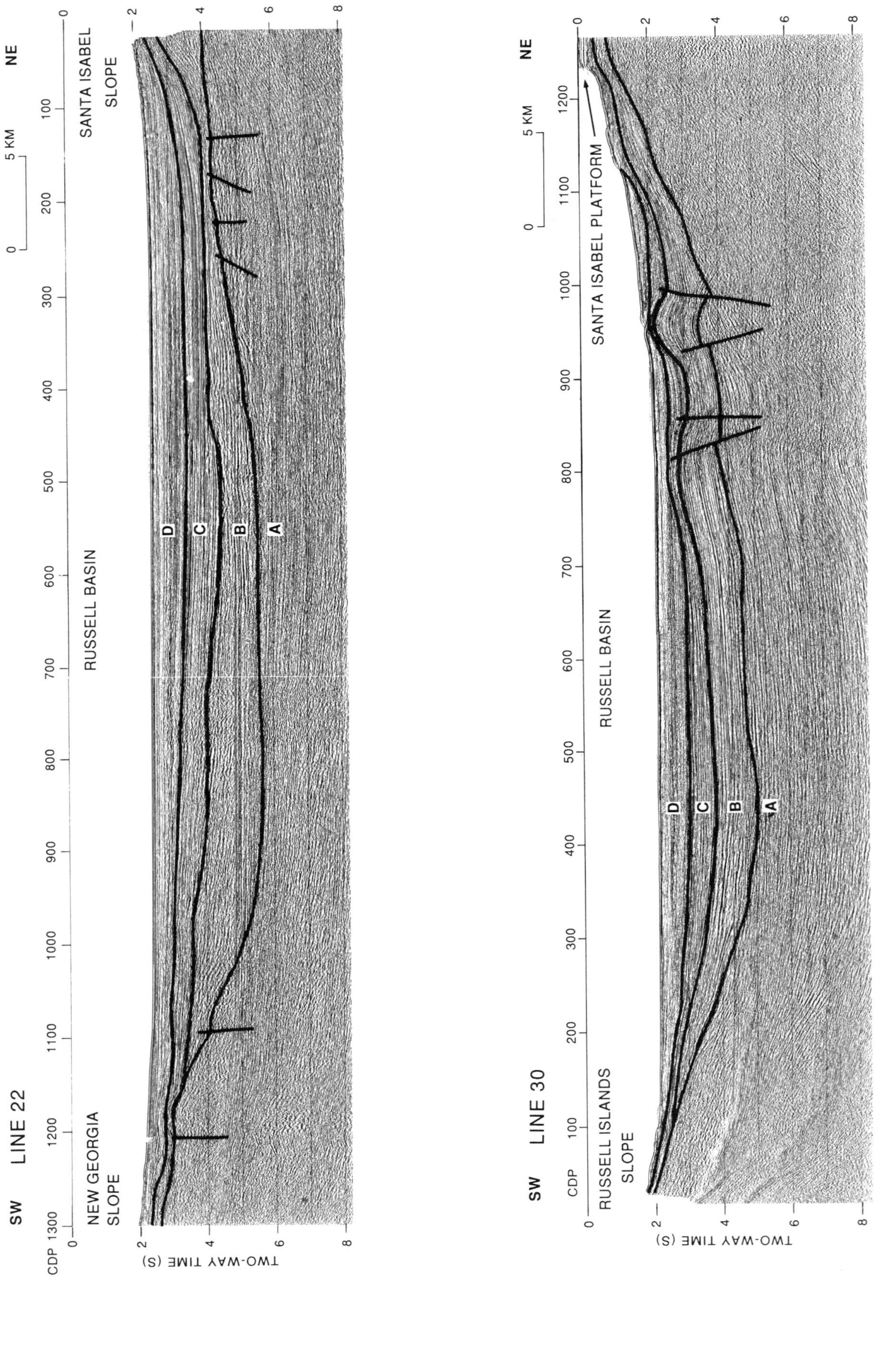
LINE 22
SW
NE
SANTA ISABEL SLOPE
RUSSELL BASIN
NEW GEORGIA SLOPE
CDP 1300
1200
1100
1000
900
800
700
600
500
400
300
200
100
0
5 KM
D
C
B
A
TWO-WAY TIME (S)
2
4
6
8
LINE 30
SW
NE
SANTA ISABEL PLATFORM
RUSSELL BASIN
RUSSELL ISLANDS SLOPE
CDP
100
200
300
400
500
600
700
800
900
1000
1100
1200
0
5 KM
D
C
B
A
TWO-WAY TIME (S)
0
2
4
6
8

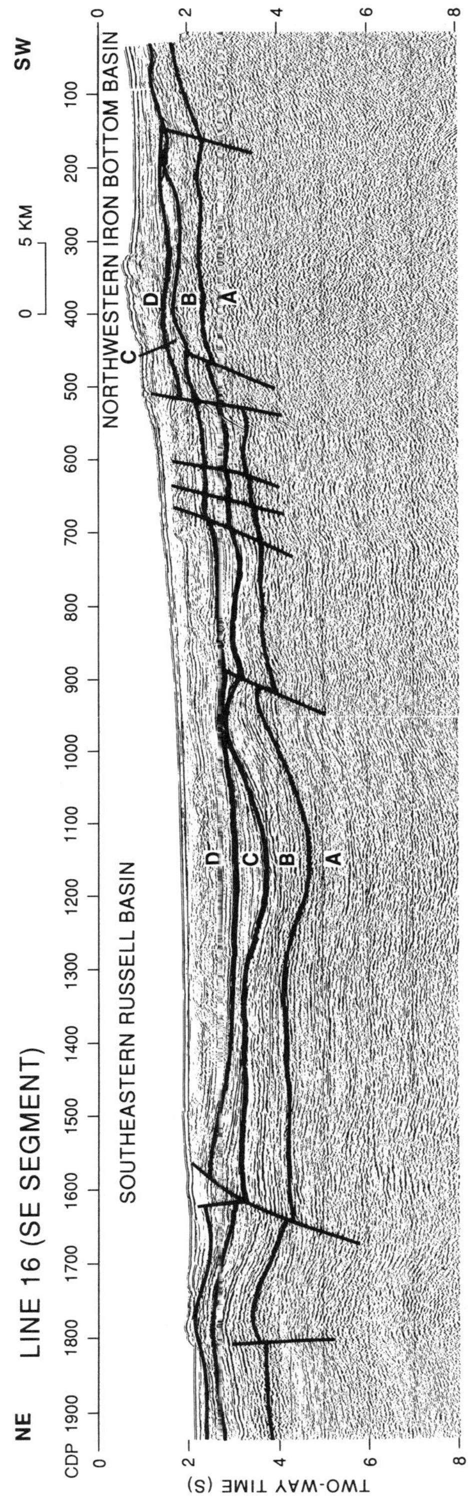

Figure 8. Interpreted multichannel seismic-reflection profiles, Russell basin, Central Solomons Trough. Well-defined unconformities divide the sequence into three seismic-stratigraphic units that correlate with those shown in Figure 5. Lines 22 and 30 are normal to the basin axis; line 16 is parallel to the axis. Line sites are shown on Figure 1.

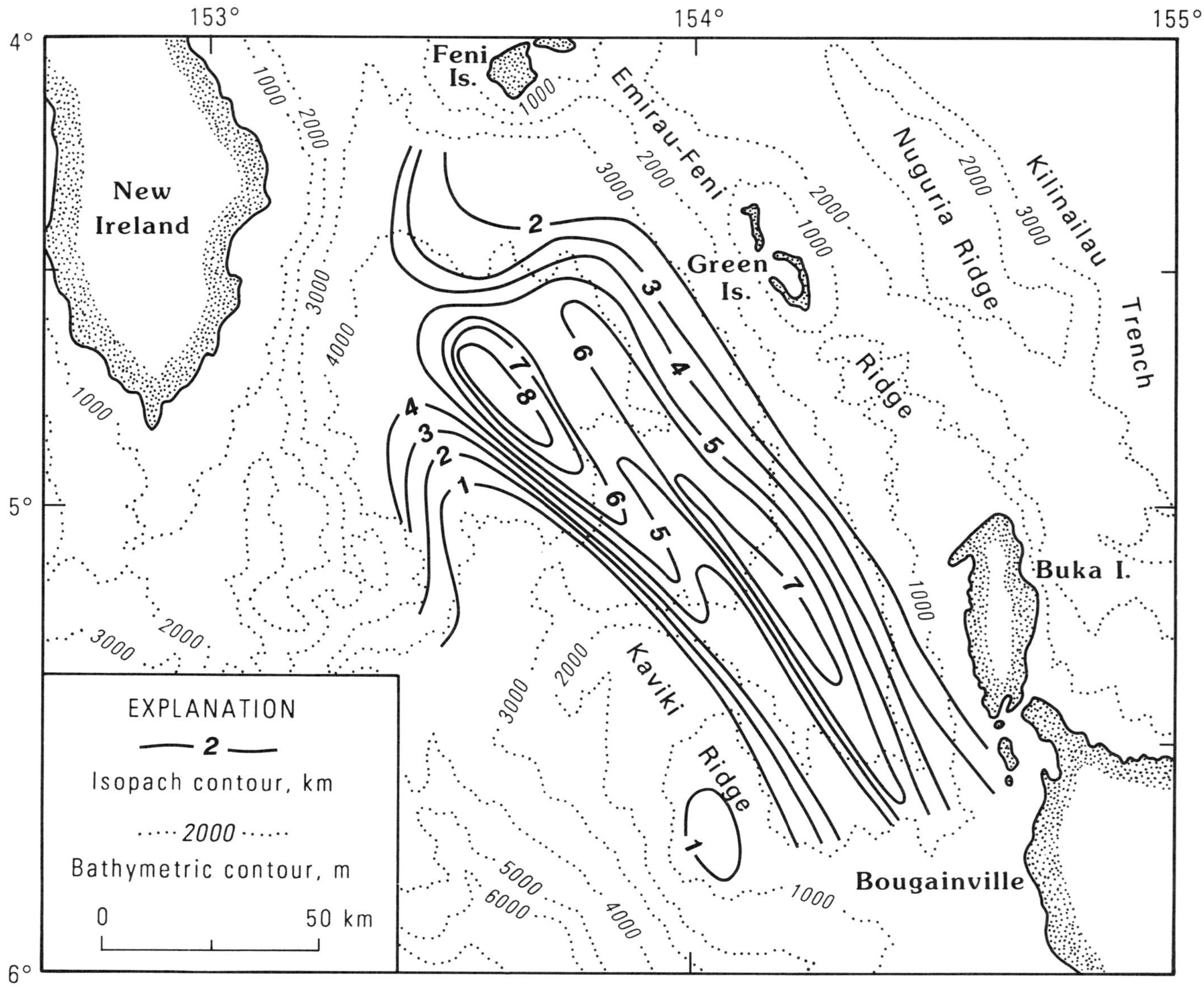

Figure 9. Isopach map showing total sediment thickness above acoustic basement, Queen Emma basin, eastern Papua New Guinea. The sedimentary sequence is believed to be Oligocene and younger in age. Adapted from Vedder, Bruns, and Cooper (this volume).

Iron Bottom Basin

Contiguous to the Russell basin on the southeast is Iron Bottom basin, the smallest of the three sub-basins that form the Central Solomons Trough. Seismic-reflection profiles indicate that the sedimentary sequence in Iron Bottom basin has a maximum thickness of about 4.5 km in an area directly north of Honiara, Guadalcanal (Figure 4). Although profiles through the passages north and south of Savo Island reveal that the seismic-stratigraphic units are continuous into Russell basin (Bruns et al, 1986), the Quaternary sequence must have been disrupted locally by the now-dormant Savo volcano. Presumably the strata in Iron Bottom basin were derived mainly from mafic igneous and metamorphic rocks on Guadalcanal with secondary contributions from similar rocks on the Florida Islands. Quaternary volcaniclastic material from Savo Island, detrital carbonate, and reef limestone probably are subordinate constituents in the sequence. Possible reefal buildups were reported by Maung and Coulson (1983) in the nearshore area northwest of Honiara, but such features were not detected on profiles farther offshore (Bruns et al, 1986).

A pronounced north-trending gravity high, the Tetere Anomaly of Coleman and Day (1965; Beyer, 1986, Figure 2), as well as seismic-reflection data (Bruns et al,

1986) indicate the presence of a shallow basement ridge that closes the southeastern end of the basin between north-central Guadalcanal and the Florida Islands. Minor low-amplitude folds and Pleistocene(?) normal faults occur locally in Iron Bottom basin (Figure 8, Line 16, SE end; Bruns et al, 1986), but their orientation is unknown

STRATAL SEQUENCES AND STRUCTURES IN OTHER BASINS

Queen Emma Basin

A northwest-trending structural and depositional intra-arc basin, named Queen Emma basin by Vedder, Bruns, and Cooper (1987; this volume), separates New Ireland and Bougainville. This basin is about 200 km long and 125 km wide. Oligocene(?) and older volcanic and metamorphic(?) rocks form the basement of the basin, upon which an Oligocene and younger sequence of strata as much as 8.2 km thick was deposited (Figure 9). Seismic-reflection profiles suggest that the lower part of this sequence (unit E2, Figure 10) originated as a forearc apron of sediments derived from a ridge to the southwest. The sequence thickens northeastward from Kaviki Ridge into the central part of the basin; from there it thins onto the Emirau-Feni Ridge. The upper part (units E3 and E4) generally is restricted to the trough of the basin and was derived mainly from the southeast (Figure 10). Although most of these strata are inferred to be volcaniclastic, local limestone bodies similar to the early Miocene Keriaka Limestone on Bougainville (Blake and Miezitis, 1967; Hilyard and Rogerson, this volume) may be present on the flanks of the basin and on Kaviki Ridge.

Major folds that deform strata in Queen Emma basin include a buried anticline in the central part and a broad antiform that forms the bordering Emirau-Feni Ridge (Figures 10 and 11). These folds are nearly symmetrical except for a local southwest vergence on the buried anticline. Small normal and reverse faults that cut strata as young as Quaternary are common on Kaviki and Emirau-Feni ridges. Pre-Pliocene(?) normal and reverse faults associated with the buried anticline suggest the "flower" structures that typically develop along strike-slip faults (Figure 11). A zone of faults southwest of the northwestern end of Kaviki Ridge probably is part of the offshore extension of the Weitin fault (Figure 9)—a strike-slip fault on southeastern New Ireland (Hohnen, 1978; Vedder, Bruns, and Cooper, this volume).

Kilinailau Trench

The Kilinailau Trench, which extends about 650 km along the northeastern side of Bougainville and the Emirau-Feni Ridge, separates the island chain northeast of New Ireland from the Ontong Java Plateau. Three seismic stratigraphic units constitute the sequence beneath the trench (Bruns, Vedder, and Culotta, this volume). These units (Figure 12) are believed to represent oceanic basalt of the Ontong Java Plateau (unit Kl), an overlying 700- to 900-m-thick sequence of pelagic sedimentary rocks (unit K2), and a 3- to 5-km-thick succession of hemipelagic trench-fill strata (unit K3). The pelagic strata may range in age from Late Cretaceous to Miocene, and the hemipelagic trench fill may be late Miocene and younger.

Folds and faults that deform the trench-fill strata are as young as Quaternary. Although the folds are nearly symmetrical, the faults generally verge to the northeast and show reverse separation along planes that flatten southeastward (Figure 12).

Trench-slope basins along the southwestern wall of the Kilinailau Trench contain as much as 1.8 km of late(?) Miocene and younger strata (units B2 and B3, Figure 12). In contrast to the structurally complex underlying rocks, these slope-basin strata are only slightly deformed. Some of the warped strata are as young as Quaternary.

Indispensable Basin

East of the Central Solomons Trough, the narrow Indispensable basin separates the Santa Isabel-Florida Islands-Guadalcanal platform from Malaita (Figure 1) and may represent either a relict forearc basin or a downwarped segment of the oceanic plateau. The length of the basin is not known; geomorphically, it is at least 200 km long in its axial part and about 55 km wide at its widest part west of northern Malaita (Table 3). A discontinuous bathymetric ridge that separates Malaita and San Cristobal may mark the southeastern end of the basin.

Seismic-reflection data indicate that as much as 4.5 km of strata occur above acoustic basement in Indispensable basin (Figure 4; Bruns et al, 1986). This basement (unit I_A, Figure 13) is provisionally correlated with the Cretaceous and early Tertiary igneous rocks and pelagic limestone exposed on Malaita and northeastern Santa Isabel (Coulson and Vedder, 1986). The thickest sedimentary sequences are concentrated along the southwestern margin of the basin—an area bounded by faults (Figure 13). The lowermost seismic-stratigraphic unit in the sedimentary sequence (unit I_B) is characterized by abrupt thickness changes and by draping over some high-standing structures. This folded and faulted Miocene(?) unit is overlain by less-deformed uniformly bedded strata (unit I_{CD}) of probable Pliocene and Quaternary age. Conspicuous unconformities separate the three seismic-stratigraphic units (Figure 13).

Some strands of the major fault zone that parallels the southwestern margin of Indispensable basin cut the youngest strata discernible in seismic-reflection profiles across the northeastern slope of the Florida Islands platform. This 5- to 10-km-wide fault zone (Figure 13) may be a continuation of the Kia-Korigole-Kaipito fault zone on Santa Isabel (Coulson and Vedder, 1986). Northeast of Guadalcanal, highly deformed beds occur in a

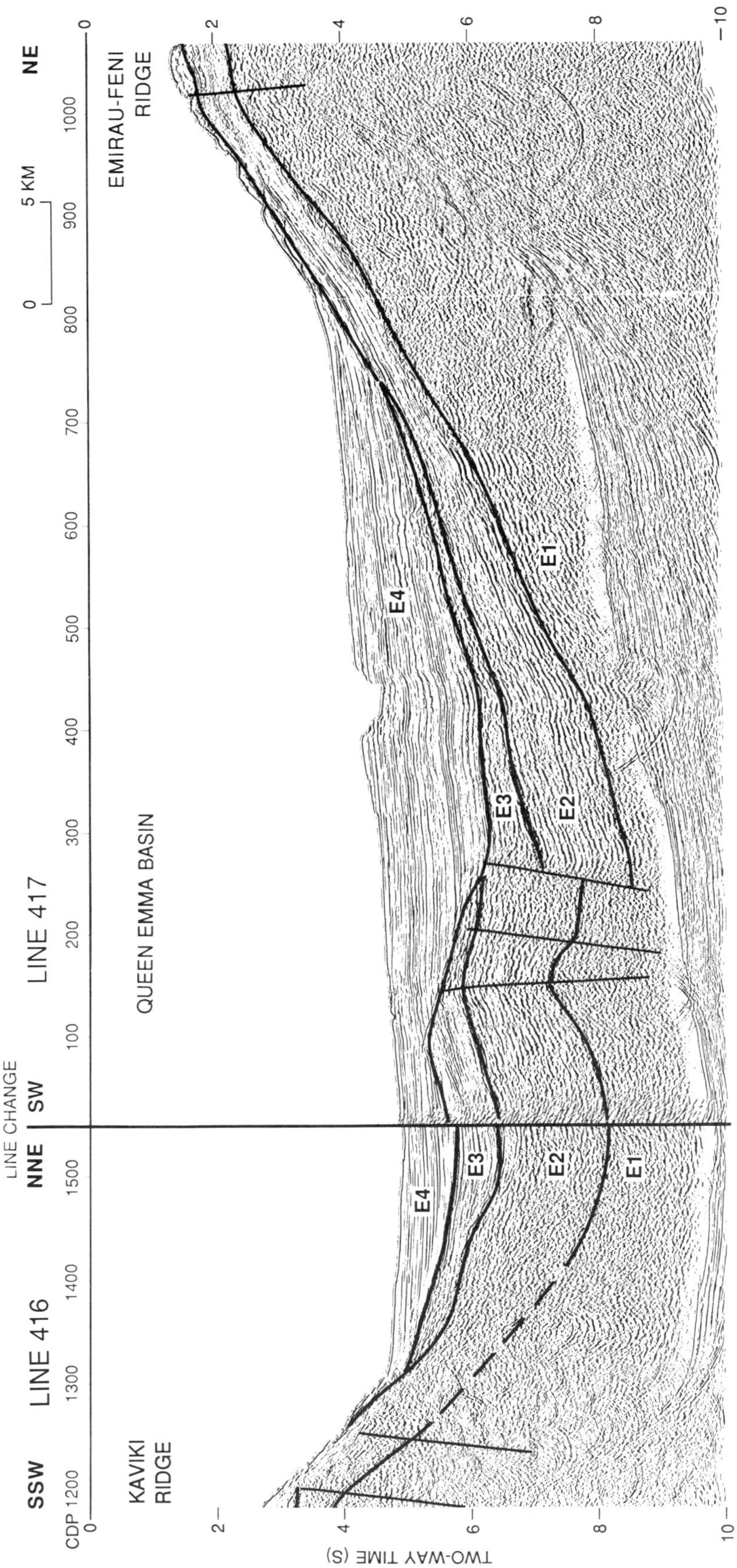
SSW
KAVIKI RIDGE
LINE 416
NNE
LINE CHANGE
SW
LINE 417
QUEEN EMMA BASIN
0
5 KM
EMIRAU-FENI RIDGE
NE
CDP 1200
1300
1400
1500
100
200
300
400
500
600
700
800
900
1000
E4
E3
E2
E1
TWO-WAY TIME (S)

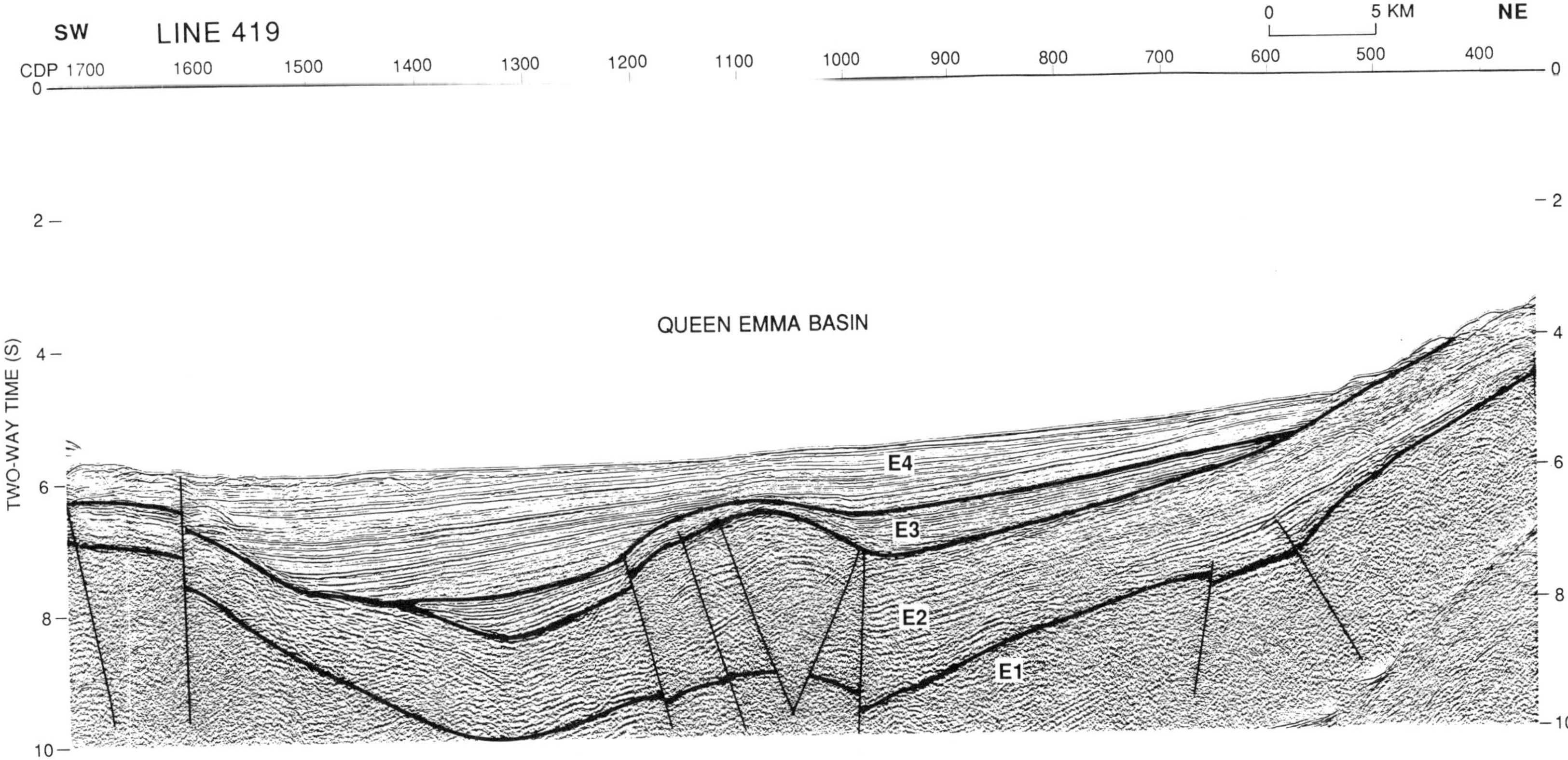

Figure 10. Interpreted multichannel seismic-reflection profiles, Queen Emma basin, eastern Papua New Guinea. Sediment wedges are bounded by unconformities. Unit El is acoustic basement; unit E2, late Eocene(?) to early Miocene strata; unit E3, middle(?) and late(?) Miocene strata; and E4, late Miocene(?) to Holocene. Line sites are shown on Figure 1.

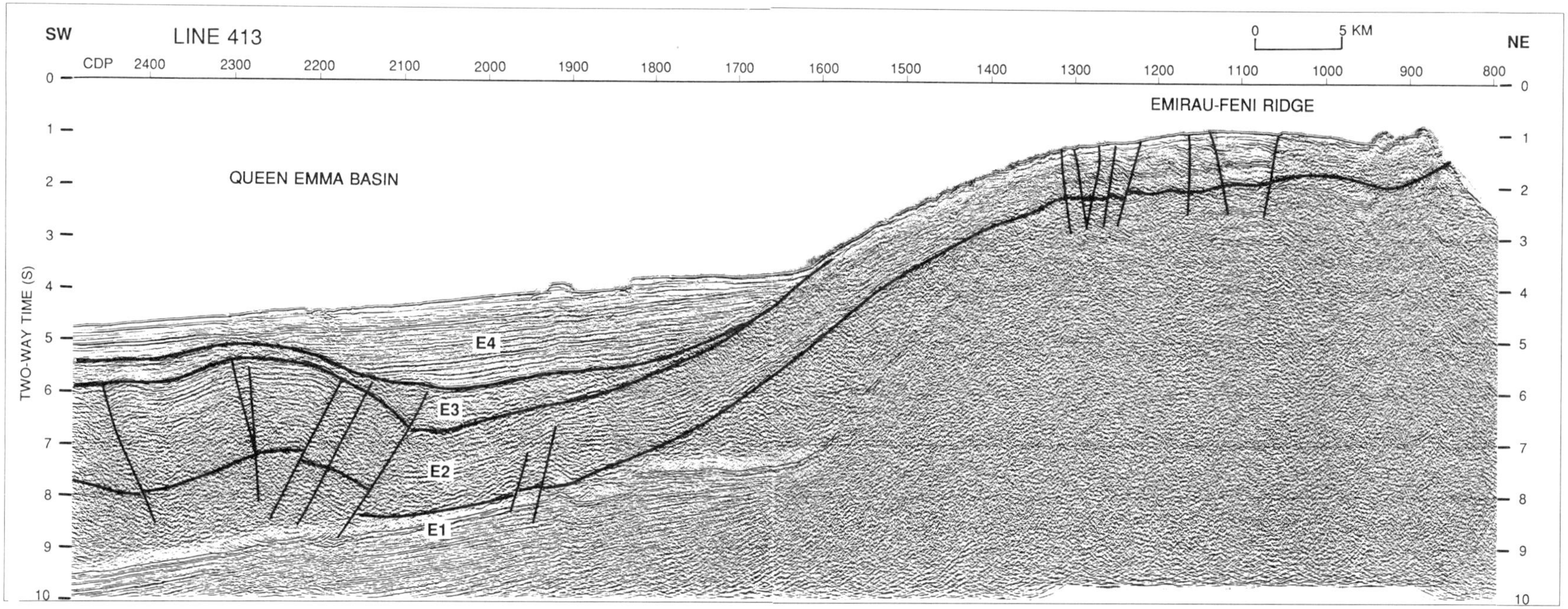

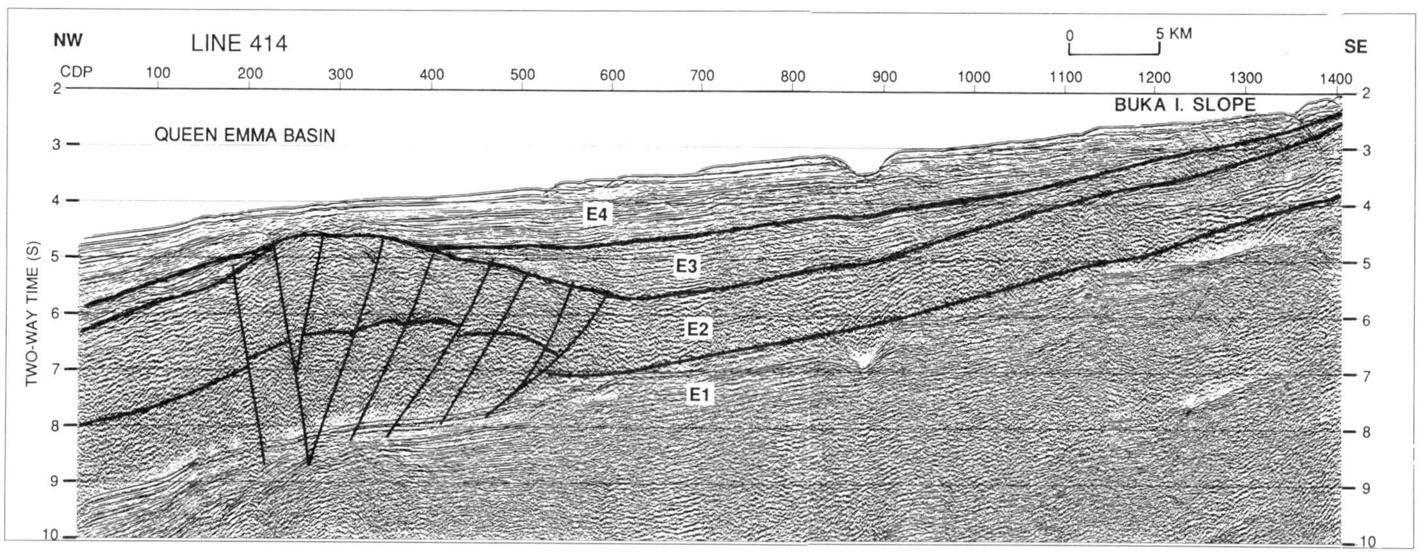

Figure 11. Interpreted multichannel seismic-reflection profiles showing folds and "flower" structure, Queen Emma basin, eastern Papua New Guinea. Seismic-stratigraphic units are the same as those shown on Figure 10. Line sites are shown on Figure 1.

10- to 15-km-wide belt on the southwest side of the offshore fault zone. Although some folds affect all but the youngest (Quaternary?) strata in the basin sequence, these structures are more obvious in the older (Pliocene? and older) parts of the sequence. Normal and reverse faults cut the pre-Pliocene(?) sequence in the basin; and numerous small, mainly basin-side-down, normal faults transect the northern margin of the basin. Presumably, most of these small faults have west and northwest trends, parallel to the dominant structural trends on Malaita. Data are not available to determine the location and nature of the southeastern end of Indispensable basin.

Mbokokimbo and Nudha Basins

Onshore in northeastern Guadalcanal, a basin containing a thick sedimentary sequence was recognized by the presence of a gravity anomaly and by the surface mapping of Coleman and Day (1965). It was designated Mbokokimbo basin by Hackman (1980, Figures 4, 22, and 24). Both studies indicated that the basin and its late Miocene and Pliocene sedimentary fill extend eastward beneath the island platform toward the island of Rua Sura, and that the Tetere basement high borders the northwestern end of the basin (Figure 14). Coleman and Day (1965) estimated that late Cenozoic strata at least 6,000 m thick lie between Taivu Point on Guadalcanal and Rua Sura. Conglomerate dredged on the slope near Nudha Island (Crook and Niem, this volume) resembles the late Pliocene and Quaternary conglomerate described by Hackman (1980) as the Vatumbulu Beds. The presence of this conglomeratic unit near Nudha Island suggests that part of the basin sequence extends eastward at least 30 km. Stratigraphic and structural relations in the offshore part of Mbokokimbo basin are little known. The northern edge of the basin is believed to be truncated by the major fault zone that borders the southwestern side of Indispensable basin, and the southern edge probably merges with the southwestern margin of Nudha basin (Bruns et al, 1986).

Nudha basin is a small narrow northwest-trending basin that lies directly north of easternmost Guadalcanal (Figure 14). Seismic-stratigraphic units in Nudha basin correlate with those in Indispensable basin (Bruns et al, 1986) and are 1.5 to 2.0 km thick (Figure 14, line 18). Whether these units were once contiguous with those in Indispensable basin and Mbokokimbo basin is uncertain. Nudha Ridge, which forms the northern margin of Nudha basin, resulted from at least 1.5 km of Quaternary uplift along the southwestern side of the major fault zone that parallels Indispensable basin. Along the southern flank of this ridge, the sedimentary section is nearly as thick as that in the central trough of the basin and is deformed by folds and north-verging reverse faults (Figure 14, line 18).

STRATAL SEQUENCES AND STRUCTURES BENEATH PLATFORMS AND RIDGES

Island platforms and submarine ridges adjoining the Bougainville-Buka area and the Central Solomons Trough generally are underlain by acoustic basement thinly covered by clastic sediment and reef-related carbonate deposits. However, there are exceptions to this general rule. One noteworthy exception is the broad platform southwest of Bougainville where a folded and faulted sequence of early Mocene and younger strata is as much as 7.6 km thick (Shaw, 1985; Stewart, Francis, and Pederson, 1987). The name Bougainville basin was applied to this depositional feature which extends onshore and is believed to contain both clastic and carbonate sequences (Shaw, 1985). The dominant trend of both the folds and faults in the offshore part of the basin is northwest and west-northwest; a secondary set of faults trends northeast (Figure 15). Although most of the faults show normal throw, opposing reverse faults occur in the southeastern part of the basin (Shaw, 1985; Stewart, Francis, and Pederson, 1987). Most of the fault activity was pre-Pliocene. Bougainville basin apparently is separated from Queen Emma basin by a basement ridge.

The sedimentary sequence that arches over the crest of Emirau-Feni Ridge between Buka and the Green Islands generally is less than 1.5 km thick (Figures 9; 10, line 417; and 11, line 413); there, the underlying acoustic basement has velocities (5.2 to > 7.5 km/s) that suggest the presence of igneous and metamorphic rocks, probably representing both early Cenozoic and late Cenozoic arc magmatism (Vedder, Bruns, and Cooper, this volume). Kaviki Ridge, south of Queen Emma basin, at most places is capped by sedimentary rocks (chiefly limestone?) less than 1.0 km thick (Figure 9), under which lie high-velocity rocks that are believed to be part of the early Cenozoic magmatic arc.

Katz (1980) noted the possibility of a substantial depositional sequence under Bougainville Strait. Landmesser (1977) predicted at least 3 km of late Miocene and Pliocene strata beneath Manning Strait, but Katz (1980) and Maung and Coulson (1983) expressed the opinion that such a thickness is excessive. Seismic-reflection profiles through Manning and Bougainville straits indicate that as much as 2 to 3 km of gently deformed strata may occur at places in each of these areas (Bruns et al, 1986; their Figure 11, line 8; and Figure 23, line 12), although the sequence in Bougainville Strait generally is less than 1.5 km thick. Unpublished data indicate that 1 to 2 km of sedimentary section is present locally on the northern flank of Choiseul. The size and orientation of structures beneath both of these straits and north of Choiseul are uncertain.

Very high velocity rocks (5.8 to >7.5 km/s) occur beneath the platform southwest of Choiseul, where gravity data imply large masses of dense basement rocks at shallow depth (Beyer, 1986). These dense masses may be thrust slabs or tectonically intruded bodies of ultramafic

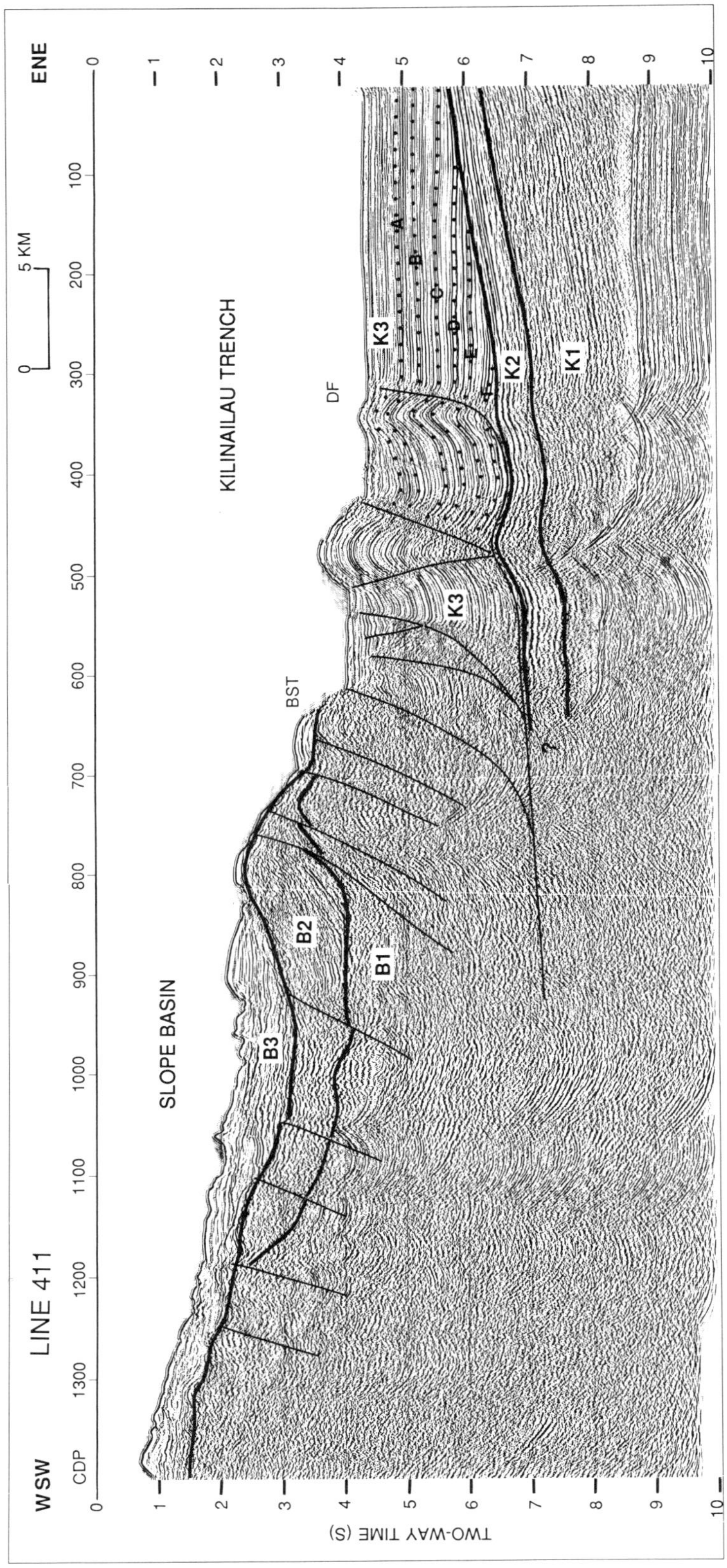
LINE 411
WSW
ENE
CDP
KILINAILAU TRENCH
SLOPE BASIN
5 KM
DF
BST
K1
K2
K3
B1
B2
B3
A
B
C
D
E
TWO-WAY TIME (S)

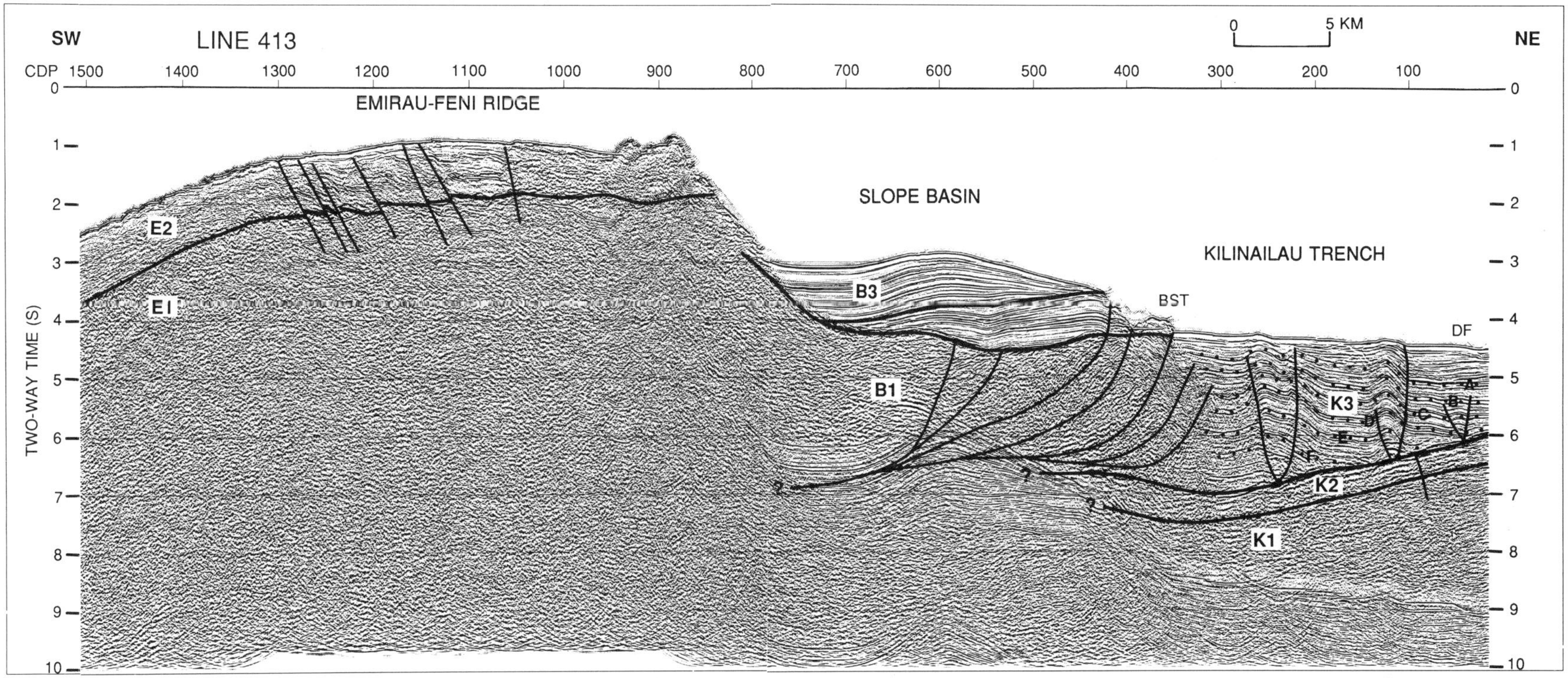

Figure 12. Interpreted time-migrated multichannel seismic-reflection profiles, Kilinailau Trench, eastern Papua New Guinea. Seismic-stratigraphic unit B1 is acoustic basement; B2 and B3, late(?) Miocene and younger trench-slope basin strata. Units El and E2 are the same as those shown on Figure 10. Unit K1 is believed to be oceanic basalt; K2, Cretaceous(?) to Miocene(?) pelagic strata; K3, late(?) Miocene and younger trench-fill strata. Line sites are shown on Figure 1.

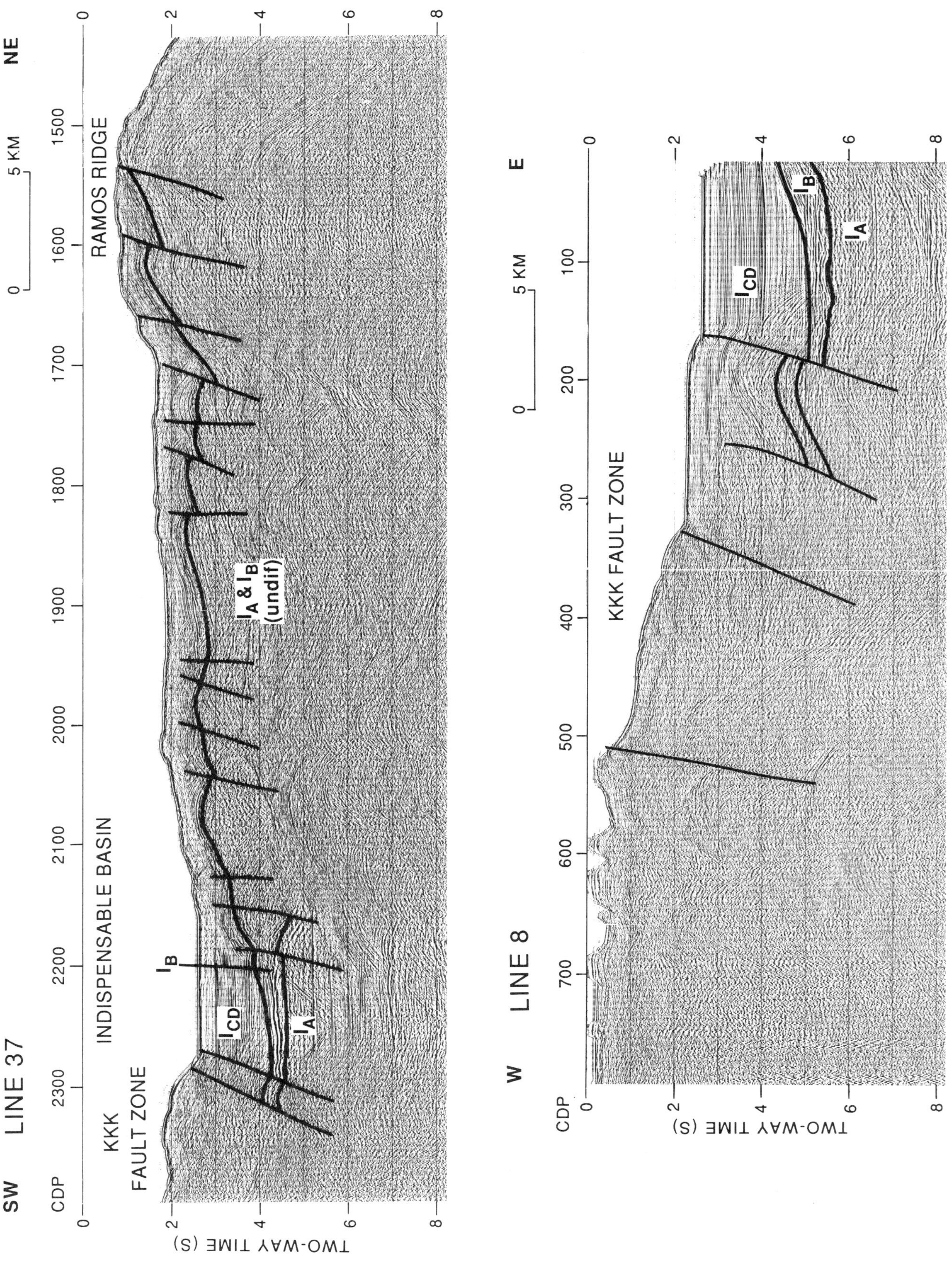

SW
LINE 37
NE
CDP
KKK
FAULT ZONE
INDISPENSABLE BASIN
RAMOS RIDGE
5 KM
I_{CD}
I_B
I_A
I_A & I_B (undif)
TWO-WAY TIME (S)
W
LINE 8
E
KKK FAULT ZONE
I_{CD}
I_B
I_A

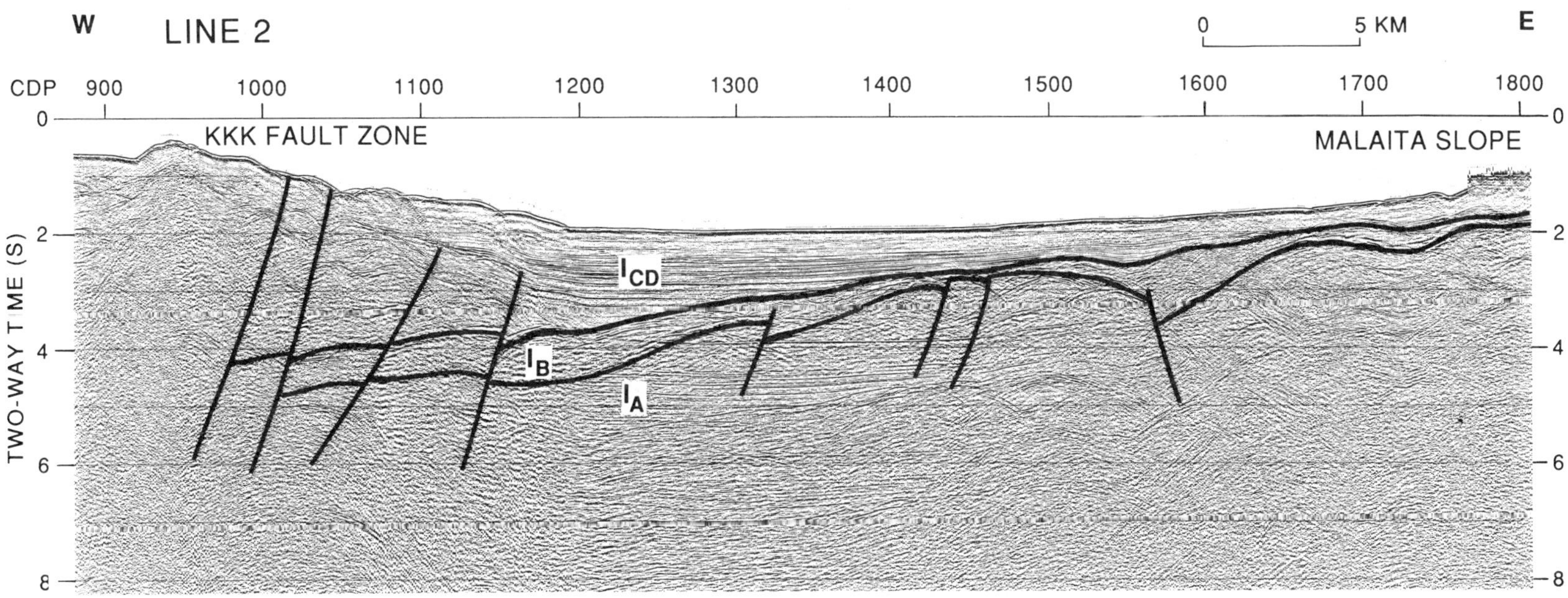

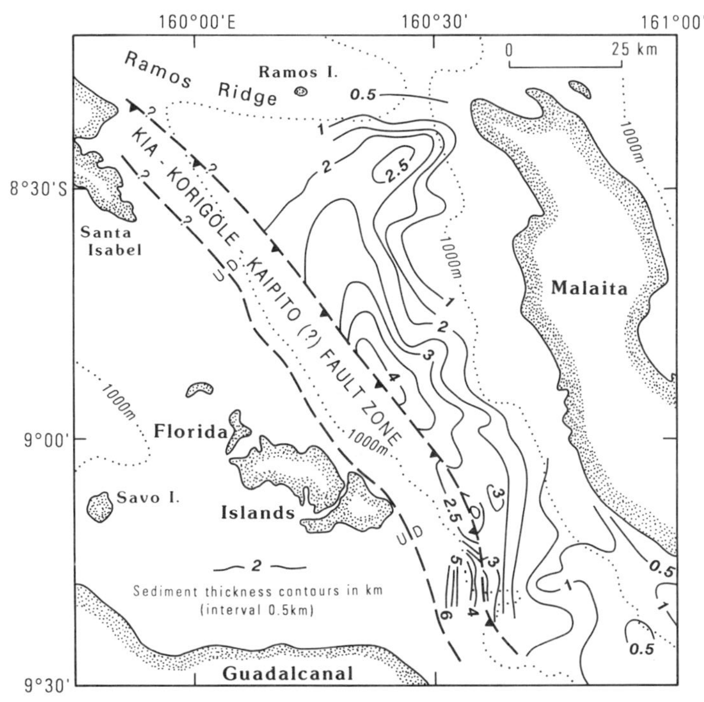

Figure 13. Contours showing sediment thickness and interpreted multichannel seismic-reflection profiles, Indispensable basin, central Solomon Islands. Seismic-stratigraphic unit I_A is believed to correlate with Cretaceous and early Tertiary rocks on Malaita; unit I_B, Miocene(?) strata; and unit I_{CD}, Pliocene(?) and Quaternary strata. Line sites are shown on Figure 1. Modified from Bruns et al (1986).

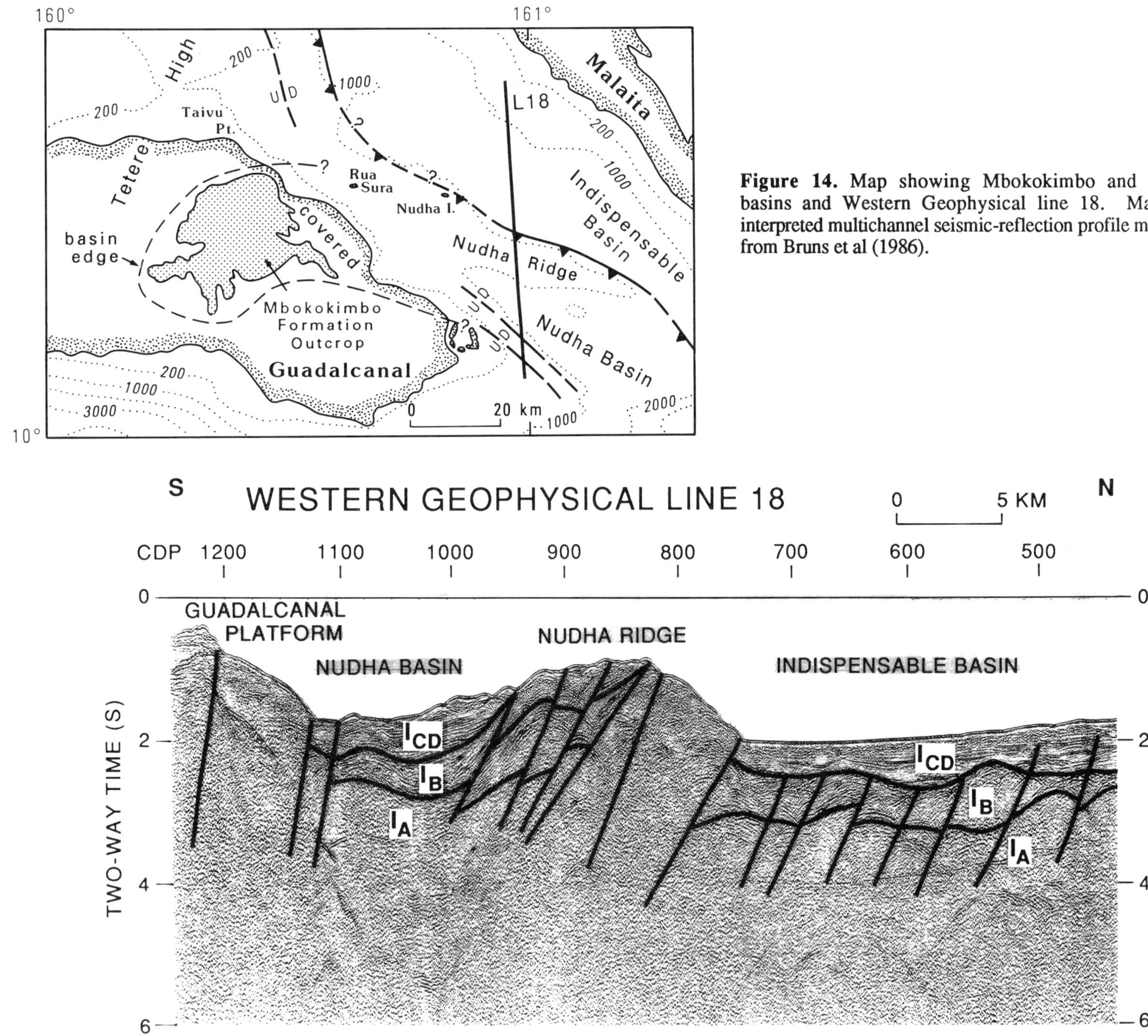

Figure 14. Map showing Mbokokimbo and Nudha basins and Western Geophysical line 18. Map and interpreted multichannel seismic-reflection profile modified from Bruns et al (1986).

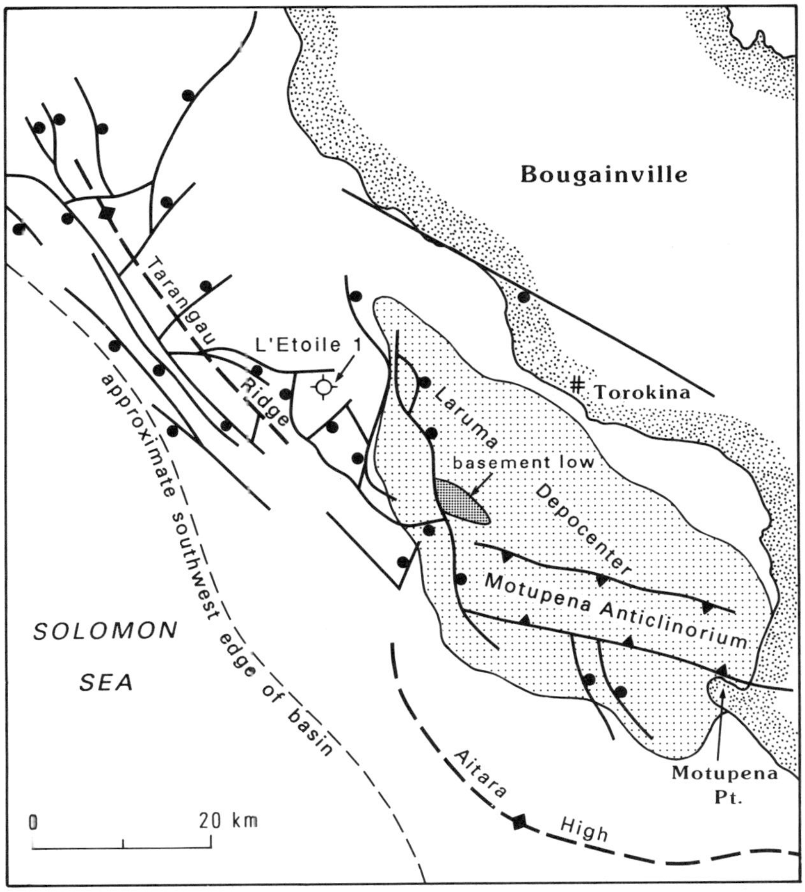

Figure 15. Structural features in Bougainville basin, eastern Papua New Guinea. Modified from Shaw (1985). Structure contours on acoustic basement shown by Hilyard and Rogerson, this volume.

rocks similar to those exposed on Choiseul (Coulson and Vedder, 1986).

Vella Ridge, between Vella Lavella and Mono Island, is underlain by as much as 3.5 km of intercalated late Miocene(?) and younger volcaniclastic, pyroclastic, and flow rocks that thin northwestward. Acoustic basement beneath this ridge is believed to consist mainly of late Cenozoic volcanic rocks (Cooper, Cochrane, and Bruns, this volume; Bruns, Vedder, and Cooper, this volume).

Between New Georgia and the Russell Islands is the Mborokua basin (Bruns et al, 1986)—a small forearc basin perched on the connecting seafloor ridge. Late Miocene(?) and younger strata in this basin are mildly deformed and are as much as 1 km thick. Small normal faults cut the Pliocene(?) and older(?) parts of the sequence. On the submergent ridge connecting Guadalcanal and the Russell Islands, sonobuoy data indicate the presence of volcanic and/or intrusive rocks (velocity range, 4.9 to 5.7 km/s) beneath a carbonate and volcaniclastic sedimentary sequence less than 2.0 km thick (Cooper, Cochrane, and Bruns, this volume). Seismic-reflection profiles across the same ridge farther west, between the Russell Islands and New Georgia, show acoustic basement, presumably volcanic, at or near the surface (Bruns et al, 1986).

The Santa Isabel-Florida Islands platform has a discontinuous capping of Quaternary reef limestone that contains pockets and lenses of layered rocks as much as 1.0 km thick above an irregular surface of acoustic basement. This acoustic basement, which locally protrudes through the layered rocks, probably consists of pre-Miocene mafic and ultramafic rocks similar to those exposed on Santa Isabel and the Florida Islands (Coulson and Vedder, 1986). Sonobuoy data indicate velocities as high as 6.4 km/s and 7.5 to 8.0 km/s in this area (Cooper, Cochrane, and Bruns, this volume). Very high velocity rocks (6.7 km/s) also lie at shallow depths just east of the Sealark Channel between the Florida Islands and Guadalcanal and represent the offshore part of the Tetere basement ridge (Bruns et al, 1986).

Beneath the inner edge of the shelf near Honiara, Guadalcanal, a sequence of Miocene and younger strata as much as 2.7 km thick may be present as indicated by gravity anomalies and stratigraphic relations (Coleman and Day, 1965). The reefal buildups reported by Maung and

Coulson (1983) in the southern part of Iron Bottom basin may extend into this shelf area as well as into the platform area surrounding the Sealark Channel.

PETROLEUM RESOURCE PROSPECTS

Despite encouraging reports by previous workers concerning offshore petroleum prospects in the Solomon Islands-Bougainville region, insufficient detailed geophysical surveys, inadequate bedrock sampling programs, and lack of subsurface stratigraphic information make realistic assessments difficult. Until offshore sections are drilled and directly correlated with island geology and offshore seismic stratigraphy, and until basin histories are resolved, only rudimentary evaluations are possible. Furthermore, water depths that are generally more than 800 m in basin areas deter exploratory drilling. Unless promising targets are identified in shallow water along the basin edges or on the island platforms, exploratory drilling throughout the investigated region should not be expected in the near future. Nevertheless, interest has been expressed for additional exploration in other offshore areas of eastern Papua New Guinea, particularly in the Cape Vogel and New Ireland basins (Exon and Tiffin, 1984; Exon et al, 1986; Stewart, Francis, and Pederson, 1987).

For the generation and accumulation of petroleum, certain conditions are necessary; among these are sufficient organic content, time and heat to form source rocks, and the presence of reservoir rocks and traps. Some of the conditions essential for generation and entrapment of hydrocarbons seem to be present in all of the basins and beneath some of the platforms and ridges. However, the most critical condition, the presence of potential source beds is not yet established. Moreover, the occurrence of suitable reservoir rocks has not been substantiated, and seeps have not been reported, either on the islands or in offshore areas.

Only one exploratory well (Oceanic Exploration Co., L'Etoile No. 1) has been drilled in the region. This well was drilled in 1975 on the island platform southwest of Bougainville, but the results were disappointing because the primary target of reef carbonates turned out to be much thinner and less porous than anticipated. However, the well was located on a structural high; consequently much of the thick flanking sedimentary sequence remains untested (Shaw, 1985; Stewart, Francis, and Pederson, 1987).

Likely areas for future exploratory drilling are described below in order of size; these include the Central Solomons Trough and the contiguous Bougainville and Manning straits, Queen Emma basin, Indispensable basin and its subsidiary basins, and Bougainville basin. Omitted from the discussion are the Kilinailau Trench and the New Britain-San Cristobal trench areas, because they are deemed to be less likely areas for commercial hydrocarbon accumulation.

Central Solomons Trough

Source Rocks and Thermal Maturation

Strata described as containing substantial organic material occur on Santa Isabel (Coleman, 1965; Landmesser, 1977; Hughes, 1982), Choiseul (Hughes, 1982; Ridgway and Coulson, 1987), and Guadalcanal (Hughes, 1977, 1982; Hackman, 1980). However, Cretaceous to Pleistocene carbonate and mudstone samples from a number of stratigraphic units exposed on Guadalcanal, Malaita, and the Florida Islands have a low total organic carbon content and are considered to be poor quality for source rocks. Buchbinder and Halley (1986) analyzed fifteen samples for total organic carbon content and found only two that gave values (0.94 and 0.48 percent) near or above the generally accepted minimum (0.50 percent) for source rocks. Even so, the organic matter in these two samples is of a chemical type that is gas prone and unlikely to yield liquid hydrocarbons, and vitrinite-reflectance analysis indicated that the organic matter contained in them is immature (Buchbinder and Halley, 1986). Other samples of Miocene and Pliocene rocks from Guadalcanal are thermally immature, based upon independent vitrinite reflectance measurements made by van Deventer (*in* Katz, 1980).

Small concentrations of hydrocarbon gases were found in samples from six stations in the Central Solomons Trough. However, these gases probably are the product of low-level microbial activity within the sediment. Although one sample had some characteristics associated with thermogenic gas, evidence for a petroleum-related source is lacking (Kvenvolden and Niem, this volume). Diapir-like acoustic voids and acoustically "transparent" layers in Quaternary sequences in Russell basin may be attributable to gas-charged sediment (Colwell and Tiffin, 1986).

Nevertheless, the foregoing analyses do not necessarily represent the source-rock potential of deeply buried strata offshore, where late Cenozoic sedimentary sequences with unknown organic carbon content are at least 5.0 km thick. High sedimentation rates prevailed during deposition of late Miocene and younger strata, and such rates in fine-grained material ordinarily decrease the chance for high organic content in arc-related basins. However, exceptions are possible in some marine turbidites and deltaic sediments. Quaternary sediment samples from the Central Solomons Trough and the flanking ridges generally have low values (0.2 to 0.5 percent) of total organic carbon (Colwell and Vedder, 1986; Colwell, Niem, and Niem, this volume). These low values, however, are not necessarily representative of older strata.

According to Hobart and Weissel (1987), heat flow in the Central Solomons Trough is low (~25-30mW/m^2). They state that even after correction for the effects of sedimentation and basin age, and using an average value of 1.0W/m-°K for the sediment conductivity, the temperature gradient could be as low as 18 to 22°C/km. They concluded that the deepest parts of the basins are not yet capable of generating hydrocarbons. However, they estimate the age of strata in the basin to be only 5 to 10 Ma, whereas we believe that these strata may be as old as 25 to 30 Ma (Bruns et al, 1986). Consequently, the sedimentation correction would be substantially less than that cited by Hobart and Weissel (1987), and the temperature at the base of the sequence as well as the geothermal gradient would be correspondingly higher. Thus, a more reasonable temperature gradient estimate of 30° to 40°C/km indicates that maturation of kerogens and the onset of hydrocarbon generation, which can begin at about 100°C in Cenozoic strata, is possible at sub-seafloor depths as shallow as 2.0 to 3.3 km in the Central Solomons Trough. Making reliable estimates of past and present temperatures in the offshore basins is difficult because sedimentation, subsidence, and uplift rates have been high, and heat flux probably has varied. Nevertheless, heat flow near the New Georgia island group is twice as great as elsewhere (Hobart and Weissel, 1987) and may indicate higher gradients in that area.

Reservoir Rocks

Strata that include potential reservoir rocks are uncommon on the islands surrounding the Central Solomons Trough, and the reservoir character of the offshore basin sequences is purely inferential. As a general rule, porosity and permeability are high in Tertiary reef deposits on the islands, but the influence of compaction and diagenesis on such deposits in the intra-arc basins is incompletely known. Early diagenetic carbonate cementation is widespread and common offshore (Colwell and Vedder, 1986; Crook and Niem, this volume; Niem, this volume). Furthermore, reef deposits may be limited to the basin flanks; our work has not demonstrated the presence of buried reefs in sequences in the deeper parts of the trough. Because uplifted volcanogenic rocks on both sides of the trough were the chief contributing sources (Colwell, Niem, and Niem, this volume; Crook and Niem, this volume), clay-rich volcaniclastic strata undoubtedly constitute much of the sedimentary sequence. Diagenesis of unstable mineral components in volcanogenic detritus ordinarily leads to rapidly decreasing porosity and permeability with burial.

Although the porosity and permeability of selected samples of clastic rocks from Guadalcanal are low (Katz, 1980), values may be higher in shallowly buried (<2 km) nearshore strata of Miocene and younger age that have not been appreciably affected by diagenesis. Miocene and Pliocene reef-limestone deposits on Guadalcanal generally are recrystallized but may contain fracture and vugular porosity. Maung and Coulson (1983) reported late Oligocene to early Miocene biostromal limestone deposits on the Shortland Islands, Choiseul, and Santa Isabel that might serve as reservoir rocks if they extend into the Central Solomons Trough, and if they are sufficiently porous and permeable; associated calcarenite beds possibly constitute other reservoirs. Samples of Pliocene calcarenite from the Florida Islands show intergranular spaces and unfilled body cavities of fossils in some beds, and a few samples of Miocene volcaniclastic sandstone are fairly well sorted and friable and possibly could be reservoir-quality rocks. Seafloor samples of late Pliocene and Pleistocene reef limestone, calcarenite, and volcaniclastic sandstone and mudstone dredged near Vella Lavella, New Georgia, and the Russell Islands all have low porosity (Colwell and Vedder, 1986). According to Niem (this volume), some of the higher porosity and permeability values from Quaternary reef rubble are local and attributable to surficial secondary processes and probably do not represent the reservoir potential of forereef limestone in the subsurface. Furthermore, reduced porosity and permeability would be expected in older, buried reef limestone.

The widespread Oligocene to Pliocene clay-rich, poorly sorted volcaniclastic strata on the islands and in the Central Solomons Trough probably would impede migration of hydrocarbons and would be considered unproductive by most petroleum geologists. Cogenetic mudstone beds, however, are common and could act as effective cap rocks and seals.

Traps

Katz (1980) and Maung and Coulson (1983) interpreted and predicted a variety of trap-forming features in the Central Solomons Trough, albeit several of their interpretations are speculative because of the sparsity and poor quality of some of the data available to them at that time. Seismic-reflection profiles obtained during the 1982 and 1984 R/V *S.P. Lee* cruises confirm the presence of many of the previously identified features; conversely, these newer profiles indicate that other features are not present.

The trough of the intra-arc basin generally was subsiding during the late Cenozoic, and large compressional anticlinal folds are uncommon. However, two sets of folds are evident in some areas; one set is pre-Pliocene, the other is Pliocene to Holocene. These folds may be transpressional and transtensional structures related to regional left shear, or they may be simple compressional features related to nearly orthogonal plate convergence (Coulson and Vedder, 1986; Bruns et al, 1986; Bruns, Vedder, and Cooper, this volume). These structures are potential traps for hydrocarbons migrating updip from sedimentary sequences deeper in the basins.

Among possible trapping structures, the most promising may be the anticlines on the northeastern and southwestern flanks of Russell and Iron Bottom basins, where strata as young as Holocene are arched and draped over increasingly steeply folded beds down section (Figure 8, lines 16 and 30). The folded sedimentary sequence above acoustic basement in each of these areas ranges in thickness from 3.0 to 4.5 km, but more importantly, the strata are updip from the basin depocenters and may be buried deeply enough to generate hydrocarbons. The most conspicuous anticline on the northeastern flank of Russell basin persists along strike for at least 20 km. Detailed surveys will be required to determine the orientations, dimensions, and closures on all of these structures. The mapping by Maung and Coulson (1983, their Figure 12, Tables III, IV, V, and VI), was based upon widely spaced reconnaissance surveys and as a result was largely inferential.

Low-amplitude folds in Shortland basin affect strata as young as Quaternary. Southeast of Fauro Island, one such anticline, or set of anticlines (Figure 5, lines 49, 50, and 407), has about 0.5 km of structural relief and extends more than 50 km along the northeastern flank of the basin. A half-graben near the basin axis extends for at least 65 km, is buried by Pliocene and younger strata (Figure 5, lines 49, 50, and 407), and trends northwestward beneath the Shortland Islands platform. Strata in the half-graben onlap the south flank and are deformed into an anticline.

Anticlines were reported to be present beneath Manning Strait by Landmesser (1977) and Maung and Coulson (1983), but the dominant structure under the strait is a shallow syncline (Bruns et al, 1986, their Figure 23). Anticlines are present on the high flanks of the syncline, but closure cannot be demonstrated with the available information. Similarly, the main structure beneath Bougainville Strait seems to be a broad syncline that is marked on its northern flank by an antiformal structure under the Bougainville-Shortland Islands platform (Bruns et al, 1986, their Figure 11, line 8). It is noteworthy that both straits are underlain by high-standing basement rocks.

Faults that may act as trapping barriers are distributed along both the northeastern and southwestern margins of the Central Solomons Trough but are more common along the northeastern margin. Other potential fault traps occur along the flanks of anticlines or uplifted blocks such as Fatu O Moana (Figure 7). Possible stratigraphic traps in the form of sedimentary wedges and buttress unconformities are present along both flanks of the Shortland and Russell basins (Figure 5, lines 49, 50, and 407; Figure 8) and along the southern and southeastern flanks of Iron Bottom basin. In combination with faults and upturned sedimentary sequences on the basin margins, such stratigraphic discontinuities may provide excellent traps for hydrocarbons.

Katz (1980) reported karst-like reef structures in the basin sequence between Vella Lavella and Choiseul, but these Pleistocene(?) structures are not discernible on seismic-reflection profiles examined for this paper. Maung and Coulson (1983) described, and we have observed on the 1982 seismic reflection data, local reef buildups of Miocene-Pliocene and Quaternary age close to shore on the southern flank of Iron Bottom basin. These buried reef-like features may contain excellent traps. Quaternary reef platforms are conspicuous along the shallow-water fringes of most of the islands and locally may form impermeable barriers or seals over potential reservoirs.

Queen Emma Basin

Source Rocks and Thermal Maturation

No potential source rocks have been reported from islands adjoining Queen Emma basin. Offshore correlatives of Miocene and Pliocene shallow-water to bathyal carbonate sequences on southeastern New Ireland presumably extend into Queen Emma basin and may include possible source rocks. Likewise, calcareous siltstone and mudstone beds incorporated in largely volcaniclastic Miocene and Pliocene sequences could be potential source rocks, but the nature and amount of organic carbon in these strata and their thermal history are unknown. Descriptions of the New Ireland rocks are given by Hohnen (1978) and Stewart and Sandy (1988), and provisional correlations with seismic-stratigraphic units in Queen Emma basin are outlined by Vedder, Bruns, and Cooper (this volume). Strata suitable as source rocks have not been recognized on Buka Island and northwestern Bougainville, although early Miocene platform limestone deposits may extend northwestward into the basin from central Bougainville.

Reservoir Rocks

As in the case of source rocks, potential reservoir rocks have not been described from islands adjoining the basin. Porous and permeable strata may occur within some of the Miocene and Pliocene volcaniclastic units on New Ireland, and some of the shallow-water limestone units may contain cavity and fracture porosity. Oligocene and early Miocene mudstone and volcaniclastic sandstone and conglomerate on Buka Island and nearby small islands are clayey, indurated, and cemented with zeolites. Younger coarse clastic strata offshore, however, may contain potential reservoirs. Unless early Miocene platform limestone, like that exposed in central Bougainville, extends northwestward in the subsurface, the likelihood of carbonate reservoirs being present in the southern part of the basin seems remote, even though Pliocene or Quaternary limestone is present on the crest of Kaviki Ridge at the south edge of the basin (Crook and Niem, this volume).

Traps

Structures that may be favorable for entrapment of hydrocarbons include a buried faulted anticlinal feature that trends northwest through the central part of the basin (Figures 10 and 11). The widely spaced profiles do not indicate whether the feature represents closely spaced en echelon anticlines or a single anticline more than 70 km long. This anticlinal feature deforms late Miocene(?) and older strata and seems to warp early Pliocene(?) strata locally, suggesting growth into Pliocene time. A submarine unconformity truncates strata along the crest of part of the same feature (Figure 10), but this feature presumably was never shallow enough to support the growth of reefs. Stratigraphic traps in the form of wedge-outs and onlaps are present in Miocene and Pliocene sequences on both flanks of the basin and over the anticlinal feature in the central part of the basin; examples are shown in Figures 10 and 11. Wedge-outs and onlaps are not evident in the early Miocene(?) and older sequences.

The broad anticlinorium that underlies the southeastern segment of Emirau-Feni Ridge displays local structures that may constitute traps. However, the largely volcanogenic composition of the rocks probably hinders migration of any liquid hydrocarbons into the structure because alteration and clay content are likely to have caused reduction of porosity and permeability. Although limestone bodies may be present locally, their extent and thickness are unknown.

Bougainville Basin

Most of the information given here for Bougainville basin was summarized from Shaw (1985) and Stewart, Francis, and Pederson (1987). The well data were provided by the Department of Minerals and Energy, Geological Survey of Papua New Guinea.

Source Rocks and Thermal Maturation

According to Shaw (1985) source-rock studies have not been made on samples from Bougainville basin; however, some of the Miocene fine-grained volcaniclastic sequences as well as an early Miocene limestone unit are considered to have moderate source-rock potential. Shaw (1985) reported that a low geothermal gradient (16°C/km) was calculated from bottom-hole temperatures in the L'Etoile No. 1 well (described below) and that the oil-generation zone would be at a depth of about 4,700 m if thermal gradients in the past were similar to those now present.

Reservoir Rocks

Although information on porosity and permeability of samples from this basin are not available, Shaw (1985) and Stewart, Francis, and Pederson (1987) suggest that reservoir rocks may be present in an early Miocene carbonate-platform limestone unit and its clastic bathyal counterparts as well as in carbonate buildups in Miocene and Pliocene clastic sequences. Pliocene reef deposits occur along structural highs on the northwestern and southeastern margins of the basin. Other potential reservoir rocks may be included in early and middle Miocene volcaniclastic turbidite sequences and in winnowed, redeposited, well-sorted pyroclastic beds. Shaw (1985) noted that fine-grained tuffaceous beds may act as intraformational seals in the Miocene sequences and that fine-grained clastic strata might form adequate seals over Pliocene carbonate buildups.

Traps

Favorable structural traps in Bougainville basin include faulted anticlines and fault-block closures (Figure 15); stratigraphic traps include truncations by unconformities, onlap wedges, and carbonate reef buildups (Shaw, 1985; Stewart, Francis, and Pederson, 1987). The major anticlinal structure and associated faults in the southeastern part of the basin formed in the late Miocene, possibly as a result of oblique convergence (Shaw, 1985), during the process of arc reversal. A fault-block closure northwest of the anticlinal structure formed chiefly during the Pliocene, although it seems to have been a positive feature as early as the middle Miocene (Shaw, 1985). The carbonate-reef buildups were constructed chiefly in Pliocene through Quaternary time, but some carbonate platform deposits that may serve as traps are as old as early Miocene.

Oceanic Exploration Co., L'Etoile No. 1 Well

Only one exploratory well has been drilled in the entire New Ireland-Bougainville-Solomon Islands region. This well, Oceanic Exploration Co., L'Etoile No. 1, was drilled in 1975 on the shallow-water platform southwest of Bougainville to test a structural high in the Bougainville basin (Figure 15). The results of this drilling were discouraging, because the primary target of expected carbonate reef deposits proved to be much thinner and less porous than anticipated, and thicker parts of the Miocene section remained untested because they lie downflank on the structure. Most of the section between the depths of 409 and 1,336 m below sea level consists of interbedded mudstone, siltstone, and volcanic rocks; below 1,336 m, the section is largely tuffaceous sandstone and siltstone to the total depth of 1,682 m below sea level. The Pleistocene-Pliocene boundary is between the depths of 458 m and 503 m, and the Pliocene-Miocene boundary lies in the interval between 1,038 and 1,065 m. A K/Ar age of 18.7 ± 0.7 Ma was obtained from a whole-rock sample of tuff taken at a depth of 1,534 m.

Indispensable Basin and Contiguous Basins

Source Rocks and Thermal Maturation

With the possible exception of a single marlstone sample of unknown age from Malaita (Buchbinder and Halley, 1986), potential source rocks have not been reported from any onshore areas around the perimeter of Indispensable basin, including the Mbokokimbo basin. However, a systematic sampling program is required before all of the sedimentary sequences are discounted as source rocks. To date, sampling for source rocks has been extremely limited. Moreover, outcrop samples, which generally are collected from weathered exposures, probably do not accurately represent the source-rock quality of the buried sequences offshore. Carbonate-cemented chimneys composed of silty mudstone from the slope southeast of the Florida Islands may have developed as a result of the venting of methane through near-surface sediment; the origin of the methane is uncertain, but it probably is biogenic (Niem, this volume0.

Reservoir Rocks

In the region of Indispensable Strait, reservoir rocks are not expected, because such strata have not been reported from either onshore or offshore areas. However, strata of favorable reservoir quality may occur in the Miocene and Pliocene sedimentary sequence in Mbokokimbo basin and its offshore extension, as well as in Nudha basin, even though porosity and permeability data have not been reported. Elsewhere, the prospects for finding reservoir rocks are less favorable. According to Katz (1980), the pelagic sedimentary rocks exposed on Malaita generally are highly indurated, dense, and "tight" Much of southeastern Santa Isabel consists of ultramafic and volcanogenic rocks, some of which are metamorphosed. The Miocene and younger sequence consists largely of volcaniclastic strata (Coleman, 1965). Pliocene and younger calcarenite and reef limestone deposits in the Florida Islands could possibly serve as reservoirs, but deformation along the slope northeast of the islands precludes correlation of these strata with the basin sequence to the east. Interior samples of Quaternary reef limestone dredged from the slope southeast of Santa Isabel have intermediate porosity and permeability that may be representative of shallowly buried fore-reef bodies (Niem, this volume). On the eastern part of the Florida Islands, exposed Miocene and Pliocene volcaniclastic strata generally are poorly sorted and appear to be of poor reservoir quality, although interbedded Miocene calcarenite beds retain some of their original porosity (Neef, 1979).

Traps

Structures that may favor entrapment of hydrocarbons occur throughout the Indispensable basin area. Buried faulted anticlines are evident in the early Pliocene(?) and older sequences beneath the central part of the basin (Figure 13, line 2) but these structures barely deform the overlying late Pliocene(?) and younger strata. Stratigraphic traps may be present where these younger strata gently drape and thin across the pre-existing folds (Figure 13, lines 2, 3, and 37). Similar structural and stratigraphic relations occur in Nudha basin (Figure 14, line 18), where they may constitute traps.

SUMMARY OF PETROLEUM POTENTIAL

Intra-arc basins and related features in the central and western Solomon Islands and the New Ireland-Bougainville region of eastern Papua New Guinea contain sequences of Oligocene and younger strata that range in maximum thickness from about 5 to nearly 8 km. These basins vary in size and shape. The Central Solomons Trough, a composite intra-arc feature nearly 450 km long, is made up of the Shortland, Russell, and Iron Bottom basins. Together these basins span a segment of the arc that extends from Bougainville to Guadalcanal. Queen Emma basin between New Ireland and Bougainville is at least 200 km long and about 125 km wide at its widest. Bougainville basin, which underlies the platform southwest of Bougainville, is about 150 km long and 50 km wide at its widest. Indispensable basin and its subsidiaries, offshore Mbokokimbo basin and Nudha basin, underlie a bathymetric depression 30 to 55 km wide and more than 200 km long between the Santa Isabel-Florida Islands-Guadalcanal platform and Malaita. The areal extent, thickness of strata, and structure make each of these basins a target for future exploration, even though the conditions necessary for the generation and accumulation of petroleum are poorly known. Other smaller basins and trench depocenters are not believed to have significant petroleum potential. Bougainville and Manning straits require further study before their potential can be assessed. Too little is known about the area northeast of Choiseul and Santa Isabel to predict its suitability for petroleum exploration, but post-Oligocene stratigraphic sequences presumably are too thin at most places to warrant optimistic forecasts. Similarly, very little is known about the area northeast of Malaita and north of San Cristobal, and consequently, detailed exploration there is unlikely in the near future.

Most petroleum geologists consider island arcs and their associated depositional basins as poor prospects for exploration because of the volcanogenic nature and low organic content of most of the sedimentary rocks. Nevertheless, some of the conditions necessary for the generation and accumulation of petroleum probably occur in the region of eastern Papua New Guinea and the central and western Solomon Islands. Although suitable source rocks for the generation of hydrocarbons have not been positively identified, potential reservoir rocks as well as structural and stratigraphic traps probably are present within the depositional sequences of the Central Solomons Trough and in the Queen Emma, Bougainville, and Indispensable basins. The thickness of the sedimentary sequences in these basins guarantees that at least some strata are buried deeply enough to be within the thermal regime required for hydrocarbon generation.

Until stratigraphic test wells are drilled into each of the basins, the likelihood of discovering petroleum accumulations in this offshore region will remain a matter of conjecture. A deterrent to exploratory drilling in the basins is the depth of water, which generally is more than 800 m over prospective targets in the Solomon Islands and more than 2,000 m between New Ireland and Bougainville. Exceptions are the island platform southwest of Bougainville and the shallow areas in Bougainville and Manning straits, where water depths generally are less than 200 m.

ACKNOWLEDGMENTS

We express our appreciation to K.W. Doble, Chief Government Geologist, and R. Rogerson of the Geological Survey, Department of Minerals and Energy, Papua New Guinea, for furnishing significant stratigraphic information on the Oceanographic Exploration Co. L'Etoile No. 1 well. We are indebted to K.J. Bird and M.A. Fisher, who made suggestions for improvement on a preliminary version of the manuscript. D.L. Moore prepared the manuscript for publication, and P.A. Swenson and S.C. Vath drafted the illustrations.

REFERENCES CITED

Beyer, L.A., 1986, Gravity anomalies of the Solomon Islands between 156° and 161°E longitude, *in* Vedder, J.G., Pound, K.S., and Boundy, S.Q., eds., Geology and offshore resources of Pacific island arcs—central and western Solomon Islands: Houston, Texas, Circum-Pacific Council for Energy and Mineral Resources Earth Science Series, v. 4, p. 177-223.

Blake, D.H., and Miezitis, Y., 1967, Geology of Bougainville and Buka Islands, New Guinea: Bureau of Mineral Resources, Geology, and Geophysics, Australia, Bulletin 93, PNG 1, 56 p.

Bruns, T.R., Cooper, A.K., Mann, D.M., and Vedder, J.G., 1986, Seismic stratigraphy and structure of sedimentary basins in the Solomon Islands region, *in* Vedder, J.G., Pound, K.S., and Boundy, S.Q., eds., Geology and offshore resources of Pacific island arcs—central and western Solomon Islands: Houston, Texas, Circum-Pacific Council for Energy and Mineral Resources Earth Science Series, v. 4, p. 177-223.

Buchbinder, B., and Halley, R.B., 1986, Source-rock evaluation of outcrop samples from Guadalcanal, Malaita, and the Florida island group, Solomon Islands, *in* Vedder, J.G., Pound, K.S., and Boundy, S.Q., eds., Geology and offshore resources of Pacific island arcs—central and western Solomon Islands: Houston, Texas, Circum-Pacific Council for Energy and Mineral Resources Earth Science Series, v. 4, p. 267-276.

Chivas, A.R., Andrew, A.S., Sinha, A.K., and O'Neil, J.R., 1982, Geochemistry of a Pliocene-Pleistocene oceanic-arc plutonic complex, Guadalcanal: Nature, v. 300, p. 139-142.

Coleman, P.J., 1965, Stratigraphical and structural notes on the British Solomon Islands with reference to the first geological map: British Solomon Islands Geological Record (1959-1962), v. 2, Report no. 29, p. 17-31.

Coleman, P.J., and Day, A.A., 1965, Petroleum possibilities and marked gravity anomalies in north-central Guadalcanal: British Solomon Islands Geological Record (1959-1962), v. 2, p. 112-119.

Coleman, P.J., and Kroenke, L.W., 1981, Subduction without volcanism in the Solomon Islands arc: Geo-Marine Letters, v. 1, p. 129-134.

Colwell, J.B., and Tiffin, D.L., 1986, Recent depositional patterns in the Central Solomons Trough of the Solomon Islands, *in* Vedder, J.G., Pound, K.S., and Boundy, S.Q., eds., Geology and offshore resources of Pacific island arcs—central and western Solomon Islands: Houston, Texas, Circum-Pacific Council for Energy and Mineral Resources Earth Science Series, v. 4, p. 243-254.

Colwell, J.B., and Vedder, J.G., 1986, Description and interpretation of dredged rocks, Solomon Islands, *in* Vedder, J.G., Pound, K.S., and Boundy, S.Q., eds., Geology and offshore resources of Pacific island arcs—central and western Solomon Islands: Houston, Texas, Circum-Pacific Council for Energy and Mineral Resources Earth Science Series, v. 4, p. 225-236.

Cooper, A.K., Bruns, T.R., and Wood, R.A., 1986, Crustal structure of the Solomon Islands intra-arc basins from sonobuoy seismic studies, *in* Vedder, J.G., Pound, K.S., and Boundy, S.Q., eds., Geology and offshore resources of Pacific island arcs—central and western Solomon Islands: Houston, Texas, Circum-Pacific Council for Energy and Mineral Resources, Earth Science Series v. 4, p. 135-156.

Coulson, F.I., and Vedder, J.G., 1986, Geology of the central and western Solomon Islands, *in* Vedder, J.G., Pound, K.S., and Boundy, S.Q., eds., Geology and offshore resources of Pacific island arcs—central and western Solomon Islands: Houston, Texas, Circum-Pacific Council for Energy and Mineral Resources Earth Science Series, v. 4, p. 59-88.

de Broin, C.E., Aubertin, F., and Ravenne, C., 1977, Structure and history of Solomons-New Ireland region: International symposium on geodynamics in southwest Pacific: Paris, Editions Technip, p. 37-50.

Dunkley, P.N., 1983, Volcanism and the evolution of the ensimatic Solomon Islands Arc, *in* Shinozura, D., and Yokoyama, I., eds., Arc volcanism: physics and tectonics: Terrapub, Tokyo, p. 225-241.

Exon, N.F., Stewart, W.D., Sandy, M.J., and Tiffin, D.L., 1986, Geology and petroleum prospects of the eastern New Ireland Basin, northeastern Papua New Guinea: BMR Journal of Australian Geology and Geophysics, n. 10, p. 39-51.

Exon, N.F., and Taylor B., 1984, Seafloor spreading, ridge subduction, volcanism, and sedimentation in the offshore Woodlark-Solomons region and Tripartite cruise report for *Kana Keoki* cruise 82-03-16, leg 4: UN ESCAP, CCOP/SOPAC Technical Report No. 34, p. 1-42.

Exon, N.F., and Tiffin, D.L., 1984, Geology and petroleum prospects of offshore New Ireland Basin in northern Papua New Guinea, *in* Watson, S.T., ed. , Transactions of the Third Circum-Pacific Energy and Minerals Resources Conference, 22-28 August, 1982, Honolulu, Hawaii, p. 623-630.

Grover, J.C., 1955, The prospects of oil accumulations, *in* Geology, mineral deposits, and prospects of mining development in the British Solomon Islands Protectorate: Interim Memoir, Geological Survey of the British Solomon Islands, no. 1, 108 p.

Hackman, B.D., 1980, The geology of Guadalcanal, Solomon Islands: Overseas Memoirs of the Institute of Geological Science, Her Majesty's Stationery Office, London, no. 6, 115 p.

Hobart, M.A., and Weissel, J.K., 1987, Geothermal surveys in the Solomon Islands-Woodlark Basin region, *in* Taylor, B., and Exon, N.F., eds., Marine geology, geophysics, and geochemistry of the Woodlark Basin-Solomon Islands: Houston, Texas, Circum-Pacific Council for Energy and Mineral Resources Earth Science Series, v. 7, p. 49-66.

Hohnen, P.D., 1978, Geology of New Ireland, Papua New Guinea: Department of National Development, Bureau of Mineral Resources, Geology and Geophysics, Bulletin 194, DNG 12, Australian Government Publishing Service, 39 p.

Hughes, G.W., 1977, The geology and foraminiferal micropaleontology of the Lungga and Itina areas, western Guadalcanal, Solomon Islands: Unpublished Ph.D. thesis, University College of Wales, Aberystwyth.

Hughes, G.W. , 1982, Stratigraphic correlation between sedimentary basins of the ESCAP region, Volume VIII, Solomon Islands, ESCAP Atlas of Stratigraphy III: New York, United Nations Mineral Resources Development Series, n. 48, p. 115-130.

Hughes, G.W., Varol, O., and Dunkley, P.N., 1984, Foraminiferal and calcareous nannofossil evidence for Quaternary vertical tectonics in the Solomon Islands (abs.), International symposium on recent crustal movements of the Pacific Region: Wellington, New Zealand, Royal Society of New Zealand and Inter-Union Committee on the Lithosphere, p. 23.

Katz, H.R., 1980, Basin development in the Solomon Islands and their petroleum potential: UN ESCAP, CCOP/SOPAC Technical Bulletin No. 3, p. 59-75.

Katz, H.R., 1982, Oil and gas developments in New Zealand and southwest Pacific Islands in 1981: American Association of Petroleum Geologists Bulletin, v. 67, p. 1689-1694.

Kroenke, L.W., 1984, Solomon Islands: San Cristobal to Bougainville and Buka, *in* Kroenke, L.W., ed., Cenozoic tectonic development of the southwest Pacific: UN ESCAP, CCOP/SOPAC Technical Bulletin, no. 6, p. 47-61.

Kroenke, L.W., Resig, J.M., and Cooper, P.A., 1986, Tectonics of the southeastern Solomon Islands: formation of the Malaita anticlinorium, *in* Vedder, J.G., Pound, K.S., and Boundy, S.Q., eds., Geology and offshore resources of Pacific island arcs—central and western Solomon Islands: Houston, Texas, Circum-Pacific Council for Energy and Minerals Resources Earth Science Series, v. 4, p. 109-116.

Landmesser, C.W., 1977, Evaluation of potential hydrocarbon occurrence in the Solomon Islands: UN ESCAP, CCOP/SOPAC, Technical Report, v. 1, no. 5, p. 47- 53.

Maung, T.U., and Coulson, F.I., 1983, Assessment of petroleum potential of the central Solomons Basin: UN ESCAP, CCOP/SOPAC Technical Report No. 26, 68 p.

Neef, G., 1979, Cenozoic stratigraphy of Small Nggela Island, Solomon Islands—early Miocene deposition in a forearc basin followed by Pliocene patch reef deposition: New Zealand Journal of Geology and Geophysics, v. 22, n. 1, p. 53-70.

Resig, J.M., 1986, Foraminiferal stratigraphy and paleobathymetry of dredged rock, R/V *S.P. Lee* cruise, Solomon Islands, *in* Vedder, J.G., Pound, K.S., and Boundy, S.Q., eds., Geology and offshore resources of Pacific island arcs—central and western Solomon Islands: Houston, Texas, Circum-Pacific Council for Energy and Mineral Resources Earth Science Series, v. 4, p. 255-260.

Ridgway, J., and Coulson, F.I.E., 1987, The geology of Choiseul and the Shortland Islands, Solomon Islands: British Geological Survey Overseas Memoir 8, 134 p.

Shaw, R.D., 1985, Hydrocarbon potential of the Bougainville basin, Papua New Guinea: 14th BMR Symposium, Extended Abstracts—Petroleum Geology of Pacific Island Countries, Marine Studies Workshop, Canberra, Australia, 22-23 October, p. 19-29.

Stewart, W.D., Francis, G., and Pederson, S.L., 1987, Hydrocarbon potential of Papua New Guinea's Bougainville, southeastern New Ireland basins: Oil and Gas Journal, v. 85, n. 47, p. 83-87.

Stewart, W.D., and Sandy, M.J., 1988, Geology of New Ireland and Djaul Island, northeastern Papua New Guinea, *in* Marlow, M.S., Dadisman, S.V., and Exon, N.F., eds., Geology and offshore resources of Pacific island arcs—New Ireland and Manus region, Papua New Guinea: Houston, Texas, Circum-Pacific Council for Energy and Mineral Resources Earth Science Series, v. 9, p. 13-30.

Strange, P.J., 1981, The geology of the Komboro and Rob Roy Island area, Choiseul, an explanation of the 1:50,000 Map Sheets CH 11: British Technical Cooperation Western Solomons Mapping Project, Report n. 11, 72 p.

Taylor, B., and Exon, N.F., 1983, 1982 R/V *Kona Keoki* cruise in the woodlark-Solomons Region, [abs.]: Basic Geo-Scientific Marine Research required for assessment of minerals and hydrocarbons in the South Pacific, A workshop, Suva, Fiji, October, 1983.

Taylor, B., and Exon, N.F., 1987, An investigation of ridge subduction in the Woodlark-Solomons region: introduction and overview, *in* Taylor, B., and Exon, N.F., eds., Marine geology, geophysics, and geochemistry of the Woodlark Basin-Solomon Islands: Houston, Texas, Circum-Pacific Council for Energy and Mineral Resources Earth Science Series, v. 7, p. 1-24.

Vedder, J.G., and Bruns, T.R., 1986, Offshore petroleum potential, Solomon Islands, *in* Vedder, J.G., Pound, K.S., and Boundy, S.Q., eds., Geology and offshore resources of Pacific island arcs—central and western Solomon Islands: Houston, Texas, Circum-Pacific Council for Energy and Mineral Resources Earth Science Series, v. 4, p. 277-294.

Vedder, J.G., Bruns, T.R., and Cooper, A.K., 1987, Geologic framework of Queen Emma basin, eastern Papua New Guinea, *in* Vedder, J.G., Bruns, T.R., and Colwell, J.B., compilers, Aspects of regional geology, Solomon Islands and Papua New Guinea, Joint Cruise Report, R/V *S.P. Lee,* 1984, Suva, Fiji, CCOP/SOPAC Technical Secretariat, p. El-21.

Vedder, J.G., and Bruns, T.R., editors, 1989, Geology and offshore resources of Pacific island arcs—Solomon Islands and Bougainville, Papua New Guinea Regions: Houston, Texas, Circum-Pacific Council for Energy and Mineral Resources Earth Science Series, v. 12.

MULTICHANNEL SEISMIC-REFLECTION PROFILES ACROSS THE SOLOMON ISLANDS ARC: GUADALCANAL-MALAITA, VELLA LAVELLA-CHOISEUL, AND BOUGAINVILLE-BUKA REGIONS

T.R. Bruns, J.G. Vedder, P.E. Hart, and D.M. Mann
U.S. Geological Survey, Menlo Park, California 94025

INTRODUCTION

During 1982 and 1984, geological and geophysical investigations of the offshore areas of the Solomon Islands and Bougainville, Papua New Guinea, were carried out from the U.S. Geological Survey research vessel R/V *S.P. Lee*. Results of these cruises are reported in Vedder, Pound, and Boundy (1986) and in this volume. In this note, we present three regional multichannel seismic-reflection profiles across the island arc (Figure 1). Each regional profile is a composite of two or more seismic sections joined together to provide the best continuous coverage across the arc. The profiles are shown as Figures 2, 3, and 4 on the separate plate that accompanies this volume. Detailed line and common depth point (CDP) locations are shown in Figures 5, 6, and 7.

Data acquisition and processing used to prepare the seismic sections followed standard U.S. Geological Survey methods as outlined by Mann (1986) and discussed in this volume in papers on the Queen Emma basin (Vedder, Bruns, and Cooper, this volume), the Kilinailau Trench (Bruns, Vedder, and Culotta, this volume), and the Shortland basin (Bruns, Vedder, and Cooper, this volume). All profiles are migrated time sections; migration velocities were selected from a series of post-stack, constant-velocity migrated sections. The vertical exaggeration at the seafloor on all profiles is about 4:1.

The purpose of presenting the cross sections is to allow interested readers to examine seismic sections across the entire arc. Such cross sections give a better view of the structure and stratigraphy of the Solomon Islands and Bougainville regions than can be gained by seeing only short sections, commonly reduced to a small scale for publication. The lines are deliberately left uninterpreted. With one exception described below, interpretation of each line is either shown elsewhere in this volume or in Vedder, Pound, and Boundy (1986). These publications also provide information on regional geology and tectonics, and therefore little discussion is included here. Because multiple interpretations of seismic-reflection data are possible, we urge readers to consult cited publications, then to make their own interpretations of the cross sections, and to compare their interpretations with the published versions.

GUADALCANAL-MALAITA CROSS SECTION

From southwest to northeast, the Guadalcanal-Malaita cross section (Figures 2 and 5) transects the active San Cristobal Trench subduction zone, the Guadalcanal-Russell Island segment of the late Cenozoic arc, the Central Solomons Trough at the east end of Russell basin, the Santa Isabel-Florida Island segment of the early Cenozoic arc, the Kia-Korigole-Kaipito fault system, Indispensable basin, the antiforms and synforms of the Malaita Anticlinorium, and the North Solomon Trench. This cross section is identical to the one presented by Bruns, Cooper, and Vedder (1986), except that the profile reproduced here has less vertical exaggeration and is migrated. All names are the same as on the previously published profile. Kroenke (this volume) shows an interpretation of the Malaita Anticlinorium portion of the line. Interpretations of other parts of the line are in Bruns et al (1986) and Bruns, Cooper, and Vedder (1986).

VELLA LAVELLA-CHOISEUL CROSS SECTION

From south to north, the Vella Lavella-Choiseul cross section (Figures 3 and 6) transects the active New Britain trench and adjacent forearc, the actively subsiding Vella Ridge segment of the late Cenozoic arc, the Central Solomons Trough at the west end of Shortland basin, the

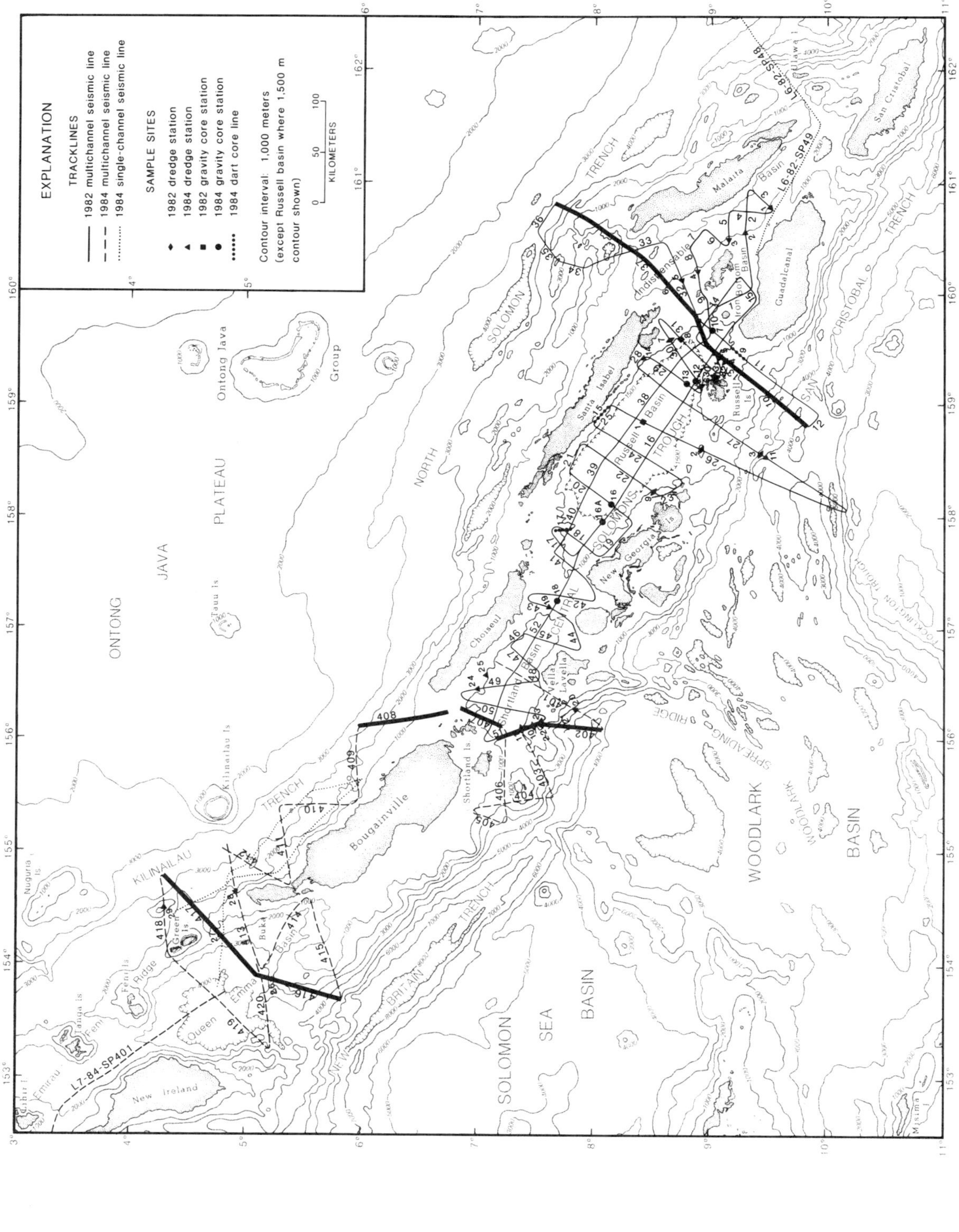

Figure 1. Regional location map for seismic-reflection lines in the Solomon Islands and Bougainville regions. Multichannel seismic cross sections are shown in heavy lines. The Guadalcanal-Malaita cross section is at the east end of the region (lines 13, 32, and 37), the Vella Lavella-Choiseul section is in the center (lines 402, 51, 407, and 408), and the Bougainville-Buka section is at the west end of the region (lines 416-417). Detailed line locations and common depth point (CDP) locations for the cross sections are shown on Figures 5-7.

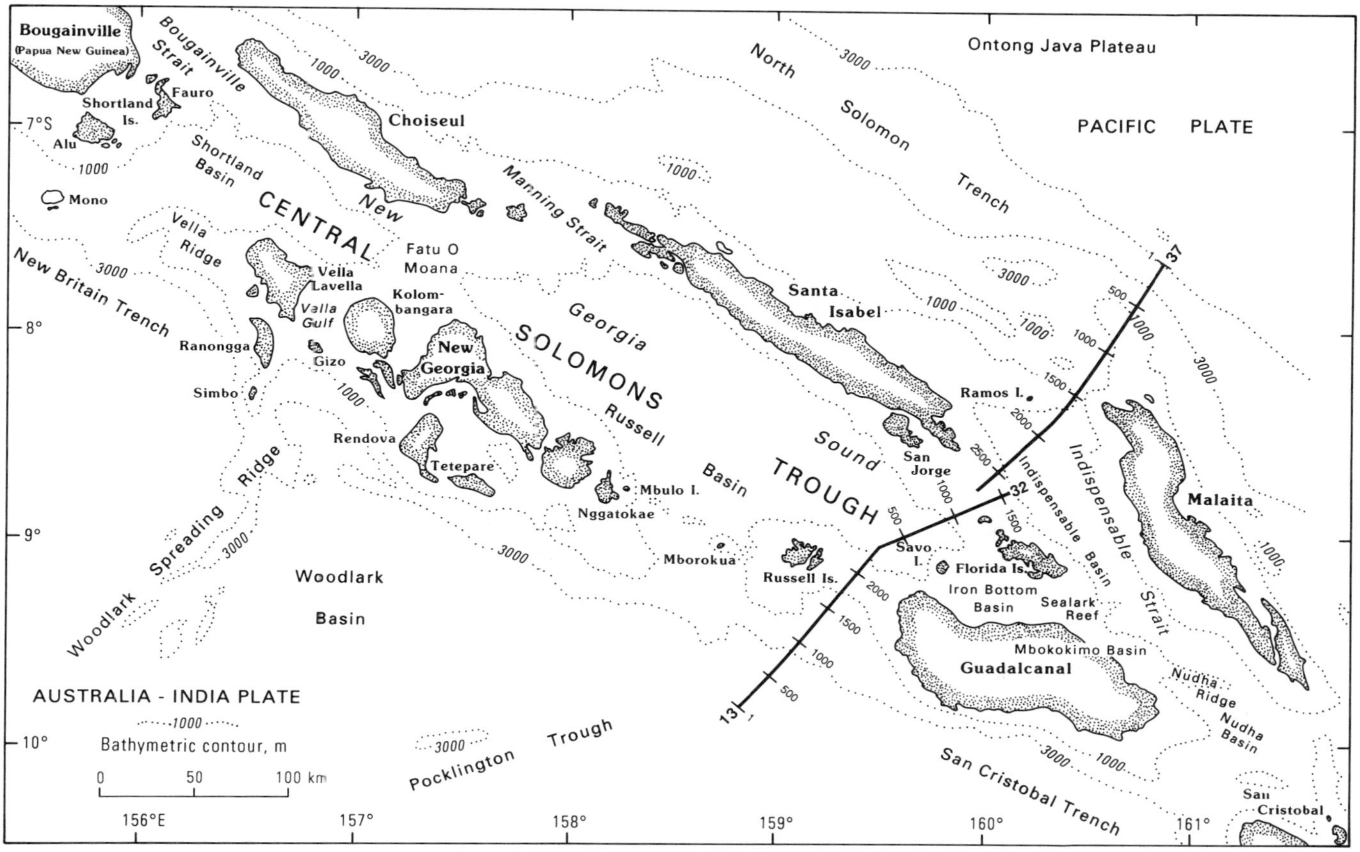

Figure 5. Index map of Guadalcanal-Malaita cross section region showing place names, and seismic line and common depth point (CDP) locations. Line segments used in cross section are shown by heavy lines. For further discussion and description, see Bruns et al (1986) and Bruns, Cooper, and Vedder (1986).

Bougainville Strait segment of the early Cenozoic arc, the reactivated Kilinailau Trench, and the southern edge of the Ontong Java Plateau. This profile is discontinuous because a line change created a small data gap in Shortland basin, and shallow water and reefs prevented data acquisition in Bougainville Strait. Additional small data gaps on the Kilinailau Trench slope portion of line 408 resulted from data loss after acquisition.

Interpretations of the Vella Ridge, Shortland basin, and Kilinailau Trench segments of the profile are presented in Bruns et al (1986), Bruns, Vedder, and Cooper (this volume), and Bruns, Vedder, and Culotta (this volume). The New Britain Trench segment of this cross section has not been illustrated previously, although a companion line with similar features is shown in Bruns, Vedder, and Cooper (this volume). The main features shown on the trench segment are an accretionary wedge at the base of the slope, a sediment apron derived from Shortland basin that thins out about half-way down the slope, and ocean plate reflectors that can be traced for 30 km beneath the slope to the water-bottom multiple.

BOUGAINVILLE-BUKA CROSS SECTION

From southwest to northeast, the Bougainville-Buka cross section begins midway up the inner trench wall of the New Britain Trench, then crosses the subsided remnant of the early Cenozoic arc beneath Kaviki Ridge, the Queen Emma intra-arc basin, the Emirau-Feni Ridge segment of the late Cenozoic arc northwest of Buka Island, the Nuguria Ridge, and the Kilinailau Trench. Note that the early Cenozoic arc is now southwest of the late Cenozoic arc, in marked contrast to the positions of the arcs on the Guadalcanal-Malaita and Vella Lavella-Choiseul cross sections. The two arcs cross on Bougainville. The Queen Emma basin and adjacent regions are described by Vedder, Bruns, and Cooper (this volume), and the Kilinailau Trench segment is described by Bruns, Vedder, and Culotta (this volume).

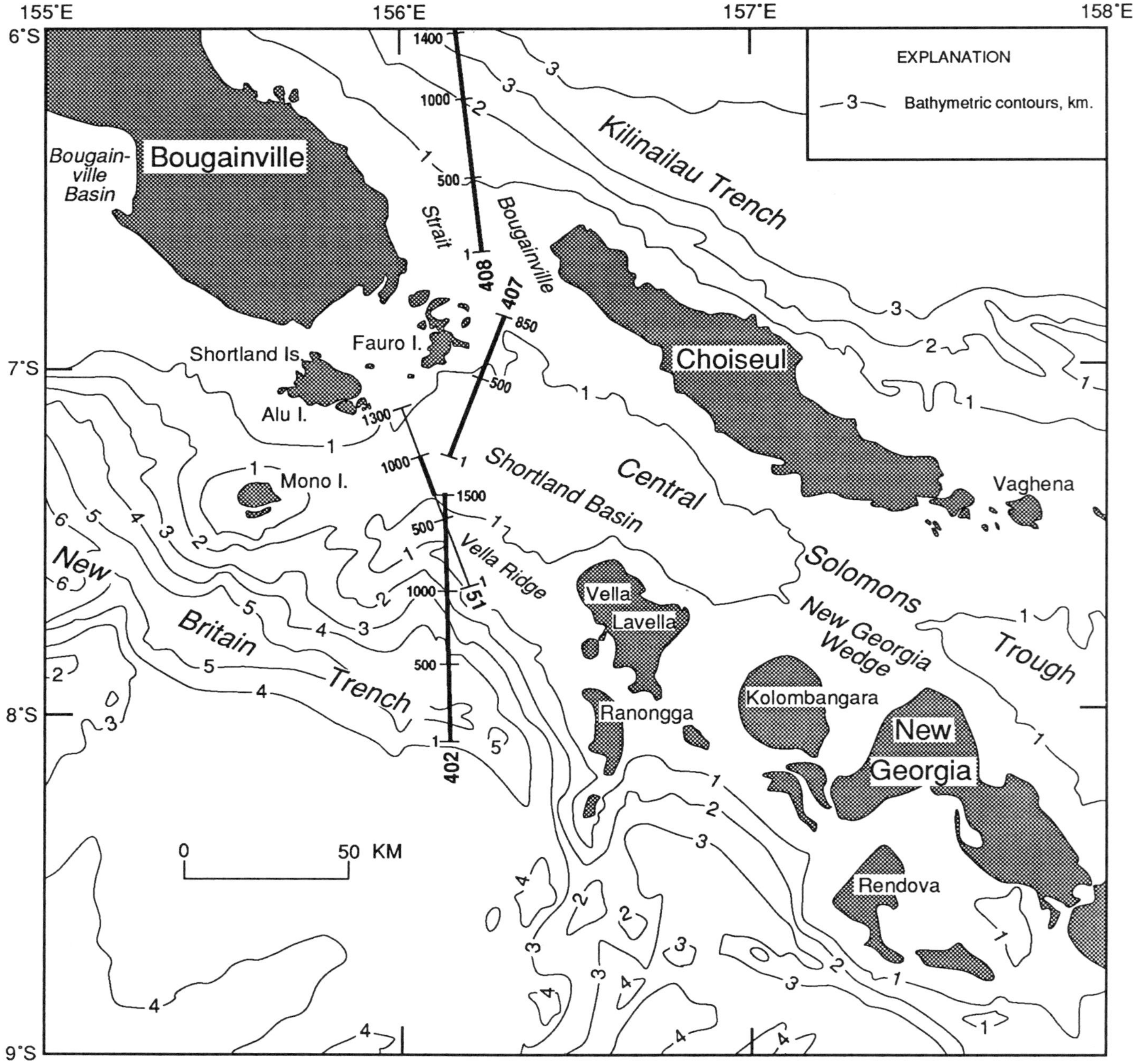

Figure 6. Index map of Vella Lavella-Choiseul cross section region showing place names, and seismic line and common depth point (CDP) locations. Line segments used in cross section are shown by heavy lines. For further discussion and description, see Bruns, Vedder, and Cooper (this volume) and Bruns, Vedder, and Culotta (this volume).

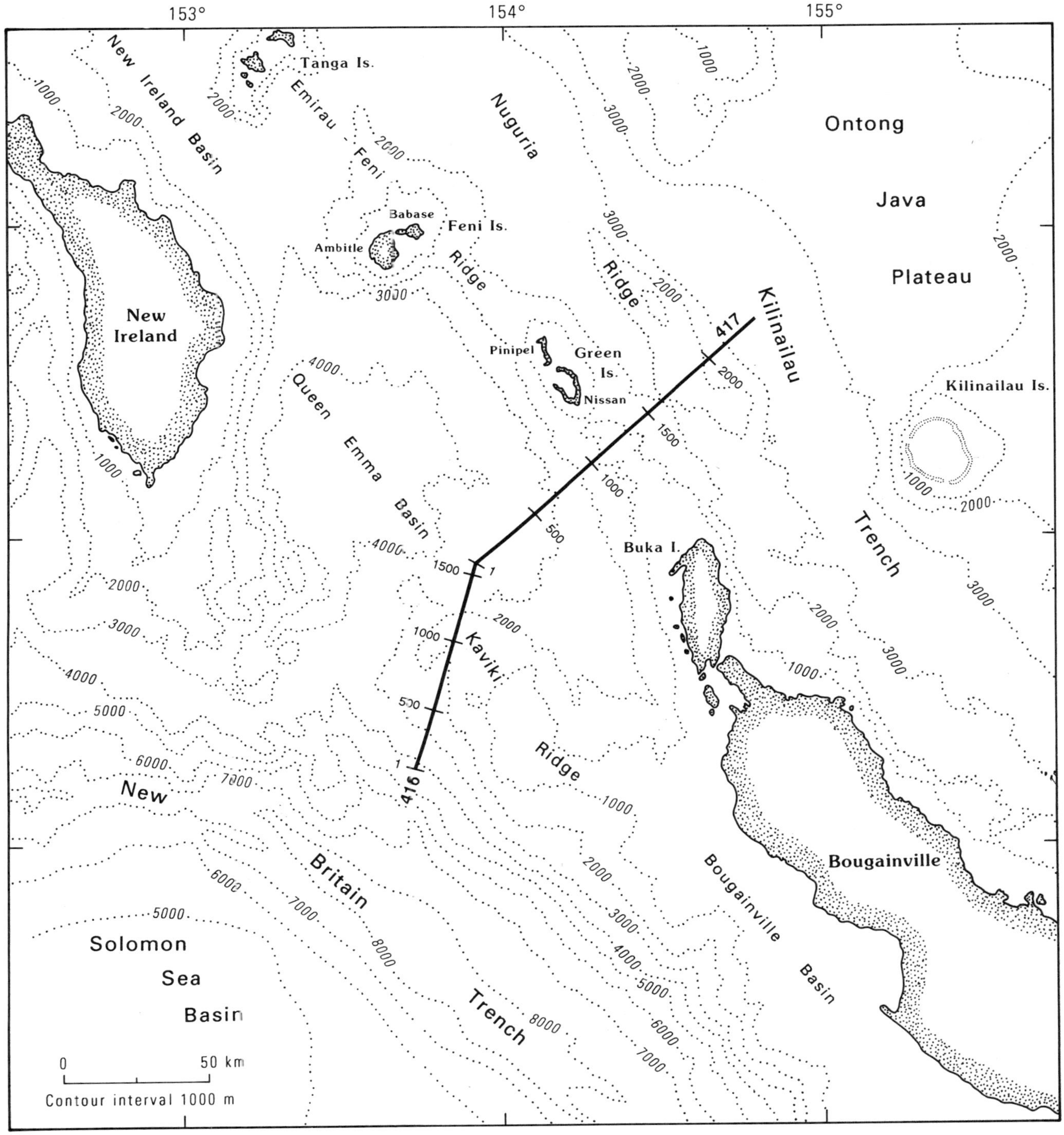

Figure 7. Index map of Bougainville-Buka cross section region showing place names, and seismic line and common depth point (CDP) locations. Line segments used in cross section are shown by heavy lines. For further discussion and description, see Vedder and Bruns (this volume) and Bruns, Vedder, and Culotta (this volume).

REFERENCES

Bruns, T.R., Cooper, A.K., Mann, D.M., and Vedder, J.G., 1986, Seismic stratigraphy and structure of sedimentary basins in the Solomon Islands region, *in* Vedder, J.G., Pound, K.S., and Boundy, S.Q., eds., Geology and offshore resources of Pacific island arcs—central and western Solomon Islands: Houston, Texas, Circum-Pacific Council for Energy and Mineral Resources Earth Science Series, v. 4, p. 177-214.

Bruns, T.R., Cooper, A.K., and Vedder, J.G., 1986, A multichannel seismic-reflection profile across the Solomon Islands arc, *in* Vedder, J.G., Pound, K.S., and Boundy, S.Q., eds., Geology and offshore resources of Pacific island arcs—central and western Solomon Islands: Houston, Texas, Circum-Pacific Council for Energy and Mineral Resources Earth Science Series, v. 4, p. 215-224.

Mann, D.M., 1986, Multichannel seismic operations for CCOP/SOPAC Cruise, Leg 3, Solomon Islands, *in* Vedder, J.G., Pound, K.S., and Boundy, S.Q., eds., Geology and offshore resources of Pacific island arcs—central and western Solomon Islands: Houston, Texas, Circum-Pacific Council for Energy and Mineral Resources Earth Science Series, v. 4, p. 35-40.

Vedder, J.G., Pound, K.S., and Boundy, S.Q., eds., 1986, Geology and offshore resources of Pacific island arcs—central and western Solomon Islands: Houston, Texas, Circum-Pacific Council for Energy and Mineral Resources Earth Science Series, v. 4, 306 p.

Vedder, J.G., and Bruns, T.R., editors, 1989, Geology and offshore resources of Pacific island arcs—Solomon Islands and Bougainville, Papua New Guinea Regions: Houston, Texas, Circum-Pacific Council for Energy and Mineral Resources Earth Science Series, v. 12.

APPENDIX

SUBMARINE TOPOGRAPHY OF THE CENTRAL AND WESTERN SOLOMON ISLANDS AND EASTERN PAPUA NEW GUINEA

T.E. Chase, B.A. Seekins, J.D. Young
U.S. Geological Survey, Menlo Park, California 94025

INTRODUCTION

A series of marine topographic maps, physiographic diagrams, and bathymetric profiles were prepared for the Solomon Islands and adjacent regions as a result of the 1982 cruise (L7-82-SP) of the R/V. *S.P. Lee* (Chase, Seekins, and Lund, 1986). These products were updated and modified after integration of new data from the 1984 cruise (L6-84-SP). Examples of the revised submarine topographic maps for the central and western Solomon Islands and eastern Papua New Guinea are shown in the accompanying figures. Figure 1 is a regional map that shows the main bathymetric features adjoining the island arc. Figures 2, 3, 4, and 5 are detailed maps that show segments of the arc investigated during the 1984 cruise (L6-84-SP).

The methods of acquisition and processing of navigation and trackline data for both the 1982 and 1984 cruises are described in Steele and Kinoshita (1986). Additional information regarding procedures used in preparing the maps, diagrams, and profiles can be obtained from the authors.

REFERENCES CITED

Chase, T.E., Seekins, B.A., and Lund, K.E., 1986, Submarine topography of the Solomon Islands region, *in* Vedder, J.G., Pound, K.S., and Boundy, S.Q., eds., Geology and offshore resources of Pacific island arc—central and western Solomon Islands: Houston, Texas, Circum-Pacific Council for Energy and Mineral Resources, Earth Science Series, v. 4, p. 21-31.

Steele, W.C., and Kinoshita, K.L., 1986, Navigation for CCOP/SOPAC Cruise Leg 3, Solomon Islands, *in* Vedder, J.G., Pound, K.S., and Boundy, S.Q., eds., Geology and offshore resources of Pacific island arcs—central and western Solomon Islands: Houston, Texas, Circum-Pacific Council for Energy and Mineral Resources Earth Science Series, v. 4, p. 19-20.